D1192331

Groundwater
Resources

Sustainability, Management, and Restoration

Neven Kresic

New York   Chicago   San Francisco
Lisbon   London   Madrid   Mexico City
Milan   New Delhi   San Juan
Seoul   Singapore   Sydney   Toronto

Cooper Union Library
MAY 0 8 2009

The **McGraw·Hill** Companies

Library of Congress Cataloging-in-Publication Data

Krešić, Neven.
  Groundwater resources : sustainability, management, and restoration /
Neven Kresic.
      p. cm.
  Includes bibliographical references and index.
  ISBN 978-0-07-149273-7 (alk. paper)
  1. Groundwater—Management.   2. Water conservation.   3. Water quality.
  4. Aquifer storage recovery.   I. Title.
  TD403.K74   2008
  333.91'04—dc22   2008025923

McGraw-Hill books are available at special quantity discounts to use as premiums and
sales promotions, or for use in corporate training programs. To contact a representative
please visit the Contact Us pages at www.mhprofessional.com.

Groundwater Resources

Copyright © 2009 by The McGraw-Hill Companies, Inc. All rights reserved. Printed in
the United States of America. Except as permitted under the United States Copyright
Act of 1976, no part of this publication may be reproduced or distributed in any form
or by any means, or stored in a data base or retrieval system, without the prior written
permission of the publisher.

1 2 3 4 5 6 7 8 9 0 DOC/DOC 0 1 4 3 2 1 0 9 8

ISBN 978-0-07-149273-7
MHID 0-07-149273-9

Sponsoring Editor	**Proofreader**
Larry S. Hager	Prakash Sharma
Acquisitions Coordinator	**Indexer**
Alexis Richard	Kevin Broccoli
Editorial Supervisor	**Production Supervisor**
David E. Fogarty	Pamela A. Pelton
Project Manager	**Composition**
Sandhya Joshi	Aptara, Inc.®
Copy Editor	**Art Director, Cover**
Sunita Dogra	Jeff Weeks

Information contained in this work has been obtained by The McGraw-Hill Companies, Inc. ("McGraw-Hill") from
sources believed to be reliable. However, neither McGraw-Hill nor its authors guarantee the accuracy or completeness of
any information published herein, and neither McGraw-Hill nor its authors shall be responsible for any errors, omissions,
or damages arising out of use of this information. This work is published with the understanding that McGraw-Hill
and its authors are supplying information but are not attempting to render engineering or other professional services.
If such services are required, the assistance of an appropriate professional should be sought.

Cooper Union Library
MAY 0 8 2009

To Joanne and Miles,
without whose help and understanding
this book would not have been possible

CROWN PUBLISHING

About the Author

Dr. Neven Kresic is the vice president of the American Institute of Hydrology for International Affairs. He is a member of the International Association of Hydrogeologists, the International Water Association, and the National Ground Water Association. Dr. Kresic is author of four books and numerous papers on the subject of groundwater. He is professional hydrogeologist, and professional geologist with over twenty five years of teaching, research and consulting experience working for a variety of clients including industry, water utilities, government agencies, and environmental law firms.

Contents

Preface x

1   Global Freshwater Resources and Their Use 1
1.1 World's Water Resources 1
1.2 Freshwater Availability 10
1.3 Water Use—Trends and Examples 11
    1.3.1 Use of Water in the United States 17
    1.3.2 Use of Water in Europe 22
    1.3.3 Use of Water in Africa 24
    1.3.4 Use of Water in China and India 26
1.4 Water Scarcity 30
1.5 Water Disputes 36
1.6 Economics of Water 39
    1.6.1 Price and Value of Water 44
    1.6.2 Virtual Water and Global Water Trade 55
1.7 Sustainability 61

2   Groundwater System 75
2.1 Definitions 75
2.2 Groundwater System Geometry 82
2.3 Groundwater Storage 83
    2.3.1 Porosity and Effective Porosity 83
    2.3.2 Specific Yield and Coefficient of Storage 88
    2.3.3 Groundwater Storage and Land Subsidence 92
2.4 Water Budget 97
2.5 Groundwater Flow 109
    2.5.1 Darcy's Law 109
    2.5.2 Types and Calculations of Groundwater Flow 114
    2.5.3 Groundwater Velocity 125
    2.5.4 Anisotropy and Heterogeneity 126
2.6 Initial and Boundary Conditions 132
    2.6.1 Initial Conditions and Contouring of Hydraulic Head 132
    2.6.2 Boundary Conditions 137
2.7 Aquifer Types 149
    2.7.1 Sand and Gravel Aquifers 150
    2.7.2 Sandstone Aquifers 160
    2.7.3 Carbonate (Karst) Aquifers 167
    2.7.4 Basaltic and Other Volcanic Rock Aquifers 174
    2.7.5 Fractured Rock Aquifers 175
    2.7.6 Withdrawals from Principal Aquifers in the United States . . 184

2.8 Aquitards   186
2.9 Springs   194
   2.9.1 Types and Classifications of Springs   196
   2.9.2 Thermal and Mineral Springs   200
   2.9.3 Spring Hydrograph Analysis   201
2.10 Groundwater in Coastal Areas and Brackish Groundwater   213
   2.10.1 Saltwater Intrusion   218
   2.10.2 Inland Brackish Water   220

3 Groundwater Recharge **235**
3.1 Introduction   235
3.2 Rainfall-Runoff-Recharge Relationship   236
3.3 Evapotranspiration   241
3.4 Infiltration and Water Movement Through Vadose Zone   245
   3.4.1 Soil Water Retention and Hydraulic Conductivity   247
   3.4.2 Darcy's Law   247
   3.4.3 Equations of Richards, Brooks and Corey, and van Genuchten   251
3.5 Factors Influencing Groundwater Recharge   253
   3.5.1 Climate   253
   3.5.2 Geology and Topography   257
   3.5.3 Land Cover and Land Use   259
3.6 Methods for Estimating Groundwater Recharge   262
   3.6.1 Lysimeters   264
   3.6.2 Soil Moisture Measurements   266
   3.6.3 Water Table Fluctuations   266
   3.6.4 Environmental Tracers   268
   3.6.5 Baseflow Separation   278
   3.6.6 Numeric Modeling   282

4 Climate Change **295**
4.1 Introduction   295
4.2 Natural Climatic Cycles   296
   4.2.1 Droughts   304
4.3 Anthropogenic Climate Change   313
   4.3.1 Impacts on Surface Water and Groundwater Resources . . .   317

5 Groundwater Quality **327**
5.1 Introduction   327
5.2 Natural Groundwater Constituents   328
   5.2.1 Total Dissolved Solids, Specific Conductance, and Salinity . .   329
   5.2.2 Hydrogen-Ion Activity (pH)   331
   5.2.3 Reduction-Oxidation (Redox) Potential (Eh)   332
   5.2.4 Primary Constituents   333
   5.2.5 Secondary Constituents   338
5.3 Groundwater Contamination and Contaminants   340
   5.3.1 Health Effects   342
   5.3.2 Sources of Contamination   347

5.3.3 Naturally Occurring Contaminants 355
5.3.4 Nitrogen (Nitrate) 359
5.3.5 Synthetic Organic Contaminants 362
5.3.6 Agricultural Contaminants 369
5.3.7 Microbiological Contaminants 374
5.3.8 Emerging Contaminants 377
5.4 Drinking Water Standards 380
5.4.1 Primary Drinking Water Standards 380
5.4.2 Secondary Drinking Water Standards 381
5.5 Fate and Transport of Contaminants 381
5.5.1 Dissolution 394
5.5.2 Volatilization 398
5.5.3 Advection 401
5.5.4 Dispersion and Diffusion 403
5.5.5 Sorption and Retardation 409
5.5.6 Biodegradation 419
5.5.7 Analytical Equations of Contaminant Fate and Transport . . 422

6   Groundwater Treatment **437**
6.1 Introduction 437
6.2 Oxidation 439
6.3 Clarification 441
6.4 Filtration 443
6.4.1 Rapid Sand Filters 443
6.4.2 Slow Sand Filters 444
6.4.3 Pressure Filters 445
6.4.4 Precoat Filters 446
6.4.5 Bag and Cartridge Filters 446
6.4.6 Ceramic Filters 446
6.5 Membrane Filtration 447
6.6 Carbon Adsorption 450
6.7 Ion Exchange and Inorganic Adsorption 450
6.8 Biological Treatment 451
6.9 Distillation 452
6.10 Disinfection 452
6.10.1 Chlorine 452
6.10.2 Chloramines 454
6.10.3 Ozone 455
6.10.4 Chlorine Dioxide 455
6.10.5 Ultraviolet Light Disinfection 455
6.11 Corrosion Control 456
6.12 Removal of Specific Constituents from Groundwater 456
6.12.1 Iron and Manganese Removal 457
6.12.2 Hardness 459
6.12.3 Nitrates 461
6.12.4 Total Dissolved Solids 462

6.12.5 Radionuclides 465
6.12.6 Hydrogen Sulfide 466
6.12.7 Volatile Organic Compounds and Synthetic
         Organic Compounds 466
6.12.8 Total Organic Carbon 467
6.12.9 Arsenic 467
6.12.10 Trace Metals and Inorganic Compounds 473
6.13 Drinking Water Treatment Costs 474

7   Groundwater Development **483**
7.1 Introduction 483
7.2 Water Wells 487
    7.2.1 Vertical Wells 488
    7.2.2 Collector Wells 515
7.3 Subsurface Dams 524
7.4 Spring Development and Regulation 530

8   Groundwater Management **539**
8.1 Introduction 539
8.2 Concept of Groundwater Sustainability 549
    8.2.1 Nonrenewable Groundwater Resources 557
8.3 Regulatory Framework 560
    8.3.1 Groundwater Quantity 560
    8.3.2 Groundwater Quality 562
    8.3.3 Transboundary (International) Aquifers 571
8.4 Integrated Water Resources Management 575
8.5 Monitoring 578
    8.5.1 Ambient Monitoring 581
    8.5.2 Compliance Monitoring 583
    8.5.3 Performance Monitoring 586
    8.5.4 Detecting Contamination 589
8.6 Data Management and GIS 594
    8.6.1 Data Management 596
    8.6.2 Types of Data 600
    8.6.3 Data Management Tools 604
    8.6.4 Geographic Information Systems (GIS) 607
    8.6.5 Data Quality 609
    8.6.6 Metadata 610
8.7 Protection of Groundwater Resources 610
    8.7.1 Groundwater Vulnerability Maps 615
    8.7.2 Delineation of Source Water Protection Zones 619
    8.7.3 Management Strategies 627
8.8 Modeling and Optimization 634
    8.8.1 Numeric Groundwater Models 635
    8.8.2 Time Series Models 657
8.9 Artificial Aquifer Recharge 663
    8.9.1 Methods of Artificial Aquifer Recharge 667

8.9.2 Aquifer Storage and Recovery 674
8.9.3 Source Water Quality and Treatment 675

9 Groundwater Restoration **695**
9.1 Introduction 695
9.2 Risk Assessment 701
    9.2.1 Data Collection and Evaluation 703
    9.2.2 Exposure Assessment 704
    9.2.3 Toxicity Assessment 705
    9.2.4 Risk Characterization 706
9.3 Remedial Investigation and Feasibility Study 708
    9.3.1 Development and Screening of Remedial Alternatives 713
    9.3.2 Detailed Analysis of Remedial Alternatives 716
    9.3.3 Treatability Studies 717
9.4 Source-Zone Remediation 718
    9.4.1 NAPLs Problem 724
    9.4.2 Physical Containment 732
    9.4.3 Fluid Removal Technologies 739
    9.4.4 In Situ Chemical Oxidation 750
    9.4.5 Enhanced Bioremediation 756
    9.4.6 Thermal Technologies 757
9.5 Dissolved Phase (Plume) Remediation 762
    9.5.1 Pump and Treat 765
    9.5.2 Permeable Reactive Barriers 773
    9.5.3 Bioremediation 778
    9.5.4 Monitored Natural Attenuation 787
9.6 Measuring Success of Remediation 792

A **Values of $W(u)$ for Fully Penetrating Wells in a Confined,
Isotropic Aquifer** **807**

B1 **Unit Conversion Table for Length, Area, and Volume** **811**

B2 **Unit Conversion Table for Flow Rate** **813**

B3 **Unit Conversion Table for Hydraulic Conductivity and
Transmissivity** **814**

    Index **815**

Preface

Over the last decade or so, few words have caused more debate and concern within the international scientific community and general public alike than the following two terms: global warming and climate change. And if one wanted to play with these four words to make an even bigger point, a third term would come to mind immediately: global change. At the time of this writing, the Intergovernmental Panel on Climate Change (IPCC) has already released eagerly anticipated reports, which present to all of us some rather alarming facts and projections of the impacts of climate change. In the era of instantaneous communication, across all continents, scientific and other facts cannot stay hidden for long. Nevertheless, there are a few skeptics, including some governments, left behind as the rest of us are rapidly learning to live with many fundamental questions we never bothered asking just a few years ago. What is arguably the common thread in various discussions about our globally changing future is the word *sustainability*. When we think of energy, transportation, food production, forests, wildlife, cities, rivers, country side, and many other things, and if we are concerned with any of them, most likely we would be able to formulate our concerns with one simple question: Is our related activity sustainable? Is using fossil fuels the way we do now sustainable? Is driving cars every day sustainable? Is deforestation sustainable? Is the population growth sustainable? Do we have the right to destroy natural habitat of other species, and indeed the species themselves? And we can add many more questions to this list.

Inseparable from the question of sustainability is the ethical question. Can we deny the rights of someone else, now or in future, to have or do the same things we ourselves have done in the past and continue to do now? Whatever our individual answer to this and other similar questions may be, there is one single thing that we can all agree on: *without water there is no life*. Although this statement may seem misplaced here because it is "just" a common truth, it is used to make the following observation important for this book: Groundwater depletion has already caused and will continue to cause many springs, rivers, lakes, and marshes to shrink or go dry, and the flora and fauna as we knew them in many parts of the world are gone and will continue to disappear because of that. And, of course, without extracted groundwater, food production and human life in many parts of the world would not be possible in present times. Related alarming headlines in the media add fuel to the ongoing debate between those who are concerned with our current practices and common future and those who seem far less concerned because they do not see anything alarming about business as usual. Here is a sample of several such headlines in the national and international media during 2007: "Southwest forecast: expect 90 years of drought" ("Human-induced change in earth's atmosphere will leave the American Southwest in perpetual drought for 90 years"); "Australia suffers worst drought in 1000 years"; "Drought lands doubled"; "At the end of September about 43 percent of the contiguous United States was in moderate to extreme drought, the National Climate Data Center said Tuesday. Worldwide, meanwhile, the agency said the year to date has been the warmest on record for land."

Because of the drought word, and after reading the short articles and maybe recalling a few common-knowledge facts, one may become genuinely alarmed. Consider that the United States and Australia are top exporters of food in the world, and both significantly rely on groundwater for agricultural irrigation, including (and especially) in areas where aquifers are already being overexploited and are under stress due to competing demands. Now add equally, if not more alarming, news from China and India, the two most populated countries in the world, about continuing groundwater depletion for agriculture and water supply. Then, consider most of Africa and the Middle East with their chronic and increasing water scarcity. Finally, try to imagine various chain reactions on the global scale involving food security, poverty, politics, geostrategic interests, refugees, environmental degradation, unrests...

According to the United Nations Educational, Scientific and Cultural Organization (UNESCO), some 1.1 billion people in the world are estimated to lack access to few tens of liters of safe freshwater that is the minimum daily range suggested by the United Nations (UN) to ensure each person's basic needs for drinking, cooking, and sanitation. Some 26 countries, totaling over 350 million people, suffer from severe water scarcity, mainly in dry lands (arid areas), although available groundwater resources appear adequate to provide an immediate relief in many such areas (UNESCO, 2006, 2007). The number of already displaced, desperate, undernourished, and thirsty people continues to grow due to climatic variability, population growth, inadequate governance, and inappropriate water management. These same factors put a continuing pressure on surface water and especially groundwater resources in many developed countries as well. Regardless of the country's economic and political development, three major themes are common everywhere: (1) competition for groundwater resources between agriculture (farmers), growing urban population, and industry; (2) depletion of these resources by all three; and (3) contamination of the resources by all three. Although this book focuses on groundwater, the following cannot be emphasized enough: Any division between surface water and groundwater as two "separate" sources of freshwater is artificial; they are interconnected in so many ways that studying one, without understanding the other, would in all probability lead to inappropriate water management decisions. This is why a part of this book explains integrated water resources management (IWRM)—a concept that is being increasingly studied and implemented at various levels, i.e., local, state, and regional (intergovernmental).

Groundwater sustainability is discussed throughout this book from the various aspects of available resource quantity and quality, including its evaluation, engineering, management, planning, and restoration. The first five chapters of the book explain in detail what groundwater is, where it comes from, how it is naturally replenished and how much, and what the possible impacts of projected climate change on groundwater recharge and use might be. The natural quality of groundwater, sources of contamination, and fate and transport of contaminants are also explained in detail. Chapter 6 covers various traditional and innovative technologies of groundwater treatment for drinking water purposes. Chapters 7 and 8 explain engineering means of groundwater extraction and regulation, delineation of source protection zones, groundwater (aquifer) vulnerability mapping, and various topics on groundwater management including modeling, monitoring, artificial aquifer recharge, and development of databases and geographic information systems. The last chapter of the book covers restoration of contaminated groundwater for beneficial uses, including remediation of contaminant source zones and dissolved contaminant plumes.

The author is grateful to the following water resources professionals who generously contributed to this book with their knowledge and enthusiasm: Alex Mikszewski, Jeff Manuszak, Marla Miller, Dr. Alessandro Franchi, Robert Cohen, Dr. Ivana Gabric, Dr. Neno Kukuric, Nenad Vrvic, Samuel Stowe, and Farsad Fotouhi.

Neven Kresic
Rixeyville, VA

CHAPTER 1

Global Freshwater Resources and Their Use

The following words of Koïchiro Matsuura, Director General of the United Nations Educational, Scientific and Cultural Organization (UNESCO), summarize the increasing importance various countries, and the international community as a whole, see in adequately addressing water resources on the global scale:

> Water, of course, is everyone's business. Hardly a day goes by when we do not hear of another flood, another drought or another pollution spill into surface waters or groundwaters. Each of these issues has a direct or indirect impact not only on human security but also on livelihoods and development. The issues involved range from those of basic human well-being (food security and health), to those of economic development (industry and energy), to essential questions about the preservation of natural ecosystems on which ultimately we all depend. These issues are inter-related and have to be considered together in a holistic manner. It is thus entirely appropriate that some twenty-four agencies and entities within the United Nations system are involved, with a shared purpose, in producing a comprehensive and objective global report on water issues and the measures being taken to address the related challenges that beset humanity worldwide.
>
> As internationalization and rapid economic growth in many societies alter traditional socio-economic structures, it is clear that change, although virtually pervasive, is not entirely positive. Many people, especially in the developing world and especially those on urban margins and in rural areas, are left behind in poverty and mired in preventable disease.
>
> Access to secure water supplies is essential. This seems self-evident. Yet, as this Report shows, it is clear that the central role of water in development is neither well understood nor appreciated. Much more needs to be done by the water sector to educate the world at large and decision-makers in particular. (UNESCO, 2006).

1.1 World's Water Resources

The total surface area of the earth is 197 million mi^2 (510 million square kilometers), of which about 139 million mi^2 (70.8 percent) was in 1960 covered by the world's oceans, about 6.9 million mi^2 (3.4 percent) by polar icecaps and glaciers, about 330,000 mi (0.17 percent) by natural freshwater lakes, and about 270,000 mi (0.14 percent) by natural saline lakes (Nace, 1960). The total land area, including that under ice, lakes, and inland seas is about 57 million mi^2 (148 million km^2). The volume of ocean water is about 317 million cubic miles (or 1320 million km^3). The estimated water volume of polar icecaps and glaciers on the continents is about 7.3 million mi^3 (30.4 million km^3). Freshwater lakes contain about 30,000 mi³ (125,000 km³) of water, and saline lakes and inland seas contain about 25,000 mi³ (104,000 km³).

Lake Baikal in Siberia, Russia, the deepest in the world (1620 m), contains about 23,000 km³ of water (Bukharov, 2001; USGS, 2007a) or close to 20 percent of all freshwater stored in the world's natural lakes. The volume of Lake Baikal is approximately equal to that of all Great Lakes on the North American Continent combined—Superior, Michigan, Huron, Erie, and Ontario—22,684 km³ (USEPA, 2007a). Most freshwater lakes are located at high latitudes, with nearly 50 percent of the world's lakes in Canada alone. Many lakes, especially those in arid regions, become salty because of evaporation, which concentrates the dissolved salts. The Caspian Sea, the Dead Sea, and the Great Salt Lake are among the world's major salt lakes. Reservoirs, or artificial lakes, contain estimated 4286 km³ of freshwater worldwide (Groombridge and Jenkins, 1998).

Wetlands, which include swamps, bogs, marshes, mires, lagoons, and floodplains, cover an estimated total global area of about 2.9 million km² (Groombridge and Jenkins, 1998). Most wetlands range in depth from 0 to 2 m. Estimating the average depth of permanent wetlands at about 1 m, the global volume of wetlands could range between 2300 and 2900 km³ (UNEP, 2007). The average amount of water in stream channels at any one time is on the order of 280 mi³ or 1166 km³ (Nace, 1960).

The main root zone (the upper 3 ft or 1 m) of the soil probably contains at least about 6000 mi³ of water (25,000 km³). The estimated additional amount of water in the rock crust of the earth is about 1 million mi³ (4.17 million km³) to a depth of half a mile (800 m) and an equal amount at the depth between 1/2 and 2 mi (Nace, 1960).

The total world supply of water is somewhat more than 326 million mi³ (1358 million km³), with about 97 percent in the oceans. The total volume of water (fresh and saline) on the land and beneath its surface is only about 9.4 million mi³ (39 million km³). About 78 percent is locked up in icecaps and glaciers, and about 0.27 percent is in inland saline lakes and seas. Most of the water stored in icecaps and glaciers is concentrated in Greenland and Antarctica, far from human habitation and not readily available for use. Approximately 43,212 mi³ (180,000 km³) of frozen freshwater on continents outside Greenland and Antarctica is stored in glaciers and mountainous icecaps spread worldwide over 212,000 mi² or 550,000 km² (UNEP, 1992, 2007; Untersteiner, 1975). Much of the groundwater at depths greater than half a mile (800 m) is economically inaccessible at present or is saline. Thus, less than 3 percent of the world's freshwater supply is available on the continents, and only little more than 11 percent of the water on the continents, actually is usable or accessible. Furthermore, the yearly renewal and continued availability of this relatively minute supply of water depend wholly on precipitation from a tenuous bit of water vapor in the atmosphere (Nace, 1960). Figure 1.1 shows volumes and percentage of usable freshwater types on the continents outside polar regions.

Although the above estimates are inexact by default, they help to define the magnitude of the problem of freshwater management. The conversion of saltwater to fresh is a great and intriguing challenge. Even though conversion processes are becoming more and more economically feasible, the cost of transportation may prohibit the use of converted seawater by inland areas for a long time. Conversion of locally available saltwater may become feasible in some inland areas and may resolve problems that are locally serious. On the whole, however, the available amount of such water is not sufficient to add materially to regional or national water supplies. Therefore, for an indefinitely long future period, inland areas will receive water from the sea only indirectly and in the same manner that they always have—vapor carried inland in the air and dropped as rain and

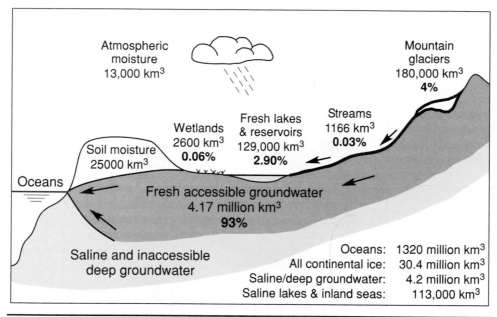

FIGURE 1.1 Volumes, in cubic kilometers, and percentages of usable freshwater on the continents outside polar regions. (Data from Nace, 1960; Groombridge and Jenkins, 1998; UNEP, 2007.)

snow. The estimated volume of moisture in the atmosphere is equivalent to only about 3100 mi³ (13,000 km³) of water, or enough to cover the entire earth to a depth of only about 1 in. (2.5 cm; Nace, 1960).

Of about 50 million mi² (130 million km²) of continental and insular dry land, somewhat more than 18 million mi² (about 36 percent) is arid to semiarid. Figure 1.2 shows distribution of world's nonpolar arid land. Such areas largely, or often entirely, depend on groundwater resources for irrigation and water supply.

Figure 1.3 is a photographic image of the Issaouane Erg (sand sea) located in eastern Algeria between the Tinrhert Plateau to the north and the Fadnoun Plateau to the south. Ergs are vast areas of moving sand with little to no vegetation cover. Part of the Sahara Desert, the Issaouane Erg covers an area of approximately 38,000 km². These complex dunes form the active southwestern border of the sand sea (NASA, 2007a). Because of prolonged droughts and desertification, in many places along the southern edge of the Sahara Desert, sand dunes advancement and rocky desert expansion are continuously taking place.

In the United States as a whole, the quantity of water in underground storage, within half a mile of the land surface, is several times that in all the large lakes of the North American Continent. Although the volume of groundwater in storage is large, its natural rate of replenishment is small in comparison. The following discussion by Nace (1960) illustrates this point:

Precipitation on the 48 States averages about 30 inches (about 2.5 feet or 762 millimeters) yearly, and the total yearly volume is about 1370 mi³ (5707 km³). Natural annual recharge of ground water may average a fourth of the precipitation, or about 340 mi³ (1416 km³) yearly. This is a liberal estimate,

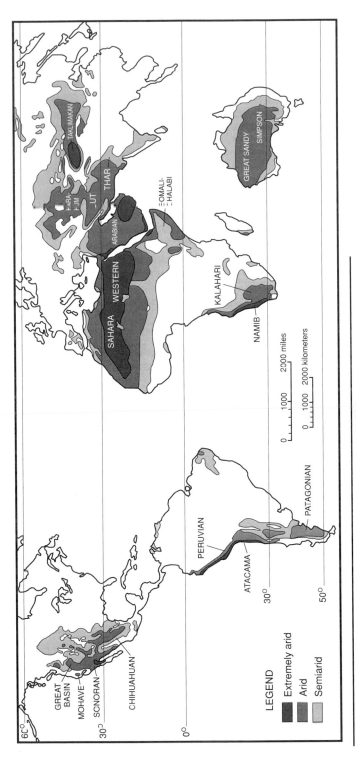

Figure 1.2 Distribution of nonpolar arid land. (Source: Meigs, 1953.)

4

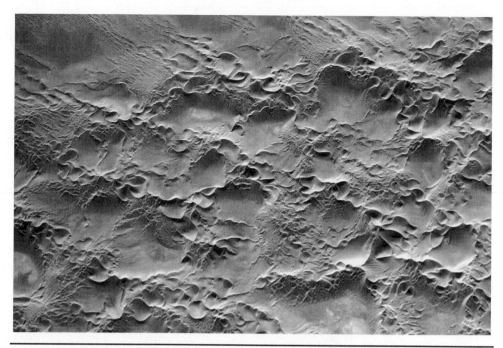

FIGURE 1.3 Astronaut photograph of the Issaouane Erg (sand sea) located in eastern Algeria. The photograph was acquired on January 16, 2005, with a Kodak 760C digital camera with an 800-mm lens, and is provided by the ISS Crew Earth Observations Experiment and the Image Science & Analysis Group, Johnson Space Center. (Available at http://earthobservatory.nasa.gov.)

which many hydrologists would dispute. However, it indicates the order of magnitude of groundwater recharge.

On the basis of the above estimates, the volume of ground water in storage above a depth of half a mile evidently is equivalent to the total of all recharge during the last 160 years. This estimate is very crude, but whether the true figure is 50, 100, or 200 years is unimportant. The significant fact is that a reserve of water has been accumulating in the groundwater bank for generations. This is the only real water reserve we have. Annual recharge in any one year is proportionately a very small increment to the total reserve. Now, by pumping, we are placing heavy drafts on the local "branch banks" in some parts of the United States—enough that the manager of the "main bank" must look to the total reserve and estimate how long the drafts can continue. This is the job for water management.

The southern high plains of Texas and New Mexico in the United States are an outstanding example of large storage and small replenishment. Overall, the groundwater in storage in the Texas region is about 200 million acre-ft (247 km^3), but if exhausted it would take considerably more than 1000 years to replace (U.S. Senate Select Committee on National Water Resources, 1960, p. 15).

Arguably, the hottest topic in the water industry at the time of this writing is the question of possible impacts of climate change on the availability of water resources, including their quantification for the planning purposes. It cannot be overemphasized, however, that any such impacts will, by default, be superimposed on the past and the ongoing impacts of extraction (exploitation) of both surface water and groundwater. These existing impacts would have to be well understood and quantified before any attempts of predictions for the future are made. Without immersing themselves into the heated political debate whom to blame for the climate change (nature, humans, or both), and what the consequences of doing (or not doing) something about it would be, water professionals are already facing the realities of redistribution of available freshwater on continents, on both local and global scales. Some of the better-known examples are reduction of snow pack in mountainous regions, melting of mountainous glaciers (Fig. 1.4), and shrinking of large natural lakes (Figs. 1.5 and 1.6). Popular and daily press are increasingly covering such phenomena. For example, *National Geographic News* for August 3, 2007, offers this lead into a story: "Lake Superior, the world's largest freshwater lake, has been shrinking for years – and now it appears to be getting hotter." The article goes on to include some attention-grabbing narrative such as: "Beachgoers at the lake, which is bounded by Wisconsin, Michigan, Minnesota, and Ontario, Canada, must walk up to 300 feet (100 yards) farther to reach shorelines. Some docks are unusable because of low water, and once-submerged lake edges now grow tangles of tall wetland plants" and "Researchers are also starting to suspect that the shrinking and heating are related—and that both are spurred by rising global temperatures and a sustained local drought" (Minard, 2007). More on climate change and its projected impacts on groundwater resources is given in Chapter 4. To set the stage, a short discussion on Aral Sea in Asia (Fig. 1.5) and Lake Chad in Africa (Fig. 1.6) illustrates the close interconnectedness of human activity, climate change, surface water and groundwater resources.

Once the fourth largest lake on earth, the Aral Sea has shrunk dramatically over the past few decades. The Aral Sea basin, covering the territories of Tajikistan, Uzbekistan, Turkmenistan, some parts of Afghanistan, Kyrgyzstan, and Kazakhstan, is located in the heart of the Euro-Asian continent. The two main rivers, the Amu-Darya and the Syr-Darya, together with some 30 primary tributaries, feed the basin, which has an areal extent of about 1.8 million km^2. In the early 1960s, the former Soviet Union launched efforts to divert almost all water from the Amu-Darya and the Syr-Darya. The diversion of millions of cubic meters of water to irrigate cotton fields and rice paddies through massive infrastructure development helped increase the irrigated area from 5 mha (million hectare) in the 1950s to 8 mha in the 1990s (Murray-Rust et al., 2003).

The water development system of the region is described as "one of the most complicated human water development systems in the world" (Raskin et al., 1992) because human interventions have gradually modified the natural water flow and the environment along the rivers' banks. The Aral Sea basin system now has highly regulated rivers with 20 medium- and large-sized reservoirs and around 60 diversion canals of different sizes. In all, the two rivers have some 50 dams of varying sizes (Murray-Rust et al., 2003).

The large, slightly brackish inland sea moderated the region's continental climate and supported a productive fishing industry. As recently as 1965, the Aral Sea received about 50 km^3 of freshwater per year—a number that fell to zero by the early 1980s. Consequently, concentrations of salts and minerals began to rise in the shrinking body of water, eventually reaching 33 g/L, up from the initial 10 to 12 g/L. This change in chemistry has led to staggering alterations in the lake's ecology, causing precipitous drops in the Aral Sea's fish population and elimination of the commercial fishing industry (NASA, 2007b; Glazovsky, 1995).

The shrinking Aral Sea has also had a noticeable affect on the region's climate. Summer and winter air temperatures at stations near the shore increased by 1.5 to 2.5°C, whereas diurnal temperatures increased by 0.5 to 3.3°C. At coastal stations, the mean annual relative air humidity decreased by 23 percent, reaching 9 percent in spring and

Figure 1.4   (Top) Photograph (northwest direction) taken several hundred meters up a steep alluvial fan located in a side valley on the east side of Queen Inlet, Glacier Bay National Park and Preserve, Alaska. (Photo by Charles W. Wright, 1906). (Bottom) Photograph (north) taken on Triangle Island, Queen Inlet (Photo by Bruce F. Molnia, 2003). Images published by the National Snow and Ice Data Center, Boulder, CO, courtesy of the U.S. Geological Survey. (Available at http://nsidc.org/cgi-bin/gpd_run_pairs.pl.)

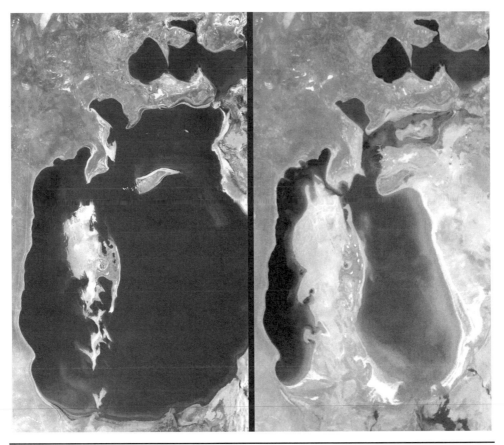

FIGURE 1.5 Satellite images showing change in the Aral Sea size between 1989 (left image) and 2003 (right image). Once part of the same lake spread over 68,300 km², the northern and southern half of the sea had already become virtually separated in 1989. The image at right shows the rapid retreat of the sea's southern half, now separated into a western and eastern half. (Images from NASA, 2007b.)

summer. Recurrence of drought days increased by 300 percent. The last spring frosts shifted to later dates and the first autumn ones occur some 10 to 12 days earlier (Glazovsky, 1995). The shorter growing season is causing many farmers to switch from cotton to rice, which demands even more diverted water.

A secondary effect of the reduction in the Aral Sea's overall size is the rapid exposure of the lakebed. Strong winds that blow across this part of Asia routinely pick up and deposit tens of thousands of tons of now exposed soil every year. This process has not only contributed to significant reduction in breathable air quality for nearby residents, but has also appreciably affected crop yields because of the heavily salt-laden particles falling on arable land. As the agricultural land becomes contaminated by the salt, the farmers try to combat it by flushing the soil with huge volumes of freshwater. What water makes its way back to the sea is increasingly saline and polluted by pesticides and fertilizers (NASA, 2007b).

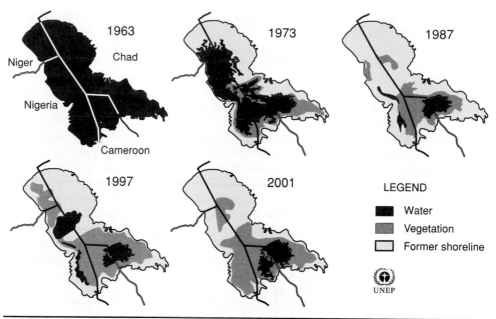

FIGURE 1.6 Change in Lake Chad surface area between 1963 and 2001. (Maps are created from the series of satellite images taken in January each year and provided by the NASA Goddard Space Center. From UNEP/GRID-Arendal, 2002; maps by Phillippe Rekacewicz, UNEP/GRID-Arendal.)

Groundwater levels rose in many regions because of irrigation. For example, in the Tashauz region, land area with a groundwater level less than 2 m below ground surface was 20 percent from 1959 to 1964, whereas from 1978 to 1982 this area increased to 31.5 percent. Over the whole of Turkmenia, 87 percent of the irrigated land has groundwater levels that have risen by at least 2.5 m. Because of this rise in groundwater levels, the area subject to soil salinization due to evaporation from a shallow water table has dramatically increased in many regions. In contrast, as the levels of the Aral Sea and the inflowing rivers dropped, the adjacent and the nonirrigated areas subject to desertification experienced a decrease in groundwater levels by 10 to 15 m (Glazovsky, 1995).

Lake Chad (Fig. 1.6) is Africa's fourth largest lake by surface area and has the largest drainage basin of any lake in the world (2.5 million km^2; Isiorho and Matisoff, 1990; Hernerdof, 1982). For thousands of years, it has been a center of trade and cultural exchange between people living north of the Sahara and people to the south. Located at the intersection of four different countries in West Africa (Chad, Niger, Nigeria, and Cameroon), Lake Chad has been the source of water for massive irrigation projects. In addition, the region has suffered from an increasingly dry climate, experiencing a significant decline in rainfall since the early 1960s. The most dramatic decrease in the size of the lake was in the 15 years between January 1973 and January 1987. Beginning in 1983, the amount of water used for irrigation began to increase. Ultimately, between 1983 and 1994, the amount of water diverted for purposes of irrigation quadrupled from the amount used in the previous 25 years. According to a study by University of Wisconsin-Madison researchers, working with NASA's Earth Observing System program, the lake is now 1/20th of the size it was 35 years ago (NASA, 2007b).

Although Lake Chad is a closed-basin, shallow lake in a semiarid region (annual precipitation of about 30 cm) with a high evaporation rate (200 cm/yr), and therefore expected to have highly alkaline and saline waters, its water is surprisingly fresh (120 to 320 mg/L; Isiorho and Matisoff, 1990). This has been attributed to several factors: low volume-to-surface ratio that ensures significant dilution by rainfall, low salinities of the input rivers (42 to 60 mg/L), seepage through the lake bottom to the phreatic (unconfined) aquifer, and biogeochemical regulations (Isiorho and Matisoff, 1990).

The upper zone of the Chad formation, consisting of Quaternary sediments, is generally unconfined and is recharged by the lake through seepage. The lake is at a higher hydraulic elevation than the aquifer, and field investigations, including direct seepage measurements, confirm groundwater flow away from the lake toward the southwest (Isiorho and Matisoff, 1990). The unconfined aquifer is tapped mostly by hand-dug wells for irrigation and domestic use throughout the southwest portion of the Chad Basin. It is this portion of the basin that would be affected by the possible disappearance of Lake Chad and an ultimate cessation of most of the aquifer recharge if current practices continue. This would lead to severe water shortages for the rural population depending upon shallow wells as well as water shortages over a huge land area already suffering from natural and social disasters.

1.2   Freshwater Availability

Efforts to characterize the volume of water naturally available to a given nation have been ongoing for several decades. The primary input for many of these estimates is an information database (AQUASTAT) that has historically been developed and maintained by FAO (Food and Agriculture Organization of the United Nations). It is based on data related to the quantity of water resources and uses a water-balance approach for each nation (FAO, 2003). This database has become a common reference tool used to estimate each nation's renewable water resources. The FAO has compiled an index of Total Actual Renewable Water Resources (TARWR). This index reflects the water resources theoretically available for development from all sources within a country. It is a calculated volume expressed in km^3/yr; divided by the nation's population and adjusted to m^3/yr, it is expressed as a per capita volume, which allows a relative evaluation of the resource available to its inhabitants. The index estimates the total available water resources per person in each nation, taking into account a number of individual component indicators by

- Adding all internally generated surface water annual runoff and groundwater recharge derived from precipitation falling within the nation's boundaries
- Adding external flow entering from other nations, which contributes to both surface water and groundwater
- Subtracting any potential resource volumes shared by the same water, which comes from surface and groundwater system interactions
- Subtracting, where one or more treaty exists, any flow volume required by that treaty to leave the country (FAO, 2003; FAO-AQUASTAT, 2007)

TARWR gives the maximum theoretical amount of water actually available for the country on a per capita basis. Beginning in about 1989, it has been used to make evaluations of water scarcity and water stress. It is important to note that the FAO estimates are maximum theoretical volumes of water renewed annually as surface water runoff and groundwater recharge, taking into consideration what is shared in both the surface and groundwater settings. However, as discussed by UNESCO (2006), these volumes do not factor in the socioeconomic criteria that are potentially and differentially applied by societies, nations, or regions to develop those resources. Costs can vary considerably when developing different water sources. Therefore, whatever the reported "actual" renewable volume of water, it is a theoretical maximum, and the extent to which it can be developed will be less for a variety of economic and technical reasons. Following are some of the factors that should be considered when using the TARWR index (UNESCO, 2006):

- Approximately 27 percent of the world's surface water runoff occurs as floods, and this water is not considered a usable resource. However, floods are counted in the nation's TARWR as part of the available, renewable annual water resource.

- Seasonal variability in precipitation, runoff, and recharge, which is important to regional and basin-level decision making and water storage strategies, is not well reflected in annualized quantities.

- Many large countries have several climatic regions as well as disparate population concentrations and the TARWR does not reflect the ranges of these factors that can occur within nations.

- There is no data in TARWR that identifies the volume of "green" water that sustains ecosystems—the volume that provides water resources for direct rain-fed agriculture, grazing, grasslands, and forests.

As already indicated, not all of the internally renewable freshwater resources (IRWR) can be controlled by the population of a country. It is estimated that even with the most feasible technical, social, environmental, and economic means, only about one-third of the IRWR can be potentially controlled. The global potentially useable water resources (PUWR) of the IRWR are estimated to be around 9000 to 14,000 km^3 (UN, 1999; Seckler, 1993). At present, about 2370 km^3 of the global PUWR are developed and are being diverted as the primary water supply (PWS) or the "virgin" or the "first" water supply for human use (IWMI, 2000). A part of the PWS is evaporated in its first use. The other part returns to rivers, streams, and aquifers as return flows and in many instances this part is again withdrawn for human use. This is known as the recycled portion of PWS. The PWS and the recycled water supply, about 3300 km^3, constitute the water diverted for use in different sectors (agriculture, industry, public supply).

1.3 Water Use—Trends and Examples

Water use is a general term that refers to water used for a specific purpose, such as for domestic water supply, irrigation, or industrial processing. Water use pertains to human interaction with and influence on the hydrologic cycle and includes elements such as water withdrawal from surface water and groundwater sources, water delivery to irrigated land, homes, and businesses, consumptive use of water, water released from wastewater-treatment plants, water returned to the environment, and in-stream uses such as production of electricity in hydropower plants. Consumptive use, or consumed water, is that part of water withdrawn that is evaporated, transpired by plants, incorporated into products or crops, consumed by humans or livestock, or otherwise removed from the immediate water environment (USGS, 2007b). It is very important to make distinction between water withdrawal and water consumption during resource evaluation. For example, not all water withdrawn for irrigation purpose and applied to a farmland will be consumed. Depending on the irrigation method, more or less diverted water will return to its original source or another body of water (e.g., surface streams and aquifers) because of drainage, runoff, and infiltration. This portion of the withdrawn water, called return flow, becomes available for further use.

The following is a list of terms commonly used by the water industry and regulators in the United States (USGS, 2007b; USEPA, 2007b):

Public supply. Water withdrawn by public governments and agencies, such as a county water department, and by private companies, which is then delivered to users. Public suppliers provide water for domestic, commercial, thermoelectric power, industrial, and public water users. Most household water is delivered by a public water supplier.

Municipal (public) water system. A water system that has at least five service connections (such as households, businesses, or schools) or which regularly serves 25 individuals for at least 60 days out of the year.

Water supply system. The collection, treatment, storage, and distribution of potable water from source to consumer.

Water purveyor. A public utility, mutual water company (including privately owned), county water district, or municipality that delivers drinking water to customers.

Potable water. Water that is safe for drinking and cooking.

Water quality criteria. Levels of water quality expected to render a body of water suitable for its designated use. Criteria are based on specific levels of pollutants that would make the water harmful if used for drinking, swimming, farming, fish production, or industrial processes.

Water quality standards. State-adopted and EPA-approved ambient standards for water bodies. The standards prescribe the use of the water body and establish the water quality criteria that must be met to protect designated uses.

Public water use. Water supplied from a public water supply and used for such purposes as firefighting, street washing, and municipal parks and swimming pools.

Domestic water use. Water used for household purposes, such as drinking, food preparation, bathing, washing clothes, dishes, and pets, flushing toilets, and watering lawns and gardens. About 85 percent of domestic water is delivered to homes by a public-supply facility, such as a county water department. About 15 percent of the nation's population supply their own water, mainly from wells.

Commercial water use. Water used for motels, hotels, restaurants, office buildings, other commercial facilities, and institutions. Water for commercial uses comes both from public-supplied sources, such as a county water department, and self-supplied sources, such as local wells.

Industrial water use. Water used for industrial purposes in such industries as steel, chemical, paper, and petroleum refining. Nationally, water for industrial uses comes mainly (80 percent) from self-supplied sources, such as local wells or withdrawal points in a river, but some water comes from public-supplied sources, such as the county/city water department.

Irrigation water use. Water application on lands to assist in growing crops and pastures or to maintain vegetative growth in recreational lands, such as parks and golf courses.

Livestock water use. Water used for livestock watering, feedlots, dairy operations, fish farming, and other on-farm needs.

Sanitation. Control of physical factors in the human environment that could harm development, health, or survival.

Sanitary water (also known as gray water). Water discharged from sinks, showers, kitchens, or other nonindustrial operations, but not from commodes.

Wastewater. The spent or used water from a home, community, farm, or industry that contains dissolved or suspended matter.

Water pollution. The presence in water of enough harmful or objectionable material to damage the water's quality.

Treated wastewater. Wastewater that has been subjected to one or more physical, chemical, and biological processes to reduce its potential of being health hazard.

Reclaimed wastewater. Treated wastewater that can be used for beneficial purposes, such as irrigating certain plants.

Publicly owned treatment works (POTWs). A waste-treatment works owned by a state, unit of local government, or Indian tribe, usually designed to treat domestic wastewaters.

Wastewater infrastructure. The plan or network for the collection, treatment, and disposal of sewage in a community. The level of treatment will depend on the size of the community, the type of discharge, and/or the designated use of the receiving water.

Groundwater, with 93 percent of the total, is by far the most abundant and readily available source of freshwater on continents outside polar regions, followed by mountainous ice caps and glaciers, lakes, reservoirs, wetlands, and rivers (Fig. 1.1). About 1.5 billion people depended upon groundwater for their drinking water supply at the end of the twentieth century (WRI, 1998). The amount of groundwater withdrawn annually is roughly estimated at about 20 percent of global water withdrawals (WMO, 1997).

According to the United Nations Environment Programme (UNEP), annual global freshwater withdrawal has grown from 3790 km^3 (of which consumption accounted for 2070 km^3 or 61 percent) in 1995 to about 4430 km^3 (of which consumption accounted for 2304 km^3 or 52 percent) in 2000. In 2000, about 57 percent of the world's freshwater withdrawal, and 70 percent of its consumption, took place in Asia, where the world's major irrigated lands are located. In the future, annual global water withdrawal is expected to grow by about 10 to 12 percent every 10 years, reaching approximately 5240 km^3 by year 2025 (an increase of 1.38 times since 1995). Water consumption is expected to grow at a slower rate of 1.33 times. In the coming decades, the water withdrawal is projected to increase by 1.5 to 1.6 times in Africa and South America, while the smallest increase of 1.2 times is expected to occur in Europe and North America (UNEP, 2007; Harrison et al., 2001; Shiklomanov, 1999).

Agriculture is by far the biggest user of water accounting for 67 percent of the world's total freshwater withdrawal, and 86 percent of its consumption in the year 2000 (UNEP, 2007). In the United States, agriculture accounts for some 49 percent of the total freshwater use, with 80 percent of this volume being used for irrigation. In Africa and Asia, an estimated 85 to 90 percent of all the freshwater used is for agriculture. By 2025, agriculture is expected to increase its water requirements by 1.2 times, and the world's irrigation areas are projected to reach about 330 mha, up from approximately 253 mha in 1995 (Shiklomanov, 1999).

Industrial uses account for about 20 percent of global freshwater withdrawals. Of this, 57 to 69 percent is used for hydropower and nuclear power generation, 30 to 40 percent for industrial processes, and 0.5 to 3 percent for thermal power generation (Shiklomanov, 1999). In the industrial sector, the biggest share of freshwater is stored in artificial reservoirs for electrical power generation and irrigation. However, the volume of water evaporated from reservoirs is estimated to exceed the combined freshwater needs of industry and domestic consumption. This greatly contributes to water losses around the world, especially in the hot tropical and arid regions.

Domestic water supply accounted for about 13 percent of global water withdrawal in year 2000. Domestic water use in developed countries is on average about 10 times more than in developing countries. UNESCO estimates that on average a person in developed countries uses 500 to 800 L/d (300 m^3/yr) for all purposes compared to 60 to 150 L/d (20 m^3/yr) in developing countries. In developing countries of Asia, Africa, and Latin America, public water withdrawal represents just 50 to 100 L/person/d. In regions with insufficient water resources, this figure may be as low as 20 to 60 L/d. In large cities with a centralized water supply and an efficient canalization system, domestic consumption does not usually represent more than 5 to 10 percent of the total water withdrawal (UNEP, 2007).

In most regions of the world, the annual withdrawal or use of water is a relatively small part (less than 20 percent) of the total annual internally renewable water resources (Table 1.1). However, in water-scarce regions, as in the case of the Middle East and North Africa, this share averages 73 percent of the total water resources (Pereira et al., 2002; from The World Bank, 1992). The relevance of the problems of water scarcity is made clear when considering that estimates for the average annual growth of the population are the world's highest in the same regions (Table 1.2).

Agriculture has the highest share among water user sectors in low- and middle-income countries, while industry is the most important user in developed countries with temperate and humid climates.

Water supply and sanitation face different problems in urban and rural settings. At present, more than one-third of the rural population is estimated to have no access to safe drinking water supply and a significant number of people do not have access to the minimum required levels (The World Bank, 2000; WRI, 1998). Moreover, almost 80 percent of the rural population is estimated to have no access to adequate sanitation, totaling 1.3 billion people in rural India and China alone (UNESCO-WWAP, 2003). This means that about 1.8 billion people in rural areas have yet to receive new and increased domestic water supply and sanitation facilities over the next few decades, which will require substantial increases in domestic withdrawals.

With growing urbanization across the world, water supply and sanitation will become an increasingly urban issue and the main challenge for the water industry worldwide. In 2007, the urban population worldwide has reached an estimated 50 percent, and this increasing trend will continue. The following megacities (defined as having more than 10 million inhabitants) currently depend upon groundwater for water supply to varying degrees: Mexico City, Kolkata (formerly Calcutta), Shanghai, Buenos Aires, Tcheran, London, Jakarta, Dhaka, Manila, Cairo, Bangkok, and Beijing (Morris et al., 2003). It is projected that in 2015 the number of megacities will reach 23 worldwide, with only 4 being in developed countries. The combined population of all these megacities will reach

Country Group	Total Annual Internal Renewable Water Resources (10^6 m^3)	Total Annual Water Withdrawal (10^6 m^3)	Annual Withdrawal as a Share of Total Water Resources (%)	Per Capita Annual Internal Renewable Water Resources (m^3)	Sectorial Withdrawal as a Share of Total Water Resources		
					Agriculture (%)	Domestic (%)	Industry (%)
Low and middle income	28,002	1,749	6	6,732	85	7	8
Sub-Saharan Africa	3,713	55	1	7,488	88	8	3
East Asia and Pacific	7,915	631	8	5,009	86	6	8
South Asia	4,895	569	12	4,236	94	2	3
Europe	574	110	19	2,865	45	14	42
Middle East and North Africa	276	202	73	1,071	89	6	5
Latin America and the Caribbean	10,579	173	2	24,390	72	16	11
High income	8,368	893	11	10,528	39	14	47
OECD members	8,365	889	11	10,781	39	14	47
Other	4	4	119	186	67	22	12
World	40,856	3,017	7	7,744	69	9	22

From Pereira et al., 2002; source of data: The World Bank, 1992.

TABLE 1.1 World Availability of Water Resources

Country Group	Population (Millions)						Average Annual Growth (Percent)				
	1973	1980	1990	2000	2030	1965–73	1973–80	1980–90	1990–00	2000–30
Low and middle income	2,923	3,383	4,146	4,981	7,441	2.5	2.1	2.0	1.9	1.4
Sub-Saharan Africa	302	366	495	668	1,346	2.7	2.8	3.1	3.0	2.4
East Asia and Pacific	1,195	1,347	1,577	1,818	2,378	2.6	1.7	1.6	1.4	0.9
South Asia	781	919	1,148	1,377	1,978	2.4	2.4	2.2	1.8	1.1
Europe	167	182	200	217	258	1.1	1.2	1.0	0.8	0.6
Middle East and North Africa	154	189	256	341	674	2.7	3.0	3.1	2.9	2.3
Latin America and the Caribbean	299	352	433	516	731	2.6	2.4	2.1	1.8	1.2
High income	726	766	816	859	919	1.0	0.8	0.6	0.5	0.2
OECD members	698	733	111	814	863	0.9	0.7	0.6	0.5	0.2
World	3,924	4,443	5,284	6,185	8,869	2.1	1.8	1.7	1.6	1.2

From Pereira et al., 2002; source of data: The World Bank, 1992.

TABLE 1.2 Population and Average Annual Growth

9.6 percent of the world's urban population, accounting for slightly more than 374 million people (UN HABITAT, 2003). Growing slums and informal settlements surround most of the megacities and other large cities in the developing countries, which are of particular concern. There is widespread water and environmental contamination from human waste in these areas because of lack of adequate drinking water supplies, sanitation, and sewage treatment services (CSD, 2004). At the beginning of the twenty-first century, an estimated 924 million people lived in slums around the world (UN HABITAT, 2003).

International Water Management Institute (IWMI) projects, under one scenario, that most developing regions will more than double their water withdrawals for domestic and industrial uses between years 1995 and 2025. Except for the African region, this level of increase would ensure an average per capita domestic supply above the basic water requirement (BWR) of 50 L/person/d. The BWR is the recommended volume of water, independent of climate, technology, and culture, needed to satisfy domestic needs— drinking, sanitation, bathing, and cooking (Gleick, 1996). In Africa, however, per capita domestic water withdrawals at present are significantly below the BWR. To raise average per capita domestic supply even to the level of BWR, Africa will have to increase its total domestic water supply by 140 percent (Molden et al., 2001).

The world's rural population is projected to grow at a slower rate over the coming decades, due to rapid urbanization and industrialization, particularly in developing countries. However, the rural population in Africa (AFR), South Asia (SA), and Middle East and North Africa (MENA) regions is projected to grow by 56 percent, 18 percent, and 20 percent, respectively. At the same time, the rural population in the East Asia and Pacific (EAP), Latin America and the Caribbean (LAC), and Europe and Central Asia (ECA) regions is projected to decline by 12 percent, 4 percent, and 27 percent, respectively. Even with the assumed slow growth in the overall rural population, more than 3.3 billion people are projected to live in rural areas by 2025, with more than 90 percent of them in South Asia, EAP, Africa, and the MENA regions (Molden et al., 2001).

Despite the expected large growth in water withdrawals for the domestic and industrial sectors, agriculture will still remain the dominant water user in developing countries. The level of water use in the agricultural sector will be influenced by goals of self-sufficiency and food security at local, regional, and national levels. In the past, food self-sufficiency has been the major goal of most developing countries. This has helped developing countries increase food production, improve overall food availability for rural households, and reduce rural unemployment and has had overall positive effects in terms of reducing poverty (Molden et al., 2001).

1.3.1   Use of Water in the United States

Estimates of water use in the United States indicate that about 408 billion gallons per day (abbreviated Bgal/d; 1000 million gallons per day; note that 1 gal equals 3.8 L) were withdrawn for all uses during year 2000. This total has varied less than 3 percent since 1985 as withdrawals have stabilized for the two largest uses—thermoelectric power and irrigation. Fresh groundwater withdrawals (83.3 Bgal/d) during 2000 were 14 percent more than that during 1985. Fresh surface water withdrawals during 2000 were 262 Bgal/d, varying less than 2 percent since 1985 (Hutson et al., 2004). Figure 1.7 shows withdrawal of surface water and groundwater between 1950 and 2000, together with the population trend. Figure 1.8 shows water withdrawals for different uses, for the same period.

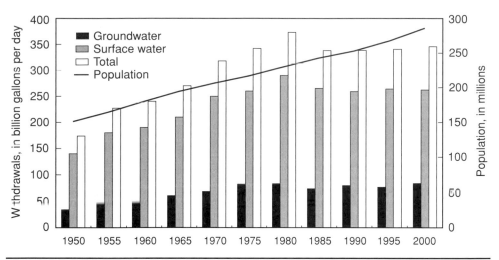

FIGURE 1.7   Trends in population and freshwater withdrawals by source, 1950–2000. (From Hutson et al., 2004.)

About 195 Bgal/d, or 48 percent, of all freshwater and saline water withdrawals during 2000 were used for thermoelectric power. Most of this water was derived from surface water and used for once-through cooling at power plants. About 52 percent of fresh surface water withdrawals and about 96 percent of saline water withdrawals were for thermoelectric power use. Withdrawals for thermoelectric power have been relatively stable since 1985.

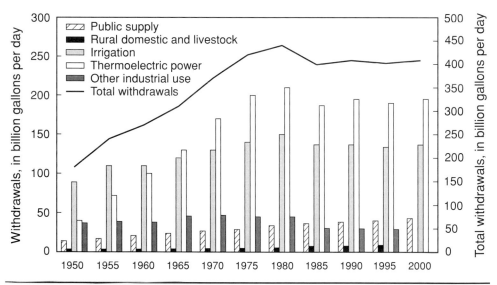

FIGURE 1.8   Trends in total water withdrawals by water use category, 1950 to 2000. Total withdrawals for rural domestic and livestock and for "other industrial use" are not available for 2000. (From Hutson et al., 2004.)

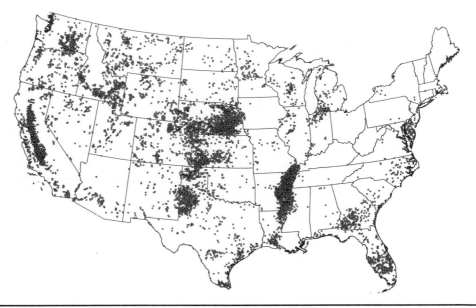

FIGURE 1.9   Distribution of irrigated land in farms. One dot represents 5000 irrigated acres. (From Gollehon and Quinby, 2006.)

Irrigation remained the largest use of freshwater in the United States and totaled 137 Bgal/d for 2000. Since 1950, irrigation has accounted for about 65 percent of total water withdrawals, excluding those for thermoelectric power. Historically, more surface water than groundwater has been used for irrigation. However, the percentage of total irrigation withdrawals from groundwater has continued to increase—from 23 percent in 1950 to 42 percent in 2000. Total irrigation withdrawals were 2 percent more in 2000 compared to that in 1995 because of a 16 percent increase in groundwater withdrawals and a small decrease in surface water withdrawals. Irrigated acreage more than doubled between 1950 and 1980, then remained constant before increasing nearly 7 percent between 1995 and 2000 in response to drought in some states, especially in the southwest. In recent years, national irrigated area reached a plateau at about 55 million acres as the continuing growth in eastern states has been offset by declines in western irrigation. The number of acres irrigated with sprinkler and microirrigation systems has continued to increase and now comprises more than one-half the total irrigated acreage in the United States (Hutson et al., 2004).

In general, there is an increasing reliance on irrigation in the humid East, and a northward redistribution of irrigation in the West (Fig. 1.9). During 1990s and early 2000s, large concentrations of irrigation have emerged in humid areas—Florida, Georgia, and especially in the Mississippi River Valley, primarily Arkansas and Mississippi (Gollehon and Quinby, 2006). Groundwater supplied most of the irrigation water in the eastern 37 states—the area experiencing the largest irrigation growth in the last decade of the twentieth century. Table 1.3 shows agricultural withdrawals for different regions in year 2000. Most withdrawals occur in the arid western states where irrigated production is concentrated. In 2000, about 85 percent of total agricultural withdrawals occurred in a 19-state area encompassing the plains, mountain, and Pacific regions. In the mountain

Region Share	Number of States	Agricultural Water Withdrawals		Components of Agricultural Withdrawals			Source of Agricultural Withdrawals	
		Percent of Total Withdrawals (%)	Quantity (1000 Acre-Feet Per Year)	Irrigation (%)	Livestock and Aquaculture (%)		Ground Water (%)	Surface Water (%)
Pacific	5	80	45,879	98	2		34	66
Mountain	8	91	64,209	96	4		20	80
Plains	6	49	25,901	97	3		80	20
South	7	30	19,054	95	5		73	27
North-Central & East	24	3	4,409	81	19		72	28
U.S. total[1]	50	41	159,558	96	4		41	59

[1] Excludes water withdrawals in the U.S. Virgin Islands, Puerto Rico, and the District of Columbia. From Gollehen and Quinby, 2006.

TABLE 1.3 Agricultural Withdrawals in Different Regions of the United States, in 2000

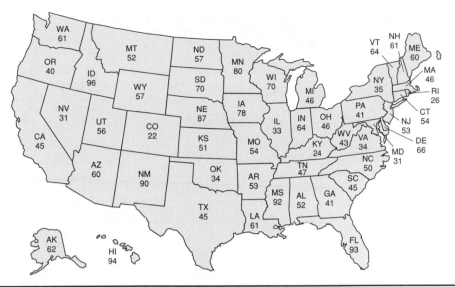

FIGURE 1.10 Estimated percentage of population in a state using groundwater as drinking water in 1995. (From USGS, 1998.)

region, more than 90 percent of the water withdrawn is used by agriculture, almost all (96 percent) for irrigation.

Public supply withdrawals were more than 43 Bgal/d in 2000. Public supply withdrawals during 1950 were 14 Bgal/d. During 2000, about 85 percent of the population in the United States obtained drinking water from public suppliers, compared to 62 percent during 1950. The percentage of groundwater use for public supply increased from 26 percent in 1950 to 40 percent in 1985 and has remained at slightly less than 40 percent since (Hutson et al., 2004). Figure 1.10 shows that groundwater is an important source of drinking water for every state.

According to the USEPA 2001 data on public water supply for federal fiscal 2001 year, the total population served by community and noncommunity water systems using groundwater as the primary water source was 101,820,639 (Williams and Fenske, 2004). The total number of such water systems was 150,793. Sixteen states had more than 1000 systems, and all but one had more than 100 systems (Rhode Island had 59). Twenty-six states had more than 1 million population served by public water systems using mostly groundwater, with the top five being Florida (>14 million), California (>9 million), Texas (>7 million), New York (>5 million), and Michigan (>3 million). Ten additional states had more or very close to three million population served primarily by groundwater: Illinois, Indiana, Louisiana, Minnesota, Mississippi, New Jersey, Ohio, Pennsylvania, Washington, and Wisconsin.

The estimated withdrawals for self-supplied domestic use increased by 71 percent between 1950 and 2000. The self-supplied domestic population was 57.5 million people for 1950, or 38 percent of the total population. During 2000, 43.5 million people, or 15 percent of the total population, were self-supplied (Hutson et al., 2004).

Self-supplied industrial withdrawals totaled nearly 20 Bgal/d in 2000, or 12 percent less than that in 1995. Compared to 1985, industrial self-supplied withdrawals declined by 24 percent. Estimates of industrial water use in the United States were largest during the years from 1965 to 1980, but during 2000, estimates were at the lowest level since reporting began in 1950. Combined withdrawals for self-supplied domestic, livestock, aquaculture, and mining were less than 13 Bgal/d for 2000 and represented about 3 percent of total withdrawals.

California, Texas, and Florida accounted for one-fourth of all water withdrawals for 2000. States with the largest surface water withdrawals were California, which had large withdrawals for irrigation and thermoelectric power, and Texas, which had large withdrawals for thermoelectric power. States with the largest groundwater withdrawals were California, Texas, and Nebraska, all of which had large withdrawals for irrigation (Hutson et al., 2004).

Total withdrawals have remained about 80 percent surface water and 20 percent groundwater during the 1950 to 2000 period. The portion of surface water withdrawals that was saline increased from 7 percent in 1950 to 20 percent in 1975 and has remained about 20 percent since. The percentage of groundwater that was saline never exceeded about 2 percent. The percentage of total withdrawals that was saline water increased from a minor amount in 1950 to as much as 17 percent during 1975 to 1990.

Water withdrawals are not the only measure of water use. Consumptive use—the water not returned to the immediate water environment—is much greater for agriculture than any other sector, both in total and as a share of water withdrawn. Estimates available from 1960 through 1995 show that agriculture accounts for more than 80 percent of the nation's consumptive use because a high share of applied irrigation water is used by plants for evapotranspiration (building of biomass), with little returning to surface or groundwater. Water diverted for cooling thermoelectric plants tends to be used as a thermal sink, with much of it returned to rivers and streams. Greater irrigation withdrawals do not necessarily translate into greater consumptive use per irrigated acre. The difference between withdrawals and consumptive use highlights the importance of losses, runoff, and return flows (Gollehon and Quinby, 2006).

1.3.2   Use of Water in Europe

The principal source of extracted freshwater in Europe is surface water with the remainder coming from groundwater sources and only minor contributions from desalination of seawater such as in Spain. According to a 1995 survey, of the total water abstracted in the European Union (EU), about 29 percent was groundwater (Krinner et al., 1999; from EEA, 1995). However, in many EU countries groundwater is the main source for public water supply because it is readily available and generally of high quality, resulting in the relative low cost of treatment and supply compared to surface water (Nixon et al., 2000; EEA, 1998). The proportion of groundwater use for public water supply in different EU countries is given in Table 1.4 (Krinner et al., 1999).

In the period from 1990 to 2001, the most marked change in total water extraction occurred in the southeastern European countries (Turkey, Cyprus, and Malta) where total water withdrawals increased by 40 percent, whereas in the northern, central, and eastern countries it decreased by 40 percent. Total water extraction in the EU-15 Member States fell by 8 to 9 percent both in the northern and in the southern countries (EEA, 2005). It appears that the drop in water extraction is a result of droughts in recent years, which have increased public awareness that water is a finite resource. The apparent downturn can also be attributed to a shift in water management strategies, moving toward demand

Country	Surface Water	Groundwater
Austria	0.7	99.3
Belgium		
Brussels	100.0	0.0
Flanders	48.5	51.5
Denmark	0.0	100.0
Finland	44.4	55.6
France	43.6	56.4
Germany	28.0	72.0
Greece	50.0	50.0
Ireland	50.0	50.0
Italy	19.7	80.3
Luxembourg	31.0	69.0
Netherlands	31.8	68.2
Portugal	20.1	79.9
Spain	77.4	21.4
Sweden	51.0	49.0
United Kingdom	72.6	27.4
Norway	87.0	13.0
Iceland	15.9	84.1
Switzerland	17.4	82.6
Czech Republic	56.0	44.0

Simplified from Krinner et al., 1999.

TABLE 1.4   Apportionment of Public Water Supply, in Percent, Between Groundwater and Surface Water

management, reducing losses, using water more efficiently, and recycling (Krinner et al., 1999).

The economic transition in central and eastern European countries during the 1990s had a large impact on water consumption in the region. The decrease in industrial activity, especially in water-intensive heavy industries, such as steel and mining, led to decreases of up to 70 percent in water extraction for industrial use. The amount of water extraction for agriculture also decreased by a similar percentage. Abstraction for public water supply declined by 30 percent after the fees were increased to reflect water costs and water meters were installed in houses (EEA, 2005).

On average, 37 percent of total water use in the EU countries is for agriculture, 33 percent for energy production (including cooling), 18 percent for urban use, and 12 percent for industry (excluding cooling). Total combined water withdrawal for agriculture remained almost unchanged over the period, while those for urban use and energy decreased by 11 percent and for industry by 33 percent (EEA, 2005). Tourism, one of the fastest increasing socioeconomic activities in Europe, places severe, often seasonal, pressures on water resources, especially in southern Europe.

The scale and importance of irrigation in Europe is most significant in regions that have semiarid climates. In these countries, such as Cyprus, Malta, Greece, parts of Spain,

Region	% Share of World Population (as of 2001)	Population Density (pop/km²)	Average Annual Per Capita Availability of Water Resources (m³)
Africa	13	27	5,157
Asia	61	117	3,159
Europe	12	32	9,027
Latin America & the Caribbean	9	26	27,354
North America	5	15	16,801
Oceania	1	4	53,711
World	—	45	7,113

From Vordzorgbe, 2003; source of data: UN, 2001.

TABLE 1.5 Population and Water Resource Features of Regions of the World

Portugal, Italy, and Turkey, irrigation accounts for more than 60 percent of water use, a large share coming from groundwater extraction. In the more humid and temperate EU member countries irrigation is carried out mainly to complement natural rainfall, and its share of total water use is generally less than 10 percent (EEA, 2005).

1.3.3 Use of Water in Africa

Africa is home to about 13 percent of the world's population, but has only about 9 percent of the world's water resources (UNEP, 2002). Average annual per capita availability of water resources in Africa is lower than the world average and higher than that of only Asia (see Table 1.5). This low level of water availability in Africa is due to three basic factors (Vordzorgbe, 2003):

1. *A significant decline in the average rainfall since the late 1960s.* In recent times, most of the continent has experienced increased aridity as mean annual rainfall has reduced by 5 percent to 10 percent between 1931–1960 and 1968–1997. The decline in Sahelian rainfall has been the largest sustained decline recorded anywhere in the world since instrumental measurements began, while deviations from the trend have been larger than in other arid regions of the world.

2. *Low runoff due to high evaporative losses.* Total runoff as a percentage of precipitation is the lowest in the world, at about 20 percent, compared to 35 percent for South America and about 40 percent for Asia, Europe, and North America.

3. *High variability of supply, due to highly variable rainfall.* For example, precipitation ranges from almost zero over some desert areas in Namibia and parts of the Horn to very high levels in the western equatorial areas. The major outcome of these extremes of rainfall is a high frequency of floods and droughts on the continent. The high variability of rainfall and river flow also reduces runoff and exacerbates vulnerability to erosion and desertification. This extreme variability of climate and hydrological conditions imposes high costs on livelihoods and raises the risks of development interventions.

Only about 4 percent of the nearly 4 million km^3 of renewable water available annually is used in Africa. Except for the northern Africa, the total amount of water withdrawn in all other subregions for use in agriculture, public water supply, and industry show that at both continental and subregional levels the withdrawals are rather low in relation to both rainfall and internal renewable resources. This may reflect a low level of development and use of water resources in the continent. However, variability in rainfall results in frequent bouts of water scarcity and, during these times, demand exceeds supply (UN Water/Africa, 2006).

In 2000, because of inadequate water storage, processing, and distribution systems, about 36 percent of the population did not have access to potable water, but the deprivation is higher in rural areas, where as much as 50 percent lacked access to safe water. Also, because of low investment in water supply and distribution infrastructure, increasing demand and weak water management policies, access to water is highly skewed in favor of urban consumers and some agricultural and industrial users (Vordzorgbe, 2003).

It is estimated that more than 75 percent of the African population uses groundwater as the main source of drinking water supply. This is particularly so in North African countries, such as Libya, Tunisia, and parts of Algeria and Morocco, as well as in southern African countries, including Botswana, Namibia, and Zimbabwe. However, groundwater accounts for only about 5 percent of the continent's total renewable water resources and the groundwater withdrawal is mainly from the nonrenewable aquifer storage. In South Africa, for example, groundwater accounts for only 9 percent of the renewable water (UN Water/Africa, 2006).

Owing to the highly variable levels of rainfall in Africa, large numbers of people are dependent on groundwater as their primary source of freshwater for various uses (UNEP, 2002). For example, in Libya and Algeria 95 percent and more than 60 percent of all withdrawals, respectively, are from groundwater. Algeria, Egypt, Libya, Mauritius, Morocco, South Africa, and Tunisia are increasingly looking at the use of desalinated water to assist in meeting their withdrawal requirements (UNEP, 2002).

Water Use in Lake Chad Basin

Water uses in eight countries of the Lake Chad Basin are typical of nonindustrialized, undeveloped, and developing countries in Africa and around the world. The majority of freshwater consumed in the region is used for agriculture followed by domestic use (Fig. 1.11). In Africa, Nigeria is the sixth largest user of water by volume (4 billion m^3/yr; Revenga and Cassar, 2002).

In the Sudan sector of the basin (West Darfur), more than 50 percent of water is obtained from dug wells with bucket collection (The World Bank, 2003). Women have to travel great distances in order to gather water for drinking, cooking, and other everyday activities. Reservoirs formed by the Tiga and Challawa Gorge dams of the Kano City Water Supply (KCWS) supply the large Nigerian urban centre of Kano City for domestic and industrial purposes (GIWA, 2004).

Traditional agriculture in the basin is predominantly rain-fed. The rivers in the Chari-Logone and Komadugu-Yobe subsystems support flood farming and recessional farming. Farmers in downstream areas therefore depend largely on river flow because rainfall is low and variable. The many large irrigation projects are located predominantly in the Komadugu-Yobe Basin.

According to GIWA (2004), there is little information concerning groundwater, but it is considered to be abundant, especially in the unconfined regional aquifers. However,

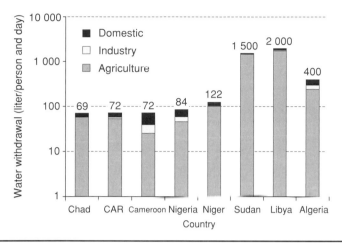

FIGURE 1.11 Freshwater withdrawal per person per day by economic sector in countries of the Chad Lake Basin, Africa. (From GIWA, 2004; source of data The World Bank, 2002.)

because of the recent declines in aquifer recharge due to prolonged droughts and reduction in river flows, aquifers are currently vulnerable to overextraction exceeding their safe yield. Surface water scarcity during these droughts as well as adaptation strategy increased the extraction of groundwater for human, agricultural, and pastoral purposes (GIWA, 2004; Thieme et al., 2005). There has been an indiscriminate drilling of wells that has led to a decrease in groundwater reserves. Groundwater drawdowns of several tens of meters have been reported in the Maiduguri area of Nigeria due to overpumping. Isiorho et al. (2000) estimate that 10 to 25 percent of water in the region is used inefficiently and attempts to improve the situation have achieved little. The droughts of the 1980s triggered the mass drilling of 537 wells between 1985 and 1989. This rapid development resulted in unsatisfactory logging of wells by several contractors who were not supervised. Most of these deep wells are uncapped and free-flowing. Normally, the local authorities cap artesian wells, but local people uncap them and allow the water to flow out and cool so that their animals can use it. This free flow of water is very inefficient and results in vast amounts of water being lost due to the high rates of evaporation in the region (Isiorho et al., 2000). Water points at Ala near Marte (Nigeria), monitored on a routine basis by the Lake Chad Basin Commission, have shown a sharp decline of about 4.5 m within a period of 1 year attributable to the general decline in the artesian pressure within the basin. Most desert species have also disappeared due to the declining water table (GIWA, 2004).

1.3.4 Use of Water in China and India

China and India have an estimated population of 1.32 and 1.13 billion people respectively or, combined, more than one-third of the world's total population of about 6.6 billion in 2007. The two countries have some of the highest rates of development growth in just about every category, which puts enormous pressure on natural resources, including water. It is therefore not surprising that the international community closely watches development trends in China and India, including their impacts on global economy, politics, and the environment.

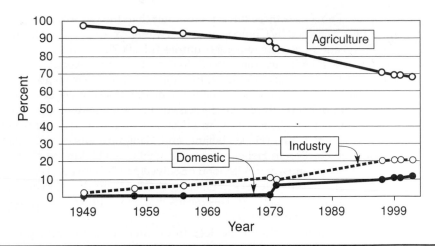

FIGURE 1.12 Percentage of water use in China for agriculture, industry, and domestic water supply between 1949 and 2002. (Data from Jin et al., 2006; original data from the Ministry of Water Resources, and Liu and Chen, 2001.)

Total water use in China has increased fivefold since 1950s. It was 103 km^3 in 1949 and increased to about 550 km^3 in recent years (Jin et al., 2006). As illustrated in Fig. 1.12, the percentage of water use in agriculture has decreased from 97 percent in 1949 to 68 percent in 2002, as domestic and industrial uses have substantially increased. In 2004, agricultural uses consumed about 359 billion m^3 (359 km^3) of water, accounting for 65 percent of total national water use. Of this, some 323 km^3, or 90 percent, went to farmland irrigation (Li, 2006).

China has experienced serious water shortages over the past two decades. According to a recent report on the country's status of water resources funded by the Chinese Ministry of Science and Technology, water shortages in China cause direct economic losses averaging 280 billion yuan (35 billion U.S. dollars (USD)) each year, which is 2.5 times more than the loss caused by floods (Li, 2006). Reportedly, some 350 million people lack access to reliable water supply.

Recognizing the seriousness of water availability, use, and management, the Chinese government passed a new regulation in 2006, updating its system of use permits and stipulating charges for water consumption in agriculture. According to officials in the State Council (China's parliament), this move is expected to enforce water-saving measures in irrigation and motivate farmers to economize on water use. Improving water efficiency in agriculture is considered the most effective way to achieve significant water savings in China. Most of China's fields use flood irrigation methods that can result in significant waste, with 1 ha of farmland typically requiring 20,000 to 30,000 m^3 of water a year (Li, 2006).

At the beginning of 2007, China's National Development and Reform Commission (NDRC), Ministry of Water Resources, and Ministry of Construction jointly released a water-saving plan to cut the nation's water use per unit of gross domestic product (GDP) by 20 percent within 5 years. The ambitious plan is expected to save China a total of 69 km^3 of water, mainly in the agriculture and industry sectors (Li, 2007). China's industrial water efficiency lags far behind many other countries. To generate 10,000 yuan (1250

USD) in GDP, China uses three times more water than the world average. Only 60 to 65 percent of the water used by Chinese industries was recycled or reused in 2004, compared to 80 to 85 percent in most developed countries (Li, 2007).

Groundwater plays a key role in the China's water supply. About 70 percent of drinking water and 40 percent of agricultural irrigation water come from groundwater (Zhan, 2006). Available groundwater, however, is not evenly distributed: about two-thirds are in the south and only one-third is in the semiarid north, where most of the agricultural irrigation is taking place. This has led to a dramatic groundwater overexploitation, particularly in northern China (Hebei, Beijing, and Tianjin provinces). For example, actual groundwater abstraction in Hebei province in 1999 was 14.9 km^3 (including 2.2 km^3 of slightly brackish groundwater with total dissolved solids between 1 and 5 g/L), but its recharge of fresh groundwater is estimated at 13.2 km^3 and the allowable yield of fresh groundwater is only 9.95 km^3/yr. This means that about 1.8 km^3 of fresh groundwater is overexploited every year (Jin et al., 2006).

Groundwater level depth for the deep freshwater in North China provinces was in the order of 20 to 100 m in 2001, but it was near the surface or artesian in the 1960s. Rates of groundwater level (hydraulic head) decrease for the deep confined freshwater aquifers are 1 to 2 m/yr. Cangzhou, a coastal city in eastern Hebei plain, is one of the cities with most serious hydraulic head decline of deep confined aquifers: 100 m (330 ft) since the 1960s. The hydraulic head decline has resulted in land subsidence, degradation of water quality, and increased costs of pumping (Jin et al., 2006).

A 2006 study by the Hebei Bureau of Hydrology and Water Resources Survey estimates that the shallow groundwater table in China's central Hebei Plain, south of Beijing, will drop an additional 16.2 m (more than 50 ft) on average by 2030, while the hydraulic head in the confined aquifers will fall additional 39.9 m (more than 130 ft) on average. These projections are based on the ongoing rates of groundwater depletion in the vast area where groundwater accounts for 90 percent of the regional water supply. This severe groundwater overexploitation has led to the shortfall between water supply and rapidly rising demands from agriculture, industry and urban residents, including two "megacities" Beijing and Tianjin, with a combined population of 26 million people (Liu, 2006).

According to the Indian Ministry of Water Resources, the total renewable groundwater resources of India have been estimated at about 433 billion m^3 (433 km^3), whereas this volume for surface water is 690 km^3 (Ministry of Water Resources, 2007a, 2007b). Table 1.6 shows that the total estimated renewable water resources (1123 km^3) will not be sufficient to satisfy India's projected water requirements of 1447 km^3 in year 2050.

Currently around 85 percent of all water use in India is for agriculture. Groundwater is the source of irrigation for about 57 percent of the irrigated area. Most of the groundwater development (about 70 percent) has been concentrated in the Indus basin, the basin of the westerly flowing rivers in Kutch and Saurashtra, and in the western parts of the Ganga basin (Amarasighne et al., 2005). Small domestic farms ("minor irrigation sector") are mostly dependent on groundwater for irrigation and cover about two-thirds of the country's total irrigation capacity. Total estimated annual groundwater withdrawal for agriculture, domestic, and industrial purposes was 231 km^3 as of February 2004 (Ministry of Water Resources, 2007b). According to The World Bank, 70 percent of India's irrigation water and 80 percent of its domestic water supply come from groundwater (The World

Sector	Year		
	2000	**2025**	**2050**
Domestic	42	73	102
Irrigation	541	910	1072
Industry	8	23	63
Energy	2	15	130
Other	41	72	80
Total	**634**	**1093**	**1447**

From Central Water Commission, 2007.

TABLE 1.6 Estimated Annual Water Requirements in India for Different Uses, in Cubic Kilometers

Bank, 2005). It therefore appears that there is a discrepancy between the various estimates of current and future water requirements, the actual groundwater withdrawal, and the renewable groundwater availability.

Keeping a provision for about 71 km^3/yr of groundwater for other uses, the Indian Government estimates that 361 km^3/yr of groundwater is available for irrigation. The current net groundwater withdrawal for irrigation is estimated at 150 km^3/yr (Ministry of Water Resources, 2007a), which is about 40 percent of the available renewable groundwater. Based on these numbers, it is estimated that India as a whole has about 211 km^3 of renewable groundwater resources available for additional growth in agriculture. Most projected water requirements in years 2025 and 2050 would therefore have to be met using the surface water and nonrenewable groundwater resources.

There are large differences in surface water and groundwater availability and utilization between regions of the country. Currently, there are 5723 groundwater assessment units (called blocks, mandals, or talukas), of which almost 30 percent are "nonsafe" in terms of groundwater extraction (i.e., the units are semicritical, critical, or overexploited). The number of overexploited and critical units is the highest in states of Andhra Pradesh, Delhi, Gujarat, Haryana, Karnataka, Punjab, Rajasthan, and Tamil Nadu (CGWB, 2006).

Recognizing the seriousness of water supply issues across the country, the Indian Government has planned and implemented various measures, including constitution of the Central Ground Water Authority with a mandate to regulate and control groundwater development and management. The ongoing activities include registration of "ground water structures" (e.g., water wells), registration of water well drilling agencies to develop a microlevel database on groundwater development and to control indiscriminate drilling activity in the country, regulation of groundwater development by the industry, promotion of artificial aquifer recharge, and general education and outreach. In addition, the Ministry of Water Resources is engaged in the planning and implementation of huge interbasin transfers of surface water aimed at curbing water shortages primarily for public water supply.

Despite all these efforts, it appears that the water supply problems in India are persisting, if not growing, as noted by various international agencies and reported by the media. In a draft report on India's water economy, written by a number of India's eminent consultants, The World Bank (2005) warns that "India faces a turbulent water future. The current water development and management system is not sustainable: unless dramatic changes are made—and made soon—in the way in which government manages water, India will have neither the cash to maintain and build new infrastructure, nor the water required for the economy and for people." World news media, following the report's release, have been much less diplomatic, often including disturbing stories:

> "There's virtually no country in the world that lives with a system as bad as you have here," John Briscoe, author of the Bank's draft country report on India, told a media conference in New Delhi. (AFP, 2005)

> "What has happened in the last 20 or 30 years is a shift to self-provision. Every farmer sinks a tubewell and every house in Delhi has a pump pumping groundwater," said Briscoe, an expert on water issues at the World Bank. "Once that water stops you get into a situation where towns will not be able to function." (AFP, 2005)

> The report says that India has no proper water management system in place, its groundwater is disappearing and river bodies are turning into makeshift sewers. (AFP, 2005)

> "Estimates reveal that by 2020, India's demand for water will exceed all sources of supply," the report says. "There is no question that the incidence and severity of conflicts (over water) has increased sharply in recent times . . . There is a high level of vitriol in the endemic clashes between states on inter-state water issues." (AFP, 2005)

> It is a rare morning when water trickles through the pipes. More often, not a drop will come. So Prasher will have to call a private water tanker, wait for it to show up, call again, wait some more and worry about whether there are enough buckets filled in the bathroom in case no water arrives. Prasher has the misfortune of living in a neighborhood on Delhi's poorly served southern fringe. As the city's water supply runs through an 8,960-kilometer network of battered public pipes, an estimated 25 to 40 percent leaks out. By the time it reaches Prasher, there is hardly enough. On average, she gets no more than 13 gallons a month from the tap and a water bill that fluctuates from $6 to $20, at its whimsy, she complains, since there is never a meter reading anyway. That means she has to look for other sources, scrimp and scavenge to meet her family's water needs. She buys 265 gallons from private tankers, for about $20 a month. On top of that she pays $2.50 toward the worker who pipes water from a private tube-well she and other residents of her apartment block have installed in the courtyard.
>
> Her well water has long turned salty. The water from the private tanker is mucky brown. Still, Prasher said, she can hardly afford to reject it. "Beggars can't be choosers," she said. "It's water." (Sengupta, *The New York Times*, 2006)

1.4 Water Scarcity

Water scarcity exists when the amount of water withdrawn from lakes, rivers, or groundwater is such that water supplies are not adequate to satisfy all human or ecosystem requirements, resulting in increased competition between water users and demands. An area is under water stress when annual water supply is below 1700 m^3 (450 thousands gallons) per person. When annual water supply is below 1000 m^3 (264,000 gallons) per person, the population experiences water scarcity (UNEP, 2007).

As water use increases, water is becoming scarce not only in arid and drought-prone areas, but also in regions where rainfall is relatively abundant. The concept of water scarcity is now viewed under the perspective of the quantities available for economic and social uses as well as in relation to water requirements for natural and human-made ecosystems. The concept of scarcity also embraces the quality of water because degraded water resources are unavailable or at best only marginally available for use in human and natural systems (Pereira et al., 2002).

Water scarcity may be permanent or temporary, may be caused by natural conditions (e.g., aridity, drought), or it may be human-induced (desertification, overexploitation of water resources). Worldwide, agriculture is the sector that has the highest demand for water. Because of its large water use, irrigated agriculture is often considered the main cause for water scarcity. Irrigation is accused of misuse of water, of producing excessive water wastes, and of degrading water quality. However, irrigated agriculture provides the livelihood of an enormous part of the world's rural population and supplies a large portion of the world's food (Pereira et al., 2002). Many countries also regard agricultural production as a matter of national security or geopolitical strategy, and thus support irrigation, including in desert areas where depletion of nonrenewable groundwater resources is the only option (Fig. 1.13).

Some 460 million people—more than 8 percent of the world's population—live in countries using so much of their freshwater resources that they can be considered highly water stressed. A further 25 percent of the world's population lives in countries approaching a position of serious water stress (UNEP, 2007; UNCSD, 1999; WMO, 1997).

Many African countries are facing alarming water shortages, which will affect nearly 200 million people. By the year 2025, it is estimated that nearly 230 million Africans will be facing water scarcity and 460 million will live in water-stressed countries (Falkenmark, 1989). As discussed by Vordzorgbe (2003),

> the future is not salutary: water stress will increase in Africa due to the influence of climate factors (increasing frequency of flood and drought and water system stress) and anthropogenic causes of increasing use (from rising population, expanding urbanization, increasing economic development, unplanned settlement patterns), inadequate storage and recycling, lack of knowledge to address concerns and weak governance of the water sector. Africa has the highest population growth rate and the fastest rate of increase in urban population in the world. This has implications for demand, quality and sustainability of water resources.

Because of the lack of renewable freshwater resources, Bahrain, Kuwait, Saudi Arabia, and the United Arab Emirates have resorted to the desalinization of seawater from the Gulf. Bahrain has virtually no freshwater (Riviere, 1989). Three-quarters of Saudi Arabia's freshwater comes from nonrenewable groundwater, which is reportedly being depleted at an average of 5.2 km^3/yr (Postel, 1997). Lester Brown, in his book *Plan B 2.0 Rescuing a Planet Under Stress and a Civilization in Trouble* (2006) writes: "When the Saudis turned to their large fossil aquifer for irrigation, wheat production climbed from 140,000 tons in 1980 to 4.1 million tons in 1992. But with rapid depletion of the aquifer, production dropped to 1.6 million tons in 2004. It is only a matter of time until irrigated wheat production ends."

According to Population Action International, based upon the UN Medium Population Projections of 1998, more than 2.8 billion people in 48 countries will face water stress or scarcity conditions by 2025. Of these countries, 40 are in West Asia, North Africa, or Sub-Saharan Africa. Over the next two decades, population increases and growing demands are projected to push all the West Asian countries into water scarcity conditions. By 2050, the number of countries facing water stress or scarcity could rise to 54, with their combined population being 4 billion people—about 40 percent of the projected global population of 9.4 billion (UNEP, 2007; from Gardner-Outlaw and Engleman, 1997;

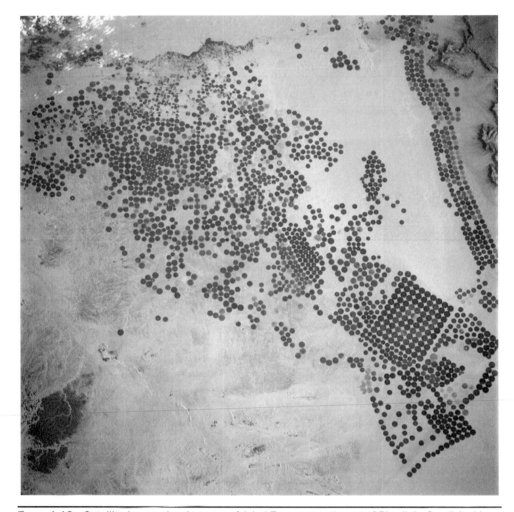

Figure 1.13 Satellite image showing area of Jabal Tuwayq, southwest of Riyadh in Saudi Arabia. Circles are fields with center-pivot irrigation systems, drawing groundwater via drilled wells in the centers. The diameter of circular fields varies from several hundred feet (tens of meters) to over a mile (2 km). (Space Shuttle astronaut photograph number STS032-096-032 taken in January 1990. Image courtesy of the Image Science & Analysis Laboratory, NASA Johnson Space Center; http://eol.jsc.nasa.gov.)

UNFPA, 1997). The numbers do not imply that the billions of people living in these countries will be without water. What they do imply, however, is that these 54 countries will most likely encounter serious constraints in their capacity to meet the demands of individual people and businesses, agriculture, industry, and the environment. Meeting these demands will require extensive planning and careful management of water supplies (CSIS and Sandia Environmental Laboratories, 2005).

Molden et al. (2001) group countries into three categories of water scarcity: physical water scarcity, economic water scarcity, and little or no water scarcity:

1. Physical water scarcity is defined in terms of the magnitude of primary water supply (PWS) development with respect to potentially utilizable water resources (PUWR). The physical water scarce condition is reached if the PWS of a country exceeds 60 percent of its PUWR. This means that even with the highest feasible efficiency and productivity, the PUWR of a country is not sufficient to meet the demands of agriculture, domestic, and industrial sectors while satisfying its environmental needs. Countries in this category will have to transfer water from agriculture to other sectors and import food or invest in costly desalinization plants.

2. Economic water scarcity is present in countries that have sufficient water resources to meet their additional PWS needs, but they require increasing their PWS through additional storage and distribution systems by more than 25 percent. Most of these countries face severe problems related to both finance and the capacity for development for increasing PWS to those levels.

3. The third category includes countries with little or no water scarcity. These countries are not physically water scarce but need to develop less than 25 percent of additional PWS to meet their 2025 needs.

Although an individual country may face physical water scarcity as a whole, substantial variations can exist within the country. For example, half of the Indian population lives in the arid northwest and southeast where groundwater is seriously overexploited, while the other half lives in regions with more abundant water resources. Substantial variations also exist between north and south China. Some parts of Mexico are physically water scarce, while others are not (Barker et al., 2000). Another important aspect is temporal variation. Some countries, especially those in monsoonal Asia, receive most of their rainfall in a few months in the wet season. These countries face severe water-scarce conditions in the other period of the year (Molden et al., 2001; Amarasinghe et al., 1999; Barker et al., 2000).

Researchers at Keele University, Great Britain, have developed the Water Poverty Index (WPI) as an interdisciplinary measure that links household welfare with water availability and indicates the degree to which water scarcity impacts on human populations (Lawrence at al., 2002). Such an index makes it possible to rank countries and communities within countries taking into account both physical and socioeconomic factors associated with water scarcity. This enables national and international organizations concerned with water provision and management to monitor both the resources available and the socioeconomic factors that have impact on access and use of those resources.

As argued by Lawrence et al. (2002), there is a strong link between "water poverty" and "income poverty" (Sullivan, 2002). A lack of adequate and reliable water supplies leads to low levels of output and health. Even where water supply is adequate and reliable, people's income may be too low to pay the user costs of clean water and drive them to use inadequate and unreliable sources of water supply. This fact was one of the drivers behind developing the WPI. The index encompasses five components:

1. Water availability, which includes both surface and groundwater that can be drawn upon by communities and countries. The available water resources are further divided into internal and external to the nation (or region).

WPI Component	Data Used
Resources	• internal freshwater flows • external inflows • population
Access	• % population with access to clean water • % population with access to sanitation • % population with access to irrigation adjusted by per capita water resources
Capacity	• log GDP per capita income (PPP) • under-five mortality rates • education enrolment rates • Gini coefficients of income distribution
Use	• domestic water use in liters per day • share of water use by industry and agriculture adjusted by the sector's share of GDP
Environment	indices of: • water quality • water stress (pollution) • environmental regulation and management • informational capacity • biodiversity based on threatened species

From Lawrence et al., 2002.

TABLE 1.7   Structure of the Water Poverty Index (WPI) and Data Used for Its Determination

2. Access to water, which means not simply safe water for drinking and cooking, but water for irrigating crops and for nonagricultural uses as well.

3. Capacity, in the sense of income to allow the purchase of improved water, and education and health which interact with income and indicate a capacity to lobby for and manage a water supply.

4. Use, which includes all domestic, agricultural, and nonagricultural uses.

5. Environmental factors, which are likely to impact regulation and affect capacity.

This conceptual framework for the WPI was developed as a consensus of opinion from a range of physical and social scientists, water practitioners, researchers, and other stakeholders in order to ensure that all relevant issues were included in the index. The five main components of the WPI include various subcomponents (Table 1.7), all expressed in a relative form and combined in the final measure, i.e., the rank of a nation as it compares with other 146 nations included in the survey. Figure 1.14 shows 10 countries with the highest and lowest scores ("water-richest" and "water-poorest" respectively), together with the United States, India, and China.

The authors of the WPI offer the following explanation of its usefulness: "However imperfect a particular index, especially one which reduces a measure of development to a single number, the purpose is political rather than statistical." As Streeten (1994, p. 235) argues: "...such indices are useful in focusing attention and simplifying the problem.

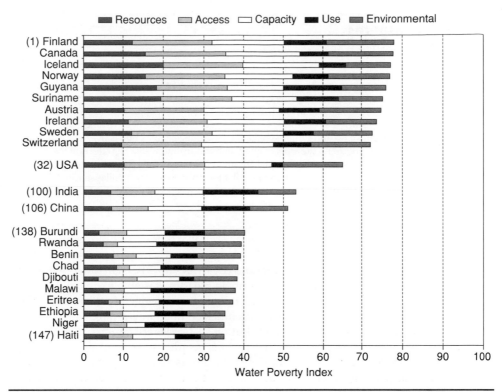

Figure 1.14 Selected national values for the Water Poverty Index. Number in parentheses is the country's rank, with Finland (1) being the water-richest. (From Lawrence et al., 2002.)

They have considerable political appeal. They have a stronger impact on the mind and draw public attention more powerfully than a long list of many indicators, combined with a qualitative discussion. They are eye-catching."

According to the authors, the results show few surprises. Of the 147 countries with relatively complete data, most of the countries in the top half are either developed or richer developing. There are a few notable exceptions: Guyana scores high on resources, access, and use, to get fifth position, while Belgium is 56th in the list, with low scores on resources and on the environment. New Zealand and the United States score very low on use, mainly due to low efficiency of water use in the agricultural and industrial sectors, which puts them 15th and 32nd on the list respectively.

Lawrence et al. (2002) compare the WPI with the Falkenmark Index Measure, expressed as availability of water resources per capita per year. The correlation between the Falkenmark Index of Water Stress and the WPI is only 0.35, which suggests that the WPI does add to the information available in assessing progress toward sustainable water provision. The Falkenmark Index indicates water stress when per capita water availability is between 1000 and 1600 m^3, and chronic water scarcity when this availability is 500 to 1000 m^3. Per capita water availability below 500 m^3 indicates a country or region beyond the "water barrier" of manageable capability (Falkenmark and Widstrand, 1992).

1.5 Water Disputes

Emily Wax of *The Washington Post* writes

> In Somalia, a well is as precious as a town bank, controlled by warlords and guarded with weapons. During the region's relentless three-year drought, water has become a resource worth fighting and dying over. The drought has affected an estimated 11 million people across East Africa and killed large numbers of livestock, leaving carcasses of cows, goats and even hearty camels rotting in the sun. The governments of Kenya and Ethiopia have mediated dozens of conflicts over water in their countries, even sending in police and the army to quell disputes around wells. The effects of the drought are most pronounced in Somalia, which has lacked an effective government and central planning, including irrigation projects, since the government of Mohamed Siad Barre collapsed in 1991. Since then, a hodgepodge of warlords and their armies have taken control of informal taxation systems, crops, markets and access to water. (Wax, 2006)

Unfortunately, violent disputes over water like these are as old as human history. As pointed out by Priscolli (1998), the *Book of Genesis* describes struggles over water wells in the Negev with the Philistines. Herodotus describes how Persian towns were subdued by filling their wells and water supply tunnels. Saladin was able to defeat the Crusaders at the Horns of Hattin in 1187 by denying them access to water. More recently, we have seen irrigation systems and hydroelectric facilities bombed in warfare. During the Gulf war, desalinization plants and water distribution systems were targeted; and the list goes on.

Arguably, however, it seems that various peoples and countries throughout the history have learned to appreciate the right for water as a basic human right, almost transcendent to life itself. FAO has identified more than 3600 treaties related to nonnavigational water use between years 805 and 1984. Since 1945, approximately 300 treaties dealing with water management or allocations in international basins have been negotiated. None of the various and extensive databases on causes of war can turn up water as a *casus belli*. Even in the highly charged Middle East, perhaps the world's most prominent meeting place for high politics and high water tension, arguably only one incident can be pointed to where water was the cause of conflict (Priscolli, 1998). Nevertheless, water scarcity and poor water quality have the potential to destabilize isolated regions within countries, whole countries, or entire regions sharing limited sources of water. There is an increasing likelihood of social strife and even armed conflict resulting from the pressures of water scarcity and mismanagement (CSIS and Sandia Environmental Laboratories, 2005).

Water disputes internal to a country are not reserved just for the Third World, or undeveloped and developing countries. Even the most developed nations are increasingly witnessing various forms of water disputes involving both water quantity and quality. Over 30 years ago, Smith (1985) warns about the competition for freshwater between individual states of the United States and the related legal consequences:

> The role of the federal government in groundwater regulation is likely to increase over the next few decades. A combination of events—including recent federal court decisions, past federal intervention in state groundwater utilization, the public pronouncements by a variety of federal actors, and increasing concern over the inability of states to control overdrafting—all suggest that the federal role in groundwater management may be increasing. These events are examined here, and it is argued that, without change in state groundwater management practices to mitigate the negative effects of state competition for groundwater, federal intervention in groundwater management seems likely.

The practices that followed indeed confirmed these words. Just recently, the federal government intervened in the case of the Colorado River water utilization in the west by reducing the ongoing overwithdrawal by California to its allocated portion. It has been instrumental in mediating disputes between Georgia, Alabama, and Florida caused by decreases in surface water flows due to the exploding water supply needs of Atlanta, or decreases of river base flows due to groundwater withdrawals in southern Georgia. Nevertheless, as reported by the *Washington Post* on October 21, 2007 (National News, p. A15):

> With water supplies rapidly shrinking during a drought of historic proportions, Gov. Sonny Perdue (R) declared a state of emergency for the northern third of Georgia and asked President Bush to declare it a major disaster area. Perdue asked the president to exempt Georgia from complying with federal regulations that dictate amount of water to be released from Georgia's reservoirs to protect federally protected mussel species downstream. On Friday, Perdue's office asked a federal judge to force the Army Corps of Engineers to curb the amount of water it drains from Georgia reservoirs into streams in Alabama and Florida.

Likely in response to this action by the Georgia governor, the federal government reacted by organizing the highest-level meeting between the three states, as reported by Associated Press from Washington, DC:

> Three Southeastern governors who are in Washington to lobby for water rights amid a potentially catastrophic drought are likely to put the Bush administration on the spot. If the administration decides to bolster Georgia's drinking supply, Alabama and Florida may claim it's crippling their economies to satisfy uncontrolled growth around Atlanta. If it continues releasing water downstream to Alabama and Florida, Georgia could argue that one of the nation's largest cities is being hung out to dry.
>
> "If it were easy it would have been solved 18 years ago," Kempthorne said. "There have been good-faith efforts, but there's also been millions of dollars spent in the courts and we do not have a solution. . . . There needs to be something where everyone says we gained here while we know we may have had to give up something else."
>
> Georgia officials have argued that the corps is turning a blind eye to a potential humanitarian crisis in Atlanta by ignoring warnings that the city's main water source, Lake Lanier, could have just a few months' worth of water remaining. The state sued the corps last month, arguing that Georgia has sacrificed more than other states and that the federal government is putting mussels before people.
>
> Alabama Gov. Bob Riley accused Georgians last week of "watering their lawns and flowers" all summer and expecting Washington to "bail them out." Florida Gov. Charlie Crist wrote Bush to say his state was "unwilling to allow the unrealistic demands of one region to further compromise the downstream communities."
>
> At a speech in Montgomery on Tuesday, Riley held up a poster-size map of Alabama and Georgia and showed that the exceptional drought area in Alabama is much larger than in Georgia. He said the state's economic prosperity was at stake. "This is about whether Alabama gets its fair share and whether we are going to have to lay off people in Alabama," he said."
>
> According to the National Drought Mitigation Center, almost a third of the Southeast is covered by an exceptional drought, the worst category. (CNNPolitics.com, 2007)

The government has also been active in mediating disputes between Georgia and South Carolina arising from saltwater intrusion at Hilton Head Island due to aquifer overpumping in the Savannah area. These are just some of the examples of the increasing frequency of government involvement in water disputes at various levels. More on regulations and legal framework for groundwater management in the United States is presented in Chapter 8.

In 1994, the Council of Australian Governments (COAG) established a strategic framework for water industry reform to ensure the efficient and sustainable use of Australia's water resources, and to use integrated and consistent approaches to water management. The framework emphasizes environmental flows, water quality, integrated watershed (catchment) management, water trading and pricing, viable and sustainable water use, and separation of responsibility for service delivery from regulation (Environment Australia, 2000).

Australia's Murray-Darling Basin Commission, established in 1985 following one of the worst droughts in Australian modern history, is an example of interstate cooperation in managing water resources on a large watershed basis. The Murray–Darling Basin produces 40 percent of Australian agricultural products and uses 75 percent of Australia's irrigation water. Because of increasing water extraction, which resulted in diminishing river flows, a moratorium on growth in water use was introduced in 1995 to ensure the reliability of supply while protecting environmental flows. This was confirmed as a permanent Cap on Diversions in 1997, but its implementation across four states, one territory and many industries continues to provide serious management challenges (Environment Australia, 2000).

In the spring of 2007, as the worst-ever, multiyear drought in Australia continued, the Commission acknowledged an overallocation of water resources and instructed further implementation of contingency measures and emergency planning (Murray-Darling Basin Commission, 2007). These measures, in some cases referred to as "severe," are raising tensions between upstream and downstream users, as well as between farmers, urban centers, and industry. As often in similar situations, the environmental users seem to be shortchanged first—the Commission has approved disconnecting flows to several wetlands in order to reduce evaporative water losses. Still more controversial is the decision by some state governments to give priority to urban users over irrigation and livestock farming. This has been reported extensively by major news media around the globe, such as BBC (British Broadcasting Corporation): "the New South Wales government actually took away water from the farmers which they'd already paid for, to cope with a shortfall in the cities. Eventually, it offered compensation, but only a third of the price paid by the farmers."

"Robbery," says Andrew Tully. "When the government comes in and steals your irrigation water that you have legally stored away as part of a good drought management strategy, that really makes you lose confidence in the whole system. It's pretty gutless." (Bryant, BBC, 2007).

Another growing source of conflicts over water resources is caused by a skyrocketing demand for bottled water worldwide. Multinational beverage and water bottling companies are facing many challenges from local communities, which are protecting groundwater resources, and resisting withdrawal by the "newcomers." An illustrative example is an outcry in India over the operations of the Coca-Cola Company, including lawsuits at a state court and the Supreme Court of India. In efforts to mitigate these and other possible negative actions worldwide, the company has elevated the strategic priority of water in its operations and the surrounding communities (Sandia National Laboratories, 2005; from Reilly and Babbit, 2005). It has surveyed 850 facilities in more than 200 countries to document and consider water issues and has begun working with conservation groups to address watershed management options around the globe. All of this serves to maintain the image of Coca-Cola as well as build rapport with local communities. Such strategies are increasingly adopted by corporations involved in the water industry, as the backlash against globalization and multinational corporations continue in many parts of the world (Sandia National Laboratories, 2005).

Cooperation in management of transboundary (internationally shared) surface waters, based on available international law and hydraulic engineering, is evident on all continents. However, the hidden nature of transboundary groundwater and lack of legal frameworks invites misunderstandings by many policymakers. Not surprisingly, transboundary aquifer management is still in its infancy since its evaluation is difficult, and it suffers from a lack of institutional will as well as finances to collect the necessary information. Although there are reliable estimates of the resources of rivers shared by two or more countries, no such estimates exist for transboundary aquifers (Salman, 1999). Unlike transboundary surface water and river basins, transboundary aquifers are not well known to policymakers. Present international law does not adequately address the issues concerning spatial flow of groundwater and has limited application in conditions where impacts from neighboring countries can be slow to develop (Puri, 2001).

After a clear consensus for an international initiative on shared groundwater resources was reached among groundwater professionals, the International Association of Hydrogeologists (IAH) established a commission to investigate the issue in 1999. Activities of the commission, over the next several years, resulted in the establishment of an international program supported by UNESCO, FAO, and UNECE (United Nations Economic Commission for Europe). One of the main drivers of the program, named the Internationally Shared/Transboundary Aquifer Resources Management (ISARM or TARM), is to support cooperation among countries so as to develop their scientific knowledge and to eliminate potential for conflict, particularly where conceptual differences might create tensions. It aims to train, educate, inform, and provide input for policies and decision making, based on good technical and scientific understanding (Puri, 2001).

1.6 Economics of Water

The concept of water as an economic good emerged during preparatory meetings for the Earth Summit of 1992 in Rio de Janeiro. It was brought forward and discussed extensively during the Dublin conference on Water and the Environment (ICWE, 1992), and became one of the four Dublin Principles, listed below:

Principle No. 1—*Freshwater is a finite and vulnerable resource, essential to sustain life, development, and the environment.* Since water sustains life, effective management of water resources demands a holistic approach, linking social and economic development with protection of natural ecosystems. Effective management links land and water uses across the whole of a catchment (drainage) area or groundwater aquifer.

Principle No. 2—*Water development and management should be based on a participatory approach, involving users, planners, and policymakers at all levels.* The participatory approach involves raising awareness of the importance of water among policymakers and the general public. It means that decisions are taken at the lowest appropriate level, with full public consultation and involvement of users in the planning and implementation of water projects.

Principle No. 3—*Women play a central part in the provision, management, and safeguarding of water.* This pivotal role of women as providers and users of water and guardians of the living environment has seldom been reflected in institutional arrangements for the development and management of water resources. Acceptance and implementation of this principle requires positive policies to address women's specific needs and to equip and empower women to participate at all levels in water resources programs, including decision making and implementation, in ways defined by them.

Principle No. 4—*Water has an economic value in all its competing uses and should be recognized as an economic good.* Within this principle, it is vital to recognize first the basic right of all human beings to have access to clean water and sanitation at an affordable price. Past failure to recognize the economic value of water has led to wasteful and environmentally damaging uses of the resource. Managing water as an economic good is an important way of achieving efficient and equitable use and of encouraging conservation and protection of water resources.

As pointed out by van der Zaag and Savenije (2006), the interpretation of the concept "water as an economic good" has continued to cause confusion and heated debate between various interpretations (advocates) of its "true" meaning. Two main schools of thought can be distinguished. The first school, the pure market proponents, maintains that water should be priced through the market. Its economic value would arise spontaneously from the actions of willing buyers and willing sellers. This would ensure that water is allocated to uses that are valued highest. The second school interprets "water as an economic good" to mean the process of integrated decision making on the allocation of scarce resources, which does not necessarily involve financial transactions (e.g. McNeill, 1998). The latter school corresponds with the view of Green (2000) who postulates that economics is about "the application of reason to choice." In other words, making choices about the allocation and use of water resources on the basis of an integrated analysis of all the advantages and disadvantages (costs and benefits in a broad sense) of alternative options (van der Zaag and Savenije, 2006).

One seemingly compelling argument in favor of the market pricing of water comes from an estimate of The World Water Council (WWC), which argues that to meet global water supply and sanitation demands, investments in water infrastructure need to increase from the current annual level of $75 billion to $180 billion (Cosgrove and Rijsberman, 2000). The development and long-term sustainability of the necessary infrastructure will require the identification of additional sources of financing and the introduction of market principles such as appropriate water-pricing mechanisms or private sector participation. Without an adequate pricing mechanism, consumers have no incentive to use water more efficiently, as they receive no signal indicating its relative value on the market. If water service providers are unable to recover the costs to adequately fund their operation, systems will inevitably deteriorate and the quality of service will suffer. This deterioration of water systems can be seen worldwide, particularly in developing countries, and partially explains the exorbitant funding needed (CSIS and Sandia National Laboratories, 2005).

Proponents of pure market principles also argue that rationalized global use of water is not possible when prices are subsidized and seriously distorted. In general, the agricultural sector is characterized by high subsidies worldwide, which includes below-market cost for irrigation water delivered through government-run irrigation projects, which are often highly inefficient. In some countries, governments subsidize the cost of energy needed to operate irrigation equipment, such as in India where installation of wells and well pumps in rural areas has been continuously promoted, including providing free energy to the farmers. Domestic and industrial water users commonly pay more than

100 times as much per unit of water as agricultural users (Cosgrove and Rijsberman, 2003). Introducing higher, more rational pricing schemes to farmers could provide the incentive for some of the water-saving measures and provide utility companies with the capital and incentive to improve infrastructure. A downside to the application of this incentive might be the agglomeration of smaller farms into larger farms, the loss of farm jobs leading to more migration to the cities, the increasing industrialization and corporatization of agriculture, and increases in food prices that affect the poor and possibly entire economies (CSIS and Sandia National Laboratories, 2005).

Market proponents argue that the poor often pay the highest price for water in developing countries and that real pricing could actually improve and expand access: unregulated vendors sell water by the container at a significant markup to families not connected to distribution infrastructure. For example, the unit pricing for household connection in Phnom Penh, Cambodia, is 1.64 USD compared to 9 cents for water provided by an informal vendor; these prices in Ulanbataar, Mongolia, are 1.51 and 0.04 USD, respectively (Clarke and King, 2004).

In recent years, private sector participation has been introduced into a number of water markets around the world, based on the belief that the private sector can deliver growth and efficiency more effectively than the public sector. However, market principles imply provision of services that are based on ability to pay, which does not fair well for poor people. For this and other reasons, developing countries express concerns over the push for privatization or changing the economic framework applied to water, led by international financial institutions. One of the key concerns is that the poor will be excluded if rich individuals or companies are allowed to buy up all the rights and establish monopolies on a universally required resource, which is both a human right and an economic commodity (CSIS and Sandia National Laboratories, 2005).

Perry et al. (1997) offer the following related discussion:

> The question is not whether water is an economic good or not—it certainly is an economic good in most cases, like almost everything else we have to worry about. Rather the question is whether it is a purely *private* good that can reasonably be left to free market forces, or a *public* good that requires some amount of extra-market management to effectively and efficiently serve social objectives. The answer to this question lies not so much in lofty principles but in value judgments, and their application to different conditions of time and place. Thus we find ourselves favoring the private good side of the argument in some cases and the public good side in others. The task is to define precisely what these cases, value judgments, and specific conditions of time and place, are. This definition is, we believe, important for two reasons: First, dogmatic posturing by the proponents of each perspective is a waste of intellectual talent. Second, and more importantly, water is far too important to its users to be the basis for socioeconomic experiments. Much is already known about the nature of successful policies and procedures for allocating water; understanding and incorporating the implications of this knowledge will avoid some potentially enormous financial, economic, environmental, and social costs.

These authors present very detailed analyses of economic theories based on both market and social principles as they apply to the water sector, focusing on water use in agriculture.

Van der Zaag and Savenije (2006) argue that water is a "special economic good" because of its unique characteristics and that contradiction exists between the first and the fourth Dublin principle, if the latter is interpreted in a narrow market sense. Water is a good that is essential, nonsubstitutable, and too bulky to be easily traded over large distances. The consequence is that water is used when and where it is available. Except for a handful of cases (such as with bottled water), and unlike for most other goods, water characteristics make it unattractive for large-scale trading. As a result, water markets can only function on a local scale and must take into account that the water flows in a downward direction ("from upstream to downstream"). It can be argued, however, that even when a combination of economic, market, and social principles in water pricing is attempted at the watershed (catchment) scale, it is rarely clear to all the existing and potential stakeholders which principles should be applied and which principles are fair or not fair. As stated by van der Zaag and Savenije (2006),

> The water "market" is not homogeneous. Different sub-sectors (agriculture, industry, power, transport, flood protection) have different characteristics. There are important water uses that have a high societal relevance but a very limited ability to pay, particularly the environmental, social and cultural requirements. Yet most if not all societies respect these interests. Decisions on water allocation appear to be taken seldom on purely "economic" (using the word in the interpretation of the first school) grounds. On the contrary: governments generally make decisions on the basis of political considerations; sometimes, and in our view more often then not, governments are sensitive to and concerned with social and cultural and, admittedly less frequently, also environmental interests. Of course, economic and financial considerations are an integral part of these decisions but these seldom are the overriding decision variable. This pragmatic approach is in agreement with the second school of thought. Sometimes governments fail to allocate the water in accordance with societal needs. This is exemplified by the lack of access to safe drinking water in many rural areas in Africa. In this example "less government more market" is unlikely to solve the problem because of a limited ability to pay of those affected.

One key and unique aspect of water is that it always belongs to a system, be it local, regional, or global, and it cannot be separated from that system or divided. For example, what happens with water in one part of the system (e.g., a watershed) always impacts users miles or hundreds of miles away. Upstream users, water diversions, and wastewater discharges will affect downstream users, water availability, and water quality. Withdrawal of groundwater may impact surface water flows and vice versa. Surface water may become groundwater at some point, and the same water may again emerge as surface water after flowing through the groundwater system. Temporal and spatial variability of water resources constantly change due to natural climatic cycles, but there also may be permanent impacts due to long-term climate and land-use changes. Water can have negative value in case of flash floods or reoccurring monsoonal floods. All this makes it difficult to establish the value of external effects on any type of water use. The following two points illustrate just some of the complexities when considering water "only" as an economic good, in the narrowest sense (van der Zaag and Savenije, 2006):

> Consider, for example, farmers in an upstream catchment area of a river basin who produce rain-fed crops and who have managed to triple yields due to prudent agronomic measures, soil husbandry practices, and nutrient management. It is known that the increase in crop yields decreases water availability downstream in the river. Do these rain-fed farmers therefore require a water right or permit for increasing their yields? If so, is it known by any measure of precision how much the additional water consumption is, compared to which baseline situation?

> Economic analysts can easily demonstrate that the future hardly has any value (in monetary terms). The discount rate makes future benefits (or costs) further than, say, 20 years ahead negligible and irrelevant. The market, by itself, will therefore ignore long-term benefits. This, like the previous aspect,

illustrates that market thinking in this limited sense goes against stated policy objectives, and that additional state control is always likely to be necessary.

Expanding on the latter point, one could envision the following scenario in the absence of any third interested parties (including a state or local government):

- A corporation purchases water rights from a rancher, somewhere in the western United States.
- The corporation pumps the aquifer and sells water (possibly including bottled water).
- The aquifer happens to be a "slowly-renewable" source, with most of the water being pumped from the storage.
- The corporation is about to close its operation after 20 years (that is, after the aquifer storage has been mostly depleted).
- In the meantime, all springs in a 5-mile radius from the pumping center dried out, and dozens of creeks are now dry year round.
- Also in the meantime, the farmer may or may not have fully enjoyed his retirement somewhere in Florida, depending on the availability and reliability of water supply in his retirement community (incidentally, almost all domestic water supply in Florida is based on groundwater).

For those who are closely following water supply issues across the America's West, the preceding scenario is not so far-fetched. In any case, there is still a significant lack of understanding when it comes to the value of environmental water uses, and any associated "price" of water. It appears that environmental groups are often left to their own means when battling water market forces and various inherited water rights historically developed for a very limited user base. For example, the Center for Biological Diversity (2007) has the following posted on its web site:

Now two cities north of Phoenix, Prescott and Prescott Valley, intend to take more than 8,717 acre-feet of water per year—nearly 8 million gallons per day—from the Big Chino aquifer and transport it through 45 mi of pipeline into new, thirsty developments. U.S. Geological Survey hydrologists have calculated that between 80 and 86 percent of the waters in the upper Verde River come from the Big Chino aquifer—and predict that in time this project will appreciably dry up the first 24 mi of the river. Meanwhile, the nearby town of Chino Valley is also ramping up groundwater-dependent development and buying up "water ranches" to feed growth.

If the upper Verde is robbed of such a significant source of its flow, the entire river will be adversely affected. The river's already imperiled species will be especially hard hit and may not survive the loss of streamflow. The threat to the Verde is so imminent that it has drawn national attention: a 2006 American Rivers report recently named it one of America's Most Endangered Rivers. But in spite of the concern expressed by a growing number of citizens, Prescott and Prescott Valley have stubbornly refused to protect the river.

Because the cities have not provided plans to protect streamflow for listed species along the Verde River, the Center in December 2004 filed a notice of intent to sue the two cities for Endangered Species Act violations. If the cities continue to refuse to develop comprehensive conservation plans, the Center will be forced to move forward with the lawsuit.

As inevitable in any similar adversarial situation, the two opposing groups of stakeholders have different interpretations of the available hydrologic and hydrogeologic information, including reports completed by the U.S. Geological Survey, arguably an independent agency. Various dueling hydrologists continue to produce studies and counter-studies, and the opposing sides are preparing for lawsuits while waiting for "state and federal governments to finish a regional computer groundwater study, which isn't expected to be complete for at least 18 months." (Davis, 2007).

1.6.1   Price and Value of Water

The price of water, in a narrow sense, is defined as the price water users are paying for the volume of water delivered per unit of time (e.g., cubic meters or gallons per month). This definition applies mostly to customers receiving water from third parties, such as homeowners or businesses supplied by public water utilities, or farmers paying for off-farm water delivered by centralized irrigation systems. In many cases, which in general cover the majority of freshwater withdrawals worldwide, the users are self-providers. Some examples include individual farmers and rural homeowners using water wells, large agricultural complexes (corporations), and industrial facilities diverting water from surface streams or using their own water wells, or power plants withdrawing water from surface water reservoirs. In most, if not all cases, neither of the two large groups of water users—customers and self-providers—pays the full ("real") price of water, which should theoretically equate to the real value of water and include all of the following (Fig. 1.15):

1. Capital cost of building water withdrawal and distribution system

2. Cost of operating and maintaining the system ("operations and maintenance" or O&M cost) including water treatment, water source and infrastructure maintenance, staff, and administrative costs

3. Capital for future major investments for augmenting the existing or finding new sources of water, and expanding the distribution system

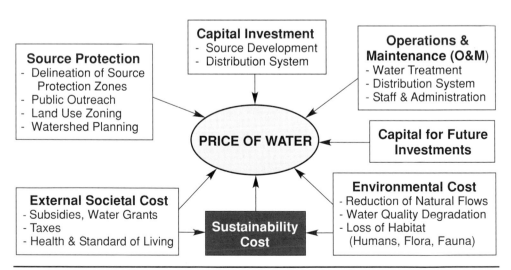

Figure 1.15   Components of the full water price, theoretically equal to its full value.

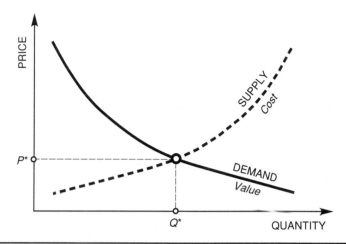

Figure 1.16 Supply and demand: price relative to value and cost (Raucher et al., 2005; copyright 2005 American Water Works Association Research Foundation).

4. Source protection cost reflecting its intrinsic value (water quality, reliability)

5. External societal cost

6. Environmental cost

7. Sustainability cost

In a market, the price of a commodity (good) is determined where the supply and demand curves intersect (Fig. 1.16). In this depiction of an economic market, P^* is the price that would clear the market. In other words, the quantity demand would equal the quantity supplied at that price. The market clearing quantity is depicted with Q^*. To the left of Q^*, the values of water (embedded in the demand curve) are all higher than the costs (depicted on the supply curve), and value also exceeds P^* to the left of Q^*. For quantities beyond Q^*, however, value is less than cost, and application of the resource for this market is not efficient beyond that level (Raucher et al., 2005).

In markets where there are many suppliers and consumers, price is often viewed as a good estimate of the marginal value of the good to both consumers and producers. It is thus considered economically efficient to allow the market process (i.e., prices determined in competitive market) to dictate the allocation of resources (Raucher et al., 2005). However, as discussed earlier, water is a special economic good, with unique characteristics that preclude open competition for virtually all uses (bottled water market being an exception). For example, there is not a single urban center where a household can chose between, say, three water utilities each providing its services through its own infrastructure ("pipes").

The value of water has many different components and they often mean different things to different people, and indeed societies. Market principles applicable to, say television sets, do not fully function in the water arena. As a result, all water users today, i.e., water utilities, agriculture, industry, are subsidized either directly or indirectly. Direct subsidies should be easily identifiable and may include factors such as reduced energy cost for certain users (such as electricity and fuel required to operate irrigation systems), and reduced O&M costs for a public water supplier subsidized by the "public," including by those that are not necessarily directly served by the same system (e.g., through reduced or waived taxes). International financial institutions have been very active in analyzing various subsidies in the water sector and are increasingly conditioning their lending to the developing world governments and utilities on the adaptation of market principles in water pricing. Recent study by The World Bank (Komives et al., 2005) shows that even in the most developed countries (OECD), only 51 percent of the water utilities surveyed charge water prices that cover O&M cost and provide for partial capital. In the same study, which compares subsidies in the electricity and water sectors, the authors present the following analysis (Komives et al., 2005):

> The water supply sector has a much lower degree of cost recovery and metering coverage than the electricity sector, leading to more untargeted and implicit subsidies in the water sector. It is also more common in the water sector to charge different prices to industrial and residential customers and to apply increasing block tariff structures that subsidize all but the very highest levels of residential consumption.
>
> The idea of subsidizing water and electricity services (the latter particularly in cold climates) has widespread support among politicians, policy makers, utility managers, and the public at large. Subsidies for basic services—particularly subsidy mechanisms such as increasing block tariffs—are considered fair and even necessary for ensuring that poor households enjoy the use of those services. They are also seen as an alternative instrument of social policy, as a way to increase the purchasing power of the poor.
>
> Utility services are characterized by a high degree of capital intensity and by long asset lives...— in the network components of the electricity and water services 70 percent to 90 percent of costs can be capital costs. Such assets typically last for much longer than 20 years. High capital intensity and long asset lives make it possible to get away without covering the full capital costs of service provision— at least for some period of time. This opens the door to unfunded subsidies of the type described above. The problem is more severe in the case of water utilities than electric utilities because water networks and their associated services deteriorate quite gradually, without necessarily threatening the continuity of provision. Power systems, however, are more sensitive. Inadequate maintenance can lead relatively quickly to outright failure and prolonged blackouts—which are, moreover, politically unpopular. For this reason, it is easier for politicians to underfinance water and sewerage services than electricity services.

Among various drawbacks to subsidized water prices charged to customers is over-consumption, which in turn creates the need for large, expensive wastewater treatment plants. In fact, water and sewerage utilities are often the same entity, charging the same customers for both services. In the past decade or so, low water prices have been targeted by utilities worldwide, regardless of the level of economic development of the population served. For example, as reported by the official Chinese news agency Xinhua, the most outstanding example of water price hikes in China is Beijing. In August of 2004, the capital raised its water price from 2.9 to 3.7 yuan/m^3 (0.48 USD; 1 yuan is approximately 0.13 USD). It was the ninth water price hike for the city in the past 14 years, making Beijing's water the most expensive in the country where average urban per capita water price was 2 yuan/m^3. Since such low water prices cannot reflect the country's severe water shortage, it is predicted that they will continue to be raised significantly in the future (China Daily, 2004).

In a 2005–2006 international cost survey, the NUS Consulting Group (2006) shows water prices increasing in 12 of the 14 countries surveyed (Table 1.8). Denmark and the United States remain the most expensive and the least expensive country surveyed

2006 Rank	Country Cost	(US cents)/m³	2005/2006 Change	5-year Trend (2001/2006)
1	Denmark	224.6	−4.6%	+1.9%
2	Germany	224.5	+1.6%	−2.7%
3	United Kingdom	190.3	+7.8%	+32.3%
4	Belgium	172.3	+1.9%	+51.1%
5	France	157.5	+3.5%	+11.8%
6	Netherlands	149.0	+1.0%	+0.3%
7	Italy	114.7	+2.0%	+23.2%
8	Finland	103.3	+9.7%	+30.2%
9	Australia	100.5	+13.8%	+45.4%
10	Spain	93.0	+3.1%	+5.2%
11	South Africa	91.8	+8.8%	+50.2%
12	Sweden	85.9	−2.4%	+10.7%
13	Canada	78.9	+8.9%	+58.0%
14	United States	65.8	+4.4%	+27.0%

From NUS Consulting Group, 2006.

TABLE 1.8   2006 International Water Cost Comparison

respectively. Australia experienced the single largest year-on-year increase in pricing at 17.9 percent, with other countries including Canada, Finland, South Africa, and the United Kingdom also showing significant increases. Australia's price increase is mainly attributable to an extended period of drought, and searching for ways to reduce overall water consumption as well as increasing supply for its growing population. In Europe, tighter EU regulations coupled with below normal rainfall levels have led some countries to adopt higher prices. Canada for the second year in a row experienced water price increases well beyond the country's annual inflation rate. Further increases are expected as the nation invests more in volume-based pricing. Given these developments around the world, the NUS Consulting Group advises that medium to large business consumers of water no longer rely on cheap and abundant supplies.

In its latest report, the Water Services Association of Australia (WSAA) concludes that

> urban water utilities must obtain a return on their investment in water infrastructure and this will flow through to water prices. This pricing approach provides urban water utilities with funds to invest in new infrastructure and repair existing infrastructure without having to rely on government. It also enables the utilities to pay a dividend to their shareholder governments. The use of inclining block tariffs can reduce the burden of higher water prices on those less well off in the community as the first increment of water used can be priced to make it affordable to all.

Figure 1.17 is an example of inclining block tariffs implemented by the Australian city Perth, which is supplied almost exclusively by groundwater.

Environment Canada (2007) reports that all surveys since at least 1991 indicate that, both nationally and provincially, Canadians use more water when they are charged a flat

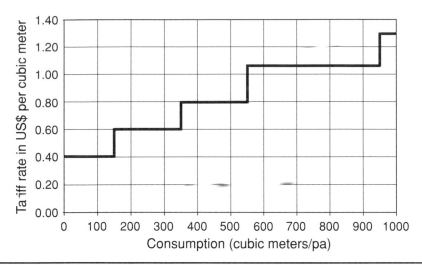

FIGURE 1.17   Perth, Australia, water prices as of June 27, 2007. (Data from WSAA, 2007.)

rate. The 2004 survey shows that in municipalities that charged according to the volume of water used, the average daily consumption rate was 266 L (70 gal) per person. In communities that charged a flat or fixed rate, the corresponding figure was 76 percent higher (467 L or 123 gal/person). These findings continue to suggest that metering and volume-based pricing can be valuable demand-management tools for promoting the responsible use of water resources. Even when water prices rise substantially it is likely that the cost of water compared to other utility services such as electricity will remain a relatively small proportion of total household expenditure. According to the Australian Bureau of Statistics data, household expenditure on water and sewage is only 0.7 percent of disposable income in Australia, compared to 2.7 percent for electricity and gas, 3.6 percent for alcohol and cigarettes, and 5 percent for household furniture (WSAA, 2007).

In summary, following are some of the strongest drivers for the increasing trend in water pricing worldwide:

- The need for capital required to meet new water supply and sanitation demands of the increasing urban population
- The need for water conservation, where increased prices are used as a tool for demand management
- Water scarcity, the existing and the expected impacts of climate change (droughts, floods, seasonal water re-distribution), which require new O&M practices, additional water storage, and development of alternative water sources

Results of an analysis of prices charged to residential customers by more than 200 water utilities in the United States are shown in Figs. 1.18 and 1.19. The raw data are from a survey conducted by Raftelis Financial Consultants for the American Water Works Association (AWWA, 2007). For this analysis, the utilities were grouped based on the predominant source of water extracted by the utility itself (>50 percent; purchased water was not considered) into groundwater-based and surface water-based. Of the 10 largest

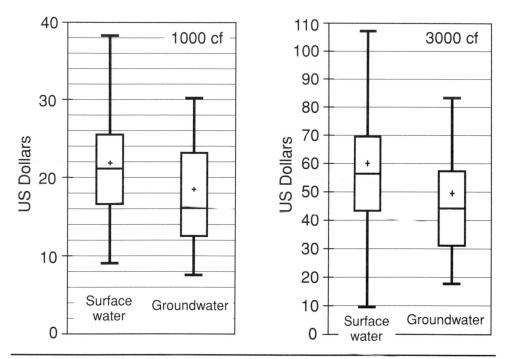

FIGURE 1.18 Box-whiskers diagrams of water prices charged to residential customers by utilities in the United States. (Left) Monthly usage of 1000 ft^3. (Right) Monthly usage of 3000 ft^3. The median price is denoted by the horizontal line inside the box, and the average price is denoted by the cross inside the box. The upper and lower box limits correspond to the 75th and 25th percentile respectively. The range of prices, excluding outliers, is shown by the vertical lines extending from the box. The analysis is performed for 123 surface water-based and 66 groundwater-based utilities. (Raw data from AWWA, 2007.)

utilities surveyed, only one (Miami) is groundwater-based. However, regardless of the size, the groundwater-based utilities on average charge less to their customers than the surface water-based utilities: 23 percent less for the monthly usage of 1000 ft^3 and 22 percent less for the monthly usage of 3000 ft^3, based on the median prices (Fig. 1.18). This may be the result of lower water treatment costs and more favorable (closer) locations of groundwater extraction relative to the user base, although a more detailed related analysis was not conducted.

Overwhelmingly, both groups have inclined water prices—they charge more for more water used. Only three utilities (Chicago, Sacramento and Juneau, AK) out of 221 surveyed have flat water prices—they charge the same amount regardless of how much water is used. In both groups, larger utilities tend to charge less for the volume unit of water delivered, which is likely a combination of two main factors: economies of scale and higher subsidies. This is illustrated in Fig. 1.19 for the 10 largest and 10 smallest utilities surveyed, for both groups. For example, the average monthly price of 1000 ft^3 of water charged to residential customers by all utilities selling more than 75 million gallons of water daily (MGD) is 19.69 USD, compared to 22.42 USD for the utilities selling less than 25 MGD.

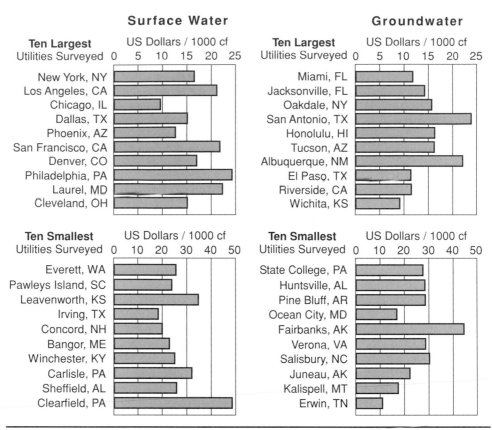

FIGURE 1.19   Monthly charges to residential customers, in USD, for 1000 ft^3 (28 m^3) of water produced by 10 largest and 10 smallest utilities based on the predominant source water type. (Raw data from AWWA, 2007.)

Economies of scale in water supply, particularly in the areas of source development and treatment, make it difficult for smaller water utilities to perform as well as larger water utilities. Declining unit costs of production indicate scale economies; as the volume of water produced (that is, withdrawn and treated) increases, the cost per gallon or cubic foot decreases. At lower unit costs, production is less costly in the aggregate and more efficient at the margin. Importantly, the economies of scale in water production are associated with the *volume* of water produced (not simply the number of service connections). Even smaller systems that are fortunate enough to have one or two large-volume customers will enjoy some economies of scale. Two utilities can have a comparable level of investment per customer and cost-of-service for the same number of residential customers, but if one also serves a large industrial firm and economies of scale are achieved, everyone in that community will enjoy lower water bills (USEPA and NARUC, 1999).

The intrinsic value of a water source is not necessarily always reflected in the price of water charged to customers, although certain aspects of it are easily quantifiable such as cost of water treatment. For example, treating water for water supply from a major river flowing through various urban and industrial areas is much more expensive than a simple preventive microbiological treatment (distribution system disinfection) of water extracted from a well-protected confined aquifer. Groundwater of high quality can also be blended with surface water thus reducing treatment cost. In that respect, groundwater used for water supply in general has higher intrinsic value than surface water. A good example is utilization of groundwater in alluvial basins of large European rivers where utilities routinely extract groundwater from the alluvial aquifers and take advantage of river bank filtration, rather than using river water. Accidental toxic spills and floods can take surface water supply out of service sometimes in the matter of days or even hours, whereas this is much less likely to happen with a groundwater-based water supply. In emergencies, including "unexpected" droughts, surface water-based utilities are often forced to look at groundwater as a last resort. In case of large urban centers, and when hydrologic and hydrogeologic characteristics are favorable, water utilities that develop both surface water and groundwater supplies are in a much better position to manage marginal cost of water and set stable, more realistic prices. For example, using supplemental groundwater during periods of peak demand, such as dry and hot summer months, can eliminate needs for large surface water storage reservoirs and treatment plants, which become unused capacity during most of the year. Artificially storing water in the subsurface (aquifer storage) has two key advantages over surface reservoirs: virtually no evaporation loss and no surface land and habitat losses due to impoundment (see Chapter 8). At the same time, in many coastal, low-lying areas around the world geomorphologic conditions preclude the building of large surface water reservoirs and groundwater is the only reliable source of water supply year round. Accumulated experience on the use of groundwater in irrigation has shown that its intrinsic value is higher than that of surface water in most cases. Groundwater can be applied to the fields in the immediate area where it is withdrawn from the aquifer, which eliminates needs for extensive network of canals and reservoirs where conveyance and evaporation losses are significant by default.

The full cost of using a high-value water source and protecting it is seldom charged directly to customers or even assessed. For example, in a study of the benefits of wellhead protection programs (WHPPs) in the United States, Williams and Fenske (2004) found that the average division of WHPP development expenditures is 2 percent from federal contributions, 62 percent from state contributions, and 36 percent from local contributions. Implementation cost is generally solely borne by the local utility or local wellhead protection authority, which in most cases do not have adequate financial, legislative (political), or technical resources to implement and enforce a program truly protective of water supplies. Together with the methodology and results of the analysis of monetary and nonmonetary benefits of WHPPs for nine diverse utilities, Williams and Fenske (2004) present a number of recommendations, including the following:

- In addition to public participation during the development of a WHPP, the authors recommend that a wellhead program contain an aggressive, on-going public awareness and education component. Effort and dollars spent on education are worthwhile because they build community support for the WHPP, will help allay public fears or concerns over potential groundwater contamination, calm concerns over potential property values and economic development effects, and instill groundwater-friendly attitudes and practices throughout economic and social sectors in the community.

- Adequate funding needs to be provided at all levels of government to encourage and support development and implementation of WHHPs. Communities should consider a wellhead protection fund similar to that administered in Dayton, Ohio. A wellhead protection fund can provide the resources to (1) administer the WHPP, (2) hire staff specialists to educate the public on practices that protect the quality of a raw water supply, (3) place land under public ownership in order to control land uses or activities that have an inherent risk of contamination to a raw water supply, (4) purchase easements or otherwise limit development rights on land owned by others to allow only environmentally-friendly development, or (5) establish a revolving fund or grant system to enable local businesses to install equipment or implement practices that are protective of groundwater quality.

- Wellhead protection often extends beyond community or water utility boundaries into neighboring jurisdictions, thus becoming a regional issue of aquifer protection beyond the defined wellhead protection area boundaries. Wellhead protection administrators need to recognize the significance of protecting aquifers on a regional basis for the benefit of all that use the aquifer for water supply. The authors encourage WHHP administrators to develop coalitions and cooperative arrangements with neighboring jurisdictions for the regional protection of water supplies.

The Dayton, Ohio, water supply example mentioned above shows that the cost of development and implementation of WHHP, using the 20-year present worth of the annual costs with the sum converted to 2003 USD, is 2.78 USD per capita annually and 0.03 USD per thousand gallons. The average water demand of the customers served by the Dayton utility is 70.62 million gallons per day (0.27 million m^3/d).

The monetary benefits of a WHHP can be demonstrated by calculating avoided costs or loss of commodity value. Avoided costs consist of the potential cost of contamination that is prevented by implementing the WHHP. The cost of contamination includes raw water treatment before delivery, contaminant source remediation, containment and/or intercept of contaminated groundwater in the aquifer, other groundwater remediation costs, and well or well-field replacement. Loss of commodity value is the potential loss of income from not being able to extract and sell contaminated groundwater. One measure of the avoided-cost benefit is the ratio of total contamination cost and the total cost of the WHHP (USEPA, 1995). If a WHPP is viewed as an investment in the future security of a community's water supply, then the avoided-cost benefit shows that the payback on the investment is at least 2.3 to 1, and as much as 13.4 to 1 in the 2004 Williams and Fenske study; the 1995 USEPA study found this same payback ranged between 5:1 and nearly 200:1.

Prices paid for irrigation water are of considerable policy interest due to their importance as a cost of production and their impact on water demand. Increasingly, adjusting the water "price" in agriculture is viewed as a mechanism to improve the economic efficiency of water use. However, as discussed earlier, water price adjustments to achieve socially desired outcomes can be difficult because prices paid for water by agricultural users are rarely set in the marketplace and generally do not reflect water scarcity. In the United States, individual states generally administer water resources and grant (not auction) rights of use to individuals without charge, except for minor administrative fees. As a result, expenditures for irrigation water usually reflect water's access and delivery costs alone—thus, costs to irrigators usually do not reflect the full social cost of water use. By contrast, those without an existing state-allocated water right—whether an irrigator, municipality, industry, or environmental group—that purchase annual water allocations or permanent water rights from existing users pay prices that more closely reflect the scarcity value of the resource (Gollehon and Quinby, 2006).

Costs of supplying irrigation water vary widely, reflecting different combinations of water sources, suppliers, distribution systems, and other factors such as field proximity to water, topography, aquifer conditions, and energy source. To generalize, groundwater is usually pumped on-farm with higher energy expenses than surface water, which is often supplied from off-farm sources through extensive storage and canal systems. Gollehon and Quinby (2006) used data from the Farm and Ranch Irrigation Survey (USDA, 2004b) to examine the cost determinants for groundwater and surface water irrigation sources in the United States. Their main findings are as follows:

Groundwater is used on nearly half of U.S. irrigated farms, with the pumped groundwater supplying over 32 million acres (Table 1.9). Energy costs in 2003 ranged from $7 per acre in Maryland to $79 per acre in California, $92 in Arizona, and over $175 per acre in Hawaii. Average costs nationwide were almost $40 per acre, and total expenditures for the sector exceeded $1.2 billion.

Surface water energy costs reflect pumping and pressurization requirements for conveyance and field application. Over 10.5 million surface-supplied acres incurred these costs in 2003, at an average cost of $26 per acre. Costs ranged from $10 per acre in Missouri to $36 in California, $41 in Washington, and $82 in Massachusetts. In general, energy costs are less for pumping surface water than groundwater since less vertical lift is required.

Nearly 40 percent of irrigated farms received water from off-farm water supplies, accounting for nearly 14 million irrigated acres. Irrigators paid an average of $42 per acre for water from off-farm suppliers, including about 20 percent of farms reporting water at zero cost (Table 1.9). Average costs ranged from $5 per acre in Minnesota to $46 in Washington, $72 in Arizona, and $86 in California. Much of the off-farm water is used in California, with over 30 percent of the Nation's acres served by off-farm sources.

It should be noted that this analysis does not address or compare the water efficiency and the return on investment, measured through crop yield, for groundwater-based versus surface water-based irrigation.

More difficult, and currently often not feasible, is to quantify that portion of the water cost attributable to external factors (Fig. 1.15). This may include societal cost such as cost of health and wealth improvements of the poor population served by the subsidized water supply systems or the cost of political stability/instability when favoring a certain group of users (e.g., urban versus agricultural users, or upgradient versus downgradient users). Environmental cost is the cost of water delivered to or, more often, diverted from environmental users (flora and fauna) to the extent that the water source cannot meet demands of all current and future non-anthropogenic users. Together, this environmental cost and those portions of the societal cost that impact future water users (generations) comprise the cost of sustainability of the water resource, sometimes referred to as water scarcity cost.

Closely related to the sustainability cost is the opportunity cost, defined as the cost of using the unit of water in its next best competing use. This cost accounts for the fact that in limited supplies, a unit of water applied to a specific use, say residential uses, cannot be applied to other uses such as industrial sector for example. If the productivity of water in the alternative sector would have been higher, then there is a real lost opportunity (Raucher et al., 2005). It is likely that the opportunity cost of using an additional volume of water is low in water-rich regions because after the senior user (e.g., water utility or water rights owner) extracts what it needs, an abundant renewable supply remains instream or in the aquifer to meet the demands of competing uses of the resource. In

Cost Category	Acres Incurring the Cost		State-Level Cost Range (Dollars/Acre)	National Average Cost (Dollars/Acre)	Total National Cost ($ Million)
	Million	Percent			
Energy expenses for pumping groundwater	32.34	61.5	7–176	39.50	1,277.54
Energy expenses for lifting or pressurizing surface water	10.56	20.1	10–82	26.35	278.72
Water purchased from off-farm sources	13.87	26.4	5–86	41.73	578.75
Maintenance/repair expenses	40.01	76.1	4–80	12.29	491.77
Total variable costs				49.87	2,622.37
Average variable cost (including acres with no cost)					
Capital investment expenses[1] incurred in 2003	26.67	50.7	16–187	42.18	1,125.13

[1] Over $13,000 per farm, distributed based on average farm size to compute per-acre expenses. From Gollehon and Quinby, 2006.

TABLE 1.9 Cost of Irrigation Water in the United States by Source and Category, 2003

contrast, in water-scarce regions every unit of water used by the senior user is taken away from an alternative use such as crop irrigation or ecological preservation. The opportunity cost of water allocated to one type use, such as agriculture, can be quite high because municipalities and industry can pay much higher prices for water than farmers. Opportunity cost in such case can include millions of dollars in foregone cropland yields. This, however, may result in the loss of local agricultural communities, lifestyles, and associated support sectors. The opportunity cost of extracting large volumes of water for municipal and industrial uses might also include the loss of important fish species and the associated value of ecological consequences or foregone recreational and/or commercial opportunities (Raucher et al., 2005).

When a current user does not pay water sustainability cost, this cost will have to be paid by other current users of the same resource as well as by future users. Unsustainable mining of an aquifer, which creates an increasingly large regional cone of depression and drawdown, or extraction from an un-renewable groundwater resource ("fossil aquifer") are typical examples—other current users and future generations are left with diminishing yields and higher prices. Prompted by the newly created global awareness regarding climate change, environmental sustainability, and population growth, the sustainability cost of water has recently started to draw attention from various stakeholders—regulators, politicians, economists, scientists, environmental organizations, and users at large. Participation of all these stakeholders in decision making can assure that the full price of water use would be assessed and that all available water resources—rain water, surface water, soil water, groundwater, and wastewater—at any given user-scale (e.g., community, watershed, country, region of the world) would be managed in an integrated and sustainable manner. Those users that are ready to innovate and change their current practices quickly can only benefit in the future as the water becomes scarcer and retains more of its true value on the emerging open water market.

It is conceivable that in a not so distant future, in its recognition as an irreplaceable and vital resource, water supplied to customers, as well as various food and other products, will be labeled as "produced using environmentally sound and sustainable water practices." This may be the ultimate means of fully valuing water.

1.6.2 Virtual Water and Global Water Trade

The water that is used in the production process of an agricultural or industrial product is called the virtual water contained in the product (Allan, 1996). For example, for producing a kilogram of grain, grown under rain-fed and favorable climatic conditions, 1 to 2 m^3 of water (1000 to 2000 kg of water) is used. For the same amount of grain, but growing in an arid country, where the climatic conditions are not favorable (high temperature, high evapotranspiration), this water use is as high as 3000 to 5000 kg (Hoekstra and Hung, 2002).

The water footprint of an individual is an indicator of that individual's total water use. It is equal to the summed virtual water content of all the products consumed by the individual. Some consumption patterns, for instance a meat diet, imply much larger water footprints than others. Awareness of one's individual water footprint may stimulate a more careful use of water (IHE Delft and World Water Council, 2003). An illustrative related web site developed by the IHE Delft ("Waterfootprint") includes an interactive calculator of the individual footprint, which takes into consideration specific living conditions and lifestyle as well as the overall standard of the residing country (http://www.waterfootprint.org/?page=cal/waterfootprintcalculator_indv).

The virtual water contents of some selected crop and livestock products for a number of selected countries are presented in Table 1.10. As can be seen, livestock products in general have higher virtual water content than crop products. This is because a live animal consumes a lot of feed crops, drinking water, and service water in its lifetime before it produces some output.

The global average virtual water content of some selected consumer goods expressed in water volumes per unit of product is given in Table 1.11. In the United States, industrial products take nearly 100 L/USD. In Germany and the Netherlands, average virtual water content of industrial products is about 50 L/USD. Industrial products from Japan, Australia, and Canada take only 10 to 15 L/USD. In world's largest developing nations, China and India, the average virtual water content of industrial products is 20 to 25 L/USD. The global average virtual water content of industrial products is 80 L/USD (Chapagain and Hoekstra, 2004).

Of particular concern in coping with the existing and the inevitable future water scarcity in many developing regions of the world is, paradoxically, the increasing standard of living and the related changes in the population diet. Patterns of food consumption are becoming more similar throughout the world, incorporating higher-quality, more expensive, and more water-intensive foods such as meat and dairy products. As presented by FAO (2002), these changes in diet have had an impact on the global demand for agricultural products and will continue to do so. Meat consumption in developing countries, for example, has risen from only 10 kg/person/yr in 1964 to 1966 to 26 kg/person/yr in 1997 to 1999. It is projected to rise still further, to 37 kg/person/yr in 2030. Milk and dairy products have also seen rapid growth, from 28 kg/person/yr in 1964 to 1966 to 45 kg/person/yr recently, and could rise to 66 kg/person/yr by 2030. The intake of calories derived from sugar and vegetable oils is expected to increase. However, average human consumption of cereals, pulses, roots, and tubers is expected to level off (FAO, 2002). These trends are partly due to simple preferences. Partly, too, they are the result of increased international trade in foods, the global spread of fast food chains, and the exposure to North American and European dietary habits.

With increasing global water shortages and awareness of the environmental impacts associated with irrigation, the concept of global trading in virtual water is receiving increased attention. However, as pointed out by IHE Delft and World Water Council (2003), trade in virtual water has been taking place unconsciously for a very long time and has steadily increased over the last 40 years. For example, during the 1990s, the U.S. Department of Agriculture and the European Community exported to the Middle East and North Africa (MENA) region as much water as it flowed down the Nile into Egypt and was used for agriculture each year: 40 billion tons (40 km^3), embedded in 40 million tons of grain (Allan, 1998).

Chapagain and Hoekstra (2004) introduce a concept of water footprint of a country, defined as the volume of water needed for the production of the goods and services consumed by the inhabitants of the country. The internal water footprint is the volume of water used from domestic water resources; the external water footprint is the volume of water used in other countries to produce goods and services imported and consumed by the inhabitants of the country. As emphasized by the authors, knowing the virtual water flows entering and leaving a country can shed an entirely new light on the actual water scarcity of a country. Jordan, as an example, imports about 5 to 7 billion m^3 of

Agricultural Product	USA	China	India	Russia	Indonesia	Australia	Brazil	Japan	Mexico	Italy	Netherlands	World Average
Rice (paddy)	1275	1321	2850	2401	2150	1022	3082	1221	2182	1679		2291
Rice (husked)	1656	1716	3702	3118	2793	1327	4003	1586	2834	2180		2975
Rice (broken)	1903	1972	4254	3584	3209	1525	4600	1822	3257	2506		3419
Wheat	849	690	1654	2375		1588	1616	734	1066	2421	619	1334
Maize	489	801	1937	1397	1285	744	1180	1493	1744	530	408	909
Soybeans	1869	2617	4124	3933	2030	2106	1076	2326	3177	1506		1789
Sugarcane	103	117	159		164	141	155	120	171			175
Cotton seed	2535	1419	8264		4453	1887	2777		2127			3644
Cotton lint	5733	3210	18,694		10,072	4268	6281		4812			8242
Barley	702	848	1966	2359		1425	1373	697	2120	1822	718	1388
Sorghum	782	863	4053	2382		1081	1609		1212	582		2853
Coconuts		749	2255		2071		1590		1954			2545
Millet	2143	1863	3269	2892		1951		3100	4534			4596
Coffee (green)	4864	6290	12,180		17,665		13,972		28,119			17,373
Coffee (roasted)	5790	7488	14,500		21,030		16,633		33,475			20,682
Tea (made)		11,110	7002	3002	9474		6592	4940				9205
Beef	13,193	12,560	16,482	21,028	14,818	17,112	16,961	11,019	37,762	21,167	11,681	15,497
Pork	3946	2211	4397	6947	3938	5909	4818	4962	6559	6377	3790	4856
Goat meat	3082	3994	5187	5290	4543	3839	4175	2560	10,252	4180	2791	4043
Sheep meat	5977	5202	6692	7621	5956	6947	6267	3571	16,878	7572	5298	6143
Chicken meat	2389	3652	7736	5763	5549	2914	3913	2977	5013	2198	2222	3918
Eggs	1510	3550	7531	4919	5400	1844	3337	1884	4277	1389	1404	3340
Milk	695	1000	1369	1345	1143	915	1001	812	2382	861	641	990
Milk powder	3234	4648	6368	6253	5317	4255	4654	3774	11,077	4005	2982	4602
Cheese	3457	4963	6793	6671	5675	4544	4969	4032	11,805	4278	3190	4914
Leather (bovine)	14,190	13,513	17,710	22,575	15,929	18,384	18,222	11,864	40,482	22,724	12,572	16,656

From Chapagain and Hoekstra, 2004.

TABLE 1.10 Average Virtual Water Content (m³/ton) of Some Selected Products for a Number of Selected Countries

Product	Virtual Water Content (Liters)	Product	Virtual Water Content (Liters)
1 glass of beer (250 ml)	75	1 glass of wine (125 ml)	120
1 glass of milk (200 ml)	200	1 glass of apple juice (200 ml)	190
1 cup of coffee (125 ml)	140	1 glass of orange juice	170
1 cup of tea (250 ml)	35	1 bag of potato crisps (220 g)	185
1 slice of bread (30 g)	40	1 egg (40 g)	135
1 slice of bread with cheese (30 g)	90	1 hamburger (150 g)	2400
1 potato (100 g)	25	1 tomato (70 g)	13
1 apple (100 g)	70	1 orange (100 g)	50
1 cotton T-shirt (500 g)	4100	1 pair of shoes (bovine leather)	8000
1 sheet of A4-paper (80 g/m^2)	10	1 microchip (2 g)	32

From Chapagain and Hoekstra, 2004.

TABLE 1.11   Global Average Virtual Water Content of Some Selected Products, per Unit of Product

virtual water per year, which is in sheer contrast with the 1 billion m^3 of annual water withdrawal from domestic water sources. As another example, Egypt, with water self-sufficiency high on the political agenda and with a total water withdrawal inside the country of 65 billion m^3/yr, still has an estimated net virtual water import of 10 to 20 billion m^3/yr.

Hoekstra and Hung (2002) present a detailed study of the volumes of virtual water trade flows between nations in the period from 1995 to 1999. The authors also calculate the virtual water balances of nations within the context of national water needs and water availability. The results of the study show that about 13 percent of the water used for crop production in the world was not consumed domestically, but exported in virtual form. Wheat accounts for 30 percent of the crop-related virtual water trade, followed by soybean (17 percent), rice (15 percent), maize (9 percent), raw sugar (7 percent), and barley (5 percent).

The virtual water trade situation strongly varies between countries, as shown in Table 1.12. World regions with a significant net virtual water import are central and south Asia, western Europe, North Africa, and the Middle East. Two other regions with net virtual water import, but less substantial, are southern and central Africa. Regions with significant net virtual water export are North America, South America, Oceania, and Southeast Asia. Three other regions with net virtual water export, but less substantial, are the FSU (former Soviet Union), Central America, and Eastern Europe. North America (the United States and Canada) is by far the biggest virtual water exporter in the world, while central and south Asia is by far the biggest virtual water importer.

As discussed by IWMI (2007), trade in food and virtual water results in "real" water savings only when the water saved can be reallocated to other uses, such as domestic, industrial, or environmental. Many traded crops are grown under rain-fed conditions. Rainwater usually cannot be allocated to other uses besides alternative rain-fed crops. Reductions in irrigation water withdrawals result in real water savings. For example, importing paddy rather than growing it can result in irrigation water savings, though not necessarily. In Asia, during the monsoon, the combination of abundant rain, floods, and limited storage capacity means that there is no alternative use for the water "saved" by importing paddy rather than growing it. Some countries where water resources are very scarce often have no option but to import. Egypt, for example, cannot grow all the cereal that it currently imports because it does not have the necessary water resources at its disposal. Thus, it is misleading to hold up Egypt as an example of water savings through global trade since, to begin with, it has little or no water to save (IWMI, 2007).

In summary, the idea of food trade as an answer to water shortages is appealing. Growing food where water is abundant and trading it to water-short areas is being recognized, in theory, as having a large potential to save water and minimize new investment in irrigation infrastructure. However, under the prevailing political and economic climate, it is unlikely that food trade alone will solve problems of water scarcity in the near term (IWMI, 2007). Many factors contribute to uncertainties water-scarce countries face when considering radical changes in food trade patterns, including food security, food sovereignty, and employment of their rural populations. On the global, geopolitical level, the fact remains that some of the largest food-exporting countries are in the group of most developed nations, but at the same time they have some of the highest subsidies in the agricultural sector (e.g., United States, Canada, Australia, and France). Some of the largest exporters also face serious environmental and societal problems due to overexploitation of water resources for irrigation, including depletion of

Rank	Country	Net Export/Import (km³)	Rank	Country	Net Export/Import (km³)	Rank	Country	Net Export/Import (km³)
Top 30 Virtual Water Export Countries								
1	United States	758.3	11	Paraguay	42.1	21	Sudan	5.8
2	Canada	272.5	12	Kazakhstan	39.2	22	Bolivia	5.3
3	Thailand	233.3	13	Ukraine	31.8	23	Saint Lucia	5.2
4	Argentina	226.3	14	Syria	21.5	24	United Kingdom	4.8
5	India	161.1	15	Hungary	19.8	25	Burkina Faso	4.5
6	Australia	145.6	16	Myanmar	17.4	26	Sweden	4.2
7	Vietnam	90.2	17	Uruguay	12.1	27	Malawi	3.8
8	France	88.4	18	Greece	9.8	28	Dominica	3.1
9	Guatemala	71.7	19	Dominican R.	9.7	29	Benin	3.0
10	Brazil	45.0	20	Romania	9.1	30	Slovakia	3.0
Top 30 Virtual Water Import Countries								
1	Sri Lanka	428.5	11	Belgium	59.6	21	Morocco	27.7
2	Japan	297.4	12	Saudi Arabia	54.4	22	Peru	27.1
3	Netherlands	147.7	13	Malaysia	51.3	23	Venezuela	24.6
4	Korea Rep.	112.6	14	Algeria	49.0	24	Nigeria	24.0
5	China	101.9	15	Mexico	44.9	25	Israel	23.0
6	Indonesia	101.7	16	Taiwan	35.2	26	Jordan	22.4
7	Spain	82.5	17	Colombia	33.4	27	South Africa	21.8
8	Egypt	80.2	18	Portugal	31.1	28	Tunisia	19.3
9	Germany	67.9	19	Iran	29.1	29	Poland	18.8
10	Italy	64.3	20	Bangladesh	28.7	30	Singapore	16.9

From Hoekstra and Hung, 2002.

TABLE 1.12 The Top 30 Virtual Water Export Countries and the Top 30 Virtual Water Import Countries in the Period 1995–1999

nonrenewable groundwater (e.g., India, Australia, and the United States). A joint effort by governments, international finance institutions, and research organizations is needed to analyze the geopolitical importance of virtual water. This should include the opportunities and threats involved and the associated political processes underlying decision making on application of the virtual water trade concept (IHE Delft and World Water Council, 2003).

1.7 Sustainability

The term "sustainable development" was popularized by the World Commission on Environment and Development in its 1987 report titled *Our Common Future*. The report, published as a book, is also known as the Brundtland Report, after the Chair of the Commission and former Prime Minister of Norway, Gro Harlem Brundtland. The aim of the World Commission was to find practical ways of addressing the environmental and developmental problems of the world. In particular, it had three general objectives:

- To re-examine the critical environmental and developmental issues and to formulate realistic proposals for dealing with them
- To propose new forms of international cooperation on these issues that will influence policies and events in the direction of needed changes
- To raise the level of understanding and commitment to action of individuals, voluntary organizations, businesses, institutes, and governments

Our Common Future was written after 3 years of public hearings and over 500 written submissions. Commissioners from 21 countries analyzed this material, with the final report submitted to and accepted by the United Nations General Assembly in 1987 (UNESCO, 2002). In various publications, debates, interpretations, and reinterpretations over the course of years, the findings of the commission and the final document (resolution) of the United Nations General Assembly were in many cases stripped down to the following widely cited single sentence which states that development is sustainable when: "meeting the needs of the present without compromising the ability of future generations to meet their own needs." Since this sentence seems to focus only on "human generations", it has been criticized by some as too narrow and failing to address the natural environment. However, the commission and the assembly did address the human and natural environments as a whole and in a holistic manner, which can be seen from key related statements of the official UN resolution 42/187 (DESA, 1999). For example, the General Assembly

- Is concerned about the accelerating deterioration of the human environment and natural resources and the consequences of that deterioration for economic and social development
- Believes that sustainable development, which implies meeting the needs of the present without compromising the ability of future generations to meet their own needs, should become a central guiding principle of the United Nations, governments and private institutions, and organizations and enterprises

- Recognizes, in view of the global character of major environmental problems, the common interest of all countries to pursue policies aimed at sustainable and environmentally sound development

- Is convinced of the importance of a reorientation of national and international policies toward sustainable development patterns

- Agrees with the commission that while seeking to remedy existing environmental problems, it is imperative to influence the sources of those problems in human activity, and economic activity in particular, and thus to provide for sustainable development

- Agrees further that an equitable sharing of the environmental costs and benefits of economic development between and within countries and between present and future generations is a key to achieving sustainable development

- Concurs with the commission that the critical objectives for environment and development policies, which follow from the need for sustainable development, must include preserving peace, reviving growth and changing its quality, remedying the problems of poverty and satisfying human needs, addressing the problems of population growth and of conserving and enhancing the resource base, reorienting technology and managing risk, and merging environment and economics in decision making

- Decides to transmit the report of the commission to all governments and to the governing bodies of the organs, organizations, and programs of the United Nations system, and invites them to take account of the analysis and recommendations contained in the report of the commission in determining their policies and programs

- Calls upon all governments to ask their central and sectoral economic agencies to ensure that their policies, programs, and budgets encourage sustainable development and to strengthen the role of their environmental and natural resource agencies in advising and assisting central and sectoral agencies in that task

Twenty years after this UN Resolution, it seems little has changed in the practice of most governments and their "agencies" at various levels, as indicated by examples described earlier. They are either unable or unwilling to fully address and then act to start solving the many problems of unsustainable development. This is partly because of the political price they are afraid to pay, anticipating that many of the required urgent measures may be unpopular with the general public. At the same time the "public," to which politicians and bureaucrats often give little credit, is navigating between sensationalist headlines and various contradicting scientific and technical reports while trying to formulate its own opinion. Educating the public (which by default consists of many "stakeholders" and many opinions) about various choices including tough ones is therefore the first but also the crucial step on the path of achieving sustainable development. Groundwater is a perfect example of many misunderstandings, by both the public and the bureaucrats (politicians), of the meaning of sustainability. This may be because groundwater is mysterious by definition: as soon as we can see it, it is not groundwater anymore. It is troubling, however, when water (groundwater) professionals including those working for government agencies announce certain groundwater policies and qualify them as being "sustainable," even though there was no public debate or involvement of independent groundwater professionals to speak of. The following example illustrates this point. It is a paraphrased paper of a "water specialist" working for a government agency in one of the Midwestern states in the United States. The article appeared in 2007, in the official journal of a nonprofit organization ostensibly working to both educate the public about and protect groundwater. The specialist was explaining how his state is blessed with precious groundwater resources that were used very successfully to better lives of its farmers through crop irrigation. The specialist also stated that in some parts of the state the aquifer water would last at least another 250 years and in some 25 years. And because it is readily available, it should be used to better lives of the farmers and whole communities even more by enabling additional irrigation of more corn fields. Those vast old and new corn fields would be used for production of ethanol in many plants that are being built and many more that would be built, all resulting in great benefits for the state rural communities, the state itself, the country, and indeed the whole world. More corn and more ethanol used for production of car fuel mean less burning of oil, less production of carbon, and it slows down global warming. What could be better? The specialist only forgot to speculate what would happen to those parts of his state where the aquifer runs dry after 25 years.

Some may call the above view of groundwater sustainability simply gambling with groundwater. There is, however, only one large-scale example in which gambling with groundwater has arguably paid off so far. It is the city of Las Vegas in Nevada, United States. The city grew in the middle of the desert thanks to the readily available significant reserves of groundwater beneath it and was known for its artesian wells. It first served as a regional railroad and mining center. The growth accelerated when surface water arrived from Lake Mead on the Colorado River in 1960s, and gambling became the dominant industry. In the meantime, groundwater withdrawal lowered groundwater levels in the regional aquifer, artesian wells stopped flowing, and all springs dried up. Las Vegas is still growing faster than any other large city in the United States (see Fig. 1.20) although the Colorado River may prove to be an unreliable source because its use is also heavily committed to southern California and Arizona. During this time countless visitors from all over the world have left hundreds of billions of dollars in the city, enabling its government to implement some of the most advanced water management practices available. Las Vegas has one of the largest deep artificial aquifer recharge operations in the world, including aquifer storage and recovery wells. It uses treated wastewater to maintain lush landscaping, numerous golf courses, and many water fountains (Fig. 1.21). Finally, it hosts offices of various state and federal agencies, consulting companies, research laboratories, and institutes many of which are directly or indirectly working on water resources–related projects. It is, of course, possible that some other communities, cities, or even entire societies may try to emulate the luck of Las Vegas in gambling with groundwater.

The multiple aspects of groundwater sustainability are addressed in the Alicante Declaration, which since its initiation has gained wide recognition among groundwater professionals worldwide. The declaration is the action agenda that resulted from the debates held in Alicante, Spain, on January 23rd to 27th, 2006, during the *International Symposium on Groundwater Sustainability* (ISGWAS). This call for action for responsible use, management, and governance of groundwater is reproduced below in its entirety (ISGWAS, 2006; available at http://aguas.igme.es/igme/ISGWAS):

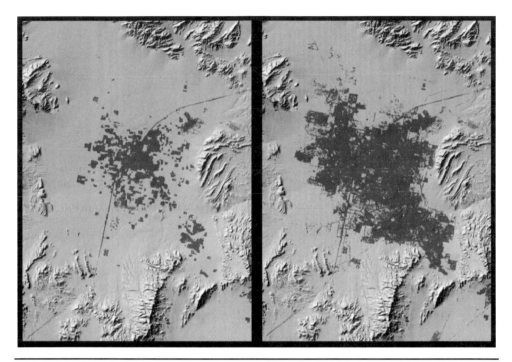

FIGURE 1.20   These two images of Las Vegas, NV, show urban extent as it was in 1973 and 1992. Between these years, the urban area grew dramatically throughout the level basin. Visible in each image are local landforms, such as the surrounding mountain ranges. The width of the area shown is approximately 48 mi or 77 km. (From Auch et al., 2004.)

Water is essential for life. Groundwater—that part of all water resources that lie underneath land surface—constitutes more than ninety five percent of the global, unfrozen freshwater reserves. Given its vast reserves and broad geographical distribution, its general good quality, and its resilience to seasonal fluctuations and contamination, groundwater holds the promise to ensure current and future world communities an affordable and safe water supply. Groundwater is predominantly a renewable resource which, when managed properly, ensures a long-term supply that can help meet the increasing demands and mitigate the impacts of anticipated climate change. Generally, groundwater development requires a smaller capital investment than surface water development and can be implemented in a shorter timeframe.

Groundwater has provided great benefits for many societies in recent decades through its direct use as a drinking water source, for irrigated agriculture and industrial development and, indirectly, through ecosystem and streamflow maintenance. The development of groundwater often provides an affordable and rapid way to alleviate poverty and ensure food security. Further, by understanding the complementary nature of ground and surface waters, thoroughly integrated water-resources management strategies can serve to foster their efficient use and enhance the longevity of supply.

Instances of poorly managed groundwater development and the inadvertent impact of inadequate land-use practices have produced adverse effects such as water-quality degradation, impairment of aquatic ecosystems, lowered groundwater levels and, consequently, land subsidence and the drying of wetlands. As it is less costly and more effective to protect groundwater resources from degradation than to restore them, improved water management will diminish such problems and save money.

To make groundwater's promise a reality requires the responsible use, management and governance of groundwater. In particular, actions need to be taken by water users, who sustain their

FIGURE 1.21   (Left) One of many golf courses in Las Vegas, NV, using treated wastewater for irrigation. (Photograph by Lynn Betts, 2000; courtesy of National Resources Conservation Service, United States Department of Agriculture). (Right) Treated wastewater is used for spectacular Fountains of Bellagio designed by Claire Kahn of WET Design. (Photograph by Claire Kahn, http://www.wetdesign.com/client/bellagio/index.html.)

well-being through groundwater abstraction; decision makers, both elected and non elected; civil society groups and associations; and scientists who must advocate for the use of sound science in support of better management. To this end, the undersigners recommend the following actions:

- *Develop more comprehensive water-management, land-use and energy-development strategies that fully recognize groundwater's important role in the hydrologic cycle.* This requires better characterization of groundwater basins, their interconnection with surface water and ecosystems, and a better understanding of the response of the entire hydrologic system to natural and human-induced stresses. More attention should be given to non-renewable and saline groundwater resources when such waters are the only resource available for use.

- *Develop comprehensive understanding of groundwater rights, regulations, policy and uses.* Such information, including social forces and incentives that drive present-day water management practices, will help in the formulation of policies and incentives to stimulate socially- and environmentally-sound groundwater management practices. This is particularly relevant in those situations where aquifers cross cultural, political or national boundaries.

- *Make the maintenance and restoration of hydrologic balance a long-term goal of regional water-management strategies.* This requires that water managers identify options to: minimize net losses of water from the hydrologic system; encourage effective and efficient water use; and ensure the fair allocation of water for human use as well as ecological needs, taking into account long-term sustainability. Hydrological, ecological, economic and socioeconomic assessments should be an integral part of any water-management strategy.

- *Improve scientific, engineering and applied technological expertise in developing countries.* This requires encouraging science-based decision-making as well as "north-south" and "south-south" cooperation. Further, it is important that funds be allocated for programs that encourage the design and mass-dissemination of affordable and low-energy consuming water harnessing devices for household and irrigation.

- *Establish ongoing coordinated surface water and groundwater monitoring programs.* This requires that data collection become an integral part of water-management strategies so that such strategies can be adapted to address changing socio-economic, environmental, and climatic conditions. The corresponding data sets should be available to all the stakeholders in a transparent and easy way.

- *Develop local institutions to improve sustainable groundwater management.* This requires that higher-level authorities become receptive to the needs of local groups and encourage the development and support of strong institutional networks with water users and civic society.

- *Ensure that citizens recognize groundwater's essential role in their community and the importance of its responsible use.* This requires that science and applied technology serve to enhance education and outreach programs in order to broaden citizen understanding of the entire hydrologic system and its global importance to current and future generations. (ISGWAS, 2006; available at http://aguas.igme.es/igme/ISGWAS)

References

ADB (the Asian Development Bank), 2004. Water in Asian cities. Utilities' performance and civil society views. 97 p. Available at: http://www.adb.org/Documents/Books/Water_for_All_Series/Water_Asian_Cities/water_asian_cities.pdf.

AFP (Agence France-Presse), 2005. India faces severe water crisis in 20 years: World Bank. New Delhi (AFP), Oct 05, 2005. Available at: http://www.terradaily.com/news/water-earth-05ze.html.

Allan, J.A., 1996. Policy responses to the closure of water resources. In: *Water Policy: Allocation and Management in Practice.* Howsam, P., Carter, R., editors. Chapman and Hall, London.

Allan, J.A., 1998. Virtual water: A strategic resource. Global solutions to regional deficits. *Ground Water*, vol. 36, no. 4, pp. 545–546.

Amarasinghe, U.A., Mutuwatta, L., and Sakthivadivel, R., 1999. *Water Scarcity Variations Within a Country: A Case Study of Sri Lanka.* International Water Management Institute, Colombo, Sri Lanka.

Amarasinghe, U.A., Sharma, B.R., Aloysius, N., Scott, C., Vladimir Smakhtin, V., and de Fraiture, C., 2005. Spatial variation in water supply and demand across river basins of India. Research Report 83. International Water Management Institute, Colombo, Sri Lanka, 37 p.

Anderson, M.T., and Woosley, L.H., Jr., 2005. Water Availability for the Western United States—Key scientific challenges. U.S. Geological Survey Circular 1261, Reston, VA, 85 p.

Auch, R., Taylor, J., and Acevedo, W., 2004. Urban growth in American cities: Glimpses of U.S. urbanization. U.S. Geological Survey Circular 1252, Reston, VA, 52 p.

Available at: http://www.iwmi.cgiar.org/waterpolicybriefing/index.asp. Accessed September 3, 2007.

Barker, R., et al., 2000. Global water shortages and the challenge facing Mexico. *International Journal of Water Resources Development*, vol. 16, no. 4, pp. 525–542.

Brown, L.R., 2006. *Plan B 2.0 Rescuing a Planet Under Stress and a Civilization in Trouble.* Earth Policy Institute, W.W. Norton & Co., New York, 365 p.

Bryant, N. (BBC), 2007. Big dry takes toll on Australia's farmers. BBC News, New South Wales, Australia, May 22, 2007. Available at: http://news.bbc.co.uk/2/hi/asia-pacific/6679845.stm.

Bukharov, A.A., 2001. *Baikal in Numbers.* Baikal Museum, Siberian Branch of the Russian Academy of Sciences, Moscow, 72 p.

CBD (Convention of Biological Diversity), 1992. *United Nations Convention on Biological Diversity.* Basic text. Secretariat of the Convention of Biological Diversity, Montreal, Canada.

Center for Biological Diversity, 2007. Save the verde. Available at: http://savetheverde.org/verde/action.html/. Accessed September 5, 2007.

Central Water Commission, 2007. Water data book: Water sector at a glance. Central Water Commission, Ministry of Water Resources, Government of India. Available at: http://cwc.nic.in/cwc_website/main/webpages/dl_index.html. Accessed August 14, 2007.

CGWB (Central Ground Water Development Board), 2006. Dynamic ground water resources of India (as on March 2004). Ministry of Water Resources, Government of India, 120 p. Available at: http://cgwb.gov.in/download.htm. Accessed August 4, 2007.

Chapagain, A.K., and Hoekstra, A.Y., 2004a. Water footprints of nations. Volume 1: Main report. Value of Water Research Report Series No. 16. UNESCO-IHE Institute for Water Education, Delft, the Netherlands, 76 p.

Chapagain, A.K., and Hoekstra, A.Y., 2004b. Water footprints of nations. Volume 2: Appendices. Value of Water Research Report Series No. 16. UNESCO-IHE Institute for Water Education, Delft, the Netherlands.

Chapagain, A.K., Hoekstra, A.Y., and Savenije, H.H.G., 2005. Saving water through global trade. Value of Water Research Report Series No. 17. UNESCO-IHE Institute for Water Education, Delft, the Netherlands, 36 p.

China Daily, 2004. Cities raise water price. Updated: 2004-12-25 09:35. Xinhua News Agency. Available at: http://www.chinadaily.com.cn/english/doc/2004-12/25/content_403306.htm.

Clarke, R. and King, J. 2004. *The Water Atlas.* The New Press, New York, 128 p.

CNNPolitics.com, 2007. Accessed November 1, 2007. Available at: http://www.cnn.com/2007/POLITICS/11/01/drought.politics.ap/index.html.

Cosgrove, W.J., and Rijsberman, F.R., 2000. *World Water Vision: Making Water Everybody's Business.* World Water Council, Earthscan Publications Ltd., London, 108 p.

CSD (Commission on Sustainable Development, United Nations), 2004. *Sanitation: Policy Options and Possible Actions to Expedite Implementation.* UN Economic and Social Council, New York.

CSIS (Center for Strategic and International Studies) and Sandia National Laboratories, 2005. *Global Water Futures. Addressing Our Global Water Future.* Center for Strategic and International Studies, Washington, DC, 133 p.

Davis, T., 2007. The battle for the Verde. *High Country News,* vol. 39, no. 9. Available at: http://www.hcn.org/servlets/hcn.PrintableArticle?article_id=17001. Accessed June 4, 2007.

EEA (European Environment Agency), 1995. *Europe's Environment.* The Dobris Assessment. European Environment Agency, Copenhagen.

EEA (European Environment Agency), 1998. *Europe's Environment*. The Second Assessment, European Environment Agency, Copenhagen.

EEA (European Environment Agency), 1999. *Groundwater Quality and Quantity in Europe*. European Topic Centre on Inland Waters (ETC/IW). European Environment Agency, Copenhagen, 123 p.

EEA (European Environment Agency), 2005. Sustainable use and management of natural resources. EEA Report No. 9. European Environment Agency, Copenhagen, 68 p.

Environment Australia, 2000. Water in a dry land – Issues and challenges for Australia's key resource. Environment Australia, Canberra, 6 p. Available at: http://www.environment.gov.au/water/publications/index.html#groundwater.

Environment Canada, 2007 Municipal water use 2004 statistics. Environment Canada, Ottawa, Ontario, 12 p.

ETC/IW (European Topic Centre on Inland Waters), 1998 The reporting directive· Report on the returns for 1993 to 1995. (France S) ETC/IW Report to DGXI, PO28/97/1.

Eurostat, 1997. Water abstractions in Europe. Internal working document, Water/97/5. Luxembourg.

Falkenmark, M., 1989. Freshwater as a factor in strategic policy and action. In: *Population and Resources in a Changing World*. Davis K., Bernstam M., and Sellers H., editors. Morrison Institute for Population and Resource Studies, Stanford, CA.

Falkenmark, M., and Widstrand, C., 1992. Population and water resources: A delicate balance. *Population Bulletin*. Population Reference Bureau, Washington, DC.

FAO (Food and Agriculture Organization of the United Nations), 2002. World agriculture: Towards 2015/2030. Summary Report. FAO, Rome, 97 p.

FAO (Food and Agriculture Organization of the United Nations), 2003. Review of world water resources by country.Water Reports 23. FAO, Rome, 110 p. Available at: ftp://ftp.fao.org/agl/aglw/docs/wr23e.pdf.

FAO-AQUASTAT, 2007. AQUASTAT main country database. Available at: www.fao.org/ag/agl/aglw/aquastat/main/.

Gardner-Outlaw, T., and Engleman, R., 1997 *Sustaining Water, Easing Scarcity: A Second Update*. Population Action International, Washington, DC.

GIWA (Global International Waters Assessment), 2004. Lake Chad Basin. Regional Assessment 43. University of Kalmar, Kalmar, Sweden. Published on behalf of United Nations Environment Programme, 129 p.

Glazovsky, N.F., 1995. The Aral Sea basin. In: Kasperson, J.X., Kasperson, R.E., and B.L. Turner II, editors. United Nations University Press, Tokyo. Available at: http://www.unu.edu/unupress/unupbooks/uu14re/uu14re00.htm#Contents

Gleick, P.H., 1994. Water, war, and peace in the Middle East. *Environment*, vol. 36, no. 3, p. 6.

Gleick, P.H., 1996. Basic water requirements for human activities: Meeting basic needs. *Water International*, vol. 21, no. 2, pp. 83–92.

Gleick, P.H., 2006. *Water Conflict Chronology*. Pacific Institute for Studies in Development, Environment, and Security, Oakland, CA. Available at: www.worldwater.org/conflictchronology.pdf.

Gollehon, N., and Quinby, W., 2006. Irrigation resources and water costs. In: *Agricultural Resources and Environmental Indicators*. Wiebe, K., and N. Gollehon, editors. United States Department of Agriculture, Economic Research Service, Economic Information Bulletin 16, p. 24–32. www.ers.usda.gov.

Green, C., 2000. If only life were that simple; optimism and pessimism in economics. *Physics and Chemistry of the Earth*, vol. 25, no. 3, pp. 205–212.

Groombridge, B., and Jenkins, M., 1998. *Freshwater Biodiversity: A Preliminary Global Assessment.* World Conservation Monitoring Centre (UNEP-WCMC) – World Conservation Press, Cambridge, UK, 104 p. and various maps.

Harrison, P., Pearce, F., and Raven, P.H., 2001. *AAAS Atlas of Population and Environment.* American Association for the Advancement of Science and the University of California Press, Berkeley, CA, 215 p.

Herdendorf, C.E., 1982. Large lakes of the world. *Journal of Great Lakes Research,* vol. 8, pp. 379–412.

Hoekstra, A.Y., 2006. The global dimension of water governance: Nine reasons for global arrangements in order to cope with local water problems. Value of Water Research Report Series No. 20. UNESCO-IHE Institute for Water Education, Delft, the Netherlands, 33 p.

Hoekstra, A.Y., and Hung, P.Q., 2002. Virtual water trade. A quantification of virtual water flows between nations in relation to international crop trade. Value of Water Research Report Series No. 11. IHE Delft, Delft, the Netherlands, 66 p. and Appendices. http://www.epa.gov/waterinfrastructure/pricing/index.htm

Hutson, S.S., Barber, N.L., Kenny, J.F., Linsey, K.S., Lumia, D.S., and Maupin, M.A., 2004. Estimated use of water in the United States in 2000. U.S. Geological Survey Circular 1268. Reston, VA, 46 p.

ICWE, 1992. The Dublin Statement and report of the conference. In: *International Conference on Water and the Environment: Development Issues for the 21st Century,* January 26–31, 1992, Dublin, Ireland.

IHE Delft and World Water Council, 2003. Session on Virtual water trade and geopolitics (brochure). In: *3rd World Water Forum,* March 16–23, 2003, Kyoto, Japan. UNESCO-IHE Institute for Water Education, Delft, the Netherlands, 8 p.

Isiorho, S.A., and Matisoff, G., 1990. Groundwater recharge from Lake Chad. *Limnology and Oceanography,* 35(4), p. 931–938.

Isiorho, S.A., Oguntola, J.A., and Olojoba, A., (2000). Conjunctive water use as a solution to sustainable economic development in Lake Chad Basin, Africa. In: *10th World Water Congress,* Melbourne, March 12–17, 2000.

IWMI (International Water Management Institute), 2000. *World Water Supply and Demand: 1995 to 2025.* IWMI, Colombo, Sri Lanka.

IWMI (International Water Management Institute), 2007. Does food trade save water? The potential role of food trade in water scarcity mitigation. Water Policy Briefing Series, Issue 25. IWMI, Colombo, Sri Lanka, 8 p.

Jin, M., Liang, X., Cao, Y., and Zhang, R., 2006. Availability, status of development, and constraints for sustainable exploitation of groundwater in China. In: Sharma, B.R., Villholth, K.G., and Sharma, K.D., editors, 2006. *Groundwater Research and Management: Integrating Science into Management Decisions. Groundwater Governance in Asia Series – 1.* International Water Management Institute, Colombo, Sri Lanka, pp. 47–61. Proceedings of IWMI-ITP-NIH International Workshop on Creating Synergy Between Groundwater Research and Management in South and Southeast Asia, Roorkee, India, February 8–9, 2005.

Komives, K., Foster, V., Halpern, J., and Wodon, Q., 2005. *Water, Electricity, and the Poor. Who Benefits from Utility Subsidies?* The World Bank, Washington, DC, 283 p.

Krinner, W., et al., 1999. Sustainable water use in Europe; Part 1: Sectoral use of water. European Environment Agency, Environmental Assessment Report No. 1, Copenhagen, 91 p.

Lawrence, P., Meigh, J., and Sullivan, C., 2002. The water poverty index: An international comparison. Keele Economics Research Papers, KERP 2002/19. Department of Economics Keele University, Keele, United Kingdom, 17 p.

LCBC (Lake Chad Basin Commission), 1998. Integrated and sustainable management of the international waters of the Lake Chad Basin. Lake Chad Basin Commission, Strategic Action Plan.

Li, L., 2007. China sets water-saving goal to tackle looming water crisis. Worldwatch Institute, Washington, DC. Available at: http://www.worldwatch.org/node/4936. Accessed August 14, 2007.

Li, Z., 2006. China issues new regulation on water management, sets fees for usage. Worldwatch Institute, Washington, DC. Available at: http://www.worldwatch.org/node/3892. Accessed August 14, 2007.

Liu, Y., 2006. Water table to drop dramatically near Beijing. Worldwatch Institute, Washington, DC. Available at: http://www.worldwatch.org/node/4407. Accessed August 14, 2007.

McNeill, D., 1998. Water as an economic good. *Natural Resources Forum*, vol. 22, no. 4, pp. 253–261.

Meigs, P., 1953. World distribution of arid and semi-arid homoclimates. In: *Reviews of Research on Arid Zone Hydrology*. United Nations Educational, Scientific, and Cultural Organization, Arid Zone Programme-1, Paris, pp. 203–209. Available at: http://pubs.usgs.gov/gip/deserts/what/world.html. Accessed July 17, 2007.

Minard, A., 2007. Shrinking Lake Superior Also Heating Up. *National Geographic News*, August 3, 2007. Available at: http://news.nationalgeographic.com/news/2007/08/070803-shrinking-lake.html. Accessed August 7, 2007.

Ministry of Water Resources, Government of India, 2007a. National water resources at a glance. Available at: http://wrmin.nic.in. Accessed August 28, 2007.

Ministry of Water Resources, Government of India, 2007b. Annual report, 2005–2006. Available at: http://wrmin.nic.in. Accessed August 28, 2007.

Molden, D., Amarasinghe, U., and Hussain, I., 2001. Water for rural development: Background paper on water for rural development prepared for the World Bank. Working Paper 32. International Water Management Institute (IWMI), Colombo, Sri Lanka, 89 p.

Morris, B.L., et al., 2003. Groundwater and its susceptibility to degradation: A global assessment of the problem and options for management. Early Warning and Assessment Report Series, RS. 03-3. United Nations Environment Programme, Nairobi, Kenya, 126 p.

Murray-Darling Basin Commission, 2007. Murray River System drought update no. 9 August 2007, 3 p. Available at: http://www.mdbc.gov.au/_data/page/1366/Drought_Update9_August071.pdf.

Murray-Rust, H., et al., 2003. Water productivity in the Syr-Darya river basin. Research Report 67. International Water Management Institute (IWMI), Colombo, Sri Lanka, 75 p.

Nace, R.L., 1960. Water management: Agriculture, and groundwater supplies. Geological Circular 415. United States Geological Survey, Washington, DC, 12 p.

NASA (National Aeronautic and Space Administration), 2007a. Earth observatory. Available at: http://earthobservatory.nasa.gov. Accessed December 10, 2006.

NASA (National Aeronautic and Space Administration), 2007b. Visible earth. A catalog of NASA images and animations of our home planet. Available at: http://visibleearth.nasa.gov. Accessed August 8, 2007.

Nixon, S.C., et al., 2000. Sustainable use of Europe's water? State, prospects and issues. European Environment Agency, Environmental Assessment Series No 7, Copenhagen, 36 p.

NUS Consulting Group, 2006. 2005–2006 International Water Report & Cost Survey: Excerpt. Available at: http://www.nusconsulting.com/p_surveys_detail.asp?PRID=33.

Pereira, L.S., Cordery, I., and Iacovides, I., 2002. Coping with water scarcity. IHP-VI, Technical Documents in Hydrology No. 58. UNESCO, Paris, 269 p.

Perry, C.J., Rock, M., and Seckler, D., 1997. Water as an economic good: A solution, or a problem? Research Report 14. International Irrigation Management Institute, Colombo, Sri Lanka.

Pigram, J.J., 2001. Opportunities and constraints in the transfer of water technology and experience between countries and regions. *Water Resources Development*, vol. 17, no. 4, pp. 563–579.

Postel, S., 1997. *Last Oasis: Facing Water Scarcity*, 2nd ed. W.W. Norton & Company, New York, 239 p.

Priscolli, J.D., 1998. Water and civilization: Conflict, cooperation and the roots of a new eco realism. A Keynote Address for the 8th Stockholm World Water Symposium, August 10–13, 1998. Available at: http://www.genevahumanitarianforum.org/docs/Priscoli.pdf.

Puri, S., editor, 2001. Internationally shared (transboundary) aquifer resources. Management, their significance and sustainable management. A framework document. International Hydrological Programme, IHP-VI, IHP Non Serial Publications in Hydrology. UNESCO, Paris, 71 p.

Raskin, P., Hansen, E., and Zhu, Z., 1992. Simulation of water supply and demand in the Aral Sea region. *Water International*, vol. 17, pp. 15–30.

Raucher, R.S., et al., 2005. *The Value of Water: Concepts, Estimates, and Applications for Water Managers*. American Water Works Association Research Foundation (AwwaRF), Denver, CO, 286 p.

Reilly, W.K., and Babbit, H.C., 2005. *A Silent Tsunami: The Urgent Need for Clean Water and Sanitation*. The Aspen Institute, Washington, DC.

Revenga, C., and Cassar, A., 2002. Freshwater trends and projections: Focus on Africa. World Wide Fund for Nature (WWF), Global Network/World Wildlife Fund.

Riviere, J.W.M., 1989. Threats to the world's water. *Scientific American*, vol. 261, no. 9, pp. 80–94.

Salman, S.M.A., editor, 1999. Groundwater, legal and policy perspectives. In: *Proceedings of a World Bank Seminar*, November 1999. Washington, World Bank. WBTP 456.

Sandia National Laboratories, 2005. *Global Water Futures. Addressing Our Global Water Future*. Center for Strategic and International Studies (CSIS), Sandia National Laboratories, Sandia, NM, 133 p.

Seckler, D., 1993. Designing water resources strategies for the twenty-first century. Discussion Paper 16. Water Resources and Irrigation Division, Winrock International, Arlington, VA.

Sengupta, S., (The New York Times), 2006. Water crisis grows worse as India gets richer. *International Herald Tribune/Asia-Pacific*, September 28, 2006. Available at: http://www.iht.com/articles/2006/09/28/news/water.php.

Shiklomanov, I.A., editor, 1999. World water resources: Modern assessment and outlook for the 21st century. (Summary of *World water resources at the beginning of the 21st century*, prepared in the framework of the IHP UNESCO). Federal Service of Russia for Hydrometeorology & Environment Monitoring, State Hydrological Institute, St. Petersburg. Available at: http://espejo.unesco.org.uy/index.html.

SIWI, IFPRI, IUCN, IWMI, 2005. Let it reign: The new water paradigm for global food security. Final Report to CSD-13. SIWI (Stockholm International Water Institute), Stockholm, Sweden, 40 p.

Smith, Z.A., 1985. Federal intervention in the management of groundwater resources: Past efforts and future prospects. *Publius*, vol. 15, no. 1, pp. 145–159.

Streeten, P., 1994. Human development: Means and ends. *American Economic Review*, vol. 84, no. 2, pp. 232–237.

The World Bank, 1992. *World Development Report 1992: Development and the Environment*. Oxford University Press, New York.

The World Bank, 2000. *Entering the 21st Century, World Development Report 1999/2000*. Oxford University Press, New York.

The World Bank, 2002. *World Development Indicators. Development Data Group*.

The World Bank, 2003. Sudan: stabilization and reconstruction – Country economic memorandum. Prepared jointly by the Government of Sudan and Poverty Reduction and Economic Management 2, Africa Region. Report No. 24620-SU.

The World Bank, 2005. India's water economy. Draft report. 95 p. Available at: http://www.worldbank.org.in/WBSITE/EXTERNAL/COUNTRIES/SOUTHASIAEXT/INDIAEXTN. Accessed August 5, 2007.

Thieme, M.L., et al., editors. 2005. *Freshwater Ecoregions of Africa: A Conservation Assessment*. World Wildlife Fund-, Island Press, Washington, DC. 483 p.

U.S. Department of Agriculture, National Agricultural Statistics Service, 2004b. *Farm and Ranch Irrigation Survey (2003)*. Volume 3, Part 1, Special Studies of 2002 Census of Agriculture, AC-02-SS-1.

U.S. Senate Select Committee on National Water Resources, 1960. National water resources and problems: Corom. Print 3, 42 p.

UN (United Nations) and WAPP (World Water Assessment Programme), 2003. *UN World Water Development Report: Water for People, Water for life*. UNESCO and Berghahn Books, Paris, 688 p. Available at: http://www.unesco.org/water/wwap/wwdr1/table_contents/index.shtml

UN (United Nations), 1999. World population prospect: 1998 revision. UN Department of Policy Coordination and Sustainable Development, New York.

UN (United Nations), 2001. Population, environment and development 2001. Population Division, Department of Economic and Social Affairs, New York.

UN HABITAT, 2003. Slums of the World: The face of urban poverty in the new millennium? Monitoring the Millennium Development Goal, Target 11 – World-wide slum dweller estimation, Working Paper, Nairobi, Kenya, 90 p.

UN Water/Africa, 2006. African water development report 2006. Economic Commission for Africa, Addis Ababa, Ethiopia, 370 p.

UNCSD (United Nations Commission for Sustainable Development), 1999. *Comprehensive Assessment of the Freshwater Resources of the World*. UN Division for Sustainable Development, New York.

UNEP (United Nations Environment Programme), 1992. Glaciers and the Environment, 1992. UNEP/GEMS Environment Library No. 9, p. 8. UNEP, Nairobi, Kenya.

UNEP (United Nations Environment Programme), 2002. *Africa Environment Outlook – Past, Present and Future Perspectives*. UNEP, Nairobi, Kenya, 448 p.

UNEP (United Nations Environment Programme), 2007. Vital water graphics. An overview of the state of the world's fresh and marine waters. Available at: http://www.unep.org/dewa/assessments/ecosystems/water/vitalwater/index.htm. Accessed August 2, 2007.

UNEP/GRID-Arendal, 2002. Chronology of change: Natural and anthropogenic factors affecting Lake Chad. UNEP/GRID-Arendal Maps and Graphics Library. Available at: http://maps.grida.no/go/graphic/. Accessed August 4, 2007.

UNESCO (United Nations Educational, Scientific and Cultural Organization), 2006. Water, a Shared Responsibility. The United Nations World Water Development Report 2. UNESCO, World Water Assessment Programme (WAPP), Paris, and Berghahn Books, New York, 584 p.

UNESCO-WWAP (World Water Assessment Program), 2003. *Water for People, Water for Life: The UN World Water Development Report*. UNESCO and Berghahn Books, Barcelona, Spain.

UNFPA (United Nations Population Fund), 1997. *Population and Sustainable Development: Five Years After Rio*. UNFPA, New York, pp. 1–36.

Untersteiner, N., 1975. Sea Ice and Ice Sheets and their Role in Climatic Variations in the Physical Basis of Climate and Climatic Modelling. Global Atmospheric Research Project (GARP). World Meteorological Organisation/International Council of Scientific Unions. Publication Series 16, pp. 206–224.

USEPA (U.S. Environmental Protection Agency) and NARUC (National Association of Regulatory Utility Commissioners), 1999. Consolidate water rates: Issues and practices in single-tariff pricing. U.S. Environmental Protection Agency, Office of Water, Washington, DC., 110 p.

USEPA (U.S. Environmental Protection Agency), 1995. Benefits and costs of prevention: Case studies of community wellhead protection. Volume 1. EPA/813/B-95/005. Washington, DC.

USEPA (U.S. Environmental Protection Agency), 2007a. Great Lakes fact sheet. Available at: http://www.epa.gov/glnpo/statsrefs.html. Accessed August 7, 2007.

USEPA (U.S. Environmental Protection Agency), 2007b. Terms of environment: Glossary, abbreviations and acronyms. Available at: http://www.epa.gov/OCEPAterms/. Accessed July 14, 2007.

USEPA (U.S. Environmental Protection Agency), 2007c. Water and wastewater pricing. Available at: USGS (United States Geologic Survey), 2007b. Water science glossary of terms. Available at: http://ga.water.usgs.gov/edu/dictionary.html. Accessed March 2, 2007.

USGS (United States Geologic Survey), 2007c. Earthshots: Riyadh, Saudi Arabia. Available at: http://earthshots.usgs.gov/Riyadh/Riyadh. Accessed January 27, 2007.

USGS (United States Geological Survey), 1998. Strategic directions for the U.S. Geological Survey Ground-Water Resources Program—A report to Congress. November 30, 1998, 14 p.

USGS (United States Geological Survey), 2001. Earthshots: satellite images of environmental change, Lake Chad, West Africa. Available at: http://edc.usgs.gov/earthshots/slow/LakeChad/LakeChad. Accessed January 2003.

USGS (United States Geological Survey), 2007a. Lake Baikal – A touchstone for global change and rift studies, USGS Fact Sheet. Available at: http://pubs.usgs.gov/fs/baikal/index.html. Accessed August 7, 2007.

Van der Zaag, P., and Savenije, H.H.G., 2006. Water as an economic good: The value of pricing and the failure of markets. Value of Water Research Report Series No. 19. UNESCO-IHE Institute for Water Education, Delft, the Netherlands, 28 p.

Vordzorgbe, S.D., 2003. Managing water risks in Africa. In: *Reports and Proceedings of the Pan-African Implementation and Partnership Conference on Water (PANAFCON)*, December 8–13, 2003, Addis Ababa. UN-Water/Africa, Economic Commission of Africa, Addis Ababa, Ethiopia, pp. 3–27.

Wax, E., 2006. Dying for water in Somalia's Drought. Amid anarchy, warlords hold precious resource. *Washington Post Foreign Service*, April 14, 2006. Available at: http://www.washingtonpost.com/wp-dyn/content/article/2006/04/13/AR2006041302116.html.

Williams, M.B., and Fenske, B.A., 2004. *Demonstrating Benefits of Wellhead Protections Programs.* AWWA Research Foundation and American Water Works Association, Denver, CO, 90 p. and appendices on CD-ROM.

WMO (World Meteorological Organization), 1997. *Comprehensive Assessment of the Freshwater Resources of the World.* WMO, Geneva, p. 9.

WRI (World Resources Institute), 1998. *World resources 1998–99. A Guide to the Global Environment.* Oxford University Press, New York, 384 p.

WSAA (Water Services Association of Australia), 2007. The WSAA Report Card for 2006/07. Melbourne, Sidney, 20 p. Available at: www.wsaa.asn.

WSAA (Water Services Association of Australia), 2007. The WSAA report card for 2006/07—performance of the Australian urban water industry and projections for the future. Melbourne, Sidney, 20 p. Available at: www.wsaa.asn.

Zan, Y., 2006. China's groundwater future increasingly murky. Worldwatch Institute, Washington, DC. Available at: http://www.worldwatch.org/node/4753. Accessed August 14, 2008.

CHAPTER 2

Groundwater System

The primary requirement from the resource management and restoration perspective is to consider available groundwater as part of an interconnected system. Traditional hydrogeology has usually been focused on one aquifer at a time as a study unit, and less attention is directed to interactions between aquifers, aquitards, and surface water features in the area of interest. However, large withdrawals of water from a single aquifer may affect adjacent aquifers and surface water and change water balance. Figure 2.1 illustrates how cessation of pumping from the Upper Floridan aquifer for the industrial water supply of Durango Paper Company at St. Marys, GA, caused a significant rebound (recovery) of groundwater levels not only in the Upper Floridan aquifer but also in the overlying shallower aquifers and the underlying Lower Floridan aquifer, indicating interaquifer leakage (Fig. 2.2). The shutdown resulted in decreased groundwater withdrawal in Camden County, GA, by 35.6 million gal/d (134.7 million mL/d). Figure 2.3 shows that the reduction in withdrawal affected water levels in the Upper Floridan aquifer more than 15 mi from the center of pumping. As a result, many wells in the St. Marys area began to flow for the first time since the early 1940s when the mill's operations began (Fig. 2.4).

Water levels in the Upper Floridan aquifer during early October of 2002, just prior to shutdown, were about 162 ft below sea level at the center of pumping. After the mill ceased operations during October 2002, water levels in all aquifers rose, changing vertical hydraulic gradients and the direction of flow between the surficial and upper Brunswick aquifers. The head in the upper Brunswick aquifer rose above the head in the confined surficial aquifer, reversing the vertical hydraulic gradient between the two aquifers (Fig. 2.2). The lower head in the upper Brunswick aquifer relative to both the surficial and the Upper Floridan aquifers before the shutdown is attributed to its role of a hydrologic sink because it likely pinches out (Peck et al., 2005).

In many cases, historic and ongoing withdrawals have an unknown effect on the overall groundwater system in the absence of long-term monitoring in various parts of the system. It is therefore very important to (1) define all major components of such a system and the ongoing interactions between them, (2) quantify the system in terms of volumes of water stored in its individual components, (3) quantify the rates of groundwater flow between those components, and (4) ascertain the overall conditions and rates of system recharge and discharge.

2.1 Definitions

Aquifer, the basic unit of a groundwater system, is defined as a geologic formation, or a group of hydraulically connected geologic formations, storing and transmitting

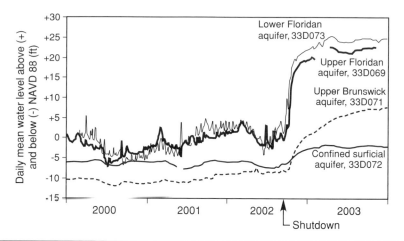

Figure 2.1 Hydrographs for the St. Marys well cluster (33D071, 33D072, and 33D073) and the nearby National Park Service well (33D069), Camden County, GA, 2000–2003. (From Peck et al., 2005.)

significant quantities of potable groundwater. The word comes from two Latin words: *aqua* (water) and *affero* (to bring, to give). The two key terms "significant" and "potable" in this definition are not easily quantifiable. The common understanding is that an aquifer should provide more than just several gallons or liters per minute to individual wells and that water should have less than 1000 mg of dissolved solids. For example, a well

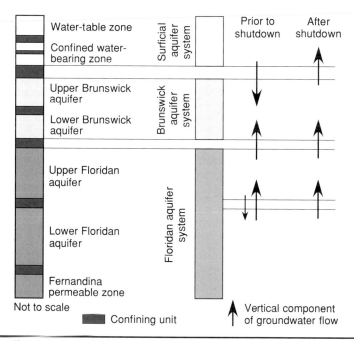

Figure 2.2 Aquifer systems at St. Marys with interpretation of vertical groundwater flow components prior to and after the shutdown of groundwater pumpage at the Durango Company (based on well hydrographs in Fig. 2.1).

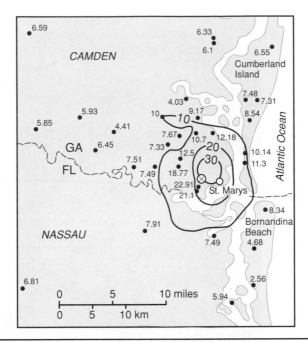

FIGURE 2.3 Observed water-level change, in feet, from September 2001 through May 2003 in wells completed in the Upper Floridan aquifer in Camden County, GA, and Nassau County, FL. Contour lines show equal water-level change in feet. The center of pumping at the Durango Company paper mill is shown with the crossed circle. (Modified from Peck et al., 2005.)

yielding 2 gal/min may be enough for an individual household. However, if this quantity is at the limit of what the geologic formation could provide via individual wells, such an "aquifer" is not considered as a source of significant public water supply. Another issue is groundwater quality. If the groundwater has naturally high total dissolved solids, say 5000 mg/L, it is traditionally disqualified from consideration as a significant source of potable water, regardless of the groundwater quantity. However, with water-treatment technologies, such as reverse osmosis (RO), aquifers with brackish groundwater are increasingly interesting for development.

Aquitard, which is closely related to aquifer, is derived from the Latin words: *aqua* (water) and *tardus* (slow) or *tardo* (to slow down, hinder, delay). An aquitard does store water and is capable of transmitting it, but at a much slower rate than an aquifer and so cannot provide significant quantities of potable groundwater to wells and springs. Determining the nature and the role of aquitards in groundwater systems is very important in both water supply and hydrogeologic contaminant studies. When the available information suggests that there is a high probability for water and contaminants to move through the aquitard within a timeframe of less than 100 years, it is called a leaky aquitard. When the potential movement of groundwater and contaminants through the aquitard is estimated in hundreds or thousands of years, such aquitard is called *competent*. More details on aquifers and aquitards is presented later in this chapter.

Aquiclude is another related term, generally much less used today in the United States but still in relatively wide use elsewhere (Latin word *claudo* means to confine, close, make inaccessible). Aquiclude is equivalent to an aquitard of very low permeability,

FIGURE 2.4 Flowing wells such as this one were common throughout the Georgia coastal area prior to large-scale groundwater development. (From Barlow, 2003; photograph by USGS, 1942.)

which, for all practical purposes, acts as an impermeable barrier to groundwater flow (note that there still is some groundwater stored in aquiclude, but it moves "very, very slowly"). A smaller number of professionals and some public agencies in the United States (such as the USGS, see Lohman et al., 1972) prefer to use the term confining bed instead of aquitard and aquiclude. Accordingly, *semiconfining bed* would correspond to a leaky aquitard. USGS suggests additional qualifiers be specified to more closely explain the nature of a confining layer (i.e., aquitard, aquiclude) of interest, such as "slightly permeable" or "moderately permeable."

Figure 2.5 illustrates major aquifer types in terms of the character and position of the hydraulic head (fluid pressure) in the aquifer, relative to the upper aquifer boundary. The top of the saturated zone of an unconfined aquifer is called the *water table*. The hydraulic head at the water table equals the atmospheric pressure. The thickness of the saturated zone and therefore the position of the water table may change in time due to varying recharge, but the hydraulic head at the water table is always equal to atmospheric

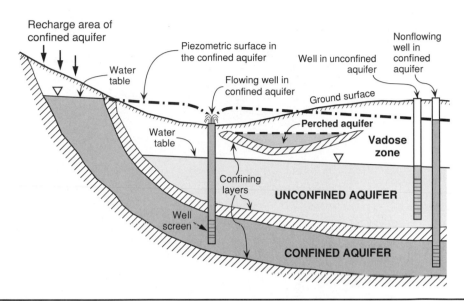

FIGURE 2.5 Schematic presentation of main aquifer types based on position of the hydraulic head (i.e., piezometric surface or water table) in the aquifer. (Modified from USBR, 1977.)

pressure. Note that there may be a low permeable layer, such as clay, somewhere between the ground surface and the water table, but as long as there is an unsaturated (vadose) zone above the water table, the aquifer is unconfined.

An impermeable or low-permeable bed of limited extent above the main water table may cause accumulation of groundwater and formation of a relatively thin saturated zone called *perched aquifer*. Groundwater in the perched aquifer may eventually flow over the edges of the impermeable bed due to recharge from the land surface and continue to flow downward to the main water table, or it may discharge through a spring or seep if the confining bed intersects the land surface.

A confined aquifer is bound above by an aquitard (confining bed), and its entire thickness is completely saturated with groundwater. The hydraulic head in the confined aquifer, also called *piezometric level*, is above this contact. The top of the confined aquifer is at the same time the bottom of the overlying confining bed. Groundwater in a confined aquifer is under pressure, such that static water level in a well screened only within the confined aquifer would stand at some distance above the top of the aquifer. If the water level in such a well rises above ground surface, the well is called a *flowing* or *artesian well* and the aquifer is sometimes called an *artesian aquifer*. The imaginary surface of the hydraulic head in the confined aquifer can be located based on measurements of the water level in wells screened in the confined aquifer. A water table of unconfined aquifers, on the other hand, is not an imaginary surface—it is the top of the aquifer and, at the same time, the top of the saturated zone below which all voids are completely filled with water. A s*emiconfined aquifer* receives water from, or loses water to, the adjacent aquifer from which it is separated by the leaky aquitard.

Hydrogeologic structure is the term used to define discharge and recharge zones of a groundwater system. Discharge and recharge are considered relative to both ground surface and subsurface. Following are the four basic types shown in Fig. 2.6:

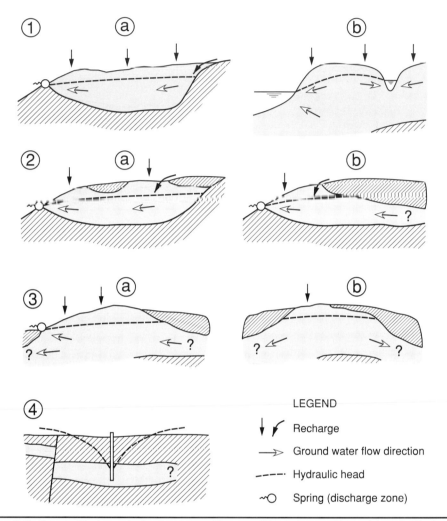

Figure 2.6 Types of hydrogeologic structures: (1) open; (2) semiopen; (3) semiclosed; (4) closed.

1. *Open hydrogeologic structure.* Recharge and discharge zones are fully defined (known). Recharge takes place over the entire areal extent of the system (aquifer), which is directly exposed to the land surface. Discharge of the system is either at the contact with the impermeable base (case 1a) or along a main erosional basis such as a large permanent river or coastal line (case 1b).

2. *Semiopen hydrogeologic structure.* The discharge zone is fully defined and the groundwater system is partially isolated from the land surface by low-permeable or impermeable cover. The recharge zones are mostly or partially known (cases 2a and 2b, respectively).

3. *Semiclosed hydrogeologic structure.* Recharge zones are known or partially known, whereas discharge zones are only partially known (case 3a) or unknown (case 3b).

4. *Closed hydrogeologic structure.* The groundwater system (aquifer) is completely isolated by impermeable geologic units and does not receive recharge. In practice, such a system can only be discovered by drilling, and the absence of any significant recharge (from the land surface or from adjacent aquifers) is manifested by large, continuingly increasing drawdowns during pumping.

In some cases, an aquifer may indeed be completely isolated from the "rest of the world." The presence of freshwater in it is testimony to a very different hydrogeologic past when the aquifer was receiving natural recharge from one or more sources such as precipitation, surface water bodies or adjacent aquifers. Various subsequent geologic processes, including faulting and folding, may have resulted in its complete isolation. Such aquifers are called *fossil aquifers* or *nonrenewable aquifers.*

In general, any aquifer that is part of a groundwater system not receiving natural recharge, regardless of the hydrogeologic structure in which it is formed, is called nonrenewable. Typical examples are aquifers and groundwater systems in arid regions with little or no precipitation and without surface water, such as the one shown in Fig. 2.7.

FIGURE 2.7   On August 25, 2000, the Moderate-Resolution Imaging Spectroradiometer (MODIS) acquired this image of a region in Africa's Sahara Desert, including the southern part of the border between Algeria and Libya. Arrows indicate some of the ancient riverbeds. (NASA Photo Library, 2007; image courtesy of Luca Pietranera, Telespazio, Rome, Italy.)

The dendritic structures of ancient riverbeds are clearly visible in this satellite image of the Acacus-Amsak region in Africa's Sahara Desert. Multidisciplinary studies (including paleoclimatology and paleobotany) suggest that this area was wet during the last glacial era, covered by forests and populated by wild animals. On the area's rocks, archaeologists have found a large number of rock paintings and engravings, faint traces of one of the most ancient civilizations of the world. Starting about 12,000 years ago, hunters rapidly learned domestication of buffalo and goat and developed one of the first systems of symbolic art. Extremely dry weather conditions began here about 5000 years ago, resulting in disappearance of surface streams and the civilization itself (NASA, 2007).

In addition to aquifers and aquitards, which can be simply referred to as porous media, a groundwater system comprises many other components that need to be defined and quantified for its successful management. Their brief description is as follows (detailed discussions are given later in the chapter).

- *System geometry*. Extent and thickness of all aquifers and aquitards in the system, including recharge and discharge zones
- *Water storage*. Types of porosity enabling storage of water, volumes of water stored, volumes available for extraction, and volumes available for addition
- *Water budget*. All natural and artificial inputs (recharge) and outputs (discharge) of water, including changes in storage over time
- *Groundwater flow*. Flow directions, velocities, and flow rates
- *Boundary and initial conditions*. Hydraulic and hydrologic conditions along external and internal boundaries of the system, including three-dimensional distribution of the hydraulic heads in the system and their fluctuations in time
- *Water quality*. Natural quality of groundwater stored in and flowing through the system, and chemical characteristics of any anthropogenic contaminants introduced into the system
- *Fate and transport of contaminants*. Movement of contaminants through porous media and various processes affecting their concentration in groundwater
- *System vulnerability*. Risk of water quality degradation and storage depletion due to groundwater extraction, risk of groundwater contamination from anthropogenic sources, and risks related to climatic changes

2.2 Groundwater System Geometry

Figure 2.8 shows key spatial features of a groundwater system. *Recharge area* is the actual land surface through which the system receives water via percolation of precipitation and surface runoff, or directly from surface water bodies such as streams and lakes. When part of a groundwater system, an aquifer may receive water from adjacent aquifers, including through aquitards, but such contact between adjacent aquifers is usually not referred to as the recharge (or discharge) area. Rather, they are indicated as lateral or vertical recharge (discharge) zones from adjacent systems. *Discharge area* is where the system loses water to the land surface, such as via direct discharge to surface water bodies (streams, lakes, wetlands, oceans) or discharge via springs. In an unconfined

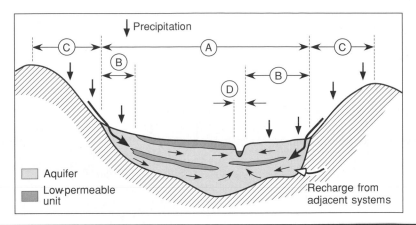

FIGURE 2.8 Key spatial elements of a groundwater system. A, extent; B, recharge area;
C, contributing (drainage) area; D, discharge area

aquifer with a shallow water table, water is also lost via direct evaporation and plant root uptake, which may be significant if riparian vegetation is abundant. An area that gathers surface water runoff, which eventually ends up recharging the system, is called *drainage* or *contributing area*. *System extent* is simply the envelope of its overall limits. It is very important to understand that system geometry is always three-dimensional by definition, and it should be presented as such, including with cross sections and two-dimensional maps for varying depths. Ideally, a three-dimensional computer model of the system geometry is generated as part of the hydrogeologic study, and serves as the basis for subsequent development of a numeric groundwater model for system evaluation and management (Fig. 2.9).

Except in cases of some simple alluvial unconfined aquifers developed in an open hydrogeologic structure, the system contributing (drainage) area, the extent, and the recharge area are usually not equal, and can all have different shapes depending upon the geology and presence of confining layers. Some large regional confined aquifers may have rather small recharge areas compared to the aquifer extent, or they may lack their own recharge area at the land surface and are receiving limited recharge through an overlying aquitard or unconfined aquifer.

In summary, defining the geometric elements of an aquifer or a groundwater system is the first and most important step in the majority of hydrogeologic studies. It is finding the answers to the following questions regarding the groundwater: "where is it coming from?" (contributing area), "where is it entering the system?" (recharge area), "where is it flowing?" (throughout the aquifer extent), and "where is it discharging from the system?" (discharge area).

2.3 Groundwater Storage

2.3.1 Porosity and Effective Porosity

The nature of the *porosity* of porous media (sediments and all rocks in general) is the single most important factor in determining the storage and movement of groundwater

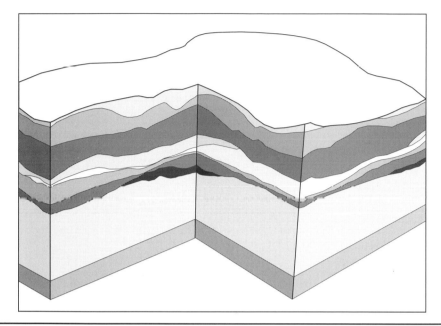

Figure 2.9 Three-dimensional conceptual site model showing different stratigraphic layers and ready for transfer into a numeric groundwater model. (Modified from Oostrom et al., 2004.)

in the subsurface. Many quantitative parameters describing "life cycle" of water and contaminants (when present) within a groundwater system directly or indirectly depend on porosity. Here are just a few: infiltration of precipitation into the subsurface, rock (sediment) permeability, groundwater velocity, volume of water that can be extracted from the groundwater system, and diffusion of contaminants into the porous media solids.

Porosity (n) is defined as the percentage of voids (empty space occupied by water or air) in the total volume of rock, which includes both solids and voids:

$$n = \frac{V_v}{V} \times 100\% \tag{2.1}$$

where V_v = volume of all rock voids and V = total volume of rock (in geologic terms, rock refers to all the following: soils, unconsolidated and consolidated sediments, and any type of rock in general). Assuming the specific gravity of water equals unity, total porosity, as a percentage, can be expressed in four different ways (Lohman, 1972):

$$n = \frac{V_i}{V} = \frac{V_w}{V} = \frac{V - V_m}{V} = 1 - \frac{V_m}{V} [\times 100\%] \tag{2.2}$$

where n = porosity, in percent per volume
   V = total volume
   V_i = volume of all interstices (voids)
   V_m = aggregate volume of mineral (solid) particles
   V_w = volume of water in a saturated sample

Porosity can also be expressed as:

$$n = \frac{\rho_m - \rho_d}{\rho_m} = 1 - \frac{\rho_d}{\rho_m}[\times 100\%] \tag{2.3}$$

where ρ_m = average density of mineral particles (grain density) and ρ_d = density of dry sample (bulk density).

The shape, amount, distribution, and interconnectivity of voids influence the permeability of rocks. Voids, on the other hand, depend on the depositional mechanisms of unconsolidated and consolidated sedimentary rocks, and on various other geologic processes that affect all rocks during and after their formation. *Primary porosity* is the porosity formed during the formation of rock itself, such as voids between the grains of sand, voids between minerals in hard (consolidated) rocks, or bedding planes of sedimentary rocks. *Secondary porosity* is created after the rock formation mainly due to tectonic forces (faulting and folding), which create micro- and macrofissures, fractures, faults, and fault zones in solid rocks. Both the primary and secondary porosities can be successively altered multiple times, thus completely changing the original nature of the rock porosity. These changes may result in porosity decrease, increase, or altering of the degree of void interconnectivity without a significant change in the overall void volume.

The following discussion by Meinzer (1923), and the figure that accompanies it (Fig. 2.10) is probably the most cited explanation of rock porosity, and one can hardly add anything to it:

The porosity of a sedimentary deposit depends chiefly on (1) the shape and arrangement of its constituent particles, (2) the degree of assortment of its particles, (3) the cementation and compacting to which it has been subjected since its deposition, (4) the removal of mineral matter through solution by percolating waters, and (5) the fracturing of the rock, resulting in joints and other openings. Well-sorted deposits of uncemented gravel, sand, or silt have a high porosity, regardless of whether they consist of large or small grains. If, however, the material is poorly sorted small particles occupy the spaces between the larger ones, still smaller ones occupy the spaces between these small particles, and so on, with the result that the porosity is greatly reduced (A and B). Boulder clay, which is an unassorted mixture of glacial drift containing particles of great variety in size, may have a very low porosity, whereas outwash gravel and sand, derived from the same source but assorted by running water, may be highly porous. Well-sorted uncemented gravel may be composed of pebbles that are themselves porous, so that the deposit as a whole has a very high porosity (C). Well-sorted porous gravel, sand, or silt may gradually have its interstices filled with mineral matter deposited out of solution from percolating waters, and under extreme conditions it may become a practically impervious conglomerate or quartzite of very low porosity (D). On the other hand, relatively soluble rock, such as limestone, though originally dense, may become cavernous as a result of the removal of part of its substance through the solvent action of percolating water (E). Furthermore hard, brittle rock, such as limestone, hard sandstone, or most igneous and metamorphic rocks, may acquire large interstices through fracturing that results from shrinkage or deformation of the rocks or through other agencies (F). Solution channels and fractures may be large and of great practical importance, but they are rarely abundant enough to give an otherwise dense rock a high porosity.

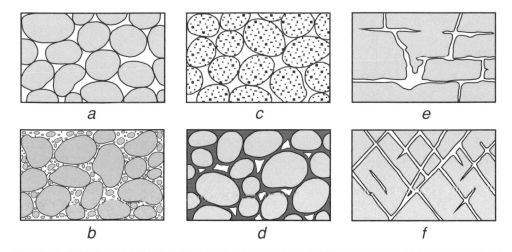

Figure 2.10 Diagram showing several types of rock interstices and the relation of rock texture to porosity. (*a*) Well-sorted sedimentary deposit having high porosity; (*b*) poorly sorted sedimentary deposit having low porosity; (*c*) well-sorted sedimentary deposit consisting of pebbles that are themselves porous and thus the deposit as a whole has a very high porosity; (*d*) well-sorted sedimentary deposit whose porosity has been diminished by the deposition of mineral matter in the interstice's; (*e*) rock rendered porous by solution; (*f*) rock rendered porous by fracturing (Meinzer, 1923).

The porosity of unconsolidated sediments (gravel, sand, silt, and clay) is often called *intergranular porosity* because the solids are loose detritic grains. When such rocks become consolidated, the former intergranular porosity is called *matrix porosity*. In general, the term matrix porosity is applied to primary porosity of all consolidated (hard) rocks, such as porosity between mineral grains (minerals) in granite, gneiss, slate, or basalt. Some unconsolidated or loosely consolidated (semiconsolidated) rocks may contain fissures and fractures, in which case the nonfracture portion of the overall porosity is also called matrix porosity. Good examples are fractured clays and glacial till sediments, or residuum deposits, which have preserved the fabric of the original bedrock in the form of fractures and bedding planes. Sometimes, microscopic fissures in rocks are also considered part of the matrix porosity as opposed to larger fissures and fractures called *macroporosity*. In general, rocks that have both the matrix and the fracture porosity are referred to as *dual-porosity* media. This distinction is important in terms of groundwater flow, which has very different characteristics in fractures and conduits compared to the bulk of the rock. It is also important in contaminant fate and transport analysis, especially when contaminant concentrations are high causing its diffusion into the rock matrix where it can remain for long periods of time. Plots of average total porosity and porosity ranges for various rock types are shown in Figs. 2.11 and 2.12.

When analyzing porosity from the groundwater management perspective, it is very important to make a very clear distinction between the total porosity and the *effective porosity* of the rock. Effective porosity is defined as the volume of interconnected pore space that allows free gravity flow of groundwater. The following anthological discussion by Meinzer (1932) explains why it is important to make this distinction between the total and the effective porosity:

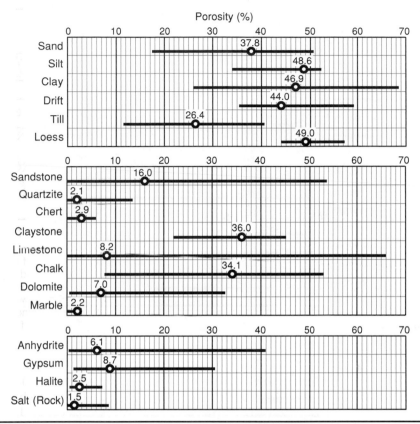

Figure 2.11   Porosity range (horizontal bars) and average porosities (circles) of unconsolidated and consolidated sedimentary rocks. (Kresic, 2007a; copyright Taylor & Francis Group, LLC; printed with permission.)

To determine the flow of ground water, however, a third factor, which has been called the *effective porosity*, must be applied. Much of the cross section is occupied by rock and by water that is securely attached to the rock surfaces by molecular attraction. The area through which the water is flowing is therefore less than the area of the cross section of the water-bearing material and may be only a small fraction of that area. In a coarse, clean gravel, which has only large interstices, the effective porosity may be virtually the same as the actual porosity, or percentage of pore space; but in a fine-grained or poorly assorted material the effect of attached water may become very great, and the effective porosity may be much less than the actual porosity. Clay may have a high porosity but may be entirely impermeable and hence have an effective porosity of zero. The effective porosity of very fine grained materials is generally not of great consequence in determinations of total flow, because in these materials the velocity is so slow that the computed flow, with any assumed effective porosity, is likely to be relatively slight or entirely negligible. The problem of determining effective porosity, as distinguished from actual porosity, is, however, important in studying the general run of water-bearing materials, which are neither extremely fine nor extremely coarse and clean. Hitherto not much work has been done on this phase of the velocity methods of determining rate of flow. No distinction has generally been made between actual and effective porosity, and frequently a factor of 33 1/3 per cent has been used, apparently without even making a test of the porosity. It is certain that the effective porosity of different water-bearing materials ranges between wide limits and that it must be at least roughly determined if reliable results as to rate of flow are to be obtained. It would seem that each

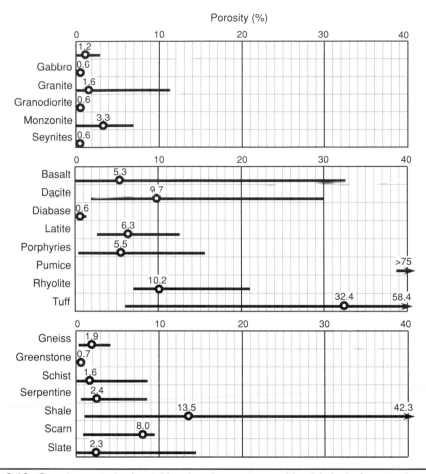

Figure 2.12 Porosity range (horizontal bars) and average porosities (circles) of magmatic and metamorphic rocks. (Kresic, 2007a; copyright Taylor & Francis Group, LLC; printed with permission.)

field test of velocity should be supplemented by a laboratory test of effective porosity, for which the laboratory apparatus devised by Slichter (1905) could be used.

2.3.2 Specific Yield and Coefficient of Storage

Two very different mechanisms are responsible for groundwater release from storage in unconfined and confined aquifers. Respectively, they are explained with two quantitative parameters: specific yield and coefficient of storage.

The *specific yield* of the porous material is defined as the volume of water in the pore space that can be freely drained by gravity due to lowering of the water table. The volume of water retained by the porous media, which cannot be easily drained by gravity, is called *specific retention*. Together, the specific yield and the specific retention are equal to the total porosity of the porous medium (rock). This is schematically shown in Fig. 2.13, in the case

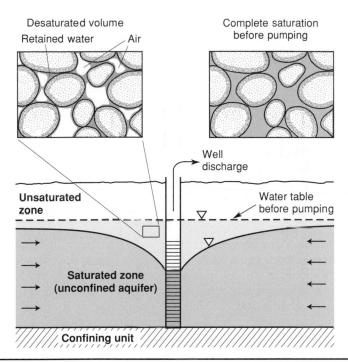

Desaturated volume
Retained water   Air

Complete saturation
before pumping

Well
discharge

**Unsaturated
zone**

Water table
before pumping

**Saturated zone
(unconfined aquifer)**

Confining unit

FIGURE 2.13 During pumping of an unconfined aquifer, water is released due to gravity drainage. Within the cone of depression (volume of aquifer affected by drawdown) not all water is released rapidly because of delayed gravity drainage, and some may be retained permanently. (Modified from Alley et al., 1999.)

of groundwater withdrawal from an unconfined aquifer. Since drainage of pore space by gravity may take long periods of time, especially in fine-grained sediments, values of specific yield determined by various laboratory and field methods are likely somewhat lower than the true values because of limited testing times. A long-term aquifer pumping test or a continuous monitoring of the hydraulic head increase due to a known recharge are arguably the only reliable methods of determining the value of specific yield, which is one of the key parameters for defining quantities of extractable groundwater. These tests provide a long-term lumped hydrodynamic response to pumping (or recharge) by all porous media present in the groundwater system. Consequently, the value of specific yield obtained from such tests cannot be explicitly related to values of effective porosity, even though these two parameters have been often equated by working professionals. The main problem in using specific yield and effective porosity interchangeably is that values of effective (and total) porosity are almost always determined in the laboratory for small samples, and have to be extrapolated (upscaled) to real field conditions, i.e., to a much larger aquifer volume. One important distinction between the specific yield and the effective porosity concepts is that the specific yield relates to volume of water that can be freely extracted from an aquifer, while the effective porosity relates to groundwater velocity and flow through the interconnected pore space. In any case, using total (instead of effective) porosity for calculations of extractable volumes of water would be completely erroneous, as pointed out by Meinzer (1923) in one of his classic publications:

The importance of water that a saturated rock will furnish, and hence its value as a source of water supply, depends on its specific yield—not on its porosity. Clayey or silty formations may contain vast amounts of water and yet be unproductive and worthless for water supply, whereas a compact but fractured rock may contain much less water and yet yield abundantly. To estimate the water supply obtainable from a given deposit for each foot that the water table is lowered, or to estimate the available supply represented by each foot of rise in the water table during a period of recharge, it is necessary to determine the specific yield. Estimates of recharge or of available supplies based on porosity, without regard to the water-retaining capacity of the material, may be utterly wrong.

The presence of fine-grained sediments such as silt and clay, even in relatively small quantities, can greatly reduce specific yield (and effective porosity) of sands and gravels. Values of specific yield for unconfined aquifers generally range between 0.05 and 0.3, although lower or higher values are possible, especially in cases of finer grained and less uniform material (lower values), and uniform coarse sand and gravel (higher values).

One additional mechanism contributing to changes in storage of unconfined aquifers is the compressibility of the water and aquifer solids in the saturated zone. In most cases, the changes in water volume due to unconfined aquifer compressibility are minor and can be ignored for practical purposes. On the other hand, storage of confined aquifers is entirely dependent on compression and expansion of both water and solids, or its elastic properties. Figure 2.14 shows the forces interacting in a confined aquifer: total load exerted on a unit area of the aquifer (σ_T), part of the total load borne by the confined water (ρ), and part borne by the structural skeleton (solids) of the aquifer (σ_e). Assuming that the total load exerted on the aquifer is constant, and if σ is reduced because of pumping, the load borne by the skeleton of the aquifer will increase. This will result in a slight compaction (distortion) of the grains of material, which means that they will encroach somewhat on pore space formerly occupied by water and water will be squeezed out (Fig. 2.15). At the same time, the water will expand to the extent permitted by its elasticity. Conversely, if ρ increases, as in response to cessation of pumping, the hydraulic (piezometric) head builds up again, gradually approaching its original value, and the water itself undergoes slight contraction. With an increase in ρ there is an accompanying decrease in σ_e and the grains of material in the aquifer skeleton return to their former shape. This releases pore space that can now be reoccupied by water moving into the part of the formation that was previously influenced by the compression (Ferris et al., 1962).

Storage properties (storativity) of confined aquifers are defined by the *coefficient of storage*. Although rigid limits cannot be established, the storage coefficients of confined aquifers may range from about 0.00001 to 0.001. In general, denser aquifer materials have smaller coefficient of storage. It is important to note that the value of coefficient of storage in confined aquifers may not be directly dependent on void content (porosity) of the aquifer material (USBR, 1977). *Specific storage* (S_s) of confined aquifers is the volume of water released (or stored) by the unit volume of porous medium, per unit surface of the aquifer, due to unit change in the component of hydraulic head normal to that surface. The unit of specific storage is length^{-1} (e.g., m^{-1} or ft^{-1}) and so when the specific storage is multiplied by aquifer thickness (b), it gives the coefficient of storage (S), which is a dimensionless number: $S = S_s b$. The specific storage is given as

$$S_s = \rho_w g(\alpha + n\beta) \tag{2.4}$$

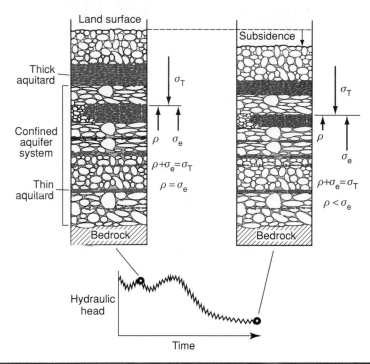

Figure 2.14 Left: In a confined aquifer system, the total weight of the overlying rock and water (σ_T) is balanced by the pore-fluid pressure (ρ) and the intergranular or effective stress (σ_e). Right: Groundwater withdrawal reduces fluid pressures (ρ). As the total stress (σ_T) remains nearly constant, a portion of the load is shifted from the confined fluid to the skeleton of the aquifer system, increasing the effective stress (σ_e) and causing some compression (reduction in porosity). Extended periods of lowered hydraulic head may result in irreversible compaction of the skeleton and land subsidence. Most of the land subsidence occurs as a result of the permanent compaction of the aquitards, which may be delayed due to their slow drainage. (Modified from Galloway et al., 1999.)

where ρ_w = density of water

g = acceleration of gravity

α = compressibility of the aquifer skeleton

n = total porosity

β = compressibility of water

All other things being equal, such as the well pumping rate, the regional nonpumping hydraulic gradient, the initial saturated aquifer thickness, and the hydraulic conductivity, the radius of well influence in a confined aquifer would be significantly larger than in an unconfined aquifer. This is because less water is actually withdrawn from the same aquifer volume in the case of confined aquifers due to the elastic nature of water release from the voids. In other words, in order to provide the same well yield (volume of water), a larger aquifer area would be affected in a confined aquifer than in an unconfined aquifer, assuming they initially have the same saturated thickness. This is illustrated in Figs. 2.16 and 2.17. The initial saturated thickness of the unconfined aquifer is 90 m, and the 90-m thick confined aquifer remains fully saturated throughout the modeled 10-year

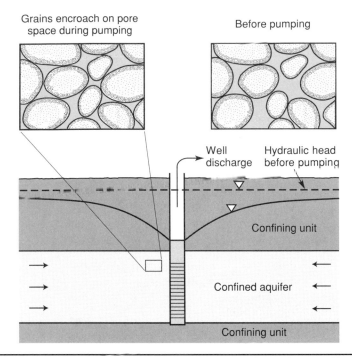

FIGURE 2.15   As the hydraulic head decreases during pumping of a confined aquifer, the fluid pressure of the stored water decreases as well and water is squeezed out of the pore space by the encroachment of solid grains (reduction of porosity). The aquifer remains fully saturated, while its skeleton (solid grains) undergoes slight compression since it has to bear a larger portion of the overburden load.

period. Neither aquifer receives any recharge, either laterally or vertically. The hydraulic conductivity of both aquifers is 5 m/d. The results of computer modeling show that the time-dependent radius of influence (map view in Fig. 2.16) of the unconfined aquifer is incomparably smaller than that of the confined aquifer with the same transmissivity and the same groundwater withdrawal rate. Figure 2.17 shows that, over time and in the absence of any recharge, the drawdown in the confined aquifer increases linearly and at a much faster rate than in the unconfined aquifer. However, this difference is not immediately apparent, as shown by the comparison between the 1-year and the 10-year drawdown. This analysis demonstrates the sensitivity and importance of the storage parameters in estimating impacts of groundwater withdrawal, and the importance of long-term monitoring in groundwater management.

2.3.3   Groundwater Storage and Land Subsidence

Groundwater is always withdrawn from storage in the porous media, regardless of the conditions of a groundwater system recharge. In other words, prior to its extraction from the subsurface, water had to be stored in the porous media voids, i.e., in the storage. It is misleading to associate "storage depletion" only with "unsustainable groundwater extraction practices," or with groundwater pumpage during long periods without significant aquifer recharge, such as multiyear droughts. Figure 2.18 shows some key

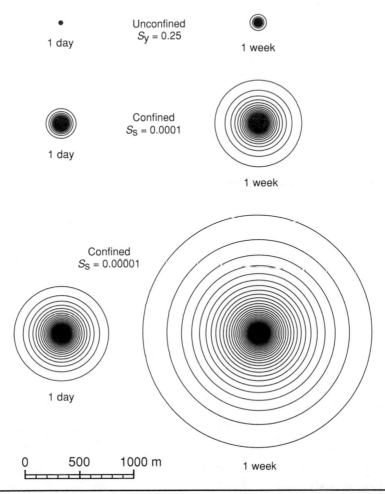

FIGURE 2.16 Development of the radius of influence from a single, fully penetrating well extracting 3600 m^3/d (660 gal/min or 42 L/s) from an unconfined aquifer (top row) and a confined aquifer (two bottom rows) having the same hydraulic conductivity (5 m/d) and the same initial saturated thickness (90 m). The specific yield of the unconfined aquifer is 0.25, and the specific storage of the confined aquifer is 0.0001 m^{-1} (middle) and 0.00001 m^{-1} (bottom). Contour interval of the hydraulic head contour lines is 0.1 m.

concepts of natural groundwater storage. The portion of the saturated zone that changes its thickness in response to natural recharge patterns represents *dynamic storage*. For confined aquifers the dynamic storage is indicated by natural variations in the piezometric surface. This storage volume can vary widely in time depending on seasonal and long-term fluctuations of precipitation and other sources of recharge. Over a multiyear period spanning several natural cycles of wet and dry years, and in the absence of artificial (anthropogenic) groundwater withdrawals, this part of the storage can be considered as fully renewable. The portion of the saturated zone below the multiyear low water table has constant volume of stored groundwater and is therefore referred to as long-term or *static storage*, even though groundwater in it is constantly flowing. In the presence of

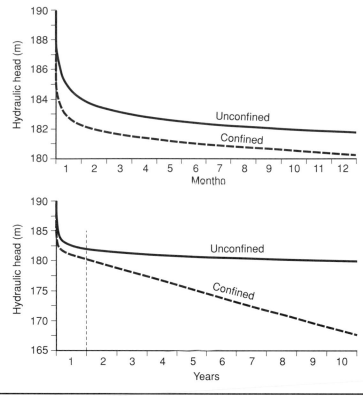

FIGURE 2.17 Drawdown versus time at the pumping well shown in Fig. 2.16, for confined and unconfined conditions. Specific yield of the unconfined aquifer is 0.25; specific storage of the confined aquifer is 0.0001 m^{-1}. Top: Drawdown development during first year of pumping. Bottom: 10 years of pumping (note different vertical scales for the two graphs).

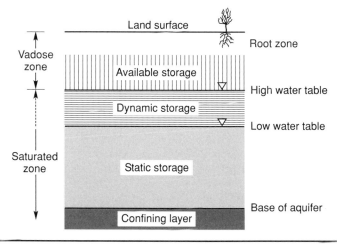

FIGURE 2.18 Schematic presentation of different groundwater storage components in an unconfined aquifer. Note that groundwater is flowing in both the dynamic and the static parts of the storage.

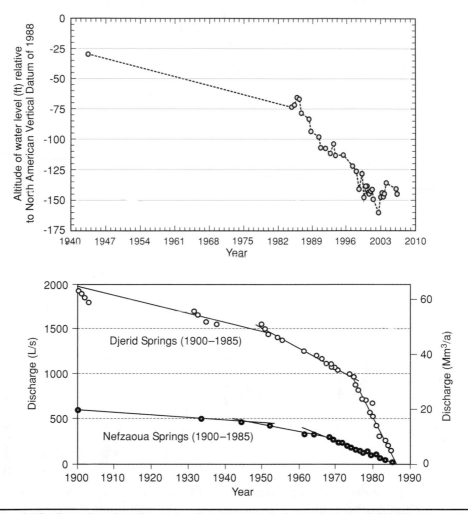

Figure 2.19 Examples of aquifer mining. Top: Water levels for monitoring well SM Df1 screened in the Aquia aquifer at Naval Air Station Patuxent River, MD, 1943–2006, showing response to groundwater withdrawals in excess of 1.0 million gal/d from at least 1946 through 1974, about 1.0 Mgal/d from 1975 through 1991, about 0.8 Mgal/d from 1992 through 1999, and about 0.7 Mgal/d from 2000 through 2005. (Modified from Klohe and Kay, 2007.) Bottom: Progressive elimination of major springflows in southern Tunisia during the twentieth century. (From Margat et al., 2006; copyright UNESCO.)

artificial groundwater withdrawals, the long-term static storage can decrease if the extracted volume of water exceeds the dynamic storage. This is called aquifer mining and is evidenced by the continuing excessive decline of the hydraulic heads or decrease of spring flows (Fig. 2.19). The static storage remains unchanged if the withdrawals equal the dynamic storage. In contrast, the renewable dynamic storage can also increase in cases of induced natural recharge caused by groundwater pumpage near surface water bodies for example. Such pumping may reverse the hydraulic gradients and result in inflow of surface water into the groundwater system as shown in Fig. 2.20. At the same

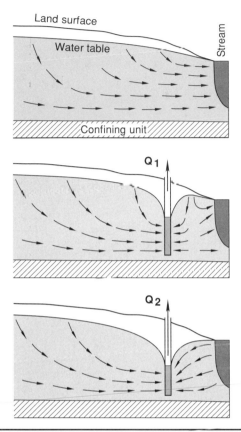

FIGURE 2.20   Induced aquifer recharge due to groundwater withdrawal near a surface water body. (Modified from Alley et al., 1999.)

time, the system does not discharge into the surface stream, which also increases the dynamic storage.

It is obvious that any meaningful quantitative assessment of different storage components is dependent on the availability of long-term monitoring data of the hydraulic head change, as well as of various system inputs (recharge) and outputs (discharge), which cause these changes and the related changes in storage.

Storage capacity of a groundwater system may be irreversibly affected by extensive groundwater withdrawals. As shown in Fig. 2.14, because of the hydraulic head decline in the aquifer system due to pumping, some of the support for the overlying material previously provided by the pressurized water filling the sediment pore space shifts to the granular skeleton of the aquifer system. This increases the intergranular pressure (load). Sand and gravel deposits are relatively incompressible, and the increased intergranular load has small effect on these aquifer materials. However, clay and silt layers comprising confining units and interbeds can be very compressible as water is squeezed from these layers in response to the hydraulic gradient caused by pumping. When long-term declines in the hydraulic head increase the intergranular load beyond the previous maximum load, the structure of clay and silt layers may undergo significant rearrangement, resulting in irreversible aquifer system compaction and land subsidence. The amount of compaction is a function of the thickness and vertical hydraulic conductivity of the clay and silt layers, and the type and structure of the clays and silts. Because of the low hydraulic conductivity of clay and silt layers, the compaction of these layers can continue for months or years after water levels stabilize in the aquifer. In confined aquifer systems that contain significant clay and silt layers and are subject to large-scale groundwater withdrawals, the volume of water derived from irreversible compaction commonly can range from 10 to 30 percent of the total volume of water pumped. This represents a one-time mining of stored groundwater and a permanent reduction in the storage capacity of the aquifer system (Alley et al., 1999; Galloway et al., 1999).

The first recognized land subsidence in the United States from aquifer compaction as a response to groundwater withdrawals was in the area of Santa Clara Valley (now known as "Silicon Valley") in California. Some other well-known areas experiencing significant land subsidence due to groundwater mining include the basin-fill aquifers of south-central Arizona, Las Vegas Valley in Nevada, and the Houston-Galveston area of Texas. Worldwide, the land subsidence in Mexico City, Mexico, is one of the most cited examples of negative impacts caused by aquifer mining. Nothing, however, compares with the example of overexploitation of confined aquifers and the related consequences illustrated in Fig. 2.21. Mining groundwater for agriculture has enabled the San Joaquin Valley of California to become one of the world's most productive agricultural regions, while simultaneously contributing to one of the single largest alterations of the land surface attributed to humankind. In 1970, when the last comprehensive surveys of land subsidence were made, subsidence in excess of 1 ft had affected more than 5200 mi^2 of irrigable land—one-half the entire San Joaquin Valley. The maximum subsidence, near Mendota, was more than 28 ft (9 m). As discussed by Galloway et al. (1999), the economic impacts of land subsidence in the San Joaquin Valley are not well known. Damages directly related to subsidence have been identified, and some have been quantified. Other damages indirectly related to subsidence, such as flooding and long-term environmental effects, merit additional assessment. Some of the direct damages have included decreased storage in aquifers, partial or complete submergence of canals and associated bridges and pipe crossings, collapse of well casings, and disruption of collector drains and irrigation ditches. Costs associated with these damages have been conservatively estimated at 25 million US dollars (EDAW-ESA, 1978). These estimates are not adjusted for changing valuation of the dollar, and do not fully account for the underreported costs associated with well rehabilitation and replacement. When the costs of lost property value due to condemnation, regarding irrigated land, and replacement of irrigation pipelines and wells in subsiding areas are included, the annual costs of subsidence in the San Joaquin Valley soar to $180 million per year in 1993 dollars (G. Bertoldi and S. Leake, USGS, written communication, March 30, 1993; from Galloway et al., 1999).

2.4 Water Budget

Healy et al. (2007) explain in detail the importance and various aspects of quantitative water budget analysis at local and global scales, including interactions between groundwater and surface water within their common water cycle:

> Water budgets provide a means for evaluating availability and sustainability of a water supply. A water budget simply states that the rate of change in water stored in an area, such as a watershed, is

FIGURE 2.21 Approximate location of maximum subsidence in the United States identified by research efforts of Joseph Poland of the USGS (pictured). Signs on pole show approximate altitude of land surface in 1925, 1955, and 1977. The pole is near benchmark S661 in the San Joaquin Valley southwest of Mendota, CA (Galloway et al., 1999).

balanced by the rate at which water flows into and out of the area. An understanding of water budgets and underlying hydrologic processes provides a foundation for effective water-resource and environmental planning and management. Observed changes in water budgets of an area over time can be used to assess the effects of climate variability and human activities on water resources. Comparison of water budgets from different areas allows the effects of factors such as geology, soils, vegetation, and land use on the hydrologic cycle to be quantified. Human activities affect the natural hydrologic cycle in many ways. Modifications of the land to accommodate agriculture, such as installation of drainage and irrigation systems, alter infiltration, runoff, evaporation, and plant transpiration rates. Buildings, roads, and parking lots in urban areas tend to increase runoff and decrease infiltration. Dams reduce flooding in many areas. Water budgets provide a basis for assessing how a natural or human-induced change in one part of the hydrologic cycle may affect other aspects of the cycle.

The most general equation of water budget that can be applied to any water system has the following form:

$$\text{Water Input} - \text{Water Output} = \text{Change in Storage} \tag{2.5}$$

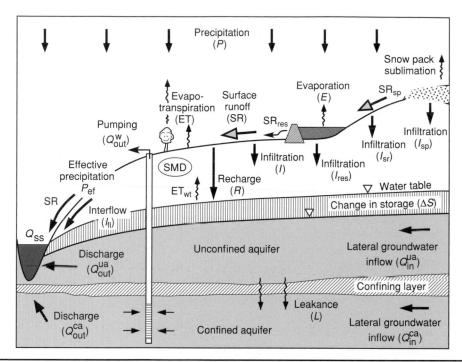

FIGURE 2.22 Elements of water budget of a groundwater system.

Water budget equations can be written in terms of volumes (for a fixed time interval), fluxes (volume per time, such as cubic meters per day or acre-feet per year), and flux densities (volume per unit area of land surface per time, such as millimeters per day). Figure 2.22 shows a majority of the components that contribute to water budget of a groundwater system. Groundwater recharge, which is usually the focus of water supply studies, as well as various methods of its quantification, is explained in detail in Chap. 3. Common to most components of water budget, including groundwater recharge, is that they cannot be measured directly and are estimated from measurements of related quantities (parameters), and estimates of other components. Exceptions are direct measurements of precipitation, stream flows, spring discharge rates, and well pumping rates. Other important quantities that can be measured directly and used in water budget calculations as part of various equations are the hydraulic head (water level) of both groundwater and surface water, and soil moisture content.

Water budget terms are often used interchangeably, sometimes causing confusion. In general, *infiltration* refers to any water movement from the land surface into the subsurface. This water is sometimes called *potential recharge* indicating that only a portion of it may eventually reach the water table (saturated zone). The term a*ctual recharge* is being increasingly used to avoid any possible confusion: it is the portion of infiltrated water that reaches the aquifer and it is confirmed based on groundwater studies. The most obvious confirmation that actual groundwater recharge is taking place is a rise in water table (hydraulic head). *Effective (net) infiltration*, or *deep percolation* refer to water movement below the root zone and are often equated to actual recharge. In hydrologic studies, the term *effective rainfall* describes portion of precipitation that reaches surface streams via direct overland flow or near-surface flow (*interflow*). *Rainfall excess* describes part of rainfall that generates surface runoff and it does not infiltrate into the subsurface. *Interception* is the part of rainfall intercepted by vegetative cover before it reaches ground surface and it is not available for either infiltration or surface runoff. The term *net recharge* is being used to distinguish between the following two water fluxes: recharge reaching the water table due to vertical downward flux from the unsaturated zone, and evapotranspiration from the water table, which is an upward flux ("negative recharge"). *Areal* (or *diffuse*) *recharge* refers to recharge derived from precipitation and irrigation that occur fairly uniformly over large areas, whereas *concentrated recharge* refers to loss of stagnant (ponded) or flowing surface water (playas, lakes, recharge basins, streams) to the subsurface.

The complexity of the water budget determination depends on many natural and anthropogenic factors present in the general area of interest, such as climate, hydrography and hydrology, geologic and geomorphologic characteristics, hydrogeologic characteristics of the surficial soils and subsurface porous media, land cover and land use, presence and operations of artificial surface water reservoirs, surface water and groundwater withdrawals for consumptive use and irrigation, and wastewater management. Following are some of the relationships between the components shown in Fig. 2.22, which can be used in quantitative water budget analyses of such a system:

$$I = P - \text{SR} - \text{ET}$$
$$I = I_{sr} + I_{res} + I_{sp}$$
$$R = I - \text{SMD} - \text{ET}_{wt}$$
$$P_{ef} = \text{SR} + I_{fl}$$
$$Q_{ss} = P_{ef} + Q_{out}^{ua} + Q_{out}^{ca}$$
$$Q_{out}^{ua} = R + Q_{in}^{ua} - L$$
$$Q_{out}^{ca} = Q_{in}^{ca} + L - Q_{out}^{w}$$
$$\Delta S = R + Q_{in}^{ua} - L - Q_{out}^{ua}$$

$$(2.6)$$

where
I = infiltration in general
SR = surface water runoff
ET = evapotranspiration
I_{sr} = infiltration from surface runoff
I_{res} = infiltration from surface water reservoirs
I_{sp} = infiltration from snow pack and glaciers
R = groundwater recharge
SMD = soil moisture deficit
ET_{wt} = evapotranspiration from water table
P_{ef} = effective precipitation
I_{if} = interflow (near-surface flow)
Q_{ss} = surface stream flow
Q_{out}^{ua} = direct discharge of the unconfined aquifer
Q_{out}^{ca} = direct discharge of the confined aquifer
Q_{in}^{ua} = lateral groundwater inflow to the unconfined aquifer
L = leakance from the unconfined aquifer to the underlying confined aquifer
Q_{in}^{ca} = lateral groundwater inflow to the confined aquifer
Q_{out}^{w} = well pumpage from the confined aquifer
ΔS = change in storage of the unconfined aquifer

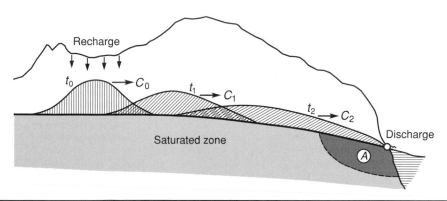

Figure 2.23 Formation and movement of a groundwater "wave" due to localized recharge event. Velocity of the wave is C_0 at time t_0, C_1 at time t_1, and C_2 at time t_2, where $C_0 \geq C_1 \geq C_2$ due to decreasing hydraulic gradients. A is the volume of "old" water discharged under pressure at the spring due to the recharge event. (Modified from Yevjevich, 1981.)

If the area is irrigated, yet two more components would be added to the list: infiltration and runoff (return flow) of the irrigation water.

Ideally, most of the above relationships would have to be established to fully quantify the processes governing the water budget of a groundwater system, including volumes of water stored in, and flowing between three general reservoirs—surface water, vadose zone, and saturated zone. By default, changes in one of the many water budget components cause a "chain reaction" and thus influence all other components. These changes take place with more or less delay, depending on both the actual physical movement of water and the hydraulic characteristics of the three general reservoirs. Figure 2.23 is an example showing how localized recharge in one part of the system can cause a rapid response far away, followed by a more gradual change between the areas of recharge and discharge as the newly infiltrated water starts flowing. The rapid response is due to propagation of the hydrostatic pressure through the system, and although this particular example illustrates behavior of large fractures and conduits in karst aquifers, the same mechanism is to a certain extent applicable to other aquifer types as well. In any case, it is very important to always consider groundwater systems as dynamic and constantly changing in both time and space.

As mentioned earlier, hydraulic head is one of the few parameters used in water budget calculations that can be measured directly. It is also the key parameter in calculations of groundwater flow rates and velocities. Figures 2.24 to 2.27 illustrate how changes in the hydraulic head (water table) can be used to calculate changes in aquifer storage and the available volume of groundwater, if the saturated aquifer thickness and the specific yield are known (estimated). The saturated aquifer thickness at any given time is determined from the water table map (Fig. 2.24) and the aquifer base map (Fig. 2.25), while the change in saturated thickness over time (Fig. 2.26) is calculated using data from individual monitoring wells with long-term hydraulic head (water table) observations such as the one shown in Fig. 2.27. The volume of groundwater stored in an aquifer for any given time is calculated by multiplying the saturated thickness for that time with the specific yield.

The example shown is from the southern portion of the Ogallala aquifer, the United States, one of the most utilized and most studied aquifers in the world. The aquifer,

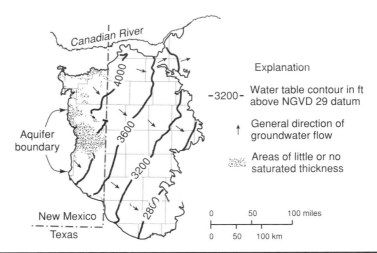

FIGURE 2.24   Predevelopment water table map for the southern portion of the High Plains aquifer. (Modified from McGuire et al., 2003.)

also known as the High Plains aquifer, is unconfined and generally composed of unconsolidated alluvial deposits. It underlies a 111-million-acre area (173,000 mi²) in parts of eight states—Colorado, Kansas, Nebraska, New Mexico, Oklahoma, South Dakota, Texas, and Wyoming. The area that overlies the aquifer varies between a semiarid to arid environment and a moist subhumid environment with gently sloping plains, fertile soil, abundant sunshine, few streams, and frequent winds. Although the area can receive a moderate amount of precipitation, precipitation in most parts is generally inadequate to provide economically sufficient yield of typical crops—alfalfa, corn, cotton, sorghum, soybeans, and wheat. The 30-year average annual precipitation ranges from about 14 in.

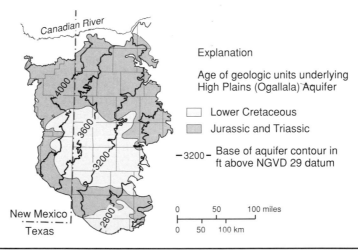

FIGURE 2.25   Elevation of the aquifer base in the southern portion of the High Plains aquifer. (Modified from McGuire et al., 2003.)

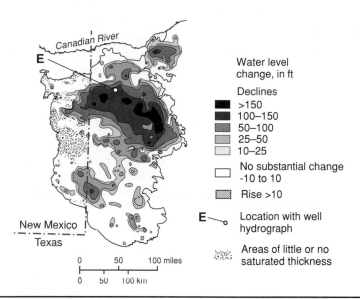

FIGURE 2.26 Water-level changes in the southern portion of the High Plains aquifer, predevelopment to 2000, and location of selected well with hydrograph. (Modified from McGuire et al., 2003.)

in the western part of the area to about 32 in. in the eastern part. Through irrigation of crops with pumped groundwater, the area overlying the aquifer has become one of the major agricultural producing regions of the world. Studies that characterize the aquifer's available water and the water chemistry begun in the early 1900s and continue to the present day. Additional studies have been conducted in selected areas to estimate the effect of water-level declines and to evaluate methods to increase the usable water in the aquifer. In the area that overlies the High Plains aquifer, farmers began extensive

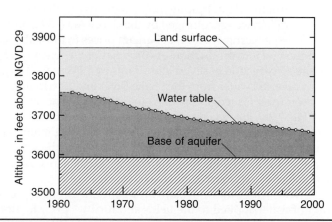

FIGURE 2.27 Hydrograph of monitoring well at location E in Castro County, TX, shown in Fig. 2.26. (Modified from McGuire et al., 2003.)

use of groundwater for irrigation in the 1940s. The estimated irrigated acreage increased rapidly from 1940 to 1980 but did not change greatly between 1980 and 1997 (McGuire et al., 2003). The change in the volume of water in storage in the High Plains aquifer from predevelopment (prior to 1940) to 2000 shows a decrease of about 200 million acre-ft, which represents 6 percent of the total volume of water in storage in the aquifer during predevelopment (McGuire et al., 2003). This change varies greatly by region and state depending upon aquifer recharge conditions (which include return flow from irrigation), rates of groundwater withdrawal, and specific yield. In Nebraska, the storage increased by 4 million acre-ft due to induced recharge from surface streams flowing over the aquifer and more favorable climatic conditions, while in Texas the storage decreased by 124 million acre-ft. In portions of some states, such as Kansas and Texas, the saturated thickness of the aquifer decreased by more than 50 percent, with the remaining thickness and storage inadequate to support feasible withdrawal of groundwater for irrigation and public supply.

The following case study illustrates changes in the water budget of another large aquifer in the United States, caused by changing agricultural and irrigation practices during the last century. The highly productive Snake River Plain aquifer in the state of Idaho has been declared a sole source aquifer by the U.S. Environmental Protection Agency, due to the nearly complete reliance on the aquifer for drinking water supplies of over 300,000 people in the area. The aquifer, developed in flood basalts and related interbedded sediments, occupies roughly 10,000 mi^2. It is a prime example of close interactions between surface water and groundwater, and a growing awareness by all stakeholders that only an integrated management of both surface water and groundwater resources can solve growing tensions between various users of a common, limited resource.

Case Study: Water Budget of the Eastern Snake River Plain Aquifer, Idaho, the United States

The information presented in this case study is from the Idaho National Laboratory (INL), Radiation Control Division (2006), and a report published by Idaho Water Resources Research Institute, University of Idaho (2007).

The history of the Eastern Snake River Plain aquifer is inexorably tied to the history of irrigation in this vast semiarid desert area (less than 10 in. or 250 mm of annual precipitation), now one of the agriculturally most productive in the world. Development of the arid Snake River Plain was encouraged by the Carey Act (1894) and other federal legislation that provided government land at bargain prices to those that could bring that land under irrigation and into production. A combination of private and federal investments resulted in the construction of seven large dams by 1938 as well as an elaborate network of canals that diverted water from the Snake River and its tributaries. Although groundwater had been used for irrigation since the 1920s in some areas on the Eastern Snake River Plain (Fig. 2.28), the development of powerful and efficient electric pumps allowed significant and ever increaseing groundwater use, causing many farmers to switch from surface water to groundwater. Currently, surface water is the source for irrigation of 1.23 million acres and groundwater the source for 930,000 acres. A combination of surface and groundwater is used to irrigate 110,000 acres.

Flooding fields with water is a relatively inefficient means of providing water to crops. The amount of water applied to the fields and furrows prior to more modern irrigation methods was sometimes as much as seven times what the crop could use. However, all of that excess water (sometimes as much as 12 ft) applied during the course of an

FIGURE 2.28   Irrigation wells at Artesian City in Twin Falls County, around 1910–1920. (Idaho State Historical Society, Bisbee collection. Printed with permission.)

irrigation season recharged the aquifer. This water became stored for later use and water levels rose substantially in some areas. For example, groundwater levels rose from 60 to 70 ft during 1907 to 1959 in areas near Kimberly and Bliss, and as much as 200 ft in areas near Twin Falls. Across the entire aquifer, the average rise was about 50 ft. This rise in aquifer levels became most evident by the increase in discharges from the major springs along the Snake River. With the transition to irrigation with groundwater along with more efficient means of applying surface water to fields, less water was added to groundwater storage and more was taken from it, resulting in a decrease of spring flows and water levels in the aquifer (Figs. 2.29 and 2.30).

Natural aquifer recharge occurs mainly in the northern and eastern portions of the plain, resulting in a generally south to southwest trending groundwater flow toward the Snake River (Fig. 2.30). Following are the primary sources of this recharge:

1. Tributary basin underflow, or groundwater that flows to the aquifer from the tributary valleys along the margins of the plain. This includes recharge from Henry's Fork and the South Fork of the Snake River, and the valleys of Birch Creek, Big and Little Lost Rivers, Big and Little Wood Rivers, Portneuf and Raft River valleys, and other smaller valleys. The Big Lost River is an example of a river directly feeding an aquifer. The river flows out of a mountain valley on the northwest margin of the Snake River Plain and entirely disappears through seepage into the permeable lava of the Plain.

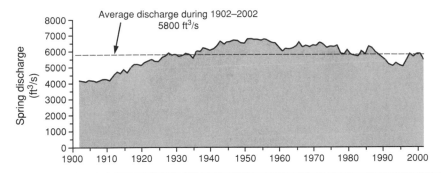

Figure 2.29   Changes in discharge of Thousand Springs between 1900 and 2000. (Modified from INL, 2006.)

2. Water infiltrating from the bed of the Snake River along some reaches north of Idaho Falls where the riverbed is above the aquifer level and water from the river seeps through the river bed to recharge the aquifer. Depending upon the seasonal hydrologic conditions in the river and diversions for irrigation, some reaches of the river can lose water during times of the year when the aquifer level is lower, and gain water when the aquifer level is above the bed of the river. During the growing season, and especially during dry years, the Snake River may nearly dry up before it reaches the famous Shoshone Falls, about 30 mi downstream of Milner Dam, due to irrigation diversions (Figs. 2.31 and 2.32).

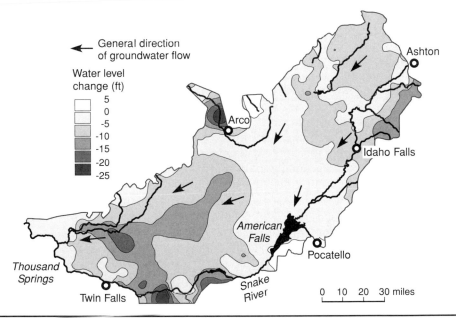

Figure 2.30   Changes in the Eastern Snake River Plain aquifer levels between 1980 and 2002. (Modified from INL, 2006.)

FIGURE 2.31 Photograph of Shoshone Falls taken in 1871, before the beginning of surface water irrigation in the 1880s. (Photograph possibly by Timothy O'Sullivan, USGS, Wheeler Survey 1871 Expedition; U.S. Geological Survey Photographic Library; http://libraryphoto.cr.usgs.gov.)

FIGURE 2.32 Photograph of Shoshone Falls taken in 2006. (Photograph courtesy of Denise Tegtmeyer.)

FIGURE 2.33   Thousand Springs discharging in Hagerman Valley along the Snake River, ID, circa 1910–1920. (Idaho Historical Society, Bisbee Collection. Printed with permission.)

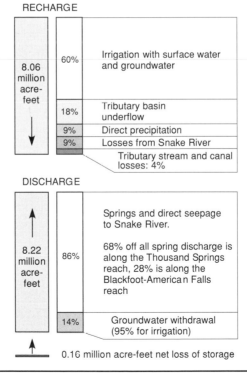

FIGURE 2.34   Water budget of the Eastern Snake River Plain aquifer for 1980. (Data from INL, 2006.)

Natural discharge from the aquifer occurs primarily along two reaches of the Snake River: (1) near American Falls Reservoir, in which the spring discharge a total of about 2600 cfs and (2) from Kimberly to King Hill (Thousand Springs reach; see Fig. 2.33), where the collective discharge is about 5200 cfs. During summer, spring flows provide a majority of the flow in the Snake River below the irrigation diversions of Milner Dam. The Snake River Basin contains 15 of the nation's 65 first magnitude springs (discharge greater than 100 cfs). Many of the springs are utilized for power generation, water supply, and aquaculture (for example, the largest trout farm in the world is fed by springs discharging from the aquifer).

Figure 2.34 illustrates the aquifer water budget for 1980, indicating that the largest source of water recharging the aquifer was irrigation. Discharge from the aquifer that year exceeded all the recharge, thus depleting the storage by about 0.16 million acre-ft, a trend that continues to this day.

2.5 Groundwater Flow

Groundwater in the saturated zone is always in motion, and this flow takes place in a three-dimensional space. When one or two flow directions appear dominant, quantitative analyses may be performed using one or two-dimensional flow equations for the purposes of simplification. When it is important to accurately analyze the entire flow field, which is often the case in contaminant fate and transport studies, a three-dimensional groundwater modeling may be the only feasible quantitative tool since three-dimensional analytical equations of groundwater flow are rather complex and often cannot be solved in a closed form.

2.5.1   Darcy's Law

The three main quantities that govern the flow of groundwater are as follows: *hydraulic gradient*, which is the driving force, *hydraulic conductivity*, which describes both the transmissive properties of the porous media and the hydraulic properties of the flowing fluid (water), and the *cross-sectional area* of flow. Their relationship is described by *Darcy's law* (Darcy was a French civil engineer who was first to quantitatively analyze the flow of water through sands as part of his design of water filters for the city of Dijon; his findings, published in 1856, are the foundation of all modern studies of fluid flow through porous media):

$$Q = KA\frac{\Delta h}{L} \; [\text{m}^3/\text{s}] \tag{2.7}$$

This linear law states that the rate of fluid flow (Q) through porous medium is directly proportional to the cross-sectional area of flow (A) and the loss of the hydraulic head between two points of measurements (Δh), and it is inversely proportional to the distance between these two points of measurement. K is the proportionality constant of the law called *hydraulic conductivity* and has units of velocity. This constant is arguably the most important quantitative parameter characterizing the flow of groundwater. Following are the other common forms of Darcy's equation:

$$v = K\frac{\Delta h}{L} \; [\text{m/s}] \tag{2.8}$$

$$v = Ki \; [\text{m/s}] \tag{2.9}$$

where v = the so-called Darcy's velocity and i = hydraulic gradient.

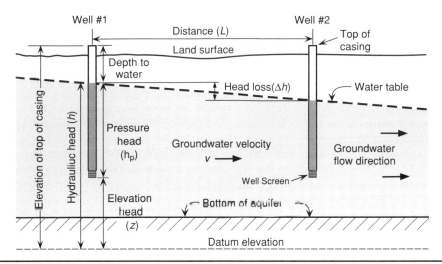

FIGURE 2.35   Schematic presentation of key elements for determining the hydraulic head and the hydraulic gradient in an unconfined aquifer. (Kresic, 2007a; copyright Taylor & Francis Group, LLC, printed with permission.)

Hydraulic Head and Hydraulic Gradient

The principle of the hydraulic head and the hydraulic gradient is illustrated in Fig. 2.35. At the bottom of monitoring well #1, where the well screen is open to the saturated zone, the total energy (H) or the driving force for water flow at that point in the aquifer is

$$H = z + h_p + \frac{v^2}{2g} \tag{2.10}$$

where z = elevation above datum (datum is usually mean sea level, but it could be any reference level)
h_p = pressure head due to the pressure of fluid (groundwater) above that point
v = groundwater velocity
g = acceleration of gravity

Since the groundwater velocity in most cases is very low, the third factor on the right-hand side may be ignored for practical purposes and the Eq. (2.10) becomes

$$H = h = z + h_p \tag{2.11}$$

where h = *hydraulic head*, also called *piezometric level*. The pressure head represents the pressure of fluid (p) of constant density (ρ) at that point in aquifer:

$$h_p = \frac{p}{\rho g} \tag{2.12}$$

In practice, the hydraulic head is determined in monitoring wells or piezometers by subtracting the measured depth to the water level from the surveyed elevation of the top of the casing:

$$h = \text{elevation of top of casing} - \text{depth to water in the well} \qquad (2.13)$$

As the groundwater flows from well #1 to well #2 (Fig. 2.35), it loses energy due to friction between groundwater particles and the porous media. This loss equates to a decrease in the hydraulic head measured at the two wells:

$$\Delta h = h_1 - h_2 \qquad (2.14)$$

The *hydraulic gradient* (i) between the two wells is obtained when this decrease in the hydraulic head is divided by the distance (L) between the wells:

$$i = \frac{\Delta h}{L} \text{ [without dimension]} \qquad (2.15)$$

Groundwater flow always takes place from the higher hydraulic head toward the lower hydraulic head (just as in the case of surface water: "water cannot flow uphill"). It is also important to understand that, except in case of a very limited portion of an aquifer, there is no such thing as strictly horizontal groundwater flow. In an area where aquifer recharge is dominant, the flow is vertically downward and laterally toward the discharge area; in a discharge area, such as surface stream, this flow has an upward component (Fig. 2.36).

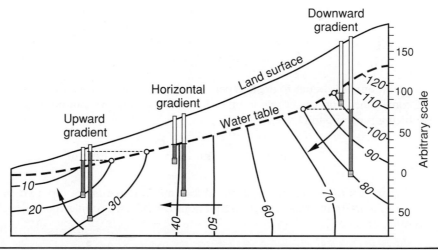

Figure 2.36 Movement of groundwater in an unconfined aquifer showing the importance of both vertical and horizontal hydraulic gradients. (Modified from Winter et al., 1998.)

Hydraulic Conductivity and Permeability

In addition to hydraulic conductivity, another quantitative parameter called *intrinsic permeability* (or simply *permeability*) is also used in studies of fluid flow through porous media. It is defined as the ease with which a fluid can flow through a porous medium. In other words, permeability characterizes the ability of a porous medium to transmit a fluid (water, oil, gas, etc.). It is dependent only on the physical properties of the porous medium: grain size, grain shape and arrangement, or pore size and interconnections in general. On the other hand, hydraulic conductivity is dependent on the properties of both the porous medium and the fluid. The relationship between the permeability (K_i) and the hydraulic conductivity (K) is expressed by the following formula:

$$K_i = K \frac{\mu}{\rho g} \ [\text{m}^2] \tag{2.16}$$

where $\mu = $ *absolute viscosity* of the fluid (also called *dynamic viscosity* or simply *viscosity*)
  $\rho = $ *density* of the fluid
  $g = $ *acceleration of gravity*

The viscosity and the density of the fluid are related through the property called *kinematic viscosity* (υ):

$$\upsilon = \frac{\mu}{\rho} \ [\text{m}^2/\text{s}] \tag{2.17}$$

Inserting the kinematic viscosity into Eq. (2.16) somewhat simplifies the calculation of the permeability since only one value (that of υ) has to be obtained from tables or graphs (note that, for most practical purposes, the value of the acceleration of gravity (g) is 9.81 m/s^2, and is often rounded to 10 m/s^2):

$$K_i = K \frac{\upsilon}{g} \ [\text{m}^2] \tag{2.18}$$

Although it is better to express permeability in units of area (m^2 or cm^2) for reasons of consistency and easier use in other formulas, it is more commonly given in darcys (which is a tribute to Darcy):

$$1 \text{ darcy} = 9.87 \times 10^{-9} \text{ cm}^2 = 9.87 \times 10^{-13} \text{ m}^2$$

When laboratory results of permeability measurements are reported in darcys (or meters squared), the following two equations can be used to find the hydraulic conductivity:

$$K = K_i \frac{g}{\upsilon} \quad \text{or} \quad K = K_i \frac{\rho g}{\mu} \ [\text{m/s}] \tag{2.19}$$

Water temperature influences both water density and viscosity and, consequently, the hydraulic conductivity is strongly dependent on groundwater temperature. Kinematic viscosity of water at temperature of 20°C is approximately 1×10^{-6} m^2/s, and rounding gravity acceleration to 10 m/s^2, gives the following conversion between permeability (given in m^2) and hydraulic conductivity (given in m/s):

$$K \,[\text{m/s}] = K_i \,[\text{m}^2] \times 10^7 \tag{2.20}$$

Since effective porosity, as the main factor influencing the permeability of a porous medium, varies widely by rock types, the hydraulic conductivity and permeability also have wide ranges as shown in Fig. 2.37. As is the case with porosity, limestones have the widest range of hydraulic conductivity of all rocks. Vesicular basalts can have very high hydraulic conductivity, but they are on average less permeable than medium to coarse sand and gravel, which are rock types with the highest average hydraulic conductivity. Pure clays and fresh igneous rocks generally have the lowest permeability, although some field-scale bedded salt bodies were determined to have permeability of zero (Wolff, 1982). This is one of the reasons why salt domes are considered as potential depositories of high radioactivity nuclear wastes in some countries.

Except in rare cases of uniform and nonstratified, homogeneous unconsolidated sediments, hydraulic conductivity and permeability vary in space and in different directions

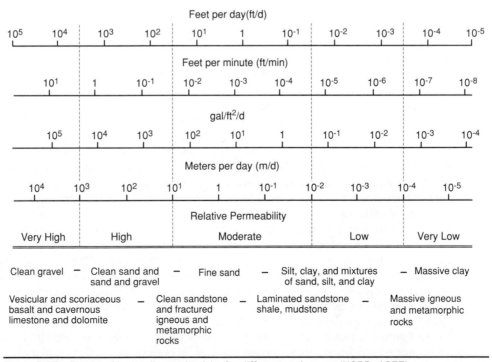

FIGURE 2.37   Range of hydraulic conductivity for different rock types (USBR, 1977).

within the same rock mass due to its heterogeneity and anisotropy. Most practitioners tend to simplify these inherent characteristics of porous media by dividing the complex three-dimensional hydraulic conductivity tensor into just two main components: horizontal and vertical hydraulic conductivity. Unfortunately, it seems common practice to apply some "rules of thumb" indiscriminately, such as vertical conductivity is ten times lower than the horizontal conductivity, without trying to better characterize the underlying hydrogeology. This difference in the two hydraulic conductivities can vary many orders of magnitude in highly anisotropic rocks and, in many cases, it may be completely inappropriate to apply the concept altogether: a highly transmissive fracture or a karst conduit may have any shape and spatial extent, at any depth.

2.5.2   Types and Calculations of Groundwater Flow

There are three general factors for determining types of groundwater flow and the equations for its quantification: (1) hydraulic conditions in the aquifer, (2) space (cross-sectional area) in which the flow is taking place, and (3) time. Flow can be confined or unconfined (hydraulic conditions) and this may change along the flow direction, in both space and time. For confined conditions, the cross-sectional area of flow at any given location in the aquifer remains constant, regardless of time, which is why equations describing confined flow are generally less complex. In contrast, the position of the water table and therefore the thickness of the saturated zone in an unconfined aquifer usually varies in time due to varying recharge conditions. This also means that the cross-sectional area of flow changes in time at any given location by default. If the hydraulic head and the hydraulic gradient do not change in time, the flow is in *steady state* (for both confined and unconfined conditions). Steady-state conditions are rarely completely satisfied, except in the case of some nonrenewable aquifers where the natural long-term balance is not disturbed by artificial groundwater withdrawal. Over short-term periods, groundwater flow is often described with steady-state equations for reasons of simplification. The term quasi-steady-state is used to describe an apparent stabilization of the hydraulic head after an initial response to some external stress, such as stabilization of drawdown at an extraction well. This stabilization may be the result of additional inflow of water into the system, such as from a nearby surface stream or due to drainage of the porous media from the constantly increasing radius of influence of the extraction well. When the rate of groundwater withdrawal significantly exceeds additional inflow of water (recharge from any direction), it is obvious that quasi-steady-state calculations are not applicable. The same is true when the porous media storage provides significant volumes of withdrawn water. In fact, the "safest" way to distinguish between a steady state and a transient groundwater flow equation is the presence of storage parameters (specific yield for unconfined, and storage coefficient for confined conditions). If these parameters are present, the flow is *transient* (time-dependent). When it is important to fully describe groundwater flow in a system, such as for resource management or aquifer restoration purposes, the only valid approach is to apply transient equations, which incorporate time-dependent flow parameters including full description of the hydraulic head (hydraulic gradient) changes in time.

Two simple cases of estimating steady-state groundwater flow rates in unconfined and confined conditions are shown in Fig. 2.38. In both cases the flow is planar, through

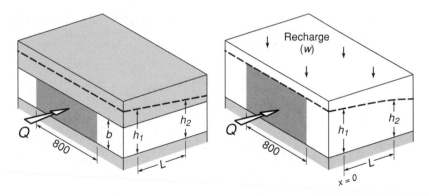

Figure 2.38 Confined (left) and unconfined (right) planar flow in steady state. In both cases, the groundwater flow rate is calculated for an aquifer width of 800 units (feet, meters) using Eqs. (2.21) and (2.22), respectively.

a constant rectangular cross-section, and over an impermeable horizontal base. The hydraulic conductivity is spatially constant (aquifers are homogeneous), and the hydraulic gradient is also constant. Equations (2.21) and (2.22) describe these simple conditions. For the confined flow case, the relationship is linear since the saturated aquifer thickness (b) does not change along the flow path. Equation (2.22), which describes unconfined conditions and includes a possible recharge rate (w), is nonlinear because the saturated thickness (position of the water table) does change between h_1 and h_2.

$$Q = 800 \times bK\frac{h_1 - h_2}{L} \tag{2.21}$$

$$Q = 800 \times K\frac{h_1^2 - h_2^2}{2L} + w\left(x - \frac{L}{2}\right) \quad \text{for} \quad x > 0 \tag{2.22a}$$

$$Q = 800 \times K\frac{h_1^2 - h_2^2}{2L} + w\frac{L}{2} \quad \text{for} \quad x = 0 \tag{2.22b}$$

In reality, flow conditions are almost always more complicated, including changing aquifer thickness, possible transition between confined and unconfined flow, nonhorizontal base, heterogeneous porous media, and time-dependent (changing in time) recharge from different directions. Various analytical equations have been developed for different flow conditions, as described in hydrogeology textbooks (e.g., see Freeze and Cherry, 1979; Domenico and Schwartz, 1990; Kresic, 2007a). Although such equations are still used to quickly estimate groundwater flow rates, numeric groundwater flow models have replaced them as a quantitative tool of choice for describing groundwater systems.

One key parameter for various calculations of groundwater flow rates is *transmissivity* of porous media. For practical purposes, it is defined as the product of the aquifer thickness (b) and the hydraulic conductivity (K):

$$T = b \times K \tag{2.23}$$

It follows that an aquifer is more transmissive (more water can flow through it) when it has higher hydraulic conductivity and when it is thicker. Although there are many laboratory and field methods for determining hydraulic conductivity and the transmissivity of aquifers, the most reliable are long-term field pumping tests, which register hydraulic response of all porous media present in the system. Aquifer testing is not a focus of this book, and the reader can consult various general and special publications on designing and analyzing aquifer tests including pumping tests, such as guidance documents by the United States Geological Survey (USGS: Ferris et al., 1962; Stallman, 1971; Lohman, 1972; Heath, 1987), the United States Environmental Protection Agency (USEPA: Osborne, 1993), the United States Bureau of Reclamation (USBR, 1977), and American Society for Testing and Materials (ASTM, 1999a, 1999b), and books by Driscoll (1989), Walton (1987), Kruseman et al. (1991), Dawson and Istok (1992), and Kresic (2007a).

Figure 2.39 shows the simplest case of steady-state radial groundwater flow toward a fully penetrating pumping well in a homogeneous confined aquifer with a constant thickness and horizontal impermeable base, and without any vertical recharge (leakage) from adjacent aquifers or aquitards. The rate of groundwater flow is calculated using the Thiem equation (Thiem, 1906), named after a German engineer who developed it in 1906 based on field experiments conducted as part of an investigation to find additional water supply for the city of Prague, Czech Republic (then part of Austrian Empire):

$$Q = \frac{2\pi T s_{\text{w}}}{\ln R/r_{\text{w}}} \tag{2.24}$$

(the symbols are explained in Fig. 2.39). A detailed description of the Thiem equation and its application is given in a work by Wenzel (1936).

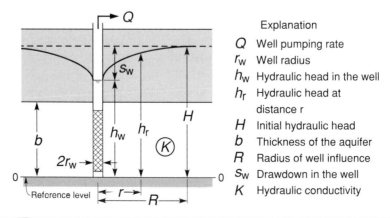

Explanation

Q Well pumping rate
r_{w} Well radius
h_{w} Hydraulic head in the well
h_{r} Hydraulic head at distance r
H Initial hydraulic head
b Thickness of the aquifer
R Radius of well influence
s_{w} Drawdown in the well
K Hydraulic conductivity

Figure 2.39 Elements of groundwater flow toward a fully penetrating well in a confined aquifer.

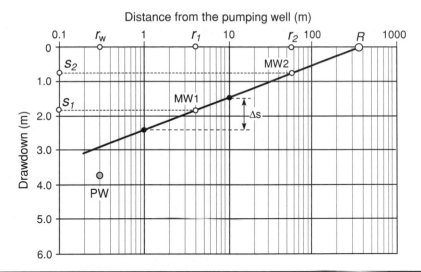

Figure 2.40 Semilog graph drawdown versus distance for a steady-state pumping test in a confined aquifer, showing data for two monitoring wells, MW1 and MW2, and pumping well PW.

Although these conditions are seldom satisfied in reality, there are several situations when a steady-state approach to calculating well pumping rates or analyzing well pumping test results may be justified for a preliminary assessment, such as when the drawdown and the radius of well influence do not change in time. This includes pumping near a large stream or a lake that is hydraulically connected with the aquifer, or at a locality partly surrounded and completely hydraulically influenced by a large river. The radius of well influence reaches the boundary relatively soon after the beginning of pumping and the drawdown remains constant afterwards.

The Thiem equation can be used to estimate aquifer transmissivity if steady-state ("stabilized") drawdown measurements are available for at least two monitoring wells placed at different distances from the pumping well. Figure 2.40 shows a semilogarithmic graph where the one-log cycle difference in drawdown (Δs) along the straight line formed by the data from two monitoring wells is noted and used in the following equation:

$$T = \frac{0.366Q}{\Delta s} \tag{2.25}$$

Monitoring wells MW1 and MW2 are located at distances r_1 and r_2 from the pumping well, respectively, and their recorded drawdowns are s_1 and s_2, respectively. Equation (2.25) is derived from the initial Eq. (2.24), including convenient conversion from natural to decadal logarithms, and taking advantage of the following relation: $\log 10 = 1$.

The graph in Fig. 2.40 shows that the drawdown recorded in the pumping well does not fall on the straight line connecting the monitoring well data; it is below the straight line indicating that there is an additional drawdown in the well because of the *well loss*. The well loss, which is inevitable for any well, is explained in detail in Chap. 7. In short, it is a consequence of various factors such as disturbance of porous medium near the well during drilling, improper (insufficient) well development, poorly designed gravel pack and/or well screen, and/or turbulent flow through the screen. Because of the well loss, at least two monitoring wells are needed in order to apply the Thiem equation properly. Using the pumping well drawdown and drawdown in one monitoring well would give erroneous results.

The steady-state radius of well influence (R) is the intercept of the straight line connecting the monitoring well data and zero drawdown. The hydraulic conductivity of the aquifer porous media is found by dividing the transmissivity with the aquifer thickness ($K = T/b$). Aquifer storage cannot be found using the steady-state approach.

Similarly to the unconfined planar flow, the radial flow toward a well in an unconfined aquifer is described with a somewhat more complicated equation since the top of the aquifer corresponds to the pumping water table. In other words, the saturated aquifer thickness increases away from the pumping well. Assuming that the aquifer impermeable base is horizontal, and when the reference level is set at the base, the hydraulic head equals the water table, which gives the following flow equation:

$$Q = \pi K \frac{h_2^2 - h_1^2}{\ln(r_2/r_1)} \qquad (2.26)$$

where h_2 = steady-state (stabilized) hydraulic head at a monitoring well farther away from the pumping well, at distance r_2, and h_1 = hydraulic head at a monitoring well closer to the pumping well, at distance r_1. The hydraulic conductivity (K) can be found using a procedure similar to that for confined aquifers and the following equation:

$$K = \frac{0.733 Q}{\Delta(H^2 - h^2)} \qquad (2.27)$$

Note that instead of drawdown (s), the y axis of the semilog graph shown in Fig. 2.40 would represent values of $H^2 - h^2$, where H = hydraulic head farther away from the pumping well and h = head closer to the pumping well.

Theis Equation

Theis equation (Theis, 1935), which describes transient (time-dependent) groundwater flow toward a fully penetrating well in a confined aquifer, is the basis for most methods of transient pumping test analysis. It is also often used to calculate pumping rates of a well when assuming certain values of drawdown, aquifer transmissivity, and storage coefficient. The equation enables the determination of aquifer parameters from drawdown measurements without drawdown stabilization. In addition, data from only one observation well are sufficient, as opposed to steady-state calculations where at least two observation wells are needed. Theis equation gives drawdown (s) at any time after the beginning of pumping:

$$s = \frac{Q}{4\pi T} W(u) \qquad (2.28)$$

where Q = pumping rate kept constant during the test
$\qquad\quad$ T = transmissivity
$\qquad\quad$ $W(u)$ = *well function of u*, also known as *the Theis function*, or simply *well function*

Dimensionless parameter u is given as

$$u = \frac{r^2 S}{4Tt} \tag{2.29}$$

where r = distance from the pumping well where the drawdown is recorded
S = storage coefficient
t = time since the beginning of pumping

Values of $W(u)$ for various values of the parameter u are given in Appendix A and can be readily found in groundwater literature. *Theis type curve* is a log-log graph of $W(u)$ versus $1/u$, such as the one shown in Fig. 2.41, and is used to match data observed in the field.

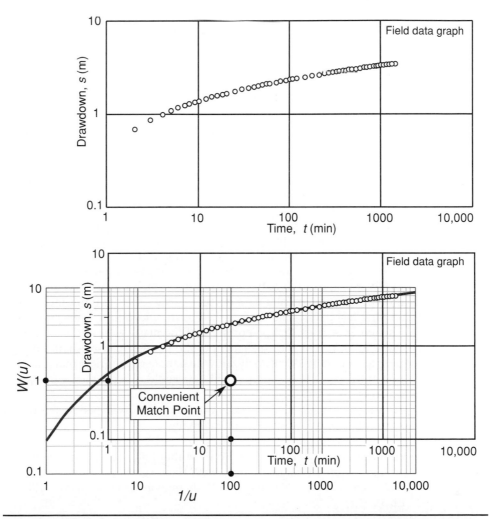

FIGURE 2.41 Field data of drawdown versus time for a monitoring well plotted on a log-log graph paper with the same scale as the theoretical Theis type curve, and superimposed on it.

Equation (2.28) has no explicit solution and Theis introduced a graphical method, which gives T and S if other terms are known. Field data of drawdown (s) versus time (t) for a monitoring well is plotted separately on a log-log graph with the same scale as the theoretical curve. Keeping coordinate axes of the curves parallel, the field data is matched to the type curve. Once a satisfactory match is found, a *match point* on the overlapping graphs is selected. The match point is defined by four coordinates, the values of which are read on two graphs: $W(u)$ and $1/u$ on the type curve graph, s and t on the field graph. The match point can be any point on the overlapping graphs, i.e., it does not have to be on the matching curve. Figure 2.41 shows a match point chosen outside the curves to obtain convenient values of $W(u)$ and $1/u$: 1 and 100, respectively. The transmissivity is calculated using Eq. (2.28) and the match point coordinates s and $W(u)$:

$$T = \frac{Q}{4\pi s} W(u) \qquad (2.30)$$

The storage coefficient is calculated using Eq. (2.29), the match point coordinates $1/u$ and t, and the previously determined transmissivity value:

$$S = \frac{4T t u}{r^2} \qquad (2.31)$$

Theis derived his equation based on quite a few assumptions and it is very important to understand its limitations. If the aquifer tested and the test conditions significantly deviate from these assumptions (which, in fact, is very often the case in reality), other methods of analysis applying appropriate analytical equations should be used. The Theis equation assumes that the aquifer is confined, homogeneous, and isotropic; it has uniform thickness; the pumping never affects its exterior boundary (the aquifer extent is considered infinite); the aquifer does not receive any recharge; the well discharge is derived entirely from aquifer storage; the pumping rate is constant; the pumping well is fully penetrating (it receives water from the entire thickness of the aquifer); it is 100 percent efficient (there are no well losses); the water removed from storage is discharged instantaneously when the head is lowered; the radius of the well is infinitesimally small (the storage in the well can be ignored); and the initial potentiometric surface (before pumping) is horizontal. When pumping test data cannot be matched to the theoretical Theis curve because of an "odd" shape, it is likely that one or more of the many assumptions is not satisfied. In such cases, a hydrogeologic assessment of the possible causes should be made and the pumping test data should be analyzed with a more appropriate method. Figure 2.42 shows some of the possible cases why field data could differ from the theoretical Theis curve (dashed line).

Various analytical methods have been continuously developed to account for these and other complex situations such as the following:

- Presence of leaky aquitards, with or without storage and above or below the pumped aquifer (Hantush, 1956, 1959; 1960; Hantush and Jacob, 1955; Cooper, 1963; Moench, 1985; Neuman and Witherspoon, 1969; Streltsova, 1974; Boulton, 1973).

- Delayed gravity drainage in unconfined aquifers (Boulton, 1954; Boulton, 1963; Neuman, 1972; Neuman, 1974; Moench, 1996).

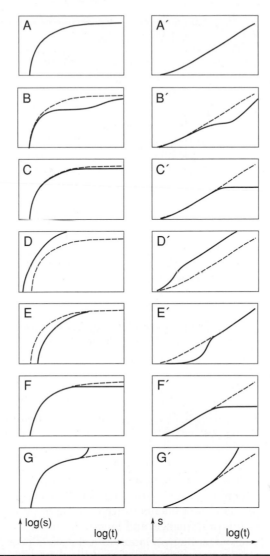

Figure 2.42 Log-log and semilog curves of drawdown versus time. A and A′, confined aquifer; B and B′, unconfined aquifer; C and C′, leaky (or semiconfined) aquifer; D and D′, effect of partial penetration; E and E′, effect of well-bore storage (large diameter well); F and F′, effect of recharge boundary; G and G′, effects of an impervious boundary. (From Griffioen and Kruseman, 2004.)

- Other "irregularities" such as large-diameter wells and presence of bore skin on the well walls (Papadopulos and Cooper, 1967; Moench, 1985; Streltsova, 1988).
- Aquifer anisotropy (Papadopulos, 1965; Hantush, 1966a; 1966b; Hantush and Thomas, 1966; Boulton, 1970; Boulton and Pontin, 1971; Neuman, 1975; Maslia and Randolph, 1986).

Attempts have also been made to develop analytical solutions for fractured aquifers, including dual-porosity approach and fractures with skin (e.g., Moench, 1984; Gringarten

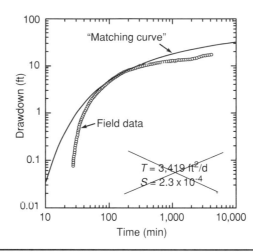

Figure 2.43 Example of inadequate curve fitting of aquifer pumping test data using an automated procedure within a computer program.

and Witherspoon, 1972; Gringarten and Ramey, 1974). However, because of the inevitable simplicity of analytical solutions, all such methods are limited to regular geometric fracture patterns such as orthogonal or spherical blocks, and single vertical or horizontal fractures.

With some minor changes and corrections, the Theis equation has also been applied to unconfined aquifers and partially penetrating wells (e.g., Hantush, 1961a, 1961b; Jacob, 1963a; Jacob, 1963b; Moench, 1993, 1996), including when monitoring wells are placed closer to the pumping well where the flow is not horizontal (e.g., Stallman, 1961; Stallman, 1965).

It cannot be emphasized enough that an appropriate interpretation of any aquifer pumping test, and especially a long-term one, is crucial for groundwater resource management and restoration. Aquifer transmissivity (hydraulic conductivity) and storage parameters are the key components for calculating optimum pumping rates of a well or a well field, radius of influence, and long-term impacts of groundwater withdrawal on the available resource. Interpretation of test data is rarely unique, and requires an experienced hydrogeologist with a thorough understanding of the overall hydrogeologic characteristics of the groundwater system. Figure 2.43 is an example of how test data should not be interpreted because the selected theoretical model does not explain the majority of the data, and does not explain any of the data later in the test even though this data is much more representative of a possible long-term response of the system.

As mentioned before, numeric groundwater models are being increasingly utilized not only for quantification of groundwater flow in a system, but also for the analysis of aquifer pumping tests because they can simulate heterogeneity, anisotropy, and the varying geometry of the system, as well as the presence of any boundaries to groundwater flow. Various hydrogeologic assumptions can be changed and tested in a numeric model until the field data is matched, and the final conceptual model is selected. Some software programs for analytical evaluation of aquifer tests offer a variety of theoretical models for automated curve matching, including manual matching. One such program, widely utilized, verified, and constantly updated is AQTESOLV by HydroSOLVE (2002).

Flow in Fractured Rock and Karst Aquifers

Characterization and quantification of groundwater flow in fractured rock, and especially karst aquifers, is very difficult because of the nature of their porosity. The flow is taking place in rock matrix, small and large fractures and, in the case of karst, in conduits and underground channels, all of which have very different values for the parameters required for the calculation of flow rates: hydraulic conductivity, transmissivity, storage properties, and hydraulic heads. In fact, the concept of hydraulic conductivity developed for intergranular porous media is not applicable to flow in fractures and conduits (channels). Despite this "hydraulic" fact, in an attempt to provide some quantification of these complex systems, many professionals use a so-called *equivalent porous medium* (*EPM*) approach, which appears to be the predominant one in hydrogeologic practice. This approach assumes that all porous media in the aquifer, at some representative scale, behave similarly and the overall flow can be approximated by Darcy's equation. However, in many practical applications, the EPM approach fails to give correct answers and cannot be used as a sound basis for groundwater management and restoration. For example, its inadequacy is evident when trying to predict discharge at a large karstic spring, or change in hydraulic heads after precipitation events, or when predicting contaminant fate and transport, including contaminant pathways in the subsurface and arrival times at points of interest (such as at a well used for public water supply).

Various equations, analytical and numeric modeling approaches have been proposed and applied to problems of groundwater flow, and contaminant fate and transport in fractured rock and karst aquifers. In the analytically most complicated, but at the same time the most realistic case, the groundwater flow rate is calculated by integrating equations of flow through the rock matrix (Darcy's flow) with the hydraulic equations of flow through various sets of fractures, pipes, and channels. This integration, or interconnectivity between the four different flow components, can be deterministic, stochastic, or some combination of the two. Deterministic connectivity is established by a direct translation of actual field measurements of the geometric fracture parameters such as dip and strike (orientation), aperture, and spacing between individual fractures in the same fracture set, and then doing the same for any other fracture set. Cavities (caves) are connected in the same way, by measuring the geometry of each individual cavity. Finally, all of the discontinuities (fractures and cavities) are connected based on the field measurements and mapping. This approach will include many uncertainties and assumptions ("you have walked and measured this cave, but what if there is a very similar one somewhere in the vicinity you don't know anything about?"). Stochastic interconnectivity is established by randomly generating fractures or pipes using some statistical and/or probabilistic approach based on field measures of the geometric fracture (pipe) parameters. An example of combining deterministic and stochastic approaches is when computer-generated (probabilistic or random) fracture sets are intersected by a known major preferential flow path such as a fault or a cave.

Except for relatively simple analytical calculations that use a homogeneous, isotropic, equivalent porous medium approach, most other quantitative methods for fractured rock and karst groundwater flow calculations include some type of modeling. Extensive reviews of various analytical equations and modeling approaches, including detailed quantitative explanations, can be found in Bear et al. (1993), Faybishenko et al. (2000),

FIGURE 2.44   Example of dual-porosity media in a karst aquifer. Arrows illustrate the exchange of water between fractures and rock matrix. Fractures are enlarged by dissolution. Limestone landscape in northwest Ireland. (Photograph by George Sowers; printed with kind permission of Francis Sowers.)

and Kovács and Sauter (2007). Figure 2.44 illustrates some of the complexities facing groundwater professionals when trying to quantify groundwater flow in karst aquifers, starting with the recharge rates ("how quickly and how much of the rain water reaches the saturated zone?"), continuing with the calculation of flow rates through the matrix and through irregular, "rough" fractures widened by dissolution, and finally trying to calculate the rate of water exchange between the matrix and the fractures.

The capacity of a karst aquifer to transmit groundwater flow ranges from very low to very high, depending upon its location and heterogeneity. A good example is the analysis of carbonate-rock aquifers in southern Nevada by Dettinger (1989). Coyote Spring Valley aquifer transmissivity at one of the major production wells is extremely high (about 200,000 ft^2/d or 18,600 m^2/d) providing for a well yield of 3400 gal/min (214 L/s) with only 12 ft (4 m) of drawdown. However, transmissivities elsewhere in the Central Corridor region, based on tests at 33 other water wells, are between 5000 and 11,000 ft^2/d, and the average well capacity is about 455 gal/min with 85 ft of drawdown. The same study shows that within 10 mi of regional springs, aquifers are an average of 25 times more transmissive than they are farther away. These are areas where flow is converging, flow rates are locally high, and the conduit-type of flow likely plays a significant, if not predominant, role.

When a fractured rock or karst aquifer is drained by a large spring, the spring flow rate would be the best point of reference for any regional flow calculations using common hydrogeologic parameters. Simple quantitative analysis of spring flow hydrographs, including autocorrelation of flow and cross correlation of flow and precipitation (or other water inputs), can also give some clues about likely types of flow and storage in the aquifer.

2.5.3 Groundwater Velocity

One common, basic relationship connects the flow rate (Q), the velocity (v), and the cross-sectional area of flow (A) in virtually all equations describing flow of fluids, regardless of the scientific (engineering) field of study:

$$Q = v \times A \tag{2.32}$$

One form of Darcy's law states that the velocity of groundwater flow is the product of the hydraulic conductivity of the porous medium (K) and the hydraulic gradient (i):

$$v = K \times i \tag{2.33}$$

However, this velocity, called Darcy's velocity, is not the real velocity at which water particles move through the porous medium. Darcy's law, first derived experimentally, assumes that the groundwater flow occurs through the entire cross-sectional area of a sample (porous medium) including both voids and grains (adequately, Darcy's velocity is called "smeared velocity" in Russian literature). Since the actual cross-sectional area of flow is smaller than the total area (water moves only through voids), another term is introduced to account for this reduction—linear groundwater velocity (v_L). From Eq. (2.32) it follows that that the linear velocity must be greater than Darcy's velocity: $v_L \geq v$. One handy parameter that can be used to describe the reduced cross-sectional area of flow is effective porosity (n_{ef}), defined as that portion of the overall rock porosity which allows free flow of groundwater (see Section 2.3.1). Accordingly, linear groundwater velocity is expressed by the following equation:

$$v_L = \frac{K \times i}{n_{ef}} \tag{2.34}$$

The linear groundwater velocity is appropriate when used to estimate the average travel time of groundwater, and Darcy's velocity is appropriate for calculating flow rates. Neither, however, is the real groundwater velocity, which is the time of travel of a water particle along its actual convoluted path through the voids. It is obvious that, for practical purposes, the real velocity cannot be measured or calculated.

Two main forces act upon individual water particles that move through the porous medium: friction between the moving water particles and friction between the water particles and the solids surrounding the voids. This results in uneven velocities of individual water particles: some travel faster and some slower than the overall average velocity of a group of particles (Fig. 2.45). This phenomenon is called *mechanical dispersion* and it is very important when quantifying the transport of contaminants dissolved in groundwater (more on fate and transport of contaminants is given in Chap. 5). Because of mechanical dispersion, the spreading of individual water (or dissolved contaminant) particles is in all three main directions with respect to the overall groundwater flow direction: longitudinal, transverse and vertical. Accurate calculation of travel times and arrival times of water and contaminant particles therefore has to include the phenomenon

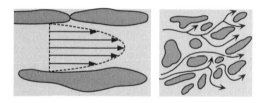

Figure 2.45 Schematic presentation of mechanical dispersion caused by varying velocity of water particles and tortuous flow paths between porous medium grains. (Franke et al., 1990.)

of dispersivity. At the same time, quantifying dispersivity accurately, without extensive field testing, including tracing, is virtually impossible.

As explained earlier, the nature of groundwater flow in fractured rock and karst aquifers differs from that in intergranular porous media. Large fractures and conduits, filled with water, do not behave as "Darcian continuum" and the concept of hydraulic conductivity and effective porosity does not apply. Groundwater velocity in such cases cannot be calculated in a meaningful way without extensive field investigations specifically targeting particular fractures or conduits—a very expensive proposition for any project type. Dye tracing and tracing with environmental isotopes remain investigative techniques of choice when assessing groundwater flow velocities in fractured rock and karst aquifers (see Benischke et al., 2007; Geyh, 2000).

Because of the unique nature of porous media in karst, groundwater velocity can vary over many orders of magnitude even within the same aquifer system. One should therefore be very careful when making a (surprisingly common) statement such as "groundwater velocity in karst is generally very high." Although this may be true for flow taking place in karst conduits and large fractures, a disproportionately greater volume of any karst aquifer has relatively low groundwater velocities (laminar flow) through small fissures and rock matrix. However, most dye tracing tests in karst are designed to analyze possible connections between known (or suspect) locations of surface water sinking and locations of groundwater discharge (springs). Because such connections involve some kind of preferential flow paths (sink-spring type), the apparent velocities calculated from the dye tracing data are usually biased toward the high end.

2.5.4 Anisotropy and Heterogeneity

Sediments and other rocks can be *homogeneous* or *heterogeneous* within some representative volume of observation. Clean beach sand made of pure quartz grains of similar size is one example of a homogeneous rock (unconsolidated sediment). If, in addition to quartz grains, there are other mineral grains but all uniformly mixed, without groupings of any kind, the sediment is still homogeneous. various possible scales, say centimeter to decameter, it is hardly ever satisfied for rock volumes representative of an aquifer or an aquitard. For simplification purposes, and when different groupings of minerals within the same rock, or sediments of different sizes within one sedimentary deposit behave similarly relative to groundwater flow, one may consider such volume as homogeneous and representative. In reality, however, all aquifers and aquitards are more or less heterogeneous, and it is only a matter of convention, or agreement between various interested stakeholders, which portion of the subsurface under investigation can be considered homogeneous. At the same time, simplification of an aquifer volume appropriate for general

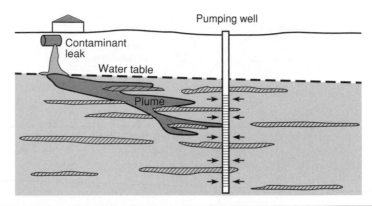

FIGURE 2.46   An aquifer consisting of predominantly gravel and sand provides water to a well through the entire screen length. At the same time, dissolved contaminants may enter the well through just a few discrete intervals.

water supply purposes may be completely inadequate for characterizing contaminant fate and transport. Figure 2.46 illustrates this point. Alluvial aquifers almost always consist of various proportions of gravel, sand, silt, and clay, deposited as layers and lenses of varying thickness. When gravel and sand dominate, with finer fractions forming thin interbeds, the aquifer may be considered as one continuum providing water to a pumping well through its entire screen. However, when the aquifer is contaminated, dissolved contaminants will move faster through more permeable porous media which may form quite convoluted preferential pathways intersecting the well at discrete intervals. Detecting such pathways, although difficult, is often the key for successful groundwater remediation, whereas it may not be of much importance when quantifying groundwater flow rates for water supply.

One important aspect of heterogeneity is that groundwater flow directions change at boundaries between rocks (sediments) of notably different hydraulic conductivity such as the ones shown in Fig. 2.47. An analogy would be refraction of light rays when they enter a medium with different density, e.g., from air to water. The refraction causes the incoming angle, or *angle of incidence,* and the outgoing angle, or *angle of refraction,* to be different (angle of incidence is the angle between the orthogonal line at the boundary and the incoming streamline; angle of refraction is the angle between the orthogonal at the boundary and the outgoing streamline). The only exception is when the streamline is perpendicular to the boundary in which case both angles are the same, i.e. -90 degrees. The mathematical relationship between the angle of incidence (α_1), angle of refraction (α_2), and the hydraulic conductivities of two porous media, K_1 and K_2, is shown in Fig. 2.47. The figure applies to both map and cross-sectional views as long as there is a clearly defined boundary between the two porous media.

Heterogeneity of the hydraulic conductivity is the main cause of macrodispersion in groundwater systems, which is of particular importance when analyzing capture zones of extraction wells, and transport of contaminants. Figure 2.48 shows two capture zones for the same well, pumping with the same rate, when the aquifer in question is modeled with a homogeneous hydraulic conductivity (right), and with a heterogeneous (spatially varying) hydraulic conductivity (left). Similarly, the shape of a plume of dissolved contaminants will be significantly influenced by the porous media heterogeneity.

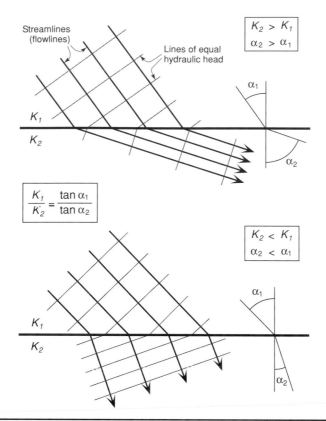

FIGURE 2.47 Refraction of groundwater flowlines (streamlines) at a boundary of higher hydraulic conductivity (top) and a boundary of lower hydraulic conductivity (bottom).

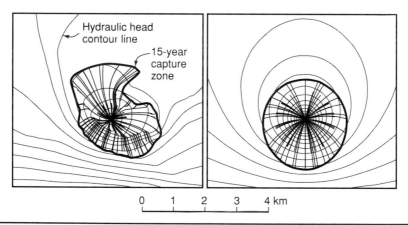

FIGURE 2.48 Right: 15-year capture zone, defined by flowlines, of a well pumping from a semiconfined aquifer modeled with uniform average hydraulic conductivity. Left: Capture zone of the same well when the aquifer is represented by spatially varying (heterogeneous) hydraulic conductivity.

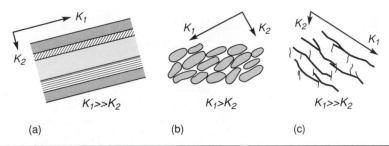

$K_1 >> K_2$

$K_1 > K_2$

$K_1 >> K_2$

(a) (b) (c)

FIGURE 2.49 Some possible reasons for anisotropy of hydraulic conductivity. (a) Sedimentary layers of varying permeability; (b) orientation of gravel grains in alluvial deposit; (c) two sets of fractures in massive bedrock. (Kresic, 2007a; copyright Taylor & Francis Group, LLC, printed with permission.)

Anisotropy of porous media is another very important factor influencing directions of groundwater flow and transport of contaminants. It is a result of the so-called *geologic fabric* of rocks comprising aquifers and aquitards. Geologic fabric refers to spatial and geometric relationships between all elements of which the rock is composed, such as grains of sedimentary rocks, and the component crystals of magmatic and metamorphic rocks. Fabric also refers to discontinuities in rocks, such as fissures, fractures, faults, fault zones, folds, and bedding planes (layering). Without elaborating further on the geologic portion of hydrogeology, it is appropriate to state that groundwater professionals lacking a thorough geologic knowledge (i.e., "nongeologists") would likely have various difficulties in understanding the many important aspects of heterogeneity and anisotropy.

In hydrogeology, a porous medium is considered anisotropic when hydraulic conductivity varies in different directions. All aquifer types are more or less anisotropic, with fractured rock and karst aquifers often exhibiting the highest degree of anisotropy; such aquifers may have zones of extremely high hydraulic conductivity with almost any shape imaginable. Figures 2.49 and 2.50 illustrate just some of many possible causes of anisotropy in various types of rocks. It is important to understand that a varying degree of anisotropy can (and usually does) exist in all spatial directions. It is for reasons of simplification and/or computational feasibility that hydrogeologists consider only three main perpendicular directions of anisotropy: two in the horizontal plane and one in the vertical plane; in the Cartesian coordinate system these three directions are represented with the X, Y, and Z axes. Figure 2.51 illustrates the importance of aquifer anisotropy in determining well capture zones.

Again, for reasons of simplicity or feasibility, one may decide that the groundwater system under consideration, or any of its parts, could be represented by a volume including "all" important aspects of heterogeneity and anisotropy of the porous media present. Such volume is sometimes called *representative elementary volume* (REV) and is defined by only one value for each of the many quantitative parameters describing groundwater flow, and fate and transport of contaminants. The REV concept is considered by many to be rather theoretical, since it is not independent of the nature of the practical problem to be solved. For example, less than 1 m^3 (several cubic feet) of rock may be more than enough for quantifying phenomena of contaminant diffusion into rock matrix, whereas this volume would be completely inadequate for calculating groundwater flow rate in a fractured rock aquifer where major transmissive fractures are spaced more than 1 meter

FIGURE 2.50   Cross-bedded sandstone of the Cutler Formation in southern Utah. The banding within the outcrop represents cross-stratification of river (fluvial) deposits. Note how cross stratification truncates underlying strata. Rock hammer for scale. (Photograph courtesy of Jeff Manuszak.)

apart. Deciding on the representative volume will also depend on the funds and time available for collecting field data and performing laboratory tests. Extrapolations and interpolations based on data from several borings or monitoring wells will by default be very different than those using data from tens of wells. Another related difficulty, which always presents a major challenge, is upscaling. This term refers to assumptions made when applying parameter values obtained from small volumes of porous media (e.g., laboratory sample) to larger, field-scale problems. Whatever the final choice of each quantitative parameter may be, every attempt should be made to fully describe and quantify the associated uncertainty and sensitivity of that parameter.

The following example illustrates how two different choices of two basic hydro-geologic parameters reflecting heterogeneity can produce very different quantitative answers, even though both selections may seem reasonable. Consider the following scenario: point of contaminant release and a potential receptor are 2500 ft apart; the regional hydraulic gradient in the shallow aquifer, which consists of "fine sands," is estimated from available monitoring well data to be 0.002. How long would it take a dissolved contaminant particle to travel between the two points, assuming that the contaminant does not degrade or adsorb to solid particles (i.e., it is "conservative" and moves at the same velocity as water)?

As shown in Fig. 2.37, fine sand can have hydraulic conductivity anywhere between a little less than 1 and about 40 ft/d. Effective porosity (specific yield) of "sand" can vary anywhere between 20 and 45 percent. Assuming the lowest values from the two ranges,

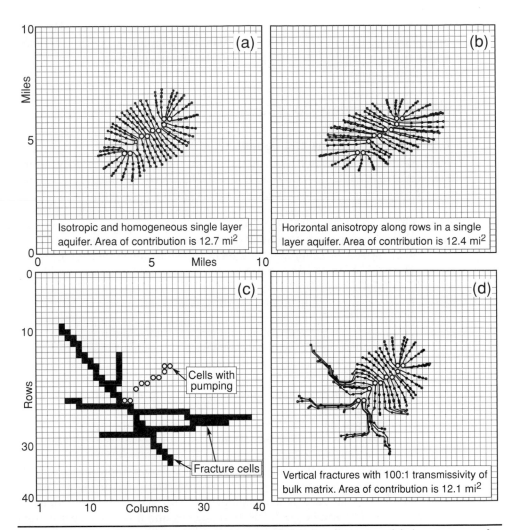

FIGURE 2.51 The influence of aquifer anisotropy and heterogeneity on a modeled capture zone for the Central Swamp region, Cypress Creek well field near Tampa, FL. (*a*) Isotropic and homogeneous 1-layer aquifer; (*b*) anisotropic hydraulic conductivity with five times greater value along rows; (*c* and *d*) simulation of vertical fractures with "fracture" cells where transmissivity is 100 times greater than in the surrounding "matrix" cells. (Modified from Knochenmus and Robinson, 1996.)

the linear velocity of a groundwater particle, using Eq. (2.34), is:

$$v_{\text{L}} = \frac{K \times i}{n_{\text{ef}}} = \frac{0.8\,(\text{ft/d}) \times 0.002}{0.2} = 0.016\,[\text{ft/d}]$$

Based on this velocity, the time of groundwater (and dissolved contaminant) travel between the two points of interest would be 156,250 days or about 428 years (2500 ft-distance is divided by the velocity of 0.016 ft/d). Using the highest values from the two ranges (40 ft/d and 45 percent), the time of travel is calculated to be about 14,045 days or

38.5 years, which is a very significant difference, to say the least. This simple quantitative example shows inherent uncertainties in quantifying groundwater flow characteristics, even when assuming that the porous medium is "homogeneous."

2.6 Initial and Boundary Conditions

2.6.1 Initial Conditions and Contouring of Hydraulic Head

In the world of groundwater modeling, the term initial conditions refers to the three-dimensional distribution of observed hydraulic heads within the groundwater system, which is the starting point for transient (time-dependent) modeling simulations. These hydraulic heads (water table of unconfined aquifers and potentiometric surface of confined aquifers) are the result of various boundary conditions acting upon the system during a certain time period. The initial distribution of the hydraulic heads for transient modeling can also be the calibrated solution of a steady-state model, which is the closest match to the field-observed heads when assuming constant boundary conditions and no change in storage. In a broad sense, any set of field-measured or calibrated hydraulic heads can serve as the starting point for further analysis, including for transient groundwater modeling. Ideally, the initial conditions should be as close as possible to the state of a long-term equilibrium between all natural water inputs and outputs from the system, or with as little anthropogenic (artificial) influences as possible: the so-called predevelopment conditions (Fig. 2.52). However, in many cases there is insufficient hydraulic head data for such natural conditions, which causes various difficulties with data interpolation and extrapolation, including uncertainties associated with any assumed predevelopment boundary conditions.

Whatever the case may be regarding the selection of initial conditions, contouring of the hydraulic head data is the first important step. Contour maps of the water table (unconfined aquifers) or the piezometric surface (confined aquifers) are made in the majority of hydrogeologic investigations and, when properly drawn, represent a very powerful tool in aquifer studies. Although commonly used for determination of groundwater flow directions only, contour maps, when accompanied with other data, allow for the analyses and calculations of hydraulic gradients, flow velocity and flow rate, particle travel time, hydraulic conductivity, and transmissivity. In addition, the spacing and the orientation (shape) of the contours directly reflect the existence of flow boundaries. When interpreting contour maps, one should always remember that it is a two-dimensional representation of a three-dimensional flow field, and as such it has limitations. If the groundwater system of interest is known to have significant vertical gradients, and enough field information is available, it is always wise to construct at least two contour maps: one for the shallow depth and one for the deeper depth. As with geologic and hydrogeologic maps in general, a contour map should be accompanied with several cross-sections showing locations and vertical points of the hydraulic head measurements with posted data, or ideally showing the contour lines on the cross sections as well. Probably the most incorrect and misleading case is developed when data from monitoring wells screened at different depths in an aquifer with vertical gradients are lumped together and contoured as one "average" data package. A perfect example would be a fractured rock or karst aquifer with thick residuum (regolith) deposits and monitoring wells screened in the residuum and at various depths in the bedrock. If data from all the wells were lumped

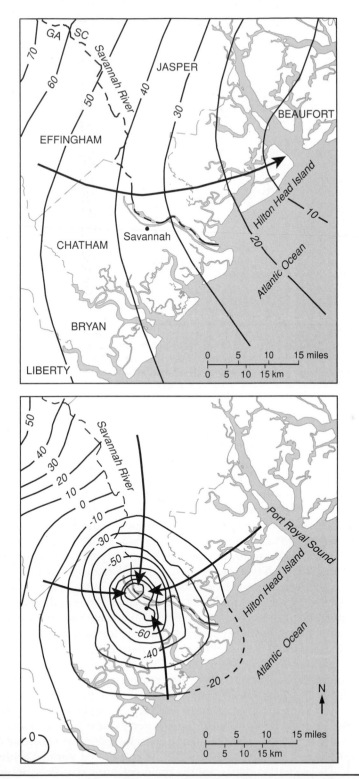

Figure 2.52 Potentiometric surface of the Upper Floridan aquifer in the area of Savannah, GA, and Hilton Head Island, SC. Top: Predevelopment conditions (datum is NGVD 29); bottom: recorded in May 1998 (datum is NAVD 88). Contour interval is 10 ft, contour lines dashed where approximate; arrows show general directions of groundwater flow. (Modified from Provost et al., 2006.)

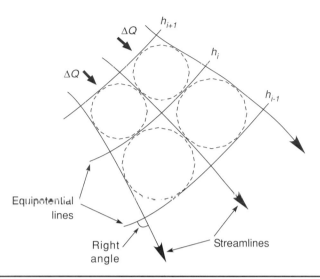

FIGURE 2.53 Flow net is a set of equipotential lines and streamlines which are perpendicular to each other. The equipotential line connects points with the same groundwater potential, i.e., hydraulic head h. The streamline is an imaginary line representing the path of a groundwater particle as it flows through an aquifer. Flow rate between adjacent pairs of streamlines, ΔQ, is the same. Equipotential lines are more widely spaced where the aquifer is more transmissive.

together and contoured, it would be impossible to interpret where the groundwater is actually flowing for the following reasons: (1) the residuum is primarily an intergranular porous medium in unconfined conditions (it has water table), and horizontal flow directions may be influenced by local (small) surface drainage features; (2) the bedrock has discontinuous flow through fractures at different depths, which is often under pressure (confined conditions), and may be influenced by regional features such as major rivers or springs. The flow in two distinct porous media (the residuum and the bedrock) may therefore be in two different general directions at a particular site, including strong vertical gradients from the residuum toward the underlying bedrock. Creating one "average" contour map for such system does not make any hydrogeologic sense (Kresic, 2007a).

The contour map of the hydraulic head is one of two parts of a *flow net*: flow net in a homogeneous isotropic aquifer is a set of streamlines and equipotential lines, which are perpendicular to each other (see Fig. 2.53). Streamline (or flow line) is an imaginary line representing the path of a groundwater particle as it flows through the aquifer. Two streamlines bound a flow segment of the flow field and never intersect, i.e., they are roughly parallel when observed in a relatively small portion of the aquifer. The main requirement of a flow net is that the flow rate between adjacent pairs of streamlines is the same (ΔQ in Fig. 2.53), which enables calculations of flow rates in various portions of the aquifer, providing that the hydraulic conductivity and the aquifer thickness are known.

Equipotential line is a horizontal projection of the equipotential surface—everywhere at that surface the hydraulic head has a constant value. Two adjacent equipotential lines (surfaces) never intersect and can also be considered parallel within a small aquifer portion. These characteristics are the main reason why a flow net in a homogeneous, isotropic aquifer is sometimes called the net of small (curvilinear) squares. In general, the following simple rules apply for graphical flow net construction in heterogeneous, isotropic systems (Freeze and Cherry, 1979):

1. Flow lines and equipotential must intersect at right angles throughout the system.
2. Equipotential lines must meet impermeable boundaries at right angles.
3. Equipotential lines must parallel constant-head boundaries.
4. The tangent law must be satisfied at geologic boundaries (see Fig. 2.47).
5. If the flow net is drawn such that squares are created in one portion of one formation, squares must exist throughout that formation and throughout all formations with the same hydraulic conductivity. Rectangles will be created in formations of different conductivity.

The last two rules make it extremely difficult to manually draw accurate quantitative flow nets in complex heterogeneous systems. If a system is anisotropic in addition, it would not be feasible to draw an adequate flow net manually in most cases. However, drawing an approximate contour map (flow net without streamlines) manually is always recommended since it allows the interpreter to incorporate the understanding of various hydrogeologic complexities. Complete reliance on contouring with computer programs could lead to erroneous conclusions since they are unable to recognize interpretations apparent to a groundwater professional such as presence of geologic boundaries, varying porous media, influence of surface water bodies, or principles of groundwater flow. Thus, manual contouring and manual reinterpretation of computer-generated maps are essential and integral parts of hydrogeologic studies. Although some advocates of computer-based contouring argue that it is the most "objective" method since it excludes possible "bias" by the interpreter, little can be added to the following statement: if something does not make hydrogeologic sense, it does not matter who or what created the senseless interpretation.

The ultimate tool for creating contour maps, tracking particles as they flow through the system, and calculating flow rates for any part of a groundwater system, is a numeric model, which can incorporate and test all known or suspected heterogeneities, boundaries, and anisotropy, in all the three dimensions. Figures 2.54 to 2.56 show output from a model used to test influence of varying hydraulic conductivity and anisotropy on tracks of particles released at certain locations in the aquifer.

When analyzing initial conditions, several synoptic data sets collected in different time periods should be used in order to better understand the system and select what appears to be a "representative" spatial distribution of the hydraulic heads. In addition to recordings from piezometers, monitoring wells, and other water wells, every effort should be made to record elevations of water levels in the nearby surface streams, lakes, ponds, and other surface water bodies. Information about hydrometeorologic conditions (e.g., rainfall) prior to the time of hydraulic head measurements is also important for understanding possible influence of recharge episodes on groundwater flow directions and fluctuations of the hydraulic heads. All this information is essential for making a correct contour map.

One of the most important aspects of constructing contour maps in alluvial aquifers is to determine the relationship between groundwater and surface water. In hydraulic

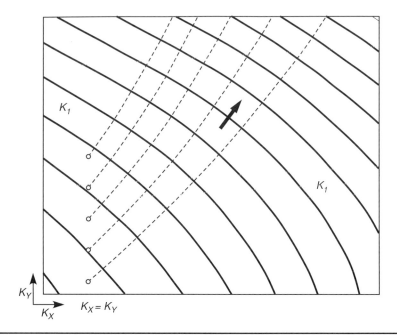

FIGURE 2.54 Hydraulic head contour lines and particle tracks (dashed) in an isotropic, homogeneous aquifer of uniform hydraulic conductivity (K_1).

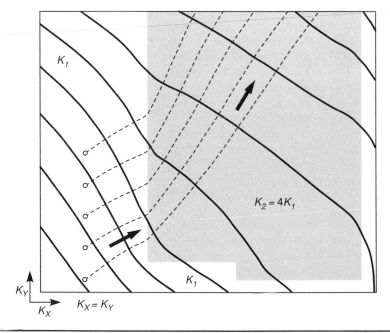

FIGURE 2.55 Influence of a geologic boundary (heterogeneity) on contour lines and particle tracks. The shaded area has four times higher hydraulic conductivity than the rest of the flow field. Aquifer is isotropic (the hydraulic conductivity is same in X and Y directions).

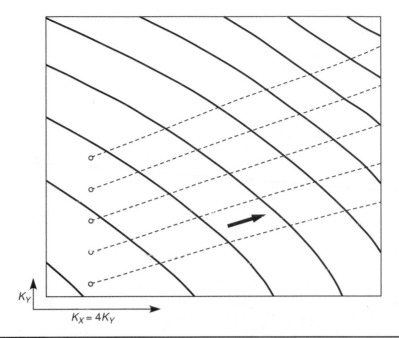

FIGURE 2.56 Influence of anisotropy on particle tracks (dashed lines). The hydraulic conductivity in X direction is four times higher than in Y direction.

terms, the contact between an aquifer and a surface water body is an equipotential boundary. In case of lakes and wetlands, this contact can be approximated with the same hydraulic head. In case of flowing streams, the hydraulic head along the contact decreases in the downgradient direction. If enough measurements of a stream stage are available, it is relatively easy to draw the water table contours near the river and to finish them along the river-aquifer contact. However, often little or no precise data is available on river stage and, at the expense of precision, it has to be estimated from a topographic map or from the monitoring well data by extrapolating the hydraulic gradients until they intersect the river. Figure 2.57 shows some of the examples of surface water-groundwater interaction represented with the hydraulic head contour lines.

In highly fractured and karst aquifers, where groundwater flow is discontinuous (it takes place mainly along preferential flow paths such as fractures and karst conduits), Darcy's Law does not apply and flow nets are not an appropriate method for the flow characterization. However, contour maps in such aquifers are routinely made by many professionals who often find themselves excluding certain "anomalous" data points while trying to develop a "normal-looking" map. Contour maps showing regional (say, on a square-mile scale) flow-pattern in a fractured rock or karst aquifer may be justified since groundwater flow generally is from recharge areas toward discharge areas and the regional hydraulic gradients will reflect this simple fact. The problems usually arise when interpreting local flow patterns, as schematically shown in Fig. 2.58.

2.6.2 Boundary Conditions

It has become standard practice in hydrogeology and groundwater modeling to describe the inflow and outflow of water from a groundwater system with three general boundary

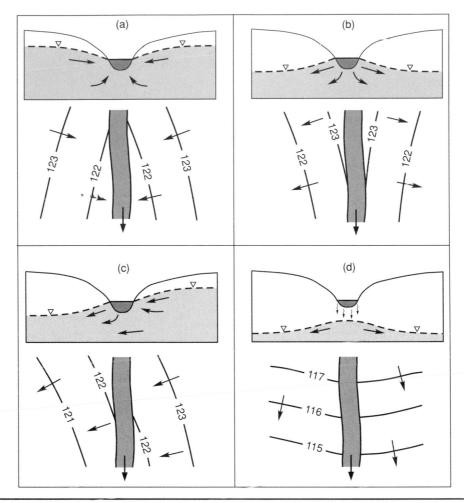

Figure 2.57 Basic hydraulic relationships between groundwater and surface water shown in cross-sectional views (top), and map views using hydraulic head contour lines. (a) Perennial gaining stream; (b) perennial losing stream; (c) perennial stream gaining water on one side and losing water on the other side; (d) losing stream disconnected from the underlying water table, also called ephemeral stream. (From Kresic, 2007a; copyright Taylor & Francis Group, LLC, printed with permission.)

conditions: (1) known flux, (2) head-dependent flux, and (3) known head, where "head" refers to the hydraulic head. These conditions are assigned to both external and internal boundaries, that is to all locations and surfaces where water is entering or leaving the system. One example of an external system boundary, sometimes overlooked as such, is the water table of an unconfined aquifer that receives recharge from percolating precipitation or irrigation return. This estimated or measured flux of water into the system is applied as recharge rate over certain land surface area. It is usually expressed in inches (millimeters) per time unit of interest (e.g., day, month, year), which, when multiplied by the area, gives the flux of water as volume per time. A large spring with a measured

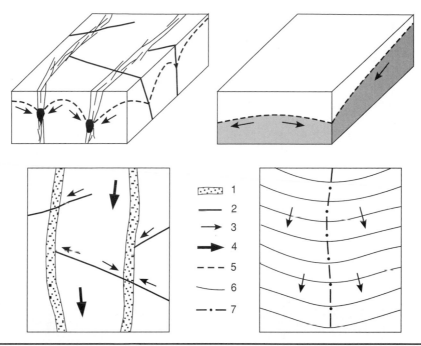

FIGURE 2.58 Groundwater flow and its map presentation for a fractured rock or karst and an intergranular aquifer. (1) Preferential flow path (e.g., fracture or fault zone or karst conduit/channel); (2) fracture/fault; (3) local flow direction; (4) general flow direction; (5) position of the hydraulic head (water table in the intergranular aquifer); (6) hydraulic head contour line; (7) groundwater divide. (From Kresic, 1991.)

discharge rate, draining an aquifer, is another example of an external boundary with a known flux. An example of an internal boundary with a known flux, where water is leaving the system, is a water well with the recorded pumping rate expressed in gallons per minute (gal/min) or liters per second (L/s). It is obvious that water can enter or leave a groundwater system in a variety of natural and artificial ways, depending upon hydrogeologic, hydrologic, climatic, and anthropogenic conditions specific to the system of interest. In many cases, these water fluxes cannot be measured directly and have to be estimated or calculated using different approaches and parameters (see Section 2.4 and Chap. 3). The simplest boundary condition is one that can be assigned to a contact between an aquifer and a low-permeable porous medium, such as "aquiclude." Assuming that there is no groundwater flow across this contact, it is called a zero-flux boundary. Although this no-flow boundary condition may exist in reality, it is very important not to assign it indiscriminately just because it is convenient. For example, contact between unconsolidated alluvial sediments and surrounding "bedrock" is often modeled as a zero-flux boundary, even though there may be some flow across this boundary in either direction. Without site-specific information on the underlying hydrogeologic conditions, a zero-flux assumption may lead to erroneous conclusions (calculations) regarding groundwater flow, or fate and transport of contaminants.

Recording hydraulic heads at external or internal boundaries, and using them to determine water fluxes indirectly, rather than assigning them directly, is very common in

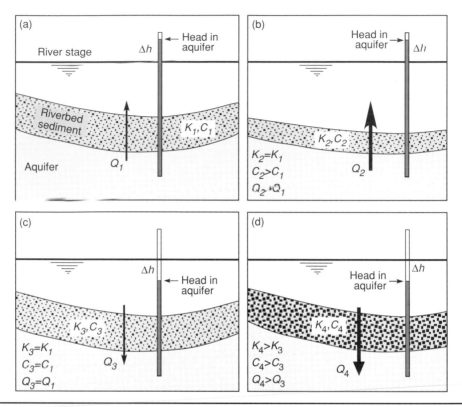

FIGURE 2.59 River boundary represented with a head-dependent flux. K is hydraulic conductivity, C is riverbed conductance, Q is flow rate between the aquifer and the river, and Δh is hydraulic gradient between the aquifer and the river (same in all four cases). (a and b) Gaining stream; (c and d) losing stream. Lower hydraulic conductivity of the riverbed sediments, and their greater thickness result in lower conductance, and lower flow rate.

hydrogeologic practice. The hydraulic heads provide for determination of the hydraulic gradients which, together with the hydraulic conductivity and the cross-sectional area of the boundary, give the groundwater flow entering or leaving the system across that boundary. This boundary condition, expressed by the hydraulic heads on either side of the boundary, and the hydraulic conductance of the boundary (i.e., the transmissivity of the boundary) is called head-dependent flux. One example of the head-dependent flux boundary would be a river having riverbed sediments of the hydraulic conductivity different than that of the underlying aquifer. As illustrated in Fig. 2.59, the rate of flow between the aquifer and the river will depend on the difference between the hydraulic heads in the aquifer adjacent to the river and the river stage (hydraulic head of the river), as well as on the riverbed conductance. Lower conductance corresponds to more fines (silt) in the riverbed sediment and a lower hydraulic conductivity, resulting in a lower water flux between the aquifer and the river (boundary). Thicker riverbed sediments will have the same effect.

When not much is known about the real physical characteristics of a boundary, or for reasons of simplification, the boundary may be represented only by its hydraulic head:

so-called known-head, or fixed-head, or equipotential boundary. River or lake stages, without considering riverbed (lakebed) conductance, are examples of such a boundary. The flux of water across the boundary (Q) is calculated using Darcy's equation: $Q = AKi$, where A = cross-sectional area of the boundary, i is the hydraulic gradient between the boundary (river or lake) and the aquifer, and K is the hydraulic conductivity of the aquifer porous media. One potential problem with such interpretations of boundary conditions is that, as the hydraulic head in the aquifer decreases, the flow entering the system from the boundary also increases due to the increased hydraulic gradient since the head at the boundary is fixed. This is of particular concern when performing transient modeling which takes into account time-dependent changes that can affect the system. If one still prefers to model a certain boundary with a fixed head condition, the hydraulic head should be adjusted in the model for different time periods, based on available field information.

The ultimate reason for selecting any of the three general boundary types is the determination of the overall water budget of a groundwater system. The sum of all water fluxes entering and leaving the system through its boundaries has to be equal to the change in water storage inside the system. When using groundwater models for system evaluation or management, the user has to determine (measure, calculate) flux to be assigned to the known-flux boundaries. In case of the other two boundary types (head-dependent flux and fixed-head), the model calculates the flux across the boundaries using other assigned parameters—hydraulic heads at the boundary and inside the system, boundary conductance, and hydraulic conductivity of the system's porous media.

An illustration of how various boundary conditions can affect groundwater flow and groundwater withdrawal is shown in Figs. 2.60 to 2.62 with an example of a basin-fill basin. Such basins, common in the semiarid western United States, may have permanent (perennial) or intermittent surface streams and may be recharged by surface water runoff

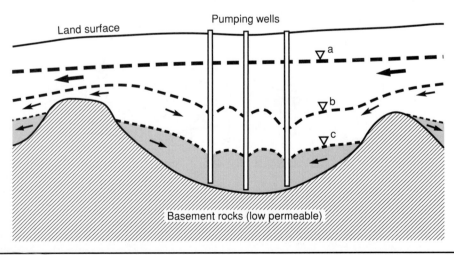

Figure 2.60 Schematic longitudinal cross section along a simplified basin-fill basin (no heterogeneities shown), connected to an upgradient and a downgradient basin. (*a*) Predevelopment hydraulic head; (*b*) hydraulic head resulting from early stages of groundwater extraction; (*c*) hydraulic head resulting from excessive groundwater extraction in all three basins, which causes cessation of groundwater flow between the basins.

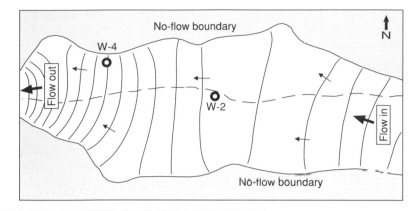

Figure 2.61 Hydraulic head contour lines in the alluvial-fill basin illustrating no hydraulic connection with the surface stream flowing through it and no water inputs from the basin margins. Arrows indicate general directions of groundwater flow.

and underflow from the surrounding mountain fronts. They can also be connected with adjacent basins, thus forming rather complex groundwater systems with various local and regional water inputs and water outputs. Assigning representative, time-dependent boundary conditions in both surface and subsurface areas (zones) may therefore be quite difficult, but it is necessary for an appropriate groundwater management.

When deciding on boundary conditions, it is essential to work with as many hydraulic head observations as possible, in both space and time, because fluctuations in the shape and elevations of the hydraulic head contour lines directly reflect various water inputs and outputs along the boundaries. For example, the cross-section in Fig. 2.60 shows one basin connected with an upgradient and a downgradient basin, with all three basins being pumped for water supply. Depending on the rates of groundwater withdrawal

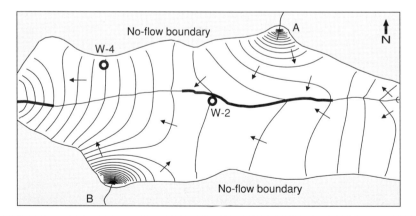

Figure 2.62 Hydraulic head contour lines showing influence of two surface streams (*A* and *B*) flowing into the basin from the surrounding bedrock areas and losing all water to the underlying aquifer short distance from the contact. The main surface stream flowing through the basin is in hydraulic connection with the underlying aquifer. Gaining reaches are shown with thick lines. Arrows indicate general directions of groundwater flow.

and recharge, more or less groundwater may be flowing between the basins, including a complete cessation of the interflows. Availability of the hydraulic head data at various locations within the basin, and at various times, will determine the accuracy of the hydraulic head contours which therefore may or may not show existence or influence of various boundary conditions. Figure 2.61 indicates general inflow of groundwater from the upgradient basin in the east and outflow to the downgradient basin in the west, with no other water inputs (i.e., all other basin boundaries are assumed to be zero-flux), and no connection between the basin aquifer and the surface stream flowing through the basin. Figure 2.62 includes influence of two streams (A and B) entering the basin and losing all water to the aquifer a short distance from the boundary. It also shows hydraulic connection between the aquifer and the surface stream flowing through the basin, including river reaches that lose water to, or gain water from, the aquifer.

As mentioned earlier, accurate representation of surface water-groundwater interactions is often the most critical when selecting boundary conditions in alluvial basins and flood plains of surface streams. A river may be intermittent or perennial, it may lose water to the underlying aquifer in some reaches and gain water in others, and the same reaches may behave differently depending on the season. The hydraulic connection between a river and "its" aquifer may be complete, without any interfering influence of riverbed sediments. In some cases, however, a well pumping close to a river may receive little water from it because of a thick layer of fine silt along the river channel or simply because there is a low-permeable sediment layer separating the aquifer and the river. In these situations it would be completely erroneous to represent the river as a constant-head (equipotential) boundary directly connected to the aquifer. Such a boundary in a quantitative model would essentially act as an inexhaustible source of water to the aquifer (or a water well) regardless of the actual conditions, as long as the hydraulic head in the aquifer is lower than the river stage.

It is obvious that any number of combinations between the two extreme cases shown in Figs. 2.61 and 2.62 is possible in real-world situations which, by default, include time-dependent (changing) boundary conditions. In our case this may simply mean that the river is not perennial. An assumption that boundary conditions do not change in time for any reason, including "simplification" or "screening," will in all likelihood result in erroneous conclusions and a false quantitative basis for groundwater management decisions. Moreover, changing boundary conditions cause changes in groundwater storage which has to be taken into account in any quantitative analysis of available groundwater resources. In other words, simulating groundwater systems with steady-state models, which exclude storage parameters by default, will in all likelihood also result in erroneous conclusions and a false quantitative basis for groundwater management decisions.

The importance of boundary conditions for estimating long-term well yield is illustrated in Fig. 2.63. The curves for hydraulic head versus time are for 1 year of simulated pumping, with and without the river acting as a complete equipotential boundary respectively. In the presence of a constant-head boundary, the hydraulic head at well W-2 stabilizes after two weeks of pumping, whereas in the absence of the boundary it continues to decrease. However, this decrease is at a considerably slower rate than at well W-4 which is located closer to the impermeable boundary. The drawdown at W-4 is about 17 m after 1 year of pumping, compared to about 10 m at W-2, with both wells pumping at the same rate, and assuming the same hydraulic conductivity. Note that W-4 is not influenced by the river boundary during the first year of pumping, showing the exactly same curve for both boundary conditions.

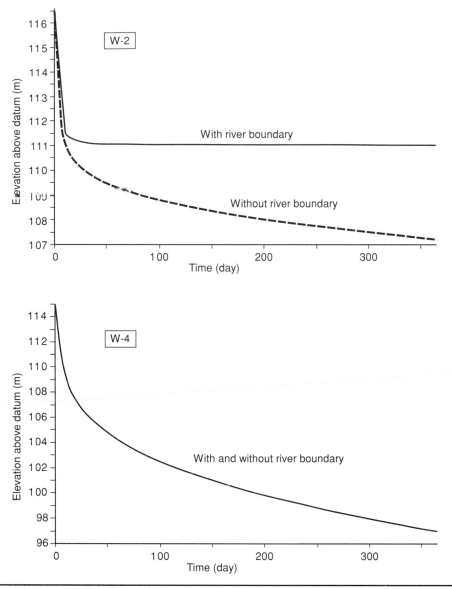

Figure 2.63 Hydraulic head versus time at wells W-2 and W-4 during the first year of simulated pumping. Both wells are pumping at the same rate, assuming same hydraulic conductivity. The location of the wells is shown in Figs. 2.61 and 2.62.

Faults often form hydraulic boundaries for groundwater flow in both consolidated and unconsolidated rocks. They may have one of the following three roles: (1) conduits for groundwater flow, (2) storage of groundwater due to increased porosity within the fault (fault zone), and (3) barriers to groundwater flow due to decrease in porosity within the fault. The following discussion by Meinzer (1923) illustrates this point:

Faults differ greatly in their lateral extent, in the depth to which they reach, and in the amount of displacement. Minute faults do not have much significance with respect to ground water except, as they may, like other fractures, serve as containers of water. But the large faults that can be traced over the surface for many miles, that extend down to great depths below the surface, and that have displacements of hundreds or thousands of feet are very important in their influence on the occurrence and circulation of ground water. Not only do they affect the distribution and position of aquifers, but they may also act as subterranean dams, impounding the ground water, or as conduits that reach into the bowels of the earth and allow the escape to the surface of deep-seated waters, often in large quantities. In some places, instead of a single sharply defined fault, there is a fault zone in which there are numerous small parallel faults or masses of broken rock called fault breccia. Such fault zones may represent a large aggregate displacement and may afford good water passages.

The impounding effect of faults is caused by following main mechanisms:

- The displacement of alternating permeable and impermeable beds in such manner that the impermeable beds are made to abut against the permeable beds.
- Presence of clayey gouge along the fault plane produced by the rubbing and mashing during displacement of the rocks. (The impounding effect of faults is most common in unconsolidated formations that contain considerable clayey material.)
- Cementation of the pore space by precipitation of material, such as calcium carbonate, from the groundwater circulating through the fault zone.
- Rotation of elongated flat clasts parallel to the fault plane so that their new arrangement reduces permeability perpendicular to the fault.

Mozley et al. (1996) discuss reduction in hydraulic conductivity associated with high-angle normal faults that cut poorly consolidated sediments in the Albuquerque Basin, New Mexico. Such fault zones are commonly cemented by calcite, and their cemented thickness ranges from a few centimeters to several meters, as a function of the sediment grain size on either side of the fault. Cement is typically thickest where the host sediment is coarse grained and thinnest where it is fine grained. In addition, the fault zone is widest where it cuts coarser-grained sediments. Extensive discussion on deformation mechanisms and hydraulic properties of fault zones in unconsolidated sediments is given by Bense et al. (2003). Various aspects of fluid flow related to faults and fault zones are discussed by Haneberg et al. (1999).

An example of major faults in unconsolidated alluvial-fill basins in southern California acting as impermeable barriers for groundwater flow is shown in Fig. 2.64. The Rialto-Colton Basin, which is heavily pumped for water supply, is almost completely surrounded by impermeable fault barriers, receives negligible recharge from precipitation, and very little lateral inflow in the far northwest from the percolating Lytle Creek waters. In contrast, the Bunker-Hill Basin to the north, which is also heavily pumped for water supply, receives most of its significant recharge from numerous losing surface streams, and runoff from the mountain front. As a result, the hydraulic heads in the Rialto-Colton Basin (not shown in Fig. 2.64) are hundreds of feet lower than in the Bunker-Hill Basin.

As repeatedly discussed earlier, one of the most important aspects of boundary conditions is that they change in time. Which time interval will be used for their inevitable averaging depends upon the goals of every particular study. Seasonal or perhaps annual time period may be adequate for a long-term water supply evaluation when

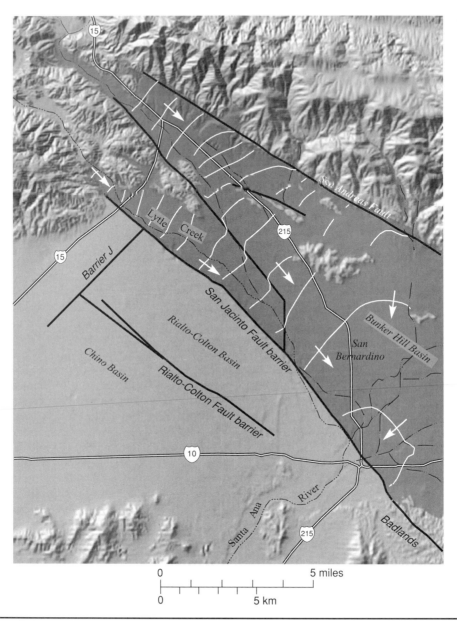

FIGURE 2.64 Groundwater basins in southern California separated by impermeable faults developed in unconsolidated alluvial-fill sediments. White lines are contours of hydraulic head; white arrows are general directions of groundwater flow; dashed lines are surface streams; bold black lines are major faults. (Modified from Danskin et al., 2006.)

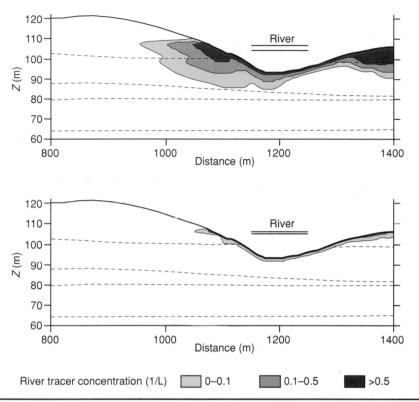

FIGURE 2.65   River water tracer concentrations at end of simulation. Two-dimensional numeric model of interaction between aquifer, vadose zone, and the Columbia River in the Hanford 300 area, WA. Top: Hourly boundary conditions; bottom: monthly boundary conditions. (Modified from Waichler and Yabusaki, 2005.)

considering recharge from precipitation. When a boundary is quite dynamic hydrauli-cally and the required accuracy of predictions is high, the time interval for describing changing boundary conditions may have to be much shorter. For example, Fig. 2.65 shows a comparison of two time intervals used to model the interaction between a large river and a highly transmissive alluvial aquifer. The Columbia River stage at this site is dominated by higher frequency diurnal fluctuations that are principally the result of water released at Priest Rapid Dam to match power generation needs. The magnitude of these diurnal river-stage fluctuations can exceed the seasonal fluctuation of monthly average river stages. During the simulation period, the mean 24-hour change (difference between minimum and maximum hourly values) in river stage was 0.48 m, and the maximum 24-hour change was 1.32 m. Groundwater levels are significantly correlated with river stage, although with a lag in time and decreased amplitude of fluctuations. A two-dimensional, vertical, cross-sectional model domain was developed to capture the principal dynamics of flow to and from the river as well as the zone where groundwater and river water mix (Waichler and Yabusaki, 2005).

Forcing the model with hourly boundary conditions resulted in frequent direction and magnitude changes of water flux across the riverbed. In comparison, the velocity fluctuations resulting from averaging the hourly boundary conditions over a day were

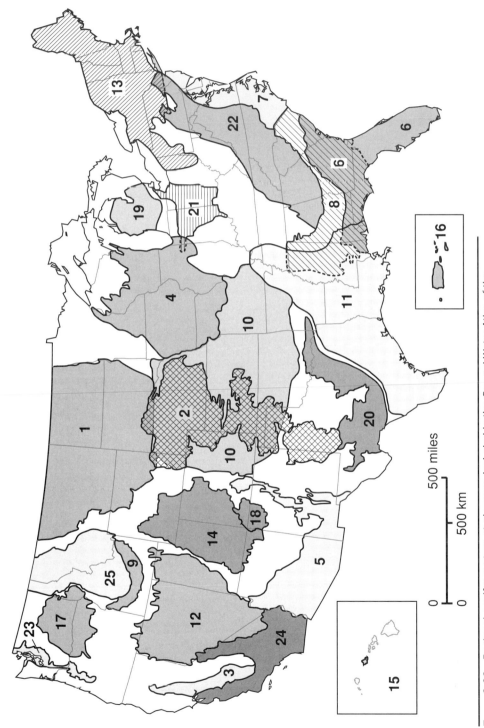

Figure 2.66 Regional aquifer system study areas included in the *Ground Water Atlas of the United States*. (1) Northern great plains; (Continued)

considerably attenuated and for the month average boundary conditions were nonexistent. A similar pattern was held for the river tracer, which could enter the aquifer and then return to the river later. Simulations based on hourly water level boundary conditions predicted an aquifer-river water mixing zone that reached 150 m inland from the river based on the river tracer concentration contours. In contrast, simulations based on daily and monthly averaging of the hourly water levels at the river and interior model boundaries were shown to significantly reduce predicted river water intrusion into the aquifer, resulting in underestimation of the volume of the mixing zone. The relatively high-frequency river-stage changes associated with diurnal release schedules at the dams generated significant mixing of the river and groundwater tracers, and flushing of the subsurface zone near the river. This mixing was the essential mechanism for creating a fully developed mixing zone in the simulations. Although the size and position of the mixing zone did not change significantly on a diurnal basis, it did change in response to seasonal trends in river stage. The largest mixing zones occurred with the river-stage peaks in May–June and December–January, and the smallest mixing zone occurred in September when the river stage was relatively low (Waichler and Yabusaki, 2005).

In conclusion, the availability and interpretation of both hydraulic head data and boundary conditions of a groundwater system are the most critical components of its quantitative evaluation. At the same time, without a thorough geologic and hydrogeologic knowledge of the underlying conditions, any quantitative analysis of the system has a high chance of failing.

2.7 Aquifer Types

The most common classification of aquifers is based on the lithology of the porous media in which they are developed. Three main groups are (1) unconsolidated sediments, (2) sedimentary rocks, and (3) fractured rock (bedrock) aquifers. They are further subdivided based on specific depositional environments (for sediments) and their general geologic origin into various aquifer types that behave similarly in terms of groundwater flow and storage. Rocks and deposits with minimal permeability, which are not considered to be aquifers, consist of unfractured intrusive igneous rocks, metamorphic rocks, shale, siltstone, evaporite deposits, silt, and clay.

An excellent overview of different types of aquifers and their main characteristics is *The Ground Water Atlas of the United States*. The atlas provides a summary of the most important information available for each principal aquifer, or rock unit that will yield usable quantities of water to wells, throughout the 50 States, Puerto Rico, and the U.S. Virgin Islands. The atlas is an outgrowth of the Regional Aquifer-System Analysis (RASA) program of the USGS—a program that investigated 24 of the most important aquifers and aquifer systems of the Nation and one in the Caribbean Islands (Fig. 2.66). The objectives

---

FIGURE 2.66   (*Continued*) (2) High plains; (3) Central Valley, California; (4) Northern Midwest; (5) Southwest alluvial basins; (6) Floridan; (7) Northern Atlantic coastal plain; (8) Southeastern coastal plain; (9) Snake River plain; (10) Central Midwest; (11) Gulf costal plain; (12) Great Basin; (13) Northeast glacial aquifers; (14) Upper Colorado River basin; (15) Oahu, Hawaii; (16) Caribbean islands; (17) Columbia Plateau; (18) San Juan basin; (19) Michigan basin; (20) Edwards–Trinity; (21) Midwestern basins and arches; (22) Appalachian valleys and Piedmont; (23) Puget–Willamette lowland; (24) Southern California alluvial basins; (25) Northern Rocky Mountain intermontane basins. (Modified from Miller, 1999).

of the RASA program were to define the geologic and hydrologic frameworks of each aquifer system, to assess the geochemistry of the water in the system, to characterize the groundwater flow system, and to describe the effects of development on the flow system. Although the RASA studies did not cover the entire Nation, they compiled much of the data needed to make the National assessments of groundwater resources presented in the *Ground Water Atlas of the United States*. The atlas, however, describes the location, extent, and geologic and hydrologic characteristics of all the important aquifers in the United States, including those not studied by the RASA program. The atlas is written in such a manner that it can be understood even by readers who are not hydrogeologists and hydrologists. Simple language is used to explain the principles that control the presence, movement, and chemical quality of groundwater in different climatic, topographic, and geologic settings. The atlas also provides an overview of groundwater conditions for consultants who need information about an individual aquifer. Finally, it serves as an introduction to regional and national groundwater resources for lawmakers, and personnel of local, state, or federal agencies. The entire atlas is available online, and detailed printed sections with color maps can be ordered, at nominal cost, from the USGS. Excerpts from the atlas are included throughout this chapter (Miller, 1999), together with other information on aquifer types and examples from around the world.

Regional aquifer system study areas included in the *Ground Water Atlas of the United States* are shown in Fig. 2.66.

2.7.1 Sand and Gravel Aquifers

Unconsolidated sand and gravel aquifers can be grouped into four general categories: (1) stream-valley aquifers, located beneath channels, floodplains, and terraces in the valleys of major streams; (2) basin-fill aquifers, also referred to as valley-fill aquifers since they commonly occupy topographic valleys; (3) blanket sand and gravel aquifers; and (4) glacial-deposit aquifers. All the unconsolidated sand and gravel aquifers are characterized by intergranular porosity and have interbeds and layers of finer sediments (silt and clay) that vary in thickness and spatial distribution depending on depositional environments. Overall, they are the most prolific and utilized aquifer type in the United States as well as worldwide because of three main reasons: (1) groundwater is stored in sand and gravel deposits (Fig. 2.67) which, overall, have the highest total and effective porosity of all aquifer types; (2) geologically the youngest, they are exposed at the land surface and receive direct recharge from precipitation; and (3) they are often in direct hydraulic connection with surface water bodies, which may serve as additional sources of recharge. For all these reasons, some of the world's largest well fields for public and industrial water supply are located in flood plains of major rivers. They are often designed to induce additional recharge from the river and take advantage of bank filtration, a natural process which improves quality of the infiltrating surface water as it flows through aquifer porous media. For example, the photograph in Fig. 2.68 shows a part of the Sava River flood plain underlain by a thick alluvial aquifer utilized for water supply of Belgrade, Serbia. The well field, one of the largest of its kind in the world, has 99 collector wells and tens of vertical wells stretching along the river banks for almost 50 km.

Figure 2.67 Details of sorting in gravel of the Provo formation east of Springville, Utah County, UT, circa 1940. (Photograph courtesy of USGS Photographic Library, 2007.)

Basin-Fill Aquifers

Basin-fill aquifers consist of sand and gravel deposits that partly fill depressions which were formed by faulting or erosion or both (Fig. 2.69). Fine-grained deposits of silt and clay, where interbedded with sand and gravel, form confining units that retard the movement of groundwater, particularly in deeper portions. In basins that contain thick sequences of deposits, the sediments become increasingly more compacted and less permeable with depth. The basins are generally bounded by low-permeability igneous, metamorphic, or sedimentary rocks.

The sediments that comprise the basin-fill aquifers mostly are alluvial deposits eroded by streams from the rocks in the mountains adjacent to the basins. They may locally include windblown sand, coarse-grained glacial outwash, and fluvial sediments deposited by streams that flow through the basins. Coarser sediment (boulders, gravel, and sand) is deposited near the basin margins and finer sediment (silt and clay) is deposited in the central parts of the basins. Some basins contain lakes or playas (dry

FIGURE 2.68 Alluvial flood plain underlain by thick prolific sand and gravel aquifer utilized for water supply of the city of Belgrade, Serbia, located at the confluence of two major European rivers—the Sava and the Danube. The Sava River is on the left and an aquifer recharge basin is on the right; the Danube is at the top of the photograph. (Photograph courtesy of Vlado Marinkovic.)

lakes) at or near their centers. Windblown sand might be present as local beach or dune deposits along the shores of the lakes. Deposits from mountain, or alpine, glaciers locally form permeable beds where the deposits consist of outwash transported by glacial melt-water. Sand and gravel of fluvial origin are common in and adjacent to the channels of through-flowing streams. Basins in arid regions might contain deposits of salt, anhydrite,

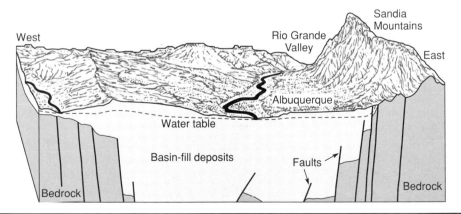

FIGURE 2.69 Schematic block diagram of the basin-fill aquifer utilized for water supply of the city of Albuquerque, New Mexico. (Modified from Robson and Banta, 1995.)

gypsum, or borate produced by evaporation of mineralized water, in their central parts (Miller, 1999).

Recharge to basin-fill aquifers is primarily by infiltration of streamflow that originates as precipitation which falls on the mountainous areas that surround the basins. This recharge, called mountain-front recharge, is mostly intermittent because the streamflow that enters the valleys is also mostly intermittent. As the streams exit their bedrock channels and flow across the surface of the alluvial fans, the streamflow infiltrates the permeable deposits on the fans and moves downward to the water table. In basins which are located in arid climates, much of the infiltrating water is lost by evaporation or as transpiration by riparian vegetation (plants on or near stream banks).

Open basins contain through-flowing streams and commonly are hydraulically connected to adjacent basins. Some recharge might enter an open basin as surface flow and underflow (groundwater that moves in the same direction as streamflow) from an upgradient basin, and recharge occurs as streamflow infiltration from the through-flowing stream. Before development, water discharges from basin-fill aquifers largely by evapotranspiration within the basin but also as surface flow and underflow into down-stream basins. After development, most discharge is through withdrawals from wells. As illustrated by examples in preceding sections of the book, during early groundwater development stages in the western United States many wells in such basins were artesian and high-yielding (e.g., see the one shown in Fig. 2.70). These days are long gone in basins with urban development or intensive irrigation for agriculture, but flowing wells are for now "doing just fine" in undeveloped basins such as the one shown in Fig. 2.71.

Figure 2.70 Flowing well on Antill Tract, San Bernardino Valley, San Bernardino County, CA, 1905. (Photograph courtesy of USGS Photographic Library, 2007.)

Figure 2.71 Flowing artesian water well at the Bonham Ranch, southern Smoke Creek Desert, NV. (Photograph taken in 1994 by Terri Garside.)

Many basin-fill aquifers in southwestern alluvial basins, Great Basin (also known as Basin and Range physiographic province), basins in Southern California (see Fig. 2.72), and northern Rocky Mountain intermontane basins, are utilized for water supply and irrigation (numbers 5, 12, 24, and 25 in Fig. 2.66). Current groundwater extraction is usually from deeper, more protected portions of basins, although there are examples of unwanted effects of such extraction due to induced upconing (vertical upward migration) of highly mineralized saline groundwater. This water is residing at greater depths where there is no flushing by fresh meteoric water. Another negative effect of groundwater extraction from basin-fill aquifers in arid and semiarid climates is aquifer mining because of the lack of significant present-day natural aquifer recharge (Fig. 2.73).

Blanket Sand and Gravel Aquifers

Thick widespread sheet-like deposits that contain mostly sand and gravel form unconsolidated and semiconsolidated aquifers called blanket sand and gravel aquifers. They largely consist of alluvial deposits brought in from mountain ranges and deposited in lowlands. However, some of these aquifers, such as the High Plains aquifer in the United States (Ogallala aquifer), include large areas of windblown sand, whereas others, such as the surficial aquifer system of the southeastern United States, contain some alluvial deposits but are largely composed of beach and shallow marine sands (Miller, 1999). The High Plains aquifer extends over about 174,000 mi^2 in parts of eight states (number 2 in Fig. 2.66). The principal water-yielding geologic unit of the aquifer is the Ogallala

Figure 2.72   Image of Southern California, from the desert at Mojave to the ocean at Ventura (distant left); Tehachapi Mountains are in the right foreground. The elevation data used in this image was acquired by the Shuttle Radar Topography Mission (SRTM) aboard the Space Shuttle Endeavour and designed to collect three-dimensional measurements of the earth's surface. The image is combination of the SRTM topography map and Landsat bands 1, 2, & 4. View width is 27 mi (43 km), vertical exaggeration 3×. (Image courtesy of NASA, 2007.)

Formation of Miocene age, a heterogeneous mixture of sand, gravel, silt, and clay deposited by a network of braided streams, which flowed eastward from the ancestral Rocky Mountains. Permeable dune sand is part of the aquifer in large areas of Nebraska and smaller areas in the other states. The Ogallala aquifer is principally unconfined and in direct hydraulic connection with the alluvial aquifers along the major rivers which flow over it.

The origin of water in the Ogallala aquifer is mainly from the last ice age, and the rate of present-day recharge is much lower. This has resulted in serious long-term water table decline in certain portions of the aquifer due to intensive groundwater extraction for water supply and irrigation. Decreases in saturated thickness result in a decrease in well yields and an increase in pumping costs because the pumps must lift the water from greater depths—conditions occurring over much of the Ogallala aquifer. Despite this, the aquifer can still be described by the following quote: "The whole world depends on the Ogallala. Its wheat goes, in large part, to Russia, China, and Africa's Sahel. Its pork

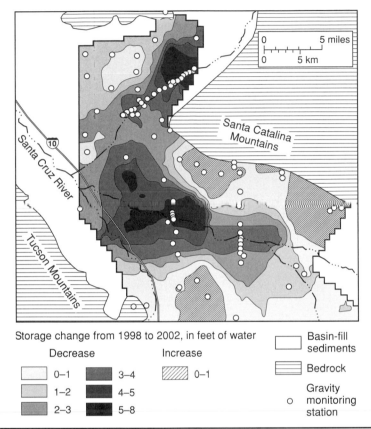

Storage change from 1998 to 2002, in feet of water

Decrease | Increase

0–1	3–4	0–1
1–2	4–5	
2–3	5–8	

Basin-fill sediments

Bedrock

o Gravity monitoring station

FIGURE 2.73 Example of groundwater storage change in part of the Tucson Basin, AZ, determined by microgravity measurements. (From Anderson and Woosley, 2005; source: Don Pool, USGS, written communication, 2003.)

ends up in Japanese and American supermarkets. Its beef goes everywhere ... " (Opie, 2000, from McGuire et al., 2003).

Other major blanket sand and gravel aquifers in the United States include the Seymour aquifer of Texas which, like the High Plains aquifer, was deposited by braided, eastward flowing streams but has been dissected into separate pods by erosion; the Mississippi River Valley alluvial aquifer, which consists of sand and gravel deposited by the Mississippi River as it meandered over an extremely wide floodplain; and the Pecos River Basin alluvial aquifer, which is mostly stream-deposited sand and gravel, but locally contains dune sands (Miller, 1999).

Semiconsolidated Sand Aquifers

Sediments that primarily consist of semiconsolidated sand, silt, and clay, interbedded with some carbonate rocks, underlie the coastal plains that border the Atlantic Ocean and the Gulf of Mexico. The sediments extend from Long Island, New York, southwestward to the Rio Grande, and generally form a thick wedge of strata that dips and thickens seaward from a featheredge at its updip limit. Coastal plain sediments are water-laid and were deposited during a series of transgressions and regressions of the sea. Depositional environments ranged from fluvial to deltaic to shallow marine, and the exact location of each environment depends upon the relative position of land masses, shorelines, and streams at a given point in geologic time. Complex interbedding and variations in lithology result from the constantly-changing depositional environments. Some beds are thick and continuous for tens to hundreds of miles, whereas others are traceable only for short distances. Consequently, the position, shape, and number of the bodies of sand and gravel that form aquifers in these sediments vary greatly from place to place (Miller, 1999).

The semiconsolidated sand aquifers have been grouped into several major aquifer systems interfingering with and grading into each other (numbers 7, 8, and 11 in Fig. 2.66). The Northern Atlantic Coastal Plain aquifer system extends from North Carolina through Long Island, NY, and locally contains as many as 10 aquifers. Figures 2.74 and 2.75 show two generalized cross-sections of this system. The Mississippi Embayment aquifer system consists of six aquifers, five of which are equivalent to aquifers in the Texas coastal uplands aquifer system to the west. The coastal lowlands aquifer system extends from Rio Grande River in Texas across southern and central Louisiana, southern Mississippi, southern Alabama, and the western part of the Florida panhandle. It contains five thick, extensive permeable zones and has been used extensively for water supply throughout the region. Its heavy pumping in the Houston metropolitan area caused one of better-known cases of major land subsidence in the Nation. The Southeastern Coastal Plain aquifer system consists of predominantly clastic sediments that crop out or are buried at shallow depths in large parts of Mississippi and Alabama, and in smaller areas of Georgia and South Carolina. Toward the coast, the aquifer system is covered either by shallower aquifers or confining units. Some of the aquifers and confining units of the Southeastern Coastal Plain aquifer system grade laterally into adjacent clastic aquifer systems in North Carolina, TN, and Mississippi and adjacent States to the west; some also grade vertically and laterally southeastward into carbonate rocks of the Floridan aquifer system. Within each aquifer system, the numerous local aquifers have been grouped into regional aquifers that are separated by regional confining units consisting primarily of silt and clay, but locally are beds of shale or chalk. The rocks that comprise these aquifer systems are of cretaceous and tertiary age. In general, the older rocks crop out farthest inland, and successively younger rocks are exposed coastward (Miller, 1999).

Coastal plains aquifers have enabled continuing urban growth and development of the Atlantic and Gulf coasts for decades. However, they have been overexploited in many locations, and are at a continuously increasing risk of saltwater encroachment in coastal areas with heavy pumping.

Glacial-Deposit Aquifers

Large areas of the north-central and northeastern United States are covered with sediments that were deposited during several advances and retreats of continental glaciers. The massive ice sheets planed off and incorporated soil and rock fragments during advances and redistributed these materials as ice-contact or meltwater deposits or both during retreats. Thick sequences of glacial materials were deposited in former river valleys cut into bedrock, whereas thinner sequences were deposited on the hills between the valleys. The glacial ice and meltwater derived from the ice laid down several types of deposits, which are collectively called glacial drift. Till, which consists of unsorted and unstratified material that ranges in size from boulders to clay, was deposited directly by

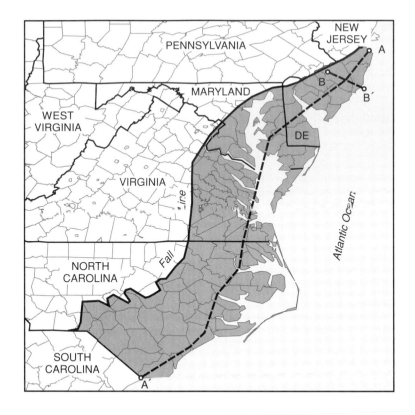

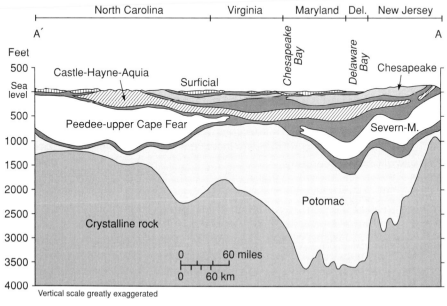

FIGURE 2.74 Map and hydrogeologic cross section AA′ showing principal aquifers in the North Atlantic Coastal Plain aquifer system (Severn-M. is Severn-Magothy aquifer). (Modified from Trapp, 1992.)

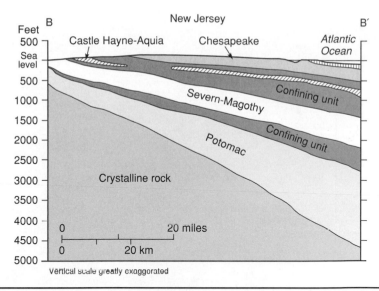

FIGURE 2.75 Hydrogeologic cross section BB' showing principal aquifers in the North Atlantic Coastal Plain aquifer system. The line of cross section is shown in Fig. 2.74. (Modified from Trapp, 1992.)

the ice. Outwash, which is mostly stratified sand and gravel (Fig. 2.76), and glacial-lake deposits consisting mostly of clay, silt, and fine sand, were deposited by meltwater. Ice-contact deposits consisting of local bodies of sand and gravel were deposited at the face of the ice sheet or in cracks in the ice.

The distribution of the numerous sand and gravel beds that make up the glacial-deposit aquifers and the clay and silt confining units that are interbedded with them is extremely complex. The multiple advances of lobes of continental ice originated from different directions and different materials were eroded, transported, and deposited by the ice, depending upon the predominant rock types in its path. When the ice melted, coarse-grained sand and gravel outwash was deposited near the ice front, and the meltwater streams deposited successively finer material farther and farther downstream. During the next ice advance, heterogenous deposits of poorly permeable till might be laid down atop the sand and gravel outwash. Small ice patches or terminal moraines dammed some of the meltwater streams, causing large lakes to form. Thick deposits of clay, silt, and fine sand accumulated in some of the lakes and these deposits form confining units where they overlie sand and gravel beds. The glacial-deposit aquifers are either localized in bedrock valleys or are in sheet-like deposits on outwash plains (Miller, 1999).

The glacial sand and gravel deposits form numerous local but highly productive aquifers. Yields of wells completed in aquifers formed by continental glaciers are as much as 3000 gal/min, where the aquifers consist of thick sand and gravel. Locally, yields of 5000 gal/min have been obtained from wells completed in glacial-deposit aquifers that are located near rivers and can obtain recharge from the rivers. Aquifers that were formed by mountain glaciers yield as much as 3500 gal/min in Idaho and Montana, and wells completed in mountain-glacier deposits in the Puget Sound, Washington area yield as much as 10,000 gal/min (Miller, 1999).

Figure 2.76 Stratified glacial sand and gravel, 1 mi north of Asticou, northeast of Lower Hadley Pond. (Acadia National Park, Maine, September 14, 1907. Photograph courtesy of USGS Photographic Library, 2007.)

2.7.2 Sandstone Aquifers

Sandstone aquifers in the United States are more widespread than those in all other types of consolidated rocks. Although generally less permeable, and usually with a lower natural recharge rate than surficial unconsolidated sand and gravel aquifers, sandstone aquifers in large sedimentary basins are one of the most important sources of water supply both in the United States and worldwide. Loosely cemented sandstone retains significant primary (intergranular) porosity, whereas secondary fracture porosity may be more important for well-cemented and older sandstone (Fig. 2.77). In either case, storage capacity of such deposits is high because of the thickness of major sandstone basins.

Sandstone aquifers are highly productive in many places and provide large volumes of water for all uses. The Cambrian-Ordovician aquifer system in the north-central United States is composed of large-scale, predominantly sandstone aquifers that extend over parts of seven states. The aquifer system consists of layered rocks that are deeply buried where they dip into large structural basins. It is a classic confined, or artesian, system and contains three aquifers. In descending order, these are the St. Peter-Prairie du Chien-Jordan aquifer (sandstone with some dolomite), the Ironton-Galesville aquifer (sandstone), and the Mount Simon aquifer (sandstone). Confining units of poorly permeable sandstone and dolomite separate the aquifers. Low-permeability shale and dolomite compose the Maquoketa confining unit that overlies the uppermost aquifer and is considered to be part of the aquifer system. Wells that penetrate the Cambrian-Ordovician aquifer system commonly are open to all three aquifers, which are collectively called the sandstone aquifer in many reports.

Figure 2.77 Conglomerate of the lower sandstone of the Dakota group about 2 mi north of Bellvue, resting unconformably on Morrison shale. (Larimer County, CO, 1922. Photograph courtesy of USGS Photographic Library, 2007.)

The rocks of the aquifer system are exposed in large areas of northern Wisconsin and eastern Minnesota. Regionally, groundwater in the system flows from these topographically high recharge areas eastward and southeastward toward the Michigan and Illinois Basins. Subregionally, groundwater flows toward major streams, such as the Mississippi and the Wisconsin Rivers, and toward major withdrawal centers, such as those at Chicago, IL, and Green Bay and Milwaukee, WI. One of the most dramatic effects of groundwater withdrawals known in the United States is shown in Fig. 2.78. Withdrawals from the Cambrian-Ordovician aquifer system, primarily for industrial use in Milwaukee, WI, and Chicago, IL, caused declines in water levels of more than 375 ft in Milwaukee and more than 800 ft in Chicago from 1864 to 1980, with the pumping influence extending over 70 mi. Beginning in the early 1980s, withdrawals from the aquifer system decreased as some users, including the city of Chicago, switched to Lake

FIGURE 2.78 Decline of water levels in the Cambrian-Ordovician aquifer system from 1864 to 1980, as a result of large groundwater withdrawals centered at Chicago and Milwaukee. Contour lines are in feet; dashed where approximately located. (Modified from Miller, 1999, and Young, 1992.)

Michigan as a source of supply. Water levels in the aquifer system began to rise in 1985 as a result of decreased withdrawals (Miller, 1999).

The chemical quality of the water in large parts of the aquifer system is suitable for most uses. The water is not highly mineralized in areas where the aquifers crop out or are buried to shallow depths, but mineralization generally increases as the water moves downgradient toward the structural basins. The deeply buried parts of the aquifer system contain saline water.

Other large layered sandstone aquifers that are exposed adjacent to domes and uplifts or that extend into large structural basins or both are the Colorado Plateau aquifers, the Denver Basin aquifer system, Upper and Lower Cretaceous aquifers in North and South Dakota, Wyoming, and Montana, the Wyoming Tertiary aquifers, the Mississippian aquifer of Michigan, and the New York sandstone aquifers (Miller, 1999).

Examples of continental-scale sandstone aquifers include the Guaraní aquifer system in South America, the Nubian Sandstone Aquifer System in Africa, and the Great Artesian Basin in Australia. The Guaraní Aquifer System (also called Botucatu aquifer) includes areas of Brazil, Uruguay, Paraguay, and Argentina. Water of very good quality is exploited for urban supply, industry, and irrigation as well as for thermal, mineral, and tourist purposes. This aquifer is one of the most important fresh groundwater reservoirs in the world, due to its vast extension (about 1,200,000 km^2), and volume (about 40,000 km^3). The aquifer storage volume could supply a total population of 5.5 billion people for 200 years at a rate of 100 L/d/person (Puri et al., 2001). The gigantic aquifer is located in the Paraná and Chaco-Paraná Basins of southern South America. It is developed in consolidated aeolian and fluvial sands (now sandstones) from the Triassic-Jurassic, usually covered by thick basalt flows (Serra Geral Formation) from the Cretaceous, which provide a high confinement degree. Its thickness ranges from a few meters to 800 m. The specific capacities of wells vary from 4 m^3/h/meter of drawdown to more than 30 m^3/h/m. The total dissolved solids (TDS) contents are generally less than 200 mg/L. The production costs per cubic meter of water from wells of depths between 500 and 1000 m and yielding between 300 and 500 m^3/h vary from US$ 0.01 to US$ 0.08, representing only 10 to 20 percent of the cost of storing and treating surface water sources (Rebouças and Mente, 2004).

The rocks of the Nubian Sandstone Aquifer System in northern Africa (NSAS; Fig. 2.79), which is shared by Egypt, Libya, Sudan, and Chad, vary in thickness from zero in outcrop areas to more than 3000 meters in the central part of the Kufra and Dakhla Basins, and range in age from Cambrian to Neogene. The main productive aquifers, separated by regional confining units, are (from land surface down) Miocene sandstone, Mesozoic (Nubian) sandstone, Upper Paleozoic-Mesozoic sandstone, and Lower Paleozoic (Cambrian-Ordovician) sandstone (Salem and Pallas, 2001). In some locations, the confined portions of the system provide water to high-yielding artesian wells such as the one shown in Fig. 2.80. The groundwater of the Nubian Basin is generally of high quality. TDS range from 100 to 1000 parts per million, with an increased salinity northward toward the Mediterranean sea where the freshwater saline water interface passes through the Qattara depression in Egypt. In Libya, the TDS of the deep Nubian aquifers ranges from 160 to 480 mg/L and from 1000 to 4000 mg/L in the shallow aquifers (Khouri, 2004).

Because of the semiarid to arid climate, the present-day natural recharge of the NSAS is negligible and countries in the region have formed a joint commission for assessment and management of this crucial, nonrenewable source of water supply. Table 2.1 shows comparison of the total recoverable freshwater resources stored in the system and the present-day annual withdrawals (Bakhbakhi, 2006).

Groundwater withdrawals from the system have been increasing each year for the past 40 years. During this time period over 40 billion m^3 of water has been extracted from the system in Libya and Egypt. This has produced a maximum drawdown of about 60 m. All but 3 percent of the free flowing wells and springs have been replaced by deep wells. Until recently, almost all the water extracted was used for agriculture, either for development projects in Libya or for private farms located in old traditional oases in Egypt. With completion of the first phase of the so-called "Great Manmade River"

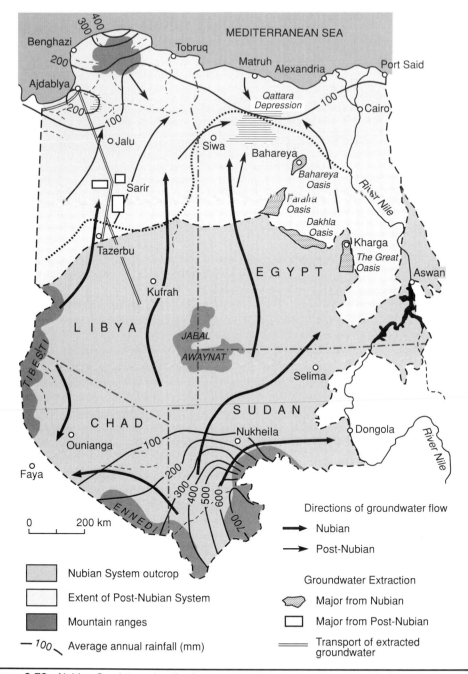

FIGURE 2.79   Nubian Sandstone Aquifer System (NSAS) in North Africa. (Modified from Bakhbakhi, 2006.)

Figure 2.80   Flowing well at Kharga Oasis, Egypt, 1961. (Photograph courtesy of USGS Photographic Library, 2007.)

project in Libya, 2 million m^3 of groundwater extracted from NSAS is transported daily via buried large-diameter reinforced concrete pipes for some 2000 km to the coastal cities in the north. Other phases of the project include groundwater extraction from another nonrenewable aquifer west of NSAS called North Western Sahara Aquifer System (NWSAS). The combined groundwater extraction from the two aquifer systems in Libya for centralized water supply of various users along the coast is reportedly 6.5 million m^3/d. It has been estimated that the cost of this megaproject is more than 25 billion US dollars (Wikipedia, 2007). The estimated remaining volume of freshwater that can be extracted from the entire NSAS in all four countries (Egypt, Libya, Chad, and Sudan) is about 14,500 km^3 (Bakhbakhi, 2006).

The Great Artesian Basin in Australia covers 1.7 million km^2 and is one of the largest groundwater basins in the world. It underlies parts of Queensland, New South Wales, South Australia, and the Northern Territory. The basin is up to 3000 m thick and contains a multilayered confined aquifer system, with the main aquifers occurring in Mesozoic sandstones interbedded with mudstone (Jacobson et al., 2004).

Groundwater in the Great Artesian Basin has been exploited from flowing wells since artesian water was discovered in 1878, allowing an important pastoral industry to be established. Wells are up to 2000 m deep, but average about 500 m. Artesian flows

Region	Nubian System		Post-Nubian System		Total Volume in Storage (km³)	Total Recoverable Volume (km³)	Present Extraction from Nubian (km³)	Present Extraction from Post-Nubian (km³)	Total Present Extraction from the NSAS (km³)
	Area (km²)	Volume in Storage (km³)	Area (km²)	Volume in Storage (km³)					
Egypt	815,670	154,720	494,040	35,867	190,587	5,367	0.200	0.306	0.506
Libya	754,088	136,550	426,480	48,746	185,296	4,850	0.567	0.264	0.831
Chad	232,980	47,810	—	—	47,810	1,630	0.000	—	0.000
Sudan	373,100	33,880	—	—	33,880	2,610	0.833	—	0.833
Total	2,175,838	372,960	920,520	84,614	457,570	14,470	1.607	0.570	2.170

From Bakhbakhi, 2006.

TABLE 2.1 Freshwater Stored in the Nubian Sandstone Aquifer System (NSAS) and Present-Day Groundwater Withdrawals per Region

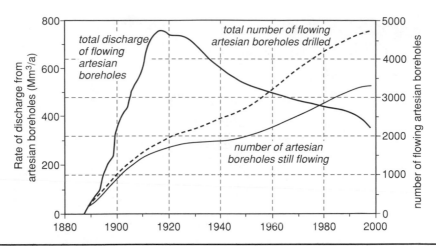

Figure 2.81 Trends in artesian overflow from wells in the Great Artesian Basin of Australia. (From Habermehl, 2006.)

from individual wells exceed 10 million L/d (more than 100 L/s), but the majority have much smaller flows. About 3100 of the 4700 flowing artesian wells drilled in the basin remain flowing. The accumulated discharge of these wells (including water supply wells in about 70 towns, as in most cases the artesian groundwater supply is the only source of water) is about 1200 million L/d, compared to the maximum flow rate of about 2000 million L/d from about 1500 flowing artesian wells around 1918 (Fig. 2.81). Nonflowing artesian wells, about 20,000, are generally shallow—several tens to hundreds of meters deep. It is estimated that these generally windmill-operated pumped wells supply on average 0.01 million L/d per well and produce a total of about 300 million L/d. High initial flow rates and pressures of artesian wells have diminished as a result of the release of water from elastic storage in the groundwater reservoir. Exploitation of the aquifers has caused significant changes in the rate of natural aquifer discharge. Spring yields have declined as a result of well development in many parts of the basin during the last 120 years, and in some areas springs have ceased to flow (Habermehl, 2006).

2.7.3 Carbonate (Karst) Aquifers

As explained earlier, what sets karst aquifers apart from any other aquifer type is their unique porosity, often referred to as dual (or even triple) porosity which consists of matrix porosity and then porosity of fractures and solutional openings (karst conduits). As a consequence, groundwater flow in karst does not conform to relatively straight forward principles governing flow in intergranular porous media and based on Darcy's law. Probably the only common characteristic karst aquifers share with other aquifer types on a regional scale is that the groundwater has to move from the areas of aquifer recharge to the areas (points) of aquifer discharge due to the hydraulic gradient in between, aquifer transmissivity and the effective porosity. In karst, this flow can often take quite unexpected turns defying the expectations of professionals not used to working in such a complex groundwater environment. It is not uncommon to have a couple of wells screened in the same interval, and only a few hundred meters apart, with completely

Figure 2.82 Large-diameter core from the Miami oolitic limestone (Biscayne aquifer), Miami, FL. Hydraulic conductivity $\geq$1000 ft/d. (Photograph by George Sowers; printed with kind permission of Francis Sowers.)

different yields. In many instances this "behavior" of karst aquifers makes it very difficult to design and reliably predict effectiveness of well fields in karst.

The original texture and porosity of carbonate deposits are highly variable because of the wide range of environments in which the deposits form. The primary porosity of the deposits can range from 1 to more than 50 percent. Compaction, cementation, dolomitization, and dissolution are diagenetic processes, which act on the carbonate deposits to change their porosity and permeability. For example, the Biscayne aquifer in south Florida is developed in young Pliocene and Pleistocene limestones and is highly productive thanks both to its very high primary porosity and karstification (Fig. 2.82). It is extensively used for water supply including for the city of Miami, which is the largest metropolitan area in the United States relying solely on groundwater. The aquifers in older carbonate rocks of Cretaceous to Precambrian age yield water primarily from solution openings (Fig. 2.83). The Ozark Plateaus aquifer system, the Silurian-Devonian aquifers, the Ordovician aquifers, the Upper Carbonate aquifer of southern Minnesota, the Arbuckle-Simpson aquifer of Oklahoma, and the New York carbonate-rock aquifers are all in layered limestones and dolomites of Paleozoic age, in which solution openings are locally well developed. The Blaine aquifer in Texas and Oklahoma likewise yields water from solution openings, some of which are in carbonate rocks and some of which are in beds of gypsum and anhydrite interlayered with the carbonate rocks (Miller, 1999).

The three most important characteristics of regional groundwater flow in karst aquifers are (1) natural groundwater divides in karst aquifers often do not coincide with surface water (topographic) divides as a result of loosing, sinking, and dry streams;

FIGURE 2.83   Large cavity at a construction site in Paleozoic limestone, Hartsville, TN. The matrix porosity is measured at 2 percent. (Photograph by George Sowers; printed with kind permission of Francis Sowers.)

(2) actual groundwater velocities in some portions of the aquifer could be extremely high, on the order of hundreds of meters or more per day; and (3) aquifer discharge is often localized through large karst springs, which have subsurface drainage areas commonly larger than the topographic ones.

A good starting point for characterizing available groundwater in a karst aquifer is to determine its regional carbonate litostratigraphy and geologic setting. Several major karst types are closely associated with the thickness of carbonate sediments, their age, and the position of regional erosional basis for both the surface streams and the groundwater flow (e.g., see Cvijic, 1893, Cvijic, 1918, Cvijic, 1926; Grund, 1903, 1914; Herak and Stringfield, 1972; Milanovic, 1979, 1981; White, 1988; Ford and Williams, 1989). Large epicontinental carbonate shelf platforms, at the scale of hundreds to thousands of kilometers, and thousands of meters thick, together with isolated carbonate platforms in open-ocean basins, have developed through the Paleozoic and Mesozoic eras, with the trend continuing into the Cenozoic era. At the same time, smaller (tens to hundreds of kilometers wide) carbonate platforms and build-ups associated with intracratonic basins have also developed (James and Mountjoy, 1983). Both types have been redistributed and reshaped during various phases of plate tectonics and can presently be found throughout the world, both adjacent to oceans and seas or deep inside the continents. The region of classic Dinaric karst in the Balkans, Europe, (the word karst comes from this area) is an example of a tectonically disturbed large thick Mesozoic carbonate platform where the Adriatic Sea is the regional erosional basis for groundwater discharge (Kresic, 2007b).

The sequence of carbonate sediments is thousands of meters thick and deep borings have encountered paleokarstification more than 2000 meters below ground surface.

The utilization of karst aquifers developed in the Mesozoic platforms of Euro-Asia, originally through the use of springs, has been documented since the beginnings of civilizations in the Old World (Mediterranean and Middle East) and Europe, and it is still irreplaceable in most European and Middle Eastern countries. In addition to modern well fields, public water supply is still heavily based on the use of springs by small and large water utilities alike (Kresic and Stevanovic, in preparation).

Like the Euro-Asian Mesozoic platforms, the Floridan aquifer in the southeast United States (North and South Carolinas, Georgia, and Florida) is also developed on a thick epicontinetal platform but with gently sloping undisturbed carbonate layers mostly covered with less permeable clastic sediments. The aquifer consists primarily of limestone and dolomite of Paleocene to Miocene age. Regional flow directions are from the inland outcrops toward the Atlantic Ocean and the Gulf of Mexico, with submerged discharge zones along the continental shelf. Florida has over 20 well-documented large offshore springs and a number of undocumented ones. These springs provided resources to prehistoric people and wildlife when the sea level was lower. Evidence for occupation of offshore sites has been discovered by researchers from the Florida State University Department of Anthropology, which have conducted surveys at and near some offshore springs and have recovered an abundance of chert tools (Scott et al., 2004; from Faught, in preparation). Although some of the offshore springs may be discharging brackish to saline water today, they almost certainly discharged freshwater during times of lower sea levels when prehistoric human occupation occurred at these sites.

Areas where limestone is exposed at the surface, as in north-central Florida, karst features such as sinkholes, large springs, and caves are fully developed. Many large caves are now completely filled with groundwater. Quite a few of such caves providing water to large karst springs have been explored by cave divers. Figure 2.84 shows a large submerged karst passage in the Yucatan karst, Mexico, which has very similar characteristics to Florida. It is estimated that there are nearly 700 springs in Florida of which 33 are the first magnitude springs with an average discharge greater than 100 ft^3/s (2.83 m^3/s). Florida represents perhaps the largest concentration of freshwater springs on the earth.

Karst of the Caribbean islands, such as Puerto Rico and Jamaica, is an example of relatively small carbonate platforms laying on a low permeable, usually magmatic base (Fig. 2.85). Surface karst features are fully developed including sinkholes and cone-shaped hills ("mogote karst" in Puerto Rico and "cockpit karst" in Jamaica). Regional groundwater flow direction is from the upland-highland recharge areas toward the coast line, with the submerged discharge zones along the freshwater-saltwater interface.

The majority of karst areas in the United States constitute portions of carbonate platforms now located away from the coast lines. Examples include Edwards aquifer in Texas and karst of Kentucky, Tennessee, West Virginia, Indiana, and Missouri. Regional groundwater flow in such cases is directed toward large karst springs located at the lowest contact between karst and non-karst, or toward the lowest large permanent surface stream intersecting the carbonates. Groundwater discharge along streams is commonly through large springs, often naturally submerged or due to river damming (see Fig. 2.86).

Probably the best-known karst terrain in the United States is in the Mammoth Cave area of central Kentucky. Recharge water enters the aquifer through sinkholes, swallow holes, and sinking streams, some of which terminate at large depressions called blind

FIGURE 2.84 Divers exploring submerged passages in Ponderosa Cave, Yucatan, Mexico. (Photograph courtesy of David Rhea, Global Underwater Explorers.)

valleys. Surface streams are scarce because most of the water is quickly routed underground through solution openings. In the subsurface, most of the water moves through caverns and other types of large solution openings, which riddle the Mississippian limestones that underlie the Mammoth Cave Plateau in Kentucky and the Pennyroyal Plain to the south and southwest of it. Some of these cavities form the large, extensive passages of Mammoth Cave, one of the world's largest and best studied cave systems.

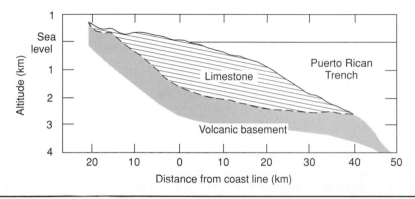

FIGURE 2.85 Section of the north coast limestone belt, Puerto Rico. (Giusti, 1978, modified from Shubert and Ewing, 1956.)

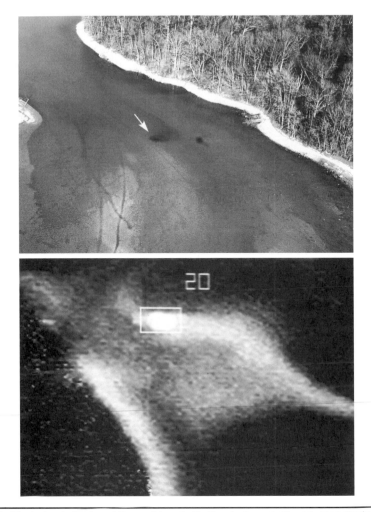

FIGURE 2.86 Photograph (top) and thermal image (bottom) of warm groundwater discharging into cold reservoir water, Tennessee, the United States. Spring #20 shown with the arrow on the top photograph. (Photograph courtesy of Frank Bogle.)

In addition to the classic Dinaric karst, the karst of southwest China is best known in terms of its spectacular development of surface and subsurface karst topography (Fig. 2.87). Carbonate rocks are widely spread and occupy about one-third of the total area of 500,000 km^2 in which mountains and hills are dominant morphologic features. Altitude decreases from 2500 m in northwest to less than 200 m in southeast. Annual precipitation is more than 1000 mm/yr. Karst features are extensively developed and the rainfall infiltration rate is generally about 30 to 70 percent. It is estimated that karst aquifers store 40 to 70 percent of total groundwater resources in southwest China (Table 2.2). Large portion of discharge of karst aquifers is via springs; there are 1293 registered significant springs with yields greater than 50 L/s (Table 2.3).

FIGURE 2.87   Tower karst near Guilin in southwestern China. (Photograph by George Sowers; printed with kind permission of Francis Sowers.)

Because of a humid climate and abundant surface water, groundwater has not been as extensively utilized for irrigation as it has been in northern China, but it is still important for urban and domestic water supply. Examples of significant use of karst aquifers and springs are in the provincial capitals Kunming and Guiyang, and the city of Tianjin (Zhaoxin and Chuanmao, 2004).

When of considerable thickness, young coastal carbonate sediments may constitute important aquifers for both local and centralized water supply. Examples can be found in Jamaica, Cuba, Hispaniola and numerous other islands in the Caribbean, the Yucatán peninsula of Mexico, Bermuda, the Cebu limestone of the Philippines, the Jaffna limestone in Sri Lanka, and some low-lying coral islands of the Indian oceans such as the Maldives (Morris et al., 2003). The high infiltration capacity of young carbonate sediments in coastal areas and islands means that there are few streams or rivers, and groundwater

Province	Karst Water Resources	Total Groundwater Resources	Ratio (%)
Yunnan	3,250	7,420	43.7
Guizhou	1,680	2,290	73.2
Guangxi	4,840	7,760	62.3
Sichuan	2,940	6,300	46.6

From Zhaoxin and Chuanmao, 2004.

TABLE 2.2   Karst Water Resources (in km^3/yr) in Four Provinces of Southwest China

Province	Flow Rate Range (L/s)				
	50–500	**500–1000**	**1000–2000**	**>2000**	**Total**
Yunnan	648	45	35	3	731
Guizhou	231	20	11	1	26
Guangxi	284	13	2	0	229
Total	1,163	78	48	4	1,293

From Zhaoxin and Chuanmao, 2004.

TABLE 2.3   Large Karst Springs with Different Flow Rates in Three Provinces in Southwest China

may be the only available source of water supply. This source is often very vulnerable to salt (sea) water encroachment due to overpumping of fresh groundwater.

Surface water can enter karstified subsurface rapidly through a network of large fractures and dissolutional openings that extend through the entire vadose zone. Consequently, any contaminants in the infiltrating water can quickly reach the water table and spread through a karst aquifer via conduits and karst channels faster than in any other porous media. The exceptions are aquifers developed in coarse uniform gravel and some fractured rocks where infiltration rates and groundwater velocities can also be very high. For this reason, karst, fractured rock, and gravel aquifers are subject to new regulation in the United States aimed at protecting vulnerable public water supplies. This regulation, named *Groundwater Rule*, and promulgated in 2006 by the United States Environmental Protection Agency (USEPA), is explained in more detail in Chap. 8.

In the United States, sandstone and carbonate rock deposits are often interbeded over large areas and form aquifers of a mixed type called sandstone and carbonate-rock aquifers. This aquifer type is present mostly in the eastern half of the Nation, but also occur in Texas and in Oklahoma, Arkansas, Montana, Wyoming, and South Dakota. The carbonate rocks generally yield more groundwater than do the sandstone rocks because of dissolution and larger open-pore space. Water in aquifers of this type may exist under confined and unconfined conditions (Maupin and Barber, 2005).

2.7.4   Basaltic and Other Volcanic Rock Aquifers

In the United States, aquifers in basaltic and other volcanic rocks are widespread in Washington, Oregon, Idaho, and Hawaii and extend over smaller areas in California, Nevada, and Wyoming. Volcanic rocks have a wide range of chemical, mineralogical, structural, and hydraulic properties. The variability of these properties is due largely to rock type and the way the rock was ejected and deposited. Pyroclastic rocks, such as tuff and ash deposits, might have been placed by flowing of a turbulent mixture of gas and pyroclastic material, or might form as windblown deposits of fine-grained ash. Where they are unaltered, pyroclastic deposits have porosity and permeability characteristics like those of poorly sorted sediments; however, where the rock fragments are very hot as they settle, the pyroclastic material might become welded and almost impermeable. Silicic lavas, such as rhyolite or dacite, tend to be extruded as thick, dense flows and have low permeability except where they are fractured. Basaltic lavas tend to be fluid and form thin flows that have a considerable amount of primary pore space at the tops and bottoms of the flows. Numerous basalt flows commonly overlap and the flows commonly are separated by soil zones or alluvial material that form permeable zones. Basalts are the most productive aquifers of all volcanic rock types.

The permeability of basaltic rocks is highly variable and depends largely on the following factors: the cooling rate of the basaltic lava flow, the number and character of interflow zones, and the thickness of the flow. The cooling rate is most rapid when a basaltic lava flow enters water. The rapid cooling results in pillow basalt, in which ball-shaped masses of basalt form, with numerous interconnected open spaces at the tops and bottoms of the balls. Large springs that discharge thousands of gallons per minute issue from pillow basalt in the wall of the Snake River Canyon at Thousand Springs, ID (see Fig. 2.33).

The Snake River Plain regional aquifer system in southern Idaho and southeastern Oregon is an example of an aquifer system in basaltic rocks. Pliocene and younger basaltic-rock aquifers are the most productive aquifers in the Snake River Plain. The saturated thickness of these rocks is locally greater than 2500 ft in parts of the eastern Snake River Plain, but is much less in the western plain. Aquifers in Miocene basaltic rocks underlie the Pliocene and younger basaltic-rock aquifers. They are used as a source of water supply only near the margins of the plain. Unconsolidated-deposit aquifers are interbedded with the basaltic-rock aquifers, especially near the boundaries of the plain (Miller, 1999).

Other basalt aquifers in the United States are the Hawaii volcanic-rock aquifers, the Columbia Plateau aquifer system, the Pliocene and younger basaltic-rock aquifers, and the Miocene basaltic-rock aquifers. Volcanic rocks of silicic composition, volcaniclastic rocks, and indurated sedimentary rocks compose the volcanic- and sedimentary-rock aquifers of Washington, Oregon, Idaho, and Wyoming. The Northern California volcanic-rock aquifers consist of basalt, silicic volcanic rocks, and volcaniclastic rocks. The Southern Nevada volcanic-rock aquifers consist of ash-flow tuffs, welded tuffs, and minor flows of basalt and rhyolite (Miller, 1999).

Worldwide, extensive lava flows occur in west-central India, where the Deccan basalts occupy an area of more than 500,000 km^2. Other extensive volcanic terrains occur in Central America, Central Africa, while many islands are entirely or predominantly of volcanic origin, such as Hawaii, Iceland, and the Canaries. Some of the older, more massive lavas can be practically impermeable (such as the Deccan) as are the dykes, sills, and plugs which intrude them (Morris et al., 2003).

Highly permeable but relatively thin rubbly or fractured lavas act as excellent conduits but have themselves only limited storage. Leakage from overlying thick, porous but poorly permeable, volcanic ash may act as the storage medium for this dual system. The prolific aquifer systems of the Valle Central of Costa Rica and of Nicaragua and El Salvador are examples of such systems (Morris et al., 2003).

2.7.5 Fractured Rock Aquifers

This category includes aquifers developed in crystalline igneous and metamorphic rocks. Most such rocks are permeable only where they are fractured, and generally yield small amounts of water to wells through several water-bearing discontinuities (e.g., fractures, foliation; see Fig. 2.88) usually associated with a certain rock type. However, because fractured rocks can extend over large areas, significant volumes of groundwater may be withdrawn from them and, in many places, they are the only reliable source of

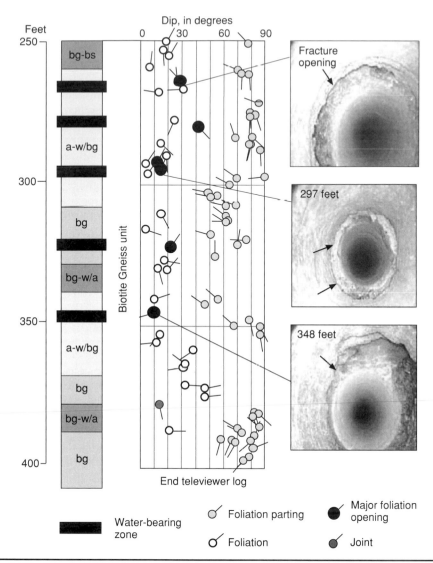

Figure 2.88 Subsurface lithologic characteristics and water-bearing zones tapped by a well in a fractured rock aquifer, Lawrenceville, GA. Tadpole with each circle shows azimuth direction; Images are obtained with a downhole camera (televiewer). (Modified from Williams et al., 2005.)

water supply. Examples in the United States include the crystalline rocks of the northern Minnesota, northeastern Wisconsin, and Appalachian and Blue Ridge regions of the eastern United States.

In some cases, the bedrock has disintegrated into a layer of unconsolidated highly weathered rock with a clayey residue of low permeability ("regolith," "saprolite," or "residuum"). Below this zone, the rock becomes progressively less weathered and more consolidated, transitioning into fresh fractured bedrock.

Case Study: Evaluating Groundwater Supplies in Fractured Metamorphic Rock of the Blue Ridge Province in Northern Virginia

Courtesy of Robert M. Cohen, Charles R. Faust, and David C. Skipp, GeoTrans Inc., Sterling, VA

Introduction   Loudoun County, which is located in northern Virginia approximately 30 mi west of Washington, DC, has been one of the fastest growing counties in the United States since the 1980s. In its western portion, the primary source of municipal, commercial, and individual domestic water supplies is groundwater pumped from thousands of wells drilled into the fractured metamorphic rock of the Blue Ridge Geologic Province.

The north-northeast trending Bull Run Fault (Fig. 2.89) separates the Blue Ridge Province anticlinorium to the west from the Culpeper Basin in Loudoun County. The anticlinorium is cored by weakly to strongly foliated high-grade Mesoproterozoic granitic and nongranitic gneisses, which were deformed and metamorphosed during the Grenville orogeny. Nine granitic gneiss (metagranite) types, which compose more than 90 percent of the basement rock volumetrically, and three nongranitic basement units were mapped by Southworth et al. (2006). A cover sequence of Late Proterozoic to

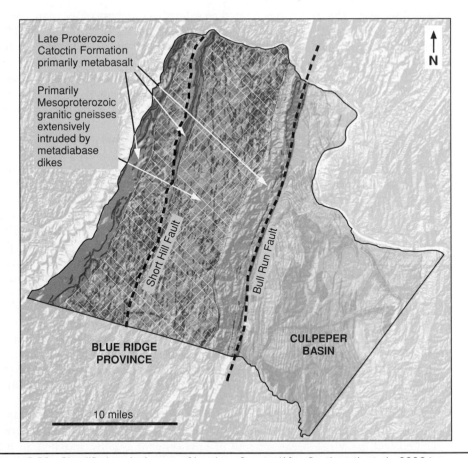

FIGURE 2.89   Simplified geologic map of Loudoun County. (After Southworth et al., 2006.)

Early Cambrian metasedimentary and metavolcanic rocks unconformably overlies the basement gneisses along ridges where it has not been eroded. The metavolcanic rocks (primarily Catoctin Formation metabasalt) were fed by a northeast-trending swarm of tabular late Proterozoic metadiabase dikes that intruded the basement rocks during continental rifting. The cover rocks were later deformed and metamorphosed to greenschist facies during the Alleghenian orogeny.

Concern about the adequacy of available groundwater quantity and quality to meet the needs of the growing County led to the implementation of hydrogeologic testing requirements in 1987, development of an extensive wells database, and creation of an integrated water monitoring program. The current hydrogeologic testing requirements at residential subdivision sites underlain by the Blue Ridge rocks include drilling test wells on 50 percent of proposed lots, conducting controlled 8-hour aquifer tests with observation wells at each test well, performing detailed quality analyses on groundwater sampled from each test well, and related data analysis. For proposed community water-supply systems, requirements include fracture fabric mapping (if outcrops are present), lineament analysis, performance of a surface geophysical survey to site wells, conduct of a minimum 72-hour aquifer test with observation wells, and related data analysis.

Wells Database   The county has created an extensive database of location, construction, and yield information for approximately 19,000 water wells, including 11,500 wells in the Blue Ridge Province of which 1800 are hydrogeologic study test wells (Figs. 2.90 and 2.91). Statistical analyses were performed using the database to evaluate differences in well yield characteristics as a function of rock type, proximity to lineaments, proximity to streams, proximity to faults, and other factors. Using GIS methods, well data were attributed to specific rock units, and well distances were calculated to lineaments, streams, and faults. Reported yield data are primarily based on "air-lift" well discharge measurements made by drillers during well construction and are of variable accuracy.

Well yield distribution curves for bedrock types in the Blue Ridge Province are shown in Figs. 2.92 and 2.93. Overall, yield distributions in the metagranites are similar and tend to be higher than yield distributions associated with nongranitic rocks, the Catoctin metabasalt, and the Harpers phyllite/metasiltstone. Approximately 5 to 10 percent of the wells are reported to yield less than 1 gal/min, which is the minimum acceptable yield for domestic wells in Loudoun County. Less than 5 percent of the wells have very high yields ($\geq$50 gal/min) desirable for community water-supply use. Given the cost of drilling, this emphasizes the need to use scientific methods (e.g., lineament analysis and surface geophysical surveys) to increase the probability of drilling high-yield wells for community water-supply development.

Mean and median well depths in the Blue Ridge of Loudoun County have increased from approximately 300 to 500 ft between 1980 and 2007. Well drillers generally continue drilling to greater depths until a satisfactory yield is achieved at a particular location. Thus, there is a negative (weak) correlation between well yield and well depth. Table 2.4 presents the mean yield per 100 ft of drilling interval for approximately 1800 hydrogeologic study test wells in the Blue Ridge based on reported yield zone depth and rate information.

Monitoring Program   Loudoun County, in conjunction with the U.S. Geological Survey (USGS) and the Virginia Department of Environment Quality, initiated an integrated Water Resource Monitoring Program (WRMP) in 2002 to provide a scientific basis for

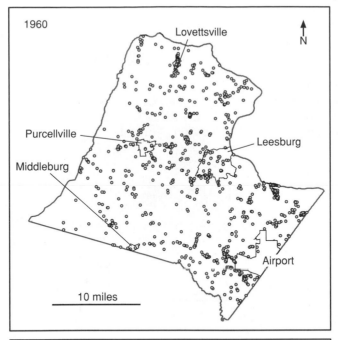

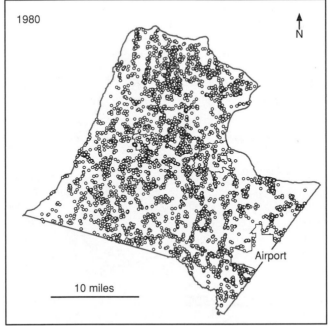

FIGURE 2.90 Number of wells in Loudoun County in 1960 and 1980.

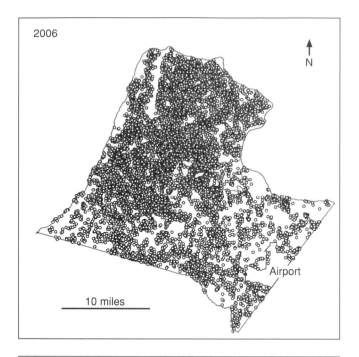

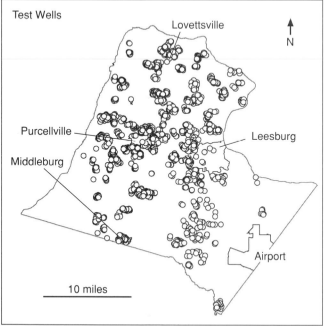

Figure 2.91 Number of wells in Loudoun County in 2006 (top) and the hydrogeologic study wells (bottom).

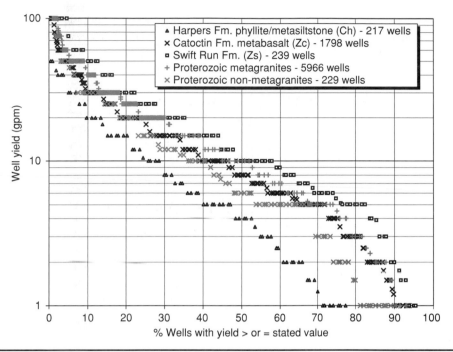

FIGURE 2.92 Well yield distributions by rock type.

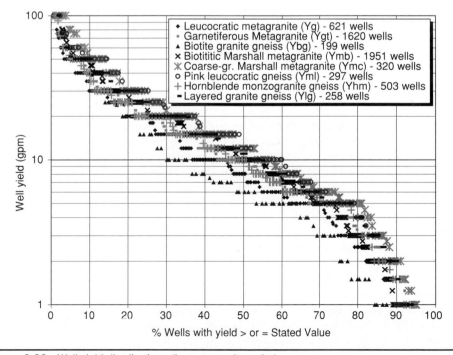

FIGURE 2.93 Well yield distributions (in metagranite units).

Interval (ft bgs)	Feet Drilled	Total Yield (gal/min)	Mean Yield (gal/min) per Interval
100 to <200	182,950	5,537	3.03
200 to <300	160,032	6,932	4.33
300 to <400	107,997	4,789	4.43
400 to <500	73,663	2,532	3.44
500 to <600	45,050	872	1.93
600 to <700	24,660	559	2.27
700 to <800	12,059	123	1.02

TABLE 2.4   Reported Yield (gal/min) Versus Depth Interval in Approximately 1800 Test Wells

making land use decisions that affect water resources. Currently, stream levels and flows are monitored at 10 stations, groundwater levels are being recorded in 11 wells, and precipitation is recorded at a few stations. Ultimately, the county plans to establish a network of 20 to 30 monitoring wells distributed throughout the county.

Groundwater levels have been monitored in one well located in the Catoctin Formation on Short Hill Mountain since the 1960s and in six other bedrock wells in the Blue Ridge Province. Review of this data show that (1) hydraulic heads in bedrock fluctuate less than 10 ft/yr, (2) heads generally rise due to recharge in late fall to early spring and during heavy precipitation events at other times, and (3) there is no evidence of a long-term hydraulic head trend at the monitored locations. Streamflow has been monitored at stations on Goose Creek and Catoctin Creek since 1930 and 1971, respectively. Streamflow appears well-correlated with precipitation rate.

Streamflow data and watershed information have been used to estimate recharge rates in the Blue Ridge province. Calculations of recharge rate based on the streamflow recession curve displacement method (USGS RORA program) were made for the periods between 1973 and 2006 for the Catoctin Creek watershed and between 2002 and 2006 for seven smaller watershed areas in the Blue Ridge Province of Loudoun County. Estimated recharge rates are typically between 10 and 13 in./yr, but range from less than 5 in/yr to more than 20 in/yr during periods of drought and extreme precipitation, respectively. These recharge rates greatly exceed groundwater pumping rates associated with rural residential subdivisions with large lots (≥3 acres each) that are present in western Loudoun County. For example, domestic water use of 300 gal/d on a 4-acre lot is equivalent in volume to a recharge rate of 1.0 in/yr. The net effective withdrawal rate is much less than 300 gal/d because a substantial portion of pumpage is returned to the groundwater system as recharge from onsite septic drainfields. Water removed by pumping is balanced by (1) a lowered hydraulic head locally in the aquifer (removal of water from storage), (2) an increase in recharge to the aquifer from above, (3) a decrease in the rate of natural discharge from the aquifer to streams, or, most likely, (4) a combination of all three sources.

High-Yield Well Siting   The capacity of crystalline metamorphic rock to transmit groundwater is highly dependent on the density and interconnectivity of open fractures in the rock. Lineament (fracture trace) analysis (Lattman and Parizek, 1964) and surface geophysical surveys have been used to site high-yielding wells in fractured metamorphic rock with varied success at many sites in Loudoun County.

A review of prior studies on the relationship of well yields in crystalline metamorphic rock to lineaments and topographic setting reveals mixed findings (Yin and Brook, 1992; Mabee, 1999; Henriksen, 2006; Mabee et al., 2002). Lineament analysis has been performed in Loudoun County using a variety of imagery platforms, including black and white aerial photographs, color and color infrared aerial photographs, shaded-relief digital elevation maps (DEMs), and topographic contour maps. GIS analysis of well yield data in the Blue Ridge of Loudoun County suggests that siting high yield wells can benefit from lineament analysis and "lay-of-the-land" methods. However, most wells drilled near lineaments and/or in valley settings will have low-to-moderate yields ($\leq$20 gal/min).

A comparative study was performed in New Hampshire by the USGS (Degnan et al., 2001) to examine the efficacy of several surface geophysical methods to locate major water-bearing fracture zones in metamorphic rock. Of the methods studied, two-dimensional electrical resistivity (ER) surveys provided the most quantitative information on fracture-zone location and dip direction. Experience in Loudoun County is consistent with these findings and ER profile imaging has been used successfully to select drilling locations for community water-supply wells at many sites in western Loudoun County. High-yield bedrock fracture zones are inferred by low resistance anomalies ($\leq$400 Ωm).

Investigation of Anisotropy It has been hypothesized (e.g., Drew et al., 2004) that the most dominant fracture fabric features, which control groundwater flow in the Blue Ridge of Loudoun County, include (1) the pervasive northeast-striking, moderately to steeply dipping (generally to the southeast) metadiabase dikes that intrude the older metagranites, and (2) subparallel northeast-trending Paleozoic cleavage (schistosity). Northwest-trending foliation in the Mesoproterozoic basement rock, which was overprinted by dike intrusion and Paleozoic cleavage, is also observed in much of western Loudoun County.

In order to examine aquifer anisotropy in a more direct manner, automated water-level recording devices were deployed in numerous observation wells during aquifer tests conducted at seven sites in the Blue Ridge of Loudoun County. Data acquired during 22 tests where drawdown was observed at three or more observations wells were analyzed using the Papadopulous (1965) equation for nonsteady groundwater flow in an infinite anisotropic confined aquifer as implemented in the TENSOR2D (Maslia and Randolph, 1986) and AQTESOLV (beta version, Duffield, 2007) computer programs. The results shown for the analyses of 15 tests where the data reasonably fit an anisotropic solution are presented based on the AQTESOLV analysis in Fig. 2.94. The anisotropic aquifer analyses indicate that different tensor orientations are observed in different areas of 100 to 250 acre study sites and that observed anisotropy is not always consistent with mapped geologic structural features. Interpreted tensor orientations vary between N70E and N79W. Nine of the fifteen orientations are between N5E and N38W.

Conclusion Extensive hydrogeologic investigation in the Blue Ridge of northern Virginia confirms that the fractured metamorphic bedrock is very complex. Adequate groundwater supply is generally available to support low-density ($\geq$3 acres per lot) rural residential use. High-yield well development for municipal (e.g., towns with denser populations) and commercial water supplies, however, presents greater challenges related to siting and potential drawdown impacts. Bedrock fracture complexity emphasizes the need for extensive monitoring to determine the impacts of high-rate pumping.

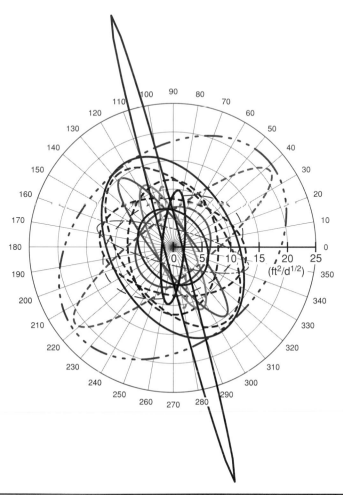

FIGURE 2.94 Anisotropic transmissivity tensor [(ft²/d)^0.5] results for 15 aquifer tests at seven sites.

2.7.6 Withdrawals from Principal Aquifers in the United States

Fresh groundwater withdrawals from 66 principal aquifers in the United States were estimated for irrigation, public-supply, and self-supplied industrial water uses for the year 2000. Total ground-water withdrawals were 76,500 million gal/d or 85,800 thousand acre-feet per year for these three uses. Irrigation used the largest amount of groundwater, 56,900 million gal/d, followed by public supply with 16,000 million gal/d, and self-supplied industrial with 3570 million gal/d. These three water uses represented 92 percent of the fresh groundwater withdrawals for all uses in the United States. The remaining 8 percent included self-supplied domestic, aquaculture, livestock, mining, and thermoelectric power uses (Maupin and Barber, 2005). Figure 2.95 compares total groundwater withdrawals for major aquifers.

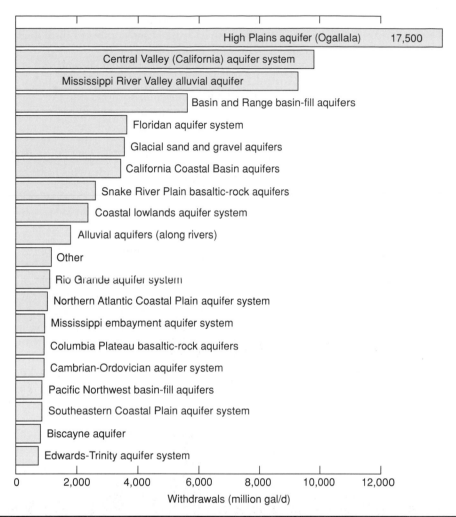

FIGURE 2.95 Aquifers that provided most of total withdrawals for irrigation, public supply, and self-supplied uses in the United States during 2000. (Modified from Maupin and Barber, 2005.)

The largest withdrawals were from unconsolidated and semiconsolidated sand and gravel aquifers, which accounted for 80 percent of total withdrawals from all aquifers. Carbonate-rock aquifers provided 8 percent of the withdrawals, and igneous and metamorphic-rock aquifers, 6 percent. Withdrawals from sandstone aquifers, from sandstone and carbonate-rock aquifers, and from the "other" aquifers category each constituted about 2 percent of the total withdrawals reported.

Fifty-five percent of the total withdrawals for irrigation, public-supply, and self-supplied industrial water uses were provided by the High Plains aquifer, California Central Valley aquifer system, the Mississippi River Valley alluvial aquifer, and the Basin and Range Basin-fill aquifers. These aquifers provided most of the withdrawals for irrigation. The High Plains aquifer was the most intensively used aquifer in the United States.

This aquifer provided 23 percent of the total withdrawals from all aquifers for irrigation, public-supply, and self-supplied industrial water uses combined and 30 percent of the total withdrawals from all aquifers for irrigation.

The primary aquifers used for public supply were the glacial sand and gravel aquifers of the Northeastern and North-Central States, the California Coastal Basin aquifers, the Floridan aquifer system, the Basin and Range Basin-fill aquifers, and the Coastal lowlands aquifer system along the Gulf Coast. These five aquifers provided 43 percent of the total withdrawals from all aquifers for public supply. The glacial sand and gravel aquifers, Coastal lowlands aquifer system, Floridan aquifer system, and Cambrian-Ordovician aquifer system were the primary sources of water for self-supplied industrial use; these aquifers provided 46 percent of the total groundwater withdrawals for that use.

2.8 Aquitards

Although aquitards play a very important role in groundwater systems, in many cases they are still evaluated qualitatively rather than quantitatively. Only relatively recently focus of field and laboratory research studies started including the role of aquitards in fate and transport of various contaminants in the subsurface. A similar effort is yet to materialize in terms of aquitards as storage of groundwater available for water supply. As illustrated with several examples in the previous sections of the book, aquitards can release ("leak") significant volumes of water to adjacent aquifers that are being stressed by pumping; they can also transfer water from one aquifer to another, both under natural conditions and as a result of artificial groundwater withdrawal. Understanding various roles aquitards can play in a hydraulically stressed groundwater system is especially important when designing artificial aquifer recharge and predicting long-term exploitable reserves of groundwater.

One usually thinks of an aquitard, when continuous and thick, and when overlying a highly productive confined aquifer, as a perfect "protector" of the valuable groundwater resource. Some professionals, however, would argue that "every aquitard leaks" and it is only a matter of time when existing shallow groundwater contamination would enter the confined aquifer and threaten the source. Of course, it does not help anyone (i.e., interested stakeholders) if such professionals rely only on their "best professional judgment" and are much less specific in terms of the "reasonable amount of time" after which the contamination would break through the aquitard. If confronted with some field-based data, such as the thickness and the hydraulic conductivity of the aquitard porous material, they may have the "best" answer ready in hand: "But the measurements did not include flow through the fractures, and we all know that all rocks and sediments comprising an aquitard, including clay, do have some fractures, somewhere." Additionally, there may be a number of old wells screened across aquifer(s) and aquitard(s), or wells with degraded casing and seal that provide for direct hydraulic communications between various water-bearing zones in the system. And the final argument is the hardest one to address: "But how do we know that the aquitard is continuous? There must be a pathway through it, such as interconnected lenses of some "sandy" material somewhere." The truth is, as always, somewhere in between. There are perfectly protective *competent* aquitards, of high *integrity*, which would not allow migration of shallow contamination to the underlying aquifer for thousands of years or more, and there are *leaky* aquitards, of low integrity, which do not prevent such migration for more than several decades or so. Of course, if an aquitard is not continuous, or is only a few feet thick in places, all bets are off. In such cases, the site-specific conditions in the adjacent aquifers would play the key role in contaminant transport. These conditions, in a "worst" case, may include large regional drawdowns caused by pumping in the underlying confined aquifer, and the resulting steep hydraulic gradients between the two aquifers (the shallow and the confined) separated by the aquitard. Contamination with dense nonaqueous phase liquids (DNAPLs), which are denser than water, is especially difficult to assess or predict since they can move irrespective of the groundwater hydraulic gradients in an aquifer-aquitard-aquifer system. However, it is surprising how many investigations in contaminant hydrogeology fail to collect more (or any) field information on the aquitard, even though determining its role may be crucial for success of a groundwater remediation project.

The only direct method for determining if actual flow through an aquitard is taking place is dye tracing, but it is of no practical use due to normally very long travel times through aquitards. Several indirect methods, which utilize hydraulic head measurements in the system and chemical and isotopic analyses of water residing in an aquitard, can be used to reasonably accurately assess the rates of groundwater movement through it. However, caution should be used when relying on hydraulic head data collected from monitoring wells that are not completed in the aquitard itself. A difference in the hydraulic heads measured in the overlying and underlying aquifers does not necessarily mean that groundwater is moving between them at any appreciable rate. The existence of the actual flow can be indirectly confirmed only by hydraulically stressing (pumping) one of the aquifers and confirming the obvious related hydraulic head change in the other two units (i.e., including the aquitard itself). When interpreting the hydraulic head changes (fluctuations) caused by pumping, all possible natural causes such as barometric pressure changes or tidal influences should be accounted for.

Figure 2.96 is a good example of possibly misleading conclusions based on measuring the hydraulic heads at only one depth in the surficial aquifer (say, at MP-4 A, where the head is 180.07 ft), and only one depth in the confined aquifer (MP-4 F, the head is 61.77 ft). The vertical difference between these two hydraulic heads is 118.3 ft,

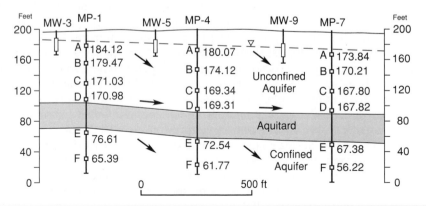

Figure 2.96 Measurements of the hydraulic head at multi-port monitoring wells screened above and below an aquitard. The confined aquifer is being pumped for water supply with an extraction well located approximately 4600 ft from MP-7. (From Kresic, 2007a; copyright Taylor & Francis Group, LLC, printed with permission.)

which may lead one to believe that there must be a significant vertical flow downward through the aquitard caused by such a strong vertical hydraulic gradient (incidentally, the confined aquifer is being pumped for water supply). However, the head difference between the last two ports in the aquifer above the aquitard, at all multi-port wells, is absent for all practical purposes: it is within one hundredth of 1 ft, upward or downward. The flow is "strictly" horizontal indicating absence of advective flow (free gravity flow) of groundwater from the unconfined aquifer into the underlying aquitard. The higher downward vertical gradients at shallow depths in the unconfined aquifer may be the result of recharge, possibly combined with the influence of some lateral pumping (boundary) in the unconfined aquifer.

When measurements of the hydraulic head are available at various depths within an aquitard, a more definitive conclusion as to the probable rates and velocities of groundwater flow through it can be made, including presence of possibly varying hydraulic head inside the aquitard caused by heterogeneities. Figure 2.97 shows recommended setup of monitoring wells for a long-term aquifer test designed to evaluate characteristics of the confined aquifer and possible interactions with the unconfined aquifer, as well as the integrity of the aquitard. Continuous measurements of the hydraulic head at different depths within the entire system can be made with cluster wells, with multi-port wells, or with their combination.

As in any porous media, the main question when attempting to quantify rates and velocities of groundwater flow through an aquitard is the selection of two critical parameters: hydraulic conductivity and effective porosity. This question becomes even more important when considering the fate and transport of contaminants across an aquitard since they can migrate through discrete pathways difficult to detect. As discussed by Cherry et al. (2006), such pathways are common in many settings and include fractures and macropores, or large openings, caused by the roots or burrowing animals. Several processes cause fractures in fine-grained unlithified aquitards. Unsaturated aquitards with lower clay content are particularly susceptible to extensive fracturing by geologic stresses and deformation. Where unlithified aquitards are subject to weathering, shrinking and drying of the sediments can cause fractures to form in the unsaturated zone above the water table. The density of fractures in these settings typically decreases significantly with depth below the weathered portion of the aquitard, but fractures can extend to depths on the order of 30 to 150 ft below the water table. Deposits with higher percentages of clay may be relatively plastic. The plasticity can promote fracture closure at depth, at some later time, if sand or silt has not been washed into the fracture (Bradbury et al., 2006).

Hydraulic properties of aquitards greatly depend on depositional environments that created them, as well as on possible exposure to the land surface at any point in their geologic past. For example, glaciolacustrine sediments, although with high proportion of clay and deposited in lakes, may have horizontal interbedded sand layers resulting in a relatively higher horizontal than vertical conductivity. Vertical fractures in clay may have been formed during some period in the past when the sediment was exposed to the land surface and subject to weathering. These fractures may be truncated by sand layers and, after a subsequent resaturation of the whole sequence, the predominantly vertical flow through the fractures may be redirected by the sand interbeds. All this may result in a quite complicated overall flow pattern in the aquitard.

The following example illustrates difficulties when attempting to calculate a representative groundwater velocity and flow rate through an aquitard that behaves like a dual porosity medium where the flow takes place in both the matrix and the fractures.

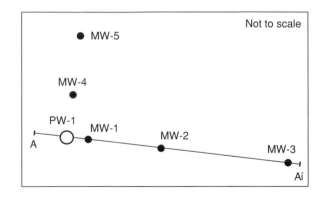

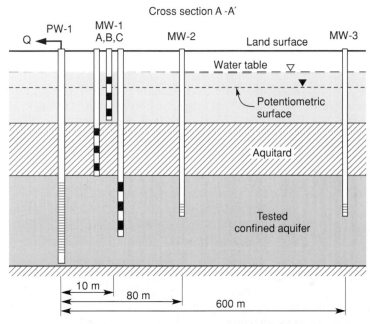

FIGURE 2.97   An example of a pumping test monitoring well network designed to determine characteristics and anisotropy of the tested confined aquifer, and nature of the aquitard including possible leakage from the aquitard and the unconfined aquifer into the underlying confined aquifer. MW-1 is a well cluster with each of the wells having multiple screens for monitoring discrete intervals. Top: map view; bottom: cross section.

Figure 2.98a shows elements for calculation of vertical flow velocity and flow rate through a 4-meter thick aquitard without fractures. The linear velocity (v_L) is calculated using Darcy's law (Eq. 2.34):

$$v_L = \frac{K_v \times i}{n_{ef}} = \frac{K_v \times \left(\frac{\Delta h}{L}\right)}{n_{ef}} = \frac{5 \times 10^{-8} \text{cm/s} \times \left(\frac{2m}{4m}\right)}{0.03}$$

$$= 8.3 \times 10^{-7} \text{cm/s}$$

$$= 26.3 \text{ cm/yr}$$

where K_v = vertical hydraulic conductivity typical for a fairly competent clay matrix
Δh = difference in the hydraulic heads between the unconfined and confined aquifers (it equals 2 m in this case)
L = thickness of the aquitard (4 m)
n_{ef} = effective porosity of clay (3 percent)

The time of travel across the aquitard is 15.2 years, and it is found by dividing the thickness of the aquitard (travel distance; $L = 4$ m) with the velocity ($v_L = 0.263$ m/yr). The flow rate through the aquitard is found by multiplying the cross-sectional area of flow ($A = 1$ m^2 in this case) with the Darcy's velocity (not the linear velocity):

$$Q = v \times A = K_v i \times A$$
$$= 5 \times 10^{-10} \text{m/s} \times \frac{2\,\text{m}}{4\,\text{m}} \times 1\,\text{m}^2$$
$$= 2.5 \times 10^{-10}\,\text{m}^3/\text{s}$$
$$= 2.16 \times 10^{-5}\,\text{m}^3/\text{d}$$

Figure 2.98b shows elements for calculating the flow velocity (v) through a single fracture of aperture $B = 5 \times 10^{-5}$ (50 μm) across an aquitard 4 m thick, using an equivalent hydraulic conductivity of the fracture (Witherspoon, 2000; $B = 2b$ in Witherspoon's notation):

$$v = K_f \times i = B^2 \frac{\rho g}{12\mu} \times \frac{\Delta h}{L} = B^2 \frac{g}{12v} \times \frac{\Delta h}{L}$$
$$= (5 \times 10^{-5}\text{m})^2 \times \frac{9.81\,\text{m/s}^2}{12 \times 0.000001\,\text{m}^2/\text{s}} \times \frac{2\,\text{m}}{4\,\text{m}} \tag{2.35}$$
$$= 1.02 \times 10^{-3}\,\text{m/s}$$
$$= 88.3\,\text{m/d}$$

where K_f = hydraulic conductivity of the fracture
v = flow velocity through the fracture
μ = dynamic viscosity
ρ = water density
g = gravity
v = kinematic viscosity

Note that dynamic viscosity and density are related through kinematic viscosity as follows: $v = \mu/\rho$. The kinematic viscosity of water at temperature of 20°C is 1×10^{-6} m^2/s (McCutcheon et al., 1993), and the acceleration of gravity is rounded to 9.81 m/s^2. The time of travel across the 4-meter thick aquitard is very short, less than one day, and it is calculated by dividing the distance of travel ($L = 4$ m) with the flow velocity ($v_L = 88.3$ m/d). The equivalent hydraulic conductivity of the fracture is here calculated as 2×10^{-3} m/s, or seven orders of magnitude greater than the hydraulic conductivity of the clay matrix (5×10^{-10} m/s).

The flow rate through this single fracture, for a one-meter width ($a = 1$ m), is found by applying the so-called cubic law, i.e., by multiplying the flow velocity with the

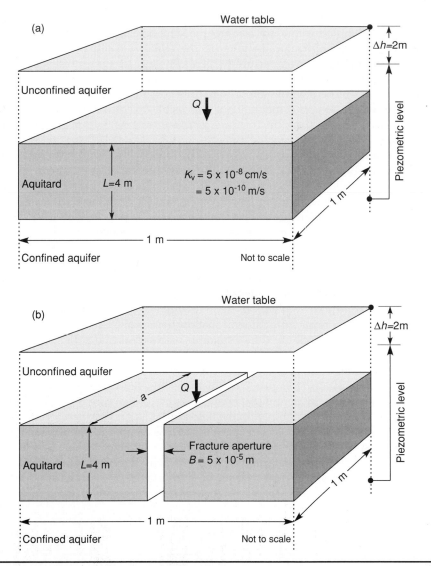

FIGURE 2.98 (a) Elements for calculating groundwater velocity and flow rate across an unfractured clay aquitard, 4-m thick. Calculation based on Darcy's law. (b) Elements for calculating groundwater velocity and flow rate through a single fracture with aperture $B = 5 \times 10^{-5}$ m, crossing an aquitard 4-m thick. (Modified from Cherry et al., 2006.)

cross-sectional area of flow (A), where $A = a\,B$:

$$Q = A \times v = a \times B \times B^2 \frac{\rho g}{12\mu} \times \frac{\Delta h}{L} = a\,B^3 \frac{g}{12v} \times \frac{\Delta h}{L}$$

$$- 1\,\mathrm{m} \times (5 \times 10^{-5}\,\mathrm{m})^3 \times \frac{9.81\,\mathrm{m/s^2}}{12 \times 0.000001\,\mathrm{m^2/s}} \times \frac{2\,\mathrm{m}}{4\,\mathrm{m}} \qquad (2.36)$$

$$= 4.4 \times 10^{-3}\,\mathrm{m^3/d}$$

Comparison of the two results shows that both the flow velocity and the flow rate through a single fracture are incomparably higher than through the matrix. It is obvious that, in case of fractured aquitards, the actual water flux through them will mostly depend on the effective aperture (which takes into account presence of asperities and fill materials), the number, the three-dimensional extent, and the interconnectivity of all fractures present within a representative volume of the aquitard. However, in many cases it would be very difficult, if not impossible, to accurately define the effective aperture and the geometry of the fractures (fracture systems) within an aquitard and various assumptions would have to be made.

Conducting field (in-situ) testing of the bulk hydraulic conductivity at various locations and depths is arguably still the only direct method that can provide answers as to the combined influence of both the matrix and the fractures on the effective hydraulic conductivity of an aquitard. Vargas and Ortega-Guerrero (2004) present results of the hydraulic conductivity tests conducted in 225 piezometers installed in a regional lacustrine clay aquitard in the metropolitan area of Mexico City. The aquitard (split into first and second sub-aquitards) has thickness between 50 and 300 meters, and covers the main aquifer used for water supply of 25 million people. The results of this study show notable differences between the matrix hydraulic conductivity determined in the laboratory, which is on the order of 1×10^{-10} to 1×10^{-11} m/s, and the field-determined hydraulic conductivity at various depths. In general, the aquitard is more heterogeneous and contains more microfractures at shallow depths of 25 to 40 meters. This is reflected in the hydraulic conductivity values spanning as much as five orders of magnitude at some locations: between 1×10^{-11} and 1×10^{-7} m/s. The range of variation generally narrows down with depth, so that field values for the second regional aquitard are between 1×10^{-11} and 1×10^{-9} m/s. Figure 2.99 illustrates this trend of decreasing hydraulic conductivity with depth, evident in the shallow aquitard as well. For example, in this general area, all 14 values determined in the field at depths greater than 15 meters are less than 1×10^{-9} m/s, which would label the aquitard as competent for all practical purposes.

Hart et al. (2005) present results of laboratory testing for shale hydraulic conductivities, a methodology for determining the vertical hydraulic conductivity (K_v) of aquitards at regional scales, and demonstrate the importance of discrete flow pathways across

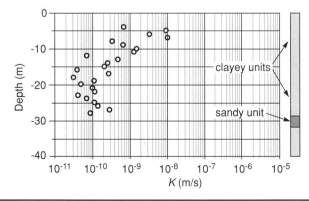

Figure 2.99 Vertical profile of hydraulic conductivity of the regional lacustrine aquitard, measured in the field at the Medical Center in Mexico City (From Vargas and Ortega-Guerrero, 2004; copyright Springer-Verlag; reprinted from Hydrogeology Journal with permission.)

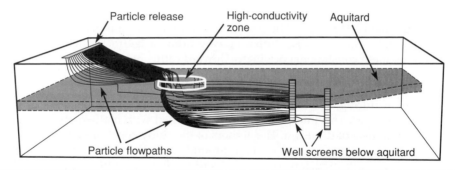

FIGURE 2.100 Results of a three-dimensional particle tracking model showing effects of a high-conductivity zone (window) in an aquitard on particle flow paths. (Modified from Chiang et al., 2002.)

aquitards. A regional shale aquitard in southeastern Wisconsin, the Maquoketa Formation, was studied to define the role that an aquitard plays in a regional groundwater flow system. Calibration of a regional groundwater flow model for southeastern Wisconsin using both predevelopment steady-state and transient targets suggested that the regional K_v of the Maquoketa Formation is 1.8×10^{-11} m/s. The core-scale measurements of the K_v of the Maquoketa Formation range from 1.8×10^{-14} to 4.1×10^{-12} m/s. Flow through some additional pathways in the shale, potential fractures or open boreholes, can explain the apparent increase of the regional-scale K_v. Based on well logs, erosional windows or high-conductivity zones seem unlikely pathways. Fractures cutting through the entire thickness of the shale spaced 5-km apart with an aperture of 50 μm could provide enough flow across the aquitard to match that provided by an equivalent bulk K_v of 1.8×10^{-11} m/s. In a similar fashion, only 50 wells of 0.1 m radius open to aquifers above and below the shale and evenly spaced 10-km apart across southeastern Wisconsin can match the model K_v (Hart et al., 2005).

Windows in aquitards can play a major role in transmitting significant quantities of water or contaminants between the adjacent aquifers. Such windows can be the result of various geologic processes and it is very important to understand the geologic history of an area under investigation. Figure 2.100 shows modeling results for a portion of a groundwater system with a high-permeability zone (window) within an aquitard, including tracks of particles released at the water table which are pulled in by two wells pumping below the aquitard.

Interpretation of chemical composition of groundwater present in an aquitard is another important element for assessing its hydraulic role relative to adjacent aquifers. Farvolden and Cherry (1988) present results of hydrogeologic investigations of thick clayey aquitards in Ontario and Quebec, Canada, conducted as part of studying possible locations for waste-disposal sites. Vertical profiles of major ions and environmental isotopes, together with the hydraulic head and conductivity profiles, were used to interpret the mechanisms of groundwater movement in the aquitards. The relatively high concentrations of major ions in and near the weathered zone are attributed to chemical weathering that took place primarily during Altithermal time when a warmer, drier climate caused the average water table to be 2 or 3 m deeper than the present-day water table. Dessication caused the fractures to form during this drier period. The vertical changes in concentrations of the analyzed constituents in the unweathered clay are primarily caused by molecular diffusion, which causes their migration due to concentration gradients. The vertical flow of groundwater (advection) has negligible effect in this respect. Cl^-, Na^+, and CH_4 diffuse upward from the bedrock where the high concentrations of these constituents originate. Upward diffusion dominates over the downward advective flow; therefore, the net movement is upward. Ca^{2+}, Mg^{2+}, HCO_3^-, and SO_4^{2-} diffuse downward from the weathered zone where they originate. ^{18}O and ^2H (deuterium) are also diffusing downward from the bottom of the weathered zone into the unweathered clayey till. This interpretation was confirmed with mathematical modeling based on Fick's Laws of diffusion. The authors of the study conclude that the clayey material beneath the weathered zone contains groundwater that is many thousands of years old and that exhibits diffusion-controlled distribution of major ions and isotopes (Farvolden and Cherry, 1988; based on studies by Desaulniers, 1986, and Desaulniers et al., 1986). Carbon-14 dating in the Sarnia district is additional evidence of the age of deep groundwater: it is between 10,000 and 14,000 years old. Based on all the information presented above, one would easily conclude that the clayey aquitard in question is competent.

A very detailed description of various field and laboratory methods of determining hydrogeologic characteristics and integrity of aquitards, including calculations of groundwater flow rates and contaminant fate and transport, is given in Cherry et al. (2006) and Bradbury et al. (2006).

2.9 Springs

Early human settlements were located near reliable sources of freshwater—springs and streams. Many ancient cities and their modern counterparts grew thanks to large permanent springs, typically located in karst regions of the Mediterranean and Middle East. Figure 2.101 shows part of a 9-km long aqueduct built in the third century AD by Roman

FIGURE 2.101   Part of an aqueduct built in third century AD by Roman emperor Diocletian for water supply of his summer residence on the Adriatic coast. The aqueduct brought water from the large karst spring shown in Fig. 2.102, which is still being used for the water supply of the Croatian port city of Split. (Photograph courtesy of Ivo Eterovic.)

FIGURE 2.102   The karst spring Jadar, initially tapped by Roman emperor Diocletian in third century AD, is still used for water supply of the Croatian port city of Split. (Photograph courtesy of Ivana Gabric, University of Split.)

emperor Diocletian for water supply of his summer residence. This residence, known as Diocletian's Palace, is the largest ever built in the Roman Empire. The palace is now the old urban core of the Croatian port city of Split on the Adriatic coast, which still uses the same karst spring tapped by Diocletian for his water supply (Fig. 2.102).

Springs of any size and of any type have been used for rural and domestic water supply throughout the world. In the United States, interest in groundwater initially focused on springs, particularly in the arid West. Emigrant trails linked many springs, which served as watering holes for people and livestock. As the eastern part of the country developed and most of the land became privately owned, the focus of groundwater-based public water supply shifted to wells. Advances in well drilling, pump technology, and rural electrification made possible the broad-scale development of groundwater in the West beginning in the early 1900s. Large-scale irrigation with groundwater, especially after World War II, spread rapidly throughout the West resulting in the cessation of spring flows at many locations. As a consequence, in both the Eastern and Western United States, the overall utilization of springs for centralized water supply is minor compared to other parts of the world. Where there are numerous large springs, for example as in the karst regions of Florida, Texas, and Missouri, many such springs are located on private land or public park land, and preserved for other uses including recreation. A paragraph from

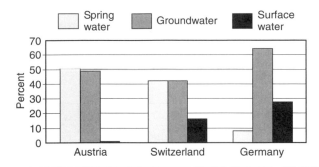

Figure 2.103   Utilization of springs for public water supply in Switzerland, Austria, and Germany. (Modified from Austrian Museum for Economic and Social Affairs, 2003.)

Meinzer's publication on large springs in the United States (Meinzer, 1927) illustrates this point for Florida springs:

> Some of the springs have become well-known resorts, but otherwise not much use is made of their water. The fascinating character of these springs is indicated by the following vivid description of Silver Spring, abbreviated from a description given in a booklet published by the Marion County Chamber of Commerce (see pl. 1.). "The deep, cool water of Silver Spring, clear as air, flows in great volume out of immense basins and caverns in the midst of a subtropical forest. Seen through the glass-bottom boats, with the rocks, under-water vegetation, and fish of many varieties swimming below as if suspended in mid-air, the basins and caverns are unsurpassed in beauty. Bright objects in the water catch the sunlight, and the effects are truly magical. The springs form a natural aquarium, with 32 species of fish. The fish are protected and have become so tame that they feed from one's hand. At the call of the guides, hundreds of them, of various glistening colors, gather beneath the glass-bottom boats."

In recent years there has been a renewed interest in spring water in the United States driven by multinational water-bottling corporations which profit from the booming consumer demand for safe drinking water. In contrast, European countries, where springs of high water quality are abundant, utilize them as preferred sources of public water supply (Fig. 2.103) and are continuously implementing various measures for their protection. The city of Vienna, Austria, is a prime example of scientific, engineering, and regulatory efforts, at all levels, aimed at protecting its famed water supply based on springs.

2.9.1   Types and Classifications of Springs

In general, a spring is any location at the land surface where groundwater discharges from an aquifer, creating a visible flow. When the flow is not visible, but the land surface is wet compared to the surrounding area, such a discharge of groundwater to the surface is called *seep*. A *seepage spring* is a term often used to indicate the discharge of water through numerous small intergranular openings of unconsolidated sediments (e.g., sand and gravel). They are usually marked by abundant vegetation and commonly occur where valleys are cut downward into the zone of saturation of a uniform water-bearing deposit. A *fracture* (or *fissure*) *spring* refers to a discharge of water along bedding planes, joints, cleavage, faults, and other breaks in the consolidated (hard) rock. *Geysers* are springs in which at more or less regular intervals hot water and steam are ejected with force

Magnitude	Discharge
First	100 ft³/s or more
Second	10 to 100 ft³/s
Third	1 to 10 ft³/s
Fourth	100 gal/min to 1 ft³/s
Fifth	10 to 100 gal/min
Sixth	1 to 10 gal/min
Seventh	1 pint per minute to 1 gal/min
Eight	less than 1 pint per minute

From Meinzer, 1923.

TABLE 2.5   Classification of Springs Based on Average Discharge

from considerable depth. Geyser springs generally emerge from tubular conduits that are lined with silica, deposited by the water, and end at the surface in a cone of similar material.

There have been various proposed classifications of springs, based on different characteristics, of which the following are the most common:

- Discharge rate and uniformity
- Character of the hydraulic head (pressure) creating the discharge
- Geologic structure controlling the discharge
- Water quality and temperature

Meinzer's classification of springs based on the average discharge expressed in U.S. units is still widely used in the United States (Table 2.5). However, the classification based solely on average spring discharge, without specifying other discharge parameters, is not very useful when evaluating the potential for spring utilization. For example, a spring may have a very high average discharge but it may be dry or just trickling most of the year. It is therefore essential that a spring is evaluated based on the minimum discharge recorded over a long period, typically longer than several hydrologic years (hydrologic year is defined as spanning all wet and dry seasons in a full annual cycle). When evaluating the availability of spring water, it is important to include a measure of spring discharge variability, which should also be based on periods of record longer than one hydrologic year. The simplest measure of variability is the ratio of the maximum and minimum discharge called the index of variability (I_v):

$$I_v = \frac{Q_{max}}{Q_{min}} \qquad (2.37)$$

Springs with the index of variability greater than 10 are considered highly variable, and those with $I_v \leq 2$ are sometimes called *constant* or *steady springs*. Meinzer (1923) proposed the following measure of variability expressed in percentage:

$$V = \frac{Q_{max} - Q_{min}}{Q_{av}} \times 100(\%) \tag{2.38}$$

where Q_{max}, Q_{min}, and Q_{av} are maximum, minimum, and average discharge respectively. Based on this equation, a constant spring would have variability less than 25 percent, and a variable spring would have variability greater than 100 percent.

Intermittent springs discharge only for a period of time, while at other times they are dry, reflecting directly the aquifer recharge pattern. *Ebb-and-flow springs*, or *periodic springs*, are usually found in limestone (karst) terrain and are explained by the existence of a siphon in the rock mass behind the spring that fills up and empties with certain regularity, regardless of the recharge (rainfall) pattern. Periodic springs can be permanent or intermittent. *Estavelle* has a dual function: it acts as a spring during high hydraulic heads in the aquifer, and as a surface water sink during periods when the hydraulic head in the aquifer is lower than in the body of surface water (estavelles are located within or adjacent to surface water features). *Secondary springs* issue from locations located away from the primary location of spring discharge, which is covered by colluvium or other debris and therefore not visible.

Springs are usually divided into two main groups based on the nature of the hydraulic head in the underlying aquifers that forces them to discharge to the land surface:

- *Gravity springs* emerge under unconfined conditions where the water table intersects land surface. They are also called *descending springs*.

- *Artesian* springs discharge under pressure due to confined conditions in the underlying aquifer, and are also called *ascending* or *rising springs*.

Geomorphology and geologic fabric (rock type and tectonic features such as folds and faults) play a key role in the emergence of springs. When site-specific conditions are rather complicated, springs of formally different types may actually appear next to each other causing confusion. For example, a lateral impermeable barrier in fracture rock, caused by faulting, may force groundwater from a greater depth to ascend and discharge at the surface. This water may have high temperature due to the normal geothermal gradient in the earth's crust—such springs are called thermal springs. At the same time, groundwater of normal temperature may issue at a spring located very close to the thermal spring. Yet a third spring may be present with water temperature varying between "hot" and "cold." All three springs are caused by the same lateral contact between the aquifer and the impermeable barrier, and can all be called *barrier springs*, although the hydraulic mechanism of groundwater discharge is quite different.

Figure 2.104 shows several common spring types. In general, when the contact between the water-bearing porous medium and the impermeable medium is sloping toward the spring, in the direction of groundwater flow, and the aquifer is above the impermeable contact, the spring is called a *contact spring* of *descending* type (Fig. 1.104a). When the impermeable contact slopes away from the spring, in the direction opposite of groundwater flow, the spring is called *overflowing* (Fig. 1.104b). *Depression springs* are formed in unconfined aquifers when topography intersects the water table, usually due to surface stream incision (Fig. 1.104c). Possible contact between the aquifer and the

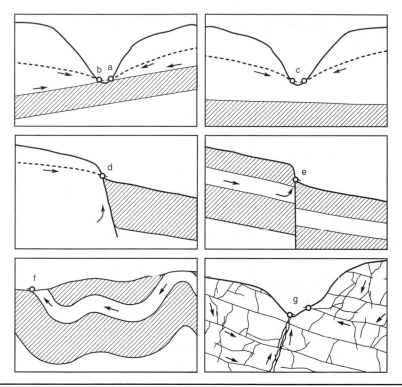

FIGURE 2.104   Different spring types based on the hydraulic head and geologic controls. (From Kresic, 2007a; copyright Taylor & Francis Group, LLC, printed with permission.)

underlying low-permeable formation is not the reason for spring emergence (this contact may or may not be known). Figure 1.104d–f shows some examples of *barrier springs*, the term generally referring to springs at steep (vertical) or hanging lateral contacts between the aquifer and the impermeable rock. When such contact forces groundwater to ascend under hydrostatic pressure, i.e., because the hydraulic head in the aquifer is higher than the land surface elevation at the spring location, the spring is called *ascending* or *artesian*. Artesian springs are usually caused by tectonic structures (faults, fractures, and folds) and often have steady temperatures and discharge, because they are not directly exposed to the atmosphere and recharge from precipitation. Thermal springs are almost always ascending. Figure 2.104g shows both ascending and descending springs in fractured rock aquifers.

Meinzer (1940) gives this account of large springs in the United States:

According to a study completed about 10 years ago, there are in the United States 65 springs of the first magnitude. Of these springs, 38 rise in volcanic rock or in gravel associated with volcanic rock, 24 in limestone, and 3 in sandstone. Of the springs in volcanic rock or associated gravel 16 are in Oregon, 15 in Idaho, and 7 in California. Of the springs in limestone, 9 rise in limestone of Paleozoic age, 8 of them in the Ozark area of Missouri and Arkansas; 4 are in Lower Cretaceous limestone in the Balcones fault belt in Texas; and 11 are in Tertiary limestone in Florida. The 3 springs that issue from sandstone are in Montana. The great discharge of these springs is believed to be due to faults or to other special features. With the additional data now available, some revision of these figures could be made but it would be of minor character.

Since this account by Meinzer and the USGS, the numbers have changed due to more precise flow measurements and contributions of other agencies and investigators across the country. In Florida alone there are 33 documented first-magnitude springs and nearly 700 other significant springs (Scott et al., 2004). Florida represents perhaps the largest concentration of freshwater springs on the earth. Other regions of the world with large springs are also located in karst areas such as the Dinarides (the Balkans), the Alps in Europe, France, Mediterranean countries, Turkey and the Middle East, and China (Kresic and Stevanovic, in preparation).

2.9.2 Thermal and Mineral Springs

Thermal springs can be divided into *warm springs* and *hot springs* depending on their temperature relative to the human body temperature of 98°F or 37°C: hot springs have a higher and warm springs a lower temperature. Warm springs have temperatures higher than the average annual air temperature at the location of discharge. Stearns et al. (1937) give a detailed description of thermal springs in the United States. Meinzer (1940) provides the following illustrative discussion regarding the occurrence and nature of thermal springs:

> An exact statement of the number of thermal springs in the United States is, of course, arbitrary, depending upon the classification of springs that are only slightly warmer than the normal for their localities and upon the groupings of those recognized as thermal springs. A recently published report lists 1059 thermal springs or spring localities. Of these 52 are in the East-Central region (46 in the Appalachian Highlands and 6 in the Ouachita area in Arkansas), 3 are in the Great Plains region (in the Black Hills of South Dakota), and all the rest are in the Western Mountain region. The States having the largest number of thermal springs, according to the listing in the report, are Idaho 203, California 184, Nevada 174, Wyoming 116, and Oregon 105. The geyser area of Yellowstone National Park, however, exceeds all others in the abundance of springs of high temperature (29). Indeed, the number of thermal springs in this area might be given as several thousand if the springs were counted individually instead of being grouped. . . . Nearly two-thirds of the recognized thermal springs issue from igneous rocks-chiefly from the large intrusive masses, such as the great Idaho batholith, which still retain some of their original heat. Few, if any, derive their heat from the extrusive lavas, which were widely spread out in relatively thin sheets that cooled quickly. Many of the thermal springs issue along faults, and some of these may be artesian in character, but most of them probably derive their heat from hot gases or liquids that rise from underlying bodies of intrusive rock. The available data indicate that the thermal springs of the Western Mountain region derive their water chiefly from surface sources, but their heat largely from magmatic sources. . . . The thermal springs in the Appalachian Highlands owe their heat to the artesian structure, the water entering the aquifer at a relatively high altitude, passing to considerable depth through a syncline or other inverted siphon and reappearing at a lower altitude; in the deep part of its course the water is warmed by the normal heat of the deep-lying rocks.

The term *mineral spring* (or *mineral water* for that matter) has very different meanings in different countries, and can be very loosely defined as a spring with water having one or more chemical characteristics different from normal potable water used for public supply. For example, the water can have an elevated content of free gaseous carbon dioxide (naturally carbonated water), or high radon content ("radioactive" water—still consumed in some parts of the world as "medicinal" water of "miraculous" effects), or high hydrogen sulfide content ("good for skin diseases" and "soft skin"), or high dissolved magnesium, or simply have total dissolved solids higher than 1000 mg/L. Some water bottlers, exploiting a worldwide boom in the use of bottled spring water, label water derived from a spring as "mineral" even when it does not have any unusual chemical or physical characteristics. In the United States, public use and bottling of spring and mineral water is under the control of the Food and Drug Administration and such water must conform to strict standards including source protection.

2.9.3   Spring Hydrograph Analysis

The analysis of spring discharge hydrographs may reveal useful information regarding the nature of the aquifer system drained by the spring, as well as the usable water quantities. In many cases, spring hydrographs represent the only available direct quantitative information about the aquifer, which is the main reason why various methods of spring hydrograph analyses have been continuously introduced. The hydrograph of a spring is the final result of various processes that govern the transformation of precipitation and other water inputs in the spring drainage area into the flow at the point of discharge. In some cases, the discharge hydrograph of a spring closely resembles hydrographs of surface streams, particularly if the aquifer is unconfined and has a high transmissivity. In relatively low permeable media, in both unconsolidated and consolidated rocks, springs are weak and usually do not react visibly to daily, weekly, or even monthly (seasonal) water inputs. On the other hand, the reaction to precipitation events is sometimes only a matter of hours in cases of large springs draining karst or intensely fractured aquifers. Although the processes that generate hydrographs of springs and surface streams are quite different, there is much that is analogous between them, and the hydrograph terminology is the same. Figure 2.105 shows a typical hydrograph of a spring that reacts rapidly to precipitation events.

An example of a spring influenced by groundwater withdrawal in its drainage area is shown in Fig. 2.106. Hydrographs of the average monthly discharge of Comal Springs

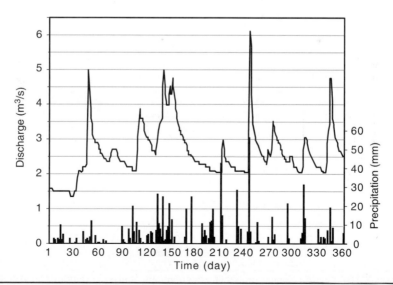

FIGURE 2.105   Characteristic hydrograph of a karst spring with fast response to recharge events.

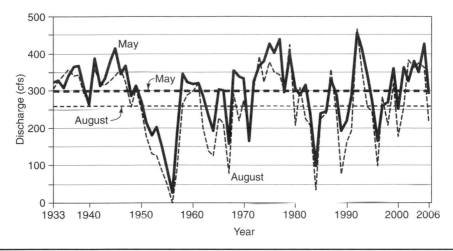

FIGURE 2.106 Average monthly discharges, in cubic feet per second, at Comal Springs, TX, for May (bold line) and August (dashed line) for the 73-year long period of record. (Source: USGS, 2008.)

in Texas for May and August, for the 73-year long period of record, show the impact of several droughts which are compounded by increased pumpage from the Edwards Aquifer. May typically has the highest recorded daily flows, and August the lowest. During the drought of the 1950s, the springs were dry from June to November of 1956. In cases like this one it would not be possible to accurately estimate the natural recharge influences on the spring hydrograph and the nature of the aquifer, without subtracting the influences of pumpage. The same figure also illustrates how using average values, even when having an unusually long period of record, could lead to erroneous conclusions about "secure" discharge rates for any given time. Probability graphs, such as the one shown in Fig. 2.107, are a much more appropriate tool for the assessment of long-term discharge records. For example, in this case the theoretical probability that the average

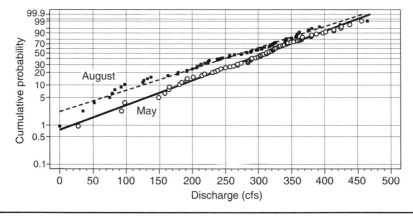

FIGURE 2.107 Extreme value probability distributions of average monthly flows in May and August for Comal Springs.

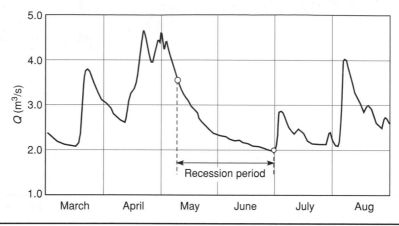

FIGURE 2.108 Part of annual spring hydrograph with a sufficiently long recession period.

spring discharge in August would be less than 50 cfs is about 4 percent and the probability that it would be 0 cfs is about 2 to 3 percent (note that we know from the record that the spring went dry in August 1956). However, it should be noted again that this probability analysis also reflects historic artifical groundwater withdrawals from the system and therefore should not be used alone for any planning purposes. In other words, such withdrawals may change in the future and their impact would have to be accounted for in some quantitative manner.

Recession Analysis

The analysis of the falling hydrograph limb shown in Fig. 2.108, which corresponds to a period without significant precipitation, is called the *recession analysis*. Knowing that the spring discharge is without disturbances caused by an inflow of new water into the aquifer, the recession analysis provides good insight into the aquifer structure. By establishing an appropriate mathematical relationship between the spring discharge and time, it is possible to predict the discharge rate after a given period without precipitation, and to calculate the volume of discharged water. For these reasons, recession analysis has been a popular quantitative method in hydrogeological studies for a long time.

The shape and characteristics of a recession curve depend upon different factors such as aquifer porosity (the most important), position of the hydraulic head, and recharge from other aquifers. The ideal recession conditions—a long period of several months without precipitation—are rare in moderate/humid climates. Consequently, summer and fall storms can cause various disturbances in the recession curve that cannot be removed unambiguously during analysis. It is therefore desirable to analyze as many recession curves from different years as possible (Kresic, 2007a). Larger samples allow for the derivation of the average recession curve as well as the envelope of long-term minimum discharges. In addition, conclusions about the porosity structure, its accumulative ability, and expected long-term minimum discharge, are more accurate.

Two well-known mathematical formulas that describe the falling limb hydrographs during recession periods are proposed by Boussinesq (1904) and Maillet (1905). Both equations give dependence of the flow at specified time (Q_t) on the flow at the beginning of recession (Q_0). The Boussinesq equation is of hyperbolic form:

$$Q_t = \frac{Q_0}{[1 + \alpha(t - t_0)]^2} \tag{2.39}$$

where t = time since the beginning of recession for which the flow rate is calculated and t_0 = time at the beginning of recession usually (but not necessarily) set equal to zero.

The Maillet equation, which is more commonly used, is an exponential function:

$$Q_t = Q_0 \cdot e^{-\alpha(t - t_0)} \tag{2.40}$$

The dimensionless parameter α in both equations represents the *coefficient of discharge* (or *recession coefficient*), which depends on the transmissivity and specific yield of the aquifer. The Maillet equation, when plotted on a semilog diagram, is a straight line with the coefficient of discharge (α) being its slope:

$$\log Q_t = \log Q_0 - 0.4343 \cdot \alpha \cdot \Delta t \tag{2.41}$$
$$\Delta t = t - t_0$$

$$\alpha = \frac{\log Q_0 - \log Q_t}{0.4343 \cdot (t - t_0)} \tag{2.42}$$

Introduction of the conversion factor (0.4343) is a convenience for expressing discharge in Eq. (2.42) in cubic meters per second and time in days. Dimension of α is day^{-1}.

Figure 2.109 is a semilog plot of time versus discharge rate for the recession period shown in Fig. 2.108. The recorded daily discharges form three straight lines which means that the recession curve can be approximated by three corresponding exponential functions with three different coefficients of discharge (α). The three lines correspond to three *microregimes of discharge* during the recession. The coefficient of discharge for the first, second, and third microregimes, using Eq. (2.42), is 0.019, 0.0045, and 0.0015, respectively.

After determining the coefficients of discharge, the flow rate at any given time after the beginning of recession can be calculated using the Maillet equation for one of the regimes. For example, discharge of the spring 35 days after the recession started, when the second microregime is active, is calculated at 2.146 m^3/s.

It is often argued that the variation of the coefficient of discharge has a physical explanation. It is commonly accepted that α on the order of 10^{-2} indicates rapid drainage of well-interconnected large fissures/fractures (or karst channels in case of karst aquifers), while milder slopes of the recession curve (α on the order of 10^{-3}) represent slow drainage of small voids, i.e., narrow fissures and aquifer matrix porosity. Accordingly, the main contribution to the spring discharge in our case is from storage in small voids.

The coefficient of discharge (α) and the volume of free gravitational groundwater stored in the aquifer above spring level (i.e., groundwater that contributes to spring

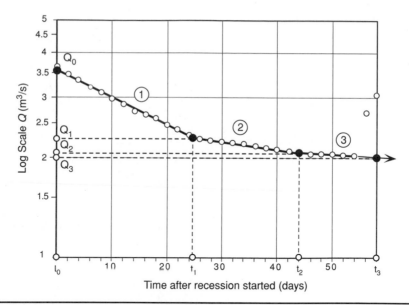

FIGURE 2.109 Semilog graph of discharge versus time for the recession period shown in Fig. 2.108. The duration of the recession period is 54 days.

discharge), are inversely proportional:

$$\alpha = \frac{Q_t}{V_t} \qquad (2.43)$$

where Q_t = discharge rate at time t and V_t = volume of water stored in the aquifer above the level of discharge (spring level). This relationship is valid only for descending gravitational springs. Equation (2.43) allows calculation of the volume of water accumulated in the aquifer at the beginning of recession, as well as the volume discharged during a given period of time. The calculated remaining volume of groundwater always refers to the reserves stored above the current level of discharge. The draining of an aquifer with three microregimes of discharge (as in our case), and the corresponding volumes of the discharged water are shown in Fig. 2.110. The total initial volume of groundwater stored in the aquifer (above the level of discharge) at the beginning of the recession period is the sum of the three volumes that correspond to three different types of storage (effective porosity):

$$V_0 = V_1 + V_2 + V_3 = [\frac{Q_1}{\alpha_1} + \frac{Q_2}{\alpha_2} + \frac{Q_3}{\alpha_3}] \times 86400\,\text{s} \qquad (2.44)$$

where discharge rates are given in cubic meters per second and the volume is obtained in cubic meters. The volume of groundwater remaining in the aquifer at the end of the third microregime is the function of the discharge rate at time t^* and the coefficient of

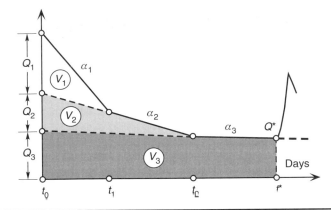

Figure 2.110   Volumes of water discharged during three microregimes of spring recession.

discharge α_3:

$$V^* = \frac{Q^*}{\alpha_3} \tag{2.45}$$

The difference between volumes V_0 and V^* is the volume of all groundwater discharged during the period $t^* - t_0$.

Recession periods of large perennial karstic springs, or springs draining highly permeable fractured aquifers, often have two or three microregimes of discharge as in our example. However, the first microregime rarely corresponds to the simple exponential expression of the Maillet type and is better explained by hyperbolic functions. Deviations from exponential dependence can be easily detected if recorded data plotted on a semilog diagram do not form straight line(s). Usually the best approximation of the rapid (and often turbulent) drainage of large groundwater transmitters at the beginning of recession is the hyperbolic relation of the Boussinesq type. Its general form is:

$$Q_t = \frac{Q_0}{(1 + \alpha t)^n} \tag{2.46}$$

In many cases this function correctly describes the entire recession curve. On the basis of 100 analyzed recession curves of karstic springs in France, Drogue (1972) concludes that among the 6 exponents studied, the best first approximations of exponent n are $1/2$, $3/2$, and 2.

Autocorrelation and Cross Correlation

In general, autocorrelation and cross correlation are analyses applied to any time series (time-dependent variable). They are also the first step in developing stochastic models of hydrologic time series such as stream (river) and spring flows, or hydraulic head fluctuations, which are dependent on some water input such as precipitation. In the case of spring hydrographs, autocorrelation and cross correlation analyses can also give clues about likely types of flow and storage in the aquifer.

Autocorrelation is the correlation between successive values of the same variable. For example, if a hydrologic variable is measured on a daily basis, for lag 1 autocorrelation we pair values recorded at days 1 and 2, days 2 and 3, days 3 and 4, and so on. The number of pairs in the autocorrelation is $N-1$, where N is the number of data. For lag 2 we pair days 1 and 3, days 2 and 4, and so on. Consequently, the number of pairs in correlation decreases again and it is now $N-2$. Autocorrelation is measured by the *autocorrelation coefficient*, also called *serial correlation coefficient*, whose estimate for any lag k is:

$$r_k = \frac{\frac{1}{N-k} \sum_{i=1}^{N-k} (x_i - x_{av})(x_{i+k} - x_{av})}{\frac{1}{N} \sum_{i=1}^{N} (x_i - x_{av})^2} \tag{2.47}$$

where N = total number of data in the sample

x_i = value of the variable (e.g., spring discharge) at time $t = i$

x_{i+k} = value of the variable at time $t = i + k$

h_{av} = average value of the data in the sample

The numerator in Eq. (2.47) is called the *autocovariance* (or just covariance, COV), and the denominator is called the *variance* (VAR) of the time series (note that the square root of the variance is called *standard deviation*). Autocorrelation coefficients are calculated for various lags and then plotted on a graph called an *autocorrelogram*. The number of lags (autocorrelation coefficients) should be approximately 10 percent of the total number of data for smaller samples. For large samples, such as daily values over one or several years, the number of lags can be up to 30 percent.

If there is some predictability based on past values of the series to its present value, the series is *autocorrelated*. Terms that are also often used to describe an autocorrelated series are *persistence* and *memory*. If a series is not autocorrelated it is called independent (i.e., persistence is absent; the series is without memory). The hypothesis that a time series is dependent (autocorrelated) is tested by various statistical tests. One of the simpler tests is proposed by Bartlett (from Gottman, 1981). To be significantly different from zero at the level of confidence 0.05 (i.e., with 95% probability), the autocorrelation coefficient must be

$$r_k > \frac{2}{\sqrt{N}} \tag{2.48}$$

where N = total number of data in the sample. This test is in hydrologic practice often performed just for the first, or the first two lags which is not recommended. It is more correct to perform a test for the entire correlogram introducing limits of confidence. This may uncover possible delayed or periodic components in the time series that would otherwise be considered as independent if, say, lag 2 was found to be not significantly different from zero. A test proposed by Anderson gives limits of confidence for the entire correlogram (Prohaska, 1981):

$$LC(r_k) = \frac{1 \pm Z_\alpha \sqrt{N - k - 2}}{N - k - 1} \tag{2.49}$$

Level of significance α	0.1	0.05	0.01	0.005	0.002
Z for one-tail test	±1.28	±1.645	±2.33	±2.58	±2.88
Z for two-tail test	±1.645	±1.96	±2.58	±2.81	±3.08

From Spiegel and Meddis, 1980.

TABLE 2.6   Values of Z for the Most Often Used Levels of Significance α

where N = the sample size
  k = lag
  Z_α = value of the normally distributed standardized variable at the α level of confidence

Values of Z for various levels of confidence can be found in statistical tables and those used most often are given in Table 2.6.

Mangin (1982) proposed that the time required for the correlogram to drop below 0.2 is called memory effect. According to the author, a high memory of a system indicates a poorly developed karst network with large groundwater flow reserves (storage). In contrast, a low memory is believed to reflect low storage in a highly karstified aquifer. However, Grasso and Jeannin (1994) analyzed the autocorrelograms of a synthetic, regular discharge time series and demonstrated that the increase in the frequency of flood events resulted in a steeper decreasing limb in the associated correlogram. They also pointed out that the sharper the peak of the flood event, the steeper a decreasing limb of the correlogram. Similarly, the decrease of the recession coefficient entails a steeper decreasing limb of the correlogram. Numerical forward simulation of spring hydrographs by Eisenlohr et al. (1997) confirmed that the shape of the resulting correlogram strongly depends on the frequency of precipitation events. These authors also showed that the spatial and temporal distribution of rainfall and the ratio between diffuse and concentrated infiltration had a strong influence on the shape of the hydrograph and subsequently on the correlogram. Consequently, the shape of the correlogram and the derived memory effect depend not only on the state of maturity of the karst system, but also on the frequency and distribution of the precipitation events under consideration (Kovács and Sauter, 2007; Kresic, 1995).

In the cross-correlation analysis, the daily spring flow, represented by the Y series, is the dependent variable influenced by daily precipitation (X series), and it lags behind X. The time-dependent relationship between the two series is analyzed by computing *coefficients of cross-correlation* for various lags and plotting the corresponding *cross-correlogram*. The cross-correlation coefficient for any lag k is given as:

$$r_k = \frac{\mathrm{COV}(x_i, y_{i+k})}{(\mathrm{VAR}x_i \cdot \mathrm{VAR}y_i)^{1/2}} \tag{2.50}$$

where   COV = covariance between the two series
  x_i and y_i = observed daily precipitation and flow respectively
  VAR = variance of each series

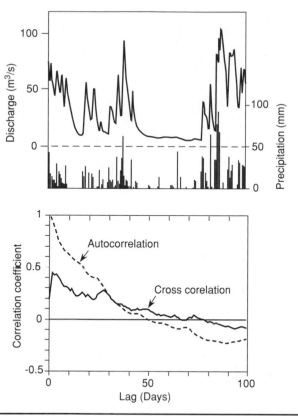

Figure 2.111 Autocorrelaton analysis of spring flow, and cross correlation analysis of spring flow and precipitation for Ombla spring in mature classic karst of the Dinarides. (From Kresic, 1995; copyright American Institute of Hydrology; printed with permission.)

In practice, the coefficient of cross-correlation for lag k is estimated from the sample using the following equation:

$$r_k = \frac{\sum_{i=1}^{N-k} x_i \cdot y_{i+k} - \frac{1}{N-k} \sum_{i=1}^{N-k} x_i \sum_{i=1}^{N-k} y_{i+k}}{\left[\sum_{i=1}^{N-k} x_i^2 - \frac{1}{N-k}\left(\sum_{i=1}^{N-k} x_i\right)^2\right]^{1/2} \left[\sum_{i=1}^{N-k} y_{i+k}^2 - \frac{1}{N-k}\left(\sum_{i=1}^{N-k} y_{i+k}\right)^2\right]^{1/2}} \tag{2.51}$$

The following examples illustrate a possible application of the autocorrelation and cross-correlation analyses. Ombla spring (Fig. 2.111), tapped for the water supply of the Croatian coastal city of Dubrovnik, drains approximately 600 km² of pure mature classic karst terrain of the Dinarides. The ratio of maximum to minimum flow (coefficient of spring nonuniformity) is more than 10 for most years, and a very rapid response to major rain events is evidenced by a short time lag of 2 to 3 days and the corresponding high coefficient of cross correlation close to 0.5 (the peak on the cross correlogram in Fig. 2.111). Statistically significant autocorrelation of flow ($r_k \geq 0.2$) lasts over 30 days due to both frequent precipitation and stable (although low) baseflow during

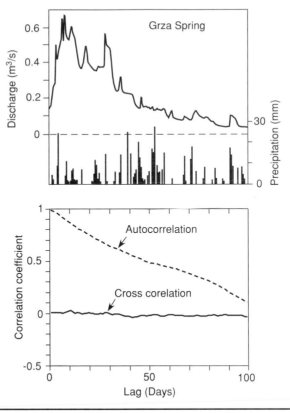

FIGURE 2.112 Autocorrelaton analysis of spring flow, and cross correlation analysis of spring flow and precipitation for Grza spring in semicovered karst of Eastern Serbia. (From Kresic, 1995; copyright American Institute of Hydrology; printed with permission.)

summer months. These facts provide for a quick preliminary assessment: the flow takes place mainly through large conduits capable of rapidly transmitting equally rapidly infiltrated rainfall. The conduit network, however, drains quickly and does not have any significant storage. Other types of porosity contribute to a very uniform regional flow (between 6 and 7 m^3/s) during long summer periods. However, knowing that the spring drainage area is 600 km^2, it appears that the effective matrix porosity of the aquifer is quite low.

Grza spring (Fig. 2.112), located in the semicovered karst of Eastern Serbia, has a very high coefficient of nonuniformity of 22.5 but at the same time a significantly higher and longer autocorrelation. The cross correlation is statistically insignificant although the precipitation in the drainage area is frequent and relatively uniformly distributed throughout the year. Preliminary assessment is that the infiltration is quite slow for a karst terrain. The conduit flow is not predominant, and other nonconduit types of effective porosity/storage are more significant. It also helps to know that the drainage area is a mountainous terrain with a significant snow cap, which melts relatively quickly during spring. This snowmelt contributes to peak flows, which are not directly related to the ongoing precipitation events.

Hydrochemical Separation of Spring Hydrographs

The simultaneous recording of spring discharge and chemical constitution of the spring water allows for a fairly accurate separation into "old" (prestorm), and "new" (rain) water. This separation is based on the assumption that the constitution of the water entering the aquifer is considerably different than that already within. When recharge by rain takes place, it is evident that the concentration of most cations characterizing groundwater, such as calcium and magnesium, is much lower in rain water. Additional preconditions for the application of a hydrochemical hydrograph separation are (after Dreiss, 1989):

- Concentrations of the chemical constituents in the rain water chosen for monitoring are uniform in both area and time.

- Corresponding concentrations in the prestorm water are also uniform in area and time.

- The effects of other processes in the hydrologic cycle during the episode, including recharge by surface waters, are negligible.

- The concentration and transport of elements are not changed by chemical reactions in the aquifer.

This last condition assumes a minor dissolution of rocks during the flow of new water through porous medium. Assuming a simple mixing of old aquifer water (Q_{old}) and newly infiltrated rain water (Q_{new}), the total recorded discharge of the spring is the sum of the two (after Dreiss, 1989):

$$Q_{total} = Q_{old} + Q_{new} \qquad (2.52)$$

If chemical reactions in the aquifer do not cause significant and rapid changes in the concentration of a selected ion in the infiltrating rain water (e.g., calcium in case of unconfined karst and intensely fractured aquifers where the flow velocity is high), the ion (e.g., calcium) balance in the spring water is:

$$Q_{total} \times C_{total} = Q_{old} \times C_{old} + Q_{new} \times C_{new} \qquad (2.53)$$

where Q_{total} = recorded spring discharge
  C_{rec} = recorded concentration of the ion in the spring water
  Q_{old} = portion of the spring flow attributed to the "old" water (i.e., water already present in the aquifer before the rain event)
  C_{old} = recorded concentration of (calcium) ion in the spring water before the rain event
  Q_{new} = portion of the spring flow attributed to the "new" water
  C_{new} = concentration of calcium ion in the new water

If C_{new} is much smaller than C_{old} the input mass of (calcium) ion is relatively small compared to its mass in the "old" aquifer water:

$$Q_{new} \times C_{new} \ll Q_{old} \times C_{old} \qquad (2.54)$$

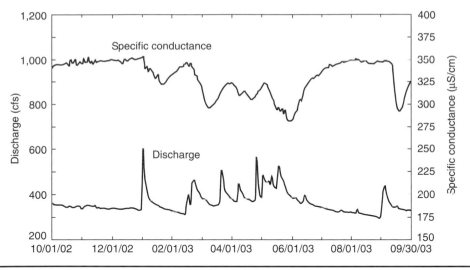

FIGURE 2.113   Big Spring, MS, mean daily discharge and specific conductance data. (Modified from Imes et al., 2007.)

From Eq. (2.54) it follows, after excluding the (small) input mass, that

$$Q_{old} = \frac{Q_{total} \times C_{total}}{C_{old}} \tag{2.55}$$

Combining Eqs. (2.53) and (2.55) gives

$$Q_{new} = Q_{total} - \frac{Q_{total} \times C_{total}}{C_{old}} \tag{2.56}$$

By applying Eq. (2.56) it is possible to estimate the discharge component formed by the inflow of new rain water if the spring discharge recordings and a continuous hydrochemical monitoring are performed before, during, and after the storm event.

Figure 2.113 is an example of changes in chemical composition of spring water that can be used to perform the above described analysis. Separation of the Big Spring hydrograph is based on specific conductance which is significantly lower for the "quick" flow component compared to the baseflow discharge component. Imes et al. (2007) explain a very similar procedure to the one described above, where the initial specific conductance of the baseflow is accounted for. The mixing of the two flow components varies for different recharge events as illustrated in Fig. 2.114.

One useful indicator of water residence in a carbonate aquifer is the calcite saturation index ($SI_{calcite}$), which can be calculated using the following formula:

$$SI_{calcite} = \log(IAP/K_T) \tag{2.57}$$

where IAP = ion activity product of the mineral (calcite) and K_T = thermodynamic equilibrium constant at a given temperature. PHREEQC, a public domain computer program

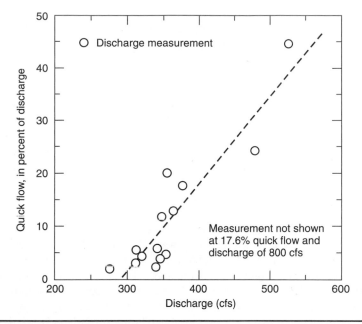

FIGURE 2.114 Discharge from Big Spring as a function of percentage quick flow, water years 2001–2004. (From Imes et al., 2007.)

for the simulation of various geochemical reactions developed at USGS (Parkhurst and Appelo, 1999), can be used to quickly perform this calculation. A value of $SI_{calcite}$ equal to 0 indicates that the water sample is saturated with calcite. A value for $SI_{calcite}$ greater than 0 indicates that the sample is supersaturated with calcite, and value of $SI_{calcite}$ less than 0 would indicate a water sample undersaturated with respect to calcite. The $SI_{calcite}$ can be used to determine the hydrogeologic characteristics of the spring water. For example, water flowing diffusely through carbonate rocks or water flowing through small fractures relatively quickly becomes saturated with respect to calcite. Conversely, water moving through large fractures and conduits requires longer flow paths and residence times to become saturated with respect to calcite (Adamski, 2000). If there is a notable, fast increase in spring discharge after major rainfall events, but the calcium concentration and $SI_{calcite}$ do not change or even increase, this would be an indicator of expulsion of old water residing in the aquifer.

Stable isotopes such as carbon-13, and deuterium and 18-oxygen and their ratio can be used to determine sources of aquifer recharge, while radiogenic isotopes such as tritium and carbon-14 and chlorofluorocarbons (CFCs) are commonly used to determine the relative age of the water discharging from a spring and mixing between "old" and "new" water (Galloway, 2004; Imes et al., 2007; see also Chap. 3).

2.10 Groundwater in Coastal Areas and Brackish Groundwater

Fresh groundwater that does not discharge into surface streams, lakes and marshes, is not evaporated from the water table or transpired by plants, and is not withdrawn artificially will eventually discharge into seas and oceans under either unconfined (shallow) or confined (deep) conditions. This direct discharge may be very significant in coastal areas with permeable surficial sediments and rocks where surface water drainage is minor or absent. For example, karst comprises 60 percent of the shoreline of the Mediterranean and is estimated to contribute 75 percent of its freshwater input, mostly via direct discharge to the sea (UNESCO, 2004).

Knowledge concerning the submarine discharge of groundwater (SDG) has existed for many centuries. The Roman geographer Strabo, who lived from 63 BC to 21 AD, mentioned a submarine fresh groundwater spring 4 km from Latakia, Syria (Mediterranean) near the island of Aradus. Water from this spring was collected from a boat, utilizing a lead funnel and leather tube, and transported to the city as a source of freshwater. Other historical accounts tell of water vendors in Bahrain collecting potable water from offshore submarine springs for shipboard and land use, Etruscan citizens using coastal springs for "hot baths" (Pausanius, ca. second century A.D.) and submarine "springs bubbling freshwater as if from pipes" along the Black Sea (Pliny the Elder, ca. first century A.D.; from UNESCO, 2004). Until relatively recently, most studies of submarine springs were driven almost exclusively by potable water supply objectives. One of the arguments for continuing efforts in that respect is that even if the captured water is not entirely fresh it may be less expensive to desalinate than undiluted seawater. Another argument is that discharge of freshwater across the sea floor may be considered a waste, especially in arid regions. In such places, the detection of SGD may provide new sources of drinking and agricultural water (UNESCO, 2004).

The groundwater flow toward the coast and its submarine discharge are driven by the hydraulic gradient between the inland recharge areas and the sea level (Fig. 2.115). If the aquifer is confined and well protected by a thick aquitard, the groundwater flow may continue well beyond the coastline with the ultimate discharge taking place along distant submarine aquifer outcrop. Figure 2.116 shows how freshwater in the Floridan aquifer in Georgia and Florida, the United States, has been detected miles away from the coast. Multiple stratified confined aquifers along the Atlantic coast of the United States contain large quantities of freshwater that extend to various distances off the coastline. These aquifers have enabled continuing development along the coast, including numerous barrier islands.

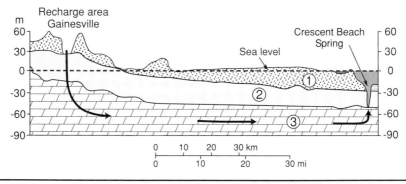

FIGURE 2.115 Idealized cross section of groundwater flow to Crescent Beach Spring, FL. (1) Post-Miocene deposits (green clay, sand, and shell); (2) confining unit (Hawthorn formation); (3) Upper Floridan aquifer (Eocene Ocala Limestone). The morphology of the spring vent and discharge characteristics were investigated in detail. (Modified from Barlow, 2003.)

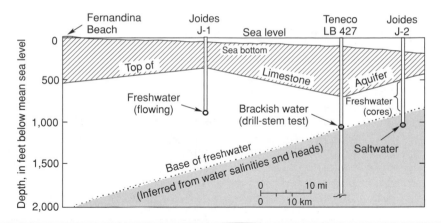

FIGURE 2.116 Inferred position of the freshwater-saltwater interface based on hydraulic testing and water analyses at offshore exploratory oil wells, Atlantic Ocean off the coast of Georgia and Florida. (From Johnston et al., 1982.)

The submarine groundwater discharge is schematically illustrated in Fig. 2.117. Depending primarily on the porous media characteristics, the interface between the freshwater and the saline water intruded naturally from the sea may be rather sharp or there may a wider transitional (mixing) zone in between. In any case, this interface has a characteristic quantifiable shape because of the density difference between freshwater and saltwater. Lighter (less dense) fresh groundwater overlies more dense saltwater and the thickness of the freshwater above the interface with saltwater can be estimated based on the ratio of their respective densities. This relationship was first recognized by Ghyben

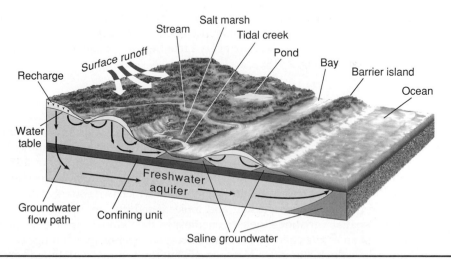

FIGURE 2.117 Shallow (unconfined) and deep (confined) submarine discharge of fresh groundwater showing flow paths in an idealized watershed along the Atlantic coast. Fresh groundwater is bounded by saline groundwater beneath the bay and ocean. Fresh groundwater discharges to coastal streams, ponds, salt marshes, and tidal creeks and directly to the bay and ocean. (From Barlow, 2003.)

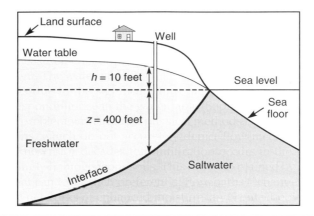

Figure 2.118 Illustration of the Ghyben-Herzberg hydrostatic relationship between freshwater and saltwater. (Modified from Barlow, 2003.)

and Herzberg, two European scientists who derived it independently in the late 1800s:

$$z = \frac{\rho_f}{\rho_s - \rho_f} h \tag{2.58}$$

where z = thickness of freshwater between the interface and the sea level
ρ_f = density of freshwater
ρ_s = density of saltwater
h = thickness of freshwater between the seal level and water table

Freshwater has a density of about 1.000 g/cm^3 at 20°C, whereas that of seawater is about 1.025 g/cm^3. Although this difference is small, Eq. (2.58) indicates that it results in 40 ft of freshwater below sea level for every 1 ft of freshwater above sea level as illustrated with the example in Fig. 2.118:

$$z = 40h \tag{2.59}$$

Although in most applications this simple equation is sufficiently accurate, it does not describe the true nature of freshwater-saltwater interface since it assumes hydrostatic conditions (no movement of either water). In reality, fresh groundwater discharges into the saltwater body (sea, ocean) with a certain velocity and through a seepage surface of certain thickness, thus creating a transition zone in which two waters of different density mix by the processes of dispersion and molecular diffusion. Mixing by dispersion is caused by spatial variations (heterogeneities) in the geologic structure, the hydraulic properties of an aquifer, and by dynamic forces that operate over a range of time scales, including daily fluctuations in tide stages, seasonal and annual variations in groundwater recharge rates, and long-term changes in sea-level position. These dynamic forces cause the freshwater and saltwater zones to move seaward at times and landward at times. Because of the mixing of freshwater and saltwater within the transition zone, a circulation of saltwater is established in which some of the saltwater is entrained within the overlying

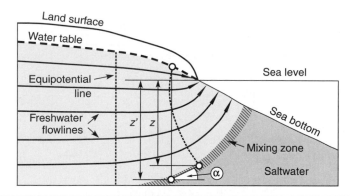

FIGURE 2.119 Hydrodynamic relationship between freshwater and saltwater in an unconfined coastal aquifer. α is the angle of the interface slope. True depth to saltwater (z') is greater than the one assumed based on the Ghyben-Herzberg relationship (z).

freshwater and returned to the sea, which in turn causes additional saltwater to move landward toward the transition zone (Barlow, 2003).

By convention, freshwater is defined as water having total dissolved solids less than 1000 mg/L and chloride concentration less than 250 mg/L. For seawater, these values are 35,000 and 19,000 mg/L, respectively. Everything in between would correspond to a mixing zone. The thickness of a mixing zone depends on local conditions in the aquifer but, in general, it is much smaller than the general vertical field scale of interest. In many cases, quantitative analyses and groundwater modeling codes are based on the assumption of a sharp interface between freshwater and saltwater.

Discharge of freshwater causes flow lines in the aquifer to deviate from horizontal as illustrated in Fig. 2.119. Because Dupuit's hypothesis about vertical equipotential lines does not apply, the true vertical thickness of freshwater is somewhat greater than the one estimated using the Ghyben-Herzberg equation, as first recognized by Hubbert (1940). The slope of the interface (α) can be calculated using the following equation (Davis and DeWiest, 1991):

$$\sin \alpha = \frac{\partial z}{\partial s} = -\left[\frac{1}{K_f} \times \frac{\rho_f}{\rho_f - \rho_s} V_f - \frac{1}{K_s} \times \frac{\rho_f}{\rho_f - \rho_s} V_s\right] \qquad (2.60)$$

where s = trace of the interface in a vertical plane
       K_f and K_s = hydraulic conductivities for freshwater and saltwater, respectively
         V_f and V_s = velocities of freshwater and saltwater along the interface

Equation (2.60) can be simplified if it is assumed that saltwater is stagnant compared to freshwater which flows over it, so that the second term in the brackets becomes zero.

Understanding submarine groundwater discharge mechanisms, as well as the dynamic nature of the freshwater-saltwater interface, is particularly important for islands. People who live on small coral islands are heavily reliant on fresh groundwater as their dominant source of potable water. On such islands, groundwater is found as a thin veneer of freshwater, called a freshwater lens, floating over saltwater in the unconfined aquifer. A typical example is Tarawa atoll in the Pacific Ocean, which consists of coral sediments

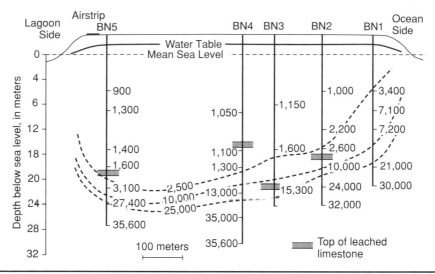

FIGURE 2.120 Depth of freshwater limits at selected cross sectional area, Tarawa atoll in the Pacific Ocean. (From Metai, 2002.)

and limestone of unknown thickness, overlying a volcanic seamount. The freshwater lenses in the islands of Tarawa atoll are up to 30-m deep (Falkland, 1992, from Metai, 2002). Two major geological layers are found within this 30-m zone, a younger layer (Holocene age) consisting largely of unconsolidated coral sediments overlying an older layer (Pleistocene age) of coral limestone. Unconformity between these two layers is at depths generally between about 10 and 15 meters below mean sea level (Jacobson and Taylor, 1981, from Metai, 2002). This unconformity is very significant to the formation of freshwater lenses. The Pleistocene limestone below the unconformity has relatively high permeability which enhances the mixing of freshwater and seawater. Mixing is less likely to occur in the relatively less permeable upper Holocene sediments. As seen in Fig. 2.120, unconformity is the main feature controlling the depth of freshwater lenses (Metai, 2002).

Island aquifers are vulnerable to any change in the delicate water balance between recharge from rainfall, evapotranspiration from the water table, mixing with the surrounding saltwater, and discharge into the open sea (ocean) water. Uncontrolled groundwater withdrawal may cause saltwater intrusion and loss of freshwater, and can have serious consequences for island people. Unfortunately, possible sea level rise caused by climate change would have similar impact on low-lying islands even if groundwater management practices were prudent.

2.10.1 Saltwater Intrusion

During the last several decades, groundwater use in coastal areas worldwide has dramatically increased due to rapid population growth. With this increase came the public recognition that groundwater supplies are vulnerable to overuse and contamination. Groundwater development depletes the amount of groundwater in storage and causes reductions in groundwater discharge to streams, wetlands, and coastal estuaries, and lowered water levels in ponds and lakes. Contamination of groundwater resources has resulted in degradation of some drinking-water supplies and coastal waters. Although overuse and contamination of groundwater are common for all types of aquifers, the proximity of coastal aquifers to saltwater creates unique challenges with respect to groundwater sustainability. Two main concerns are saltwater intrusion into freshwater aquifers and changes in the amount and quality of fresh groundwater discharge to coastal saltwater ecosystems. Saltwater intrusion is the movement of saline sea water into freshwater aquifers caused primarily by groundwater pumping from coastal wells. Because saltwater has high concentrations of total dissolved solids and certain inorganic constituents, it is unfit for human consumption and many other uses. Saltwater intrusion reduces fresh groundwater storage and, in extreme cases, leads to the abandonment of supply wells when concentrations of dissolved ions exceed drinking-water standards (Barlow, 2003). The problem of saltwater intrusion was recognized as early as 1854 on Long Island, New York (Back and Freeze, 1983), thus predating many other types of drinking-water contamination issues in the news.

When natural conditions in a coastal aquifer are altered by groundwater withdrawal, the position and shape of the freshwater-saltwater interface, as well as the mixing zone thickness, may change in all three dimensions and result in saltwater intrusion (encroachment). The presence of leaky and discontinuous aquitards, and pumping from different aquifers or different depths in the same aquifer, may create a rather complex spatial relationship between freshwater and saltwater. Figure 2.121 shows a simple case of saltwater intrusion caused by well pumpage from an unconfined homogeneous aquifer resting on an impermeable horizontal case. As the pumping rate and drawdown increase, the interface continues to move landward until it reaches the critical hydraulic condition. At this point the hydraulic head at the groundwater divide caused by pumping and the interface toe are positioned on the same vertical. Any further increase in the pumping rate or lowering of the hydraulic head will result in a rapid advance of the interface until

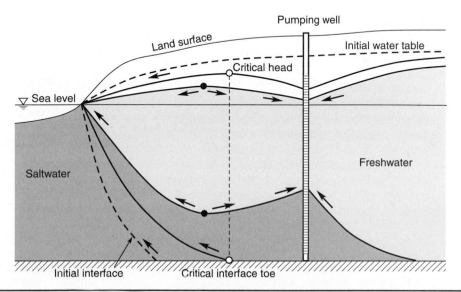

Figure 2.121 Changing freshwater-saltwater interface position resulting from a single well pumping in a coastal unconfined aquifer. (Modified from Strack, 1976; Bear, 1979.)

new equilibrium is reached, with the interface toe landward of the well (Bear, 1979). In many cases this lateral intrusion of saltwater would result in complete abandonment of the well.

Well pumping above the saltwater-freshwater interface will cause upconing of the denser saltwater, which is not necessarily always accompanied by a significant lateral landward movement of the interface. This upconing may reach the well and also result in cessation of pumping due to unacceptable concentrations of total dissolved solids and other constituents in the extracted water. However, unlike in the case of complete lateral saltwater encroachment, once the pumping stops and the hydraulic head of freshwater increases, the cone of dense saltwater dissipates relatively quickly driven by gravity.

Strack (1976), Bear (1979), Kashef (1987), and Bear et al. (1999) provide analytical solutions for calculating the positions and movement of the saltwater-freshwater interface for various cases of groundwater extraction including stratified aquifer-aquitard systems. There are several excellent commercial and public domain (free) computer programs for three-dimensional numeric modeling of density-dependent groundwater flow. SUTRA developed by the USGS (Voss and Provost, 2002) is an example of a program widely used for modeling saltwater-freshwater interactions in coastal groundwater systems.

2.10.2   Inland Brackish Water

Nonpotable groundwater, with naturally elevated concentrations of total dissolved solids (TDS) exceeding 1000 mg/L, can be found in deeper portions of sedimentary basins of all scales. Higher TDS concentrations of brackish groundwater are the result of its long residence times and slow rates or complete absence of present-day recharge with freshwater. More mineralized groundwater can sometimes also be found at shallow depths in unconfined aquifers, particularly in arid regions, where evaporation from the water table results in increased concentration of dissolved minerals. Finally, shallow groundwater may have naturally elevated TDS concentrations because of mixing with brackish groundwater migrating from deeper aquifers.

When TDS concentrations in groundwater exceed 50,000 mg/L, it is called brine. Brines are generally associated with geologic formations of marine origin rich in evaporates such as anhydrite, gypsum, or halite (jointly often referred to as salts). The origin of groundwater contained in such formations may be trapped sea water that was never flushed by fresh groundwater and became more mineralized in time. Various geologic processes including tectonics may also allow fresh groundwater to circulate into buried evaporates and dissolve them thus becoming brine. Brine groundwater can naturally migrate upward via faults or other geologic features and contaminate shallow freshwater aquifers and surface water bodies.

With the exploding demand for reliable water supply worldwide, followed by advances in technology and a decrease in desalination costs, brackish groundwater has become increasingly targeted for large-scale development. Another reason for the growing interest in brackish groundwater is the fact that the overpumping of many fresh groundwater aquifers allows them to become contaminated by brackish groundwater. The city of El Paso, Texas, the United States, is a prime example illustrating both the contamination of fresh groundwater with brackish water, and a large-scale development of brackish water for water supply.

Approximately 50 percent of the present-day water supply of El Paso is groundwater extracted from two deep basins called the Hueco and Mesilla Bolsons (Fig. 2.122), and

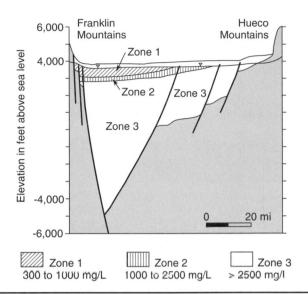

Zone 1
300 to 1000 mg/L

Zone 2
1000 to 2500 mg/L

Zone 3
> 2500 mg/l

FIGURE 2.122 Cross section of the Hueco Bolson showing three zones of total dissolved solids (TDS) concentrations in the deep basin aquifer. (From Hutchinson, 2004)

the other 50 percent is treated surface water from the Rio Grande. Until the late 1980s, groundwater provided more than 75 percent of the city's annual water supply which at its peak in 1989 was 125,215 acre-ft (Hutchinson, 2004). Concerns about groundwater level declines and changes in water quality were raised as early as 1921 (Lippincott, 1921). Figures 2.123 and 2.124 illustrate the impact of excessive groundwater pumping from the basin aquifer. Recognizing that the ongoing groundwater exctraction is not sustainable, the El Paso Water Utility (EPWU) implemented a new water management strategy which includes a reduction of groundwater pumpage, rate structure increase, expansion of reuse of reclaimed water, increased use of Rio Grande water, and treatment of brackish water for potable use (Hutchinson, 2004).

EPWU is operating the world's largest inland desalination plant jointly financed by the United States Defense Department and the local community. The plant is capable of producing 15.5 million gal/d of permeate. It uses reverse osmosis to obtain potable water from brackish groundwater pumped from the Hueco Bolson. Raw water from new and rehabilitated existing wells is pumped to the plant and filtered before being sent to reverse osmosis membranes. Approximately 83 percent of the water is recovered while the remainder is output as a concentrate. The long process of planning, designing and finally building the entire system for brackish groundwater extraction, treatment and disposal started in 1997. EPWU and the Juárez water utility, the Junta Municipal de Aqua y Saneamiento, along with other agencies on both sides of the United States–Mexico border, commissioned the United States Geological Survey to conduct a detailed analysis of the amount of fresh groundwater remaining in the Hueco Bolson, the amount of brackish groundwater available, and groundwater flow patterns. The results of the groundwater model were used for selecting the location of the desalination plant and source wells, and for characterization of possible injection well sites. The most complex analysis was directed toward the problem of concentrate disposal. A comprehensive study examined

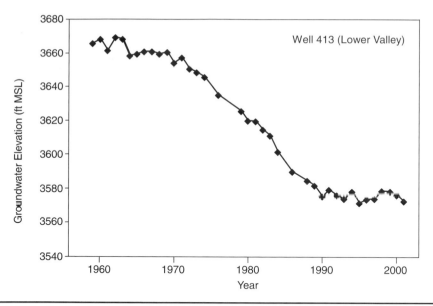

FIGURE 2.123   Declining water level at one of the El Paso Water Utility wells caused by excessive groundwater withdrawals. (From Hutchinson, 2004.)

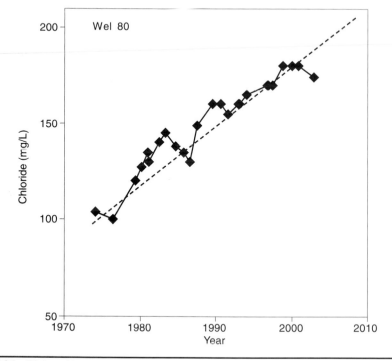

FIGURE 2.124   Rising chloride concentration at one of the El Paso water utility wells caused by aquifer overexploitation and encroachment of brackish water. (From Hutchinson, 2004.)

six alternatives for disposal resulting in selection of deep-well injection as the preferred method. The injection sites criteria include (1) confinement of the concentrate to prevent migration to fresh groundwater, (2) storage volume sufficient for 50 years of operation, and (3) meeting all the requirements of the Texas Commission on Environmental Quality (EPWU, 2007).

References

Adamski, J.C., 2000. Geochemistry of the Springfield Plateau aquifer of the Ozark Plateaus Province in Arkansas, Kansas, Missouri and Oklahoma, U.S.A. *Hydrological Processes*, vol. 14, pp. 849–866.

Alley, W.M., Reilly, T.E., and Franke O.L., 1999. Sustainability of ground-water resources. U.S. Geological Survey Circular 1186. Denver, CO, 79 p.

American Society for Testing and Materials (ASTM), 1999a. *ASTM Standards on Determining Subsurface Hydraulic Properties and Ground Water Modeling*, 2nd ed. West Conshohocken, PA, 320 p

American Society for Testing and Materials (ASTM), 1999b. *ASTM Standards on Ground Water and Vadose Zone Investigations; Drilling, Sampling, Geophysical Logging, Well Installation and Decommissioning*, 2nd ed. West Conshohocken, PA, pp. 561.

Anderson, M.T., and Woosley L.H., Jr., 2005. Water availability for the Western United States—Key scientific challenges. U.S. Geological Survey Circular 1261, Reston, VA, 85 p.

Austrian Museum for Economic and Social Affairs, 2003. *Water Ways*. Vienna, 14 p.

Back, W., and Freeze R.A., editors, 1983. Chemical hydrogeology. Benchmark Papers in Geology, 73. Hutchinson Ross Publication Company, Stroudsburg, PA, 416 p.

Bakhbakhi, M., 2006. Nubian sandstone aquifer. In: *Non-renewable Groundwater Resources; A Guidebook on Socially-Sustainable Management for Water-Policy Makers*, Foster, S. Loucks, D.P., eds. IHP-VI, Series on Groundwater No. 10, UNESCO, Paris, pp. 75–81.

Barlow, P.M., 2003. Ground water in freshwater-saltwater environments of the Atlantic coast. U.S. Geological Survey Circular 1262, Reston, VA, 113 p.

Bear, J., Tsang, C.F., and G. deMarsily, G., editors, 1993. *Flow and Contaminant Transport in Fractured Rock*. Academic Press, San Diego, pp. 548.

Benischke, R., Goldscheider, N., and Smart, C., 2007. Tracer techniques. In: *Methods in Karst Hydrogeology*. Goldscheider, N., Drew, D., editors. International Contributions to Hydrogeology 26, International Association of Hydrogeologists, Taylor & Francis, London, pp. 148–170.

Bear, J., 1979. *Hydraulics of Groundwater*. McGraw-Hill Series in Water Resources and Environmental Engineering. McGraw-Hill, New York, 567 p.

Bear, J., Cheng, A.H.-D., Sorek, S., Ouazar, D., and Herrera I., editors. 1999. *Seawater Intrusion in Coastal Aquifers—Concepts, Methods and Practices*. Kluwer Academic Publishers, Dordrecht, the Netherlands, 625 p.

Bense, V.F., Van den Berg, E.H., and Balen Van, R.T., 2003. Deformation mechanisms and hydraulic properties of fault zones in unconsolidated sediments; the Roer Valley Rift System, the Netherlands. *Hydrogeology Journal*, vol. 11, pp. 319–332.

Boulton, N.S., 1954. Unsteady radial flow to a pumped well allowing for delayed yield from storage. *International Association of Scientific Hydrology Publications*, vol. 37, pp. 472–477.

Boulton, N.S., 1963. Analysis of data from non-equilibrium pumping tests allowing for delayed yield from storage. *Proceedings of the Institution of Civil Engineers* (London), vol. 26, pp. 469–482.

Boulton, N.S., 1970. Analysis of data from pumping tests in unconfined anisotropic aquifers. *Journal of Hydrology*, vol. 10, pp. 369

Boulton, N.S., 1973. The influence of delayed drainage on data from pumping tests in unconfined aquifers: *Journal of Hydrology*, vol. 19, no. 2, pp. 157–169.

Boulton, N.S., and Pontin, J.M.A., 1971. An extended theory of delayed yield from storage applied to pumping tests in unconfined anisotropic aquifers. *Journal of Hydrology*, vol. 19, pp. 157–169.

Boussinesq, J., 1904. Recherches théoriques sur l'écoulement des nappes d'eau infiltrées dans le sol et sur les débits des sources. *Journal de Mathématiques Pures et Appliquées*, Paris, vol. 10, pp. 5–78.

Bradbury, K.R., Gotkowitz, M.B., Hart, D.J., Eaton, T.T., Cherry, J.A., Parker, B.L., and Borchardt, M.A., 2006. Contaminant transport through aquitards: Technical guidance for aquitard assessment. American Water Works Association Research Association (AwwaRF), Denver, CO, 144 p.

Cherry, J.A., Parker, B.L., Bradbury, K.R., Eaton, T.T., Gotkowitz, M.B., Hart, D.J., and Borchardt, M.A., 2006. Contaminant transport through aquitards: A state of the science review. American Water Works Association Research Association (AwwaRF), Denver, CO, 126 p.

Chiang, W.H., Chen, J., and Lin, J., 2002. 3D Master – A computer program for 3D visualization and real-time animation of environmental data. Excel Info Tech, Inc., 146 p.

Cooper, H.H., Jr., 1963. Type curves for nonsteady radial flow in an infinite leaky artesian aquifer. In: *Compiler: Shortcuts and Special Problems in Aquifer Tests*. Bentall, R., editor. U.S. Geological Survey Water-Supply Paper 1545-C, pp. C48–C55.

Cvijić, J., 1893. Das Karstphänomen. Versuch einer morphologischen Monographie. *Geographische Abhandlungen herausgegeben von Prof. Dr A. Penck, Wien,* Bd. V, Heft. 3, pp. 1–114.

Cvijić, J., 1918. Hydrographie souterraine et évolution morphologique du karst. *Receuilles Travaux de 1'Institute de Geographic Alpine*, vol. 6, no. 4, pp. 376–420.

Cvijić, J., 1924. Geomorfologija (Morphologie Terrestre). Knjiga druga (Tome Second). Beograd, 506 p.

Danskin, W.R., McPherson, K.R., and Woolfenden, L.R., 2006. Hydrology, description of computer models, and evaluation of selected water-management alternatives in the San Bernardino area, California. U.S. Geological Survey Open-File Report 2005–1278, Reston, VA, 178 p.

Davis, S.N., and DeWiest, R.J.M., 1991. *Hydrogeology*. Krieger Publishing Company, Malabar, FL, 463 p.

Dawson, K., and Istok, J., 1992. *Aquifer Testing; Design and Analysis*. Lewis Publishers, Boca Raton, FL, 280 p.

Degnan, J.R., Moore, R.B., and Mack, T.J., 2001. Geophysical investigations of well fields to characterize fractured-bedrock aquifers in southern New Hampshire. USGS Water-Resources Investigations Report 01–4183, 54 p.

Desaulniers, D.E., 1986. Groundwater origin, geochemistry, and solute transport in three major clay plains of east-central North America [Ph.D. thesis]. Department of Earth Sciences, University of Waterloo, 450 p.

Desaulniers, D.E., Kaufmann, R.S., Cherry, J.A., and Bentley, H.W., 1986. 37Cl-35Cl variations in a diffusion-controlled groundwater system. *Geochimica et Cosmochimica Acta*, vol. 50, pp. 1757–1764.

Dettinger, M.D., 1989. Distribution of carbonate-rock aquifers in southern Nevada and the potential for their development, summary of findings, 1985–88. Program for the Study and Testing of Carbonate-Rock Aquifers in Eastern and Southern Nevada, Summary Report No. 1, Carson City, NV, 37 p.

Domenico, P.A., and Schwartz, F.W., 1990. *Physical and Chemical Hydrogeology*. John Willey and Sons, New York, 824 p.

Dreiss, S.J., 1989. Regional scale transport in a karst aquifer. 1. Component separation of spring flow hydrographs. *Water Resources Research*, vol. 25, no. 1, pp. 117–125.

Drew, L.J., Southworth, S., Sutphin, D.M., Rubis, G.A., Schuenemeyer, J.H., and Burton, W.C., 2004. Validation of the relation between structural patterns in fractured bedrock and structural information interpreted from 2D-variogram maps of water-well yields in Loudoun County, Virginia. *Natural Resources Research*, vol. 13, no. 4, pp. 255–264.

Driscoll, F.G., 1989. *Groundwater and Wells*. (Third Printing). Johnson Filtration Systems Inc, St. Paul, MN, 1089 p.

Drogue, C., 1972. Analyse statistique des hydrogrammes de decrues des sources karstiques. *Journal of Hydrology*, vol. 15, pp. 49–68.

Duffield, G.M., 2007. AQTESOLV, beta version with Papadopulous (1965) equation, HydroSOLVE, Inc., Reston, VA.

EDAW-ESA, 1978, Environmental and economic effects of subsidence: Lawrence Berkeley Laboratory Geothermal Subsidence Research Program Final Report—Category IV, Project 1, various pages.

Eisenlohr, L., Kiraly, L., Bouzelboudjen, M., and Rossier, I., 1997. Numerical versus statistical modeling of natural response of a karst hydrogeological system. *Journal of Hydrology*, vol. 202, pp. 244–262.

EPWU (El Paso Water Utilities), 2007. Water in the desert. An opportunity for innovation. Available at: http://www.epwu.org/water/desal_info.html

Falkland, A.C., 1992. Review of Tarawa freshwater lenses. Hydrology and Water Resources Branch, ACT Electricity and Water, Prepared for AIDAB.

Farvolden, R.N., and Cherry, J.A., 1988. Chapter 18, Region 15, St. Lawrence Lowland. In: *The Geology of North America, Vol. 0–2, Hydrogeology*, Back, W., Rosenshein, J.S., and Seaber, P.R., editors. The Geological Society of America, Boulder, CO, pp. 133–140.

Faybishenko, B., Witherspoon, P.A., and Benson, S.M., editors, 2000. Dynamics of Fluids in Fractured Rock, Geophysical Monograph 122. American Geophysical Union, Washington, DC, 400 p.

Ferris, J.G., Knowles, D.B., Brown, R.H., and Stallman R.W., 1962. Theory of aquifer tests. U.S. Geological Survey Water Supply Paper 1536-E, Washington, DC, 173 p.

Ford, D.C., and Williams, P.W., 1989. *Karst Geomorphology and Hydrology*. Unwin Hyman, London, 601 p.

Franke, O.L., Reilly, T.E., Haefner, R.J., and Simmons, D.L., 1990. Study guide for a beginning course in ground-water hydrology: Part 1–course participants. U.S. Geological Survey Open File Report 90–183, Reston, VA, 184 p.

Freeze, R.A., and Cherry, J.A., 1979. *Groundwater*. Prentice-Hall, Englewood Cliffs, NJ, 604 p.

Galloway, J.M., 2004. Hydrogeologic characteristics of four public drinking-water supply springs in northern Arkansas. Water-Resources Investigations Report 03–4307, Little Rock, AR, 68 p.

Galloway, D., Jones, D.R., and Ingebritsen, S.E., 1999. Land subsidence in the United States. U.S. Geological Survey Circular 1182, Reston, VA, 177 p.

Geyh, M., 2000. Volume IV. Groundwater. Saturated and unsaturated zone. In: *Environmental Isotopes in the Hydrological Cycle. Principles and Applications*. Mook, W.G., editor. International Hydrological Programme, IHP-V, Technical Documents in Hydrology, No. 39, Vol. IV, UNESCO, Paris, 196 p.

Giusti, E.V., 1978. Hydrogeology of the karst of Puerto Rico. U.S. Geological Survey Professional Paper 1012, Washington, DC, 68 p.

Gottman, J.M., 1981. *Time-Series Analysis; A Comprehensive Introduction for Social Scientists*. Cambridge University Press, Cambridge, 400 p.

Grasso, D.A., and P-Jeannin, Y., 1994. Etude critique des methodes d'analyse de la reponse globale des systemes karstiques. Application au site de Bure (JU, Suisse). Bulletin d'Hydrogéologie (Neuchatel), vol. 13, pp. 87–113.

Griffioen, J., and Kruseman, G.P., 2004. Determining hydrodynamic and contaminant transfer parameters of groundwater flow. In: *Groundwater Studies: An International Guide for Hydrogeological Investigations*. Kovalevsky, V.S., Kruseman, G.P., Rushton, K.R., editors. IHP-VI, Series on Groundwater No. 3, UNESCO, Paris, France, pp. 217–238.

Gringarten, A.C., and Ramey, H.J., 1974. Unsteady state pressure distributions created by a well with a single horizontal fracture, partial penetration or restricted entry. *Society of Petroleum Engineers Journal*, pp. 413–426.

Gringarten, A.C., and Whiterspoon, P.A., 1972. A method of analyzing pump test data from fractured aquifers. In: *International Society of Rock Mechanics and International Association of Engineering Geology, Proceedings of the Symposium Rock Mechanics*, Stuttgart, vol. 3-B, pp. 1–9.

Grund, A., 1903. Die Karsthydrographie. Studien aus Westbosnien. Geograph. Abhandl. von Penck BD, VII, H.3,1–200, Leipzig.

Grund, A., 1914. Der geographische Zyklus im Karst. Z. Ges. Erdkunde, pp. 621–640.

Habermehl, M.A., 2006. The great artesian basin, Australia. In: *Non-Renewable Groundwater Resources; A Guidebook on Socially-Sustainable Management for Water-Policy Makers*. Foster, S., Loucks, D.P., editors. IHP-VI, Series on Groundwater No. 10, UNESCO, Paris, pp. 82–88.

Haneberg, W., Mozley, P., Moore, J., and Goodwin, L., editors. 1999. Faults and subsurface fluid flow in the shallow crust. *American Geophysical Union Monograph*, vol. 113, pp. 51–68.

Hantush, M.S., 1956. Analysis of data from pumping tests in leaky aquifers. *Transactions, American Geophysical Union*, vol. 37, no. 6, pp. 702–714.

Hantush, M.S., 1959. Nonsteady flow to flowing wells in leaky aquifers. *Journal of Geophysical Research*, vol. 64, no. 8, pp. 1043–1052.

Hantush, M.S., 1960. Modification of the theory of leaky aquifers. *Journal of Geophysical Research*. vol. 65, pp. 3713–3725.

Hantush, M.S., 1961a. Drawdown around a partially penetrating well. *Journal of the Hydrology Division, Proceedings of the American Society of Civil Engineers.*, vol. 87., no. HY4, pp. 83–98.

Hantush, M.S., 1961b. Aquifer tests on partially penetrating wells. *Journal of the Hydrology Division, Proceedings of the American Society of Civil Engineers,* vol. 87., no. HY5, pp. 171–194.

Hantush, M.S., 1966a. Wells in homogeneous anisotropic aquifers. *Water Resources Research,* vol. 2, no. 2, pp. 273–279.

Hantush, M.S., 1966b. Analysis of data from pumping tests in anisotropic aquifers. *Journal of Geophysical Research,* vol. 71, no. 2, pp. 421–426.

Hantush, M.S., and Jacob C.E., 1955. Nonsteady radial flow in an infinite leaky aquifer. *Transactions, American Geophysical Union,* vol. 36, no. 1, pp. 95–100.

Hantush, M.S., and Thomas R.G., 1966. A method for analyzing a drawdown test in anisotropic aquifers. *Water Resources Research,* vol. 2, no. 2, pp. 281–285.

Hart, D.J., Bradbury, K.R., and Feinstein, D.T., 2005. The vertical hydraulic conductivity of an aquitard at two spatial scales. *Ground Water,* vol. 44, no. 2, pp. 201–211.

Healy, R.W., Winter, T.C., LaBaugh, J.W., and Franke, O.L., 2007. Water budgets: Foundations for effective water-resources and environmental management. U.S. Geological Survey Circular 1308, Reston, VA, 90 p.

Heath, R.C., 1987. Basic ground-water hydrology. U.S. Geological Survey Water-Supply Paper 2220, Fourth Printing, Denver, CO, 84 p.

Henriksen, H., 2006. Fracture lineaments and their surroundings with respect to groundwater flow in the bedrock of Sunnfjord, western Norway. *Norwegian Journal of Geology,* vol. 86, pp. 373–386.

Herak, M., and Stringfield, V.T., 1972. *Karst; Important Karst Regions of the Northern Hemisphere.* Elsevier, Amsterdam, 551 p.

Hubbert, M.K., 1940. The theory of ground-water motion. *Journal of Geology,* vol. 48, no. 8, pp. 785–944.

Hutchinson, W.R., 2004. Hueco Bolson groundwater conditions and management in the El Paso area. EPWU Hydrogeology Report 04–01. Available at: http://www.epwu.org/water/hueco_bolson.html.

HydroSOLVE, Inc., 2002. *AQTESOLV for Windows, User's Guide.* HydroSOLVE, Inc., Reston, VA, 185 p.

Idaho Water Resources Research Institute, 2007. Eastern Snake River Plain surface and ground water interaction. University of Idaho. Available at: http://www.if.uidaho.edu/~johnson/ifiwrri/sr3/esna.html. Accessed September 12, 2007.

Imes, J.L., Plummer, L.N., Kleeschulte, M.J., and Schumacher, J.G., 2007. Recharge area, base-flow and quick-flow discharge rates and ages, and general water quality of Big Spring in Carter County, Missouri. U.S. Geological Survey Scientific Investigations Report 2007–5049, Reston, VA, 80 p.

INL (Idaho National Laboratory), Radiation Control Division, 2006. Our changing aquifer. The Eastern Snake River Plain aquifer. Oversight Monitor, State of Idaho, Department of Environmental Quality, 6 p.

Jacob, C.E., 1963a. Determining the permeability of water-table aquifers. In: Compiler: Methods of determining permeability, transmissibility, and drawdown. Bentall, R., editor. U.S. Geological Survey Water-Supply Paper 1536-I, pp. 245–271.

Jacob, C.E., 1963b. Corrections of drawdown caused by a pumped well tapping less than the full thickness of an aquifer. In: Compiler: Methods of determining permeability, transmissibility, and drawdown. Bentall, R., editor. U.S. Geological Survey Water-Supply Paper 1536-I, pp. 272–292.

Jacobson G., and Taylor, F.J., 1981. Hydrogeology of Tarawa atoll, Kiribati. Bureau of Mineral Resources Record No. 1981/31, Australian Government.

Jacobson, G., et al., 2004. Groundwater resources and their use in Australia, New Zealand and Papua New Guinea. In: *Groundwater Resources of the World and Their Use*. Zektser, I.S., and L.G. Everett, editors. IHP-VI, Series on Groundwater No. 6, UNESCO, Paris, France, pp. 237–276.

James, N.P., and Mountjoy, F.W., 1983. Shelf-slope break in fossil carbonate platforms: an overview. In: *The Shelfbreak: Critical Interface on Continental Margins*. Stanley D.J. Moore, G.T., editors. SEPM Spec. Pub. No. 33, pp. 189–206.

Johnston, R.H. et al., 1982. Summary of hydrologic testing in tertiary limestone aquifer, Tenneco offshore exploratory well–Atlantic OCS, lease-block 427 (Jacksonville NH 17–5). U.S. Geological Survey Water-Supply Paper 2180, Washington, DC, 15 p.

Kashef, A-A. I., 1987. *Groundwater Engineering*. McGraw-Hill International Editions, Civil Engineering Series, McGraw-Hill, Inc., Singapore, 512 p.

Khouri, J., 2004. Groundwater resources and their use in Africa. In: Groundwater resources of the world and their use, IHP-VI, Series on Groundwater No. 6, Zektser, I.S. Everett, L.G. editors. UNESCO, Paris, France, pp. 209–237.

Klohe, C.A., and Kay, R.T., 2007. Hydrogeology of the Piney Point-Nanjemoy, Aquia, and Upper Patapsco Aquifers, Naval Air Station Patuxent River and Webster Outlying Field, St. Marys County, Maryland, 2000–06. U.S. Geological Survey Scientific Investigations Report 2006–5266, 26 p.

Knochenmus, L.A., and Robinson, J.L., 1996. Descriptions of anisotropy and heterogeneity and their effect on ground-water flow and areas of contribution to public supply wells in a karst carbonate aquifer system. U.S. Geological Survey Water-Supply Paper 2475, Washington, DC, 47 p.

Kovács, A., and Sauter, M., 2007. Modelling karst hydrodynamics. In: Methods in Karst Hydrogeology. Goldscheider, N., Drew, D., editors. International Contributions to Hydrogeology 26, International Association of Hydrogeologists, Taylor & Francis, London, pp. 201–222.

Kresic, N., 1991. Kvantitativna hidrogeologija karsta sa elementima zaštite podzemnih voda (in Serbo-Croatian; Quantitative karst hydrogeology with elements of ground-water protection). Naučna knjiga, Beograd, 196 p.

Kresic, N., 1995. Stochastic properties of spring discharge. In: *Toxic Substances and the Hydrologic Sciences*, Dutton, A.R., editor. American Institute of Hydrology, Minneapolis, MN, pp. 582–590.

Kresic, N., 2007a. *Hydrogeology and Groundwater Modeling*, 2nd ed. CRC Press, Taylor & Francis Group, Boca Raton, FL, 807 p.

Kresic, N., 2007b. Hydraulic methods. In: *Methods in Karst Hydrogeology*. Goldscheider, N., and Drew, D., editors. International Contributions to Hydrogeology 26, International Association of Hydrogeologists, Taylor & Francis, London, pp. 65–92.

Kresic, N., and Stevanovic, Z. (eds.), 2009 (in preparation). *Groundwater Hydrology of Springs: Engineering, Theory, Management and Sustainability*. Elsevier, New York.

Kruseman, G.P., de Ridder, N.A., and Verweij, J.M., 1991. *Analysis and Evaluation of Pumping Test Data* (completely revised 2nd ed). International Institute for Land Reclamation and Improvement (ILRI) Publication 47, Wageningen, the Netherlands, 377 p.

Lattman, L.H., and Parizck, R.R., 1964. Relationship between fracture traces and the occurrence of ground water in carbonate rocks. *Journal of Hydrology*, vol. 2, pp. 73–91.

Lippincott, J.B., 1921. Report on available water supplies, present condition, and proposed improvements of El Paso City Water Works. Report to City Water Board.

Lohman, S.W., 1972. Ground-water hydraulics. U.S. Geol. Survey Professional Paper 708, 70 p.

Lohman, S.W., et al., 1972. Definitions of selected ground-water terms – revisions and conceptual refinements. U.S. Geological Survey Water Supply Paper 1988 (Fifth Printing 1983), Washington, DC, 21 p.

Mabee, S.B., 1999. Factors influencing well productivity in glaciated metamorphic rocks. *Ground Water*, vol. 37, no. 1, pp. 88–97.

Mabee, S.B., Curry, P.J., and Hardcastle, K.C., 2002. Correlation of lineaments to ground water inflows in a bedrock tunnel. *Ground Water*, vol. 40, no. 1, pp. 37–43.

Maillet, E., editor., 1905. *Essais D'hydraulique Souterraine Et Fluviale*. Herman Paris.

Mangin, A., 1982. *L'approche Systémique Du Karst, Conséquences Conceptuelles Et Méthodologiques*. Proc. Réunion Monographica sobre el karst, Larra. pp. 141–157.

Margat, J., Foster, S., and Droubi, A., 2006. Concept and importance of non-renewable resources. In: Non-renewable groundwater resources. A guidebook on socially-sustainable management for water-policy makers. Foster, S., Loucks, D.P., editors. IHP-VI, Series on Groundwater No. 10, UNESCO, Paris, 103 p.

Maslia, M.L., and Randolph, R.B., 1986. Methods and computer program documentation for determining anisotropic transmissivity tensor components of two-dimensional ground-water flow. U.S. Geological Survey Open-File Report 86–227, 64 p.

Maupin, M.A., and Barber, N.L., 2005. Estimated withdrawals from principal aquifers in the United States, 2000. U.S. Geological Survey Circular 1279, Reston, VA, 46 p.

McCutcheon, S.C., Martin, J.L., and Barnwell, T.O., Jr., 1993. Water quality. In: *Handbook of Hydrology*. Maidment, D.R., editor. McGraw-Hill, Inc., New York, pp. 11.1–11.73.

McGuire, V.L., Johnson, M.R., Schieffer, R.L., Stanton, J.S., Sebree, S.K., and Verstraeten, I.M., 2003. Water in storage and approaches to ground-water management, High Plains aquifer, 2000. U.S. Geological Survey Circular 1243, Reston, VA, 51 p.

Meinzer, O.E., 1923 (reprint 1959). The occurrence of ground water in the United States with a discussion of principles. Geological U.S. Survey Water-Supply Paper 489, Washington, DC, 321 p.

Meinzer, O.E., 1927. Large springs in the United States. U.S. Geological Survey Water-Supply Paper 557, Washington, DC, 94 p.

Meinzer, O.E., 1932 (reprint 1959). Outline of methods for estimating ground-water supplies. Contributions to the hydrology of the United States, 1931. U.S. Geological Survey Water-Supply Paper 638-C, Washington, DC., p. 99–144.

Meinzer, O.E., 1940. Ground water in the United States; a summary of ground-water conditions and resources, utilization of water from wells and springs, methods of scientific investigation, and literature relating to the subject. U.S. Geological Survey Water-Supply Paper 836D, Washington, DC, pp. 157–232.

Metai, E., 2002. Vulnerability of freshwater lens on Tarawa – the role of hydrological monitoring in determining sustainable yield. Presented at *Pacific Regional Consultation on Water in Small Island Countries*, July 29 to August, 3, 2002, Sigatoka, Fiji, 17 p.

Milanovic, P., 1979. Hidrogeologija karsta i metode istraživanja (in Serbo-Croatian; Karst hydrogeology and methods of investigations). HE Trebišnjica, Institut za korištenje i zaštitu voda na kršu, Trebinje, 302 p.

Milanovic, P.T., 1981. *Karst Hydrogeology*. Water Resources Publications, Littleton, CO, 434 p.

Miller, J.A., 1999. Introduction and national summary. In: *Ground-Water Atlas of the United States.* United States Geological Survey, A6. Available at: http//caap.water.usgs.gov/gwa/index.html.

Moench, A.F., 1984. Double-porosity models for a fissured groundwater reservoir with fracture skin. *Water Resources Research,* vol. 21, no. 8, pp. 1121–1131.

Moench, A.F., 1985. Transient flow to a large-diameter well in an aquifer with storative semiconfining layers. *Water Resources Research,* vol. 8, no. 4, pp. 1031–1045.

Moench, A.F., 1993. Computation of type curves for flow to partially penetrating wells in water-table aquifers. *Ground Water,* vol. 31, no. 6, pp. 966–971.

Moench, A.F., 1996. Flow to a well in a water-table aquifer: an improved Laplace transform solution. *Ground Water,* vol. 34, no. 4, pp. 593–596.

Morris, B.L., Lawrence, A.R.L., Chilton, P.J.C., Adams, B., Calow, R.C., and Klinck, B.A., 2003. Groundwater and its susceptibility to degradation: a global assessment of the problem and options for management. Early Warning and Assessment Report Series, RS. 03–3. United Nations Environment Programme, Nairobi, Kenya, 126 p.

Mozley, P.S., et al., 1996. Using the spatial distribution of calcite cements to infer paleoflow in fault zones: examples from the Albuquerque Basin, New Mexico [abstract]. American Association of Petroleum Geologists 1996 Annual Meeting.

NASA (National Aeronautic and Space Administration), 2007. Visible Earth: a catalog of NASA images and animations of our home planet. Available at: http://visibleearth.nasa.gov. Accessed August 8, 2007.

NASA Photo Library, 2007. Available at: http://www.photolib.noaa.gov/brs/nuind41.htm.

Neuman, S.P., 1972. Theory of flow in unconfined aquifers considering delayed response to the water table. *Water Resources Research,* vol. 8, no. 4, pp. 1031–1045.

Neuman, S.P., 1974. Effects of partial penetration on flow in unconfined aquifers considering delayed gravity response. *Water Resources Research,* vol. 10, no. 2, pp. 303–312.

Neuman, S.P., 1975. Analysis of pumping test data from anisotropic unconfined aquifers considering delayed gravity response. *Water Resources Research,* vol. 11, no. 2, pp. 329–342.

Neuman, S.P., and Witherspoon, P.A., 1969. Applicability of current theories of flow in leaky aquifers. *Water Resources Research,* vol. 5, pp. 817–829.

Oostrom, M., Rockhold, M.L., Thorne, P.D., Last, G.V., and Truex, M.J., 2004. *Three-Dimensional Modeling of DNAPL in the Subsurface of the 216-Z-9 Trench at the Hanford Site.* Pacific Northwest National Laboratory, Richland, WA, various pages.

Opie, J., 2000. *Ogallala, Water for a Dry Land,* 2nd ed. University of Nebraska Press Lincoln, NE, 475 p.

Osborne, P.S., 1993. Suggested operating procedures for aquifer pumping tests. Ground Water Issue, United States Environmental Protection Agency, EPA/540/S-93/503, 23 p.

Parkhurst, D.L., and Appelo, C.A.J., 1999. User's guide to PHREEQC (Version 2)—a computer program for speciation, batch-reaction, one-dimensional transport, and inverse geochemical calculations. U.S. Geological Survey Water-Resources Investigations Report 99–4259, 310 p.

Papadopulos, I.S., 1965. Nonsteady flow to a well in an infinite anisotropic aquifer. In: *Proc. Dubrovnik Symposium on the Hydrology of Fractured Rocks.* International Association of Scientific Hydrology, p. 21–31.

Papadopulos, I.S., and Cooper, H.H., 1967. Drawdown in a well of large diameter. *Water Resources Research*, vol. 3, pp. 241–244.

Peck, M.F., McFadden, K.W., and Leeth, D.C., 2005. Effects of decreased ground-water withdrawal on ground-water levels and chloride concentrations in Camden County, Georgia, and ground-water levels in Nassau County, Florida, From September 2001 to May 2003. U.S. Geological Survey Scientific Investigations Report 2004–5295, Reston, VA, 36 p.

Prohaska, S., 1981. Stohasticki model za dugorocno prognoziranje recnog oticaja (Stochastic model for long-term prognosis of river flow; in Serbian). Vode Vojvodine, Posebna izdanja, Novi Sad, 106 p.

Provost, A.M., Payne, D.F., and Voss, C.I., 2006. Simulation of saltwater movement in the Upper Floridan aquifer in the Savannah, Georgia–Hilton Head Island, South Carolina, area, predevelopment–2004, and projected movement for 2000 pumping conditions. U.S. Geological Survey Scientific Investigations Report 2006–5058, Reston, VA, 132 p.

Puri, S., et al., 2001. Internationally shared (transboundary) aquifer resources management: their significance and sustainable management. A framework document. IHP-VI, Series on Groundwater 1, IHP Non Serial Publications in Hydrology, UNESCO, Paris, 76 p.

Rebouças, A., and Mente, A., 2003. Groundwater resources and their use in South America, Central America and the Caribbean. In: *Groundwater Resources of the World and Their Use*. Zektser, I.S., and Everett, L.G., editors. IHP-VI, Series on Groundwater No. 6, UNESCO, Paris, France, pp. 189–208.

Robson, S.G., and Banta, E.R., 1995. Rio Grande aquifer system. In: *Ground Water Atlas of the United States*. Arizona, Colorado, New Mexico, Utah, HA 730-C.

Salem, O., and Pallas, P., 2001. The Nubian Sandstone Aquifer System (NSAS). In: *Internationally Shared (Transboundary) Aquifer Resources Management: Their Significance and Sustainable Management; A Framework Document*. Puri, et al., editors. IHP-VI, Series on Groundwater 1, IHP Non Serial Publications in Hydrology, UNESCO, Paris, pp. 41–44.

Scott, T.M., Means, G.H., Meegan, R.P., Means, R.C., Upchurch, S.B., Copeland, R.E., Jones, J., Roberts, T., and Willet, A., 2004. *Springs of Florida*. Florida Geological Survey, Bulletin No. 66, Tallahassee, FL, 658 p.

Shubert, G.L., and Ewing, M., 1956. Gravity reconnaissance survey of Puerto Rico. *Geological Society of America Bulletin*, vol. 67, no. 4, pp. 511–534.

Slichter, C.S., 1905. Field measurements of the rate of movement of underground waters. *U.S. Geological Survey Water-Supply and Irrigation Paper* 140, Series 0, Underground waters, 43, Washington, DC, 122 p.

Southworth, S., Burton, W.C., Schindler, J.S., and Froelich, A.J., 2006. Geologic map of Loudoun County, Virginia, USGS Geologic Investigations Series Map I-2553 and pamphlet, 34 p.

Spiegel, M.R., and Meddid, R., 1980. *Probability and Statistics*. Schaum's Outline Series. McGraw-Hill, New York, 372 p.

Stallman, R.W., 1961. The significance of vertical flow components in the vicinity of pumping wells in unconfined aquifers. In: Short Papers in the Geologic and Hydrologic Sciences, *U.S. Geological Survey Professional Paper* 424-B, Washington, DC, pp. B41–B43.

Stallman, R.W., 1965. Effects of water-table conditions on water-level changes near pumping wells. *Water Resources Research*, vol. 1, no. 2, pp. 295–312.

Stallman, R.W., 1971. Aquifer-test, design, observation and data-analysis. U.S. Geological Survey Techniques of Water-Resources Investigations, book 3, chap. B1, 26 p.

Stearns, N.D., Stearns, H.T., and Waring, G.A., 1937. *Thermal Springs in the United States.* U.S. Geological Survey Water-Supply Paper 679-B.

Strack, O.D.L., 1976. A single-potential solution for regional interface problems in coastal aquifers. *Water Resources Research*, vol. 12, no. 6, pp. 1165–1174.

Streltsova, T.D., 1974. Drawdown in compressible unconfined aquifer. *Journal of the Hydrology Division, Proceedings of the American Society of Civil Engineers,* vol. 100, no. HY11, pp. 1601–1616.

Streltsova, T.D., 1988. *Well Testing in Heterogeneous formations.* John Wiley and Sons, New York, 413 p.

Trapp, H., Jr., 1992. Hydrogeologic framework of the Northern Atlantic Coastal Plain in parts of North Carolina, Virginia, Maryland, Delaware, New Jersey, and New York. U.S. Geological Survey Professional Paper 1404-G, 59 p.

Theis, C.V., 1935. The lowering of the piezometric surface and the rate and discharge of a well using ground-water storage. *Transactions, American Geophysical Union*, vol. 16, pp. 519–524.

Thiem, G., 1906. *Hydrologische methoden.* Leipzig, Gebhardt, 56 p.

UNESCO (United Nations Educational, Scientific and Cultural Organization), 2004. Submarine groundwater discharge. Management implications, measurements and effects. IHP-VI, Series on Groundwater No. 5, IOC Manuals and Guides No. 44, Paris, 35 p.

USBR, 1977. *Ground Water Manual.* U.S. Department of the Interior, Bureau of Reclamation, Washington, DC, 480 p.

USGS (United States Geological Survey), 2007. USGS Photographic Library. Available at: http://libraryphoto.cr.usgs.gov.

USGS (United States Geological Survey), 2008. National Water Information System: Web Interface, USGS Ground-Water Data for the Nation. Available at: http://waterdata.usgs.gov/nwis/gw. Accessed January 2008.

Vargas, C., and Ortega-Guerrero, A., 2004. Fracture hydraulic conductivity in the Mexico City clayey aquitard: Field piezometer rising-head tests. *Hydrogeology Journal*, vol. 12, pp. 336–344.

Voss, C.I., and Provost, A.M., 2002. SUTRA; A model for saturated-unsaturated, variable-density ground-water flow with solute or energy transport. U.S. Geological Survey Water-Resources Investigations Report 02–4231, Reston, VA, 250 p.

Waichler, S.R., and Yabusaki, S.B., 2005. Flow and transport in the Hanford 300 Area vadose zone-aquifer-river system. Pacific Northwest National Laboratory, Richland, WA, various pages.

Walton, W.C., 1987. *Groundwater Pumping Tests, Design & Analysis.* Lewis Publishers, Chelsea, MI, 201 p.

Wenzel, L.K., 1936. The Thiem method for determining permeability of water-bearing materials and its application to the determination of specific yield; results of investigations in the Platte River valley, Nebraska. U.S. Geological Survey Water Supply Paper 679-A, Washington, DC, 57 p.

White, B.W., 1988. *Geomorphology and hydrology of karst terrains.* Oxford University Press, New York, 464 p.

Wikipedia, 2007. Great Manmade River. Available at: http://wikipedia://en.wikipedia.org/wiki/Geat_Manmade_River.

Williams, L.J., Kath, R.L., Crawford, T.J., and Chapman, M.J., 2005. Influence of geologic setting on ground-water availability in the Lawrenceville area, Gwinnett County, Georgia. U.S. Geological Survey Scientific Investigations Report 2005–5136, Reston, VA, 50 p.

Witherspoon, P.A., 2000. Investigations at Berkeley on fracture flow in rocks: from the parallel plate model to chaotic systems. In: *Dynamics of Fluids in Fractured Rock*. Faybishenko, B., Witherspoon, P.A., and Benson, S.M., editors. Geophysical Monograph 122, American Geophysical Union, Washington, DC, pp. 1–58.

Wolff, R.G., 1982. Physical properties of rocks—Porosity, permeability, distribution coefficients, and dispersivity: U.S. Geological Survey Open-File Report 82–166, 118 p.

Yin, Z-Y, and Brook, G.A., 1992. The topographic approach to locating high-yield wells in crystalline rocks: Does it work? *Ground Water,* vol. 30, no. 1, pp. 96–102.

Young, H.L., 1992. Summary of ground-water hydrology of the Cambrian-Ordovician aquifer system in the northern midwest, United States. U.S. Geological Survey Professional Paper 1405–A, 55 p.

Zhaoxin, W., and Chuanmao, J., 2004. Groundwater resources and their use in China. In: Groundwater resources of the world and their use. Zekster, I.S., and Everet, L.G., editors. IHP-VI, Series on Groundwater No. 6, UNESCO, Paris, pp. 143–159.

CHAPTER 3

Groundwater Recharge

Neven Kresic

Alex Mikszewski   *Woodard & Curran, Inc., Dedham, Massachusetts*

3.1 Introduction

Together with natural groundwater discharge and artificial extraction, groundwater recharge is the most important water budget component of a groundwater system. Understanding and quantifying recharge processes are the prerequisites for any analysis of the resource sustainability. It also helps policymakers make better-informed decisions regarding land use and water management since protection of natural recharge areas is paramount to the sustainability of the groundwater resource. The first important step in recharge analysis is to consider the scale of the study area since the approach and methods of quantification are directly influenced by it. For example, it may be necessary to delineate and quantify local areas of focused recharge within several acres or tens of acres at a contaminated site where contaminants may be rapidly introduced into the subsurface. This scale of investigation would obviously not be feasible or needed for groundwater resources assessment in a large river basin (watershed). However, as illustrated further in this chapter, groundwater recharge is variable at all scales, in both space and time. It is, therefore, by default that any estimate of recharge involves averaging a number of quantitative parameters and their extrapolation-interpolation in time and space. This also means that there will always be a degree of uncertainty associated with quantitative estimates of recharge and that this uncertainty would also have to be analyzed and quantified. In other words, groundwater recharge is both a probabilistic and a deterministic process: if and when it rains (laws of probability), the infiltration of water into the subsurface and the eventual recharge of the water table will follow physical (deterministic) laws. At the same time, both sets of laws (equations) use parameters that are either measured directly or estimated in some (preferably quantitative) way but would still have to be extrapolated-interpolated in space and time. For this reason, the quantification of groundwater recharge is one of the most difficult tasks in hydrogeology. Unfortunately, this task is too often reduced to simply estimating a percentage of total annual precipitation that becomes groundwater recharge and then using that percentage as the "average recharge" rate for various calculations or groundwater modeling at all spatial and time scales. Following are some of the examples illustrating the importance of not reducing the determination of recharge to a "simple," nonconsequential task.

- Recharge is often "calibrated" as part of groundwater modeling studies where other model parameters and boundary conditions are considered to be more certain ("better known"). In such cases, the model developer should clearly discuss uncertainties related to the "calibrated" recharge rates and the sensitivity of all key model parameters; a difference of 5 or 10 percent in aquifer recharge rate may not be all that sensitive compared to aquifer transmissivity (hydraulic conductivity) when matching field measurements of the hydraulic head; however, this difference is very significant in terms of aquifer water budget and analyses of aquifer sustainability.

- Recharge rate has implications on the shape and transport characteristics of groundwater contaminant plumes. More or less direct recharge from land surface may result in a more or less diving plume, respectively. Different recharge rates also result in different overall concentrations—higher recharge results in lower concentrations (assuming that the incoming water is not contaminated).

- Rainwater that successfully infiltrates into the subsurface and percolates past the root zone may take tens, or even hundreds, of years to traverse the vadose zone and reach the water table. The effects of recharge reduction are thus abstract, as groundwater users do not face immediate consequences. As a result, land use changes are often made without consideration of impacts to groundwater recharge.

- Natural and anthropogenic climate changes also alter groundwater recharge patterns, the effects of which will be faced by future generations.

As discussed earlier in Chap. 2, water budget and groundwater recharge terms are often used interchangeably, sometimes causing confusion. In general, infiltration refers to any water movement from the land surface into the subsurface. This water is called potential recharge, indicating that only a portion of it may eventually reach water table (saturated zone). The term "actual recharge" is being increasingly used to avoid any possible confusion: it is the portion of infiltrated water that reaches the aquifer, and it is confirmed based on groundwater studies. The most obvious confirmation that actual groundwater recharge is taking place is rise in water table elevation (hydraulic head). However, the water table can also rise because of cessation of groundwater extraction (pumping), and this possibility should always be considered. Effective infiltration and deep percolation refer to water movement below the root zone and are often used to approximate actual recharge. Since evapotranspiration (ET) (loss of water to the atmosphere) refers to water at both the surface and subsurface, this should be clearly indicated.

3.2 Rainfall-Runoff-Recharge Relationship

Most natural groundwater recharge is derived directly from rainfall and snowmelt that infiltrate through ground surface and migrate to the water table. To quantify recharge from precipitation, it is critical to understand rainfall-runoff relationships. The first step is to determine the fraction of precipitation available for groundwater recharge, after subtracting what is lost to overland flow (runoff) and evapotranspiration (ET). The key to the rainfall-runoff relationship is the soil type, the antecedent moisture condition, and the land cover. Soils that are well drained generally have high effective porosities and high hydraulic conductivities, whereas soils that are poorly drained have higher total porosities and lower hydraulic conductivities. These physical properties combine with initial moisture content to determine the infiltration capacity of surficial soils. Wet, poorly drained soils will readily produce runoff, while dry, well-drained soils will readily absorb rainfall. Land cover determines the fraction of precipitation available for infiltration. Impervious, paved surface prevent any water from entering the soil column, while open, well-vegetated fields are conducive to infiltration. Regardless of soil type, antecedent moisture, or land cover, the chain of events occurring during a storm event is the same.

Available rainfall will infiltrate the subsurface until the rate of precipitation exceeds the infiltration capacity of the soil, at which point ponding and subsequent runoff will occur. Runoff either collects in discrete drainage channels or moves as overland sheet flow. It is important to understand that infiltration continues throughout the storm event, even as runoff is being produced. The infiltration rate after ponding begins to decrease and asymptotically approaches the saturated hydraulic conductivity of the soil media. Figure 3.1 illustrates typical infiltration patterns for four different rainfall events for the same soil with the saturated hydraulic conductivity of about 0.001 cm/s.

Simple calculations of runoff and water retention volumes in a watershed are possible using the U.S. Department of Agriculture Soil Conservation Service (SCS) runoff curve number (CN) method, updated in Technical Release 55 (TR-55) of the U.S. Department of Agriculture (USDA, 1986). TR-55 presents simplified procedures for estimating direct surface runoff and peak discharges in small watersheds. While it gives special emphasis to urban and urbanizing watersheds, the procedures apply to any small watershed in which certain limitations (assumptions) are met. Hydrologic studies to determine runoff

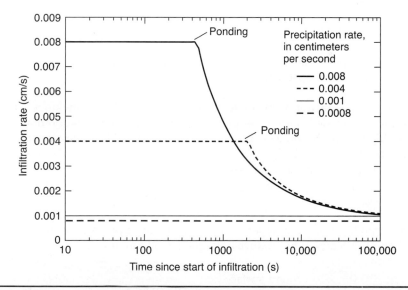

FIGURE 3.1 Infiltration in a one-dimensional soil column for four different rainfall intensities and a soil with saturated hydraulic conductivity of about 0.001 cm/s. (From Healy et al., 2007.)

and peak discharge should ideally be based on long-term stationary streamflow records for the area. Such records are seldom available for small drainage areas. Even where they are available, accurate statistical analysis of them is often impossible because of the conversion of land to urban uses during the period of record. It, therefore, is necessary to estimate peak discharges with hydrologic models based on measurable watershed characteristics (USDA, 1986).

In TR-55, runoff is determined primarily by amount of precipitation and by infiltration characteristics related to soil type, soil moisture, antecedent rainfall, cover type, impervious surfaces, and surface retention. Travel time is determined primarily by slope, length of flow path, depth of flow, and roughness of flow surfaces. Peak discharges are based on the relationship of these parameters and on the total drainage area of the watershed, the location of the development, the effect of any flood control works or other natural or manmade storage, and the time distribution of rainfall during a given storm event.

The model described in TR-55 begins with a rainfall amount uniformly imposed on the watershed over a specified time distribution. Mass rainfall is converted to mass runoff by using a runoff CN. Selection of the appropriate CN depends on soil type, plant cover, amount of impervious areas, interception, and surface storage. Runoff is then transformed into a hydrograph by using unit hydrograph theory and routing procedures that depend on runoff travel time through segments of the watershed (USDA, 1986). As pointed out by the authors, to save time, the procedures in TR-55 are simplified by assumptions about some parameters. These simplifications, however, limit the use of the procedures and can provide results that are less accurate than more detailed methods. The user should examine the sensitivity of the analysis being conducted to a variation of the peak discharge or hydrograph.

The SCS runoff equation is

$$Q = \frac{(P - I_a)^2}{(P - I_a) + S} \tag{3.1}$$

where Q = runoff (in)
$\quad P$ = rainfall (in)
$\quad S$ = potential maximum retention after runoff begins (in)
$\quad I_a$ = initial abstraction (in)

Initial abstraction is all losses before runoff begins. It includes water retained in surface depressions and water intercepted by vegetation, evaporation, and infiltration. I_a is highly variable but generally is correlated with soil and cover parameters. Through studies of many small agricultural watersheds, I_a was found to be approximated by the following empirical equation:

$$I_a = 0.2S \tag{3.2}$$

By removing I_a as an independent parameter, this approximation allows use of a combination of S and P to produce a unique runoff amount. Substituting Eq. (3.2) into Eq. (3.1) gives

$$Q = \frac{(P - 0.2S)^2}{(P + 0.8S)} \tag{3.3}$$

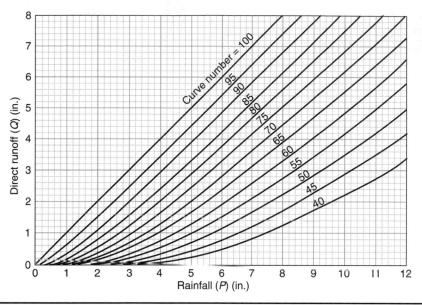

FIGURE 3.2 Solution of runoff equation. Curves are for condition $I_a = 0.2S$ and Eq. (3.3). (From USDA, 1986.)

Water retention S, which includes infiltration, interception by vegetation, ET, and storage in surface depressions, is calculated using the CN approach, taking into account antecedent soil moisture, soil permeability, and land cover. It is related to CN through the equation:

$$S = \frac{1000}{CN} - 10 \qquad (3.4)$$

CNs for watershed range from 0 to 100, with 100 being perfectly impervious. The graph in Fig. 3.2 solves Eqs. (3.2) and (3.4) for a range of CNs and rainfall events. Table 3.1 is an example of CNs for several types of agricultural land cover that can be selected based on available information. TR-55 also includes tables for various other land covers. Soils are classified into four hydrologic soil groups (A, B, C, and D) according to their minimum infiltration rate, which is obtained for bare soil after prolonged wetting. Appendix A in TR-55 defines the four groups and provides a list of most of the soils in the United States and their group classification. The hydrologic characteristics of the four groups are as follows (Rawls et al., 1993):

- Group A soils have low runoff potential and high infiltration rates even when thoroughly wetted. They consist mainly of deep, well to excessively drained sands or gravels. The USDA soil textures normally included in this group are sand, loamy sand, and sandy loam. These soils have infiltration rate greater than 0.76 cm/h.

- Group B soils have moderate infiltration rates when thoroughly wetted and consist mainly of moderately deep to deep and moderately well-drained to

Cover Type	Hydrologic Condition	Curve Numbers for Hydrologic Soil Group			
		A	B	C	D
Pasture, grassland, or range-continuous forage for grazing[1]	Poor	68	79	86	89
	Fair	49	69	79	84
	Good	39	61	74	80
Meadow—continous grass, protected from grazing, generally moved for hay	–	30	58	71	78
Brush—brush-weed-grass mixture with brush the major element[2]	Poor	48	67	77	83
	Fair	35	56	70	77
	Good	30[3]	48	65	73
Woods-grass combination (orchard or tree farm)[4]	Poor	57	73	82	86
	Fair	43	65	76	82
	Good	32	58	72	79
Woods[5]	Poor	45	66	77	83
	Fair	36	60	73	79
	Good	30[3]	55	70	77

[1] *Poor*: <50% ground cover or heavily gazed with no mulch; *fair*: 50–75% ground cover and not heavily gazed; *good*: >75% ground cover and lightly or only occasionaly grazed.

[2] *Poor*: <50% ground cover; *fair*: 50–75% ground cover; *good*: >75% ground cover.

[3] Actual curve number is less than 30; use CN = 30 for runoff computations.

[4] Computed for areas with 50% woods and 50% grass (pasture) cover.

[5] *Poor*: Forest litter, small trees, and brush are destroyed by heavy grazing or regular burning; *fair*: woods are grazed but not burned, and some forest litter covers the soil; *good*: woods are protected from grazing, and litter and brush adequately cover the soil.

From USDA, 1986.

TABLE 3.1   Runoff Curve Numbers for Selected Agricultural Lands, for Average Runoff Condition and $I_a = 0.2S$.

well-drained soils having moderately fine to moderately coarse textures. The USDA soil textures normally included in this group are silt loam and loam. These soils have an infiltration rate between 0.38 and 0.76 cm/h.

- Group C soils have low infiltration rates when thoroughly wetted and consist mainly of soils with a layer that impedes downward movement of water and soils with moderately fine to fine texture. The USDA soil texture normally included in this group is sandy clay loam. These soils have an infiltration rate between 0.13 and 0.38 cm/h.

- Group D soils have high runoff potential. They have very low infiltration rates when thoroughly wetted and consist mainly of clay soils with a high swelling potential, soils with a permanent high water table, soils with a claypan or clay layer at or near the surface, and shallow soils over a nearly impervious material. The USDA soil textures normally included in this group are clay loam, silty clay loam, sandy clay, silty clay, and clay. These soils have a very low rate of infiltration (0.0 to 0.13 cm/h). Some soils are classified in group D because of a high water table that creates a drainage problem; however, once these soils are effectively drained, they are placed into another group.

Watersheds with higher CNs generate more runoff and less infiltration. Examples include watersheds with high proportion of paved, impervious surfaces. Dense forestland and grasslands have the lowest CNs and retain high proportions of rainfall. However, it is important to understand that most urban areas are only partially covered by impervious surfaces; the soil remains an important factor in runoff estimates. Urbanization has a greater effect on runoff in watersheds with soils having high infiltration rates (sands and gravels) than in watersheds predominantly of silts and clays, which generally have low infiltration rates (USDA, 1986).

TR-55 includes tables and graphs for selection of all quantitative parameters needed to select CNs and calculate runoff for thousands of soil types and land covers in the United States. Representative land covers include bare land, pasture, western desert urban area, and woods to name a few. The soils in the area of interest may be identified from a soil survey report, which can be obtained from local SCS offices or soil and water conservation district offices.

While the SCS method enables calculation of runoff, it does not provide for exact estimation of infiltration, which is only one of the calculated overall water retention components. The calculated volume of water retained by the watershed includes terms for ET and interception by vegetation. Knowledge of the vegetative conditions of the watershed in question will help determine the distribution of rainfall retention. Vegetative interception will be more significant for a forested area than for an open field. One must also remember that runoff-producing storm events allow infiltration rates to asymptotically approach the saturated hydraulic conductivity of surficial soils. Knowledge of the physical properties of watershed soils is, therefore, necessary for estimation of infiltration rates.

3.3 Evapotranspiration

ET is often the second largest component of the water budget, next to precipitation. Approximately 65 percent of all precipitation falling on landmass returns to the atmosphere through ET, which can be defined as the rate of liquid water transformation to vapor from open water, bare soil, or vegetation with soil beneath (Shuttleworth, 1993). Transpiration is defined as the fraction of total ET that enters the atmosphere from the soil through the plants. The rate of ET, expressed in inches per day or millimeters per day, has traditionally been estimated using meteorological data from climate stations located at particular points within a region and parameters describing transpiration by certain types of vegetation (crop). There are two standard rates used as estimates of ET: *potential evaporation* and *reference crop evaporation*.

Potential evaporation E_0 is the quantity of water evaporated per unit area, per unit time from an idealized, extensive free water surface under existing atmospheric conditions. This is a conceptual entity that measures the meteorological control on evaporation from an open water surface. E_0 is commonly estimated from direct measurements of evaporation with evaporation pans (Shuttleworth, 1993). Note that E_0 is also called potential evapotranspiration (PET), even though as defined it does not involve plant activity.

Reference crop evapotranspiration E_c is the rate of evaporation from an idealized grass crop with a fixed crop height of 0.12 m, an albedo of 0.23, and a surface resistance of 69 s·m^{-1} (Shuttleworth, 1993). This crop is represented by an extensive surface of short green grass cover of uniform height, actively growing, completely shading the ground, and not short of water. When estimating actual ET from a vegetated surface it is common practice to first estimate E_c and then multiply this rate by an additional complex factor called the crop coefficient K_c.

There are many proposed and often rather complex empirical equations for estimating E_0 and E_c, using various parameters such as air temperature, solar radiation, radiation exchange for free water surface, hours of sunshine, wind speed, vapor pressure deficit (VPD), relative humidity, and aerodynamic roughness (for example, see Singh, 1993; Shuttleworth, 1993; Dingman, 1994). The main problem in applying empirical equations to a very complex physical process such as ET is that in most cases such equations produce very different results for the same set of input parameters. As pointed out by Brown (2000), even in cases of the most widely used group of equations referred to as "modified Penman equations," the results may vary significantly (Penman proposed his equation in 1948, and it has been modified by various authors ever since).

The PET from an area can be estimated from the free water evaporation assuming that the supply of water to the plant is not limited. Actual evapotranspiration E_{act} equals the potential value, E_0 as limited by the available moisture (Thornthwaite, 1946). On a natural watershed with many vegetal species, it is reasonable to assume that ET rates do vary with soil moisture since shallow-rooted species will cease to transpire before deeper-rooted species (Linsley and Franzini, 1979). A moisture-accounting procedure can be established by using the continuity equation:

$$P - R - G_0 - E_{act} = \Delta M \tag{3.5}$$

where P = precipitation
        R = surface runoff
        G_0 = subsurface outflow
        E_{act} = actual ET
        ΔM = the change in moisture storage

E_{act} is estimated as

$$E_{act} = E_0 \frac{M_{act}}{M_{max}} \tag{3.6}$$

where M_{act} = computed soil moisture stage on any date and M_{max} = assumed maximum soil moisture content (Kohler, 1958, from Linsley and Franzini, 1979).

In addition to available soil moisture, plant type is another important component of actual ET, as certain species require more water than others. A Colorado State University study in eastern Colorado examined the water requirements of different crops in 12 agricultural areas (Broner and Schneekloth, 2007). On average, corn required 24.6 in/season, sorghum required 20.5 in/season, and winter wheat required 17.5 in/season. Allen et al. (1998) present detailed guidelines for computing crop water requirements together with representative values for various crops. Native vegetation to semiarid and arid environments is more adept at surviving in low-moisture soils and generates much lower E_{act} values than introduced species. E_{act} increases with increasing plant size and canopy densities. With regard to stage of development, actively growing plants will transpire at a much greater rate than dormant plants (Brown, 2000). Wind augments PET by actively transporting heat from the air to vegetation and by facilitating the transfer of water vapor from vegetation to the atmosphere. Humidity and temperature together determine the

VPD, a measure of the "drying power" of the atmosphere. The VPD quantifies the gradient in water vapor concentration between vegetation and the atmosphere and increases with increasing temperature and decreasing humidity (Brown, 2000). A less publicized factor influencing open soil evaporation is soil type. Open soil evaporation occurs in two phases: a rapid phase involving capillary conduction followed by a long-term, energy-intensive phase involving vapor diffusion. Initial evaporation rates from coarse-grained soils are therefore very high, as these soils have high conductivities. However, over time, fine-grained soils with high porosities yield greater evaporation quantities because of greater long-term water retention (Wythers et al., 1999). In other words, the conductive phase lasts much longer for fine-grained soils than coarse-grained soils. Yet for either soil type, extended dry periods will lead to desiccation of soils through vapor diffusion.

Figure 3.3 shows regional PET patterns across the continental United States, demonstrating the significance of the above contributing factors. The highest PET values are found in areas of high temperatures and low humidity, such as the deserts of southeastern California, southwestern Arizona, and southern Texas (Healy et al., 2007). Mountainous regions exhibit low PET values because of colder temperatures and the prevalence of moist air. It is interesting that southern Florida has remarkably high PET rates rivaling those in desert environments. This may be attributable to the dense vegetative cover of the subtropical Florida landscape, or the more seasonal trends in annual precipitation (i.e., more defined rainy and dry seasons).

A very detailed study of ET rates by vegetation in the spring-fed riparian areas of Death Valley, United States, was performed by the U.S. Geological Survey (USGS). The study was initiated to better quantify the amount of groundwater being discharged annually from these sensitive areas and to establish a basis for estimating water rights

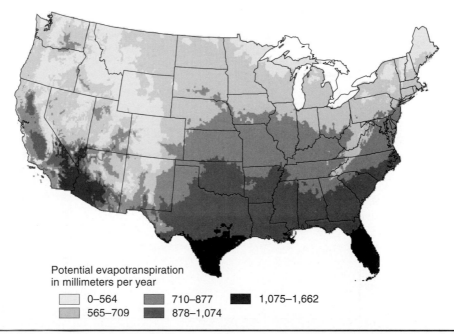

Potential evapotranspiration
in millimeters per year

☐ 0–564	▨ 710–877	■ 1,075–1,662
▨ 565–709	▨ 878–1,074	

Figure 3.3 Potential evapotranspiration map of the continental United States. (From Healy et al., 2007.)

and assessing future changes in groundwater discharge in the park (Laczniak et al., 2006). ET was estimated volumetrically as the product of ET-unit (general vegetation type) acreage and a representative ET rate. ET-unit acreage was determined from high-resolution multispectral imagery. A representative ET rate was computed from data collected in the Grapevine Springs area using the Bowen ratio solution to the energy budget or from rates given in other ET studies in the Death Valley area. The groundwater component of ET was computed by removing the local precipitation component from the ET rate.

Figure 3.4 shows instrumentation used to collect data for the ET computations at the Grapevine Springs site. The instruments included paired temperature and humidity probes, multiple soil heat-flux plates, multiple soil temperature and moisture probes, a net radiometer, and bulk rain gauge. In addition, a pressure sensor was set in a nearby

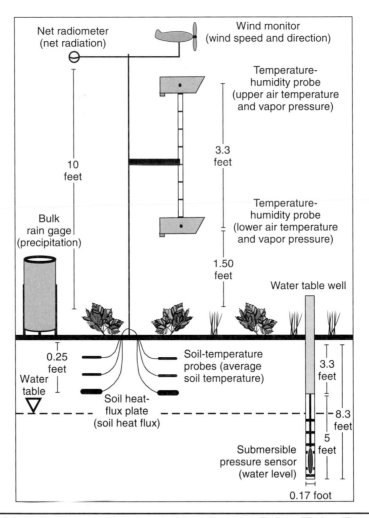

Figure 3.4   Schematic diagram of instrumentation at Grapevine Springs ET site. (From Laczniak et al., 2006.)

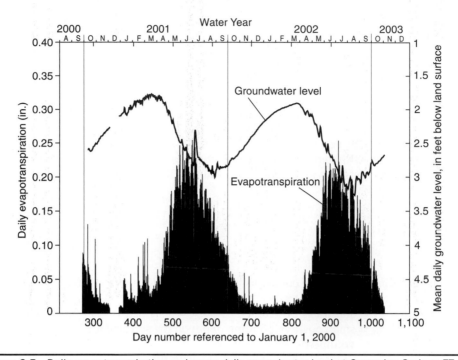

Figure 3.5 Daily evapotranspiration and mean daily groundwater level at Grapevine Springs ET site, from September 28, 2000, to November 3, 2002 (day numbers 272 and 1038, respectively). (From Laczniak et al., 2006.)

shallow well to acquire information on the daily and annual water table fluctuation. Micrometeorologic data were collected at 20-min intervals and water levels were collected at hourly intervals (Laczniak et al., 2006).

The results of the study show that ET at the Grapevine Springs site generally begins increasing in late spring and peaks in the early through midsummer period (June and July). During this peak period, daily ET ranged from about 0.18 to 0.25 in (Fig. 3.5) and monthly ET ranged from about 5.7 to 6.2 in. ET totaled about 2.7 ft in water year 2001 and about 2.3 ft in water year 2002. The difference in precipitation between the two water years is nearly equivalent to the difference in annual ET. Annual trends in daily ET show an inverse relation with water levels—as ET begins increasing in April, water levels begin declining, and as ET begins decreasing in September, water levels begin rising. The slightly greater ET and higher water levels in water year 2001, compared with water year 2002, are assumed to be a response to greater precipitation.

The groundwater component of ET at the Grapevine Springs ET site ranged from 2.1 to 2.3 ft, with the mean annual groundwater ET from high-density vegetation being 2.2 ft (Laczniak et al., 2006).

3.4 Infiltration and Water Movement Through Vadose Zone

The most significant factors affecting infiltration are the physical characteristics and properties of soil layers. The rate at which water enters soil cannot exceed the rate at which water is transmitted downward through the soil. Thus, soil surface conditions alone cannot increase infiltration unless the transmission capacity of the soil profile is adequate. Under conditions where the surface entry rate (rainfall intensity) is slower than the transmission rate of the soil profile, the infiltration rate will be limited by the rainfall intensity (water supply). Until the top soil horizon is saturated (i.e., the soil moisture deficit is satisfied), the infiltration rate will be constant, as shown in Fig. 3.1. For higher rainfall intensities, all the rain will infiltrate into the soil initially until the soil surface becomes saturated ($\theta = \theta_s$, $h \geq 0$, $z = 0$), that is, until the so-called *ponding time* (t_p) is reached. At that point, the infiltration is less than the rainfall intensity (approximately equal to the saturated hydraulic conductivity of the media) and surface runoff begins. These two conditions are expressed as follows (Rawls et al., 1993):

$$-K(h)\frac{\partial h}{\partial z} + 1 = R \quad \theta(0, t) \leq \theta_s \quad t \leq t_p \tag{3.7}$$

$$h = h_0 \quad \theta(0, t) = \theta_s \quad t \geq t_p \tag{3.8}$$

where $K(h)$ = hydraulic conductivity for given soil water potential (degree of saturation)
h = soil water potential
h_0 = small positive ponding depth on the soil surface
θ = volumetric water content
θ_s = volumetric water content at saturation
z = depth from land surface
t_p = time from the beginning of rainfall until ponding starts (ponding time)
R = rainfall intensity

The transmission rates can vary at different horizons in the unsaturated soil profile. After saturation of the uppermost horizon, the infiltration rate is limited to the lowest transmission rate encountered by the infiltrating water as it travels downward through the soil profile. As water infiltrates through successive soil horizons and fills in the pore space, the available storage capacity of the soil will decrease. The storage capacity available in any horizon is a function of the porosity, horizon thickness, and the amount of moisture already present. Total porosity and the size and arrangement of the pores have significant effect on the availability of storage. During the early stage of a storm, the infiltration process will be largely affected by the continuity, size, and volume of the larger-than-capillary ("noncapillary") pores, because such pores provide relatively little resistance to the infiltrating water. If the infiltration rate is controlled by the transmission rate through a retardant layer of the soil profile, then the infiltration rate, as the storm progresses, will decrease as a function of the decreasing storage availability above the restrictive layer. The infiltration rate will then equal the transmission rate through this restrictive layer until another, more restrictive, layer is encountered by the water (King, 1992). The soil infiltration capacity decreases in time and eventually asymptotically reaches the value of overall saturated hydraulic conductivity K_s of the affected soil column as illustrated in Fig. 3.1.

In general, the infiltration rate decreases with increasing clay content in the soil and increases with increasing noncapillary porosity through which water can flow freely under the influence of gravity. Presence of certain clays such as montmorillonite, even in relatively small quantities, may dramatically reduce infiltration rate as they become wet and swell. Runoff conditions on soils of low permeability develop much sooner and more often than on uniform, coarse sands and gravels, which have infiltration rates higher than most rainfall intensities.

The soil surface can become encrusted with, or sealed by, the accumulation of fines or other arrangements of particles that prevent or retard the entry of water into the soil. As rainfall starts, the fines accumulated on bare soil may coagulate and strengthen the crust, or enter soil pores and effectively seal them off. A soil may have excellent subsurface drainage characteristics but still have a low infiltration rate because of the retardant effect of surface crusting or sealing (King, 1992).

3.4.1 Soil Water Retention and Hydraulic Conductivity

Similar to groundwater flow in the saturated zone, the flow of water in the unsaturated zone is governed by two main parameters—the change in total potential (hydraulic head) along the flow path between the land surface and the water table and the hydraulic conductivity of the soil media. However, both parameters depend on the volumetric water content in the porous medium; they change in time and space as the soil becomes more or less saturated in response to water input and output such as infiltration and ET.

Air that fills pores unoccupied by water in the vadose zone exerts an upward suction effect on water caused by capillary and adhesive forces. This suction pressure of the soil, also called *matric potential*, is lower than the atmospheric pressure. It is the main characteristic of the vadose zone governing water movement from the land surface to the water table. The water-retention characteristic of the soil describes the soil's ability to store and release water and is defined as the relationship between the soil water content and the matric potential. The water-retention characteristic is also known as *moisture characteristic curve*, which when plotted on a graph shows water content at various depths below ground surface versus matric potential (Fig. 3.6). Again, the matric potential is always negative above the water table, since the suction pressure is lower than the atmospheric pressure. The matric potential is a function of both the water content and the sediment texture: it is stronger in less saturated and finer soils. As the soil saturation increases, the matric potential becomes "less negative." At the water table, the matric potential is zero and equals the atmospheric pressure. Other terms that are synonymous with matric potential but may differ in sign or units are soil water suction, capillary potential, capillary pressure head, suction head, matric pressure head, tension, and pressure potential (Rawls et al., 1993).

As the moisture content increases, suction pressure decreases, causing a corresponding increase in hydraulic conductivity. As a result, soils with high antecedent moisture contents will support a greater long-term drainage rate than dry soils. At saturation (i.e., at and immediately above the water table in the capillary fringe), the hydraulic conductivity of the soil media is equal to the saturated hydraulic conductivity. Figure 3.7 shows unsaturated hydraulic conductivity as the function of matric potential for the same two soil types shown in Fig. 3.6.

3.4.2 Darcy's Law

Flow in the vadose zone can be best understood through interpretation of Darcy's law for unsaturated flow in one dimension. The governing equation (known as the

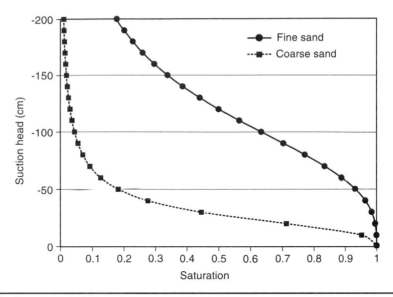

FIGURE 3.6 Two characteristic moisture curves for fine and coarse sand.

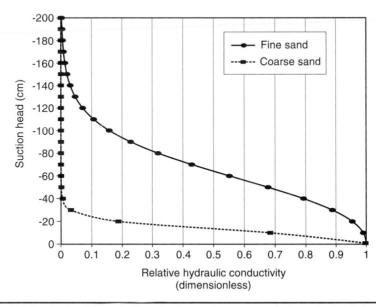

FIGURE 3.7 Relative hydraulic conductivity for fine and coarse sand shown in Fig. 3.6.

Darcy-Buckingham equation) is

$$q = -K(\Psi)\frac{\partial H}{\partial z} = -K(\theta)\frac{\partial}{\partial z}(\Psi + z) \qquad (3.9)$$

where q = flow per unit width
$K(\Psi)$ = unsaturated hydraulic conductivity given as function of matric potential
θ = soil moisture content
H = total water potential
z = elevation potential
Ψ = matric potential or pressure head

The matric and elevation potentials (heads) combine to form the total water potential (head). Pressure head is negative in the unsaturated zone (causing the "suction" effect), zero at the water table, and positive in the saturated zone. As in the saturated zone, water in the vadose zone moves from large to small water potentials (heads).

For deep vadose zones common to the American West, soil hydraulic properties can be used to simply calculate recharge flux using Darcy's law. The underlying assumption of the "Darcian method" is that moisture content becomes constant at some depth, such that there is no variation in matric potential or pressure head h with depth z and that all drainage is due to gravity alone (see Fig. 3.8). Under these conditions, the deep drainage rate is approximately equal to the measured unsaturated hydraulic conductivity (Nimmo et al., 2002). The equations describing this relationship are the following:

$$q = -K(\Psi)\left[\frac{d\Psi}{dz} + 1\right]$$
$$\frac{d\Psi}{dz} \approx 0 \qquad (3.10)$$
$$q \approx -K(\Psi)$$

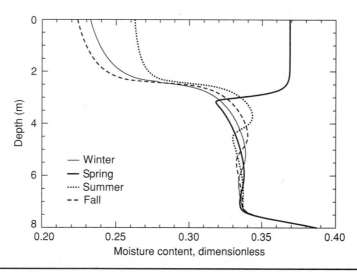

FIGURE 3.8 Hypothetical moisture-content profiles at four different times of the year. As depth increases, the variation in moisture content decreases. (From Healy et al., 2007.)

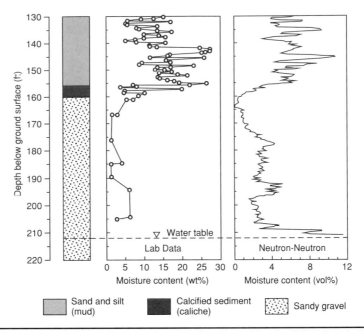

Figure 3.9   Lithology and moisture distribution as a function of depth within a borehole in a deep vadose zone. (Modified from Serne et al., 2002.)

Alternatively, knowledge of matric potential (pressure head) at different depths enables direct calculation of recharge flux from the above equation. Velocity of the recharge front can be calculated by dividing the flux by soil moisture content.

Darcy's law illustrates the complexity of unsaturated flow, as both hydraulic conductivity and pressure head are functions of the soil moisture content (θ). Another crucial factor influencing drainage rates is vadose zone lithology. Deep vadose zones often have a high degree of heterogeneity, created by distinct depositional periods. This limits the application of Eq. (3.10). For example, layers of fine-grained clays and silts are common to deep basins of the American West. As a result, moisture percolating through the vadose zone often encounters low permeability strata, including calcified sediment layers as shown in Fig. 3.9. Presence of low-permeable intervals in the vadose zone greatly delays downward migration and may cause significant lateral spreading. Therefore, the time lag for groundwater recharge may be on the order of hundreds of years or more for deep vadose zones with fine-grained sediments. Furthermore, moisture is lost to lateral spreading and storage. Detailed lithologic characterization of the vadose zone is essential in quantification of groundwater recharge as well as in planning and designing artificial recharge systems. Knowledge of site-specific geology is also helpful in qualitatively understanding the time scale of vadose zone processes.

Flow through the vadose zone is comparatively less significant in humid environments with shallow water tables where travel times from the land surface to the water table are on the order of years or less.

3.4.3 Equations of Richards, Brooks and Corey, and van Genuchten

Water flow in variably saturated soils is traditionally described with the Richards equation (Richards, 1931) as follows (van Genuchten et al., 1991):

$$C\frac{\partial h}{\partial t} = \frac{\partial}{\partial z}\left(K\frac{\partial h}{\partial z} - K\right)$$ (3.11)

where h = soil water pressure head or matric potential (with dimension L)
   t = time (T)
   z = soil depth (L)
   K = hydraulic conductivity (LT^{-1})
   C = soil water capacity (L^{-1}) approximated by the slope $d\theta/dh$ of the soil water retention curve $\theta(h)$, in which θ is the volumetric water content (L^3L^{-1}).

Equation (3.11) may also be expressed in terms of the water content if the soil profile is homogeneous and unsaturated $(h \leq 0)$:

$$\frac{\partial \theta}{\partial t} = \frac{\partial}{\partial z}\left(D\frac{\partial \theta}{\partial z} - K\right)$$ (3.12)

where D = soil diffusivity (L^2T^{-1}) defined as

$$D = K\frac{dh}{d\theta}$$ (3.13)

The unsaturated soil hydraulic functions in the above equations are the soil water retention curve, $\theta(h)$, the hydraulic conductivity function, $K(h)$ or $K(\theta)$, and the soil water diffusivity function $D(\theta)$. Several functions have been proposed to empirically describe the soil water retention curve. One of the most widely used is the equation of Brooks and Corey (van Genuchten et al., 1991; Šimnek et al., 1999):

$$\theta = \begin{cases} \theta_r + (\theta_s - \theta_r)(\alpha h)^{-\lambda} & (h < -1/\alpha) \\ \theta_s & (h \geq -1/\alpha) \end{cases}$$ (3.14)

where θ = volumetric water content
   θ_r = residual water content
   θ_s = saturated water content
   α = an empirical parameter (L^{-1}) whose inverse $(1/\alpha)$ is often referred to as the air entry value or bubbling pressure
   α = a negative value for unsaturated soils
   λ = a pore-size distribution parameter affecting the slope of the retention function
   h = the soil water pressure head, which has negative values for unsaturated soil.

Equation (3.14) may be written in a dimensionless form as follows:

$$S_e = \begin{cases} (\alpha h)^{-\lambda} & (h < -1/\alpha) \\ 1 & (h \geq -1/\alpha) \end{cases} \tag{3.15}$$

where S_e = effective degree of saturation, also called the reduced water content ($0 < S_e < 1$):

$$S_e = \frac{\theta - \theta_r}{\theta_s - \theta_r} \tag{3.16}$$

The residual water content θ_r in Eq. (3.16) specifies the maximum amount of water in a soil that will not contribute to liquid flow because of blockage from the flow paths or strong adsorption onto the solid phase (Luckner et al., 1989, from van Genuchten et al., 1991). Formally, θ_r may be defined as the water content at which both $d\theta/dh$ and K reach zero when h becomes large. The residual water content is an extrapolated parameter and may not necessarily represent the smallest possible water content in a soil. This is especially true for arid regions where vapor phase transport may dry out soils to water contents well below θ_r. The saturated water content θ_s denotes the maximum volumetric water content of a soil. The saturated water content should not be equated to the porosity of soils; θ_s of field soils is generally about 5 to 10 percent smaller than the porosity because of entrapped or dissolved air (van Genuchten et al., 1991).

The Brooks and Corey equation has been shown to produce relatively accurate results for many coarse-textured soils characterized by relatively uniform pore- or particle-size distributions. Results have generally been less accurate for many fine-textured and undisturbed field soils because of the absence of a well-defined air-entry value for these soils. A continuously differentiable (smooth) equation proposed by van Genuchten (1980) significantly improves the description of soil water retention:

$$S_e = \frac{1}{[1 + (\alpha h)^n]^m} \tag{3.17}$$

where α, n, and m are empirical constants affecting the shape of the retention curve ($m = 1 - 1/n$). By varying the three constants, it is possible to fit almost any measured field curve. It is this flexibility that made the van Genuchten equation arguably the most widely used in various computer models of unsaturated flow and contaminant fate and transport. Combining Eqs. (3.16) and (3.17) gives the following form of the van Genuchten equation:

$$\theta(h) = \theta_r + \frac{\theta_s - \theta_r}{[1 + (\alpha h)^n]^{1-\frac{1}{n}}} \tag{3.18}$$

One of the widely used models for predicting the unsaturated hydraulic conductivity from the soil retention profile is the model of Mualem (1976), which may be written in the following form (van Genuchten et al., 1991):

$$K(S_e) = K_s S_e \left[\frac{f(S_e)}{f(l)} \right]^2 \tag{3.19}$$

$$f(S_e) = \int_0^{S_e} \frac{1}{h(x)} dx \tag{3.20}$$

where S_e is given by Eq. (3.16) K_s = hydraulic conductivity at saturation, and l = an empirical pore-connectivity (tortuosity) parameter estimated by Mualem to be about 0.5 as an average for many soils.

Detailed solution of the Mualem's model by incorporating Eq. (3.17) is given by van Genuchten et al. (1991). This solution, sometimes called van Genuchten-Mualem equation, has the following form:

$$K(S_e) = K_o S_e^l \left\{ 1 - \left[1 - S_e^{n/(n-1)} \right]^{1-\frac{1}{n}} \right\}^2 \tag{3.21}$$

where K_o is the matching point at saturation, a parameter often similar, but not necessarily equal, to the saturated hydraulic conductivity (K_s). Fitting the van Genuchten-Mualem equation to soil data gives good results in most cases. It is important to note that the curve-fit parameters and those representing matching-point saturation and tortuosity tend to lose physical significance when fit to laboratory data. Hence, one must remember that these terms are best described as mathematical constants rather than physical properties of the soils in question.

Rosetta (Schaap, 1999) and RETC (van Genuchten et al., 1991) are two very useful public domain programs developed at the U.S. Salinity Laboratory for estimating unsaturated zone hydraulic parameters required by the Richards and van Genuchten equations. The programs provide models of varying complexity, starting with simple ones such as percentages of sand, silt, and clay in the soil and ending with rather complex options where laboratory data are used to fit unknown hydraulic coefficients.

3.5 Factors Influencing Groundwater Recharge

3.5.1 Climate

The simplest classifications of climate are based on annual precipitation. Strahler and Strahler (1978, p. 129, from Bedinger, 1987) refer to arid climates as having 0 to 250 mm/year precipitation, semiarid as having 250 to 500 mm/year precipitation, and sub-humid as having 500 to 1000 mm/year precipitation. The soil water balance of Thornthwaite (1948) provides a classification of climate by calculating the annual cycle of soil moisture availability or deficiency, thus providing measures of soil moisture available for plant growth. Strahler and Strahler (1978) discuss the worldwide distribution of 13 climatic types based on the average annual variations of precipitation, PET, and the consequent soil moisture deficit or surplus. Soil water balance models, which are very similar to the soil moisture model for classifying worldwide climate, have been developed to estimate recharge. These models utilize more specific data, such as soil type and moisture-holding capacity, vegetation type and density, surface-runoff characteristics, and spatial and temporal variations in precipitation. Various soil water balance models used in estimating recharge in arid and semiarid regions of the world are discussed by Bedinger (1987), who includes an annotated bibliography of 29 references. The estimated recharge rates have wide scatter, which is attributed to differences in applied methods, real differences in rates of infiltration to various depths and net recharge, and varying characteristics of soil types, vegetation, precipitation, and climatic regime.

Based on a detailed study of a wide area in the mid-continental United States spanning six states, Dugan and Peckenpaugh (1985) concluded that both the magnitude and the proportion of potential recharge from precipitation decline as the total precipitation declines (see Fig. 3.10), although other factors including climatic conditions, vegetation, and soil type also affect potential recharge. The limited scatter among the points in Fig. 3.10 indicates a close relationship between precipitation and recharge. Furthermore, the relationship becomes approximately linear where mean annual precipitation exceeds 30 in and recharge exceeds 3 in. Presumably, when the precipitation and recharge are less than these values, disproportionably more infiltrating water is spent on satisfying the moisture deficit in dry soils. The extremely low recharge in the western part of the study area, particularly Colorado and New Mexico, appears to be closely related to the high PET found in these regions. Seasonal distribution of precipitation also shows a strong relationship to recharge. Areas of high cool-season precipitation tend to receive higher amounts of recharge. Where PET is low and long winters prevail, particularly in the Nebraska and South Dakota parts of the study area, effectiveness of cool-season precipitation as a source of recharge increases. Overall, however, when cool-season precipitation is less than 5 in, recharge is minimal. Dugan and Peckenpaugh (1985) conclude that generalized patterns of potential recharge are determined mainly by climatic conditions.

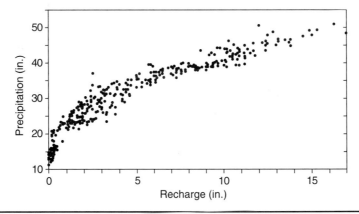

Figure 3.10 Computed mean annual recharge using soil moisture program versus mean annual precipitation, by model grid element, in mid-continental United States. (Modified from Dugan and Peckenpaugh, 1985.)

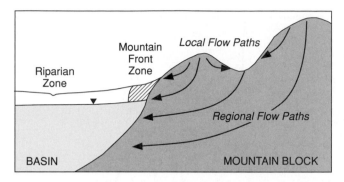

FIGURE 3.11 Schematic of mountain block contribution to recharge of a groundwater system. (From Wilson and Guan, 2004; copyright American Geophysical Union.)

Smaller variations within local areas, however, are related to differences in land cover, soil types, and topography.

A large portion of groundwater recharge is derived from winter snowpack, which provides a slow, steady source of infiltration. Topographically, higher elevations typically receive more precipitation, including snow, than valleys or basins. Coupled with low PET, they produce ideal recharge conditions. This fact is especially important in areas such as the Basin and Range province of the United States where the so-called mountain block recharge (MBR) is critical component of a groundwater system water balance (Fig. 3.11; Wilson and Guan, 2004). High-elevation recharge produces the confined aquifer conditions that millions of people rely on for potable water. Snowmelt recharge in mountain ranges occurs in a cyclical pattern, described as follows (Flint and Flint, 2006):

- During daytime, snowmelt infiltrates thin surface soils and migrates down to the soil-bedrock interface.

- The soil-bedrock interface becomes saturated, and once the infiltration rate exceeds the bulk permeability of the bedrock matrix, moisture enters the bedrock fracture system.

- At night, snow at the ground surface refreezes while stored moisture in surface soils drains into the bedrock.

This wetting-drying cycle of snowmelt recharge minimizes surface water runoff and promotes infiltration. Rapid snowmelt produces surface runoff, which also significantly contributes to basin groundwater recharge along the mountain front.

Snowmelt can provide critical recharge for low-lying areas where PET is high. The snowpack in such areas is usually smaller than that at higher elevations, but often snowmelt still provides the vast majority of recharge in a calendar year. Thanks to the presence of a major Department of Energy site, many studies have been conducted in Hanford, Washington, to quantify recharge on the Columbia Plateau. Lysimeters were installed below numerous sites of differing vegetative cover, including sand dunes, grasslands, riparian areas, and agricultural fields. The results revealed that snowmelt plays an integral role in groundwater recharge on the plateau, as shown in Figs. 3.12 and 3.13. During winter months at Hanford, temperatures cool enough to lower PET rates.

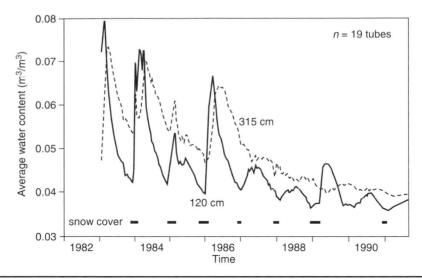

Figure 3.12 Soil water content at two depths and snow cover in Hanford, WA. (Modified from Fayer et al., 1995.)

When temperatures rise above freezing, snowmelt occurs, resulting in infiltration. Rapid snowmelt may lead to ponding of water where slopes are insignificant, creating prolonged infiltration periods (Fayer and Walters, 1995). Figure 3.12 also illustrates how years of limited snow cover produce much less recharge than years with extended durations of snow cover, a very important fact when considering the potential impact of climate change on snowpack and groundwater recharge.

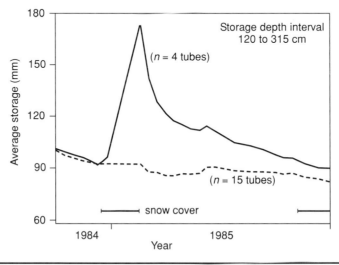

Figure 3.13 Water storage variations for two groups of lysimeters during snowmelt at Hanford Site. The two groups were installed in different soils illustrating the importance of soil type on recharge rate. (Modified from Fayer et al., 1995.)

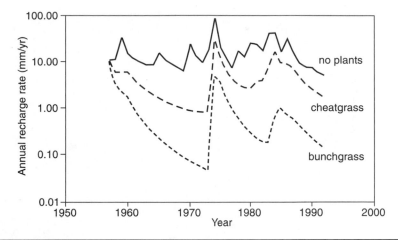

FIGURE 3.14 Simulated recharge rates for Ephrata sandy loam and three different vegetation covers, Hanford Site, WA. (From Fayer and Walters, 1995.)

Variations of precipitation and air temperature are also very important for recharge. In some years it may never rain (or snow) enough to cause any recharge in semiarid climates with high PET. Even that portion of rainfall that infiltrates into the shallow subsurface may be evapotranspired back to the atmosphere before it percolates deeper than the critical depth of ET. Higher air temperatures cause more ET from the shallow subsurface and a higher soil moisture deficit (drier soil).

The impact of vegetative cover on recharge also depends on climatic variations. During drier and warmer years, plants will uptake a higher percentage of the infiltrated water than during average years, and this will also vary for different plant species. What all this means is that the relationship between precipitation and actual groundwater recharge is not linear; simply adopting values of certain quantitative parameters measured during 1 or 2 years, as representative of the long-term groundwater recharge would be erroneous. Figure 3.14 illustrates some of the above points. Recharge rates for a 30-year period were simulated for different vegetation covers based on extensive site-specific, multiyear field analyses of various soil properties, vegetation root density, root water uptake, and infiltration rates (Fayer and Walters, 1995). The starting point for all three simulations was the same soil water content in 1957. The model-simulated recharge rates show two orders of magnitude difference in recharge for soil with bunchgrass, compared to one order of magnitude difference for nonvegetated soil. At the same time, there is much less overall variation in recharge for the vegetated than for the nonvegetated soil.

The above short discussion on the role of climate variability shows that in any given case (e.g., vegetated or nonvegetated soil, type of vegetative cover, and type of soil), it is very important to account for the type and temporal variability of precipitation, as well as the variability of air temperature, if one were to base groundwater management decisions on any time-dependent basis.

3.5.2 Geology and Topography

When soil cover is thin or absent, the lithologic and tectonic characteristics of the bedrock play dominant roles in aquifer recharge. Fractured bedrock surface and steep or vertical

FIGURE 3.15 Subvertical layers of limestone near land surface and thin residuum enhance aquifer recharge. Zlatar karst massif in western Serbia.

bedding greatly increase infiltration (Fig. 3.15), while layers of unfractured bedrock sloping at the same angle as the land surface may almost completely eliminate it.

Mature karst areas, where rock porosity is greatly increased by dissolution, generally have the highest infiltration capacity of all geologic media. For example, actual aquifer recharge rates of over 80 percent of total precipitation, even for high-intensity rainfall events of more than 250 mm/day, have been routinely recorded in classic karst areas of Montenegro. These rates were determined by measuring flow of large temporary karst springs, which would become active within only a few hours after the start of rainfall. Several temporary springs in the area have recorded maximum discharge rate of over 300 m^3/s and are among the largest such springs in the world (Fig. 3.16).

Steeper slopes, with the same vegetative cover and soil permeability, generally have more runoff and less infiltration. Depressions in land surface collect runoff and retain water longer, thus enhancing infiltration. Topography also plays a significant role on the general influence of various hydrometeorologic factors: precipitation in most cases increases with elevation, mountains create precipitation shadows on the leeward side, and snow accumulates behind surface barriers where it stays longer and can significantly

Figure 3.16   Karst spring Sopot on the Adriatic coast of Montenegro discharging over 200 m³/s within 24 h of a summer storm. Before the storm, the spring was completely dry. (Photograph courtesy of Igor Jemcov.)

contribute to increased infiltration as discussed earlier; the same is true for northern slopes.

3.5.3   Land Cover and Land Use

Because of the time lag between surface water processes, including infiltration, and the actual groundwater recharge arriving at the water table, it is very important to take into consideration historic land use and land cover changes when estimating representative recharge rates. It is equally important to consider future land use changes when making predictions or when modeling groundwater availability.

Three main trends in land use and the associated human-induced changes in land cover have been taking place worldwide and disrupting natural hydrologic cycles: (1) conversion of forests into agricultural land, mostly a practice still taking place in undeveloped countries; (2) rapid urbanization in undeveloped and developing countries converting all other land uses into urban land; and (3) rapid decentralization of cities, particularly in the United States, where urban sprawl has changed the American landscape and resulted in deeply entrenched social and environmental problems. Urban development and the creation of impervious surfaces beyond a city core inevitably result in increase of runoff and soil erosion. In turn, this reduces infiltration potential and groundwater recharge. Increased sediment load carried by surface streams often results in the formation of fine-sediment deposits along stream channels, which reduces hydraulic connectivity and exchange between surface water and groundwater in river flood plains. Clear-cutting of forests also alters the hydrologic cycle and results in increased erosion and sediment loading to surface streams. Conversion of low-lying forests into agricultural land may increase groundwater recharge, especially if it is followed by irrigation. However, it is important to remember that often this additional recharge is irrigation return, and the origin of water may be the underlying aquifer in question. In such case, irrigation return would be a small fraction of the groundwater originally pumped. In contrast, cutting of forests in areas with steeper slopes will generally decrease groundwater recharge because of the increased runoff, except in cases of very permeable bedrock, such as karstified limestone at or near land surface.

The evolution of land use in the United States follows a typical pattern outlined by Taylor and Acevedo (2006). During the eighteenth and nineteenth centuries, natural forest or grassland habitat was extensively converted to agricultural use. The industrial revolution of the late nineteenth and early twentieth centuries spurred urban development and massive migration to city centers. Agricultural land was extensively reforested during this time period, either naturally or artificially. Following World War II, development patterns took on a more "modern" approach, as people moved out of cities into peripheral suburban communities. During this phase, which is still taking place, both agricultural and forested lands are converted to residential, commercial, and industrial areas, as jobs tend to rapidly follow Americans to the suburbs. Figure 3.17 shows land use changes in central and southern Maryland, an area experiencing dramatic sprawl due to expansion from Washington, DC, and Baltimore metropolitan areas.

Using the SCS CN method, one can approximate changes in runoff over time in the study area. Data from the Cub Run Watershed in Northern Virginia illustrate how land use changes impact runoff. A weighted approach to the SCS method was applied to derive one CN for the entire watershed. The basic procedure is multiplying the percent of the watershed comprising a certain land use by the CN corresponding to that land use and then calculating the summation of all such products. The study in the Cub Run Watershed begins in 1990, well after the transition from agricultural land to forest land.

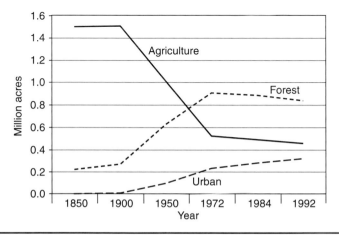

Figure 3.17 Land use changes in Central and Southern Maryland, 1850–1992. (From Taylor and Acevedo, 2006.)

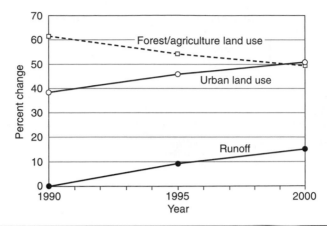

Figure 3.18 Runoff changes in Cub Run watershed, northern Virginia.

Between 1990 and 2000, land was rapidly converted to urban use, consisting of higher density residential and commercial developments (Dougherty et al., 2004). The effects of these changes result in an approximate 15 percent increase in runoff from the watershed, as shown in Fig. 3.18. The increase in runoff means less groundwater recharge in the watershed and more erosion of streambanks and impairment of water quality.

A similar analysis for watersheds in California reveals the same trend, namely, that rapid urbanization leads to exponential growth in runoff (Warrick and Orzech, 2006). Figure 3.19 shows the annual average discharge normalized by precipitation for four rivers in Southern California from 1920 to 2000. The construction of dams for flood control purposes on the Santa Ana and Los Angeles rivers only temporarily delayed dramatic increases in river discharge. A significant portion of this runoff was once groundwater recharge, placing further strain on over-allocated aquifers in the region. Figure 3.20 shows how the cumulative sediment discharge of the Santa Ana River also increased between 1970 and 2000 as a consequence of urbanization and the increase in runoff. This and many other similar studies show that city planners and water managers must promote infiltration in urban and suburban environments, both to reduce runoff and erosion and to sustain groundwater resources.

As previously discussed, urban development often causes decreases in infiltration rates and increases in surface runoff because of the increasing area of various impervious surfaces (rooftops, asphalt, and concrete). However, Table 3.2 illustrates that the infiltration rate varies significantly within an urban area based on actual land use. This is particularly important when evaluating fate and transport of contaminant plumes, including development of groundwater models for such diverse areas. For example, a contaminant plume may originate at an industrial facility, with high percentage of impervious surfaces resulting in hardly any actual infiltration, and then migrate toward a residential area where infiltration rates may be rather high because of the open space (yards) and irrigation (watering of lawns).

Agricultural activities have had direct and indirect effects on the rates and compositions of groundwater recharge and aquifer biogeochemistry. Direct effects include dissolution and transport of excess quantities of fertilizers and associated materials and hydrologic alterations related to irrigation and drainage. Some indirect effects include

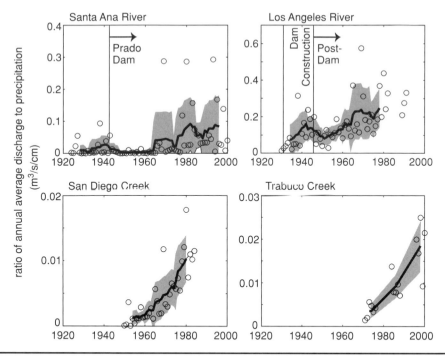

Figure 3.19   The time history of the relationship between river discharge and precipitation for southern California rivers showing increases in discharge with respect to precipitation. All records have been normalized by the annual precipitation measured at Santa Ana, CA. Solid lines show 10-year means; shadings are 1 standard deviation about the means. (From Warrick and Orzech, 2006.)

changes in water-rock reactions in soils and aquifers caused by increased concentrations of dissolved oxidants, protons, and major ions. Agricultural activities have directly or indirectly affected the concentrations of a large number of inorganic chemicals in groundwater, such as NO_3^-, N_2, Cl, SO_4^{2-}, H^+, P, C, K, Mg, Ca, Sr, Ba, Ra, and As, as well as a wide variety of pesticides and other organic compounds (Böhlke, 2002).

3.6 Methods for Estimating Groundwater Recharge

Direct quantitative measurements of groundwater recharge flux, actually arriving at the water table and determined as volume per time (e.g., ft^3/day or m^3/day) are often-cost prohibitive or not feasible. Installation, operation, and maintenance of lysimeters, which are the only devices capable of direct measurement of the recharge flux in vadose zone, are very expensive. Moreover, because of the inherent heterogeneity of soils, many lysimeters would be needed for any reliable estimate of recharge at a scale greater than the extent of one single lysimeter. In semiarid and arid regions with deep water tables, installation of lysimeters is not feasible. Most likely because of these simple facts every now and then a prominent hydrogeologists has to remind decision makers that "project designs and management strategies need to be flexible enough not to require radical change if initial predictions prove wrong, due to incorrect assumptions about recharge

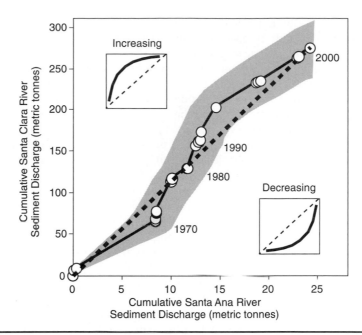

FIGURE 3.20 Cumulative sediment loads for the Santa Ana River and Santa Clara during 1965 to 2000. Dashed line represents a constant relationship in sediment loads; gray shade is the accumulated standard error of the cumulative loads. Inserts represent sediment load pattern for the Santa Ana River. (From Warrick and Orzech, 2006.)

rates and other hydrogeological factors" (Foster, 1988). Similarly, "Groundwater recharge estimation must be treated as an iterative process that allows progressive collection of aquifer-response data and resource evaluation. In addition, more than one technique needs to be used to verify results" (Sophocleous, 2004).

Indirect estimates of groundwater recharge have numerous limitations, particularly in arid environments. First and foremost, calculations are highly sensitive to changes in physical and empirical matric head-moisture curve fit parameters. This problem is

Land Use and Land Cover	Area (mi²)	Precipitation (in/yr)	Recharge (in/yr)	Recharge %
Undeveloped and nonbuilt-up	641	44.2	24.1	54.5
Residential	13	43.3	12.7	29.3
Built-up	35	45.0	13.3	29.6
Urban	99	43.7	8.1	18.5
All categories	788	44.2	21.4	48.4

Modified from Lee and Risley (2002).

TABLE 3.2 Estimates of Mean Annual Recharge on the Basis of Mean Annual Precipitation, Generalized Surficial Geology, and Land Use and Land Cover Categories from the Willamette Lowland Regional Aquifer System Analysis

exacerbated at low moisture contents in arid environments, where order of magnitude changes in flux result from small variations in physical measurements. These difficulties are especially problematic for the Darcian and numerical modeling methods, which rely on limited point measurements of pressure head, unsaturated hydraulic conductivity, and moisture content. The water table fluctuation method experiences similar problems in semiarid settings due to the significant time lag between infiltration at the ground surface and a corresponding rise in the water table (Sophocleous, 2004).

Another major problem with indirect physical methods is their reliance on idealized, theoretical equations, which do not accurately depict flow mechanisms in the vadose zone. It has been known for decades that infiltration occurs in the form of an uneven front even in seemingly homogeneous soils. This can be explained by a number of different mechanisms that can form such preferential flow paths ("channeling"): wormholes, fractures, dendritic networks of enhanced moisture, and contact points of differing soil media (Nimmo, 2007). Water velocity through these "macropores" is often an order of magnitude greater than movement through the soil matrix. The macropores issue may further complicate measurement in semiarid and arid environments, where normally dry fractures may only become activated after threshold rainfall events. In humid environments, higher moisture contents persist in surface soils and there is less sensitivity with regard to pressure head and unsaturated hydraulic conductivity. The water table fluctuation method is also more useful, as immediate changes in water table elevation are visible after recharge events (Sophocleous, 2004). To summarize, indirect physical methods are better suited in humid climates where water managers can have a better handle on the water balance. Heterogeneity and hydraulic sensitivity dominate semiarid and arid environments, where the influence of preferential flow through macropores further compromises accuracy of recharge estimates. When measured physical parameters fall in the dry range, recharge flux calculations are often in error by at least an order of magnitude (Sophocleous, 2004). Recommendations on choosing appropriate techniques for quantifying groundwater recharge are given by Scanlon et al. (2002).

3.6.1 Lysimeters

The most common procedure for direct physical measurement of recharge flux (net infiltration) involves the construction of lysimeters. Lysimeters are vessels filled with soil that are placed below land surface and collect the percolating water. The construction and design of lysimeters vary significantly depending on their purpose. Figures 3.21 and 3.22 show one of the most elaborate and expensive lysimeter facilities today, designed to directly measure various quantitative and qualitative parameters of water migrating through the vadose zone. Lysimeter stations may be equipped with a variety of automated instruments including tensiometers, which measure matric potential at different depths and instruments that measure actual flux (flow rate) of infiltrating water at different depths. Some stations may also include piezometers for recording water table fluctuations. Water quality parameters may also be measured and recorded automatically.

Worldwide, the primary use of lysimeters was traditionally in agricultural studies, although more recently their use is increasing in general groundwater studies for water supply and contaminant fate and transport. Data collected from lysimeters is often used to calibrate empirical equations or numeric models for determining other water balance elements such as evapotranspiration.

FIGURE 3.21 Array of lysimeters operated by Helmholtz Center Munich—German Research Center for Environmental Health (GmbH). (Photograph courtesy of Dr. Sascha Reth.)

FIGURE 3.22 Below ground view of one of the lysimeters shown in Fig. 3.21. Various sensors and sampling equipment collect data for multidisciplinary studies, including water budget and groundwater recharge. (Photograph courtesy of Dr. Sascha Reth.)

Lysimeters may have varying degrees of surface vegetation and may contain disturbed or undisturbed soil. The clear advantage of lysimeters is that they enable direct measurement of the quantity of water descending past the root zone over a time period of interest. Net infiltration flux is easily calculated from these measurements, eliminating much uncertainty in surficial processes such as ET and runoff. Lysimeters also capture infiltration moving rapidly through preferential flow pathways like macropores and fractures. The main disadvantages of lysimeters are their expensive construction costs and difficult maintenance requirements. These costs generally limit the total depth of lysimeters to about 10 ft, which inhibits direct correlation of net infiltration with actual groundwater recharge, since low-permeability clay layers may lie below the bottom of the lysimeter (Sophocleous, 2004). Additionally, lysimeters constructed with disturbed soils may have higher moisture contents, possibly skewing measurement results and overestimating recharge.

A major difficulty when extrapolating lysimeter data to a wider aquifer recharge area is the inevitably high variability in soil and vegetative characteristics. A recent study by the USGS (Risser et al., 2005a) illustrates this problem. Data collected from seven lysimeters installed on a 100 ft^2 plot show that the coefficient of variation between monthly recharge rates at individual lysimeters is greater than 20 percent for 6 months, with the June, July, and August values of about 50, 100, and 60 percent, respectively. Coordinating direct measurements with a water balance or water table fluctuation approximation on a larger scale may help resolve some of the scale problems associated with point measurements using lysimeters.

3.6.2 Soil Moisture Measurements

Small negative pressure heads (less than about 100 kPa) in the unsaturated zone can be measured with tensiometers, which couple the measuring fluid in a manometer, vacuum gauge, or pressure transducer to water in the surrounding partially saturated soil through a porous membrane. The pressure status of water held under large negative pressures (greater than 100 kPa) may be measured using thermocouple psychrometers, which measure the relative humidity of the gas phase within the medium and with heat dissipation probes or HDPs (Lappala et al., 1987; McMahon et al., 2003). The measuring instrumentation may be permanently installed in wells screened at different depths to measure soil moisture profile in the vadose zone, or measurements may be performed on a temporary basis using direct push methods, such as cone penetrometers equipped with tension rings. In shallow applications, soil moisture probes can be installed in trenches.

By simultaneously measuring the matric potential and the moisture content at same vertical locations during different hydrologic conditions (e.g., prior, during, and after periods of major recharge), it is possible to plot several moisture characteristic curves, which then enables accurate determination of the unsaturated hydraulic conductivity and calculations of flow rates and velocities in the unsaturated zone. This method also helps to distinguish between the relative proportions of net water loss to evaporation (upward water movement) and drainage (downward water movement) by establishing a plane of zero potential gradient, called "zero flux" plane (Fig. 3.23).

3.6.3 Water Table Fluctuations

Rise in water table after rainfall events is the most accurate indicator of actual aquifer recharge. It can also be used to estimate the recharge rate, which is assumed to be equal

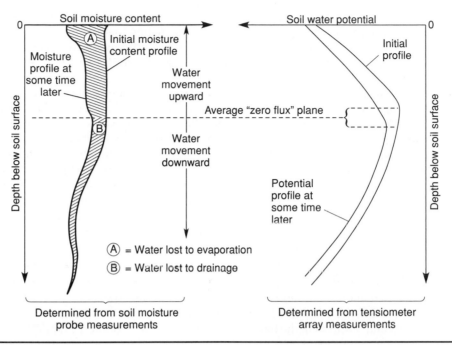

Figure 3.23 Illustration of the measurement of evaporation using soil moisture depletion supplemented with the determination of an average "zero flux" plane to discriminate between (upward) evaporation and (downward) drainage. (From Shuttleworth, 1993; copyright McGraw-Hill.)

to the product of water table rise and specific yield (Fig. 3.24):

$$R = S_y \Delta h \tag{3.22}$$

where R = recharge (L)
   S_y = specific yield (dimensionless)
   Δh = water table rise (L).

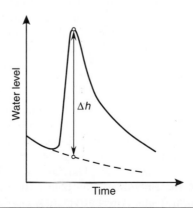

Figure 3.24 Principle of water table fluctuation method for estimating recharge.

Because of its simplicity and general availability of water-level measurements in most groundwater projects, the water table fluctuation method may be the most widely used method for estimating recharge rates in humid regions.The main uncertainty in applying this approach is the value of specific yield, which, in many cases, would have to be assumed.

An important factor to consider when applying water table fluctuation method is the frequency of water-level measurement. Delin and Falteisek (2007) point out that measurements made less frequently than about once per week may result in as much as a 48-percent underestimation of recharge based on an hourly measurement frequency.

Zaidi et al. (2007) used the double water table fluctuation method and an extensive network of observation wells to calculate water budget for a semiarid crystalline rock aquifer in India. The monsoonal character of rainfall in the area allows division of the hydrologic year into two distinct dry and wet seasons and application of the following water budget equation twice a year:

$$R + RF + Q_{in} = E + PG + Q_{out} + S_y \Delta h \qquad (3.23)$$

where two parameters, natural recharge, R, and specific yield, S_y, are unknown. Water table fluctuations, Δh, are measured, and other components of the water budget are independently estimated: RF is the irrigation return flow, Q_{in} and Q_{out} are horizontal inflows and outflows in the basin, respectively, E is evaporation, and PG is groundwater extraction by pumping. The method is called the "double water table fluctuation" because the equation is applied two times (for wet and dry season) so that both unknowns (recharge and specific yield) can be solved. This eliminates inaccuracies associated with estimating specific yield at large field scales.

3.6.4   Environmental Tracers

Environmental tracers have been irreplaceable in groundwater sustainability studies, as they provide answers about contemporary and historic recharge rates at time scales varying from days to thousands of years. They are useful in finding answers about possible mixing of groundwater of different age and origin within a groundwater system, or sources of groundwater recharge. They are also used to assess impacts and effectiveness of artificial recharge. In the unsaturated zone, environmental tracers mirror soil moisture movement, providing a sound means for estimating present-day recharge. Tracer methods are a good alternative to physical estimation in arid environments with low recharge fluxes, as tracer concentration measurements are much more precise than those of soil hydraulic properties. Environmental tracers also provide the only reliable means of quantifying the influence of preferential flow networks and macropores. Tracers are able to move together with soil moisture through such regions of preferential flow and are therefore invaluable for contaminant fate and transport analyses. As discussed by Sophocleous (2004), tracer studies are often surprising and cast doubt on the utility of physical methods of recharge assessment based on vadose zone investigations. A classic example is an attempt to quantify recharge at grassland and irrigated sites above the High Plains (Ogallala) aquifer. Using Darcian methodology based on physical vadose zone measurements, the calculated flux below the root zone ranged from 0.1 to 0.25 mm/year for a grassland site. Alternative analysis of chloride profiles led to estimates of 2.5 to 10 mm/year at the same exact location. This discrepancy of one to two orders of magnitude shows that preferential flow in some cases may be a dominant recharge mechanism and illustrates the major need for more research in this area (Sophocleous, 2004).

Environmental tracers commonly used to estimate the age of young groundwater (less than 50 to 70 years old) are the chlorofluorocarbons (CFCs) and the ratio of tritium and helium-3 ($^3H/^3He$). Because of various uncertainties and assumptions that are associated with sampling, analysis, and interpretation of the environmental tracer data, groundwater ages estimated using CFC and ^3H-^3He methods are regarded as apparent ages and must be carefully reviewed to ensure that they are geochemically consistent and hydrologically realistic (Rowe et al., 1999). Isotopes typically used for the determination of older groundwater ages are carbon-14, oxygen-18 and deuterium, and chlorine-36, with many other isotopes are increasingly studied for their applicability (Geyh, 2000).

Although reference is often made to dating of groundwater, the age actually applies to the date of introduction of the tracer substance and not the water. Unless one recognizes and accounts for all the physical and chemical processes that affect the concentrations of an environmental tracer in the aquifer, the tracer-based age is not necessarily equal to the transit time of the water (Plummer and Busenberg, 2007). The concentrations of all dissolved substances are affected, to some extent, by transport processes. For some tracers, the concentrations can also be affected by chemical processes, such as degradation and sorption during transit. For this reason, the term "age" is usually qualified with the word "model" or "apparent," i.e., "model age" or "apparent age." The emphasis on model or apparent ages is needed because simplifying assumptions regarding the transport processes are often made and chemical processes that may affect tracer concentrations are usually not accounted for (Plummer and Busenberg, 2007).

Detailed international field research on the applicability of isotopic and geochemical methods in the vadose (unsaturated) zone for groundwater recharge estimation was coordinated by the International Atomic Energy Agency (IAEA) during 1995 to 1999, with results obtained from 44 sites mainly in arid climates (IAEA, 2001). Information on the physiography, lithology, rainfall, vadose zone moisture content, and chemical and isotopic characteristics was collected at each profiling site and used to estimate contemporary recharge rates (see Table 3.3 for examples).

The best source for numerous studies on the application of various environmental isotopes in surface water and groundwater studies are the proceedings of the international conferences organized by the IAEA as well as the related monographs published by the agency, many of which are available for free download at its Web site (www.iaea.org).

Chloride

Mass balance of a natural environmental tracer in the soil pore water may be used to measure infiltration where the sole source of the tracer is atmospheric precipitation and where runoff is known or negligible. The relationship for a steady-state mass balance is (Bedinger, 1987)

$$R_q = (P - R_s)\frac{C_p}{C_z} \tag{3.24}$$

where R_q (infiltration) is a function of P (precipitation), R_s (surface runoff), C_p (tracer concentration in precipitation), and C_z (tracer concentration in the soil moisture). In an

Country	Mean Precipitation (mm/a)	Vadose Zone Depth (m)	Rainfall Cl Content (mg/L)	Mean Recharge (mm/a)
Jordan				
Jarash	480	21	10	28
Azraq	67	7	61	2
Saudi Arabia				
Qasim	133	18	13	2
Syria				
Damascus Oasis	220	21	7	2–6
Egypt				
Rafaa	300	20	16	18–24
Nigeria				
Mfi	389	10	—	<1

From Puri et al. (2006).

TABLE 3.3 Rates of Contemporary Recharge at Six Sites in Arid Climate Based on Isotopic and Geochemical Investigations of Contemporary Groundwater Recharge at Sites with Arid Climate.

ideal model, the tracer content of soil water increases with depth due to the loss of soil water to ET and conservation of tracer. The tracer content of soil water attains a maximum at the maximum depth of ET. The model postulates a constant content of tracer in the soil moisture from this point to the water table.

The mass balance Eq. (3.24) assumes that recharge occurs by piston flow. However, it has been shown that diffusion and dispersion are important components of tracer flux in the unsaturated zone. Johnston (1983, from Bedinger, 1987) proposed the following equation for mass balance where runoff is negligible:

$$PC_p = -D_s - \frac{\partial C}{\partial Z} + C_z R_q \qquad (3.25)$$

where D_s is the diffusion-dispersion coefficient; $\partial C / \partial C_z$ is the rate of change in soil moisture concentration with depth; and P, C_z, and R_q are as defined in Eq. (3.24).

Departures from the ideal model of tracer variation with depth are common and have been attributed to changes in land use, such as clearing of native vegetation, replacement of native vegetation by cropped agriculture, bypass mechanisms for infiltrating water through the soil profile, and changes in climate (Bedinger, 1987).

Chloride (Cl) is a widely used environmental tracer since it is conservative and highly soluble. Knowledge of wet and dry deposition cycles of Cl on the ground surface can be used in conjunction with vadose zone samples to determine the recharge rate at a site. Chloride is continuously deposited on the land surface by precipitation and dry fallout. The high solubility of Cl enables its transport into the subsurface by infiltrating water. Because Cl is essentially nonvolatile and its uptake by plants is minimal, it is retained in the sediment when water is removed by evaporation and transpiration. An increase in Cl within the root zone of the shallow subsurface, therefore, is proportional to the amount of water lost by ET (Allison and Hughes, 1978, from Coes and Pool, 2005). Izbicki

(2002) cites high Cl concentrations near the base of the root zone as proof that ground-water recharge is not occurring at desert sites removed from intermittent washes or arroyos.

In areas where active infiltration is occurring, an increase in Cl in the shallow subsurface will generally be absent, and concentrations will be very low through the unsaturated zone. In areas where little to no active infiltration is occurring, an increase in Cl in the shallow subsurface will be present. After reaching maximum they will stay relatively constant down to the water table. However, there are published studies showing that Cl concentrations in arid areas with little infiltration may decrease below the peak concentration in the root zone due to varying factors such as paleoclimatic variations and nonpiston flow (Coes and Pool, 2005).

If the Cl deposition rate on the land surface is known, the average travel time of Cl (t_{Cl}) to a depth in the unsaturated zone (z) can be calculated as (Coes and Pool, 2005)

$$t_{Cl} = \frac{\int_0^z Cl_{soil}\, dz}{Cl_{dep}} \tag{3.26}$$

where Cl_{soil} is chloride mass in the sample interval (M/L^3) and Cl_{dep} is chloride deposition rate ($M/L^2/t$). The above equation entails several assumptions: (1) flow in the unsaturated zone is downward vertical and piston type, (2) bulk precipitation (precipitation plus dry fallout) is the only source of Cl and there are no mineral sources of Cl, (3) the Cl deposition rate has stayed constant over time, and (4) there is no recycling of Cl within the unsaturated zone.

Figure 3.25 shows examples of chloride and tritium concentrations in a thick unsaturated zone in the semiarid climate of Arizona, the United States.

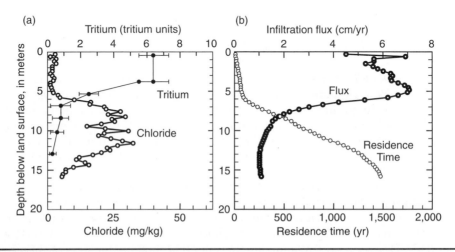

Figure 3.25 Determination of basin floor infiltration at borehole BF1, Sierra Vista sub-watershed, AZ. (*a*) Sediment chloride and pore water tritium data; tritium values include standard error of estimate bars. (*b*) Residence time and infiltration flux calculated from chloride data. (From Coes and Pool, 2005.)

Tritium

Tritium ($_3$H), a naturally occurring radioactive isotope of hydrogen with a half-life of 12.43 years, has been used extensively as a hydrologic tracer and dating tool. It is produced naturally in the upper atmosphere by bombardment of nitrogen with cosmic radiation and, although few measurements are available, it is estimated that the natural concentration of tritium in precipitation is between 5 and 20 TU (Kauffman and Libby, 1954). Large quantities of tritium were released to the atmosphere during thermonuclear-weapons testing from 1952 until the late 1960s, with maximum releases occurring in the early 1960s. As a result, the amount of tritium in precipitation sharply increased during testing as tritium was introduced into the water cycle and decreased after testing ended. The atmospheric-testing peak therefore provides an absolute time marker from which to estimate groundwater age. However, because radioactive decay and hydrodynamic dispersion have greatly reduced maximum tritium concentrations in groundwater, identification of the 1960s atmospheric-testing peak has become increasingly difficult. The interpretation of ages from tritium data alone is further complicated by the fact that monitoring and extraction wells are commonly screened over intervals that represent a wide range of groundwater ages.

The amount of tritium in subsurface water at a given time is a function of the amount of tritium in the atmosphere when infiltration occurred and the radioactive decay rate of tritium. If flow in the unsaturated zone is assumed to be downward vertical and piston type, the average infiltration flux (q_i), can be estimated by (Coes and Pool, 2005)

$$q_i = \frac{\Delta z}{\Delta t}\theta_v \tag{3.27}$$

where z = depth to maximum tritium activity (L)
   t = elapsed time between sampling and maximum historic atmospheric tritium activity (T)
   θ_v = volumetric soil water content (L^3/L^3).

Changes in the amount of tritium in precipitation through time, radioactive decay, hydrodynamic dispersion, and mixing water of different ages in the subsurface typically preclude the use of this isotope for quantitatively estimating groundwater residence times. In many cases, qualitative observations may be the best use of tritium data (Clark and Fritz, 1997, from Kay et al., 2002). The most accurate use of the data is to indicate pre- or post-1952 groundwater recharge. For a well-defined hydrologic system, analysis of the tritium input function may provide sufficient information to make quantitative age estimates for groundwater in the system. More complex flow systems generally require more complex models to quantitatively use the tritium data (Plummer et al., 1993). Time-sequence data and multidepth samples also are required to accurately date groundwater horizons using tritium data.

Assuming that piston-flow conditions (no dispersion or mixing) are applicable, Clark and Fritz (1997, from Kay et al., 2002) provide the following guidelines for using the tritium data: (1) groundwater that contains less than 0.8 TU and is underlying regions with continental climates has recharged the water table prior to 1952, (2) water with 0.8–4 TU may represent a mixture of water that contains components of recharge from before and after 1952, (3) tritium concentrations from about 5 to 15 TU may indicate recharge after about 1987, (4) tritium concentrations between about 16 and 30 TU are indicative of recharge since 1953 but cannot be used to provide a more specific time of recharge, (5) water with more than 30 TU probably is from recharge in the 1960s or 1970s, and (6) water with more than 50 TU predominately is from recharge in the 1960s. The continuing depletion of the artificial tritium in the environment will likely reduce its future usefulness in the groundwater studies.

Tritium-Helium-3 (^{3}H-^{3}He)

The ^{3}H/^{3}He method was developed to remove the ambiguity associated with tritium age estimation. Radioactive decay of tritium produces the noble gas helium-3 (^{3}He). Determination of the ^{3}H/^{3}He ratio, therefore, can be used to estimate the apparent date when a sample of water entered the aquifer as recharge. Because these substances virtually are inert in groundwater, unaffected by groundwater chemistry, and not derived from anthropogenic contamination, ^{3}H/^{3}He dating can be applied to a wide range of hydrologic investigations since the input function of ^{3}H does not have to be known (Kay et al., 2002; Geyh, 2000). Possible applications include site characterization, corroboration of the results of other age-dating methods, surface water–groundwater interaction studies, and calibration and interpretation of groundwater flow models (Aeschbach-Hertig et al., 1998; Ekwurzel et al., 1994; Solomon et al., 1995; Sheets et al., 1998; Szabo et al., 1996; Stute et al., 1997; Kay et al., 2002).

The activity of ^{3}H in the sample ($^3H_{spl}$) is given as (Geyh, 2000)

$$^3\text{H}_{spl} = {^3\text{H}_{init}}\, e^{-\lambda t} \tag{3.28}$$

where $^3H_{init}$ = activity of initial tritium
$\qquad \lambda$ = radioactive decay constant
$\qquad t$ = time since decaying started (absolute age).
The growth of ^{3}He in a sample is given by

$$^3\text{He}_{spl} = {^3\text{H}_{init}}(1 - e^{-\lambda t}) \tag{3.29}$$

By combining Eqs. (3.28) and (3.29), the unknown and variable initial ^{3}H activity ($^3H_{init}$) is eliminated and the age of water (t) is obtained from

$$^3\text{He}_{spl} = {^3\text{H}_{spl}}(e^{-\lambda t} - 1) \tag{3.30}$$

$$t = -\frac{\ln\left(1 + \frac{^3\text{He}_{spl}}{^3\text{H}_{spl}}\right)}{\lambda} \tag{3.31}$$

The ^{3}He concentration in the sample has to be corrected for admixed ^{3}He from the earth's crust and from the atmosphere. The concentration of tritiogenic ^{3}He will increase as tritium decays; thus, older waters will have higher $^3\text{He}_{trit}/^3\text{H}$ ratios.

Oxygen and Deuterium

The ^{16}O, ^{18}O, ^{1}H, and ^{2}H isotopes are the main isotopes that make up the water molecule. These isotopes of oxygen and hydrogen are stable and do not disintegrate by radioactive decay. In shallow groundwater systems with temperatures less than 50°C, the isotopic compositions of δ^2H and δ^{18}O in water are not affected by water-rock interactions (Perry et al., 1982). Since these are part of the water molecule, these isotopes can be used as natural tracers. Differences in the isotopic composition of groundwater and precipitation can be used to detect differences in the source water including recent precipitation.

Stable isotopes of oxygen and hydrogen in the water molecule are measured as the ratio of the two most abundant isotopes of a given element, $^2H/^1H$ and $^{18}O/^{16}O$, relative to a reference standard. These ratios are expressed in delta units (δ) as parts per thousand (per mil, written as ‰). The general expression for stable isotope notation is (Kay et al., 2002)

$$\delta x = \left(\frac{R_x}{R_{STD}} - 1 \right) \times 1000 \tag{3.32}$$

where R_x and R_{STD} are the $^2H/^1H$ and $^{18}O/^{16}O$ of the sample and reference standard, respectively. The delta units are given in parts per thousand (per mil, written as ‰). Ocean water has $\delta^{18}O$ and δ^2H values of ±0‰, and has been chosen as the Vienna Standard Mean Ocean Water (V-SMOW) standard. Most freshwaters have negative delta values (Geyh, 2000). For example, an oxygen sample with a $\delta^{18}O$ value of -50‰ is depleted in ^{18}O by 5 percent or 50‰ relative to the standard.

The difference in the mass of oxygen and hydrogen isotopes in water results in distinct partitioning of the isotopes (fractionation) as a result of evaporation, condensation, freezing, melting, or chemical and biological reactions. For example, δ^2H and $\delta^{18}O$ values in precipitation are isotopically lighter in areas with lower mean annual temperature. Strong seasonal variations are expected at any given location, whereas average annual values of δ^2H and $\delta^{18}O$ in precipitation show little variation at any one location (Dansgaard, 1964, from Kay et al., 2002). The IAEA provides δ^2H and $\delta^{18}O$ precipitation data measured at various locations throughout the world (accessible at: ftp://ftp.iaea.org).

The strong relationship between the $\delta^{18}O$ and δ^2H values of precipitation is reflected in the global meteoric water line (MWL). The slope is 8, and the so-called deuterium excess is $+10$‰. The deuterium excess (d) is defined as

$$d_{excess} = \delta^2H - 8\delta^{18}O \tag{3.33}$$

The deuterium excess near the coast is smaller than $+10$‰ and approximately 0‰ only in Antarctica. In areas where, or during periods in which, the relative humidity immediately above the ocean is or was below the present mean value, d is greater than $+10$‰. An example is the deuterium excess of $+22$‰ in the eastern Mediterranean. The value of d is primarily a function of the mean relative humidity of the atmosphere above the ocean water. The coefficient d can therefore be regarded as a paleoclimatic indicator (Geyh, 2000; Merlivat and Jouzel, 1979; Gat and Carmi, 1970).

The genesis of groundwater in relation to present-day and paleoclimatic conditions is an important aspect of resource characterization in the more arid regions. Determination of the proportions of the common stable isotopes (^2H and ^{18}O), together with radiometric dating (through ^{14}C, ^3H, ^3He, and other determinations), can be used for this purpose. The stable isotope composition characteristics of groundwater from several major aquifer systems in the Middle East are shown in Fig. 3.26. For example, the Umm Er Rhaduma and Neogene aquifers in Saudi Arabia, together with the Dammam aquifer in Kuwait and the Qatar aquifers, all have low excess ^2H in relation to the present meteoric line, which is a classic paleogroundwater indicator, and it has been confirmed by radiometric dating that these aquifers were mainly replenished during a humid Pleistocene episode (Yurtsever, 1999, from Puri et al., 2006).

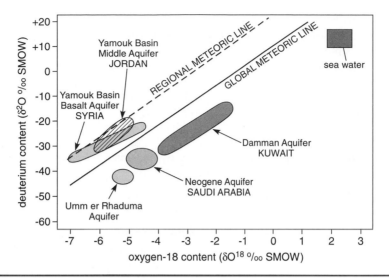

Figure 3.26 Characteristic isotopic composition of groundwater from some major Middle East aquifers (Yurtsever, 1999, from Puri et al., 2006).

Geyh (2000) discusses various processes and factors that affect local isotopic compositions of precipitation and groundwater as well as their deviations from the global and local MWLs. This includes groundwater mixing, reactions, evaporation, temperature, altitude, and continental effects.

CFCs and Sulfur Hexafluoride

Clorofluorocarbons (CFCs), together with tritium and an emerging environmental tracer sulfur hexafluoride (SF_6), can be used to trace the flow of young water (water recharged within the past 50 years) and to determine the time elapsed since recharge. Information about the age of young groundwater can be used to define recent recharge rates, refine hydrologic models of groundwater systems, predict contamination potential, and estimate the time needed to flush contaminants from groundwater systems. CFCs can also be used to trace seepage from rivers into groundwater systems, provide diagnostic tools for detection and early warning of leakage from landfills and septic tanks, and assess susceptibility of water supply wells to contamination from near-surface sources (Plummer and Friedman, 1999).

CFCs are stable, synthetic organic compounds that were developed in the early 1930s as safe alternatives to ammonia and sulfur dioxide in refrigeration and have been used in a wide range of industrial and refrigerant applications. Production of CFC-12 (dichlorodifluoromethane, CF_2Cl_2) began in 1931, followed by CFC-11 (trichlorofluoromethane, $CFCl_3$) in 1936, and then by many other CFC compounds, most notably CFC-113 (trichlorotrifluoroethane, $C_2F_3Cl_3$). CFC-11 and CFC-12 were used as coolants in air conditioning and refrigeration, as blowing agents in foams, insulation, and packing materials, as propellants in aerosol cans, and as solvents. CFC-113 was primarily used by the electronics industry in semiconductor chip manufacturing, in vapor degreasing and cold immersion cleaning of microelectronic components, and in surface cleaning. Commonly known as Freon[TM], CFCs are nontoxic, nonflammable, and noncarcinogenic, but they contribute to ozone depletion. Therefore, in 1987, 37 nations signed an agreement to limit release of CFCs and to halve CFC emissions by 2000. Production of CFCs ceased in the United States as of January 1, 1996, under the Clean Air Act. Current estimates of the atmospheric lifetimes of CFC-11, CFC-12, and CFC-113 are about 45, 87, and 100 years, respectively.

Groundwater dating with CFC-11, CFC-12, and CFC-113 is possible because (1) their amounts in the atmosphere over the past 50 years have been reconstructed, (2) their solubilities in water are known, and (3) their concentrations in air and young water are high enough that they can be measured. Age is determined from CFCs by relating their measured concentrations in groundwater back to known historical atmospheric concentrations and to calculated concentrations expected in water in equilibrium with air.

For best results, the apparent age should be determined using multiple dating techniques because each dating technique has limitations. CFC dating is best suited for groundwater in relatively rural environments without localized, nonatmospheric CFC contamination from septic systems, sewage effluent, landfills, or urban runoff. The dating method appears to work well in shallow, aerobic, sand aquifers that are low in particulate organic matter, where results can be accurate within 2 to 3 years before the study date. Even where there are problems with CFC dating of groundwater, the presence of CFCs indicates that the water sample contains at least some post-1940s water, making CFCs useful as tracers of recent recharge. Where CFC and ^{3}H/^{3}He ages agree, or where all three CFCs indicate similar ages, considerable confidence can be placed in the apparent age (Plummer and Friedman, 1999).

An example of groundwater dating with CFCs in agricultural areas on the Delmarva Peninsula of Maryland and Virginia, the United States, is shown in Fig. 3.27. The results indicate that water recharged since the early 1970s exceeds the U.S. Environmental Protection Agency drinking water maximum contaminant level (MCL) for nitrate of 10 mg/L (as N), while water recharged prior to the early 1970s, before the heavy use of nitrogen fertilizers, does not exceed the MCL (Böhlke and Denver, 1995; from Plummer

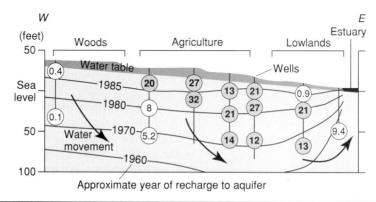

Figure 3.27 CFCs and nitrate concentrations were measured between June 1989 and January 1990 on a section of the Delmarva Peninsula, in the Fairmount watershed. Groundwater dating reveals a pattern of high nitrate concentrations moving slowly toward the estuary. Numbers within circles show nitrate concentrations, in milligrams per liter (mg/L, as N). Bold numbers indicate concentrations higher than 10 mg/L. (Modified from Plummer and Friedman, 1999.)

and Friedman, 1999). Nitrate concentrations in groundwater under woodlands were low, whereas groundwater recharged under agricultural fields had nitrate concentrations that exceeded the MCL. CFC concentrations indicate that groundwater now discharging to streams that drain agricultural areas of the Delmarva and then flow into the Chesapeake Bay or the Atlantic Ocean was recharged in nearby fields in the 1960s and 1970s (Böhlke and Denver, 1995; Focazio et al., 1998). Thus, even if the application of nitrogen fertilizers to the fields stopped today, streams, rivers, and estuaries can be expected to receive increasing amounts of nitrate from groundwater discharge until the contaminated water is flushed through the system (Modica et al., 1998). Up to 30 years may be needed to flush the high-nitrate water present in the analyzed agricultural watersheds (Plummer and Friedman, 1999).

The use of SF_6 is an emerging alternative to using CFCs in dating groundwater as atmospheric CFC concentrations continue to fall. Industrial production of SF_6 began in 1953, with the introduction of gas-filled high-voltage electrical switches. SF_6 is extremely stable and is accumulating rapidly in the atmosphere. The historical atmospheric mixing ratio of SF_6 is being reconstructed from production records, archived air samples, and atmospheric measurements. The mixing ratio is also being retrieved from concentrations measured in seawater and in previously dated groundwater. As atmospheric CFC concentrations fall, an even more sensitive dating tool will be the ratio of SF_6 to, for example, CFC-12. Although SF_6 is almost entirely of human origin, there is likely a natural, igneous source of SF_6 that will complicate dating in some environments. USGS scientists have successfully used SF_6 to date shallow groundwater on the Delmarva Peninsula, Maryland, and water from springs in the Blue Ridge Mountains of Virginia, the United States, with SF_6 (Plummer and Friedman, 1999).

Carbon-14

Radiocarbon (carbon-14 or ^{14}C) is the radioactive isotope of carbon with a half-life of 5730 years. It occurs in atmospheric CO_2, living biosphere, and the hydrosphere after its production by cosmic radiation. Underground production is negligible. The ^{14}C activity is usually given as an activity ratio relative to a standard activity, about equal to the activity of recent or modern carbon. Therefore, the ^{14}C content of carbon-containing materials is given in percent modern carbon (pMC); 100 pMC (or 100 percent modern carbon) corresponds, by definition, to the ^{14}C activity of carbon originating from (grown in) 1950 AD (Geyh, 2000). In addition to radioactive isotope ^{14}C, two other stable carbon isotopes, ^{13}C and ^{12}C, are important for understanding the origin of CO_2 involved in the dissolved carbonate–CO_2 system in groundwater and for correcting the age-dating results obtained from the ^{14}C isotope.

The ^{14}C composition of groundwater is the result of its radioactive decay and of various chemical reactions between water and porous media in the unsaturated and saturated zones. These reactions include dissolution of carbon dioxide and carbonate minerals. Recently, infiltrating water and dissolved carbon dioxide gas in the unsaturated zone have a ^{14}C composition of about 100 percent modern because carbon dioxide gas diffuses from the atmosphere and because plants respire carbon dioxide gas to the soil zone or unsaturated zone that is 100 percent modern. Water that has dissolved carbon dioxide gas and is infiltrating through the unsaturated zone or moving through the aquifer can also dissolve carbonate minerals, which increases dissolved inorganic carbon concentrations and can significantly reduce the ^{14}C composition of the water (Anderholm and Heywood, 2002). Groundwater with a long residence time in the aquifer was subject to similar processes at the time of its origin and has undergone various transformations through reactions with the aquifer porous media since that time. Because of all these factors, ^{14}C is not a conservative tracer, and its use for groundwater dating studies is not straightforward.

Groundwater dating using ^{14}C is generally considered applicable for water up to 30,000 years old, although the original dating technique proposed for organic carbon samples is applicable to the 45,000 to 50,000-year range (Libby, 1946). Before determining how long groundwater has been isolated from the atmosphere or from the modern ^{14}C reservoir, the effect of chemical reactions on the ^{14}C composition of groundwater needs to be determined. Various models have been used to adjust or estimate the ^{14}C composition of water resulting from processes in the unsaturated zone and aquifer (Mook, 1980, from Anderholm and Heywood, 2002; Geyh, 2000). These models range from simple ones that require little data, to complex models that require much information about the carbon isotopic composition, as well as geochemical reactions that occur as water moves through the vadose zone and the aquifer. The following equation can be used to estimate the apparent age of groundwater (Anderholm and Heywood, 2002):

$$t = \frac{5730}{\ln 2} \ln \left(\frac{A_0}{A_S} \right) \qquad (3.34)$$

where t = apparent age, in years
$A_0 = {}^{14}C$ composition of water before radioactive decay and after chemical reactions, in percent modern
$A_S = {}^{14}C$ composition measured in the sample, in percent modern.

Experience with various correction models shows that, using the same hydrochemical and isotope information, different models may produce corrections varying by many thousands of years (Geyh, 2000).

3.6.5 Baseflow Separation

A surface stream hydrograph is the final quantitative expression of various processes that transform precipitation into streamflow. Separation of the surface stream hydrograph is a common technique of estimating the individual components that participate in the flow formation. Theoretically, they are divided into flow formed by direct precipitation over the surface stream, surface (overland) runoff collected by the stream, near-surface flow of the newly infiltrated water (also called underflow), and groundwater inflow. However, it is practically impossible to accurately separate all these components of streamflow generated in a real physical drainage area. In practice, the problem of component separation is therefore reduced to an estimation of the *baseflow*, formed by groundwater, and *surface runoff*, which is the integration of all the other components. In natural long-term conditions and in the absence of artificial groundwater withdrawal, the rate of groundwater recharge in a drainage basin of a permanent gaining stream is equal to the rate of groundwater discharge. Assuming that all groundwater discharges into the surface stream drainage network, either directly or via springs, it follows that the stream baseflow equals the groundwater recharge in the drainage basin. Although some professionals view the hydrograph separation method as a "convenient fiction" because of its subjectivity and lack of rigorous theoretical basis, it does provide useful information in the absence of detailed (and expensive) data on many surface water runoff processes and drainage basin characteristics that contribute to streamflow generation. In any case, the method should be applied with care and regarded only as an approximate estimate of the actual groundwater recharge. In addition, geologic and hydrogeologic characteristics of the basin should be well understood before attempting to apply the method. The following examples illustrate some situations where baseflow alone should not be used to estimate groundwater recharge (Kresic, 2007):

1. Surface streamflows through a karst terrain where topographic and groundwater divides are not the same. The groundwater recharge based on baseflow may be grossly over- or underestimated depending on the circumstances.

2. The stream is not permanent, or some river segments are losing water (either always or seasonally); locations and timing of the flow measurements are not adequate to assess such conditions.

3. There is abundant riparian vegetation in the stream floodplain, which extracts a significant portion of groundwater via ET.

4. There is discharge from deeper aquifers, which have remote recharge areas in other drainage basins.

5. A dam regulates the flow in the stream.

A simple hydrograph generated by an isolated precipitation event and the principle of baseflow separation is shown in Figure 3.28. In reality, unless the surface stream is intermittent, the recorded hydrograph has a more complex shape, which reflects the influence of antecedent precipitation. Actual hydrographs are formed by the superposition of single hydrographs corresponding to separate precipitation events.

The first method of hydrograph component separation shown in Fig. 3.28 (line ABC) is commonly applied to surface streams with significant groundwater inflow. Assuming that point C represents the end of all surface runoff, and the beginning of flow generated solely by groundwater discharge, the late near-straight line section of the hydrograph is extrapolated backward until it intersects the ordinate of the maximum discharge (point

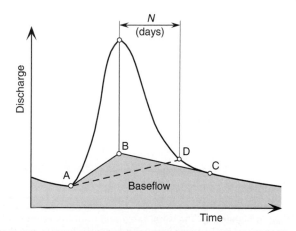

Figure 3.28 Single hydrograph formed by isolated rainfall event showing two common methods of baseflow separation.

B). Point A, representing the beginning of surface runoff after rainfall, and point B are then connected with the straight line. The area under the line ABC is the baseflow or the groundwater component of the surface streamflow.

The second graphical method of baseflow separation is used for surface streams in low-permeable terrain without significant groundwater flow. It is conditional since point D (the hydrograph falling time) is found by the following empirical formula (Linsley et al., 1975):

$$N = 0.8A^{0.2} \text{ (days)} \tag{3.35}$$

where A is the drainage area in square kilometers. In general, this method gives short falling times: for an area of 100 km^2, N is 2 days; for 10,000 km^2, N is 5 days. Thus, the method should be applied cautiously after analyzing a sufficient number of single hydrographs and establishing an adequate area-time relationship.

As illustrated in Fig. 3.29, graphical methods of baseflow separation may not be applicable at all in some cases. A stream with alluvial sediments having significant bank storage capacity may, during floods or high river stages, lose water to the subsurface so that no baseflow is occurring (Fig. 3.29a). Or, a stream may continuously receive baseflow from a regional aquifer that has a different primary recharge area than the shallow aquifer and maintain a higher head than the stream stage (Fig. 3.29b). Although one could use the same approach and "separate" either of the two hydrographs, it would

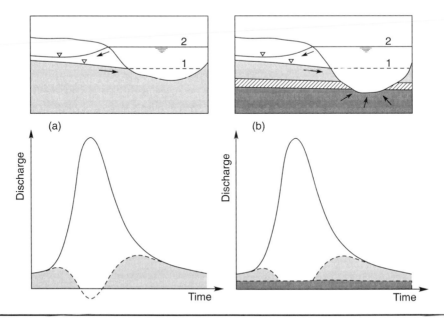

FIGURE 3.29   Stream hydrograph showing flow components after major rise due to rainfall when (a) the stream stage is higher than the water table and (b) the stream stage is higher than the water table in the shallow aquifer, but lower than the hydraulic head in the deeper aquifer that is discharging into the stream. (1) Initial stream stage before rainfall and (2) stream stage during peak flow.

not be possible to make any conclusions as to the groundwater component of the surface streamflow without additional field investigations. One such field method is hydrochemical separation of the streamflow hydrograph using dissolved inorganic constituents or environmental tracers. It is often more accurate than simple graphoanalytical techniques because surface water and groundwater almost always have significantly different chemical signatures.

Risser et al. (2005b) present a detailed application and comparison of two automated methods of hydrograph separation for estimating groundwater recharge based on data from 197 streamflow gauging stations in Pennsylvania. The two computer programs—PART and RORA (Rutledge, 1993, 1998, 2000)—developed by the USGS are in public domain and available for free download from the USGS Web site. The PART computer program uses a hydrograph separation technique to estimate baseflow from the streamflow record. The RORA computer program uses the recession-curve displacement technique of Rorabaugh (1964) to estimate groundwater recharge from each storm period. The RORA program is not a hydrograph-separation method; rather, recharge is determined from displacement of the streamflow-recession curve according to the theory of groundwater drainage.

The PART program computes baseflow from the streamflow hydrograph by first identifying days of negligible surface runoff and assigning baseflow equal to streamflow on those days; the program then interpolates between those days. PART locates periods of negligible surface runoff after a storm by identifying the days meeting a requirement of antecedent-recession length and rate of recession. It uses linear interpolation between the log values of baseflow to connect across periods that do not meet those tests. A detailed description of the algorithm used by PART is provided in Rutledge (1998, pp. 33–38).

Rorabaugh's method used by RORA is a one-dimensional analytical model of groundwater discharge to a fully penetrating stream in an idealized, homogenous aquifer with uniform spatial recharge. Because of the simplifying assumptions inherent in the equations, Halford and Mayer (2000) caution that RORA may not provide reasonable estimates of recharge for some watersheds. In fact, in some extreme cases, RORA may estimate recharge rates that are higher than the precipitation rates. Rutledge (2000) suggests that estimates of mean monthly recharge from RORA are probably less reliable than estimates for longer periods and recommends that results from RORA should not be used at time scales smaller than seasonal (3 months), because results differ most greatly from manual application of the recession-curve displacement method at small time scales. It should be noted that neither RORA nor PART computer programs can account for situations shown in Fig. 3.29 or other possible complex relationships between surface streams and groundwater.

Spring Flow Hydrograph

Although the processes that generate hydrographs of springs and surface streams are quite different, there is much that is analogous between them, and the hydrograph terminology is the same. Increase in spring flow after rainfall events is a direct indicator of the actual aquifer recharge. Knowing the exact area where this direct recharge from rainfall takes place, and the representative (average) amount of rainfall, enables relatively accurate determination of the aquifer recharge based on the measured increase of spring discharge rate. However, it is often difficult to accurately determine a spring drainage area, especially in karst. In addition, the spring hydrograph reflects the response to rainfall of all porosity types in the entire volume of the aquifer and the response to all water inputs into the aquifer; these inputs may include percolation of sheet surface runoff from less permeable areas beyond the aquifer extent and direct percolation from surface streams.

The impact of newly infiltrated water on spring discharge varies with respect to predominant type of porosity and position of the hydraulic head in the aquifer. In any case, the first reaction of karst or fractured aquifers to recharge in form of a rapid initial increase in discharge rate is in many cases the consequence of pressure propagation through karst conduits and large fractures, and not necessarily the outflow of newly infiltrated water (see also Fig. 2.23). The new water arrives at the spring with certain delay, and its contribution is just a fraction of the overall flow rate. The contribution (percentage) of the newly infiltrated water discharging at the spring can be determined using hydrochemical separation methods (Section 2.9.3) and environmental tracers.

3.6.6 Numeric Modeling

Variably Saturated Flow Models

Numerous equations and analytical mathematical models have been developed for estimating soil water movement and infiltration rates for various purposes such as irrigation and drainage, groundwater development, soil and groundwater contamination studies, managed aquifer recharge, and wastewater management, to name just a few. Ravi and Williams (1998) and Williams et al. (1998) have prepared a two-volume publication for the U.S. Environmental Protection Agency, in which they present a number of widely applied analytical methods, divided into three types: (1) empirical models, (2) Green-Ampt models, and (3) Richards equation models. These methods (except the empirical models) are based on widely accepted concepts of soil physics, and soil hydraulic and climatic parameters representative of the prevailing site conditions. The two volumes (1) categorize infiltration models presented based on their intended use, (2) provide a conceptualized scenario for each infiltration model that includes assumptions, limitations, mathematical boundary conditions, and application, (3) provide guidance for model selection for site-specific scenarios, (4) provide a discussion of input parameter estimation, (5) present example application scenarios for each model, and (6) provide a demonstration of sensitivity analysis for selected input parameters (Ravi and Williams, 1998).

Common to all analytical methods is that they describe only one-dimensional water movement through the vadose zone and make various simplifying assumptions, of which those of a homogeneous soil profile and uniform initial soil water content are the most limiting. Because of the limitations of analytical equations, numeric models of water movement through the vadose zone and direct recharge of the water table are starting to prevail in practice. In addition, they are easily linked with, or are part of, numeric models of the saturated zone, which makes their development and utilization even more attractive. There are several versatile public domain unsaturated-saturated (variably saturated) flow, and fate and transport numeric models that can be used to estimate aquifer recharge rates. Examples of models with friendly graphical user interface (GUI) include VS2DT (developed by the USGS) and HYDRUS-2D/3D (available in public domain). The latter one, although not in public domain, is a successor of HYDRUS-1D initially developed at the U.S. Salinity Laboratory of the U.S. Department of Agriculture.

The use of VS2DT is illustrated with a study of recharge through a desert wash. In semiarid and arid groundwater basins, aquifer recharge is dependent on ephemeral

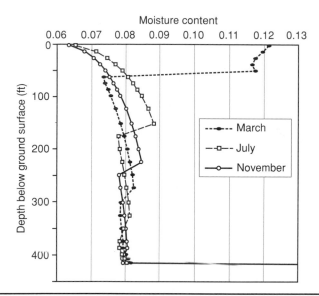

Figure 3.30 Moisture content versus depth for model simulation of recharge through a desert wash.

streams draining snowmelt or direct precipitation from surrounding highlands. Minimal recharge occurs through thick deposits on the basin floor due to low localized precipitation and high PET (Izbicki, 2002). As a result, it is very important to protect intermittent washes at the basin margins and to further understand how they drive groundwater recharge. This point is especially salient, given the increasing demand put on groundwater basins in semiarid and arid settings. Physical and hydrologic parameters and wash dimensions are largely based on work by Izbicki (2002). As flow through the wash occurs during snowmelt periods or flash flood events, it is assumed that all infiltration occurs during the month of March. The total quantity of infiltration is approximately 10 percent of the total annual flow through the wash.

To illustrate flow patterns through deep vadose zones, the model is first run with homogenous fill deposits consisting of coarse sand, from the land surface down to the 415-ft-deep water table. The model is then run for 25 years to depict the resulting steady-state moisture profile through the vadose zone. Note that all recharge enters the model through the wash throughout the month of March. The moisture profiles for select months in the final year are shown in Fig. 3.30. The long-term deep drainage rate can be estimated from the model-simulated equilibrium moisture content of approximately 8 percent using the Darcy-Buckingham equation (3.9). It is important to reinforce that flow through the vadose zone does not occur as a uniform wetting front, but rather as a diffusing pulse of moisture that is dampened with depth. This diffusion causes the uniform gravity drainage below certain depth, as each recharge event is dampened to a relatively constant moisture content and pressure head.

While the above scenario is useful as a proof of concept, it is more pertinent to examine flow through a heterogeneous vadose zone with layering of finer sedimentary deposits. The same recharge conditions are simulated for a period of 20 years, at which point all recharge in the month of March is cutoff due to a hypothetical "paving" of the wash. The

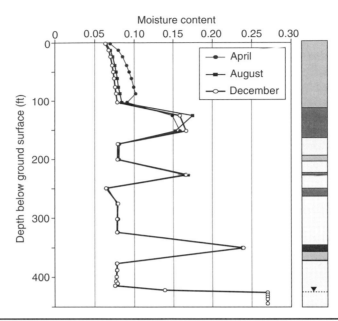

FIGURE 3.31 Model-simulated moisture content versus depth for wash recharge through a heterogeneous vadose zone with layering of finer sedimentary deposits, excluding clay and silt.

steady-state moisture profile for the 20-year duration is shown in Fig. 3.31, together with the layering of finer sedimentary deposits. As is evident from the figure, there is little variation in moisture content in time throughout the deeper vadose zone. This results in a constant drainage flux, as in the homogenous model. A measurement of unsaturated hydraulic conductivity within any layer allows approximation of the long-term drainage rate throughout the entire system (Nimmo et al., 2002). However, the sedimentary heterogeneity causes significant variation in moisture content with depth because of the increased water retention capacities of fine-grained soils. The flux reaching the water table will be less than that for the homogenous coarse-grained model because of significant moisture storage in and above fine-grained layers. Furthermore, lateral spreading of moisture will occur when a wetting front reaches deposits of lower permeability. It is intuitive that this spreading will further reduce the magnitude of downward flux through the vadose zone.

Once the recharge source is removed in year 20, moisture will continue to enter the saturated zone at the same rate for approximately 3 years, at which point the moisture content immediately above the capillary fringe will begin to decline. Figure 3.32 shows the moisture content profile in time at a point just above the capillary fringe directly below the wash. It is interesting to note that it takes approximately 20 years for any moisture to reach the water table. This number could be greater by an order of magnitude if clays or silts were present in the vadose zone. Vadose zone modeling is critical in establishing travel times of moisture through heterogeneous sediments. Another important point is that the moisture content (and thus the recharge flux) does not decrease at a rapid, uniform rate once the recharge source is cut off. Twenty-five years after the wash is paved over, recharge is still entering the water table, albeit at a decreasing rate every year. It

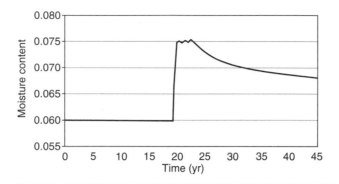

Figure 3.32 Model-simulated moisture content versus time immediately above capillary fringe.

will take many more years for the unsaturated sediments to drain all stored moisture and return to residual moisture conditions. From a management perspective, the "time lag" between infiltration reduction and recharge reduction results in long-term, abstract consequences of land use changes. Paving washes to accommodate urbanization may not result in immediate water level declines. However, long-term effects are undeniable, and the danger is that, once disrupted, the natural recharge equilibrium will take many more years to be reestablished.

Distributed-Parameter Areal Recharge Models

USGS has developed two versatile public-domain computer programs for estimating areally distributed deep percolation, or actual groundwater recharge, based on surficial processes that control various water budget elements: INFILv3 and Deep Percolation Model (DPM). A very detail report presenting the development and application of the distributed-parameter watershed model, INFILv3, for estimating the temporal and spatial distribution of net infiltration and potential recharge in the Death Valley region, Nevada and California, is given by Hevesi et al. (2003). To estimate the magnitude and distribution of potential recharge in response to variable climate and spatially varying drainage basin characteristics, the INFILv3 model uses a daily water balance model of the root zone, with a primarily deterministic representation of the processes controlling net infiltration and potential recharge (Fig. 3.33). The daily water balance includes precipitation, as either rain or snow accumulation, sublimation, snowmelt, infiltration into the root zone, ET, drainage, water content change throughout the root-zone profile (represented as a six-layered system in the Death Valley model), runoff and surface water run-on (defined as runoff that is routed downstream), and net infiltration simulated as drainage from the bottom root-zone layer. PET is simulated using an hourly solar radiation model to simulate daily net radiation, and daily ET is simulated as an empirical function of root-zone water content and PET.

The model uses daily climate records of precipitation and air temperature from a regionally distributed network of climate stations and a spatially distributed representation of drainage basin characteristics defined by topography, geology, soils, and vegetation. The model simulates daily net infiltration at all locations, including stream channels with intermittent streamflow in response to runoff from rain and snowmelt. The temporal distribution of daily, monthly, and annual net infiltration can be used to evaluate the potential effect of future climatic conditions on potential recharge.

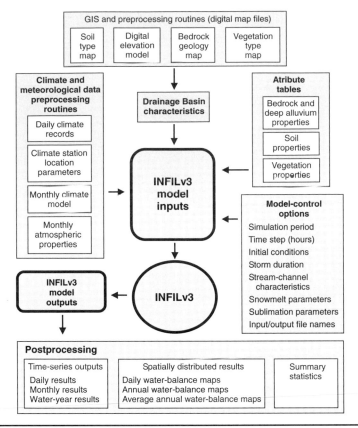

Figure 3.33 Inputs and outputs in the program structure of the INFLv3 model of the Death Valley region, Nevada and California. (From Hevesi et al., 2003.)

The INFILv3 model inputs representing drainage basin characteristics were developed using a geographic information system (GIS) to define a set of spatially distributed input parameters uniquely assigned to each grid cell of the INFILv3 model grid (Hevesi et al., 2003).

The USGS' DPM calculates, on a daily basis, the potential quantity of recharge to an aquifer via the unsaturated zone. Recharge is defined as the amount of water leaving either the active root zone (deep percolation) or, in the case of bare soils such as sand dunes, the mapped depth of the soil column (called the soil zone to distinguish it from the root zone). Recharge is derived from precipitation and irrigation. The model is physically based and, to the extent possible, was developed so that few parameters need to be calibrated. It was developed to fill the need between rigorous unsaturated flow models (or complex land surface process models) and overly simple methods for estimating groundwater recharge. The model can be applied to areas as large as regions or as small as a field plot. For a detailed description of DPM, see Bauer and Vaccaro (1987) and Bauer and Maslin (1997). DPM calculates daily PET, snow accumulation and ablation, plant interception, evaporation of intercepted moisture, soil evaporation, soil moisture changes (abstractions and accumulations), transference of unused energy, plant transpiration, and surface runoff. The residual, including any cumulative errors associated with calculations, is deep percolation (recharge). Transference is the amount of unused PET that is transferred to potential plant transpiration after abstractions from snow sublimation, evaporation of intercepted water, and soil evaporation (Vaccaro, 2007).

DPM spatially distributes input parameters to distinct areas within a modeled region, watershed, or area, or to a point that has a unit area. These distinct areas subdivide the modeled area, and they can be of any size or shape and are called hydrologic response units (HRUs). Generally, the physical properties for a HRU are such that the hydrologic response is assumed to be similar over the entire area of an HRU. The land use and land cover (LULC) can vary by HRU. For typical applications of DPM, the soil properties and LULC are the factors that define the HRU's hydrologic response. For forested mountainous terrains with winter snowpacks, a watershed model would provide better estimates of deep percolation than those calculated by DPM (Vaccaro, 2007).

One of DPM's convenient features is that the user can input observed surface runoff directly into the model. Direct surface runoff is defined as observed daily streamflow minus an estimate of daily baseflow made by a user, both in units of cubic feet per second. The use of observed streamflow allows the model to calculate improved estimates of recharge, which generally is one of the smaller components of the water budget, because at times the potential error in calculated surface runoff can be larger than the calculated recharge. Calculated runoff can be used when direct runoff data are unavailable.

References

Aeschbach-Hertig, W., Schlosser, P., Stute, M., Simpson, H.J., Ludin, A., and Clark, J.F., 1998. A ^{3}H/^{3}He study of groundwater flow in a fractured bedrock aquifer. *Ground Water*, vol. 36, no. 4, pp. 661–670.

Allen, R.G., Pereira, L.S., Raes, D., and Smith, M., 1998. Crop evapotranspiration—Guidelines for computing crop water requirements. Food and Agriculture Organization (FAO) of the United Nations. Irrigation and Drainage Paper 56, Rome, Italy, 41 p.

Allison, G.B., and Hughes, M.W., 1978. The use of environmental chloride and tritium to estimate total recharge to an unconfined aquifer. *Australian Journal of Soil Resources*, vol. 16, pp. 181–195.

Anderholm, S.K., and Heywood, C.E., 2003. Chemistry and age of ground water in the southeastern Hueco Bolson, New Mexico and Texas. U.S. Geological Survey Water-Resources Investigations Report 02-4237, Albuquerque, NM, 16 p.

Bauer, H.H., and Mastin, M.C., 1997. Recharge from precipitation in three small glacial-till mantled catchments in the Puget Sound Lowland. U.S. Geological Survey Water-Resources Investigations Report 96-4219, 119 p.

Bauer, H.H., and Vaccaro, J.J., 1987. Documentation of a deep percolation model for estimating ground-water recharge. U.S. Geological Survey Open-File Report 86-536, 180 p.

Bedinger, M.S., 1987. Summary of infiltration rates in arid and semiarid regions of the world, with an annotated bibliography. U.S. Geological Survey Open-File Report 87-43, Denver, CO, 48 p.

Böhlke, J.-K., 2002. Groundwater recharge and agricultural contamination. *Hydrogeology Journal*, vol. 10, no. 1, pp. 153–179.

Böhlke, J.K., and Denver, J.M., 1995. Combined use of groundwater dating, chemical, and isotopic analyses to resolve the history and fate of nitrate contamination in two agricultural watersheds, Atlantic coastal plain, Maryland. *Water Resources Research*, vol. 31, pp. 2319–2339.

Broner, I., and Schneekloth, J., 2007. Seasonal water needs and opportunities for limited irrigation for Colorado Crops. Colorado State University Extension. Available at: http://www.ext.colostate.edu/Pubs/crops/04718.html. Accessed August 2007.

Brown, P., 2000. Basis of evaporation and evapotranspiration. Turf Irrigation Management Series: I. The University of Arizona College of Agriculture, Tucson, AZ, 4 p.

Clark, I.D., and Fritz, P., 1997. *Environmental Isotopes in Hydrogeology*. Lewis Publishers, New York, 311 p.

Coes, A.L., and Pool, D.R., 2005. Ephemeral-stream channel and basin-floor infiltration and recharge in the Sierra Vista subwatershed of the upper San Pedro basin, Southeastern Arizona. U.S. Geological Survey Open-File Report 2005-1023, Reston, VA, 67 p.

Dansgaard, W., 1964. Stable isotopes in precipitation. *Tellus*, vol. 16, no. 4, pp. 437–468.

Delin, G.N., and J.D. Falteisek, 2007. Ground-water recharge in Minnesota. U.S. Geological Survey Fact Sheet 2007-3002, 6 p.

Dingman, S.L., 1994. *Physical Hydrology*. Macmillan, New York, 575 p.

Dougherty, M., Dymond, R.L., Goetz, S.J., Jantz, C.A., and Goulet, N., 2004. Evaluation of impervious surface estimates in a rapidly urbanizing watershed. *Photogrammetric Engineering & Remote Sensing*, vol. 70, no. 11, pp. 1275–1284.

Dugan, J.T., and Peckenpaugh, J.M., 1985. Effects of climate, vegetation, and soils on consumptive water use and ground-water recharge to the central Midwest regional aquifer system, mid-continent United States. U.S. Geological Survey Water-Resources Investigations Report 85-4236, Lincoln, NE, 78 p.

Ekwurzel, B., Schlosser, P., Smethie, W.M., Plummer, L.N., Busenberg, E., Michel, R.L., Weppernig, R., and Stute, M., 1994. Dating of shallow ground-water—Comparison of the transient tracers $^3H/^3He$, chlorofluorocarbons, and 85 Kr. *Water Resources Research*, vol. 30, no. 6, pp. 1693–1708.

Fayer, M.J., and Walters, T.B., 1995. *Estimating Recharge Rates at the Hanford Site*. Pacific Northwest Laboratory, Richland, WA, various pages.

Fayer, M.J., Rockhold, M.L., Kirham, R.R., and Gee, G.W., 1995. Appendix A: Multiyear observations of water content to characterize low recharge. In: *Estimating Recharge Rates at the Hanford Site*. Fayer, M.J., and T.B. Walters, editors. Pacific Northwest Laboratory, Richland, WA, pp. A.1–A.14.

Flint, A.L., and Flint, L.E., 2006. Modeling soil moisture processes and recharge under a melting snowpack. Proceedings, TOUGH Symposium May 15–17, 2006. Lawrence Berkeley National Laboratory, Berkeley, CA.

Focazio, M.J., Plummer, L.N., Bohlke, J.K., Busenberg, E., Bachman, L.J., and Powers, D.S., 1998. Preliminary estimates of residence times and apparent ages of ground water in the Chesapeake Bay watershed and water-quality data from a survey of springs. U.S. Geological Survey Water-Resources Investigations Report 97-4225, 75 p.

Foster, S.S.D., 1988. Quantification of ground-water recharge in arid regions—a practical view for resource development and management. In: *Estimation of Natural*

Ground-Water Recharge, NATO ASI Series C, vol. 222. Simmers, I., editor. Reidel Publishing, Dordrecht, the Netherlands, pp. 323–338.

Gat, J.R., and Carmi, I., 1970. Evolution of the isotopic composition of atmospheric waters in the Mediterranean Sea area. *Journal Geophysics Research*, vol. 75, pp. 3039–3048.

Geyh, M., 2000. Groundwater, saturated and unsaturated zone. In: *Environmental Isotopes in the Hydrological Cycle; Principles and Applications*. Mook, W.G., editor. IHP-V, Technical Documents in Hydrology, No. 39, Vol. IV. UNESCO, Paris, 196 p.

Halford, K.J., and Mayer, G.C., 2000. Problems associated with estimating ground-water discharge and recharge from stream-discharge records. *Ground Water*, vol. 38, no. 3, pp. 331–342.

Healy, R.W., Winter, T.C., LaBaugh, J.W., and Franke, O.L., 2007. Water budgets: Foundations for effective water-resources and environmental management. U.S. Geological Survey Circular 1308, Reston, VA, 90 p.

Hevesi, J.A., Flint, A.L., and Flint, L.E., 2003. Simulation of net infiltration and potential recharge using a distributed-parameter watershed model of the Death Valley region, Nevada and California. U.S. Geological Survey Water-Resources Investigations Report 03-4090, Sacramento, CA, 161 p.

IAEA (International Atomic Energy Agency), 2001. Isotope based assessment of groundwater renewal in water scarce regions. International Atomic Energy Association TECDOC-1246.

Izbicki, J.A., 2002. Geologic and hydrologic controls on the movement of water through a thick, heterogenous unsaturated zone underlying an intermittent stream in the Western Mojave Desert, Southern California. *Water Resources Research,* vol. 38, no. 3, doi: 10.1029/2000WR000197.

Johnston, C.D., 1983. Estimation of groundwater recharge from the distribution of chloride in deeply weathered profiles from south-west Western Australia. In: *Papers of the International Conference on Groundwater and Man*, vol. 1. Investigation and Assessment of Groundwater Resources, Sydney, 1983. Australian Water Resources Council, Conference Series 8, Canberra, pp. 143–152.

Kauffman, S., and Libby, W.S., 1954. The natural distribution of tritium. *Physical Review*, vol. 93, no. 6, pp. 1337–1344.

Kay, R.T., Bayless, E.R., and Solak, R.A., 2002. Use of isotopes to identify sources of ground water, estimate ground-water-flow rates, and assess aquifer vulnerability in the Calumet Region of Northwestern Indiana and Northeastern Illinois. U.S. Geological Survey Water-Resources Investigation Report 02-4213, Indianapolis, IN, 60 p.

King, R.B., 1992. Overview and bibliography of methods for evaluating the surface-water-infiltration component of the rainfall-runoff process. U.S. Geological Survey Water-Resources Investigations Report 92-4095, Urbana, IL, 169 p.

Kohler, M.A., 1958. *Meteorological Aspects of Evaporation*, vol. III. Int. Assn. Sci. Hydr. Trans., General Assembly, Toronto, pp. 423–436.

Kresic, N., 2007. *Hydrogeology and Groundwater Modeling*, 2nd ed. CRC Press, Boca Raton, FL, 807 p.

Laczniak, R.J., Smith J.L., and DeMeo, G.A., 2006. Annual ground-water discharge by evapotranspiration from areas of spring-fed riparian vegetation along the eastern margin of Death Valley, 2002–02. U.S. Geological Survey Scientific Investigations Report 2006-5145, 36 p. Available at: http://pubs.water.usgs.gov/sir2006-5145.

Lappala, E.G., Healy, R.W., and Weeks, E.P., 1987. Documentation of computer program VS2D to solve the equations of fluid flow in variably saturated porous media. U.S. Geological Survey Water-Resources Investigations Report 83-4099, Denver, CO, 131 p.

Lee, K.K., and Risley, J.C., 2002. Estimates of ground-water recharge, base flow, and stream reach gains and losses in the Willamette River Basin, Oregon. U.S. Geological Survey Water-Resources Investigations Report 01-4215, Portland, OR, 52 p.

Libby, W.F., 1946. Atmospheric helium three and radiocarbon from cosmic radiation. *Physical Review*, vol. 69, pp. 671–672.

Linsley, R.K., Kohler, M.A., and Paulhus, J.L.H., 1975. *Hydrology for Engineers*. McGraw-Hill, New York, 482 p.

Linsley, R.K., and Franzini, J.B., 1979. *Water-Resources Engineering*, 3rd ed. McGraw-Hill, New York, 716 p.

McMahon, P.B., Dennehy, K.F., Michel, R.L., Sophocleous, M.A., Ellett, K.N., and Hurlbut, D.B., 2003. Water movement through thick unsaturated zones overlying the Central High Plains Aquifer, Southwestern Kansas, 2000–2001. U.S. Geological Survey Water-Resources Investigations Report 03–4171, Reston, VA, 32 p.

Merlivat, L., and Jouzel, J., 1979. Global climatic interpretation of the deuterium-oxygen 18 relationship for precipitation. *Journal Geophysics Research*, vol. 84, pp. 5029–5033.

Modica, E., Buxton, H.T., and Plummer, L.N., 1998. Evaluating the source and residence times of ground-water seepage to headwaters streams, New Jersey Coastal Plain. *Water Resources Research*, vol. 34, pp. 2797–2810.

Mook, W.G., 1980. Carbon-14 in hydrogeological studies. In: *Handbook of Environmental Isotope Geochemistry*, Vol. 1: *The Terrestrial Environment*, A. Fritz, P., and Fontes, J.Ch., editors. Elsevier Scientific, , New York, Chap 2, pp. 49–74.

Mualem, Y., 1976. A new model for predicting the hydraulic conductivity of unsaturated porous media. *Water Resources Research*, vol. 12, pp. 513–522.

Nimmo, J.R., Deason, J.A., Izbicki, J.A., and Martin, P., 2002. Evaluation of unsaturated zone water fluxes in heterogeneous alluvium at a Mojave Basin site. *Water Resources Research*, vol. 38, no. 10, pp. 1215, doi:10.1029/2001WR000735.

Nimmo, J.R., 2007. Simple predictions of maximum transport rate in unsaturated soil and rock. *Water Resources Research*, vol. 43, W05426, doi:10.1029/2006WR005372.

Perry, E.C., Grundl, T., and Gilkeson, R.H., 1982. H, O, and S isotopic study of the ground water in the Cambrian-Ordovician aquifer system of northern Illinois. In: *Isotope Studies of Hydrologic Processes*. Perry, E.C., Jr., and Montgomery, C.W., editors. Northern Illinois University Press, DeKalb, IL, pp. 35–45.

Plummer, L.N., and Friedman, L.C., 1999. Tracing and dating young ground water. U.S. Geological Survey Fact Sheet-134-99, 4 p.

Plummer, L.N., Michel, R.L., Thurman, E.M., and Glynn, P.D., 1993. Environmental tracers for age-dating young ground water. In: *Regional Ground-Water Quality*. Alley, W.M., editor. Van Nostrand Reinhold, New York, pp. 255–294.

Plummer, L.N., and Busenberg, E., 2007. Chlorofluorocarbons. In: *Excerpt from Environmental Tracers in Subsurface Hydrology*. Cook, P., and Herczeg, A., editors. Kluwer, The Reston Chlorofluorocarbon Laboratory, U.S. Geological Survey, Reston, VA.

Puri, S., Margat, J., Yucel Yurtsever, Y., and Wallin, B., 2006. Aquifer characterization techniques. In: Non-Renewable Groundwater Resources; A Guidebook On

Socially-Sustainable Management for Water-Policy Makers. Foster, S., and Loucks, D.P. editors. IHP-VI, Series on Groundwater No. 10. UNESCO, Paris, pp. 35–47.

Ravi, V., and Williams, J.R., 1998. Estimation of infiltration rate in the vadose zone: Compilation of simple mathematical models, volume I. EPA/600/R-97/128a, U.S. Environmental Protection Agency, Cincinnati, OH, 26 p. + appendices.

Rawls, W.J., Lajpat, R.A., Brakensiek, D.L., and Shirmohammadi, A., 1993. Infiltration and soil water movement. In: *Handbook of Hydrology*. Maidment, D.R., editor. McGraw-Hill, New York, pp. 5.1–5.51.

Richards, L.A., 1931. Capillary conduction of liquids through porous mediums. *Physics*, vol. 1, no. 3, pp. 318–333.

Risser, D.W., Gburek, W.J., and Folmar, G.J., 2005a. Comparison of methods for estimating ground-water recharge and base flow at a small watershed underlain by fractured bedrock in the eastern United States. U.S. Geological Survey Scientific Investigations Report 2005-5038, Reston, VA, various pages.

Risser, D.W., Conger, R.W., Ulrich, J.E., and Asmussen, M.P., 2005b. Estimates of ground-water recharge based on streamflow-hydrograph methods: Pennsylvania. U.S. Geological Survey Open File Report 2005-1333, Reston, VA, 30 p.

Rorabaugh, M.I., 1964. Estimating changes in bank storage and ground-water contribution to streamflow. Extract of publication no. 63 of the I.A.S.H. Symposium Surface Waters, pp. 432–441.

Rowe, G.L., Jr., Shapiro, S.D., and Schlosser, P., 1999. Ground-water age and water-quality trends in a Buried-Valley aquifer, Dayton area, Southwestern Ohio. U.S. Geological Survey Water-Resources Investigations Report 99-4113. Columbus, OH, 81 p.

Rutledge, A.T., 1993. Computer programs for describing the recession of ground-water discharge and for estimating mean ground-water recharge and discharge from streamflow records. U.S. Geological Survey Water-Resources Investigations Report 93-4121, 45 p.

Rutledge, A.T., 1998. Computer programs for describing the recession of ground-water discharge and for estimating mean ground-water recharge and discharge from streamflow records—update. U.S. Geological Survey Water-Resources Investigations Report 98-4148, 43 p.

Rutledge, A.T., 2000. Considerations for use of the RORA program to estimate ground-water recharge from streamflow records. U.S. Geological Survey Open-File Report 00-156, Reston, VA, 44 p.

Scanlon, B.R., Healy, R.W., and Cook, P.G., 2002. Choosing appropriate techniques for quantifying groundwater recharge. *Hydrogeology Journal*, vol. 10, no. 1, pp. 18–39.

Schaap, M.G., 1999. *Rosetta*, Version 1.0. U.S. Salinity Laboratory, U.S. Department of Agriculture, Riverside, CA.

Serne, R.J., et al., 2002. Characterization of vadose zone sediment: Borehole 299-W23-19 [SX-115] in the S-SX Waste Management Area. PNNL-13757-2, Pacific Northwest National Laboratory, Richland, WA.

Shuttleworth, W.J., 1993. Evaporation. In: *Handbook of Hydrology*. Maidment, D.R., editor. McGraw-Hill, New York, pp. 4.1–4.53.

Šimůnek, J., Šejna, M., and van Genuchten, M.Th., 1999. *The Hydrus-2D Software Package for Simulating the Two-Dimensional Movement of Water, Heat, and Multiple Solutes in Variably-Saturated Media*, Version 2.0. U.S. Salinity Laboratory, U.S. Department of Agriculture, Riverside, CA, 227 p.

Singh, V.P., 1993. *Elementary Hydrology*. Prentice Hall, Englewood Cliffs, NJ, 973 p.

Sheets, R.A., Bair, E.S., and Rowe, G.L., 1998. Use of ^{3}H/^{3}He ages to evaluate and improve groundwater flow models in a complex buried-valley aquifer. *Water Resources Research*, vol. 34, no. 5, pp. 1077–1089.

Solomon, D.K., Poreda, R.J., Cook, P.G., and Hunt, A., 1995. Site characterization using ^{3}H/^{3}He ground-water ages, Cape Cod, MA. *Ground Water*, vol. 33, no. 6, pp. 988–996.

Sophocleous, M., 2004. Ground-water recharge and water budgets of the Kansas High Plains and related aquifers. Kansas Geological Survey Bulletin 249, Kansas Geological Survey. The University of Kansas, Lawrence, KS, 102 p.

Strahler, A.N., and Strahler, A.H., 1978. *Modern Physical Geography*. John Wiley, New York, 502 p.

Stute, M., Deak, J., Revesz, K., Bohlke, J.K., Deseo, E., Weppernig, R., and Schlosser, P., 1997. Tritium/^{3}He dating of river infiltration—an example from the Danube in the Szigetkoz area, Hungary. *Ground Water*, vol. 35, no. 5, pp. 905–911.

Szabo, Z., Rice, D.E., Plummer, L.N., Busenberg, E., Drenkard, S., and Schlosser, P., 1996. Age dating of shallow groundwater with chlorofluorocarbons, tritium/helium-3, and flowpath analyses, southern New Jersey coastal plain. *Water Resources Research*, vol. 32, pp. 1023–1038.

Taylor, J.L., and Acevedo, W., 2006. Change to urban, agricultural, and forested land in Central and Southern Maryland from 1850-1990. In: *Rates, Trends, Causes, and Consequences of Urban Land-Use Change in the United States*. Acevedo, W., Taylor, J.L., Hester, D.J., Mladinich, C.S., and Glavac, S., editors. U.S. Geological Survey Professional Paper 1726, pp. 129–137.

Thornthwaite, C.W., 1946. The moisture factor in climate. *Transactions, American Geophysical Union*, vol. 27, pp. 41–48.

Thornthwaite, C.W., 1948. An approach toward a rational classification of climate. *The Geological Review*, vol. January, pp. 55–94.

USDA (United States Department of Agriculture), 1986. Urban hydrology for small watersheds; TR-55. Natural Resources Conservation Service, Technical Release 55, Second Revised Edition, June 1986, Soil Conservation Service, Engineering Division, Washington, D.C., various pages and appendices.

Vaccaro, J.J., 2007. A deep percolation model for estimating ground-water recharge: Documentation of modules for the modular modeling system of the U.S. Geological Survey. U.S. Geological Survey Scientific Investigations Report 2006-5318, 30 p.

Van Genuchten, M.Th., 1980. A closed-form equation for predicting the hydraulic conductivity of unsaturated soils. *Soil Science Society of America Journal*, vol. 44, no. 5, pp. 892–898.

Van Genuchten, M.Th., Leij, F.J., and Yates, S.R., 1991. The RETC code for quantifying the hydraulic functions of unsaturated soils. EPA/600/2-91/065, Ada, Oklahoma, 83 p.

Warrick, J.A., and Orzech, K.M., 2006. The effects of urbanization on discharge and suspended-sediment concentrations in a Southern California river. In: *Rates, Trends, Causes, and Consequences of Urban Land-Use Change in the United States*. Acevedo, W., Taylor, J.L., Hester, D.J., Mladinich, C.S., and Glavac, S. editors. U.S. Geological Survey Professional Paper 1726, pp. 163–170.

Williams, J.R., Ouyang, Y., and Chen, J.-S., 1998. Estimation of infiltration rate in the vadose zone: Application of selected mathematical models, Volume II. EPA/600/R-97/128b, U.S. Environmental Protection Agency, Cincinnati, OH, 44 p. + appendices.

Wilson, J.L., and Guan, H., 2004. Mountain-block hydrology and mountain-front recharge. In: *Groundwater Recharge in a Desert Environment: The Southwestern United States.* Phillips, F.M., Hogan, J., and Scanlon, B., editors. American Geophysical Union, Washington, DC. Available at: http://www.utsa.edu/LRSG/Staff/Huade/publications/. Accessed September 2007.

Wythers, K.R., Lauenroth, W.K., and Paruelo, J.M., 1999. Bare soil evaporation under semiarid field conditions. *Soil Science Society of America Journal*, vol. 63, pp. 1341–1349.

Yurtsever, Y. 1999. An overview of nuclear science and technology in groundwater; Assessment/management and IAEA activities in the Gulf Region. In proceedings of 4th Gulf Intl Water Conference "Water in the Gulf, Challenges of the 21st century", Water Science and Technology Association, Bahrain, 13-19 February 1999.

Zaidi, F.K., Ahmed, S., Dewandel, B., and Maréchal, J-C., 2007. Optimizing a piezometric network in the estimation of the groundwater budget: A case study from a crystalline-rock watershed in southern India. *Hydrogeology Journal*, vol. 15, pp. 1131–1145.

CHAPTER 4

Climate Change

4.1 Introduction

The following words of Lester Snow, Director of the Association of California Water Agencies, summarize the crucial interconnection between climate, water supply, and population growth:

> The Colorado River has historically been an abundant source of supply for water users in the United States and Mexico. With growth of demands on this water supply, the time of historical abundance has ended. The previous five years of drought remain manifested in low reservoir levels. The Secretary of the Interior is beginning preparation of first-ever shortage criteria for the reservoir system. These conditions demonstrate the need for a strong scientific foundation in understanding climatic and hydrologic conditions that influence Colorado River water supplies. We know that droughts will inevitably occur in the future—a future made more uncertain by the impacts of climate change and increased hydrologic variability. Uncertainty surrounding the magnitude of water supplies will be coupled with increased competition for these supplies as the Southwest's population continues its rapid growth.
>
> I encourage those attending this conference to become informed about the uncertainties associated with our present understanding of Colorado River Basin climate and hydrology, and to incorporate them in water management decision-making. In California, we are placing increasing emphasis on integrated regional water management planning as a way to enable us to better respond to hydrologic variability through use of a diversified portfolio of resource management strategies (Snow, 2005).

Since 2005, several events took place, further reinforcing these words. Most notably, the phrases "climate change" and "global warming" became household words across the globe as the Intergovernmental Panel on Climate Change (IPCC) and the former vice president of the United States, Mr. Al Gore, both received the 2007 Nobel Peace Prize for their efforts in researching and explaining the many aspects and impacts of climate change. There are few scientific graphs, if any, that have become so widely reproduced in the media and discussed worldwide as the two shown here in Fig. 4.1. These illustrate the connection between an increasing concentration of carbon dioxide (CO_2) in the atmosphere caused by human activity such as burning of fossil fuels and an increasing global temperature. As if determined to convert a few remaining skeptics, 2007 was also the driest year on record in southern California, with a catastrophic drought affecting the southeastern United States as well. Exceptional droughts tend to galvanize politicians, economists, water professionals, and the general public in searching for answers to many drought-related problems, water supply being the most important. In that sense, 2007 may be called a "perfect year" for rethinking societal approaches to both water management and climate change.

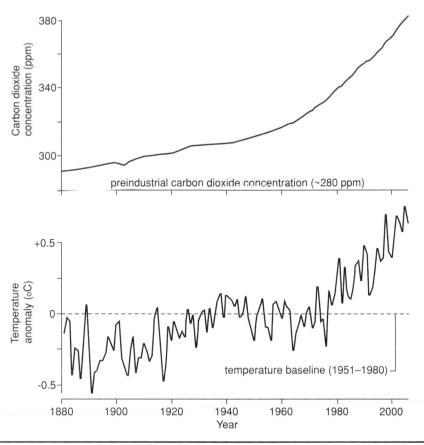

FIGURE 4.1 Global temperature anomaly and carbon dioxide concentration in parts per million since 1880. (From Riebek, 2007. NASA graphs by Robert Simmon, based on carbon dioxide data from Dr. Pieter Tans, NOAA/ESRL, and temperature data from NASA Goddard Institute for Space Studies.)

4.2 Natural Climatic Cycles

Climate is defined as an aggregate of weather conditions, representing a general pattern of weather variations at a location or in a region. It includes average weather conditions, as well as the variability of elements and information on the occurrence of extreme events (Lutgens and Tarbuck, 1995). The nature of both weather and climate is expressed in terms of basic elements, the most important of which are (1) the temperature of the air, (2) the humidity of the air, (3) the type and amount of cloudiness, (4) the type and amount of precipitation, (5) the pressure exerted by the air, and (6) the speed and direction of the wind. These elements constitute the variables by which weather patterns and climatic types are depicted (Lutgens and Tarbuck, 1995). The main difference between weather and climate is the time scale at which these basic elements change. Weather is constantly changing, sometimes from hour to hour, and these changes create almost an infinite variety of weather conditions at any given time and place. In comparison, climate changes are much more subtle and were, until relatively recently, considered important for time scales of hundreds of years or more, and usually only discussed in academic circles.

A more broad definition of climate is that it represents the long-term behavior of the interactive climate system, which consists of the atmosphere, hydrosphere, lithosphere, biosphere, and cryosphere or ice and snow that are accumulated on the earth's surface. To understand fully and to predict changes in the atmosphere component of the climate system, one must understand the sun, oceans, ice sheets, solid earth, and all forms of life (Lutgens and Tarbuck, 1995).

The most significant theory relating earth motions and long-term climate change, later confirmed with geologic and paleoclimatic evidence collected from around the globe, was developed in the 1930s by the Serbian mathematician and astrophysicist Milutin Milankovitch, professor at the University of Belgrade. His work titled *Kanon der Erdbestrahlung und seine Anwendung auf das Eiszeitenproblem* (*Canon of Insolation of the Earth and Its Application to the Problem of the Ice Ages*) was published in German in 1941 by the Royal Serbian Academy, but was largely ignored by the international scientific community. In 1969, it was translated into English and published with the title *Canon of Insolation of the Ice-Age Problem* by the U.S. Department of Commerce and the National Science Foundation, Washington, DC. In 1976, a study published in the journal *Science* examined deep-sea sediment cores and found that Milankovitch's theory did in fact correspond to periods of climate change (Hays et al., 1976). Specifically, the authors were able to analyze the record of temperature change going back 450,000 years and found that major variations in climate were closely associated with changes in the geometry (eccentricity, obliquity, and precession) of the earth's orbit; ice ages had indeed occurred when the earth was going through different stages of orbital variation. Since this study, the National Research Council of the U.S. National Academy of Sciences has embraced the Milankovitch cycle model (NRC, 1982):

> . . . orbital variations remain the most thoroughly examined mechanism of climatic change on time scales of tens of thousands of years and are by far the clearest case of a direct effect of changing insolation on the lower atmosphere of Earth.

Milankovitch was intrigued by the puzzle of climate change and studied climate records, noting differences over time. He theorized that global climate change was brought about by regular changes in earth's axis, tilt, and orbit that altered the planet's relationship to the sun, triggering ice ages. Milankovitch determined that the earth wobbles in its orbit and calculated the slow changes in the earth's orbit by careful measurement of the position of the stars and using the gravitational pull of other planets and stars. The three variables quantified by Milankovitch are now known as Milankovitch cycles:

1. Eccentricity cycle of the earth's orbit; every 90,000 to 100,000 years there is a change in the earth's orbit about the sun. Its almost circular orbit becomes more elliptical, taking earth farther from the sun.

2. The tilt of the earth's axis or obliquity cycle; on average, every 40,000 years there is a change in the tilt of the earth's equatorial plane in relation to its orbital plane, moving either the northern or the southern hemisphere farther from the sun.

3. Precession or orientation of the earth's rotational axis; on average, every 22,000 years there is a slight change in its wobble (the earth does not rotate perfectly like a wheel about an axis; it spins like a wobbling top).

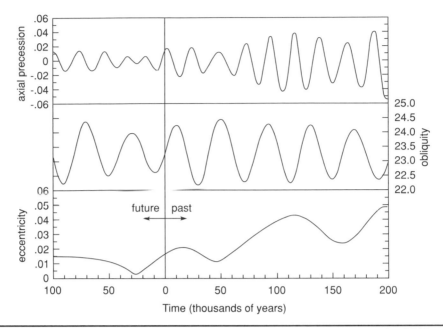

Figure 4.2   Calculated values for 300,000 years of Milankovitch cycles. (From NASA, 2007; source: Berger and Loutre, 1991.)

These cycles mean that during certain periods there is less solar energy arriving to the earth, resulting in less melting of snow and ice. Instead of melting, these cold expanses of frozen water grow. The snow and ice last longer and, over many seasons, begin to accumulate. Snow and ice reflect some sunlight back into space, which also contributes to cooling. Temperatures drop, and glaciers begin to advance (Tesla Memorial Society of New York, 2007).

The climate is influenced by all three cycles that can combine in a number of different ways, sometimes strongly reinforcing each other and sometimes working against each other. The general influence of the Milankovitch cycles on the long-term climate and their current state is presented below based on NASA (2007; see also Fig. 4.2).

The eccentricity of the earth's orbit changes slowly over time from nearly zero (circular) to 0.07 (eccentric). As the orbit becomes more eccentric (oval), the difference between the distance from the sun to the earth at perihelion (closest approach) and aphelion (furthest away) becomes greater and greater. Currently, a difference of only 3 percent (5 million km) exists between perihelion, which occurs on or about January 3, and aphelion, which occurs on or about July 4. This difference in distance amounts to about a 6 percent increase in incoming solar radiation (insolation) from July to January. The current trend of eccentricity is decreasing. When the orbit is highly elliptical, the amount of insolation received at perihelion would be on the order of 20 to 30 percent greater than at aphelion, resulting in a substantially different climate from what we experience today.

Today, the earth's axis is tilted 23.5° from the plane of its orbit around the sun. During a cycle that averages about 40,000 years, the tilt of the axis varies between 22.1° and 24.5°. Because of tilt changes, the seasons as we know them can become exaggerated. More tilt means more severe seasons—warmer summers and colder winters; less tilt means less severe seasons—cooler summers and milder winters. It is the cool summers that allow snow and ice to last from year to year in high latitudes, eventually building up into massive ice sheets. An earth covered with more snow reflects more of the sun's energy into space, causing additional cooling. The current trend in the earth's axis tilt is decreasing.

Changes in axial precession alter the dates of perihelion and aphelion and, therefore, increase the seasonal contrast in one hemisphere and decrease the seasonal contrast in the other hemisphere. If a hemisphere is pointed toward the sun at perihelion, that hemisphere will be pointing away at aphelion and the difference in seasons will be more extreme. This seasonal effect is reversed for the opposite hemisphere. Currently, the northern hemisphere summer occurs near aphelion, which means that the northern hemisphere should have somewhat less extremes between the seasons. The climatic precession is close to its peak and shows a decreasing trend.

Although the Milankovitch cycles can explain long-term climatic changes on geologic time scales (on the order of tens of thousands of years or more), their long duration makes them ineffective tools to explain or predict changes that are of significance for water resources evaluation and planning, namely, at time scales of decades to centuries. However, what we can learn from the well-established science of long term climate change and the geologic evidence of it occurring in the past is that it will inevitably occur in the future as well. A fourth cycle, not addressed by Milankovitch, may accelerate natural climate change—human activity on earth. The photograph in Fig. 4.3 is an evidence that

Figure 4.3 Cave divers in submerged cave passages with an abundance of speleothems—stalactites, stalagmites, flowstone, and columns—formed prior to submergence. Nohoch Nah Chich in the Yucatan Peninsula, 2007. (Photograph courtesy of David Rhea, Global Underwater Explorers.)

the sea level in the past was lower than it is today. One of the reasons is that the ice accumulated on the continents during the last ice age did melt to a large degree, causing a significant global sea level rise. As a result, the water table in coastal aquifers also rose, as evidenced from submerged caves in karst regions such as the Yucatan Peninsula in Mexico. Speleothems visible in the photograph could have only been formed when the cave was not submerged. Vast cave systems, many of which are now completely filled with freshwater like this one, were developed in the Yucatan when the sea level was lower than today.

Accurate and systematic measurements of weather and climate elements are paramount to fully understanding the climate of a region and anticipating future climatic changes that may impact water supplies. Unfortunately, records of air temperature and precipitation, the most important direct measures of climate, go back only several hundred years in Europe and less than that in other parts of the world. The situation is even worse with hydrologic measurements of streamflows or spring flows and worse yet with records of groundwater levels, the two most important direct measures of freshwater budget. Even though the time record of direct climatic and hydrologic measurements is increasing, it is becoming more and more evident that 100 hundred years or so is still too short to capture the statistics necessary for a more accurate probability analysis of the extreme climate events such as floods and droughts. For example, it was during a wet period in the measured hydrologic record that the 1922 Colorado River Compact established the basic apportionment of the river between the Upper and Lower Colorado River Basins in the United States. At the time of Compact negotiations, it was thought that an average annual flow volume of about 21 million acre-ft (MAF; 1 acre-ft equals 136.8 m^3) was available for apportionment. The Compact provided for 7.5 MAF of consumptive use annually for each of the basins, plus the right for the lower basin to develop 1 MAF of consumptive use annually. Subsequently, a 1944 Treaty with Mexico provided a volume of water of 1.5 MAF annually for Mexico. During the period of measured hydrology now available, the river's average annual natural flow has been about 15 MAF at Lee Ferry (ACWA and CRWUAC, 2005). This over-allocation of the Colorado River is now causing many political and societal problems in the region.

Studies in the last two decades have revealed that some climatic fluctuations once thought to be local phenomena are part of a large-scale atmospheric circulation that periodically affects global weather and contributes to long-term climate characteristics of different world regions. The best known and the most studied is ENSO (El Niño–Southern Oscillation). Centuries ago, the local residents on the coasts of Ecuador and Peru named a regular annual weather event El Niño ("the child") after the Christ child because it usually appeared during the Christmas season. During this event that lasts a few weeks, a weak, warm countercurrent flows southward along the coasts of Ecuador and Peru, replacing the cold Peruvian current. However, every 3 to 7 years this countercurrent is unusually warm and strong and is accompanied by a pool of warm ocean surface water in the central and eastern Pacific, which influences weather worldwide (Lutgens and Tarbuck, 1995).

The second strongest El Niño on record occurred in 1982 and 1983 (Fig. 4.4) and was blamed for weather extremes of a variety of types in many parts of the world. Heavy rains and flooding affected normally dry portions of Peru and Ecuador. Australia, Indonesia, and the Philippines experienced severe droughts, while one of the warmest winters on record was followed by one of the wettest springs for much of the United States. Heavy snows in the Sierra Nevada and the mountains of Utah and Colorado led to mudflows

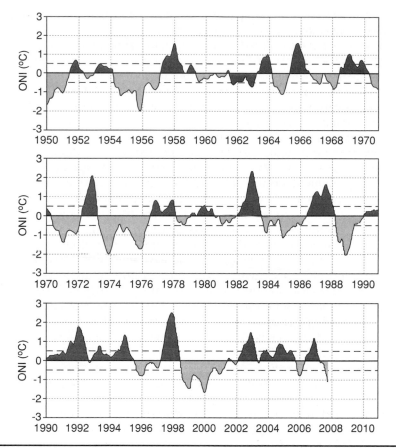

FIGURE 4.4 Oceanic Niño Index, ONI (°C), evolution since 1950. ONI is the principal measure for monitoring, assessing, and predicting ENSO. Positive values greater than +0.5 generally indicate El Niño conditions, and negative values less than –0.5 generally indicate La Niña conditions. (From CPC/NCEP, 2007b.)

and flooding in Utah and Nevada and along the Colorado River in the spring of 1983. The unusual rains brought floods to the Gulf States and Cuba. Unfortunately, as discussed by Lutgens and Tarbuck (1995), the effects of El Niño are highly variable, depending in part on the temperatures and size of the warm pools in the Pacific. During one El Niño, an area may experience flooding, only to be hit by drought during the next event. It is such extreme events that water managers both fear and are constantly preparing for. Climate Prediction Center (CPC) of the National Weather Service, National Oceanic and Aeronautic Administration (NOAA), maintains a Web page dedicated to the research and weather predictions associated with El Niño and La Niña events (CPC, 2007a).

The probability of floods and droughts is the key design element for water supply systems relying on surface water. Although systems based on groundwater are much less vulnerable to extreme weather events, they too can be stressed during prolonged droughts as a result of increased demand for water. Edwards and Redmond's discussion on the climatic conditions in the Colorado River Basin, the United States, illustrates the importance of understanding and predicting cyclical climate patterns for water supply management (Edwards and Redmond, 2005).

The waters of the Colorado River originate primarily in the high mountain basins of Colorado, Utah, and Wyoming and flow through seven states and two countries. With headwaters about 1500 and 1700 mi from the Gulf of California, the Colorado and the Green Rivers, respectively, contribute equally to about 80 percent of the flow into Lake Powell, with the remainder mostly from the San Juan River and Mountain Range. Within the 242,000 mi^2 U.S. portion of the Colorado River Basin (Basin), the highest one-seventh of the basin supplies about six-sevenths of the total flow, and many of the lower river reaches lose water under natural conditions. Most of the precipitation supply falls in winter as snow on interior mountain ranges. Spring precipitation can be important, but summer precipitation is usually nearly negligible in altering water supply, although it does influence demand. Thus, climatic influences on the interior mountain ranges are key factors governing the supply of water in the river from 1 year or decade to the next.

The warm phase of ENSO, El Niño, typically brings wet and cool winters to the southwest United States and dry and warm winters to the Pacific northwest and northern Rockies. Overall, El Niño winters tend to have more wet days, more precipitation per wet day, and more persistent wet episodes in the southwestern United States. All of these favor increased runoff. Notably, extremely high or low flow is better correlated with ENSO than is total runoff volume. The opposite cool phase of ENSO, La Niña, has been reliably associated with dry and warm winters in the Southwest for the past 75 years and, to a less reliable extent, with wet and cool winters in the northern West. The understanding of ENSO and its effects on the Basin are crucial in predicting winter snowpack. So far, western North America climate relationships to ENSO appear to be confined to the winter, with slight or ambiguous associations with summer climate. In the Colorado River Basin, the strongest relationships are seen in the lower basin, south of the San Juan Mountains of Colorado. The relationship becomes less clear farther north and begins to have the opposite effect in the upper Green River Basin and the Wind River Mountains in Wyoming.

On the basis of the analysis of the Colorado River Basin's climate, Edwards and Redmond (2005) offer the following summary:

> Through multiple re-use, the river provides water supply needs for 28 million people, a number expected to continue to grow in coming decades as the Southwest's population continues to expand at the fastest rate in the nation. The river basin has been developed through an extensive infrastructure system that was designed to buffer against the region's significant climate variability. Of note, however, the system has not been thoroughly tested by events of the magnitude that we have learned from the paleoclimate record may occur. The recent drought has provided a taste of what is possible, though not the full meal.

Figures 4.5. and 4.6 illustrate the combined impact of several recent droughts and water use on Lake Mead, one of the most important water resources in the American west. Created in the 1930s, it ensures a steady water supply for Arizona, Nevada, California, and northern Mexico by holding back the flow of the Colorado River behind the Hoover dam. It is one of the largest water reservoirs in the world. When full, the lake contains roughly the same amount of water as would have otherwise flowed through the Colorado River over a 2-year period: roughly 36 trillion liters (9.3 trillion gallons). Ninety percent of southern Nevada's water comes from Lake Mead, with releases being regulated by the

FIGURE 4.5 Lake Mead in 2004. (Photograph courtesy of Andy Pernick, the U.S. Bureau of Conservation.)

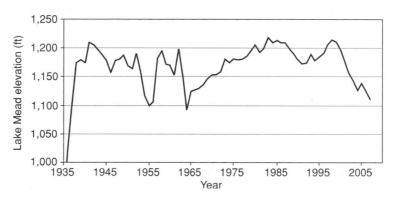

FIGURE 4.6 Lake Mead level for September, 1935–2007. (Source of raw data: U.S. Bureau of Reclamation.)

Southern Nevada Water Authority. When the water levels in the lake are declining and expected to cross below an elevation of 1145 ft, the Water Authority declares a drought watch. Once the water is below 1145 ft, the watch is shifted to a drought alert. If the lake level drops below 1125 ft, a drought emergency goes into effect. Each of the water level alert states triggers various water restrictions and practices in the area, from restrictions on watering gardens, washing cars, running fountains in civic parks, and public places to increases in the rates charged for water to encourage conservation (Allen, 2003).

In April of 2007, the water level dropped below 1125 ft for the first time since 1965 and remained below this benchmark through September 2007, the last month with data available to the author. The graph in Fig. 4.6 shows that it took about 20 years for the lake levels to recover from the 1965 low. As discussed throughout this chapter, the overallocation of the Colorado River water, combined with population growth and the impact of droughts, is putting additional stress on groundwater resources in this semiarid to arid region where natural groundwater recharge is very low.

4.2.1 Droughts

As pointed out by the National Drought Mitigation Center (NDMC, 2007), drought is a normal, recurrent feature of climate, although many erroneously consider it a rare and random event. Graphs like the one shown in Fig. 4.7 remind us of this simple fact. It is understandable, however, that every current drought may always be the hardest ever for the people affected by it, since human memory tends to block unpleasant experiences from the past. (*Note*: As opposed to the general public, water resources managers are not expected to have this characteristic.) When droughts are of historic proportions, they may trigger major societal changes and forever impact the use and management of water resources. For example, the major drought of the twentieth century in the United States, in terms of duration and spatial extent, is considered to be the 1930s Dust Bowl drought, which lasted up to 7 years in some areas of the Great Plains (Fig. 4.8). This drought, memorialized in John Steinbeck's novel *The Grapes of Wrath* was so severe, widespread, and lengthy that it resulted in a mass migration of millions of people from the Great Plains to the western United States in search of jobs and better living conditions. It also dramatically changed agricultural practices including the unprecedented large-scale use of groundwater for irrigation across the Great Plains and throughout the American west.

Although drought has scores of definitions, it originates from a deficiency of precipitation over an extended period of time, usually a season or more. This deficiency results in a water shortage for some activity, group, or environmental sector. Drought should be considered relative to some long-term average condition of balance between precipitation and evapotranspiration in a particular area, a condition often perceived as "normal."

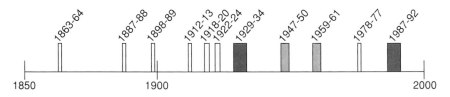

FIGURE 4.7 California's multiyear historical dry periods of statewide or major regional extent, 1850–2000. Dry periods prior to 1900 are estimated from limited data. (Source: http://watersupplyconditions.water.ca.gov/.)

Figure 4.8 A great "roller" of dust moves across the land in Colorado during the Dust Bowl of 1930s. (Photograph courtesy of National Resources Conservation Service.)

It is also related to the timing (such as principal season of occurrence, delays in the start of the rainy season, and occurrence of rains in relation to principal crop growth stages) and the effectiveness of the rains such as rainfall intensity and number of rainfall events. Other climatic factors such as high temperature, high wind, and low relative humidity are often associated with droughts in many regions of the world and can significantly aggravate their severity (NDMC, 2007).

Drought should not be viewed as merely a physical phenomenon or natural event. Its impacts on society result from the interplay between a natural event (less precipitation than expected resulting from natural climatic variability) and the demand people place on water supply. Human beings often exacerbate the impact of drought. Recent droughts in both developing and developed countries and the resulting economic and environmental impacts and personal hardships have underscored the vulnerability of all societies to this natural hazard (NDMC, 2007).

Two main drought definitions are conceptual and operational. Conceptual definitions, formulated in broad terms, help the general public understand the concept of drought. For example, "drought is a protracted period of deficient precipitation resulting in extensive damage to crops, resulting in loss of yield." Conceptual definitions may also be important in establishing drought policy. For example, Australian drought policy incorporates an understanding of normal climate variability into its definition of drought. The country provides financial assistance to farmers only under "exceptional drought circumstances," when drought conditions are beyond those that could be considered part of normal risk management. Declarations of exceptional drought are based on science-driven assessments. Previously, when drought was less well defined from a policy standpoint and less well understood by farmers, some farmers in the semi-arid Australian climate claimed drought assistance every few years (NDMC, 2007).

Operational definitions of drought help identify the beginning, end, and degree of severity of a drought. To determine the beginning of drought, operational definitions specify the degree of departure from the average of precipitation or some other climatic variable over some time period. This is usually done by comparing the current situation to the historical average, often based on a 30-year period of record. The threshold identified as the beginning of a drought (e.g., 75 percent of average precipitation over a specified time period) is usually established somewhat arbitrarily, rather than on the basis of its precise relationship to specific impacts.

An operational definition for agriculture might compare daily precipitation values to evapotranspiration rates to determine the rate of soil moisture depletion and then express these relationships in terms of drought effects on plant behavior (i.e., growth and yield) at various stages of crop development. Operational definitions can also be used to analyze drought frequency, severity, and duration for a given historical period. Developing a climatology of drought for a region provides a greater understanding of its characteristics and the probability of recurrence at various levels of severity. Information of this type is extremely beneficial in the development of response and mitigation strategies and preparedness plans (NDMC, 2007).

Although the major droughts of the twentieth century, the 1930s Dust Bowl and the 1950s droughts, had the most severe impact on the central United States., droughts regularly occur all across North America. Florida suffered from the 1998 drought along with the states of Oklahoma and Texas. Extensive drought-induced fires burned over 475,000 acres in Florida and cost $500 million in damages. In the same year, Canada suffered its fifth-highest fire occurrence season in 25 years. Starting in 1998, 3 years of record low rainfall plagued northern Mexico. The year 1998 was declared the worst drought in 70 years. It became worse as 1999 spring rainfalls were 93 percent below normal. The government of Mexico declared five northern states as disaster zones in 1999 and nine in 2000. The U.S. west coast experienced a 6-year drought in the late 1980s and early 1990s, causing Californians to take aggressive water conservation measures. Even the typically humid northeastern United States experienced a 5-year drought in the 1960s, draining reservoirs in New York City down to 25 percent of capacity (NCDC, 2007a).

The impact of the 2007 drought in southern California and the southeastern United States is yet to be assessed, although it is already apparent that it will have a major influence on water management decisions. For example, the governor of California and the democratic party-led state legislature came to an impasse over the emergency state investments in major water supply projects, with the governor favoring construction of large surface water reservoirs and the legislature favoring the use of groundwater and artificial aquifer recharge.

Drought is a natural hazard that cumulatively has affected more people in North America than any other natural hazard (Riebsame et al., 1991). In the United States, the cost of losses due to drought averages $6 to $8 billion every year but range as high as $39 billion for the 3-year drought of 1987 to 1989, which was the most costly natural disaster documented in U.S. history at the time. Continuing uncertainty in drought prediction contributes to crop insurance payouts of over $175 million per year in western Canada (NCDC, 2007a).

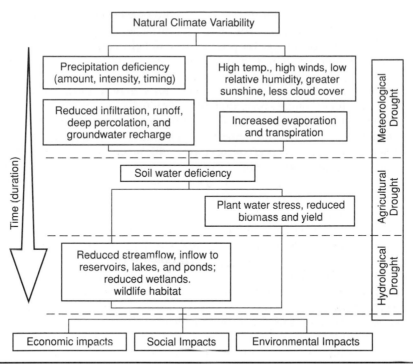

FIGURE 4.9 Three categories of drought identified by the National Drought Mitigation Center. (From NDMC, 2007.)

Figure 4.9 illustrates the concept of three different drought categories identified by the National Drought Mitigation Center as follows:

Agricultural drought links various characteristics of meteorological (or hydrological) drought to agricultural impacts, focusing on precipitation shortages, differences between actual and potential evapotranspiration, soil water deficits, reduced groundwater or reservoir levels, and so forth. Crop water demand depends on prevailing weather conditions, biological characteristics of the specific plant, its stage of growth, and the physical and biological properties of the soil. A good definition of agricultural drought should be able to account for the variable susceptibility of crops during different stages of crop development, from emergence to maturity. Deficient topsoil moisture at planting may hinder germination, leading to low plant populations per hectare and a reduction of final yield. However, if topsoil moisture is sufficient for early growth requirements, deficiencies in subsoil moisture at this early stage may not affect final yield if subsoil moisture is replenished as the growing season progresses or if rainfall meets plant water needs.

Hydrological drought is concerned with the effects of periods of precipitation (including snowfall) shortfalls on surface or subsurface water supply (such as streamflow, reservoir and lake levels, and groundwater). The frequency and severity of hydrological drought are often defined on a watershed or river basin scale. Although all droughts originate with a deficiency of precipitation, hydrologists are more concerned with how this deficiency plays out through the hydrologic system. Hydrological droughts are usually out of phase with or lag the occurrence of meteorological and agricultural droughts. It takes longer for precipitation deficiencies to show up in components of the hydrological system such as soil moisture, streamflow, and groundwater and reservoir levels. As a result, these impacts are out of phase with impacts in other economic sectors. For example, precipitation deficiency may result in rapid depletion of soil moisture that is almost immediately discernible to agriculturalists, but the impact of this deficiency on reservoir levels may not affect hydroelectric power production or recreational uses for many months. Also, water in hydrologic storage systems (e.g., reservoirs and rivers) is often used for multiple and competing purposes (e.g., flood control, irrigation, recreation, navigation, hydropower, and wildlife habitat), further complicating the sequence and quantification of impacts. Competition for water in these storage systems escalates during drought, and conflicts between water users increase significantly.

Socioeconomic drought definition associates the supply and demand of some economic good with elements of meteorological, hydrological, and agricultural drought. It differs from the aforementioned types of drought because its occurrence depends on the time and space processes of supply and demand to identify or classify droughts. The supply of many economic goods, such as water, forage, food grains, fish, and hydroelectric power, depends on weather. Because of the natural variability of climate, water supply is ample in some years but unable to meet human and environmental needs in other years. Socioeconomic drought occurs when the demand for an economic good exceeds supply as a result of a weather-related shortfall in water supply. For example, in Uruguay in 1988 to 1989, drought resulted in significantly reduced hydroelectric power production because power plants were dependent on streamflow rather than storage for power generation. Reducing hydroelectric power production required the government to convert to more expensive (imported) petroleum and enforce stringent energy conservation measures to meet the nation's power needs.

In most instances, the demand for economic goods is increasing as a result of increasing population and per capita consumption. Supply may also increase because of improved production efficiency, technology, or the construction of reservoirs that increase surface water storage capacity. If both supply and demand are increasing, the critical factor is the relative rate of change. Is demand increasing more rapidly than supply? If so, vulnerability and the incidence of drought may increase in the future as supply and demand trends converge (NDMC, 2007).

The sequence of impacts associated with meteorological, agricultural, and hydrological drought further emphasizes their differences. When drought begins, the agricultural sector is usually the first to be affected because of its heavy dependence on stored soil water. Soil water can be rapidly depleted during extended dry periods. If precipitation deficiencies continue, then people dependent on other sources of water will begin to feel the effects of the shortage. Those who rely on surface water (i.e., reservoirs and lakes) and groundwater are usually the last to be affected. A short-term drought that persists for 3 to 6 months may have little impact on these sectors, depending on the characteristics of the hydrologic system and water use requirements.

When precipitation returns to normal and meteorological drought conditions have abated, the sequence is repeated for the recovery of surface and subsurface water supplies. Soil water reserves are replenished first, followed by streamflow, reservoirs and lakes, and finally groundwater. Drought impacts may diminish rapidly in the agricultural sector because of its reliance on soil water, but linger for months or even years in other sectors dependent on stored surface or subsurface supplies. Groundwater users, often the last to be affected by drought during its onset, may be the last to experience a return to normal water levels. The length of the recovery period is a function of the intensity of the drought, its duration, and the quantity of precipitation received as the episode terminates (NDMC, 2007).

Drought Indices

Drought indices assimilate data on rainfall, snowpack, streamflow, and other water supply indicators into a comprehensible big picture. A drought index value is typically a single number, far more useful than raw data for decision making. There are several indices that measure how much precipitation for a given period of time has deviated from historically established norms. Although none of the major indices is inherently superior to the rest in all circumstances, some indices are better suited than others for certain uses. For example, the Palmer Drought Severity Index (PDSI or The Palmer) has been widely used by the U.S. Department of Agriculture to determine when to grant emergency drought assistance. The Palmer index is better when working with large areas of uniform topography. Western states, with mountainous terrain and the resulting complex regional microclimates, find it useful to supplement Palmer values with other indices such as the Surface Water Supply Index, which takes snowpack and other unique conditions into account. Detailed discussion of various drought indices, including their advantages and drawbacks, is given by Hayes (2007).

The National Drought Mitigation Center is using a newer index, the Standardized Precipitation Index (SPI), to monitor moisture supply conditions. Some distinguishing traits of this index are that it identifies emerging drought months sooner than the Palmer index and that it is computed on various time scales. The understanding that a deficit of precipitation has different impacts on groundwater, reservoir storage, soil moisture, snowpack, and streamflow led McKee, Doesken, and Kleist of Colorado State University to develop the SPI in 1993. The SPI was designed to quantify the precipitation deficit for multiple time scales. These time scales reflect the impact of drought on the availability of different water resources. Soil moisture conditions respond to precipitation anomalies on a relatively short scale. Groundwater, streamflow, and reservoir storage reflect the longer-term precipitation anomalies. For these reasons, McKee et al. (1993) originally calculated the SPI for 3-, 6-, 12-, 24-, and 48-month time scales.

The SPI calculation for any location is based on the long-term precipitation record for a desired period. This long-term record is fitted to a probability distribution, which is then transformed into a normal distribution so that the mean SPI for the location and desired period is zero (Edwards and McKee, 1997). Positive SPI values indicate greater than median precipitation, and negative values indicate less than median precipitation. Because the SPI is normalized, wetter and drier climates can be represented in the same way, and wet periods can also be monitored using the SPI. PSI values smaller than –2 indicate extreme drought and greater than 2 extreme wet conditions.

Occurrence of Droughts

Instrumental records of drought for the United States extend back approximately 100 years. These records capture the twentieth-century droughts but are too short to assess the reoccurrence of major droughts such as those of the 1930s, 1950s, and 2000s.

As droughts continue to have increasingly costly impacts on the society, economy, and environment, it is becoming even more important to put the severe droughts of the twentieth and the beginning of the twenty-first century into a long-term water management perspective. This perspective can be gained through the use of paleoclimatic records of drought, called proxies. Paleoclimate is the climate of the past, before the development of weather-recording instruments, and is documented in biological and geological systems that reflect variations in climate in their structure. Different proxies record variations in drought conditions on the order of single seasons to decadal- and century-scale changes, providing scientists with the information about both rapid and slow changes and short and long periods of drought.

Historical records, such as diaries and newspaper accounts, can provide detailed information about droughts for the last 200 (mid-western and western United States) or 300 (eastern United States) years. Tree-ring records can extend back 300 years in most areas and thousands of years in some regions. In trees that are sensitive to drought conditions, tree rings provide a record of drought for each year of the tree's growth. Geologic evidence is used for records longer than those provided by trees and historical accounts and for regions where such accounts are absent. It includes analysis of lake sediments and their paleontologic content, and sand dunes (NCDC, 2007a).

Lake sediments, if the cores of the sediments are sampled at very frequent intervals, can provide information about variations occurring at frequencies less than a decade in length. Fluctuations in lake level can be recorded from beach material sediments (geologic bath tub rings), which are deposited either high (further from the center under wetter conditions) or low (closer to the center under drier conditions) within a basin as the water depth and thus lake level change in response to drought. Droughts can increase the salinity of lakes, changing the species of small, lake-dwelling organisms that occur within a lake. Pollen grains get washed or blown into lakes and accumulate in sediments. Different types of pollen in lake sediments reflect the vegetation around the lake and the climate conditions that are favorable for that vegetation. For example, a change in the type of pollen found in sediments from an abundance of grass pollen to an abundance of sage pollen can indicate a change from wet to dry conditions (NCDC, 2007a).

Records of more extreme environmental changes can be found by investigating the layers within sand dunes. The sand layers are interspersed among layers of soil material produced under more wet conditions, between the times when the sand dune was active. For a soil layer to develop, the climate needs to be wet for an extended period of time; such layers therefore reflect slower, longer-lasting changes.

Large areas of the intermountain basins of the western United States contain sand dunes and other dune-related features, most of which are now stabilized by vegetation. Sand dunes and sand sheets were deposited by the wind in times of drought and contain a wealth of information about episodes of drought and aridity over the course of the Holocene, which is the period since the end of the most recent widespread glaciation, about 10,000 years ago. The soil layers, which are interspersed with sand layers, contain organic materials and can be dated with radiocarbon dating techniques. The dates from the soil layers between layers of sand can be used to bracket times of drought as signified by the presence of sand. Since there is a lag in time in the vegetation and dune response to climate conditions, this record is fairly coarse in terms of time scales that it can resolve (typically centuries or longer). In addition, radiocarbon dating, with a dating precision of 5 percent (or more during certain periods in the Holocene), contributes to low temporal resolution of this record. However, recent work has used optically stimulated

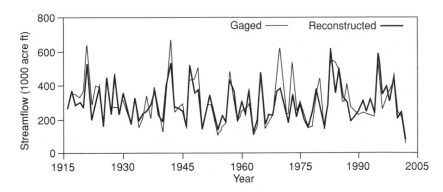

FIGURE 4.10 Calibration/verification period (1916–2002) for a tree-ring reconstruction of the South Platte River at South Platte, CO. The reconstruction explains most of the variance ($R^2 = 0.76$) of the gaged record and captures the extreme low flows, including 2002. The gaged flow record, corrected for depletions, was provided by Denver Water. Units for flow are 1000 acre-ft. (From Woodhouse and Lukas, 2005.)

luminescence techniques to date sand grains, producing records with a decadal scale resolution for the past 1000 years (Woodhouse, 2005).

Tree rings provide annually or seasonally resolved data that are precisely dated to the calendar year. Tree-ring records commonly extend 300 to 500 years into the past, and a small number are thousands of years long. Trees that are sensitive to climate reflect variations in climate in the width of their annual rings. Thus, the ring-width patterns contain records of past climate. Trees that grow in arid or semiarid areas and on open, dry, south-facing slopes are stressed by a lack of moisture. These trees can be used for reconstructing climate variables such as precipitation, streamflow, and drought. To develop a reconstruction of past climate, tree-ring data are calibrated with an instrumental record for the period of years common to both. This process yields a statistical model that is applied to the full length of the tree-ring data to generate a reconstruction of past climate. The reconstructions are only estimates of past climate, as the tree-ring-based reconstructions do not explain all the variance in the instrumental records. However, they can explain up to 60 to 75 percent of the total variance in an instrumental record (Fig. 4.10; Woodhouse and Lukas, 2005). A detailed explanation of the tree-ring paleoclimate dating including examples is given in Meko et al. (1991) and Cook et al. (1999).

A remarkably widespread and persistent period of drought in the late-sixteenth century is evident in a large number of various proxy records for the western United States. Tree-ring data document drought conditions that ranged across western North America from northern Mexico to British Columbia. Tree-ring-based streamflow reconstructions for the Sacramento River and Blue River (in the Upper Colorado River watershed) show concurrent drought conditions in both of these watersheds in the late-sixteenth century. This was one of the few periods of drought shared by both the Sacramento and the Blue River reconstructions in over 500 years and common to both records. During the period from 1580 to 1585, there were 4 years with concurrent drought conditions in both watersheds. Drought was particularly severe in the Sacramento River reconstruction, which indicated the driest 3-year period in the entire reconstruction (extending to AD 869) was 1578 to 1580. In addition to the western United States, there is also evidence of severe sustained drought in the western Great Plains about this time, with widespread mobilization of sand dunes in eastern Colorado and the Nebraska Sand Hills.

Analysis of Ni et al. (2002) is one of the cases studies confirming the late-sixteenth century megadrought and indicating similarly severe earlier droughts in Arizona and New Mexico. The authors developed a 1000-year reconstruction of cool-season (November–April) precipitation for each climate division in Arizona and New Mexico from a network of 19 tree-ring chronologies in the southwestern USA. Linear regression (LR) and artificial neural network (NN) models were used to identify the cool-season precipitation signal in tree rings. By using 1931 to 1988 records, the stepwise LR model was cross-validated with a leave-one-out procedure and the NN was validated with a bootstrap technique. The final models were also independently validated using the 1896 to 1930 precipitation data. In most of the climate divisions, both techniques can successfully reconstruct dry and normal years, and the NN seems to capture large precipitation events and more variability better than the LR. In the 1000-year reconstructions, the NN also produces more distinctive wet events and more variability, whereas the LR produces more distinctive dry events. Figure 4.11 shows the results of the combined model (LR + NN) for one of Arizona's climate divisions.

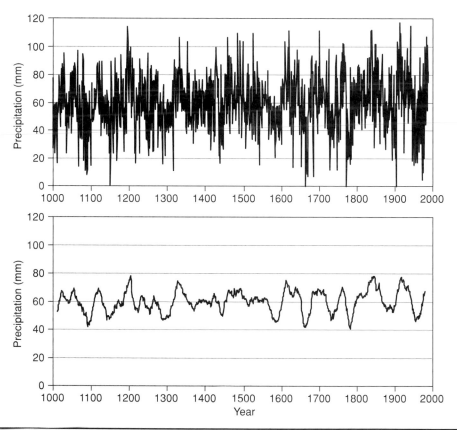

Figure 4.11 A 1000-year reconstruction of cool-season (November–April) precipitation for the Arizona Climate Division No. 5. Bottom: 21-year moving average. (Source of raw data: Ni et al., 2002.)

Several observations can be made from the bottom graph, which depicts the 21-year moving average. Among sustained dry periods comparable in duration and low precipitation to the observed 1950s drought, only two have somewhat similar wet periods preceding the droughts: 1730s and mid-1600s. In both cases, however, the sustained low precipitation is lower than the 1950s drought, and the preceding wet periods generally have lower precipitation. The fifteenth century megadrought occurred after a long period of average precipitation, without any significant wet periods. Even worse conditions are visible for the late-1200s megadrought, which was preceded by two shorter droughts during the first half of the thirteenth century. The droughts of late 1000s and mid-1100s were also likely more severe than the 1950s drought. Finally, the top graph with the model result on the annual basis shows that there were four seasons without any precipitation, the situation not recorded for the observed 1896 to 1988 period.

In the most recent study of paleo flows in the Colorado River Basin based on tree-ring records, Meko et al. (2007) show a very good agreement between the mid-1100s precipitation drought in central Arizona apparent in Fig. 4.11 and the Colorado River flow at Lee Ferry. The corresponding hydrologic drought is the most extreme low-frequency feature of the new reconstruction, covering AD 762 to 2005. It is characterized by a decrease of more than 15 percent in mean annual flow averaged over 25 years and by the absence of high annual flows over a longer period of about six decades. The drought is consistent in timing, with dry conditions inferred from tree-ring data in the Great Basin and Colorado Plateau, but regional differences in intensity emphasize the importance of basin-specific paleoclimatic data in quantifying likely effects of drought on water supply (Meko et al., 2007).

The National Climatic Data Center of the NOAA has compiled a Web site of existing hydroclimatic reconstructions (streamflow, precipitation, and drought indices) for California based on tree-ring data. The site also shows locations of existing tree-ring chronologies that could be used to generate additional reconstructions. Links are provided for similar information for the Colorado River Basin, an important source of water supply for southern California (http://www.ncdc.noaa.gov/paleo/streamflow/ca/reconstructions.html).

4.3 Anthropogenic Climate Change

There may not be any skeptical world governments left when it comes to the hard scientific evidence regarding anthropogenic causes of accelerated global warming. This may still not be true for some scientists who argue that the observed global rise in temperature is just part of the normal and well-documented, long-term Milankovitch cycles of climate change. It is very likely that the change of the official position of some governments, notably of Australia and the United States, was triggered by a series of reports produced by the IPCC and released during 2007. The reports are available for free download at the following Web site: http://www.ipcc.ch/ipccreports/assessments-reports.htm.

What has also undeniably helped in turning the attention of the general public in the United States to this scientific effort is the reality on the ground, as illustrated by the following excerpts from a report issued by the National Climatic Data Center in January of 2007 (NCDC, 2007b):

The 2006 average annual temperature for the contiguous U.S. was the warmest on record and nearly identical to the record set in 1998, according to scientists at NOAA's National Climatic Data Center in Asheville, N.C. Seven months in 2006 were much warmer than average, including December, which ended as the fourth warmest December since records began in 1895. Based on preliminary data, the 2006 annual average temperature was 55°F or 2.2°F (1.2°C) above the 20th century mean and 0.07°F (0.04°C) warmer than 1998. These values were calculated using a network of more than 1,200 U.S. Historical Climatology Network stations. These data, primarily from rural stations, have been adjusted to remove artificial effects resulting from factors such as urbanization and station and instrument changes which occurred during the period of record.

U.S. and global annual temperatures are now approximately 1.0°F warmer than at the start of the 20th century, and the rate of warming has accelerated over the past 30 years, increasing globally since the mid-1970's at a rate approximately three times faster than the century-scale trend. The past nine years have all been among the 25 warmest years on record for the contiguous U.S., a streak which is unprecedented in the historical record.

The IPCC (2007) shows that 11 of the last 12 years (1995 to 2006) rank among the 12 warmest years in the instrumental record of global surface temperature (since 1850). The temperature increase is widespread over the globe and is greater at higher northern latitudes. Land regions have warmed faster than the oceans.

Earth is naturally heated by the incoming energy from the sun. Over the long term, the amount of incoming solar radiation absorbed by the earth and atmosphere is balanced by the earth and atmosphere releasing the same amount of outgoing longwave (thermal) radiation. About half of the incoming solar radiation is absorbed by the earth's surface. This energy is transferred to the atmosphere by warming the air in contact with the surface (thermals), by evapotranspiration, and by thermal radiation that is absorbed by clouds and greenhouse gases (GHGs). The atmosphere, in turn, radiates thermal energy back to earth as well as out to space (Kiehl and Trenberth, 1997).

Thermal (longwave) radiation that leaves the earth (is not radiated back to the land surface) allows earth to cool. The portion that is reabsorbed by water vapor, CO_2, and other greenhouse in the atmosphere (called GHGs because of their heat-trapping capacity) and then re-radiated back toward the earth's surface contributes to its heating. On the whole, these reabsorption and re-radiation processes are beneficial. If there were no GHGs or clouds in the atmosphere, the earth's average surface temperature would be a very chilly −18°C (0°F) instead of the comfortable 15°C (59°F) that it is today (Riebek, 2007).

Changes in atmospheric concentrations of GHGs and aerosols, land cover, and solar radiation all can alter the energy balance of the climate system. Global atmospheric concentrations of main GHGs, CO_2, methane (CH_4), and nitrous oxide (N_2O), have increased markedly as a result of human activities since 1750 and now far exceed preindustrial values determined from ice cores spanning back 650,000 years. CO_2 is the most important anthropogenic GHG. Its annual emissions grew by about 80 percent between 1970 and 2004 (IPCC, 2007). The long-term trend of declining CO_2 emissions per unit of energy supplied reversed after 2000, causing an additional alarm as energy consumption is increasing in all countries. Global increases in CO_2 concentrations are due primarily to fossil fuel use, with land use change providing another significant but smaller contribution. It is very likely that the observed increase in CH_4 concentration is predominantly due to agriculture and fossil fuel use. Methane growth rates have declined since the early 1990s, consistent with total emission (sum of anthropogenic and natural sources) being nearly constant during this period. The increase in N_2O concentration is primarily due to agriculture (IPCC, 2007).

Observations of the CO_2 concentrations in the atmosphere and the related increasing temperature (see Fig. 4.1), as well as hundreds of scientific studies analyzed by the IPCC, univocally support the theory that GHGs are warming the world. Between 1906 and 2006, the average surface temperature of the earth rose 0.74°C. Average northern hemisphere temperatures during the second half of the twentieth century were very likely higher than during any other 50-year period in the last 500 years and likely the highest in at least the past 1300 years.

Consistent with warming, global average sea level has risen since 1961 at an average rate of 1.8 (1.3 to 2.3) mm/year and since 1993 at 3.1 (2.4 to 3.8) mm/year, with contributions from thermal expansion, melting glaciers and ice caps, and the polar ice sheets. The world's mountain glaciers and snow cover have receded in both hemispheres, and Arctic sea ice extent has been continuously shrinking by 2.7 percent per decade since 1978, with larger decreases observed during summers (average decrease of 7.4 [range 5.0 to 9.8] percent per decade; IPCC, 2007).

As the 2007 reports by the IPCC were being released, new data from satellites and ground observations indicated that the rate of ice melting is even faster than originally estimated. One of the mechanisms of particular concern is the accelerated formation of basal streams of melted water flowing at the base of continental glaciers (Fig. 4.12). Basal streams can cause accelerated melting and retreat of glaciers, as well as their increased movement and sliding into the oceans. This means that the worldwide observed sea level rise may continue at a more rapid rate than what various models of global warming have predicted. A faster shrinking of snow and ice cover will also produce faster global warming because less solar energy will be reflected back into the atmosphere.

Figure 4.12   Basal glacial stream flowing into the Gulf of Alaska near Seward, AK. Ice discharge can be accelerated by basal streams like this one. (Photograph courtesy of Jeff Manuszak.)

The emerging evidence suggests that global warming is already influencing the weather. For example, the IPCC reports the following:

- From 1900 to 2005, precipitation increased significantly in eastern parts of North and South America, northern Europe, and northern and central Asia but declined in the Saharan Africa, the Mediterranean, southern Africa, and parts of southern Asia. Globally, the area affected by drought has likely increased since the 1970s.

- It is very likely that over the past 50 years cold days, cold nights, and frosts have become less frequent over most land areas, and hot days and hot nights have become more frequent.

- It is likely that heat waves have become more frequent over most land areas, the frequency of heavy precipitation events has increased over most areas, and since 1975 the incidence of higher sea levels has increased worldwide.

- There is observational evidence of an increase in intense tropical cyclone activity in the North Atlantic since about 1970, with limited evidence of increases elsewhere.

By using best estimates from various modeled scenarios, the IPCC projects that average surface temperatures could rise between 1.8°C and 4°C by the end of the twenty-first century. Taking into consideration modeling uncertainties, the likely range of these scenarios is between 1.1°C and 6.4°C (IPCC, 2007). The lower estimates come from best-case scenarios in which environmental-friendly technologies such as fuel cells and solar panels replace much of today's fossil-fuel combustion.

As discussed by Riebek (2007), these numbers probably do not seem threatening at first glance. After all, temperatures typically change a few tens of degrees whenever a storm front moves through. Such temperature changes, however, represent day-to-day regional fluctuations. When surface temperatures are averaged over the entire globe for extended periods of time, it turns out that the average is remarkably stable. Not since the end of the last ice age 20,000 years ago, when earth warmed about 5°C, has the average surface temperature changed as dramatically as the 1.1 to 6.4°C plausible change predicted for the twenty-first century.

Regional-scale changes projected by the IPCC models are as follows:

- Warming will be greatest over land and at most high northern latitudes and least over Southern ocean and parts of the North Atlantic ocean, continuing recent observed trends in contraction of snow cover area, increases in thaw depth over most permafrost regions, and decrease in sea ice extent; in some projections, Arctic late-summer sea ice disappears almost entirely by the latter part of the twenty-first century.

- A likely increase in the frequency of hot extremes, heat waves, and heavy precipitation.

- A likely increase in tropical cyclone intensity.

- A poleward shift of extra-tropical storm tracks with subsequent changes in wind, precipitation, and temperature patterns.

- Precipitation increase in high latitudes and likely decreases in most subtropical land regions, continuing observed recent trends.

Particularly alarming is the possibility of some abrupt or irreversible changes, which cannot be simulated by various models or show less agreement between models but nevertheless cannot be excluded. According to IPCC, they are as follows:

- Partial loss of ice sheets on polar land could imply meters of sea-level rise, major changes in coastlines, and inundation of low-lying areas, with the greatest effects in river deltas and low-lying islands. Such changes are currently projected to occur over millennial time scales, but more rapid sea-level rise on century time scales cannot be excluded.

- Contraction of the Greenland ice sheet is projected to continue to contribute to sea-level rise after 2100. Current models suggest virtually complete elimination of the Greenland ice sheet and a resulting contribution to sea-level rise of about 7 m if global average warming were sustained for a millennium in excess of 1.9 to 4.6°C relative to preindustrial values. The future temperatures in Greenland will be comparable to those for the last interglacial period 125,000 years ago, when paleoclimatic information suggests a reduction in polar land ice and a 4 to 6 m of sea-level rise. However, as in the case of the Antarctic, net loss of ice mass may happen faster if dynamic ice discharge dominates the ice sheet mass balance (see Fig. 4.12).

- There is medium confidence that approximately 20 to 30 percent of species assessed so far are likely to be at increased risk of extinction if increases in global average warming exceed 1.5°C to 2.5°C (relative to 1980 to 1999). As global average temperature increase exceeds about 3.5°C, model projections suggest significant extinctions (40 to 70 percent of species assessed) around the globe.

4.3.1   Impacts on Surface Water and Groundwater Resources

Based on global climate models, the IPCC (2007) has high confidence that by mid-century, annual river runoff and water availability are projected to increase at high latitudes (and in some tropical wet areas) and decrease in some dry regions in the mid-latitudes and tropics. There is also high confidence that many semiarid areas (e.g., Mediterranean basin, western United States, southern Africa, and northeast Brazil) will suffer a decrease in water resources due to climate change. For example, Fig. 4.13 shows how the warming trend in western United States winters mirrors annual global temperature trends. Warmer-than-average winter temperatures have been evident across the region, particularly over the last 6 years. The western United States has also experienced drought over the last 6 years.

The most obvious consequence of increasing winter temperatures will be the reduction in snowpack, which is the main source of river runoff in the West. In 2004, the USGS modeled the effects of climate change on three river basins draining the Sierra Nevada Mountains (Dettinger et al., 2004). Two basins experienced a dramatic decrease in snowpack under a "business as usual" scenario (no cutbacks in carbon emissions). Figure 4.14 demonstrates these results graphically for all three basins. The American River Basin has the lowest average elevation and, therefore, the highest sensitivity to climate change (Dettinger et al., 2004). The USGS model predicts that the snow cover in the American River Basin will be negligible by the year 2100. Reduced snowpack occurs as temperature increases result in a higher percentage of precipitation falling as rainfall, as demonstrated in Fig. 4.15 for the Merced River Basin.

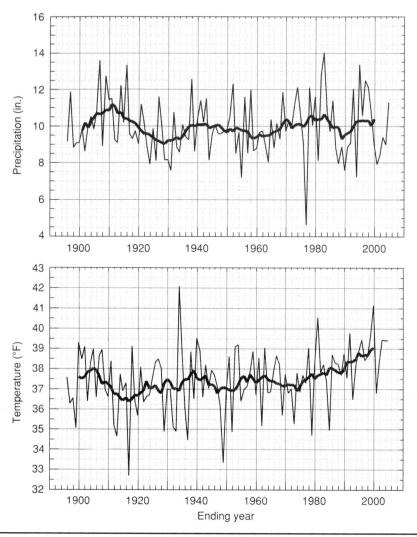

FIGURE 4.13 Western United States (11 states) October–March precipitation and temperature, 1895–2005. Bold line is the 11-year running mean. (From Edwards and Redmond, 2005; source: Western Regional Climate Center.)

The USGS notes that the most significant hydrologic change, due to global warming, will be higher streamflows in the late winter/early spring (March) due to more rainfall and earlier snowmelt. The earlier snowmelt of reduced magnitude will cause water shortages in the late spring and summer, when water demand is high (Dettinger et al., 2004).

According to IPCC (2007), there is high confidence that some hydrological systems have been already affected through increased runoff and earlier spring peak discharge in many glacier- and snow-fed rivers. Global warming has also affected the thermal structure and water quality of warming rivers and lakes.

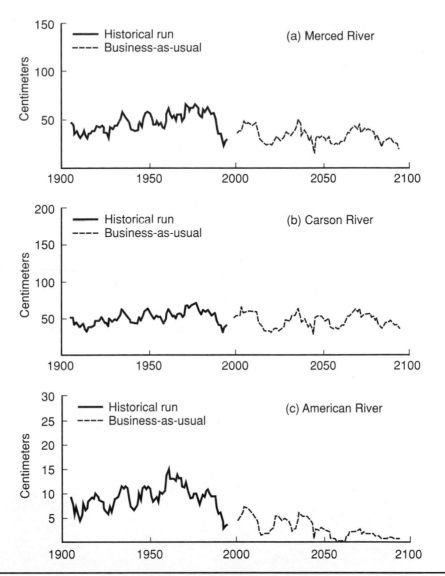

Figure 4.14   Basin average snow water contents on April 1, in three river basins draining the Sierra Nevada Mountains in the western United States. (From Dettinger et al., 2004.)

IPCC projects that the mechanism of decreasing snowpack, receding glaciers, and increasing early spring runoff will affect other mountainous regions across the globe and cause shortages of surface water supplies in regions at lower elevations. Most notably, the water supply of some of the most populated regions of India and China depends on large rivers originating in the high mountain ranges of Himalayas. Figure 4.16 is a glimpse of the receding glaciers across the globe.

The reduced winter snowpack will also have consequences for groundwater recharge of intermountain basins; less and shorter snowmelt means that the surface flow will last

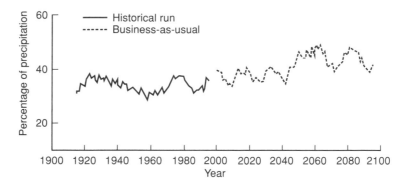

FIGURE 4.15 Rainfall as a percentage of precipitation in the Merced River Basin. (From Dettinger et al., 2004.)

for shorter periods of time, which directly translates into less groundwater recharge at basin margins and via bedrock. A greater percentage of precipitation falling on mountain ranges and at basin margins will be in the form of rainfall immediately becoming surface water runoff. Similarly unfavorable for groundwater recharge conditions is the IPCC's projection that an increase in rainfall in some regions will come in the form of bigger, wetter storms, rather than in the form of more rainy days. In between those

FIGURE 4.16 Receding alpine glaciers in the Nutzotin Mountains, eastern Alaska Range. Note the series of terminal moraines in the foreground. These glaciers feed a network of streams that will become ephemeral once the glaciers have melted completely. (Photograph courtesy of Jeff Manuszak.)

larger storms will be longer periods of light or no rain, so the frequency of drought will increase.

Rising temperatures and heat waves will result in an increased demand for water and, together with droughts, will put additional pressure on both surface water and groundwater resources. Warming of reservoirs and surface streams will result in water quality problems such as algal blooms. Higher temperatures will cause higher water losses from surface water reservoirs due to increased evapotranspiration. Increased evapotranspiration will also affect shallow subsurface soil, causing higher soil moisture deficits. A larger fraction of the infiltrated precipitation will therefore be spent on satisfying this soil moisture deficit before deep circulation and actual groundwater recharge can take place. In areas with irrigated agriculture, higher temperatures, heat waves, and droughts will put additional demand on water resources because more water will have to be diverted from streams and reservoirs or pumped from aquifers to offset higher evapotranspiration and higher soil moisture deficit.

More frequent episodes of heavy precipitation will cause deterioration of water quality in surface streams due to increased turbidity and runoff from nonpoint sources of contamination. A related increased incidence of floods will put pressure on water supply infrastructure and increase the risks of water-borne diseases.

A sea-level rise will cause salinization of irrigation water in coastal areas, estuaries, and freshwater systems, and saltwater intrusion in shallow aquifers underlying low coastal areas. In addition, population relocation from the semiarid and arid regions severely affected by prolonged droughts and from the low-lying coastal areas affected by the sea-level rise will put pressure on water resources in other regions that otherwise may not be significantly impacted by global warming.

Some of the projected continental-scale impacts of global warming are given below (IPCC, 2007).

Africa
- By 2020, between 75 and 250 million people are projected to be exposed to increased water stress due to climate change.
- By 2020, in some countries, yields from rain-fed agriculture could be reduced by up to 50 percent. Agricultural production, including access to food, in many African countries is projected to be severely compromised. This would further adversely affect the predictability of food production and exacerbate malnutrition.
- Toward the end of the twenty-first century, the projected sea-level rise will affect low-lying coastal areas with large populations. The cost of adaptation could amount to at least 5 to 10 percent of gross domestic product (GDP).
- By 2080, arid and semiarid land in Africa is projected to increase by 5 to 8 percent under a range of climate scenarios.

Asia
- By the 2050s, freshwater availability in central, south, east, and southeast Asia, particularly in large river basins, is projected to decrease.
- Coastal areas, especially heavily populated megadelta regions in south, east, and southeast Asia, will be at greatest risk due to increased flooding from the sea and, in some megadeltas, flooding from the rivers.

- Climate change is projected to compound pressures on natural resources and the environment, along with rapid urbanization, industrialization, and economic development.

- Endemic morbidity and mortality due to diarrheal disease primarily associated with floods and droughts are expected to rise in east, south, and southeast Asia due to projected changes in the hydrological cycle.

Australia and New Zealand
- By 2020, a significant loss of biodiversity is projected to occur in some ecologically rich sites including the Great Barrier Reef and Queensland Wet Tropics.

- By 2030, water availability problems are projected to intensify in southern and eastern Australia and, in New Zealand, in Northland and some eastern regions.

- By 2030, production from agriculture and forestry is projected to decline over much of southern and eastern Australia, and over parts of eastern New Zealand, due to increased drought and fire. However, in New Zealand, initial benefits are projected for some other regions.

- By 2050, ongoing coastal development and population growth in some areas of Australia and New Zealand are projected to exacerbate risks from sea-level rise and increases in the severity and frequency of storms and coastal flooding.

Europe
- Climate change is expected to magnify regional differences in Europe's natural resources and assets. Negative impacts will include an increased risk of inland flash floods, more frequent coastal flooding, and increased erosion (due to an increase in storms and sea-level rise).

- Mountainous areas will experience glacier retreat, reduced snow cover and winter tourism, and extensive species losses (in some areas up to 60 percent by 2080).

- In southern Europe, climate change is projected to worsen conditions (high temperatures and drought) in a region already vulnerable to climate variability and to reduce water availability, hydropower potential, summer tourism, and, in general, crop productivity.

- Climate change is also projected to increase the health risks due to heat waves and the frequency of wildfires.

Latin America
- By mid-century, increases in temperature and associated decreases in soil water are projected to lead to gradual replacement of tropical forest by savanna in eastern Amazonia. Semiarid vegetation will be slowly replaced by arid-land vegetation.

- There is a risk of significant biodiversity loss through species extinction in many areas of tropical Latin America.

- Productivity of some important crops is projected to decrease and livestock productivity to decline, with adverse consequences for food security. In temperate zones, soybean yields are projected to increase. Overall, the number of people at risk of hunger is projected to increase.

- Changes in precipitation patterns and the disappearance of glaciers are projected to significantly affect water availability for human consumption, agriculture, and energy generation.

North America
- Warming in the mountains of the American west is projected to cause decreased snowpack, more winter flooding, and reduced summer flows, exacerbating competition for over-allocated water resources.

- In the early decades of the century, moderate climate change is projected to increase aggregate yields of rain-fed agriculture by 5 to 20 percent, but with important variability among regions. Major challenges are projected for crops that are near the warm end of their suitable range or that depend on highly utilized water resources.

- During the course of this century, cities that currently experience heat waves are expected to be further challenged by increased number, intensity, and duration of heat waves during the course of the century, with the potential for adverse health impacts.

- Coastal communities and habitats will be increasingly stressed by the changing climate.

Polar regions
- The main projected biophysical effect is a reduction in thickness and extent of glaciers and ice sheets and sea ice, and changes in natural ecosystems with detrimental effects on many organisms including migratory birds and mammals.

- For human communities in the Arctic, impacts, particularly those resulting from changing snow and ice conditions, are projected to be mixed.

- Detrimental impacts would include those on infrastructure and traditional indigenous ways of life.

- In both polar regions, specific ecosystems and habitats are projected to be vulnerable, as climatic barriers to species invasions are lowered.

Small islands
- Sea-level rise is expected to exacerbate inundation, storm surge, erosion, and other coastal hazards, thus threatening vital infrastructure, settlements, and facilities that support the livelihood of island communities.

- By mid-century, climate change is expected to reduce water resources in many small islands, e.g., in the Caribbean and Pacific, to the point where they become insufficient to meet demand during low-rainfall periods.

Finally, the author includes the following related discussion by the National Climatic Data Center of NOAA (NCDC, 2007c):

Climate change refers to the changes in average weather conditions that generally occur over long periods of time, usually centuries or longer. Occasionally, these changes can occur more rapidly, in periods as short as decades. Such climate changes are often characterized as "abrupt".

Until recently, many scientists studying changes and variations in the climate thought that the climate system was slow to change, and that it took many thousands if not millions of years for ice ages and other major events to occur. Scientists are just beginning to formulate and test hypotheses regarding the causes of abrupt climate change, but only a handful of attempts have been made to model abrupt change using computer models. These efforts are focused not only on past events, but also on abrupt events that might occur in the future as greenhouse gases increase in the atmosphere and temperatures continue to rise.

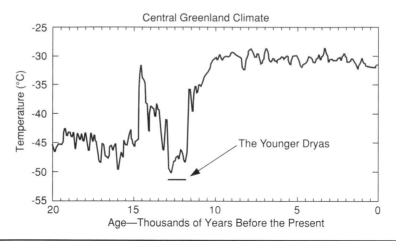

FIGURE 4.17 As the earth's climate emerged from the most recent ice age, the warming that began 15,000 years ago was interrupted by a cold period known as the Younger Dryas, which in turn ended with abrupt warming. (From NCDC, 2007c; source: Cuffey and Clow, 1997; Alley, 2000.)

One of the best studied examples of abrupt change occurred as the Earth's climate was changing from a cold glacial to a warmer interglacial state. During a brief period lasting about a century, temperatures in most of the northern hemisphere rapidly returned to near-glacial conditions, stayed there for over 1,000 years in a time called the Younger Dryas (named after a small Arctic flower) then about 11,500 years ago quickly warmed again. In some places, the abrupt changes may have been as large as 10°C, and may have occurred over a decade (Fig. 4.17). Changes this large would have a huge impact on modern human society, and there is a pressing need to develop an improved understanding and ability to predict abrupt climate change events. Indeed, this is the goal of several national and international scientific initiatives.

References

ACWA and CRWUAC (Association of California Water Agencies and Colorado River Water Users Association Conferences), 2005. Colorado River Basin climate; paleo, present, future. Special Publication for ACWA and CRWUAC, 66 p.

Allen, J., 2003. Drought lowers Lake Mead, 2003. NASA Earth Observatory. Avaialble at: http://earthobservatory.nasa.gov/Study/LakeMead/. Accessed October 2007.

Alley, R.B., 2000. The Younger Dryas cold interval as viewed from central Greenland. *Quaternary Science Reviews*, vol. 19, pp. 213–226.

Berger, A., and Loutre, M.F., 1991. Insolation values for the climate of the last 10 million years. *Quaternary Science Reviews*, vol. 10, pp. 297–317. Data available at: http://www.ncdc.noaa.gov/paleo/forcing.html#orbital. Accessed December 2007.

Cook, E.R., Meko, D.M., Stahle, D.W., and Cleaveland, M.K., 1999. Drought reconstructions for the continental United States. *Journal of Climate*, vol. 12, pp. 1145–1162.

CPC (Climate Prediction Center), 2007a. El Niño/La Niña Home. National Weather Service. Available at: http://www.cpc.ncep.noaa.gov/products/analysis_monitoring/lanina/index.html. Accessed December 2007.

CPC (Climate Prediction Center), 2007b. ENSO cycle: Recent evolution, current status and predictions, update prepared by Climate Prediction Center/NCEP, December 24, 2007. National Weather Service. Available at: http://www.cpc.ncep.noaa.gov/ products/analysis_monitoring/lanina/index.html. Accessed in December 2007.

Cuffey, K.M., and Clow, G.D., 1997. Temperature, accumulation, and ice sheet elevation in central Greenland through the last deglacial transition. *Journal of Geophysical Research*, vol. 102, pp. 26383–26396.

Dettinger, M.D., Cayan, D.R., Meyer, M.K., and Jeton, A.E., 2004. Simulated hydrologic responses to climate variations and change in the Merced, Carson, and American River Basins, Sierra Nevada, California, 1900–2099. *Climatic Change*, vol. 62, pp. 283–317.

Edwards, D.C., and McKee, T.B., 1997. Characteristics of 20th century drought in the United States at multiple time scales. Climatology Report Number 97–2. Colorado State University, Fort Collins, CO.

Edwards, L.M., and Redmond, K.T., 2005. Climate factors on Colorado River Basin water supply. In: Colorado River Basin climate; paleo, present, future. Special Publication for Association of California Water Agencies and Colorado River Water Users Association Conferences, pp. 14–22.

Hayes, M.J., 2007. What is drought? Drought indices. National Drought Mitigation Center, University of Nebraska, Lincoln. Available at: http://www.drought.unl.edu/ whatis/what.htm.

Hays, J.D., Imbrie, J., and Shackleton, N.J., 1976. Variations in the earth's orbit: Pacemaker of the ice ages. *Science*, vol. 194, no. 4270, pp. 1121–1132.

IPCC (Intergovernmental Panel on Climate Change), 2007. Summary for policymakers of the Synthesis Report of the IPCC Fourth Assessment Report; Draft Copy 16 November 2007 [subject to final copyedit]. 23 p. Available at: http://www.ipcc.ch/. Accessed November 2007.

Kiehl, J., and Trenberth, K., 1997. Earth's annual global mean energy budget. *Bulletin of the American Meteorological Society*, vol. 78, pp. 197–206.

Lutgens, F.K., and Tarbuck, E.J., 1995. *The Atmosphere*. Prentice-Hall, Englewood Cliffs, NJ, 462 p.

McKee, T.B., Doesken, N.J., and Kleist, J., 1993. The relationship of drought frequency and duration to time scales. Preprints, 8th Conference on Applied Climatology, January 17–22, 1993, Anaheim, CA, pp. 179–184.

Meko, D.M., Hughes, M.K., and Stockton, C.W., 1991. Climate change and climate variability: The paleo record. In: *Managing Water Resources in the West Under Conditions of Climate Uncertainty*. National Academy Press, Washington, DC, pp. 71–100.

Meko, D.M., Woodhouse, C.A., Baisan, C.A., Knight, T., Lukas, J.J., Hughes, M.K., and Salzer, M.W., 2007. Medieval drought in the upper Colorado River Basin. *Geophysical Research Letters*, vol. 34, L10705, doi:10.1029/2007GL029988.

Milankovitch, M., 1941. *Kanon der Erdbestrahlung und seine Anwendung auf das Eiszeitenproblem*. Königlich Serbische Akademie, Belgrad, 626 p.

NASA, 2007. On the shoulders of giants. Milutin Milankovitch (1879–1958). NASA Earth Observatory. Available at: http://earthobservatory.nasa.gov/Library/Giants/ Milankovitch/.

NCDC (National Climatic Data Center), 2007a. North American Drought: A paleo perspective. Available at: http://www.ncdc.noaa.gov/paleo/drought/drght_home. html. Accessed September 2007.

NCDC (National Climatic Data Center), 2007b. Climate of 2006 in historical perspective. Annual report, 09 January 2007. Available at: http://www.ncdc.noaa.gov/oa/climate/research/2006/ann/ann06.html. Accessed February 2007.

NCDC (National Climatic Data Center), 2007c. A paleo perspective on abrupt climate change. Available at: http://www.ncdc.noaa.gov/paleo/abrupt/story.html. Accessed December 2007.

NDMC (National Drought Mitigation Center), 2007. Understanding and defining drought. Available at: http://drought.unl.edu/whatis/concept.htm#top. Accessed September 2007.

Ni, F., et al., 2002. Southwestern USA linear regression and neural network precipitation reconstructions. International Tree-Ring Data Bank, World Data Center for Paleoclimatology, Boulder, Data Contribution Series #2002-080.

Ni, F., Cavazos, T., Hughes, M.K., Comrie, A.C., and Funkhouser, G., 2002. Cool-season precipitation in the southwestern USA since AD 1000: Comparison of linear and nonlinear techniques for reconstruction. *International Journal of Climatology*, vol. 22, no. 13, pp. 1645–1662.

NRC (National Research Council), 1982. *Solar Variability, Weather, and Climate*. National Academy Press, Washington, DC, 7 p.

Riebek, H., 2007. Global warming. NASA Earth Observatory. Available at: www.earthobservatory.nasa.gov/library/globalwarmingupdate/.

Riebsame, W.E., Changnon, S.A., and Karl, T.R., 1991. *Drought and Natural Resources Management in the United States: Impacts and Implications of the 1987–89 Drought*. Westview Press, Boulder, CO, pp. 11–92.

Snow, L.A., 2005. Foreword. In: *Colorado River Basin Climate; Paleo, Present, Future*. Special Publication for Association of California Water Agencies and Colorado River Water Users Association Conferences.

Tesla Memorial Society of New York, 2007. Milutin Milankovitch. Available at: http://www.teslasociety.com/milankovic.htm.

Woodhouse, C., 2005. Paleoclimate overview. In: *Colorado River Basin Climate; Paleo, Present, Future*. Special Publication for Association of California Water Agencies and Colorado River Water Users Association Conferences, pp. 7–13.

Woodhouse, C., and Lukas, J., 2005. From tree rings to streamflow. In: *Colorado River Basin Climate; Paleo, Present, Future*. Special Publication for Association of California Water Agencies and Colorado River Water Users Association Conferences, pp. 1–6.

CHAPTER 5

Groundwater Quality

5.1 Introduction

Water quality is the chemical, physical, biological, and radiological condition of a surface water or groundwater body. This chapter discusses natural quality and contamination of freshwater—water that contains less than 1000 mg/L of total dissolved solids (TDS); generally, more than 500 mg/L of dissolved solids is undesirable for drinking and many industrial uses. The United States Environmental Protection Agency (USEPA) defines an underground source of drinking water (USDW) as an aquifer containing water with a TDS concentration of less than 10,000 mg/L. This distinction is important, as slightly brackish and brackish groundwater is becoming increasingly of interest for development.

The evaluation of groundwater quality is a complex task. Groundwater quality can be adversely affected or degraded as a result of human activities that introduce contaminants into the environment. It can also be affected by natural processes that result in elevated concentrations of certain constituents in the groundwater. For example, elevated metal concentrations can result when metals are naturally leached into groundwater from minerals present in the rocks. High levels of arsenic and uranium are frequently found in certain types of aquifers and groundwater systems around the world. In the United States, arsenic and uranium are found at elevated concentrations primarily in some groundwater systems in the western states but occasionally may exceed drinking water standards in other parts of the country as well.

As pointed out by the USEPA (2000a), not too long ago, it was thought that soil provided a protective "filter" or "barrier" that immobilized the downward migration of contaminants released on the land surface. Soil was supposed to prevent groundwater resources from being contaminated. The detection of pesticides and other contaminants in groundwater demonstrated that these resources were indeed vulnerable to contamination. The potential for a contaminant to affect groundwater quality is dependent on its ability to migrate through the overlying soils to the underlying groundwater resource. Various physical, chemical, and biological processes and many different factors influence this potential migration of contaminants from the land surface to the water table of shallow aquifers. The same is true regarding the potential migration of contaminants from shallow to deeper groundwater systems.

Traditionally, surface water and groundwater have been treated as separate entities in the management of water resources. However, it has become apparent that they continuously interact in many different ways, as summarized in a publication by the USGS, *Ground water and surface water—a single resource* (Winter et al., 1998). In addition to precipitation, or direct recharge, water in lakes, wetlands, and streams can recharge groundwater systems, and vice versa: groundwater systems discharge into lakes, wetlands, and streams. Groundwater provides an average of 52 percent of all the streamflow in the nation; this contribution ranges from 14 to 90 percent in 24 different regions studied by the USGS (Winter et al., 1998). The water quality of both surface water and groundwater can be affected by their interactions with and transport of nutrients and contaminants. Because contamination is not restricted to either, both groundwater and surface water must be considered in water quality assessments. An understanding of their interactions is critical in water protection and conservation efforts. It is evident that protection of groundwater, as much as protection of surface water, is of major importance for sustaining their multiple uses such as drinking water supply, fish and wildlife habitats, swimming, and fishing (USEPA, 2000a).

5.2 Natural Groundwater Constituents

The British Geological Society and Environment Agency have recently produced a series of reports documenting baseline groundwater quality of major aquifers in England and Wales (e.g., Neumann et al., 2003). The two agencies defined the baseline (background) concentration of a substance as "the range in concentration (within a specified system) of a given element, species or chemical substance present in solution which is derived from natural geological, biological, or atmospheric sources." This effort was undertaken to establish a standard that serves as the scientific basis for defining natural variations in groundwater quality and whether or not anthropogenic pollution is taking place. One of the principal difficulties when attempting to define natural groundwater quality is that this baseline in many regions has been modified by humans since early times due to settlement and agricultural practices. Locating groundwater without traces of human impact, which for practical purposes may be defined as water recharged in the preindustrial era (pre-1800s), is difficult for various reasons. For example, groundwater exploitation may result in a mixing of the initially stratified system formed as the result of natural hydraulic gradients and the natural variation in the aquifer's physical and geochemical processes. Groundwater samples collected from a system that was under pumping stress for some time will, therefore, often represent mixtures of the initially stratified system. The determination of the natural baseline can be achieved by several means including the study of pristine (unaffected by anthropogenic influence) environments, the use of historical records, and the application of graphical procedures such as probability plots to discriminate different populations (Neumann et al., 2003). In addition, in order to correctly interpret the water quality variations in terms of the baseline, some knowledge of the residence time of groundwater is required. For this purpose, both inert and active chemical and isotopic tracers are essential.

As the most effective solvent of geologic materials, groundwater contains a large number of dissolved natural elements. Complete chemical analyses of groundwater (those looking for "all" possible naturally occurring constituents) would generally show more than 50 elements at levels detectable in commercial laboratories, and more at scientific laboratories capable of detecting very low concentrations. Constituents that are commonly present at concentrations greater than 1 mg/L in most geologic settings are sometimes called major or macro constituents of groundwater. Such components are analyzed by default because they most obviously reflect the type of rocks present in the subsurface and are therefore used to compare general genetic types of groundwater.

Some of the elements commonly present in groundwater at concentrations between 0.01 and 10 mg/L are significant for understanding its genesis and are also often analyzed by default. These are sometimes referred to as either minor or secondary constituents. Metallic elements that are usually found at concentrations less than 0.1 and less than 0.001 mg/L are sometimes called *minor constituents* and *trace constituents*, respectively. However, the significant concentrations and the relative importance of different groundwater constituents are site and regulations specific, and these do vary in different parts of the world, and in time. A good example is arsenic, considered for a long time to be a "minor" or "trace" constituent. However, as more and more analyses of groundwater used or considered for water supply are being conducted both in the United States and worldwide, arsenic emerges as the major groundwater constituent simply because of the new regulatory drinking water standard of 0.01 mg/L (10 ppb).

About 35 or so important natural inorganic groundwater constituents, recognizing the relativity of word "important," can be divided into the following two practical analytical groups:

1. *Primary constituents* analyzed routinely
    a. Anions: Cl^-, SO_4^{2-}, HCO_3^-, CO_3^{2-}, and NO_3^- (and other nitrogen forms)
    b. Cations: Ca^{2+}, Mg^{2+}, Na^+, K^+, and Fe^{2+} (and other iron forms)
    c. Silica as SiO_2 (present mostly in uncharged form)

2. *Secondary constituents* analyzed as needed
    a. Elements/anions: boron (B), bromine/bromide (Br), fluorine/fluoride (F), iodine/iodide (I), and phosphorus/phosphate (P)
    b. Metals, nonmetallic elements: aluminum (Al), antimony (Sb), arsenic (As), barium (Ba), beryllium (Be), cadmium (Cd), cesium (Cs), chromium (Cr), copper (Cu), lead (Pb), lithium (Li), manganese (Mn), mercury (Hg), nickel (Ni), rubidium (Rb), selenium (Se), silver (Ag), strontium (Sr), and zinc (Zn)
    c. Radioactive elements: radium (Rd), uranium (U), alpha particles, and beta particles
    d. Organic matter (total and dissolved organic carbon—TOC and DOC)
    e. Dissolved oxygen (and/or reduction-oxidation potential, *Eh*)

Almost all primary and secondary inorganic constituents listed above are included in the list of primary and secondary drinking water standards by the USEPA (see Section 5.4). Most of them, when in excess of a certain concentration, are considered contaminants, and such groundwater is not suitable for human consumption. Groundwater contamination may be the result of both naturally occurring substances and those introduced by human activities.

5.2.1 Total Dissolved Solids, Specific Conductance, and Salinity

Overall, water is the most effective solvent of geologic materials and other environmental substances—solid, liquid, and gaseous. This quality of water is the result of a unique structure of its molecule, which is a dipole—the centers of gravity and electric charges in the water molecule are asymmetric. The polarity of molecules, in general, is quantitatively expressed with the dipole moment, which is the product of the electric charge and the distance between the electric centers. Dipole moment for water is 6.17×10^{-30} Cm (kulonmeters), higher than for any other substance, and explains why water can dissolve more solids and liquids than any other liquid. Dissolution of rocks by water plays the main role in continuous redistribution of geologic materials in the environment, at and below the land surface.

Substances subject to dissolution by water (or any other liquid) are called *solutes*. Some substances are more soluble in water than others. Ionized mineral salts, such as sodium chloride, are very easily and quickly dissolved in water by its dipolar molecules. Synthetic organic substances with polarized molecules, such as methanol, are also highly soluble in water: hydrogen bonds between water and methanol molecules can readily replace the very similar hydrogen bonds between different methanol molecules and different water molecules. Methanol is, therefore, said to be miscible in water (its solubility in water is infinite for practical purposes). On the other hand, many nonpolar organic molecules, such as benzene and trichloroethylene (TCE) for example, have low water solubility.

True solutes are in the state of separated molecules and ions, all of which have very small dimensions (commonly between 10^{-6} and 10^{-8} cm), thus making a water solution transparent to light. Colloidal solutions have solid particles and groups of molecules that are larger than the ions and molecules of the solvent (water). When colloidal particles are present in large enough quantities, they give water an opalescent appearance by scattering light. Although there is no one-agreed-to definition of what exactly colloidal sizes are, a common range cited is between 10^{-6} and 10^{-4} cm (Matthess, 1982). The amount of a solute in water is expressed in terms of its concentration, usually in milligrams per liter (mg/L or parts per million—ppm) and micrograms per liter (parts per billion—ppb). It is sometimes difficult to distinguish between certain true solutes and colloidal solutions that may carry particles of the same source substance. Filtering and/or precipitating colloidal particles before determining the true dissolved concentration of a solute may be necessary in some cases. This is especially true for drinking water standards because these, for most substances, are based on dissolved concentrations. Laboratory analytical procedures are commonly designed to determine total concentrations of a substance and do not necessarily provide indication of all the individual species (chemical forms) of it. If needed, however, such speciation can be requested. For example, determination of individual chromium species, rather than the total chromium concentration, may be important in groundwater contamination studies, since hexavalent chromium or Cr(VI) is more toxic and has different mobility than trivalent chromium, Cr(III).

The total concentration of dissolved material in groundwater is called *total dissolved solids* (TDS). It is commonly determined by weighing the dry residue after heating the water sample usually to 103°C or 180°C (the higher temperature is used to eliminate more of the crystallization water). TDS can also be calculated if the concentrations of major ions are known. However, for some water types, a rather extensive list of analytes may be needed to accurately obtain the total. During evaporation, approximately one-half of the hydrogen carbonate ions are precipitated as carbonates and the other half escapes as water and carbon dioxide. This loss is taken into account by adding half of the HCO_3^- content to the evaporation (dry) residue. Some other losses, such as precipitation of sulfate as gypsum and partial volatilization of acids, nitrogen, boron, and organic substances, may contribute to a discrepancy between the calculated and the measured TDS.

Solids and liquids that dissolve in water can be divided into electrolytes and non-electrolytes. Electrolytes, such as salts, bases, and acids, dissociate into ionic forms (positively and negatively charged ions) and conduct electrical current. Nonelectrolytes, such as sugar, alcohols, and many organic substances, occur in aqueous solution as uncharged molecules and do not conduct electrical current. The ability of 1 cm^3 of water to conduct electrical current is called *specific conductance* (or sometimes simply *conductance*, although the units are different). Conductance is the reciprocal of resistance and is measured in units called *Siemen* (International System) or *mho* (1 Siemen equals 1 mho; the name mho is derived from the unit for resistance—*ohm*, by spelling it in reverse). Specific conductance is expressed as *Siemen/cm* or *mho/cm*. Since the *mho* is usually too large for most groundwater types, the specific conductance is reported in *micromhos/cm* or *microSiemens/cm* (μS/cm), with instrument readings adjusted to 25°C, so that variations in conductance are only a function of the concentration and type of dissolved constituents present (water temperature also has a significant influence on conductance). Measurements of specific conductance can be made rapidly in the field with a portable instrument, which provides for a convenient method to quickly estimate TDS and compare general types of water quality. For a preliminary (rough) estimate of TDS, in milligrams per liter, in fresh potable water, the specific conductance in micromhos/cm can be multiplied by 0.7. Pure water has a conductance of 0.055 micromhos at 25°C, laboratory distilled water between 0.5 and 5 micromhos, rainwater usually between 5 and 30 micromhos, potable groundwater ranges from 30 to 2000 micromhos, sea water from 45,000 to 55,000 micromhos, and oil field brines have commonly more than 100,000 micromhos (Davis and DeWiest, 1991).

The term *salinity* is often used for total dissolved salts (ionic species) in groundwater, in the context of water quality for agricultural uses or human and livestock consumption. Various salinity classifications, based on certain salts and their ratios, have been proposed (see Matthess, 1982). One problem with the term salinity is that a salty taste may be already noticeable at somewhat higher concentrations of sodium chloride, NaCl (e.g., 300 to 400 mg/L), even though the overall concentration of all dissolved salts may not "qualify" a particular groundwater to be called "saline." In practice, it is common to call water with less than 1000 mg/L (1 g/L) dissolved solids fresh, and water with more than 10,000 mg/L saline.

5.2.2 Hydrogen-Ion Activity (pH)

Hydrogen-ion activity, or pH, is probably the best-known chemical characteristic of water. It is also the one that either directly affects or is closely related to most geochemical and biochemical reactions in groundwater. Whenever possible, pH should be measured directly in the field, since groundwater, once outside its natural environment (aquifer), quickly undergoes several changes that directly impact pH, the most important being temperature and the CO_2-carbonate system. Incorrect values of pH may be a substantial source of error in geochemical equilibrium and solubility calculations.

Water molecules naturally dissociate into H^+ and OH^- (hydroxyl) ions. By convention, the content of the hydrogen ion in water is expressed in terms of its activity, pH, rather than its concentration in milligrams or millimoles per liter. By definition, when the number of hydrogen ions equals the number of hydroxyl ions, the solution is neutral and the hydrogen ion activity, or pH, is 7 (note that log [7] times log [7] is log [14]). Theoretically, when there are no hydrogen ions, pH is 14 and the solution is purely alkaline (base); when there are no hydroxyl ions, the solution is a pure acid with pH 1. Accordingly, when the activity (concentration) of hydrogen ions decreases, the activity of hydroxyl ions must increase because the product of the two activities is always the same, i.e., 14.

The reaction of carbon dioxide with water is one of the most important for establishing pH in natural water systems. This reaction is represented by the following three steps (Hem, 1989):

$$CO_2(g) + H_2O(l) = H_2CO_3(aq) \tag{5.1}$$

$$H_2CO_3(aq) = H^+ + HCO_3^- \tag{5.2}$$

$$HCO_3^- = H^+ + CO_3^{2-} \tag{5.3}$$

where g, l, and aq denote gaseous, liquid, and aqueous phases, respectively. The second and third steps produce hydrogen ions, which influence the acidity of solution. Other common reactions that create hydrogen ions involve dissociation of acidic solutes.

Many of the reactions between water and solid species consume H^+, resulting in the creation of OH^- and alkaline conditions. One of the most common is hydrolysis of solid calcium carbonate (calcite):

$$CaCO_3 + H_2O = Ca^{2+} + HCO_3^- + OH^- \tag{5.4}$$

Note that Eq. (5.4) explains why lime dust (calcium carbonate) is often added to acidic soils in agricultural applications to stimulate growth of crops that do not tolerate such soils.

The pH of water has a profound effect on the mobility and solubility of many substances. Only a few ions such as sodium, potassium, nitrate, and chloride remain in solution through the entire range of pH found in normal groundwater. Most metallic elements are soluble as cations in acid groundwater but will precipitate as hydroxides or basic salts with an increase in pH. For example, all but traces of ferric ions will be absent above a pH of 3, and ferrous ions diminish rapidly as the pH increases above 6 (Davis and DeWiest, 1991).

5.2.3   Reduction-Oxidation (Redox) Potential (Eh)

Reduction and oxidation can be broadly defined as a gain of electrons and loss of electrons, respectively. For a particular chemical reaction, an *oxidizing agent* is any material that gains electrons, and a *reducing agent* is any material that loses electrons. The reduction process is illustrated with the following expression (Hem, 1989):

$$Fe^{3+} + e^- = Fe^{2+} \tag{5.5}$$

where ferric iron (Fe^{3+}) is reduced to the ferrous state by gaining one electron. The symbol "e^-" represents the electron, or unit negative charge. This expression is a "half-reaction" for the iron reduction-oxidation couple; for the reduction to take place there has to be a source of electrons, i.e., another element has to be simultaneously oxidized (lose electrons). Together with the hydrogen ion activity (pH), reduction and oxidation reactions play key role in solubility of various ionic substances. Microorganisms are involved in many of the reduction-oxidation reactions, and this relationship is especially important when studying the fate and transport of contaminants subject to biodegradation.

The electric potential of a natural electrolytic solution with respect to the standard hydrogen half-cell measuring instrument is expressed (usually) in millivolts or mV. This measured potential is known as *reduction-oxidation potential* or *redox* and is denoted with Eh (*h* stands for hydrogen). Observed Eh range for groundwater is between +700 and −400 mV. This positive sign indicates that the system is oxidizing, and the negative that the system is reducing. The magnitude of the value is a measure of the oxidizing or reducing tendency of the system. Eh, just like pH, should be measured directly in the field. Concentration of certain elements is a good indicator of the range of possible Eh values. For example, a notable presence of H_2S (>0.1 mg/L) always causes negative Eh. If oxygen is present in concentrations greater than 1 mg/L, Eh is commonly between 300 and 450 mV. Generally, an increase in content of salts decreases the Eh of the solution.

The redox state is determined by the presence or absence of free oxygen in groundwater. Newly percolated (recharge) water often supplies oxygen to groundwater in the range from 6 to 12 mg/L. As groundwater moves away from the recharge zone, oxygen can be consumed in a number of different geochemical reactions, the most direct being oxidation of iron and manganese compounds. Microbial activity also consumes oxygen and may rapidly create a reducing environment in a saturated zone with an excess of dissolved organic carbon (DOC) (which is a nutrient for microbes) such as in cases of groundwater contamination with organic liquids.

Determining redox potential in an aquifer is particularly important in contaminant fate and transport and remediation studies. For example, oxidizing (*aerobic*) conditions favor biodegradation of petroleum hydrocarbons such as gasoline, while reducing (*anaerobic*) conditions favor biodegradation of chlorinated compounds such as tetrachloroethene (PCE). Based on oxygen demand of the various bacterial species, an oxygen content between 0.7 and 0.01 mg O_2/L at 8°C water temperature has been commonly defined as threshold oxygen concentration for the boundary between oxidizing and reducing conditions. However, field observations suggest that reducing conditions may appear at considerably higher oxygen contents (Matthess, 1982).

The redox potential generally decreases with rising temperature and pH, and this decrease results in an increasing reducing power of the aqueous system. Reducing systems, in addition to the absence or very much reduced oxygen content, have a noticeable content of iron and manganese; occurrence of hydrogen sulfide, nitrite, and methane; an absence of nitrate; and often a reduction or absence of sulfate (Matthess, 1982).

5.2.4 Primary Constituents

As a rule, the primary constituents dissolved in fresh groundwater always make up more than 90 percent of TDS in a sample. The following elements and ionic species, in particular, have been widely used to describe the chemical type and origin of groundwater: cations of calcium (Ca^{2+}), magnesium (Mg^{2+}), sodium (Na^+), and potassium (K^+) and anions of chloride (Cl^-), sulfate (SO_4^{2-}), hydrocarbonate (HCO_3^-), and carbonate (CO_3^-). The reason why they are the most prevalent in natural groundwaters is that the most important soluble minerals and salts occurring in relatively large quantities in rocks are calcium carbonate ($CaCO_3$), magnesium carbonate ($MgCO_3$), their combination ($CaCO_3 \times MgCO_3$), sodium chloride (NaCl), potassium chloride (KCl), calcium sulfate ($CaSO_4$), and hydrous calcium sulfate ($CaSO_4 \times 2H_2O$). The five elements (Ca, Mg, Na, K, and Cl) are abundant in various sedimentary, magmatic, and metamorphic rocks and are constantly released to the environment by weathering and dissolution.

Although aluminum and iron and are the second and the third most abundant metallic elements in the earth's crust, respectively, they rarely occur in natural groundwater in concentrations exceeding 1 mg/L (Hem, 1989). Naturally occurring forms of aluminum are especially stable under near-neutral pH, and aluminum is not considered to be a major dissolved-phase constituent. The exceptions are waters with very low pH such as acidic mine drainage. The chemical behavior of iron and its solubility in water are rather complex and depend strongly on the redox potential and pH. The forms of iron present are also strongly affected by microbial activity. Various ferrous complexes are formed by many organic molecules, and some of the complexes may be significantly more resistant to oxidation than free ferrous ions and also insoluble in groundwater. The presence of various forms of iron in groundwater is also important when evaluating the degradation of organic contaminants or deterioration of well screens by iron bacteria. For all these reasons, iron is considered one of the primary groundwater constituents and it is commonly analyzed, even though its dissolved ionic forms are found in most groundwaters in smaller concentrations compared to the major ions.

Unlike aluminum and iron, silicon, being the second only to oxygen in the earth's crust, is found in appreciable quantities in most groundwaters, usually between 1 and 30 mg/L when expressed as silica. The relative abundance of silica in natural water is due to its many different chemical forms found in minerals and rocks. This fact is contrary to the common belief that silica is not soluble in water and is therefore not present in groundwater. The most abundant forms of silica dissolved in water are thought not to form ions, although the complicated groundwater chemistry of silica is still not well understood (Hem, 1989; Matthess, 1982).

When analyzing relationships between major ions, or one of the ions to the total concentration, it is often helpful in understanding the origin of groundwater, and the similarities and differences between samples. Some useful ratios for establishing chemical types of groundwater are the ratio of calcium to magnesium for studying water from carbonate sediments (limestone and dolomite), and the ratio of silica to dissolved solids for identifying solution of different silicate minerals in magmatic rock terrains. Other ratios may be useful in different terrains, as long as the mineral contents of the aquifer porous media is well understood. Various groundwater classifications based on the presence of different ions and groups of ions have been proposed in literature and are beyond the scope of this book (e.g., see Alekin, 1953, 1962; Hem, 1989; Matthess, 1982).

A common practice in interpreting the results of chemical analyses is to present the concentrations of major ionic species graphically. Piper diagrams (Piper, 1944) are convenient for plotting the results of multiple analyses on the same graph, which may reveal grouping of certain samples and indicate their common or different origin. Box-and-Whisker plots (or simply "box" plots) are useful for comparing statistical parameters of samples known to be collected from different aquifers. For example, Figs. 5.1 and 5.2 show results of chemical analyses of groundwater from three different aquifers in coastal North Carolina, the United States. The most obvious from the Piper diagram presentation is that the surficial aquifer samples have a wide scatter, ranging from calcium bicarbonate to sodium chloride types. The source of the calcium and bicarbonate is most likely carbonate shell material in sediments of the surficial aquifer; however, the lower concentrations of these analytes compared with those in the Castle Hayne and Peedee aquifers (Fig. 5.2) are probably a result of less abundance of carbonate material in the surficial aquifer and the leaching and removal of these chemical constituents by infiltrating precipitation from the surficial deposits. The lowest pH values for recent groundwater samples were measured in the surficial aquifer and also indicate the leaching and removal of carbonate minerals (Harden et al., 2003). The pH of groundwater in the surficial aquifer was slightly

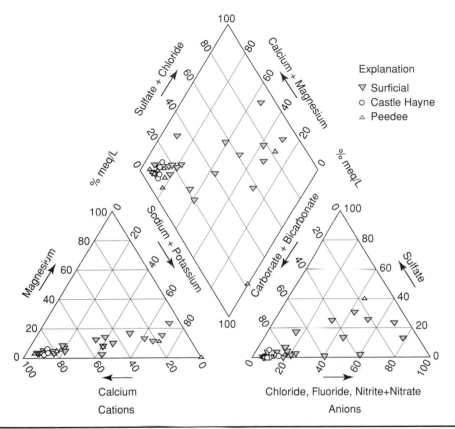

FIGURE 5.1 Piper diagram showing July–August 2000 groundwater quality in Brunswick County, NC. (Modified from Harden et al., 2003.)

acidic and ranged from 4.8 to 7.5 pH units, with a median value of about 6.9 (Fig. 5.2). The median dissolved-solids concentration (residue at 180°C) in the surficial aquifer was about 110 mg/L, almost three times less than in the Castle-Hayne aquifer, which is the deepest among the three. The highest dissolved-solids concentration of 870 mg/L was detected in a surficial aquifer well at Bald Head Island. This well also has the highest chloride concentration. Groundwater at Bald Head Island is known to be salty and is treated by reverse osmosis for supply purposes (Harden et al., 2003).

Stiff diagram (Fig. 5.3) gives an irregular polygonal shape that can help recognize possible patterns in multiple analyses and is therefore commonly used on hydrogeologic maps.

Cumulative frequency diagrams are useful in visualizing the distribution of data and may be of use in determining outlying data, certain controlling chemical mechanisms, and possible pollution. British Geologic Survey and Environment Agency provides the following discussion accompanying Fig. 5.4 (Neumann et al., 2003):

- The median and upper and lower percentile concentrations are used as a reference for the element baseline, which can be compared regionally or in relation to other elements.

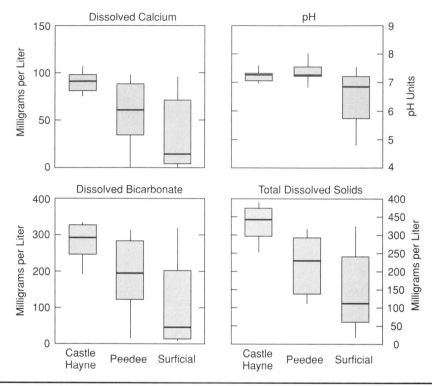

Figure 5.2 Concentrations of selected chemical constituents in groundwater samples collected during July–August 2000, Brunswick County, NC. Bottom and top of the box are 25th and 75th percentiles, respectively. Bold line inside the box is median value. Vertical lines outside the box extend to 10th and 90th percentiles. (Modified from Harden et al., 2003.)

- Normal to multimodal distributions are to be expected for many elements reflecting the range in recharge conditions, water-rock interactions, and residence time under natural conditions.

- Narrow ranges of concentration may indicate rapid attainment of saturation with minerals (e.g., Si with silica and Ca with calcite).

- A strong negative skew may indicate selective removal of an element by some geochemical process (e.g., NO_3 by in situ denitrification).

- A positive skew most probably indicates a contaminant source for a small number of the groundwaters, and this gives one simple way of separating those waters above the baseline. Alternatively, the highest concentrations may indicate waters of natural higher salinity.

An example of cumulative frequency diagrams for major constituents in the limestone aquifers in the Cotswolds, England, is shown in Fig. 5.5. The majority of the plots show a relatively narrow range, and some approach log-normal distribution with relatively steep gradients. For HCO_3 and Ca, the narrow ranges of concentrations indicate saturation with calcite in groundwaters, with the upper limit being controlled by carbonate mineral

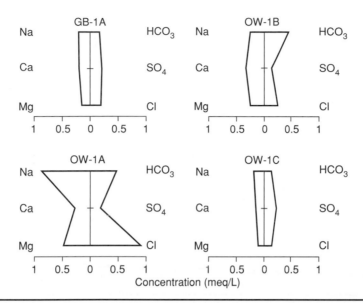

FIGURE 5.3 Stiff diagrams representing ion concentrations in water samples from public supply well GB-1A and observations wells at Green Belt Parkway well field, Holbrook, NY. (From Brown et al., 2002.)

solubility. Na and Cl concentrations are variable and show positive skew within the upper 10 percent of the data, particularly for Na. The strong negative skew in the plot for NO_3 indicates the presence of reducing waters and the removal of nitrate by in-situ denitrification. Additionally, old formation waters will exhibit low nitrate concentrations, being recharged before agricultural pollution occurred, while some groundwaters, such

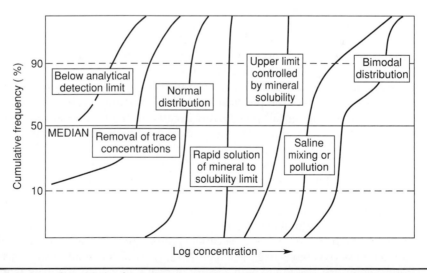

FIGURE 5.4 Use of cumulative frequency diagrams to indicate baseline characteristics in groundwaters. (From Neumann et al., 2003; copyright British Geological Survey and Environment Agency.)

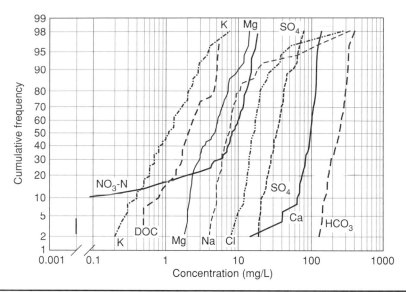

Figure 5.5   Cumulative frequency plots for major constituents in the limestone aquifers in the Cotswolds, England. (Modified from Neumann et al., 2003; copyright British Geological Survey and Environment Agency.)

as in woodland areas, might be generally unaffected by agricultural pollution (Neumann et al., 2003).

5.2.5 Secondary Constituents

Secondary constituents of importance for most natural groundwaters of drinking water quality include metals, fluoride, and organic matter. The term heavy metals (or trace metals) is applied to the group of metals and semimetals (metalloids) that have been associated with contamination and potential toxicity or ecotoxicity; it usually refers to common metals such as copper, lead, or zinc. Some define a heavy metal as a metal with an atomic mass greater than that of sodium, whereas others define it as a metal with a density above 3.5 to 6 g/cm^3. The term is also applied to semimetals (elements such as arsenic, which have the physical appearance and properties of a metal but behave chemically like a nonmetal) presumably because of the hidden assumption that "heaviness" and "toxicity" are in some way identical. Despite the fact that the term heavy metal has no sound terminological or scientific basis, it has been widely used in scientific environmental literature (van der Perk, 2006). Heavy metals commonly found in natural fresh groundwater include zinc, copper, lead, cadmium, mercury, chromium, nickel, and arsenic.

Heavy metals occur naturally as part of many primary and secondary minerals in all types of rocks. In natural waters, they are present mainly at low concentrations (usually much less than 0.1 mg/L), and as cations, although some semimetals such as arsenic may occur as oxyanions (e.g., arsenate AsO_4^{3-}). Their generally low concentrations in groundwater are due to the high affinity of heavy metals to adsorption and precipitation in soils and aquifer porous media. The maximum natural concentrations of heavy metals are usually associated with ore deposits and oxidized, low-pH water. In general, many solids control the fixation (immobilization) of heavy metals, namely, clay minerals, organic matter, iron, manganese, and aluminum oxides and hydroxides for adsorption, and poorly soluble sulfide, carbonate, and phosphate minerals for precipitation (Bourg and Loch, 1995; from van der Perk, 2006). The pH of groundwater is the most important factor controlling the fate and transport of heavy metals in the subsurface. In general, decreasing pH results in higher mobility of heavy metals and vice versa. More on general characteristics, hydrochemistry, and mobility of heavy metals in the subsurface can be found in Bourg and Loch (1995), Appelo and Postma (2005), and van der Perk (2006).

Arsenic has emerged as one of the most widespread natural contaminants in groundwater in various regions of the world. Since it can occur both naturally and as a result of anthropogenic contamination, arsenic is covered in more detail later as a groundwater contaminant.

The element fluorine is used by higher life forms in the structure of bones and teeth. The importance of fluoride; its anion, in forming human teeth; and the role of fluoride intake from drinking water in controlling the characteristics of tooth structure was recognized during the 1930s (Hem, 1989). Since that time the fluoride content of natural water has been studied extensively. Although intake of fluoride is necessary for promoting strong healthy teeth, at high concentrations it may cause bone disease and mottled teeth in children (MCL for fluoride in the United States is 4 mg/L). Although fluoride concentrations in most natural waters are small, less than 1 mg/L, groundwater exceeding this value has been found in many places in the United States, in a wide variety of geologic terrains (Hem, 1989). Fluorite and apatite are common fluoride minerals in magmatic and sedimentary rocks, and amphiboles and micas may contain fluoride that replaces part of the hydroxide. Rocks rich in alkali metals have a higher fluoride content than most other magmatic rocks. Fresh volcanic ash may be rather rich in fluoride, and ash that is interbedded with other sediments could contribute significantly to fluoride concentrations. Fluoride is commonly associated with volcanic or fumarolic gases, and, in some areas, these may be important sources of fluoride in groundwater (Hem, 1989). Fluorine is the most electronegative of all the elements, and its F^- ion forms strong solute complexes with many cations, particularly with aluminum, beryllium, and ferric iron. Anthropogenic sources of fluoride include fertilizers and discharge from ore-processing and smelting operations, such as aluminum works.

In addition to inorganic (mineral) substances, groundwater always contains natural organic substances, and almost always some living microorganisms (mainly bacteria), even at depths of up to 3.5 km in some locations (Krumholz, 2000). Organic matter in surface and groundwater is a diverse mixture of organic compounds ranging from macromolecules to low-molecular-weight compounds such as simple organic acids and short-chained hydrocarbons. In groundwater, there are three main natural sources of organic matter: organic matter deposits such as buried peat, kerogen, and coal; soil and sediment organic matter; and organic matter present in waters infiltrating into the subsurface from rivers, lakes, and marine systems (Aiken, 2002). Various components of naturally occurring hydrocarbons (oil and gas) and their breakdown products formed by microbial activity are a significant part of groundwater chemical composition in many areas throughout the world. A very large number of artificial organic chemicals have become part of groundwater reality in recent years because, if analyzed using the latest available analytical methods, they are often detected.

Organic matter in groundwater plays an important role in controlling geochemical processes by acting as proton and electron donors-acceptors and pH buffers, by affecting the transport and degradation of pollutants and by participating in mineral dissolution and precipitation reactions. Dissolved and particulate organic matter may also influence the availability of nutrients and serve as a carbon substrate for microbially mediated reactions. Numerous studies have recognized the importance of natural organic matter in the mobilization of hydrophobic ("water-hating") organic species, heavy metals, and radionuclides. Many contaminants that are commonly regarded as virtually immobile in aqueous systems can interact with dissolved organic carbon (DOC) or colloidal organic matter, resulting in migration of hydrophobic chemicals far beyond the distances predicted by the structure and activity relationships (Aiken, 2002).

A number of significant, although poorly understood, mechanisms can be responsible for the transport or retention of organic molecules in the subsurface. Once in the system, organic compounds, whether of anthropogenic or natural origin, can be truly dissolved, associated with immobile or mobile particles. Mobile particles include DOC, DOC-iron complexes, and colloids. Positively charged organic solutes are readily removed from the dissolved phase by cation exchange, which can be a significant sorption mechanism. Organic solutes that may exist as cations in natural waters include amino acids and polypeptides. Hydrophilic neutral (e.g., carbohydrates and alcohols) and low-molecular-weight anionic organic compounds (e.g., organic acids) are retained the least by aquifer solids. Hydrophobic synthetic organic compounds interact strongly with the organic matter associated with the solid phase of porous media. These interactions are controlled, in part, by the nature of the organic coatings on solid particles, especially with respect to its polarity and aromatic carbon content. Interactions of hydrophobic organic compounds with stationary particles can result in strong binding and slow release rates of these compounds (Aiken, 2002).

5.3 Groundwater Contamination and Contaminants

In general, any water that contains disease-causing or toxic substances is defined as contaminated (USEPA, 2000a). This definition does not differentiate between possible sources of contamination or types of contaminants—any substance of natural or synthetic origin that is toxic to humans or can cause disease is defined as a groundwater (water) contaminant. In the broadest sense, all sources of groundwater contamination and contaminants themselves can be grouped into two major categories: naturally occurring and artificial (anthropogenic). Some natural contaminants, such as arsenic and uranium, may have significant local or regional impacts on groundwater supplies. However, anthropogenic sources and synthetic chemical substances, in general, have much greater negative effects on the quality of groundwater resources. Almost every human activity has the potential to directly or indirectly impact groundwater to a certain extent. Figure 5.6 illustrates some of the land use activities that can result in groundwater contamination. An exponential advancement of analytical laboratory techniques in the last decade has demonstrated that many synthetic organic chemicals (SOCs) are widely distributed in the environment, including in groundwater, and that a considerable number of them can now be found in human tissue and organs of people living across the globe.

Strongly related to the ever-increasing public awareness of environmental pollution is a very rapid growth in consumption of bottled drinking water, also across the globe. Many consumers are ready to pay a premium for brands marketed as "pure spring

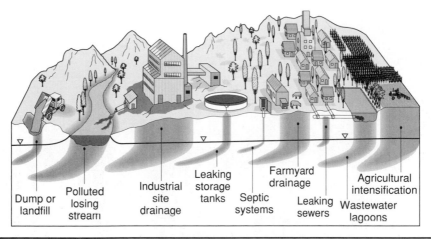

FIGURE 5.6 Land use activities commonly generating a groundwater pollution threat. (From Foster et al., 2002–2005.)

water" or "water coming from deep pristine aquifers," so that major multinational corporations are frantically looking for groundwater resources that can be marketed as such. In general, there is still a lot of truth in the following statement, very much appreciated by many groundwater professionals: groundwater in general is much less vulnerable to contamination than surface water; it is generally of better quality and thus requires less investment in water supply development. It is however also true that it usually takes more time and it is more difficult to restore a groundwater source once it becomes contaminated.

In developed industrialized countries, regulatory agencies are mostly concerned with groundwater contamination caused by organic synthetic chemicals. For example, in 1993 the USEPA reported that rapid growth of the chemical industry in the United States in the second half of the twentieth century resulted in common industrial and commercial use of at least 63,000 SOCs (synthetic organic chemicals), with 500 to 1000 being added each year. Health effects brought about by long-term, low-level exposure to these chemicals are not well known (USEPA, 1990).

In less developed countries, contamination of water supplies by organic chemicals is of minor or no concern compared to health problems related to poor sanitary conditions and diseases caused by pathogenic organisms (bacteria, parasites, and viruses). The primary health-related goal in such countries is disinfection of drinking water and development of safe water supplies. Even a simple sanitary design of water wells, such as the one shown in Fig. 5.7, can dramatically improve the health and lives of population dependent on groundwater use.

Groundwater contamination is most common in developed urban areas, agricultural areas, and industrial complexes. Frequently, groundwater contamination is discovered long after it has occurred. One reason for this is the slow movement of groundwater through groundwater systems, sometimes as little as fractions of a foot (meter) per day. This often results in a delay in the detection of groundwater contamination. In some cases, contaminants introduced into the subsurface decades ago are only now being discovered. This also means that the environmental management practices of today will have effects on groundwater quality well into the future (USEPA, 2000a).

FIGURE 5.7 Children in Northern Ghana use a hand pump to collect water for their families at a new sanitary well. (Photograph courtesy of Jenny VanCalcar.)

5.3.1 Health Effects

Various substances in drinking water can adversely affect or cause disease in humans, animals, and plants. These effects are known as toxic effects. Below are general categories of toxicity, based on the organs or systems in the body affected (USEPA, 2003a):

- Gastrointestinal—affecting the stomach and intestines.
- Hepatic—affecting the liver.
- Renal—affecting the kidneys.
- Cardiovascular or hematological—affecting the heart, circulatory system, or blood.

- Neurological—affecting the brain, spinal cord, and nervous system. In nonhuman animals, behavior changes can result in lower reproductive success and increased susceptibility to predation.

- Respiratory—affecting the nose, trachea, and lungs or the breathing apparatus of aquatic organisms.

- Dermatological—affecting the skin and eyes.

- Reproductive or developmental—affecting the ovaries or testes, or causing lower fertility, birth defects, or miscarriages. This includes contaminants with genotoxic effects, i.e., capable of altering deoxyribonucleic acid (DNA), which can result in mutagenic effects or changes in genetic materials.

Substances that cause cancer are known as *carcinogens* and are classified as such based on evidence gathered in studies. The USEPA classifies compounds as carcinogenic based on evidence of carcinogenicity, pharmacokinetics (the absorption, distribution, metabolism, and excretion of substances from the body), potency, and exposure. Based on the weight-of-evidence descriptors, USEPA has the following classification of contaminants (2005): (1) carcinogenic to humans, (2) likely to be carcinogenic to humans, (3) suggestive evidence of carcinogenic potential, (4) inadequate information to assess carcinogenic potential, and (5) not likely to be carcinogenic to humans.

Many of the SOCs commonly found in groundwater, such as most prevalent volatile organic compounds (VOCs) and pesticides (e.g., benzene, tetrachloroethene (PCE), trichloroethene (TCE), and alachlor) are carcinogens and have very low maximum contaminant levels (MCLs), which are drinking water standards legally enforceable by the USEPA.

The effects a contaminant has on various life forms depend not only on its potency and the exposure pathway but also on the temporal pattern of exposure. Short-term exposure (minutes to hours) is referred to as *acute*. For example, a person can become seriously ill after drinking only one glass of water contaminated with a pathogen (bacteria, virus, and parasite). Longer term exposure (days, weeks, months, and years) is referred to as *chronic*.

The constancy of exposure is also a factor in how the exposure affects an organism. For instance, the effects of 7 days of exposure may differ, depending on whether the exposure was on 7 consecutive days or 7 days spread over a month, a year, or several years. In addition, some organisms may be more susceptible to the effects of contaminants. If evidence shows that a specific subpopulation is more sensitive to a contaminant than the population at large, then safe exposure levels are based on that population. If no such scientific evidence exists, pollution standards are based on the group with the highest exposure level. Some commonly identified sensitive subpopulations include infants and children, the elderly, pregnant and lactating women, and immunocompromised individuals (USEPA, 2003a). The most common groups of groundwater contaminants that can cause serious health effects are heavy metals, SOCs, radionuclides, and microorganisms (pathogens).

At their natural concentrations, some heavy metals play an essential role in biochemical processes and are required in small amount by most organisms for normal, healthy growth (e.g., zinc, copper, selenium, and chromium). Other metals such as cadmium, lead, mercury, and tin, and the semimetal arsenic are not essential and do not cause deficiency disorders if absent (van der Perk, 2006). If ingested in excessive quantities, virtually all heavy metals are toxic to animals and humans. They become toxic by forming complexes with organic compounds (ligands) so that the modified molecules lose their ability to function properly, causing the affected cells to malfunction or die. In acute poisoning, large excesses of metal ions can disrupt membrane and mitochondrial function and generate free radicals. In most cases, this leads to general weakness and malaise (van der Perk, 2006). Probably the most infamous metal associated with groundwater contamination is arsenic. Exposure to naturally occurring arsenic in drinking water from groundwater sources has been widely documented in various regions of the world and has had grave health consequences for affected populations, particularly in south and southeast Asia.

According to the USEPA, human exposure to arsenic can cause both short- and long-term health effects. Short or acute effects can occur within hours or days of exposure. Long or chronic effects occur over many years. Long-term exposure to arsenic has been linked to cancer of the bladder, lungs, skin, kidneys, nasal passages, liver, and prostate. Short-term exposure to high doses of arsenic can cause other adverse health effects, but such effects are unlikely to occur from U.S. public water supplies that are in compliance with the arsenic drinking water standard, currently set at 0.01 mg/L (http://www.epa.gov/safewater/arsenic/basicinformation.html#three).

The Agency for Toxic Substances & Disease Registry (ATSDR) of the U.S. Department of Health and Human Services has, on its Web site, a very detailed discussion regarding physiologic effects of arsenic toxicity including that from drinking contaminated groundwater (http://www.atsdr.cdc.gov/csem/arsenic/physiologic_effects.html). For example: "Epidemiologic evidence indicates that chronic arsenic exposure is associated with vasospasm and peripheral vascular insufficiency. Gangrene of the extremities, known as Blackfoot disease, has been associated with drinking arsenic-contaminated well water in Taiwan, where the prevalence of the disease increased with increasing age and well-water arsenic concentration (10 to 1820 ppb). Persons with Blackfoot disease also had a higher incidence of arsenic-induced skin cancers."

Since the first publication of Rachel Carson's *Silent Spring* in 1962 (Carson, 2002), there has been increasing awareness that anthropogenic chemicals in the environment can exert profound and deleterious effects on wildlife populations and that human health is inextricably linked to the health of the environment. The last two decades, in particular, have witnessed growing scientific concern, public debate, and media attention over the possible harmful effects to humans and wildlife that may result from exposure to chemicals that have the potential to interfere with the endocrine system. These chemicals, called endocrine disruptors, are exogenous substances that act like hormones in the endocrine system and disrupt the physiologic function of endogenous hormones (Wikipedia, 2007).

On its dedicated web page, the USEPA states that "Evidence suggests that environmental exposure to some anthropogenic chemicals may result in disruption of endocrine systems in human and wildlife populations. A number of the classes of chemicals suspected of causing endocrine disruption fall within the purview of the U.S. Environmental Protection Agency's mandates to protect both public health and the environment. Although there is a wealth of information regarding endocrine disruptors, many critical scientific uncertainties still remain" (http://www.epa.gov/endocrine/).

The list of endocrine disruptors is very long, and it is constantly growing as new research results become available. They encompass a variety of chemical classes, including natural and synthetic hormones, pesticides, and compounds used in the plastics industry and in consumer products. Endocrine disruptors are often pervasive and dispersed in the environment, including in groundwater. Here are a few from the major groups of synthetic chemicals: persistent organohalogens (1,2-dibromoethane, dioxins and furans, PBBs, PCBs, and pentachlorophenol), food antioxidants (BHA), pesticides (majority, if not all; e.g., alachlor, aldrin, atrazine, chlordane, DDT, dieldrin, heptachlor, lindane, mirex, zineb, and ziram), and phatalates. Heavy metals like arsenic, cadmium, lead, and mercury are also endocrine disruptors in addition to their toxic effects.

In general, the health effects associated with endocrine-disrupting compounds include a range of reproductive problems (reduced fertility, male and female reproductive tract abnormalities, skewed male/female sex ratios, loss of fetus, and menstrual problems), changes in hormone levels, early puberty, brain and behavior problems, impaired immune functions, and various cancers (Wikipedia, 2007).

A book by Colborn et al. (1997) *Our Stolen Future* examines mechanisms with which certain synthetic chemicals interfere with hormonal messages involved in the control of growth and development, especially in the fetus. The associated Web site discusses scientific findings of the impacts of endocrine disrupters at low doses, emphasizing that new research on endocrine-disrupting compounds is revealing that these compounds have impacts at levels dramatically lower than that thought to be relevant to traditional toxicology. The site also includes numerous recent research examples with full reference details: http://www.ourstolenfuture.org/NewScience/newscience.htm.

Some of the more recent research regarding the dual effects and risks of multiple contaminants in drinking water is of particular concern. For example, in a study by the University of Wisconsin–Madison, researchers noted that common mixtures of pesticides and fertilizers can have biological effects at the current concentrations measured in groundwater. Specifically, the combination of aldicarb, atrazine, and nitrate, which are the most common contaminants detected in groundwater in agricultural areas, can influence the immune and endocrine systems as well as affect neurological health. Changes in the ability to learn and in patterns of aggression were observed. Effects are most noticeable when a single pesticide is combined with nitrate fertilizer. Research shows that children and developing fetuses are most at risk (Porter et al., 1999; from USEPA, 2000a).

On its Web site (http://www.atsdr.cdc.gov), the ATSDR has a detailed discussion on the health effects of many SOCs that have been found in groundwater supplies.

Health-Based Screening Levels

Health-based screening levels (HBSLs) are benchmark concentrations of contaminants in water that, if exceeded, may be of potential concern for human health. HBSLs are nonenforceable benchmarks that were developed by the USGS in collaboration with the USEPA and others using (1) USEPA methodologies for establishing drinking water guidelines and (2) the most recent, USEPA peer-reviewed, publicly available human-health toxicity information (Toccalino et al., 2003, 2006). HBSLs are based on health effects alone and do not consider cost or technical limitations of water treatment required to remove a contaminant (i.e., to decrease its concentration in water below detectable levels). In contrast, maximum contaminant levels (MCLs) are legally enforceable USEPA drinking water standards that set the maximum permissible level of a contaminant in water that is delivered to any user of a public water system. MCLs are set as close as feasible to the maximum level of a contaminant at which no known or anticipated adverse effects on human health would occur over a lifetime, taking into account the best available technology (BAT), treatment techniques (TTs), cost considerations, expert judgment, and public comments (USEPA, 2006).

For carcinogens, the HBSL range represents the contaminant concentration in drinking water that corresponds to an excess estimated lifetime cancer risk of 1 chance in 1 million (10^{-6}) to 1 chance in 10,000 (10^{-4}). For noncarcinogens, the HBSL represents the maximum contaminant concentration in drinking water that is not expected to cause any adverse effect over a lifetime of exposure. HBSL calculations adopt USEPA assumptions for establishing drinking water guidelines, specifically lifetime ingestion of 2 L of water per day by an adult weighing 70 kg. For noncarcinogens, it also typically is assumed that 20 percent of the total contaminant exposure comes from drinking water sources and that 80 percent comes from other sources such as food and air (Toccalino, 2007). The HBSL methodology includes the final USEPA cancer classifications (USEPA, 2005a).

HBSL for known carcinogens is calculated using the following equation (Toccalino, 2007):

$$\text{HBSL } (\mu g/L) = \frac{(70 \text{ kg body weight}) \times (\text{risk level})}{(2 \text{ L water consumed}/d) \times (\text{SF [mg/kg/d]}^{-1}) \times (\text{mg}/1000\,\mu g)}$$
(5.6)

where risk level is 10^{-6} to 10^{-4} risk range, and SF is the oral cancer slope factor, which has units of $(\text{mg}/\text{kg}/d)^{-1}$. SF is defined as an upper bound, approximating a 95 percent confidence limit, on the increased cancer risk for a lifetime exposure to a contaminant. This estimate is generally reserved for use in the low-dose region of the dose-response relationship. If the model selected for extrapolation from dose-response data is the linearized multistage model, the SF value is also known as the Q1* (carcinogenic potency factor) value.

For contaminants with suggestive evidence of carcinogenic potential, HBSLs are calculated using the following equation for calculating lifetime health advisory (lifetime HA) values:

$$\text{HBSL } (\mu g/L) = \left[\frac{(\text{RfD [mg/kg/d]}) \times (70 \text{ kg body weight}) \times (1000\,\mu g/\text{mg}) \times \text{RSC}}{(2 \text{ L water consumed}/d)} \right]$$
$$\div \text{RMF}$$
(5.7)

where RfD = reference dose in milligrams of chemical per kilogram of body weight per day
 RSC = relative source contribution (defaults to 20 percent in the absence of other data)
 RMF = risk management factor (defaults to 10 in the absence of other data)

An oral RfD is an estimate (with uncertainty spanning perhaps an order of magnitude) of a daily oral exposure to the human population (including sensitive subgroups) that is likely to be without an appreciable risk of deleterious effects during a lifetime (USEPA, 2006). Units for RfD are milligrams per kg per day (mg/kg/d).

For noncarcinogens, HBSLs are calculated using the following equation for calculating lifetime HA values:

$$\text{HBSL } (\mu g/L) = \left[\frac{(\text{RfD [mg/kg/d]}) \times (70 \text{ kg body weight}) \times (1000\,\mu g/\text{mg}) \times \text{RSC}}{(2 \text{ L water consumed}/d)} \right]$$
(5.8)

5.3.2 Sources of Contamination

Groundwater contamination can occur as relatively well-defined, localized plumes emanating from specific point sources such as leaking underground storage tanks (LUSTs), spills, landfills, waste lagoons, and industrial facilities. Nonpoint sources of pollution refer to pollution discharged over a wide land area, not from one specific location. They include forms of diffuse pollution caused by sediment, nutrients, and organic and toxic substances originating from land use activities such as agriculture or urban development. Rainwater, snowmelt, or irrigation water can wash off these substances together with soil particles and carry them with surface runoff to surface streams. A portion of this contaminant load dissolved in water can also infiltrate into the subsurface and eventually contaminate groundwater.

Results of a nationwide study of potential contaminant sources conducted by the USEPA and state environmental protection agencies is shown in Fig. 5.8. Each state was requested to indicate the 10 top sources that potentially threaten their groundwater resources. States added sources as was necessary based on state-specific concerns. When selecting sources, states considered numerous factors, including

- The number of each type of contaminant source in the state
- The location relative to groundwater sources used for drinking water purposes
- The size of the population at risk from contaminated drinking water
- The risk posed to human health and/or the environment from releases

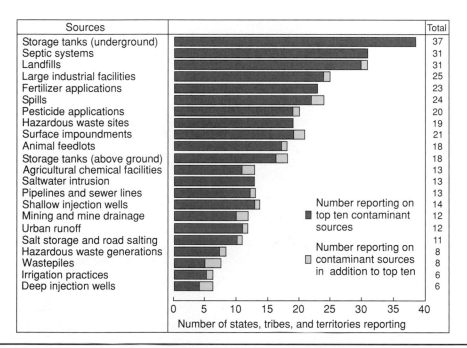

Figure 5.8 Major sources of groundwater contamination in the United States. (From USEPA, 2000a.)

- Hydrogeologic sensitivity (the ease with which contaminants enter and travel through soil and reach aquifers)
- The findings of the state's groundwater assessments and/or related studies

For each of the 10 top sources, states identified the specific contaminants that may impact groundwater quality. As seen in Fig. 5.8, the sources most frequently cited by states as a potential threat to groundwater quality are LUSTs. Septic systems, landfills, industrial facilities, and fertilizer applications are the next most frequently cited sources of concern. If similar sources are combined, five broad categories emerge as the most important potential sources of groundwater contamination: (1) fuel storage practices, (2) waste disposal practices, (3) agricultural practices, (4) industrial practices, and (5) mining operations.

Fuel Storage Practices

Fuel storage practices include the storage of petroleum products in underground and aboveground storage tanks. Underground storage tank (UST) is any system having 10 percent of the total tank volume below ground. Although tanks exist in all populated areas, they are generally most concentrated in the more heavily developed urban and suburban areas of a state. Storage tanks are primarily used to hold petroleum products such as gasoline, diesel fuel, and fuel oil. Leakages can be a significant source of groundwater contamination (Fig. 5.9). The primary causes of tank leakages are faulty installation or corrosion of tanks and pipelines. The USEPA (2000a) reports that based on information from 22 states, 57 percent of 85,000 USTs were characterized by confirmed contaminant releases to the environment and 18 percent had releases that adversely affected groundwater quality.

Petroleum products are complex mixtures of hundreds of different compounds. Over 200 gasoline compounds can be separated in the mixture. Compounds characterized

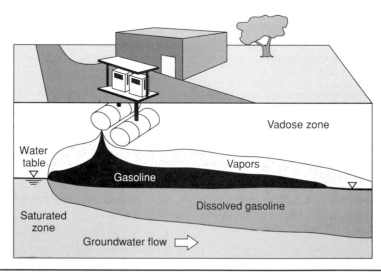

Figure 5.9 Groundwater contamination as a result of leaking underground storage tanks. (From USEPA, 2000a.)

by higher water solubility are frequently detected in groundwater. Four compounds, in particular, are associated with petroleum contamination: benzene, toluene, ethylbenzene, and xylenes, commonly named together as BTEX. Petroleum-related chemicals threaten the use of groundwater for human consumption because some, such as benzene, are known to cause cancer even at very low concentrations.

Waste Disposal Practices

Waste disposal practices include septic systems, landfills, surface impoundments, deep and shallow injection wells, dry wells, sumps, wastepiles, waste tailings, land application, and illegal disposal. Any practice that involves the handling and disposal of waste has the potential to impact the environment if protective measures are not taken. Contaminants most likely to impact groundwater include metals, volatile organic compounds (VOCs), semivolatile organic compounds (SVOCs), nitrates, radionuclides, and pathogens. As reported by the USEPA (2000a), a state survey indicates that in many instances present-day groundwater contamination is the result of historic practices at waste disposal sites.

Domestic or centralized septic systems for on-site sewage disposal are constructed using conventional, alternative, or experimental system designs. Conventional individual septic system design consists of a septic tank used to detain domestic wastes to allow the settling of solids and a leach field where liquids distributed from the septic tank (and optionally a distribution box) are allowed to infiltrate into the shallow unsaturated soil for adsorption. Septic tanks are commonly used when a sewer line is not available to carry wastewater to a treatment plant. Improperly constructed and poorly maintained septic systems can cause substantial and widespread nutrient and microbial contamination of groundwater. For example, approximately 126,000 individual on-site septic systems are used by 252,000 people in Montana, and groundwater monitoring has shown elevated nitrate levels near areas of concentrated septic systems. Nitrate contamination by individual septic systems and municipal sewage lagoons is a significant groundwater contamination problem reported by other states as well.

Leaking sewer lines in urban areas and industrial complexes can cause groundwater contamination with a variety of contaminants. Together with the leaky water supply lines, these also contribute to rising water tables under many large urban centers.

Land application is a general term for spreading of sewage (domestic and animal) and water-treatment plant sludge over tracts of land. This practice, still controversial, when improperly executed can cause widespread groundwater contamination in hydrogeologically sensitive areas.

The problem with all sewage disposal practices is the presence of various pharmaceuticals and personal care products (PPCPs) that are being continuously released to the environment including groundwater. During 1999 to 2000, the United States Geological Survey (USGS) implemented first-ever U.S. national reconnaissance of "emerging pollutants" in surface waters and groundwaters in 36 states. The objective of the study was to establish baseline occurrence data including for some commonly used PPCPs. Samples were collected from 142 streams, 55 wells, and seven wastewater-treatment effluents. The findings, published on March 15, 2002, issue of Environmental Science and Technology, show a widespread of PPCPs in surface water and groundwater. Detailed information is available at http://toxics.usgs.gov/highlights/whatsin.html.

Landfills have long been used to dispose of wastes, and, in the past, little regard was given to the potential for groundwater contamination in site selection. Landfills

FIGURE 5.10 Aerial view of the proper closure of a modern-day landfill, which is required by strict environmental regulations. The landfill is being covered by several layers of materials to prevent rainfall infiltration and leaching, and to ensure no future adversarial impact on groundwater resources. Note the highway on the left for scale. (Printed with permission of the City of Virginia Beach, VA, the United States.)

were generally sited on land considered to have no other uses. Unlined abandoned sand and gravel pits, old strip mines, marshlands, and sinkholes were often used. In many instances, the water table was at, or very near the ground surface, and the potential for groundwater contamination was high. Not surprisingly, states consistently cite landfills as a high-priority source of groundwater contamination. Generally, the greatest concern is associated with practices or activities that occurred prior to establishment of stringent construction standards to which modern landfills must adhere (Fig. 5.10).

According to the USEPA (2000a), discharges to surface impoundments such as pits, ponds, lagoons, and leach fields are generally underregulated. They usually consist of relatively shallow excavations that range in area from a few square feet to many acres and are or were used in agricultural, mining, municipal, and industrial operations for the treatment, retention, and disposal of both hazardous and nonhazardous wastes. As a consequence, they have the potential to leach metals, VOCs, and SVOCss to groundwater. For example, in Colorado, wells located downgradient from tailings ponds or cyanide heaps associated with mining operations often exhibit high concentrations of metals; in Arizona, surface impoundments and leach fields are identified as significant sources of VOCs.

During the Surface Impoundment Assessment (USEPA, 1983), more than 180,000 impoundments were located at approximately 80,000 sites. Nearly half of the sites were located over zones that are either very thin or very permeable, and more than half of these contained industrial waste. In addition, 98 percent of the sites on thick, permeable aquifers were located within a mile of potential drinking water supplies (USEPA, 1983).

Especially serious problems develop with surface impoundments in limestone terrain with extensive near-surface solution openings. In 1990, the USEPA reported that in Florida, Alabama, Missouri, and elsewhere, municipal sewage lagoons have collapsed into sinkholes draining raw influent into widespread underground openings. In some cases, the sewage has reappeared in springs and streams several miles away.

Class V injection wells are shallow disposal systems that are used to place a variety of fluids below the land surface, directly into or above shallow aquifers. They include shallow wastewater disposal wells ("dry wells"), sumps, septic systems, storm water drains, and agricultural drainage systems. Because class V injection wells did not have any specific design requirements and were not required to treat the wastewaters released through them, the USEPA revised underground injection control regulations in 2001 after recognizing their potential threat to groundwater supplies (USEPA, 2002).

Class I injection wells are defined by the USEPA as wells that inject fluids below the deepest underground source of drinking water (USDW) within a quarter mile (402 m) radius of the borehole (USEPA, 2002). In many parts of the country, these wells are used to dispose of waste fluids, primarily wastewater from municipal wastewater-treatment plants, but also landfill leachate and nonhazardous industrial wastewater. In 1983, the USEPA reported the existence of at least 188 active hazardous waste deep injection wells in the United States. Most such wells are tied to the chemical industry and their depths range from 1000 to 9000 ft. The deepest wells are in Texas and Mississippi.

Over the past 30 years, deep well injection has become an essential method for the disposal of liquid wastes in many parts of the country. For example, in Florida in 2002, approximately 1,285,000 m^3/d of liquid waste was injected in 126 active deep (Class I) injection wells in Florida (Florida Department of Environmental Protection, 2003a, 2003b; from Maliva et al., 2007). Because of improper well construction, well failure, or unforeseen hydrogeologic conditions, a deep injection site may become a source of potential groundwater contamination (Maliva et al., 2007).

Oil field brines have contaminated both surface water and groundwater in every state that produces oil (USEPA, 1990). The brine, an unwanted by-product, is produced with the oil, as well as during drilling. In the latter case, drilling fluids and brines were historically stored in reserve pits, which were filled some time after completion or abandonment of the well. Ordinarily, oil field brines are temporarily stored in holding tanks or placed in an injection well. Owing to the corrosive nature of the brine, transport pipelines and casings of the injection wells can readily corrode, causing groundwater contamination. As of 1983, the USEPA reported the existence of 24,000 Class II wells used to inject oil field brine.

Agricultural Practices

Agricultural practices that have the potential to contaminate groundwater include animal feedlots, fertilizer and pesticide applications, irrigation practices, agricultural chemical facilities, and drainage wells. Groundwater contamination can be a result of routine applications, spillage, or misuse of pesticides and fertilizers during handling and storage, manure storage and spreading, improper storage of chemicals, and irrigation return drains serving as a direct conduit to groundwater. Fields with overapplied and misapplied fertilizers and pesticides can introduce nitrogen, pesticides, cadmium, chloride, mercury, and selenium into the groundwater. As indicated by the USEPA (2000a), states report that agricultural practices continue to be a major source of groundwater contamination. Fertilizers and pesticides are applied both in rural agricultural areas on crops and orchards and in urban-suburban settings on lawns and golf courses.

Livestock is an integral component of many states' economies. As a consequence, concentrated animal feeding operations (CAFOs), where animals are kept and raised in confined areas, occur in many states. CAFOs congregate animals, feed, manure and urine, dead animals, and production operations on a small land area. Such operations can pose a number of risks to water quality and public health, mainly because of the amount of animal manure and wastewater they generate. Animal feedlots often have impoundments from which wastes may infiltrate into groundwater. Livestock waste is a source of nitrate, bacteria, TDS, and sulfates.

Shallow unconfined aquifers in many states have become contaminated from the application of fertilizer. Crop fertilization is the most important agricultural practice contributing nitrate to the environment. Nitrate is considered by many to be the most widespread groundwater contaminant. To help combat the problems associated with the overuse of fertilizers, the U.S. Department of Agriculture's Natural Resources Conservation Service assists crop producers in developing nutrient management plans.

Pesticide use and application practices are of great concern for groundwater quality nationwide. The primary route of pesticide transport to groundwater is by leaching through the vadose zone or by spills and direct infiltration through drainage controls. Pesticide infiltration is generally greatest when rainfall is intense and occurs shortly after the pesticide is applied. Within sensitive areas, groundwater monitoring has shown fairly widespread detections of pesticides, specifically the pesticide atrazine.

Human-induced salinity occurs in agricultural regions where irrigation is used extensively. Irrigation water continually flushes nitrate-related compounds from fertilizers into the shallow aquifers along with high levels of chloride, sodium, and other metals, thereby increasing the salinity of the underlying aquifers (USEPA, 2000a). Improper irrigation can cause extensive soil salinization by raising the water table above the critical depth of evaporation, resulting in precipitation of dissolved mineral salts and their accumulation at and near the land surface.

Industrial Practices

Raw materials and waste handling in industrial processes can pose a threat to groundwater quality. Industrial facilities, hazardous waste generators, and manufacturing and repair shops, all present the potential for releases. Storage of raw materials at the facility is a problem if the materials are stored improperly and leaks or spills occur. Examples include chemical drums that are carelessly stacked or damaged, and dry materials that are exposed to rainfall. Material transport and transfer operations at these facilities can also be a cause for concern.

The most common industrial contaminants are metals, VOCs, SVOCs, and petroleum compounds. VOCs are associated with a variety of activities that use them as degreasing agents. As pointed out by the USEPA (2000a), development of new technologies and new products to replace organic solvents is continuing. For example, organic biodegradable solvents derived from plants are being developed for large-scale industrial applications. Environmentally responsible dry-cleaning technologies are being developed that eliminate the need for perchloroethylene (PCE), the chemical most commonly used by dry-cleaning operations. Legislation is being considered in New York and by other local governments and states that would ban the use of PCE by the dry-cleaning industry.

In the 2000 USEPA study, accidental spills of chemicals, industrial wastes, and petroleum products from trucks, railways, aircrafts, handling facilities, and storage tanks were indicated by most states as a source of grave concern. For example, the state of Indiana reported that, in 1996, 41 million gallons of various products were spilled, with about 50 spills occurring per week. Montana reports an average of 300 accidental spills each year. On average, approximately 15 of these spills require extensive cleanup and follow-up groundwater monitoring. One of these was the 1995 derailment of railroad tanker cars in the Helena rail yard that threatened to contaminate groundwater with 17,400 gallons of fuel oil. Follow-up monitoring demonstrated that rapid response actions had prevented the majority of the contaminants from reaching local aquifers. South Carolina determined that accidental spills and leaks are the second most common source of groundwater contamination, and, as in Arizona, these releases were usually associated with petroleum-based products attributed to machinery maintenance or manufacturing.

It is virtually certain that there are leaks at any given time from tens of thousands of miles of buried pipelines carrying various petroleum products and industrial fluids somewhere in the United States. Such leaks are, however, exceedingly difficult to detect. Sometimes they may become apparent only by sudden and otherwise unexplainable changes in water quality of springs, wells, and surface streams or by dying vegetation.

Atmospheric pollutants, such as airborne sulfur and nitrogen compounds created by industrial activities and power generation using fossil fuels and by vehicle emissions, fall as dry particles or acid rain on land surface and can infiltrate into the soil column, eventually causing groundwater contamination.

Mining Operations

Mining can result in a variety of water contamination problems caused by pumping of mine waters to the surface, by leaching of the spoil material, by waters naturally discharging through the mine, and by milling wastes, among others. Literally thousands of miles of streams and hundreds of acres of aquifers have been contaminated by highly corrosive mineralized waters originating in the coal mines and dumps of Appalachia. In many western states, mill wastes and leachates from metal sulfide operations have seriously affected both surface water and groundwater (USEPA, 1990).

Many mines are deeper than the water table, and, in order to keep them dry, large quantities of water are pumped to waste. If salty or mineralized water lies at relatively shallow depths, the pumping of freshwater for dewatering purposes may cause an upward migration, which may be intercepted by pumping wells. The mineralized water most commonly is discharged into a surface stream (USEPA, 1990).

Wells as Contamination Conduits

Contamination through improperly abandoned, uncased, and failed wells, or wells with long screens and gravel packs open to several aquifers, is arguably the most problematic from the assessment standpoint (Fig. 5.11). Contamination caused by dissolved constituents or water of different density (e.g., saline water and brines) can migrate through such wells in either direction ("up" or "down" the well) depending on the differences in the hydraulic head or water density between various portions of the groundwater systems. For example, the USGS reports that thousands of deep wells were drilled into the

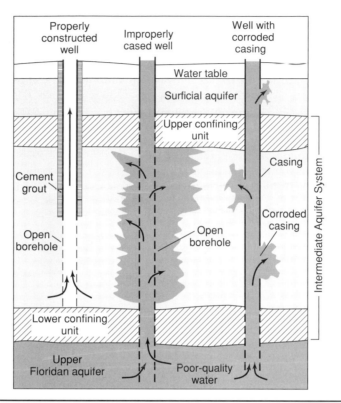

FIGURE 5.11 Contamination by saltwater upconing or by otherwise contaminated groundwater can occur through failed, uncased, or improperly constructed wells that create a conduit for flow between aquifers of differing water density or quality. A properly constructed well open to a single aquifer is shown on the left. (Modified from Metz and Brendle, 1996).

Upper Floridan aquifer in west-central Florida from 1900 to the early 1970s for irrigation, before this practice stopped due to the deterioration of water quality in the aquifer. Most of the early irrigation wells also were open to the intermediate aquifer system. Usually, the wells were completed with a short length of steel casing through the surficial aquifer and then were open to two lower aquifers. These open wells, which can be many hundreds of feet in length, provide direct conduits for water to flow upward or downward across the confining units, thus shortcutting the slower route of leakage through the confining units (Metz and Brendle, 1996). Approximately 8000 wells were reported or estimated to be open to the intermediate aquifer system and the Upper Floridan aquifer in the study area.

In a large area encompassing parts of Manatee, Sarasota, and Charlotte counties, groundwater levels were as much as 20 ft higher in the Upper Floridan aquifer than in the overlying intermediate aquifer system. It was estimated that a total of 85 Mgal/d flowed from the Upper Floridan aquifer to the overlying fresher water zones in the intermediate aquifer system through wells screened in both systems. In the majority of the area of upward flow, concentrations of chloride, sulfate, and dissolved solids in the Upper Floridan aquifer exceed recommended or permitted drinking water standards, and the upward flow is contaminating the intermediate aquifer system (Barlow, 2003). In 1974, the Southwest Florida Water Management District began the Quality of Water Improvement Program to restore hydrologic conditions altered by improperly constructed wells through a process of plugging of abandoned wells. As of October 2001, more than 5200 wells had been inspected and nearly 3000 plugged since the program began (Southwest Florida Water Management District, 2002).

5.3.3 Naturally Occurring Contaminants

Arsenic

Since 1993, the World Health Organization (WHO) lowered the drinking water standard for arsenic from 0.05 to 0.01 mg/L (10 ppb). Others followed, including the United States and the European Union. This naturally occurring element has become the most notorious and widely recognized groundwater contaminant. The following excerpts from a press release illustrate the point:

> Arsenic in drinking water is a global threat to health affecting more than 70 countries and 137 million people, according to new research presented to the annual conference at the Royal Geographical Society with IBG (The Institute of British Geographers) in London today (Wednesday 29 August 2007).
> Large numbers of people are being unknowingly exposed to unsafe levels of arsenic in their drinking water, Peter Ravenscroft from the department of geography at the University of Cambridge told the geographers' conference. At present, Bangladesh is the country worst affected, where hundreds of thousands of people are likely to die from arsenic causing fatal cancers of the lung, bladder and skin.
> Arsenic poses long-term health risks "exceeding every other potential water contaminant", according to research presented by Dr Allan Smith, of the University of California, Berkeley and adviser to the WHO on arsenic. Dr Smith added: "Most countries have some water sources with dangerous levels of arsenic, but only now are we beginning to recognise the magnitude of the problem. It is *the* most dangerous contaminant of drinking water in terms of long term health risks and we must test all water sources worldwide as soon as possible (RGS, 2007).

The most serious damage to health from drinking arsenic-contaminated water has occurred in Bangladesh and West Bengal, India. In the 1970s and 1980s, UNICEF and other international agencies helped to install more than 4 million hand-pumped wells in Bangladesh to give communities access to clean drinking water and to reduce diarrhea and infant mortality. Cases of arsenic-related diseases (generally referred to as arsenicosis) were seen in West Bengal and then in Bangladesh in the 1980s. By 1993, arsenic from the water in wells was discovered to be responsible. In 2000, a WHO report (Smith et al., 2000) described the situation in Bangladesh as: "the largest mass poisoning of a population in history ... beyond the accidents at Bhopal, India, in 1984, and Chernobyl, Ukraine, in 1986."

In 2006, UNICEF reported that 4.7 million (55 percent) of the 8.6 million wells in Bangladesh had been tested for arsenic of which 1.4 million (30 percent of those tested) had been painted red, showing them to be unsafe for drinking water: defined in this case as more than 50 ppb (UNICEF, 2006). Although many people have switched to using arsenic-free water, in a third of cases where arsenic had been identified, no action had yet been taken. UNICEF estimates that 12 million people in Bangladesh were drinking arsenic-contaminated water in 2006, and the number of people showing symptoms of arsenicosis was 40,000 but could rise to 1 million (UNICEF, 2006). Other estimates are higher still (Petrusevski et al., 2007).

Implementation of the WHO guideline value of 10 μg/L is not currently feasible for a number of countries strongly affected by the arsenic problem, including Bangladesh and India, which retain the 50 μg/L limit. Other countries have not updated their drinking water standards recently and retain the older WHO guideline of 50 μg/L. These include Bahrain, Bolivia, China, Egypt, Indonesia, Oman, Philippines, Saudi Arabia, Sri Lanka, Vietnam, and Zimbabwe. The most stringent standard currently set for acceptable arsenic concentration in drinking water is by Australia, which has a national standard of 7 μg/L (Petrusevski et al., 2007).

The disease symptoms caused by chronic arsenic ingestion, arsenicosis, develop when arsenic-contaminated water is consumed for several years. However, there is no universal definition of the disease caused by arsenic, and it is currently not possible to differentiate which cases of cancer were caused by drinking arsenic-affected water. Estimates, therefore, vary widely. Symptoms may develop only after more than 10 years of exposure to arsenic, while it may take 20 years of exposure for some cancers to develop. Long-term ingestion of arsenic in water can first lead to problems with kidney and liver function, and then to damage of the internal organs including lungs, kidney, liver, and bladder. Arsenic can disrupt the peripheral vascular system leading to gangrene in the legs, known in some areas as black foot disease. This was one of the first reported symptoms of chronic arsenic poisoning observed in China (province of Taiwan) in the first half of twentieth century. A correlation between hypertension and arsenic in drinking water has also been established in a number of studies (Petrusevski et al., 2007).

Elemental arsenic is a steel-gray metal-like substance rarely found naturally. As a compound with other elements such as oxygen, chlorine, and sulfur, arsenic is widely distributed throughout the earth's crust, especially in minerals and ores that contain copper or lead. Natural arsenic in groundwater is largely the result of dissolved minerals from weathered rocks and soils. Principal ores of arsenic are sulfides (As_2S_3, As_4S_4, and FeAsS), which are almost invariably found with other metal sulfides. The hydrogen form of arsenic is arsine, a poisonous gas. Arsenic also forms oxide compounds. Arsenic trioxide (As_2O_3) is a transparent crystal or white powder that is slightly soluble in water and has a specific gravity of 3.74. Arsenic pentoxide (As_2O_5) is a white amorphous solid that is very soluble in water, forming arsenic acid. It has a specific gravity of 4.32 (USEPA, 2005b).

Dissolved arsenic in groundwater exists primarily as oxy anions with formal oxidation states of III and V. Either arsenate [As(V)] or arsenite [As(III)] can be the dominant inorganic form in groundwater. Arsenate ($H_nAsO_4^{n-3}$) generally is the dominant form in oxic (aerobic, oxygenated) waters with dissolved oxygen >1 mg/L. Arsenite ($H_nAsO_3^{n-3}$) dominates in reducing conditions, such as sulfidic (dissolved oxygen <1 mg/L with sulfide present) and methanic (methane present) waters. Aqueous and solid-water reactions, some of which are bacterially mediated, can oxidize or reduce aqueous arsenic. Both anions are capable of adsorbing to various subsurface materials, such as ferric oxides and clay particles. Ferric oxides are particularly important to arsenate fate and transport, as ferric oxides are abundant in the subsurface and arsenate strongly adsorbs to these surfaces in acidic to neutral waters. An increase in the pH to an alkaline condition may cause both arsenite and arsenate to desorb, and they are usually mobile in an alkaline environment (Dowdle et al., 1996; Harrington et al., 1998; Welch et al., 2000; USEPA, 2005b). The toxicity and mobility of arsenic vary with its valency state and chemical form. As(III) is generally more toxic to humans and four to 10 times more soluble in water than As(V) (USEPA, 1997).

All arsenic compounds consumed in the United States are imported. Arsenic has been used primarily for the production of pesticides, insecticides, and chromated copper arsenate (CCA), a preservative that renders wood resistant to rotting and decay. Increased environmental regulation, along with the decision of the wood-treatment industry to eliminate arsenical wood preservatives from residential application by the end of 2003, caused arsenic consumption in the United States to decline drastically in 2004. Other industrial products containing arsenic include lead-acid batteries, light-emitting diodes, paints, dyes, metals, pharmaceuticals, pesticides, herbicides, soaps, and semiconductors. Anthropogenic sources of arsenic in the environment include mining and smelting operations, agricultural applications, and disposal of wastes that contain arsenic (USEPA, 2005b). Arsenic is a contaminant of concern at many remediation sites. Because arsenic readily changes valence states and reacts to form species with varying toxicity and mobility, effective treatment of arsenic can be challenging.

A recent study of arsenic concentrations in major U.S. aquifers by the USGS (accessible at: http://water.usgs.gov/nawqa/trace/pubs/) shows wide regional variations of naturally occurring arsenic due to a combination of climate and geology. Although slightly less than half of 30,000 arsenic analyses of groundwater in the United States were equal or less than 1 μg/L, about 10 percent exceeded 10 μg/L. At a broad regional scale, arsenic concentrations exceeding 10 μg/L appear to be more frequently observed in the western United States than in the eastern half (USGS, 2004). Interestingly, more detailed recent investigations of groundwater in New England, Michigan, Minnesota, South Dakota, Oklahoma, and Wisconsin suggest that arsenic concentrations exceeding 10 μg/L are more widespread and common than previously recognized. Arsenic release from iron oxide appears to be the most common cause of widespread arsenic concentrations exceeding 10 μg/L in groundwater. This can occur in response to different geochemical conditions, including release of arsenic to groundwater through the reaction of iron oxide with either natural or anthropogenic (i.e., petroleum products) organic carbon. Iron oxide also can release arsenic to alkaline groundwater, such as that found in some felsic volcanic rocks and alkaline aquifers of the western United States. Sulfide minerals in rocks may act both as a source and as a sink for arsenic, depending on local geochemistry. In oxic (aerobic, oxygenated) water, dissolution of sulfide minerals, most notably pyrite and arsenopyrite, contributes arsenic to groundwater and surface water in many parts of the United States. Other common sulfide minerals, such as galena, sphalerite, marcasite, and chalcopyrite, can contain 1 percent or more arsenic as an impurity.

Radionuclides

Radionuclides are naturally occurring elements that have unstable nuclei that spontaneously break down to form more stable energy and particle configurations. Energy released during this process is called radioactive energy, and such elements are called radioactive elements or radionuclides. The most unstable configurations disintegrate very rapidly, and some of them do not exist in the earth's crust anymore (e.g., chemical elements 85 and 87, astatine and francium). Other radioactive elements, such as rubidium-87, have a slow rate of decay and are still present in significant quantity (Hem, 1989). The decay of a radionuclide is a first-order kinetic process, usually expressed in terms of a rate constant (λ) given as

$$\lambda = \frac{\ln 2}{t_{1/2}}$$

(5.9)

where $t_{1/2}$ = half-life of the element, i.e., the length of time required for half the quantity present at time 0 to disintegrate.

Radioactive energy is released in various ways, with the following three types being of interest in water chemistry: (1) alpha radiation, consisting of positively charged helium nuclei; (2) beta radiation, consisting of electrons or positrons; and (3) gamma radiation, consisting of electromagnetic wave-type energy similar to X-rays (Hem, 1989). Potential effects from radionuclides depend on the number of radioactive particles or rays emitted (alpha, beta, or gamma) and not the mass of the radionuclides (USEPA, 1981). Becquerel (Bq) is the unit for radioactivity in the International System (SI) of units, defined as the radiation caused by one disintegration per second; this is equivalent to approximately 27.0270 picoCuries (pCi). The unit is named for a French physicist, Antoine-Henri Becquerel, the discoverer of radioactivity. One Curie (Ci; named after Pierre and Marie Curie, the discoverers of radium) is defined as 3.7×10^{10} atomic disintegrations per second, which is the approximate specific activity of 1 g of radium in equilibrium with its disintegration products. Maximum contaminant load (MCL) for radium and alpha and beta radiation is expressed in pCi/L in the United States. Where possible, radioactivity is reported in terms of concentration of specific nuclides, as is commonly the case with uranium, which is conveniently analyzed by chemical means (MCL of uranium is expressed in μg/L). For some elements, radiochemical analytical techniques permit detection of concentrations much lower than what can be analyzed by any current chemical method. This fact is of special significance when performing tracing with radioactive isotopes, which can be introduced into the groundwater in very small quantities.

Exposure to radionuclides results in an increased risk of cancer. Certain elements accumulate in specific organs. For example, radium accumulates in the bones and iodine accumulates in the thyroid. For uranium, there is also the potential for kidney damage. Many water sources have very low levels of naturally occurring radioactivity, usually low enough not to be considered a public health concern. In some parts of the United States, however, the underlying geology causes elevated concentrations of some radionuclides in aquifers used for water supply.

Contamination of water from anthropogenic radioactive materials occurs primarily as the result of improper waste storage, leaks, or transportation accidents. These radioactive materials are used in various ways in the production of nuclear energy, commercial products (such as television and smoke detectors), electricity, and nuclear weapons and in nuclear medicine in therapy and diagnosis.

Anthropogenic radionuclides have also been released into the atmosphere as the result of atmospheric testing of nuclear weapons and, in rare cases, accidents at nuclear fuel stations and discharge of radiopharmaceuticals. The two types of radioactive decay that carry the most health risks due to ingestion of water are alpha emitters and beta/photon emitters. Many radionuclides are mixed emitters, with each radionuclide having a primary mode of disintegration. The naturally-occurring radionuclides are largely alpha emitters, although many of the short-lived daughter products emit beta particles. Anthropogenic radionuclides are predominantly beta/photon emitters and include those that are released to the environment as the result of activities of the nuclear industry but also include releases of alpha-emitting plutonium from nuclear weapon and nuclear reactor facilities (USEPA, 2000b). The natural radionuclides involve three decay series, which start with uranium-238, thorium-232, and uranium-235, and are known collectively as the uranium, thorium, and actinium series. Each series decays through stages of various nuclides, which emit either an alpha or a beta particle as they decay and terminates with a stable isotope of lead. Some of the radionuclides also emit gamma radiation, which accompany the alpha or beta decay. The uranium series contains uranium-238 and -234, radium-226, lead-210, and polonium-210. The thorium series contains radium-228 and radium-224. The actinium series contains uranium-235 (USEPA, 2000b).

As part of the new MCL standard promulgation for the radionuclides, the USEPA, in cooperation with the USGS, issued a technical document (USEPA, 2000b), which includes sections on the fundamentals of radioactivity in drinking water, an overview of natural occurrence of major radionuclides in groundwater, and the results of a nationwide survey of selected wells in all hydrostratigraphic provinces in the United States performed by the USGS (Focazio et al., 2001).

5.3.4   Nitrogen (Nitrate)

Nitrate is believed by many to be the most widely spread groundwater contaminant worldwide, primarily as a result of agricultural activities utilizing fertilizers. Other significant and widely spread anthropogenic sources of groundwater contamination with nitrogen forms are the disposal of sewage by centralized and individual systems, leaking sewers, animal feeding operations, and acid rain. Nitrate is the most oxidized form of inorganic nitrogen. Nitrogen occurs in groundwater as uncharged gas ammonia (NH_3), which is the most reduced inorganic form, nitrite and nitrate anions (NO_2^- and NO_3^-, respectively), in cationic form as ammonium (NH_4^+), and at intermediate oxidation states as a part of organic solutes. Some other forms such as cyanide (CN^-) may occur in groundwater affected by waste disposal (Rees et al., 1995; Hem, 1989). Three gaseous forms of nitrogen may exist in groundwater: elemental nitrogen (oxidation state of zero), nitrous oxide (N_2O; slightly oxidized, $+1$), and nitric oxide (NO; $+2$). All three, when dissolved in groundwater, remain uncharged gasses (Rees et al., 1995).

Nitrogen can undergo numerous reactions that can lead to storage in the subsurface, or conversion to gaseous forms that can remain in the soil for periods of minutes to many years. The main reactions include (1) immobilization/mineralization, (2) nitrification, (3) denitrification, and (4) plant uptake and recycling (Keeney, 1990). Immobilization is the biological assimilation of inorganic forms of nitrogen by plants and microorganisms to form organic compounds such as amino acids, sugars, proteins, and nucleic acids. Mineralization is the inverse of immobilization. It is the formation of ammonia and ammonium ions during microbial digestion of organic nitrogen. Nitrification is the microbial oxidation of ammonia/ammonium ion first to nitrite, then ultimately to nitrate. Nitrification is a key reaction leading to the movement of nitrogen from the land surface to the water table because it converts the relatively immobile ammonium form (reduced nitrogen) and organic nitrogen forms to a much more mobile nitrate form. Chemosynthetic autotrophic soil bacteria of the family *Nitrobacteriaceae* is believed to be principally responsible for the nitrification process. Ammonium oxidizers, including the genera *Nitrosomonas*, *Nitrosospira*, *Nitrosolobus*, and *Nitrosvibrio*, oxidize ammonium to nitrite. The nitrite oxidizing bacteria, which oxidize nitrite to nitrate, include the genus *Nitrobacter*. Nitrification can also be carried out by heterotrophic bacteria and fungi (Rees et al., 1995). The nitrogen used by plants is largely in the oxidized form. Denitrification is the biological process that utilizes nitrate to oxidize (respire) organic matter into energy usable by microorganisms. This process converts the nitrate to more reduced forms, ultimately yielding nitrogen gas that can diffuse into the atmosphere. Uptake of nitrogen by plants also removes nitrogen from the soil column and converts it to chemicals needed to sustain the plants. Because the plants eventually die, the nitrogen incorporated into the plant tissues ultimately is released back to the environment, thus completing the cycle (Rees et al., 1995).

Ammonium cations are strongly adsorbed on mineral surfaces, whereas nitrate is readily transported by groundwater and stable over a considerable range of conditions. The nitrite and organic species are unstable in aerated water and easily oxidized. They are generally considered indicators of pollution by sewage or organic waste. The presence of nitrate or ammonium might be indicative of such pollution as well, but generally the pollution would have occurred at a site or time substantially removed from the sampling point. Ammonium and cyanide ions form soluble complexes with some metal ions, and certain types of industrial waste effluents may contain such species (Hem, 1989).

Nitrate is not directly toxic to humans. However, under strongly reducing conditions, such as those in human gut, it transforms to nitrite. Nitrite ions pass from the gut into the blood stream and bond to hemoglobin molecules, converting them to a form that cannot transport oxygen (methemoglobin). Nitrite can also react chemically with amino compounds to form nitrosamides, which are highly carcinogenic (UNESCO, 1998). Excessive consumption of nitrate in drinking water has been associated with the risk of methemoglobinemia or "blue baby syndrome," an acute effect that is accentuated under poor sanitary conditions such as sewage contamination or dirty drinking vessels (Buss et al., 2005). If left untreated, methemoglobinemia can be fatal for affected infants. The WHO and the European Union have set the standard for nitrate in drinking water at 11.3 mg/L measured as nitrogen (mg N/L) that corresponds to 50 mg NO_3/L. The standard in the United States, Canada, and Australia is 10 mg N/L.

Extensive application of nitrogen fertilizers has caused an increase in nitrate concentrations over large agricultural areas in many countries. As a worldwide average, pristine waters contain nitrate at approximately 0.1 mg N/L (Heathwaite et al., 1996). This is extremely low compared to typical modern groundwater concentrations. For example, studies of UK aquifers suggest that current natural background or baseline concentrations are more than an order of magnitude above the global average pristine concentration (Buss et al., 2005).

Nitrogen oxides, present in the atmosphere due to the combustion of fossil fuels, undergo various chemical alterations that produce H^+ and finally leave the nitrogen as nitrate. These processes can lower the pH of rain in the same way sulfur oxides do. Nonindustrially impacted rain may have a total nitrogen concentration of about 6 mg/L, and rainfall of 10 in/yr would yield a nitrogen load to the soil column of about 13 pounds per acre per year in such case. Significant evaporation of such rainwater could result in high concentrations of nitrogen in the infiltration water (Heaton, 1986; Rees et al., 1995). Industrially impacted rain may have a nitrogen concentration higher than 6 mg/L, resulting in higher nitrogen load to the subsurface.

Domestic sewage in sparsely populated areas of the United States is disposed of primarily in on-site septic systems. In 1980, 20.9 million residences (about 24 percent of the total in the United States) disposed of about 4 million acre-ft of domestic sewage in on-site septic systems (Reneau et al., 1989). Inherent in this method is the discharge of effluent to the local groundwater. To avoid contamination problems in an area, treated sewage effluent can be removed from a basin and discharged elsewhere if wastewater treatment is centralized. Unfortunately, removal is not possible with on-site septic systems, and even properly designed and constructed on-site septic systems frequently cause nitrate concentrations to exceed the MCL in the underlying groundwater (Wilhelm et al., 1994).

Total nitrogen concentrations in septic-tank effluent range from 25 mg/L to as much as 100 mg/L, and the average is in the range 35 to 45 mg/L (USEPA, 1980), of which about 75 percent is ammonium and 25 percent is organic. Wilhelm et al. (1994) report that nitrate concentrations in the effluent below a septic field can be two to seven times the MCL, and distinct plumes of nitrate-contaminated groundwater may extend from the septic system. Seiler (1996) estimates that septic systems contribution of nitrogen to groundwater in the East Lemmon subarea of Washoe County, Nevada, is between 16,500 and 42,000 kg (18 to 46 tons) of nitrogen annually.

In animal feeding lots, wastes may lose much of the nitrogen by ammonia volatilization, particularly in corrals that are not subject to water application; water can transport the nitrogen to the subsurface before substantial volatilization has occurred. The amount of nitrate from animal wastes that percolates to the groundwater depends on the amount of nitrate formed from the wastes, the infiltration rate, the frequency of manure removal, the animal density, the soil texture, and the ambient temperature (National Research Council, 1978).

The decay of natural organic material in the ground can contribute substantial amounts of nitrogen to groundwater. For example, in the late 1960s in west-central Texas, several cattle died from drinking groundwater containing high concentrations of nitrate; the source of the nitrate was determined to be naturally occurring organic material in the soil (Kreitler and Jones, 1975; from Seiler, 1996). The average nitrate concentration (as NO_3) for 230 wells was 250 mg/L, and the highest concentration exceeded 3000 mg/L. Native vegetation, which included a nitrogen-fixing plant, was destroyed by plowing of the soil for dryland farming. This increased oxygen delivery to the soil and the nitrate causing the contamination were formed by oxidation of the naturally occurring organic material in the soil.

The stable isotope composition of nitrate is known to be indicative of its source and can also be used to indicate that biological denitrification is occurring (Buss et al., 2005). The variable used is $\delta^{15}N$, which compares the fraction of $^{15}N/^{14}N$ of the sample to that of an internationally accepted standard (the air in the case of nitrogen):

$$\delta^{15}N(\permil) = \frac{\left(^{15}N/^{14}N\right)_{sample} - \left(^{15}N/^{14}N\right)_{standard}}{\left(^{15}N/^{14}N\right)_{standard}} \times 1000 \qquad (5.10)$$

When tracing the origins of contamination, some sources have characteristic isotopic signatures. For instance, the $\delta^{15}N$ values for inorganic nitrate fertilizers tend to be in the range $-7\permil$ to $+5\permil$, for ammonium fertilizers $-16\permil$ to $-6\permil$, for natural soil $-3\permil$ to $+8\permil$, for sewage, $+7\permil$ to $+25\permil$, and for precipitation $-3\permil$ (Fukada et al., 2004; Widory et al., 2004; BGS, 1999; Heaton, 1986). This approach is often combined with information from other species of interest: Barrett et al. (1999) used $\delta^{15}N$ and microbiological indicators to identify sewage nitrogen, while Widory et al. (2004) used $\delta^{15}N$, $\delta^{11}B$, and $^{87}Sr/^{86}Sr$ to discriminate between mineral fertilizers, sewage, and pig, cattle, and poultry manure. Bölke and Denver (1995) use $\delta^{15}N$ with $\delta^{13}C$, $\delta^{34}S$, chlorofluorocarbons, tritium, and major ion chemistry to determine the application history and fate of nitrate contamination in agricultural catchments (Buss et al., 2005).

Isotopic effects, caused by slight differences in the mass of two isotopes, tend to cause the heavier isotope to remain in the starting material of a chemical reaction. Denitrification, for example, causes the nitrate of the starting material to become isotopically heavier.

Volatilization of ammonia results in the lighter isotope preferentially being lost to the atmosphere, and the ammonia that remains behind becomes isotopically heavier. These isotopic effects mean that, depending on its origin, the same compound may have different isotopic compositions. Even when the stable-isotope composition of the source material is known, what reactions occur after its deposition and how they affect its isotopic composition also must be known, if the source of nitrate in groundwater is to be identified. Because fractionation after deposition blurs the isotopic signatures of the source materials, the use of ^{15}N data alone may not be sufficient to differentiate among sources.

Because of the many ways human activities influence various forms of nitrogen in the environment, and the public health concerns associated with elevated concentrations of nitrite and nitrate in potable groundwater, many scientific investigations on the sources of nitrogen, the nitrogen cycle, and related groundwater impacts are available (e.g., Feth, 1966; National Research Council, 1978; Zwirnmann, 1982; Keeney, 1990; Spalding and Exner, 1993; Puckett, 1994; Rees et al., 1995; Mueller et al., 1995; Buss et al., 2005).

5.3.5 Synthetic Organic Contaminants

Concern about SOCs in the drinking water supplies of some cities was a significant cause of the passage of the 1974 Safe Drinking Water Act, even though data were scarce. In 1981, the USEPA conducted the Ground Water Supply Survey to determine the occurrence of volatile organic chemicals (VOCs) in public drinking water supplies using groundwater. The survey showed detectable levels of these chemicals in 28.7 percent of public water systems serving more than 10,000 people and in 16.5 percent of smaller systems. Other USEPA and state surveys also revealed VOCs in public water supplies. The USEPA has used these surveys to support regulation of numerous organic chemicals, many of which are carcinogenic (Tiemann, 1996).

SOCs are human-made (anthropogenic) compounds that are used for a variety of industrial and agricultural purposes and include organic pesticides. SOCs can be divided into two groups: VOCs and nonvolatile (semivolatile) compounds.

Disinfection of drinking water is one of the major public health advances of the twentieth century. Disinfection is a major factor in reducing the typhoid and cholera epidemics that were common in American and European cities in the nineteenth century and the beginning of the twentieth century. While disinfectants are effective in controlling many microorganisms, certain disinfectants (notably chlorine) react with natural organic and inorganic matter in source water and distribution systems to form *disinfection by-products* (*DBPs*), which are almost all organic chemicals (chromate and bromate are notable exceptions). A large portion of the U.S. population is potentially exposed to DBPs through its drinking water. More than 240 million people in the United States are served by public water systems that apply a disinfectant to water to protect against microbial contaminants. Results from toxicology studies have shown several DBPs (e.g., bromodichloromethane, bromoform, chloroform, dichloroacetic acid, and bromate) to be carcinogenic in laboratory animals. Other DBPs (e.g., chlorite, bromodichloromethane, and certain haloacetic acids) have also been shown to cause adverse reproductive or developmental effects in laboratory animals. Epidemiological and toxicological studies involving DBPs have provided indications that these substances may have a variety of adverse effects across the spectrum of reproductive and developmental toxicity: early-term miscarriage, still birth, low birth weight, premature babies, and congenital birth defects (USEPA, 2003b). DBPs are of special concern when studying the potential for artificial aquifer recharge using treated wastewater. Three disinfectants and four DBPs are currently on the USEPA Primary Drinking Water Standards list (see Section 5.4).

VOCs and SVOCs

VOCs are synthetic chemicals used for a variety of industrial and manufacturing purposes. Among the most common VOCs are degreasers and solvents such as benzene, toluene, and TCE; insulators and conductors such as polychlorinated biphenyls (PCBs); dry-cleaning agents such as tetrachloroethylene (PCE); and gasoline compounds. VOCs have the potential to cause chromosome aberrations, cancer, nervous system disorders, and liver and kidney damage (USEPA, 2003a, 2003b). There are 53 organic chemicals included on the USEPA Primary Drinking Water Standards list (Section 5.4).

VOCs were detected in many aquifers across the United States in a study conducted by the USGS (Zogorski et al., 2006). The assessment of 55 VOCs in groundwater included analyses of about 3500 water samples collected during 1985 to 2001 from various types of wells, representing almost 100 different aquifer studies. This is the first national assessment of the occurrence of a large number of VOCs with different uses, and the assessment addresses key questions about VOCs in aquifers. Almost 20 percent of the water samples from aquifers contained one or more of the 55 VOCs, at an assessment level of 0.2 μg/L. This detection frequency increased to slightly more than 50 percent for the subset of samples analyzed with a low-level analytical method and for which an order-of-magnitude lower assessment level (0.02 μg/L) was applied. VOCs were detected in 90 of 98 aquifer studies completed across the country, with most of the largest detection frequencies in California, Nevada, Florida, and the New England and Mid-Atlantic States. Trihalomethanes (THMs), which may originate as chlorination by-products, and solvents were the most frequently detected VOC groups. Furthermore, detections of THMs and solvents and some individual compounds were geographically widespread; however, a few compounds, such as methyl *tert*-butyl ether (MTBE), ethylene dibromide (EDB), and dibromochloropropane (DBCP), had regional or local occurrence patterns. The widespread occurrence of VOCs indicates the ubiquitous nature of VOC sources and the vulnerability of many of the country's aquifers to low-level VOC contamination. The findings for VOCs indicate that other compounds with widespread sources and similar behavior and fate properties may also be occurring.

VOCs found in 1 percent or more tested supply wells at the assessment level of 0.2 μg/L are chloroform (THM), perchloroethene (tetrachloroethene or PCE), MTBE, trichloroethene (TCE), toluene, dichlorodifluoromethane (refrigerant), 1,1,1,-trichloroethane, chloromethane, bromodichloromethane (THM), trichlorodifluoromethane (refrigerant), bromoform (THM), dibromochloromethane (THM), *trans*-1,2-dichloromethene, methylene chloride, and 1,1-dichloromethane.

Although many VOCs were detected in the USGS study, they were typically at low concentrations and below their respective MCLs where applicable. For example, 90 percent of the total VOC concentrations in samples were less than 1 μg/L. Forty-two of the fifty-five VOCs were detected in one or more samples at an assessment level of 0.2 μg/L. Furthermore, VOCs in each of the seven VOC groups considered in this assessment were detected in the samples; these groups included fumigants, gasoline hydrocarbons, gasoline oxygenates (such as MTBE), organic synthesis compounds, refrigerants, solvents, and THMs. The finding that most VOC concentrations in groundwater are less than 1 μg/L is important because many previous monitoring programs did not use low-level analytical methods and therefore would not have detected such contamination.

The complexity of explaining VOC contamination in aquifers was confirmed in this assessment through statistical models for 10 frequently detected compounds. Factors describing the source, transport, and fate of VOCs were all important in explaining the national occurrence of these VOCs. For example, the occurrence of PCE was statistically associated with the percentage of urban land use and density of septic systems near sampled wells (source factors), depth to top of well screen (transport factor), and presence of dissolved oxygen (fate factor). National-scale statistical analyses provide important insights about the factors that are strongly associated with the detection of specific VOCs, and this information may benefit many local aquifer investigations in selecting compound- and aquifer-specific information to be considered. Continued efforts to reduce or eliminate low-level VOC contamination will require enhanced knowledge of sources of contamination and aquifer characteristics (Zogorski et al., 2006).

SVOCs are operationally defined as solvent-extractable organic compounds that can be determined by gas chromatography/mass spectrometry (GC/MS). They include polycyclic aromatic hydrocarbons (PAHs), azaarenes, nitrogenated compounds, phenols, phthalates, ketones, and quinones. Many of these SVOCs have been designated as priority pollutants by the USEPA because of their toxicity and association with industrial activities and processes. They are referenced in the Clean Water Act of 1977. SVOCs that are priority pollutants include phthalates used in plastics, phenols used as disinfectants and in manufacturing chemicals, and PAHs. PAHs and azaarenes contain fused carbon rings that form during the incomplete combustion of organic matter, including wood and fossil fuels (such as gasoline, oil, and coal). Azaarenes are distinguished from PAHs by having a nitrogen atom substituted for a carbon in the fused ring structure. Azaarenes tend to occur in association with PAHs in affected soils and streambed sediment samples because fossil fuel combustion is the primary source of both PAHs and azaarenes. However, additional sources of PAHs include natural or anthropogenic introduction of uncombusted coal and oil and industrial use of PAHs in the dye and plastic industries (Nowell and Capel, 2003; from Lopes and Furlong, 2001).

Most SVOCs are moderately to strongly hydrophobic (i.e., they have fairly low water solubility and fairly high octanol-water partition coefficients). Consequently, they tend to sorb to soil and sediment and partition to organic matter in water.

Nonaqueous-Phase Liquids

Nonaqueous-phase liquids (NAPLs) are hydrocarbons that exist as a separate, immiscible phase when in contact with water and/or air. Differences in the physical and chemical properties of water and NAPL result in the formation of a physical interface between the liquids, which prevents the two fluids from mixing. NAPLs are typically classified as either light nonaqueous phase liquids (LNAPLs), which have densities less than that of water, or dense nonaqueous phase liquids (DNAPLs), which have densities greater than that of water (Table 5.1). It is very important to make the distinction between the actual NAPL liquid in free phase, and the chemical of the same name dissolved in water. For example, most common organic contaminants such as PCE, TCE, and benzene can enter the subsurface as both free-phase NAPL and dissolved in percolating water. However, the fate and transport of free-phase NAPL and the same chemical dissolved in groundwater are quite different.

LNAPLs affect groundwater quality at a variety of sites. The most common contamination problems result from the release of petroleum products. Leaking of USTs at gas stations and other facilities is arguably the most widespread point-source contamination

IUPAC Name	Common or Alternative Name	Density
1,2,3-trichlorobenzene	1,2,6-trichlorobenzene	1.690
tetrachloroethene	perchloroethylene, tetrachloroethylene, PCE	1.623
tetrachloromethane	carbon tetrachloride	1.594
1,1,2-trichloroethene	1,1,2-trichloroethylene, TCE	1.464
1,2,4-trichlorobenzene	1,2,4-trichlorobenzol	1.450
1,1,2-trichloroethane	methyl chloroform	1.440
1,1,1-trichloroethane	methyl chloroform	1.339
1,2-dichlorobenzene	o-dichlorobenzene	1.306
cis-1,2-dichloroethene	cis-1,2-dichloroethylene	1.284
trans-1,2-dichloroethene	trans-1,2-dichloroethylene	1.256
1,2-dichloroethane	1,2-ethylidene dichloride, glycol dichloride	1.235
1,1-dichloroethene	1,1-dichloroethylene, DCE	1.213
1,1-dichloroethane	1,1-ethylidene dichloride	1.176
chlorobenzene	monochlorobenzene	1.106
pure water at 0°C		1.000
naphthalene	naphthene	0.997
chloromethane	methyl chloride	0.991
chloroethane	ethyl chloride	0.920
chloroethene	vinyl chloride, chloroethylene	0.910
stryrene	vinyl benzene	0.906
1,2-dimethylbenzene	o-xylene	0.880
benzene	—	0.876
ethylbenzene	—	0.867
1,2,4-trimethylbenzene	pseudocumene	0.876
methylbenzene	toluene	0.867
1,3-dimethylbenzene	m-xylene	0.864
1,4-dimethylbenzene	p-xylene	0.861
2-methoxy-2-methylpropane	methyl tert-butyl ether, MTBE	0.740

IUPAC, International Union of Pure and Applied Chemistry.
From Lawrence, 2006.

TABLE 5.1   Density (in g/cm^3 at 20°C) of Selected Volatile Organic Compounds

of groundwater in developed countries (see Fig. 5.9). Gasoline products are typically multicomponent organic mixtures composed of chemicals with varying degrees of water solubility. Some gasoline additives (e.g., MTBE and alcohols such as ethanol) are highly soluble. Other components (e.g., BTEX) are slightly soluble. Many components (e.g., n-dodecane and n-heptane) have relatively low water solubility under ideal conditions (Newell et al., 1995). At the end of the refining process, finished gasoline commonly contains more than 150 separate compounds; however, some blends may contain as many as 1000 compounds (Mehlman, 1990; Harper and Liccione, 1995). In addition to BTEX, which on average make about 16 percent of a typical gasoline blend, three minor components of gasoline, naphthalene, vinyl benzene (styrene), and 1,2,4-trimethylbenzene (124-TMB), are commonly detected in contaminated groundwater (Lawrence, 2006). Individual BTEX compounds are also widely used as solvents and in manufacturing (Swoboda-Colberg, 1995).

When a mixture of pure LNAPLs (e.g., fuel) is released to the subsurface, the components of the fuel may remain in the original free phase, dissolve into and migrate with any water present in the vadose zone, absorb to solid material in the soil, or volatilize into soil gas. Therefore, a three-phase system consisting of water, product, and air is created within the vadose zone. Infiltrating water dissolves the components within the LNAPL (e.g., benzene, toluene, xylene, and others) and carries them to the water table. These dissolved constituents then form a contaminant plume emanating from the area of the residual product, where LNAPL phase is immobile (trapped by the porous media). If enough product is spilled, free LNAPL will flow downward to the water table and form a pool floating on it, being lighter (less dense) than water.

Many of the components commonly found in LNAPLs are volatile and can partition into the soil air and be transported by molecular diffusion in all directions within the vadose zone and away from the area of the residual mass. These vapors may partition back into the water phase and spread contamination over a wider area. They can also diffuse across the land surface boundary into the atmosphere (Palmer and Johnson, 1989).

Fluctuations of the water table, combined with the downgradient migration of the three contaminant phases (product pool, phase dissolved in groundwater, and vapor phase), may create a complex horizontal and vertical distribution (redistribution) of the contaminant in the subsurface, especially if the porous media is heterogeneous (presence of clay lenses and layers).

Accumulations of LNAPL at or near the water table are susceptible to "smearing" from changes in water-table elevation such as those that occur due to seasonal changes in recharge and discharge, or tidal influence in coastal environments. Mobile LNAPL floating on the water table will move vertically as the groundwater elevation fluctuates. As the water table rises or falls, LNAPL will be retained in the soil pores, leaving behind a residual LNAPL "smear zone." If smearing occurs during a decline in groundwater elevations, residual LNAPL may be trapped below the water table when groundwater elevations rise. A similar situation may develop during product recovery efforts. LNAPL will flow toward a recovery well or trench in response to the gradient induced by water-table depression. LNAPL residual will be retained below the water table as the water-table elevation returns to prepumping conditions (Newell et al., 1995).

The major types of DNAPLs are halogenated solvents, coal tar, creosote-based wood-treating oils, PCBs, and pesticides. As a result of widespread production, transportation, utilization, and disposal practices, particularly since 1940s, there are numerous DNAPL contamination sites in North America and Europe. The potential for serious long-term contamination of groundwater by some DNAPL chemicals at many sites is high due to their toxicity, limited solubility (but much higher than drinking water limits), and significant migration potential in soil gas, in groundwater, and/or as a separate phase. DNAPL chemicals, especially chlorinated solvents, are among the most prevalent groundwater contaminants identified in groundwater supplies and at waste disposal sites (Cohen and Mercer, 1993).

Halogenated solvents, particularly chlorinated hydrocarbons, and brominated and fluorinated hydrocarbons, to a much lesser extent, are DNAPL chemicals encountered at numerous contaminated sites. These halocarbons are produced by replacing one or more hydrogen atoms with chlorine (or another halogen) in petrochemical precursors such as methane, ethane, ethene, propane, and benzene. Many bromocarbons and fluorocarbons are manufactured by reacting chlorinate hydrocarbon intermediates (such as chloroform or carbon tetrachloride) with bromine and fluorine compounds, respectively. DNAPL halocarbons at ambient environmental conditions include chlorination products of methane (methylene chloride, chloroform, and carbon tetrachloride), ethane (1,1-dichloroethane, 1,2-dichloroethane, 1,1,1-trichloroethane, and 1,1,2,2-tetrachloroethane), ethene (1,1-dichloroethene, 1,2-dichloroethene isomers, TCE, and tetrachloroethene), propane (1,2-dichloropropane and 1,3-dichloropropene isomers), and benzene (chlorobenzene, 1,2-dichlorobenzene, and 1,4-dichlorobenzene); fluorination products of methane and ethane such as 1,1,2-trichlorofluormethane (Freon-11) and 1,1,2-trichlorotrifluorethane (Freon-113); and bromination products of methane (bromochloromethane, dibromochloromethane, dibromodifluoromethane, and bromoform), ethane (bromoethane and 1,1,2,2-tetrabromoethane), ethene (ethylene dibromide), and propane (1,2-dibromo-3-chloropropane).

Coal tar and creosote are complex chemical mixture DNAPLs derived from the destructive distillation of coal in coke ovens and retorts. Historically, coal tar has been produced by coal tar distillation plants and as a by-product of manufactured gas plant and steel industry coking operations. Creosote blends are used to treat wood alone or diluted with coal tar, petroleum, or, to a very limited extent, pentachlorophenol. In addition to wood preservation, coal tar is used for road, roofing, and water-proofing solutions. Considerable use of coal tar is also made for fuels (Cohen and Mercer, 1993).

Creosote and coal tar are complex mixtures containing more than 250 individual compounds. Creosote is estimated to contain 85% PAHs, 10% phenolic compounds, and 5% N-, S-, and O-heterocyclic compounds. The compositions of creosote and coal tar are quite similar, although coal tar generally includes a light oil component (<5 percent of the total) consisting of monocyclic aromatic compounds such as BTEX. Consistent with the composition of creosote and coal tar, PAHs (in addition to BTEX compounds) are common contaminants detected in groundwater at wood-treating sites (Rosenfeld and Plumb, 1991; Cohen and Mercer, 1993).

PCBs are extremely stable, nonflammable, dense, and viscous liquids that are formed by substituting chlorine atoms for hydrogen atoms on a biphenyl (double benzene ring) molecule. PCBs, which are not produced any more in the United States and now have a very restricted regulated use, were historically used in oil-filled switches, electromagnets, voltage regulators, heat transfer media, fire retardants, hydraulic fluids, lubricants, plasticizers, carbonless copy paper, dedusting agents, and other products. PCBs were frequently mixed with carrier fluids prior to use. Due to their widespread historic use and persistence, PCBs are often detected in the environment at very low concentrations. The potential for DNAPL migration is greatest at sites where PCBs were produced, utilized in manufacturing processes, stored, reprocessed, and/or disposed of in quantity (Cohen and Mercer, 1993).

DNAPLs can have great mobility in the subsurface because of their relatively low solubility, high density, and low viscosity. Hydrophobic DNAPLs do not readily mix with water (they are immiscible) and tend to remain as separate phases (i.e., nonaqueous). The relatively high density of these liquids provides a driving force that can carry product deep into aquifers. DNAPL infiltrating from the land surface because of a spill or leak may encounter two general conditions: (1) in the presence of moisture (water) within the vadose zone; DNAPL exhibits viscous fingering during infiltration (when a high-density, low-viscosity fluid (DNAPL) displaces a lower-density, higher-viscosity fluid (water), the flow is unstable, resulting in the occurrence of viscous fingering); and (2) if the vadose zone is dry, viscous fingering is generally not observed (Palmer and Johnson, 1989). When a spill of DNAPL is small, it will flow through the vadose zone until it reaches residual saturation, i.e., until all the mobile DNAPL is trapped in the porous media. This residual DNAPL may still cause formation of a dissolved plume in the underlying saturated zone (aquifer) by partitioning into the vapor phase; these dense vapors may sink to the capillary fringe where they are eventually dissolved in water and transported downgradient. Infiltrating water can also dissolve the residual DNAPL and transport it down to the water table.

Once DNAPL enters the saturated zone, its further migration will depend on the amount (mass) of product and the aquifer heterogeneity, such as the presence of low-permeable lenses and layers of clay. Figure 5.12 shows examples of free-phase (mobile) DNAPL accumulation over low-permeable layers. This free-phase (pooled) DNAPL serves as a continuing source of dissolved-phase contamination, which is carried downgradient by the flow of groundwater. As it migrates, the DNAPL may leave a residual phase in the porous media of the saturated (aquifer) zone along its path. This residual phase also serves as a source of dissolved-phase contamination. Being denser than water, free-phase DNAPL moves because of gravity, not because of the hydraulic gradients normally present in natural aquifers. As a consequence, DNAPL encountering a low-permeable layer may flow along its slope, in a different (including opposite) direction from the dissolved plume, as shown in Fig. 5.12. This free-phase DNAPL would also create its own dissolved-phase plume. Migration of DNAPL in the unsaturated and saturated zones may create a rather complex pattern of multiple secondary sources of free- and residual-phase DNAPL, and multiple dissolved plumes at various depths within the

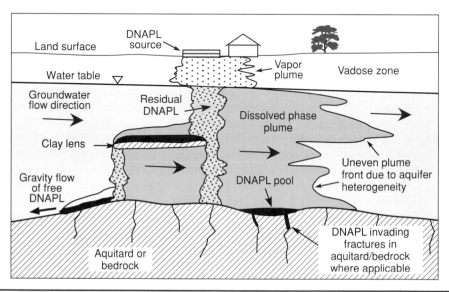

FIGURE 5.12 Schematic of possible migration pathways of free-phase DNAPL and the derived dissolved contaminant plumes in the subsurface. (From Kresic, 2007; copyright Taylor & Francis Group, LLC; printed with permission.)

aquifer. This pattern may not be definable based on the hydraulics of groundwater flow alone and, in any case, presents a great challenge when attempting to restore an aquifer to its beneficial use: "Once in the subsurface, it is difficult or impossible to recover all of the trapped residual" (USEPA, 1992).

Two invaluable technical resources for the general study and characterization of DNAPLs in the subsurface are books by Cohen and Mercer (1993) and Pankow and Cherry (1996).

5.3.6 Agricultural Contaminants

Fertilizers

As a result of population growth, the available arable land per capita has decreased and the demand for agricultural production has increased. Although the contribution of agricultural production to the gross national product (GNP) in some countries has fallen, production in absolute terms has increased greatly. This can be attributed to technological developments such as increased mechanization, intensive agriculture, irrigation, and the use of fertilizers and pesticides. Most forms of agricultural land use constitute an important diffuse or nonpoint source of contamination of soils, surface water, and groundwater (UNESCO, 1998).

Increases in soil fertility have been attained specially since 1960 by massive application of inorganic fertilizers. Organic fertilizers have been used since the early history of development, but the increased use of inorganic fertilizers began in the 1940s, reaching its peak during the "Green Revolution" (1960 to 1970). For example, Foster (2000) shows how the use of artificial fertilizers affected British agriculture between 1940 and 1980. A threefold increase in food production was accomplished by a 20-fold increase in the use of fertilizers. The unused nitrogen has, therefore, been lost to the atmosphere by denitrification and leached to surface and groundwater as nitrate, or remains stored in the unsaturated zone (Buss et al., 2005). This nitrate in the unsaturated zone will continue to serve as a long-term source of groundwater contamination even if application of fertilizers were to discontinue today.

Since the 1980s, the application of fertilizers remained at relatively the same level or has been decreasing in most European countries, South and North America (including the United States), and Australia. During the same time period, total fertilizer consumption in Asia almost doubled and is projected to more than double again by 2030 (UNESCO, 1998). Nevertheless, consumption in Asia represented only 19 percent of European consumption in 1990. The largest users of fertilizers are Europe and the United States (100 to 150 kg/ha, with a 50 to 55 percent mineral components). In comparison, most Latin American and African countries use less than 10 kg/ha (UNESCO, 1998).

Altogether, 16 mineral elements are known to be necessary for plant growth, but only three needed in large quantities—nitrogen, phosphorus, and potassium. The others, called secondary elements and microelements, are generally required for cell metabolism and enzymes and are required in very small amounts. Nitrogen is the most critical element in the fertilizer program. It is lacking in nearly all agricultural soils because it leaches readily and therefore has to be applied on a regular basis (UNESCO, 1998). Potassium also leaches readily and has to be applied at the same rate as nitrogen, whereas phosphorous accumulates in the soil and does not leach readily to the subsurface. Phosphorous is, therefore, the main nonpoint source of contamination by surface runoff, which results in eutrophication of surface water bodies.

Organic fertilizers contain essential nutrients (nitrogen, phosphorous, and potassium) and also stimulators and a considerable quantity of microbes that support biological activity and are needed for mineralizing nutrients. Worldwide, they commonly include animal manure, crop residues, municipal sewage sludge and wastewater, and a wide variety of industrial and organic wastes (UNESCO, 1998).

The first nitrogen fertilizer used commercially, Peruvian guano, formed by deposition of excreta by seawolf, was organic in nature. It is very likely that the low levels of perchlorate found in groundwater throughout agricultural areas in California and elsewhere can be attributed to widespread use of guano during the first half of the twentieth century. Perchlorate, a mineral salt of chlorine, also associated with manufacturing of rocket fuel, ammunitions, and firework, is one of the most notorious emerging contaminants.

Pesticides

If treatment is not used to protect plants, insects and fungus can destroy crops. Unfortunately, so far, the only proven efficient method for plant protection on a large scale is through the application of chemicals. Plant extracts have been used as pesticides since Roman times, nicotine since the seventeenth century and synthetic pesticides since the 1930s (Paul Muller discovered the insecticide properties of DDT in 1939). Today, new active compounds are registered in different countries every year and usually have to be handled with care because of their toxic properties (Fig. 5.13).

The word pesticide refers to any chemical that kills pests and includes insecticides, fungicides, and nematocides; it also generally includes herbicides. Extensive use of pesticides is not confined to rural agricultural areas only. They are commonly used in both urban and suburban settings on lawns, parks, and golf courses. According to UNESCO (1998), data on pesticide use are not available for most countries. It is, however, known that the use of pesticides is high in developed countries where the total amount of chemicals used per hectare varies from 1 to 3 L/year (insecticides), and from 3 to 10 kg/yr (fungicides). In developing countries, pesticides are often unavailable and beyond the financial capabilities of farmers, and their use is limited to a few crops.

The use of herbicides is increasing worldwide. Normally, herbicides are applied at a rate varying from 5 to 12 kg/ha. Preemergence herbicides are applied at lower rates (1 to 4 kg/ha).

Pesticides used in the 1950s and 1960s were generally characterized by low aqueous solubility, strong sorption by soil components, and broad-spectrum toxic effects. These pesticide properties are now known to accumulate in the environment and cause adverse impacts on aquatic ecosystems via persistence and biomagnifications. Examples of such pesticides are chlorinated hydrocarbon insecticides, including DDT and dieldrin. However, only small amounts of these types of pesticides are likely to reach groundwater systems (UNESCO, 1998). In contrast, newer pesticides are more soluble, less sorbed, and readily degradable and have more selective toxicological effects. As a result, pesticide application rates in developed countries have generally declined, but the solubility and mobility characteristics of these compounds may lead to considerable groundwater contamination. While the presence of nitrate in many aquifers in the world has been widely reported, fewer cases of contamination by pesticides have been reported so far. Reasons for this could be the potential time lag in the response of groundwater systems to this contaminant input, the high costs involved in their chemical analysis, and, in some cases, the disregard of the degradation products is the main reason (UNESCO, 1998).

FIGURE 5.13 Wearing gloves, mask, and other protection is part of handling farm chemicals safely. (Photograph courtesy of Tim McCabe, National Resources Conservation Service.)

In 1992, the USEPA issued the Pesticides in Ground Water Database (1971–1991), which showed that nearly 10,000 of 68,824 tested wells contained pesticides at levels that exceeded drinking water standards or health advisory levels. Almost all the data were from drinking water wells. The USEPA has placed restrictions on 54 pesticides found in groundwater, 28 of which are no longer registered for use in the United States but may still be present in soils and groundwater due to this widespread historic use (Tiemann, 1996).

Figure 5.14 shows the results of a nationwide study conducted during 1992 to 2002 and published by the USGS in 2006. One or more pesticides or their degradation products were detected in water more than 90 percent of the time during the year in streams draining watersheds with agricultural, urban, and mixed land uses. In addition, some

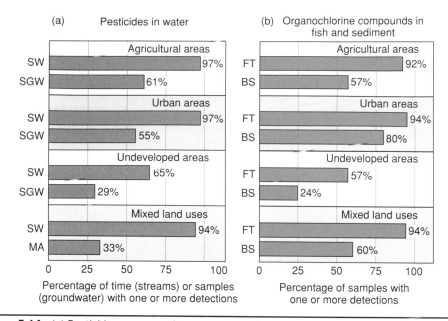

Figure 5.14 (a) Pesticide occurrence in stream water (SW), shallow groundwater (SGW), and major aquifers (MA) in the United States; most pesticides in this group are in use. (b) Organochlorine pesticides in fish tissue (FT) and streambed sediment (BS); most pesticides in this group are no longer used. (Modified from Gilliom et al., 2006.)

organochlorine pesticides that have not been used in the United States for years were detected along with their degradation products and by-products in most samples of whole fish or bed sediment from streams sampled in these land use settings. Pesticides were less common in groundwater but were detected in more than 50 percent of wells sampled to assess shallow groundwater in agricultural and urban areas.

As mentioned earlier, in addition to natural geologic sources, there are many anthropogenic sources of arsenic. The most important are derived from agricultural practices, such as the application of pesticides and herbicides. Inorganic arsenic was widely applied before it was banned for pesticide use in the 1980s and 1990s. Lead arsenate ($PbHAsO_4$) was the primary insecticide used in fruit orchards prior to the introduction of DDT in 1947. Inorganic arsenicals have also been applied to citrus, grapes, cotton, tobacco, and potato fields. For example, historic annual arsenic loading rates up to approximately 490 kg/ha (approximately 440 lb/acre) on apple orchards in eastern Washington led to arsenic concentrations in soil in excess of 100 mg/kg (Benson, 1976; Davenport and Peryea, 1991; from Welch et al., 2000). Agricultural soils in other parts of the United States also have high arsenic concentrations exceeding 100 mg/kg due to long-term application (20–40 years or more) of calcium and lead arsenate (Woolsen et al., 1971, 1973). Early studies suggested that arsenic in eastern Washington orchards was largely confined to the topsoil, although evidence for movement into the subsoil has been cited (Peryea, 1991). This apparent movement of arsenic suggests a potential for contamination of shallow groundwater. Application of phosphate fertilizers creates the potential for releasing arsenic into groundwater. Laboratory studies suggest that phosphate applied to soils contaminated with lead arsenate can release arsenic to soil water. Increased use of phosphate at relatively high application rates has been adopted to decrease the toxicity of arsenic to trees in replanted orchards. Laboratory results suggest that this practice may increase arsenic concentrations in subsoil and shallow groundwater. Application of phosphate onto uncontaminated soil may also increase arsenic concentrations in groundwater by releasing adsorbed natural arsenic (Woolsen et al., 1973; Davenport and Peryea, 1991; Peryea and Kammereck, 1997; Welch et al., 2000).

In some irrigated regions, automatic fertilizer feeders are attached to irrigation sprinkler systems. When the pump is shut off, water flows back through the pipe into the well, creating a partial vacuum that may cause fertilizer to flow from the feeder into the well. It is possible that some individuals even dump fertilizers (and perhaps pesticides) directly into the well to be picked up by the pump and distributed to the sprinkler system (USEPA, 1990).

Aurelius (1989; from USEPA, 1990) described an investigation in Texas where 188 wells were sampled for nitrate and pesticides in 10 counties where aquifer vulnerability studies and field characteristics indicated the potential for groundwater contamination from the normal use of agricultural chemicals. Nine pesticides (2,4,5-T, 2,4-DB, metolachlor, dicamba, atrazine, prometon, bromacil, picloram, and triclopyr) were found present in 10 wells, 9 of which were used for domestic supply. Also, 182 wells were tested for nitrate, and, of these, 101 contained more than the regulatory limit. Of the high nitrate wells, 87 percent were used for household purposes. In addition, 28 wells, of which 23 were domestic, contained arsenic at or above of 0.05 mg/L, which was the MCL at the time (current MCL for arsenic is 0.01 mg/L).

Concentrated Animal Feeding Operations

CAFOs result from the consolidation of small farms with animals into larger operations, leading to a higher density of animals per unit of land on CAFOs than on small farms. For example, in 2005, the United States produced over 103 million pigs at 67,000 production facilities (USDA, 2006a, 2006b; from Sapkota et al., 2007). Facilities housing over 55,000 pigs accounted for more than half of the total U.S. swine inventory, reflecting the increasing consolidation and concentration of U.S. swine production (USDA, 2006a). This trend in swine production has resulted in the concentration of large volumes of manure in relatively small geographical areas. Manure is typically stored in deep pits or outdoor lagoons and then applied to agricultural fields as a source of fertilizer. However, as a result of runoff and percolation events, components of manure, including human pathogens and chemical contaminants, can impact surface water and groundwater proximal to swine CAFOs, posing risks to human health (Anderson and Sobsey, 2006; Sayah et al., 2005). Specific swine production practices, including the use of nontherapeutic levels of antibiotics in swine feed, can exacerbate the risks associated with exposures to manure-contaminated water sources (Sapkota et al., 2007).

Elevated concentrations of nutrients, metals, bacteria, and a number of other chemicals and pathogens are observed in surface and groundwater in many agricultural areas throughout the United States. Excess nutrients may be an important contributing factor for the growth and increase in dinoflagellates such as *Pfiesteria*. Many of the infectious organisms that cause illness in animals can also cause disease in humans and can survive in water. The most common pathogens that pose a human-health risk include *Salmonella* spp., *Escherichia coli* O157:H7 (*E. coli*), *Campylobacter* spp., *Listeria monocytogenes*, as well as viruses and protozoa such as *Cryptosporidium parvum* and *Giardia*. These organisms have been found in groundwater in a number of communities (Rice et al., 2005).

In a study of surface water and groundwater situated up- and downgradient from a swine facility, Sapkota et al. (2007) found antibiotic-resistant enterococci and other fecal indicators. Collected samples were tested for susceptibility to erythromycin, tetracycline, clindamycin, virginiamycin, and vancomycin. The results of the study show that the median concentrations of enterococci, fecal coliforms, and E. coli were 4- to 33-fold higher in down- versus upgradient surface water and groundwater. Higher minimal inhibitory concentrations for four antibiotics were observed in enterococci isolated from down- versus upgradient surface water and groundwater. Elevated percentages of erythromycin- and tetracycline-resistant enterococci were detected in downgradient surface waters, and higher percentages of tetracycline- and clindamycin-resistant enterococci were detected in downgradient groundwater. The authors concluded that these findings provide additional evidence that water impacted by swine manure could contribute to the spread of antibiotic resistance.

5.3.7   Microbiological Contaminants

Microbiological contaminants are microorganisms potentially harmful to humans or animals. Jointly called pathogens, they include parasites, bacteria, and viruses. Although the previous sections illustrate the seriousness of contamination with organic chemicals and inorganic substances, pathogens are by far the most widely spread water contaminants. As pointed out by researchers from the Johns Hopkins University, water-related diseases are a human tragedy, killing millions of people each year, preventing millions more from leading healthy lives, and undermining development efforts. About 2.3 billion people in the world suffer from diseases that are linked to water, and some 60 percent of all infant mortality is linked to infectious and parasitic diseases, most of them water related (Hinrichsen et al., 1997).

Where proper sanitation facilities are lacking, waterborne diseases can spread rapidly. Untreated excreta carrying disease organisms wash or leach into freshwater sources, contaminating drinking water. Diarrheal disease, the major waterborne disease, is prevalent in many countries where sewage treatment is inadequate—human wastes are disposed of in open latrines, ditches, canals, and water courses, or they are spread on cropland. An estimated 4 billion cases of diarrheal disease occur every year, causing 3 to 4 million deaths, mostly among children (Hinrichsen et al., 1997).

Although surface water is the primary recipient and host of pathogen contamination, shallow groundwater is also greatly affected in many regions with poor or nonexistent sanitation. However, some pathogens, such as parasites Giardia and Cryptosporidium, are naturally present in surface water bodies and are not necessarily associated with poor sanitation practices. For this reason, the USEPA has instituted specific water-treatment requirements for public supply systems using "groundwater under direct influence" (GWUI) of surface water.

The increased efforts to reclaim and reuse wastewaters and gray waters pose additional public health concerns. These concerns are heightened by the fact that the indicators of the "sanitary quality" of waters, i.e., the total- and fecal-coliform bacteria, are unreliable indicators of the presence of a number of key pathogenic agents including enteric viruses and cyst-forming protozoans. Wastewater reclamation does not have a specific meaning in terms of the degree of treatment for different reclamation projects. Some reuses of wastewater are allowed with very little additional treatment beyond the conventional sewage treatment, of addressing enteric viruses and cyst-forming protozoans.

Even after what is considered to be good conventional domestic wastewater treatment and chlorination (chloramination), domestic wastewaters released to surface waters in nearby streams, lakes, estuaries, or coastal marine waters, still contain large numbers of enteroviruses and pathogenic protozoans that can readily cause human disease upon ingestion and, to a lesser extent, body contact with these waters (Lee and Jones-Lee, 1993).

In addition to nonsanitary practices at the land surface, pathogens can enter groundwater systems in natural surface water-groundwater interactions, or via artificial aquifer recharge using surface water and treated wastewater. Once in the subsurface (aquifer), their survival and transport will depend on various biogeochemical interactions with native groundwater, porous media, and native microorganisms. Whereas some bacteria and parasites cannot survive more than several weeks in the saturated zone regardless of the native conditions, some viruses are known to survive for months or even years.

Pathogens can cause an adverse effect after an acute (short-term) exposure such as ingestion of just one glass of water. They can also cause epidemics and chronic diseases. In the early 1990s, for example, raw sewage water that was used to fertilize vegetable fields caused outbreaks of cholera in Chile and Peru. In Buenos Aires, Argentina, a slum neighborhood faced continual outbreaks of cholera, hepatitis, and meningitis because only 4 percent of homes had either water mains or proper toilets, while poor diets and little access to medical services aggravated the health problems (Hinrichsen et al., 1997).

Bacteria are microscopic living organisms usually consisting of a single cell. Waterborne disease-causing bacteria include *E. coli* and *Shigella*. Protozoa or parasites are also single-cell organisms. Examples include *Giardia lamblia* and *Cryptosporidium*. A virus is the smallest form of microorganism capable of causing disease. A virus of fecal origin that is infectious to humans by waterborne transmission is of special concern for drinking water regulators. More than 120 different types of potentially harmful enteric viruses are excreted in human feces and are widely distributed in type and number in domestic sewage, agricultural wastes, and septic drainage systems (Gerba, 1999; from Banks and Battigelli, 2002). Many of these viruses are stable in natural waters and have long survival times, with half-lives ranging from weeks to months. Because they may cause disease even when just a few virus particles are ingested, low levels of environmental contamination may affect water consumers. From 1971 to 1979, approximately 57,974 people in the United States were affected by outbreaks of waterborne pathogens (Craun, 1986; from Banks and Battigelli, 2002). Outbreaks of waterborne disease attributed to enteric viruses are poorly documented, even though viruses are commonplace in natural waters contaminated with human feces. Illnesses in humans caused by waterborne viruses range from severe infections such as myocarditis, hepatitis, diabetes, and paralysis to relatively mild conditions such as self-limiting gastroenteritis. Currently, enteric viruses are included in the National Primary Drinking Water Standards issued by the USEPA, while several other groups are on the contaminant candidate list (CCL). Studies of possible groundwater contamination with viruses are still very rare (USEPA, 2003c), but because of their presence on the CCL, and the new groundwater rule promulgated by USEPA in 2006, the interest of the scientific community has increased.

Giardia (Fig. 5.15) was only recognized as a human pathogen capable of causing waterborne disease outbreaks in the late 1970s. Its occurrence in relatively pristine water as well as wastewater-treatment plant effluent called into question water system definitions of "pristine" water sources. This parasite, now recognized as one of the most common causes of waterborne disease in humans in the United States, is found in every region of

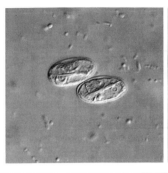 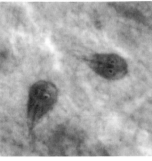

FIGURE 5.15 Left: two *Giardia intestinalis* cysts in a wet mount under DIC microscopy; image taken at 1000× magnification. The cysts are oval to ellipsoid and measure 8 to 19 μm (average 10 to 14 μm). Right: *Giardia intestinalis* trophozoites are pear shaped and measure 10 to 20 μm in length. (Photographs courtesy of the Center for Disease Control (CDC) Parasite Image Library.)

the United States and throughout the world. In 1995, outbreaks in Alaska and New York were caused by *Giardia*. The outbreak of giardiasis in Alaska affected 10 people and was associated with untreated surface water. The outbreak in New York affected an estimated 1449 people and was associated with surface water that was both chlorinated and filtered (USEPA, 2003c). The symptoms of giardiasis include diarrhea, bloating, excessive gas, and malaise.

The infectious dose for *Cryptosporidium* is less than 10 organisms, and, presumably, one organism can initiate an infection. As late as 1976, it was not known to cause disease in humans. In 1993, 403,000 people in Milwaukee, WI, became ill with diarrhea after drinking water contaminated with the parasite, resulting in the largest waterborne disease outbreak ever documented in the United States (Tiemann, 1996). For the 2-year period of 1993 to 1994, the Center for Disease Control reported that 17 states identified 30 disease outbreaks associated with drinking water. Since then, attention has been focused on determining and reducing the risk of cryptosporidiosis from public water supplies. Crypto is commonly found in lakes and rivers and is highly resistant to disinfection. Groundwater under the influence of surface water, and groundwater in highly transmissive karst and gravel aquifers, is also susceptible to contamination with parasites such as *Giardia* and *Cryptosporidium*. People with severely weakened immune systems are likely to have more severe and more persistent symptoms than healthy individuals.

In a nationwide study by the USGS, microbiological data were collected from 1205 wells in 22 study units of the National Water-Quality Assessment (NAWQA) program during 1993 to 2004. The samples of untreated groundwater were analyzed primarily for concentrations of total-coliform bacteria, fecal-coliform bacteria, and *E. coli*, and for the presence of coliphage viruses (Embrey and Runkle, 2006).

Nearly 30 percent of the 1174 wells analyzed tested positive for coliform bacteria. With at least one well in each study unit or principal aquifer testing positive, fecal-indicator bacteria were geographically widespread.

Samples were collected from 423 wells to test for the presence of coliphage viruses, which are considered indicators of the potential presence of human enteric viruses. Coliphage were present in samples from four of the 11 study units—the Central Columbia Plateau-Yakima, Georgia-Florida, San Joaquin, and Trinity, representing the Columbia Plateau, Floridan, Central Valley, and Coastal Lowlands aquifers, respectively. Overall, coliphage viruses were present in less than 4 percent of domestic and public wells used for drinking water supply.

Wells used for domestic supply made up the largest class of water use, with total-coliform concentrations analyzed in 405 wells and *E. coli* concentrations analyzed in 397 wells, followed by public supply wells and unused wells with 227 and 37 analyses of total-coliform bacteria, respectively. Total coliforms were detected in untreated water from 33 percent of domestic wells and 16 percent of public supply wells; *E. coli* were detected in 8 and 3 percent of domestic and public supply wells, respectively. Although median concentrations were <1 CFU/100 mL for all classes of water use, as defined in this report, the overall distribution of total-coliform concentrations was significantly higher in domestic wells than in public supply wells.

Generally, coliform bacteria were detected more frequently and in higher concentrations in wells completed in sandstone or shale, and in sedimentary, carbonate, and crystalline rocks than for wells in unconsolidated materials, in semiconsolidated sand, or in volcanic rocks. More than 50 percent of sampled wells completed in carbonate rocks (limestone and dolomite) or in crystalline rocks (schist and granite) tested positive for coliform bacteria. The Floridan, Piedmont and Blue Ridge, Ordovician, and Valley and Ridge aquifers, all of which had high detection rates or concentrations of coliform bacteria, are composed of these fractured and porous rocks. The lowest rates of detections (less than 5 percent) were for wells in the Basin and Range and Snake River aquifers. Materials in these aquifers are primarily unconsolidated sand, gravel, and clay, or basalt with interbeds of sand, gravel, or clay.

The depths of public supply wells (median of 427 ft below land surface) and of the wells in the Basin and Range aquifer (median depth of 400 ft) might explain, in part, the relatively low detection frequencies of the coliform bacteria observed in these samples. A thick unsaturated zone increases the potential for natural attenuation of microorganisms, preventing the transport of bacteria into the groundwater. Fifty percent of wells in principal aquifers with median depths of sampled wells ranging from 100 to 200 ft below land surface tested positive for total-coliform bacteria, whereas only 9 percent of wells in principal aquifers with median depths of sampled wells greater than 200 ft tested positive.

5.3.8 Emerging Contaminants

Thanks to the advancements of the communication age and the widespread use of the Internet, the public around the globe is becoming increasingly informed about various environmental concerns almost instantaneously. Traditional media is following this trend by also publishing on the Internet. The end result is an ever-increasing transparency when it comes to discussing the effects of environmental degradation on drinking water resources. The following quotes from an article published in the *Las Vegas Sun* illustrate this point and the role of media (October 20, 2006. *Chemicals cause changes in fish and raise concerns for humans*, by Launce Rake):

> There's something wrong with the fish. It's been confounding scientists for years: Male fish are developing female sexual characteristics in Lake Mead and other freshwater sources around the country. On Thursday, the U.S. Geological Survey released a four-page summary of more than a decade of studies linking wastewater chemicals to those changes. But a scientist who has studied the issue for years complains that the report understates the danger of those toxins at Lake Mead and elsewhere. The researcher had aired his concerns seven months ago—shortly after he was fired by the USGS.

The federal agency says the researcher was fired for failing to publish his data. The researcher says the federal agency wouldn't allow him to publish. Both sides, however, agree on the basic issue: In Lake Mead and in other freshwater sites, scientists have found traces of pharmaceuticals, pesticides, chemicals used in plastic manufacturing, artificial fragrances and other substances linked to changes in fish and animals. Thursday's report noted that the primary source for the chemicals in Lake Mead was the Las Vegas Wash, a man-made river made up almost entirely of treated wastewater from cities in the Las Vegas Valley.

Gross said the problem is acute in Lake Mead and in other freshwater sites. One element left out of the Thursday report is evidence of sperm failure in fish, he said. "On a national scale we see alterations in fish," said the scientist, who continues to research hormone-disrupting chemicals in Florida and other states. "Endocrine (a hormone) disruption is widespread across the United States and is widespread in Lake Mead." Gross said his conclusions, shared by other researchers, are not popular: "The (Southern Nevada) Water Authority doesn't want to hear it. My agency doesn't want to hear it. The Department of Interior does not want to deal with it. They want to make the argument that there is nothing to worry about, but common sense just suggests it is not that simple."

Studies documenting sexual abnormalities in fish in the Potomac River—source of drinking water for millions in the Washington, D.C., area—raised similar concerns in September. Water officials there said the studies showed no evidence that drinking water was unsafe, but the studies did not answer the question on potential impacts to human health.

The preceding quotes are an example of public concern with emerging contaminants—the constituents that are generally not regulated but whose relatively wide presence in drinking water supplies has been documented. The 1996 Amendments to the Safe Drinking Water Act specify that development of new drinking water standards requires broad public and scientific input to ensure that contaminants posing the greatest risk to public health will be selected for future regulation. A contaminant's presence in drinking water and public health risks associated with a contaminant must be considered in order to determine whether a public health risk is evident. In addition, the new contaminant selection approach explicitly takes into account the needs of sensitive populations such as children and pregnant women. Under the 1996 Amendments, the CCL guides scientific evaluation of new contaminants. Contaminants on the CCL are prioritized for regulatory development, drinking water research (including studies of health effects, treatment effects, and analytical methods), and occurrence monitoring. The Unregulated Contaminant Monitoring Rule (UCMR) guides collection of data on contaminants not included in the National Primary Drinking Water Standard. The data are used to evaluate and prioritize contaminants that the USEPA is considering for possible new drinking water standards. Currently, there are 37 SOCs on the USEPA's CCL (USEPA, 2005c).

It is important to note that the USEPA has not limited itself to making regulatory determinations for only those contaminants on the CCL. The agency can also decide to regulate other unregulated contaminants if information becomes available, showing that a specific contaminant presents a public health risk. Some of these "other" contaminants have already been regulated by the various states, which often react faster to widely expressed public concerns than the federal government. Examples include MTBE (an infamous gasoline additive), which was regulated by quite a few states before it finally made it on the last CCL, and 1,4-dioxane (solvent stabilizer), which is being increasingly detected in association with 1,1,1-TCA plumes, is regulated by some states, but it is not on the current CCL.

The main difficulty with the entire process of drinking water regulation is that humanity now lives in a chemical universe created by our diverse activities. Literally hundreds of thousands of synthetic chemicals are being widely used in manufacturing and for various other purposes, with more than 1000 new ones introduced each year. In comparison, the USEPA's CCL list has 37 SOCs under evaluation. It is simply not feasible, and it would certainly be cost prohibitive for any society, to engage in regulation of thousands of chemicals that may be present in water supplies at some minute concentrations, but of which is little known regarding their effects on the human health and the environment at such low concentrations. Instead, it is likely that the entire field of water resources management will be forced to take a holistic approach where drinking water regulations will be an integral part of much broader environmental regulations, including those of the carbon cycle. Simple examples are the questions of water treatment and the cost of it, including the required energy; is it better to "completely" treat the drinking water or the wastewater, no matter what, and how many chemical substances constitute the "complete" list. And finally, how do we estimate the true cost and benefits of our decisions with respect to the society and the environment?

Probably the only parts of the environment not yet widely influenced by the vast number of anthropogenic chemical substances are deep pristine confined groundwater systems. As such, they present an enormous treasure but are under increasing threat due to their natural connectivity with the shallow systems. As illustrated in Chap. 8, artificial aquifer recharge with surface water and treated wastewater is one of the most important aspects of water supply sustainability. Wastewater is being increasingly viewed as a true water resource and will certainly play a major role in water resources management in the very near future. In fact, in some countries the term wastewater is being replaced with the term "used water" to emphasize this trend.

With the advancement of analytical methods, which can detect concentrations at parts per trillion (ng/L) or lower, a large picture of the numerous chemicals present in water supplies has emerged only recently. Certain pharmaceutically active compounds (e.g., caffeine, aspirin, and nicotine), which have been known for over 20 years to occur in the environment, are now joined by a broad group of chemicals collectively referred to as PPCPs. It seems that this term has prevailed in practice as a synonym for emerging contaminants, although water- and wastewater-treatment industry prefers to use the term microconstituents.

PPCPs are a diverse group of chemicals comprising all human and veterinary drugs (available by prescription or over the counter, including the new genre of "biologics"), diagnostic agents (e.g., X-ray contrast media), "nutraceuticals" (bioactive food supplements such as huperzine A), and other consumer chemicals, such as fragrances (e.g., musks) and sun-screen agents (e.g., methylbenzylidene camphor); also included are "excipients," the so-called "inert" ingredients used in PPCP manufacturing and formulation (Daughton, 2007). Nanomaterials are an emerging subgroup of microcostituents considered by many as the next industrial wonder. They are already present in cosmetics, sunscreens, wrinkle-free clothing, and food products. Because of their small size, nanomaterials pose a challenge in terms of detection and treatment. Also because of their size they can enter all human organs including the brain, but very little is known of their fate and transport in the environment.

Only a subset of PPCPs, such as synthetic steroids, is known to be direct-acting endocrine disruptors. However, little is known about the individual and combined effects of long-term exposure to most PPCPs and their degradation products at very low concentrations.

The widespread use of PPCPs in the environment is a result of their unavoidable, collective discharge by humans as well as animals. Some pharmaceuticals are not completely metabolized after consumption by humans or animals and are excreted in their original form, while others are transformed into different compounds (conjugates). Almost 20 percent of prescription drugs are flushed down the toilet unused, according to some estimates (Jeyanayagam, 2008). Domestic sewage is a major source of PPCPs, and CAFOs are a major source of antibiotics and possibly steroids (Daughton, 2007).

Free excreted drugs and derivatives can escape degradation in municipal sewage-treatment facilities where their removal efficiency is a function of the drug's structure and treatment technology employed. Some conjugates can also be hydrolyzed back to the free parent drug during the treatment process. After going through the wastewater-treatment plant, PPCPs and their degradation products are discharged to receiving surface waters and can find their way to groundwater, including by direct artificial aquifer recharge. The full extent, magnitude, and ramifications of their presence in the aquatic environment are largely unknown (Daughton, 2007). Releases of PPCPs to the environment are likely to continue as the human population increases and ages; the pharmaceutical industry formulates new prescription and nonprescription drugs and promotes their use, and more wastewater is generated, which enters the hydrologic cycle and may impact groundwater resources (Masters et al., 2004).

5.4 Drinking Water Standards

5.4.1 Primary Drinking Water Standards

National Primary Drinking Water Regulations (NPDWRs or Primary Standards) are legally enforceable standards that apply to public water systems. Primary standards protect drinking water quality by limiting the levels of specific contaminants that can adversely affect public health and are known or anticipated to occur in water. They take the form of maximum contaminant levels (MCLs) or treatment techniques (TTs). These standards must be met at the point of delivery to any user of a public system (i.e., point of use, or point of discharge from the water distribution system) or, in some cases, at various points throughout the distribution system (USEPA, 2003a).

Once the USEPA has selected a contaminant for regulation, it examines the contaminant's health effects and sets an MCL goal (MCLG). This is the maximum level of a contaminant in drinking water at which no known or anticipated adverse health effects would occur, and which allows an adequate margin of safety. MCLGs do not take cost and technologies into consideration. MCLGs are nonenforceable public health goals. In setting the MCLG, the USEPA examines the size and nature of the population exposed to the contaminant, and the length of time and concentration of the exposure. Since MCLGs consider only public health and not the limits of detection and treatment technology, they are sometimes set at a level that water systems cannot meet. For most carcinogens (contaminants that cause cancer) and microbiological contaminants, MCLGs are set at zero because a safe level often cannot be determined (USEPA, 2003a, 2006).

MCLs, which are enforceable limits that finished drinking water must meet, are set as close to the MCLG as feasible. The Safe Drinking Water Act (SDWA) defines "feasible" as the level that may be achieved with the use of the BAT, TT, or other means specified by USEPA, after examination for efficacy under field conditions (i.e., not solely under laboratory conditions) and taking cost into consideration (USEPA, 2003a).

For some contaminants, especially microbiological contaminants, there is no reliable method that is economically and technically feasible to measure a contaminant at particularly low concentrations. In these cases, the USEPA establishes TTs. A TT is an enforceable procedure or level of technological performance that public water systems must follow to ensure control of a contaminant. Examples of rules with TTs are the surface water-treatment rule (aimed primarily at biological contaminants and water disinfection) and the lead and copper rule.

As of January 2008, USEPA has set MCLs or TTs for 87 contaminants included in the National Primary Drinking Water Standards list (Table 5.2).

The CCL guides scientific evaluation of new contaminants. Contaminants on the CCL are prioritized for regulatory development, drinking water research (including studies of health effects, treatment effects, and analytical methods), and occurrence monitoring. The UCMR guides collection of data on contaminants not included in the National Primary Drinking Water Standard. The data are used to evaluate and prioritize contaminants that the USEPA is considering for possible new drinking water standards.

The CCL must be updated every 5 years, providing a continuing process to identify contaminants for future regulations or standards and for prevention activities. To prioritize contaminants for regulation, USEPA considers peer-reviewed science and data to support an "intensive technological evaluation," which includes many factors: occurrence in the environment, human exposure and risks of adverse health effects in the general population and sensitive subpopulations, analytical methods of detection, technical feasibility, and impacts of regulation on water systems, the economy, and public health (USEPA, 2003a). Table 5.3 shows the contaminants on the current candidate list (CCL2) as of January 2008.

5.4.2 Secondary Drinking Water Standards

National Secondary Drinking Water Regulations (NSDWRs or Secondary Standards) are nonenforceable guidelines regarding contaminants that may cause cosmetic effects (such as skin or tooth discoloration) or have aesthetic effects (such as affecting the taste, odor, or color of drinking water). The USEPA recommends secondary standards to water systems but does not require systems to comply. However, states may choose to adopt them as enforceable standards. NSDWRs are intended to protect "public welfare" (USEPA, 2003a). There are 15 constituents included in the National Secondary Drinking Water Standards list (Table 5.4).

5.5 Fate and Transport of Contaminants

Contaminants in the vadose and saturated zones are subject to various fate and transport processes, which govern their mobility and longevity. Contaminants can enter the subsurface dissolved in the infiltrating water or as immiscible liquids, which, in time, will also dissolve into the ambient groundwater. Depending on the characteristics of the contaminant and the geochemical environment, the velocity of contaminant migration will be more or less retarded with respect to groundwater velocity as the contaminant may adsorb onto solid particles of the porous media and diffuse into the pore space of the solids and dead-end pores between the solids. The contaminant may also enter into various biogeochemical reactions with the native groundwater and solids, which can change its character and mobility. Irreversible precipitations from the solution or a

	Contaminant	MCL or TT1 (mg/L)[2]	Potential Health Effects from Exposure Above the MCL	Common Sources of Contaminant in Drinking Water	Public Health Goal
OC	Acrylamide	TT[8]	Nervous system or blood problems	Added to water during sewage/wastewater increased risk of cancer treatment	zero
OC	Alachlor	0.002	Eye, liver, kidney, or spleen problems; anemia; increased risk of cancer	Runoff from herbicide used on row crops	zero
R	Alpha particles	15 picocuries per Liter (pCi/L)	Increased risk of cancer	Erosion of natural deposits of certain minerals that are radioactive and may emit a form of radiation known as alpha radiation	zero
IOC	Antimony	0.006	Increase in blood cholesterol; decrease in blood sugar	Discharge from petroleum refineries; fire retardants; ceramics; electronics; solder	0.006
IOC	Arsenic	0.01	Skin damage or problems with circulatory systems, and may have increased risk of cancer	Erosion of natural deposits; runoff from orchards, runoff from glass & electronics production wastes	0
IOC	Asbestos (fibers >10 mm)	7 million fibers per Liter (MFL)	Increased risk of developing benign intestinal polyps	Decay of asbestos cement in water mains; erosion of natural deposits	7 MFL
OC	Atrazine	0.003	Cardiovascular system or reproductive problems	Runoff from herbicide used on row crops	0.003
IOC	Barium	2	Increase in blood pressure	Discharge of drilling wastes; discharge from metal refineries; erosion of natural deposits	2
OC	Benzene	0.005	Anemia; decrease in blood platelets; increased risk of cancer	Discharge from factories; leaching from gas storage tanks and landfills	zero

			Potential Health Effects	Sources of Contaminant	
OC	Benzo(a)pyrene (PAHs)	0.0002	Reproductive difficulties; increased risk of cancer	Leaching from linings of water	zero
IOC	Beryllium	0.004	Intestinal lesions	Discharge from metal refineries and coal-burning factories; discharge from electrical, aerospace, and defense industries	0.004
R	Beta particles and photon emitters	4 millirems per year	Increased risk of cancer	Decay of natural and man-made deposits of certain minerals that are radioactive and may emit forms of radiation known as photons and beta radiation	zero
DBP	Bromate	0.010	Increased risk of cancer	By-product of drinking water disinfection	zero
IOC	Cadmium	0.005	Kidney damage	Corrosion of galvanized pipes; erosion of natural deposits; discharge from metal refineries; runoff from waste batteries and paints	0.005
OC	Carbofuran	0.04	Problems with blood, nervous system, or reproductive system	Leaching of soil fumigant used on rice and alfalfa	0.04
OC	Carbon tetrachloride	0.005	Liver problems; increased risk of cancer	Discharge from chemical plants and other industrial activities	zero
D	Chloramines (as Cl$_2$)	MRDL = 4.0[1]	Eye/nose irritation; stomach discomfort, anemia	Water additive used to control microbes	MRDLG = 4[1]
OC	Chlordane	0.002	Liver or nervous system problems; increased risk of cancer	Residue of banned termiticide	zero

TABLE 5.2 National Primary Drinking Water Standards (Continued)

	Contaminant	MCL or TT1 (mg/L)[2]	Potential Health Effects from Exposure Above the MCL	Common Sources of Contaminant in Drinking Water	Public Health Goal
D	Chlorine (as Cl_2)	MRDL = 4.0[1]	Eye/nose irritation; stomach discomfort	Water additive used to control microbes	MRDLG = 4[1]
D	Chlorine dioxide (as ClO_2)	MRDL = 0.8[1]	Anemia; infants & young children: nervous system effects	Water additive used to control microbes	MRDLG = 0.8[1]
DBP	Chlorite	1.0	Anemia; infants & young children: nervous system effects	By-product of drinking water disinfection	0.8
OC	Chlorobenzene	0.1	Liver or kidney problems	Discharge from chemical and agricultural chemical factories	0.1
IOC	Chromium (total)	0.1	Allergic dermatitis	Discharge from steel and pulp mills; erosion of natural deposits	0.1
IOC	Copper	TT[7]; Action Level = 1.3	Short-term exposure: gastrointestinal distress; long-term exposure: liver or kidney damage; people with Wilson's Disease should consult their personal doctor if the amount of copper in their water exceeds the action level	Corrosion of household plumbing systems; erosion of natural deposits	1.3
M	*Cryptosporidium*	TT[3]	Gastrointestinal illness (e.g., diarrhea, vomiting, cramps)	Human and animal fecal waste	zero
IOC	Cyanide (as free cyanide)	0.2	Nerve damage or thyroid problems	Discharge from steel/metal factories; discharge from plastic and fertilizer factories	0.2
OC	2,4-D	0.07	Kidney, liver, or adrenal gland problems	Runoff from herbicide used on row crops	0.07

OC	Dalapon	0.2	Minor kidney changes	Runoff from herbicide used on rights of way	0.2
OC	1,2-Dibromo-3-chloropropane (DBCP)	0.0002	Reproductive difficulties; increased risk of cancer	Runoff/leaching from soil fumigant used on soybeans, cotton, pineapples, and orchards	zero
OC	o-Dichlorobenzene	0.6	Liver, kidney, or circulatory system problems	Discharge from industrial chemical factories	0.6
OC	p-Dichlorobenzene	0.075	Anemia; liver, kidney or spleen damage; changes in blood	Discharge from industrial chemical factories	0.075
OC	1,2-Dichloroethane	0.005	Increased risk of cancer	Discharge from industrial chemical factories	zero
OC	1,1-Dichloroethylene	0.007	Liver problems	Discharge from industrial chemical factories	0.007
OC	cis-1,2-Dichloroethylene	0.07	Liver problems	Discharge from industrial chemical factories	0.07
OC	trans-1,2-Dichloroethylene	0.1	Liver problems	Discharge from industrial chemical factories	0.1
OC	Dichloromethane	0.005	Liver problems; increased risk of cancer	Discharge from drug and chemical factories	zero
OC	1,2-Dichloropropane	0.005	Increased risk of cancer	Discharge from industrial chemical factories	zero
OC	Di(2-ethylhexyl) adipate	0.4	Weight loss, live problems, or possible reproductive difficulties	Discharge from chemical factories	0.4
OC	Di(2-ethylhexyl) phthalate	0.006	Reproductive difficulties; liver problems; increased risk of cancer	Discharge from rubber and chemical factories	zero
OC	Dinoseb	0.007	Reproductive difficulties	Runoff from herbicide used on soybeans and vegetables	0.007

TABLE 5.2   National Primary Drinking Water Standards (*Continued*)

	Contaminant	MCL or TT1 (mg/L)[2]	Potential Health Effects from Exposure Above the MCL	Common Sources of Contaminant in Drinking Water	Public Health Goal
OC	Dioxin (2,3,7,8-TCDD)	0.00000003	Reproductive difficulties; increased risk of cancer	Emissions from waste incineration and other combustion; discharge from chemical factories	zero
OC	Diquat	0.02	Cataracts	Runoff from herbicide use	0.02
OC	Endothall	0.1	Stomach and intestinal problems	Runoff from herbicide use	0.1
OC	Endrin	0.002	Liver problems	Residue of banned insecticide	0.002
OC	Epichlorohydrin	TT[8]	Increased cancer risk, and over a long period of time, stomach problems	Discharge from industrial chemical factories; an impurity of some water treatment chemicals	zero
OC	Ethylbenzene	0.7	Liver or kidneys problems	Discharge from petroleum refineries	0.7
OC	Ethylene dibromide	0.00005	Problems with liver, stomach, reproductive system, or kidneys; increased risk of cancer	Discharge from petroleum refineries	zero
IOC	Fluoride	4.0	Bone disease (pain and tenderness of the bones); children may get mottled teeth	Water additive which promotes strong teeth; erosion of natural deposits; discharge from fertilizer and aluminum factories	4.0
M	Giardia lamblia	TT[3]	Gastrointestinal illness (e.g., diarrhea, vomiting, cramps)	Human and animal fecal waste	zero
OC	Glyphosate	0.7	Kidney problems; reproductive difficulties	Runoff from herbicide use	0.7
DBP	Haloacetic acids (HAA5)	0.060	Increased risk of cancer	By-product of drinking water disinfection	n/a[6]
OC	Heptachlor	0.0004	Liver damage; increased risk of cancer	Residue of banned termiticide	zero

OC	Heptachlor epoxide	0.0002	Liver damage; increased risk of cancer	Breakdown of heptachlor	zero
M	Heterotrophic plate count (HPC)	TT[3]	HPC has no health effects; it is an analytic method used to measure the variety of bacteria that are common in water; the lower the concentration of bacteria in drinking water, the better maintained the water system is	HPC measures a range of bacteria that are naturally present in the environment	n/a
OC	Hexachloro-cyclopentadiene	0.05	Kidney or stomach problems	Discharge from chemical factories	0.05
IOC	Lead	TT[7]; Action Level = 0.015	Infants and children: delays in physical or mental development; children could show slight deficits in attention span and learning abilities; adults: kidney problems; high blood pressure	Corrosion of household plumbing systems; erosion of natural deposits	zero
M	Legionella	TT[3]	Legionnaire's Disease, a type of pneumonia	Found naturally in water; multiplies in heating systems	zero
OC	Lindane	0.0002	Liver or kidney problems	Runoff/leaching from insecticide used on cattle, lumber, gardens	0.0002
IOC	Mercury (inorganic)	0.002	Kidney damage	Erosion of natural deposits; discharge from refineries and factories; runoff from landfills and croplands	0.002
OC	Methoxychlor	0.04	Reproductive difficulties	Runoff/leaching from insecticide used on fruits, vegetables, alfalfa, livestock	0.04

TABLE 5.2 National Primary Drinking Water Standards (*Continued*)

	Contaminant	MCL or TT1 (mg/L)[2]	Potential Health Effects from Exposure Above the MCL	Common Sources of Contaminant in Drinking Water	Public Health Goal
IOC	Nitrate (measured as nitrogen)	10	Infants below the age of 6 months who drink water containing nitrate in excess of the MCL could become seriously ill and, if untreated, may die; symptoms include shortness of breath and blue baby syndrome	Runoff from fertilizer use; leaching from septic tanks, sewage; erosion of natural deposits	10
IOC	Nitrite (measured as nitrogen)	1	Infants below the age of 6 months who drink water containing nitrite in excess of the MCL could become seriously ill and, if untreated, may die; symptoms include shortness of breath and blue baby syndrome	Runoff from fertilizer use; leaching from septic tanks, sewage; erosion of natural deposits	1
OC	Oxamyl (Vydate)	0.2	Slight nervous system effects	Runoff/leaching from insecticide used on apples, potatoes, and tomatoes	0.2
OC	Pentachlorophenol	0.001	Liver or kidney problems; increased cancer risk	Discharge from wood preserving factories	zero
OC	Picloram	0.5	Liver problems	Herbicide runoff	0.5
OC	Polychlorinated biphenyls (PCBs)	0.0005	Skin changes; thymus gland problems; immune deficiencies; reproductive or nervous system difficulties; increased risk of cancer	Runoff from landfills; discharge of waste chemicals	zero
R	Radium 226 and Radium 228 (combined)	5 pCi/L	Increased risk of cancer	Erosion of natural deposits	zero

IOC	Selenium	0.05	Hair or fingernail loss; numbness in fingers or toes; circulatory problems	Discharge from petroleum refineries; erosion of natural deposits; discharge from mines	0.05
OC	Simazine	0.004	Problems with blood	Herbicide runoff	0.004
OC	Styrene	0.1	Liver, kidney, or circulatory system problems	Discharge from rubber and plastic factories; leaching from landfills	0.1
OC	Tetrachloroethylene (PCE)	0.005	Liver problems; increased risk of cancer	Discharge from factories and dry cleaners	zero
IOC	Thallium	0.002	Hair loss; changes in blood; kidney, intestine, or liver problems	Leaching from ore-processing sites; discharge from electronics, glass, and drug factories	0.0005
OC	Toluene	1	Nervous system, kidney, or liver problems	Discharge from petroleum factories	1
M	Total Coliforms (including fecal coliform and E. coli)	5.0%[4]	Not a health threat in itself; it is used to indicate whether other potentially harmful bacteria may be present[5]	Coliforms are naturally present in the environment as well as feces; fecal coliforms and E. coli only come from human and animal fecal waste.	zero
DBP	Total Tri-halomethanes (TTHMs)	0.080	Liver, kidney, or central nervous system problems; increased risk of cancer	By-product of drinking water disinfection	n/a[6]
OC	Toxaphene	0.003	Kidney, liver, or thyroid problems; increased risk of cancer	Runoff/leaching from insecticide used on cotton and cattle	zero

TABLE 5.2 National Primary Drinking Water Standards (*Continued*)

389

	Contaminant	MCL or TT1 (mg/L)[2]	Potential Health Effects from Exposure Above the MCL	Common Sources of Contaminant in Drinking Water	Public Health Goal
OC	2,4,5-TP (Silvex)	0.05	Liver problems	Residue of banned herbicide	0.05
OC	1,2,4-Trichlorobenzene	0.07	Changes in adrenal glands	Discharge from textile finishing factories	0.07
OC	1,1,1-Trichloroethane	0.2	Liver, nervous system, or circulatory problems	Discharge from metal degreasing sites and other factories	0.20
OC	1,1,2-Trichloroethane	0.005	Liver, kidney, or immune system problems	Discharge from industrial chemical factories	0.005
OC	Trichloroethylene (TCE)	0.005	Liver problems; increased risk of cancer	Discharge from metal degreasing sites and other factories	zero
M	Turbidity	TT[3]	Turbidity is a measure of the cloudiness of water. It is used to indicate water quality and filtration effectiveness (e.g., whether disease-causing organisms are present). Higher turbidity levels are often associated with higher levels of disease-causing micro-organisms such as viruses, parasites and some bacteria. These organisms can cause symptoms such as nausea, cramps, diarrhea, and associated headaches.	Soil runoff	n/a
R	Uranium	30 μg/L	Increased risk of cancer, kidney toxicity	Erosion of natural deposits	zero
OC	Vinyl chloride	0.002	Increased risk of cancer	Leaching from PVC pipes; discharge from plastic factories	zero
M	Viruses (enteric)	TT[3]	Gastrointestinal illness (e.g., diarrhea, vomiting, cramps)	Human and animal fecal waste	zero
OC	Xylenes (total)	10	Nervous system damage	Discharge from petroleum factories; discharge from chemical factories	10

[1] Definitions: MCLG, maximum contaminant level goal; MCL, maximum contaminant level; MRDLG, maximum residual disinfectant level goal; MRDL, maximum residual disinfectant level; TT, treatment technique.

[2] Units are in milligrams per liter (mg/L) unless otherwise noted. Milligrams per liter are equivalent to parts per million (ppm).

[3] EPA's surface water treatment rules require systems using surface water or ground water under the direct influence of surface water to disinfect their water and filter their water or meet criteria for avoiding filtration so that the following contaminants are controlled at these levels:

Cryptosporidium 99% removal; Giardia lamblia: 99.9% removal/inactivation; Viruses: 99.99% removal/inactivation

Turbidity: may never exceed 1 NTU, and must not exceed 0.3 NTU in 95% of daily samples in any month.

HPC: No more than 500 bacterial colonies per milliliter.

[4] No more than 5.0% samples total coliform-positive in a month. (For water systems that collect fewer than 40 routine samples per month, no more than one sample can be total coliform-positive per month.) Every sample that has total coliform must be analyzed for either fecal coliforms or E. coli if two consecutive TC-positive samples, and one is also positive for E. coli fecal coliforms, system has an acute MCL violation.

[5] Fecal coliform and E. coli are bacteria whose presence indicates that the water may be contaminated with human or animal wastes.

[6] Although there is no collective MCLG for this contaminant group, there are individual MCLGs for some of the individual contaminants:

Haloacetic acids: dichloroacetic acid (zero); trichloroacetic acid (0.3 mg/L)
Trihalomethanes: bromodichloromethane (zero); bromoform (zero); dibromochloromethane (0.06 mg/L). If more than 10% of tap water samples.

[7] Lead and copper are regulated by a Treatment Technique that requires systems to control the corrosiveness of their water.

To exceed the action level, water systems must take additional steps. For copper, the action level is 1.3 mg/L, and for lead is 0.015 mg/L.

[8] Each water system must certify, in writing, to the state (using third-party or manufacturers certification) that when it uses acrylamide and/or epichlorohydrin to treat water, the combination (or product) of dose and monomer level does not exceed the levels specified, as follows: Acrylamide = 0.05% dosed at 1 mg/L (or equivalent); Epichlorohydrin = 0.01% dosed at 20 mg/L (or equivalent). From USEPA. http://www.epa.gov/safewater/contaminants/index.html#listmcl. Accessed January 2008.

TABLE 5.2 National Primary Drinking Water Standards (*Continued*)

Microbial Contaminant Candidates	Chemical Contaminant Candidates	CASRN
Adenoviruses	1,1,2,2-tetrachloroethane	79-34-5
Aeromonas hydrophila	1,2,4-trimethylbenzene	95-63-6
Caliciviruses	1,1-dichloroethane	75-34-3
Coxsackieviruses	1,1-dichloropropene	563-58-6
Cyanobacteria (blue-green algae), other freshwater algae, and their toxins	1,2-diphenylhydrazine	122-66-7
Echoviruses	1,3-dichloropropane	142-28-9
Helicobacter pylori	1,3-dichloropropene	542-75-6
Microsporidia (Enterocytozoon & Septata)	2,4,6-trichlorophenol	88-06-2
Mycobacterium avium intracellulare (MAC)	2,2-dichloropropane	594-20-7
	2,4-dichlorophenol	120-83-2
	2,4-dinitrophenol	51-28-5
	2,4-dinitrotoluene	121-14-2
	2,6-dinitrotoluene	606-20-2
	2-methyl-Phenol (o-cresol)	95-48-7
	Acetochlor	34256-82-1
	Alachlor ESA & other acetanilide pesticide degradation products	N/A
	Aluminum	7429-90-5
	Boron	7440-42-8
	Bromobenzene	108-86-1
	DCPA mono-acid degradate	887-54-7
	DCPA di-acid degradate	2136-79-0
	DDE	72-55-9
	Diazinon	333-41-5
	Disulfoton	298-04-4
	Diuron	330-54-1
	EPTC (s-ethyl dipropylthiocarbamate)	759-94-4
	Fonofos	944-22-9
	p-Isopropyltoluene (p-cymene)	99-87-6
	Linuron	330-55-2
	Methyl bromide	74-83-9
	Methyl-t-butyl ether (MTBE)	1634-04-4
	Metolachlor	51218-45-2
	Molinate	2212-67-1
	Nitrobenzene	98-95-3
	Organotins	N/A
	Perchlorate	14797-73-0
	Prometon	1610-18-0
	RDX	121-82-4
	Terbacil	5902-51-2

TABLE 5.3 Contaminant Candidate List 2 as of January 2008

Microbial Contaminant Candidates	Chemical Contaminant Candidates	CASRN
	Terbufos	13071-79-9
	Vanadium	7440-62-2
	Triazines & degradation products of	21725-46-2
	triazines including, but not limited to	6190-65-4
	Cyanazine	

From USEPA. http://www.epa.gov/safewater/ccl/ccl2.html. Accessed January 2008.

TABLE 5.3 (Continued)

complete degradation (mineralization) permanently remove the contaminant from flowing groundwater. Most other processes act to decrease the concentration of the dissolved contaminant as it moves away from the source zone where it was introduced into the subsurface. Unfortunately, predicting contaminant concentrations at some distance from the source and after a certain time of travel is often very difficult because of the various contaminant fate and transport mechanisms, many (if not all) of which cannot be quantified exactly. It is, therefore, not uncommon for two or more parties ("stakeholders") to arrive at very different answers as to the expected (predicted) contaminant concentrations, using presumably the same conceptual site model. This fact is especially important when considering that most legal, engineering, and management decisions are based on

Contaminant	Secondary Standard
Aluminum	0.05–0.2 mg/L
Chloride	250 mg/L
Color	15 (color units)
Copper	1.0 mg/L
Corrosivity	noncorrosive
Fluoride	2.0 mg/L
Foaming Agents	0.5 mg/L
Iron	0.3 mg/L
Manganese	0.05 mg/L
Odor	3 threshold odor number
pH	6.5–8.5
Silver	0.10 mg/L
Sulfate	250 mg/L
Total Dissolved Solids	500 mg/L
Zinc	5 mg/L

From USEPA. http://www.epa.gov/safewater/contaminants/index.html#listsec. Accessed January 2008.

TABLE 5.4 National Secondary Drinking Water Standards

predicted contaminant concentrations, in both time and space. Some of the related questions include the following:

- When and how much contaminant was introduced into the subsurface?
- Who is responsible for groundwater contamination (whose plume is it)?
- How long would it take for the contaminant to reach point A?
- What will be the contaminant concentration when it reaches point A?
- Will this remedial technology reach the groundwater cleanup goal expressed as acceptable contaminant concentration at points A, B, C, etc.?
- How long will this remedial technology have to be implemented before cleanup goals are achieved?
- What will be the cost of implementing this remedial technology?
- What will be the life-cycle cost of this remedial alternative?

Unfortunately, in too many cases there are no unique quantitative answers to these and similar questions. This is probably the main reason why attorneys and courts are inevitable and often the most important players in resolving groundwater management and restoration issues, at least in the United States. Another possible explanation is that water resources and environmental regulations in the United States, as well as the societal framework, may not be conducive to a more holistic approach to management of water (groundwater) resources. In any case, as illustrated in the following sections, selection of "representative" quantitative parameters used in calculations of contaminant fate and transport is not a straightforward process; it always leaves enough room for criticism by those willing to practice it. One obvious answer to this problem is to collect as much site-specific (field) information as possible.

5.5.1 Dissolution

The water solubility of a given substance is the maximum amount of that substance water can dissolve and maintain in solution that is in equilibrium with the solid or liquid source of the substance. Solubilities of various inorganic and organic substances are extremely variable, from infinite (e.g., for liquid substances *miscible* with water such as alcohol ethanol) to quite low, such as for NAPLs, which are immiscible with water. The terms hydrophilic (water loving) and hydrophobic (water hating) are sometimes used in reference to water solubility and water insolubility, respectively. The water solubility of a substance is controlled by quite a few factors, including water temperature; pressure; concentrations of hydrogen (H^+) and hydroxyl (OH^-) ions, i.e., pH of water; redox potential (*Eh*); and relative concentrations of other substances in the solution. The relationships between these variables in actual field conditions are complex and constantly changing, so that exact site-specific solubilities of various substances of interest cannot be easily determined. Principles of analytic laboratory chemistry, combined with some general assumptions and geochemical modeling, can be used to establish limits of natural solubilities of common substances. Various general texts in chemistry list aqueous solubilities for inorganic and organic compounds, which can be used for initial analyses.

When a contaminant is highly soluble or completely miscible in water, such as many inorganic salts (e.g., sodium chloride and perchlorate) and a number of miscible organic compounds (e.g., ethanol and 1,4-dioxane), the rate of dissolution by water (groundwater) flowing through the source zone is not limited. The flux (concentration) of the contaminant entering the subsurface will depend on the precipitation (or applied water) infiltration rate and the mass of the contaminant, not on the contaminant's solubility. A continuous loading of the contaminant at the land surface (in the sources zone) may result in its accumulation if the water infiltration rate is lower than the contaminant dissolution rate, which may happen in some arid regions. Accumulation can also occur at the land surface and at some distance below if the evapotranspiration rate is higher than the deep percolation rate, resulting in contaminant precipitation from the water solution.

The solubility product, which is a concept of physical chemistry, is an equilibrium constant for the solution of a compound that dissociates into ions. For a saturated solution of a compound, the product of the molar concentrations of the ions is a constant at any fixed temperature. Published values of individual solubility products often are in disagreement with each other because of experimental difficulties. In addition, applying mineral solubility products for pure water (ideal solution) when estimating solubility in field conditions would be erroneous, since, to be precise in saturation calculations, it is necessary to know the chemical form (single ions, complex ions, or neutral molecules) and activity of all ionic species in a given solution, including possible chemical reactions between various ions. Therefore, simple geochemical solubility calculations based on published solubility products are only rough approximations. An additional complicating factor is that the so-called activity of an ionic species is not constant and changes with concentration of both the individual ion considered and the other ions present in the solution. In other words, for dilute solutions, it is often assumed that the activity of an ion equals its mole fraction (concentration) in the solution, but this assumption may lead to significant errors for more concentrated solutions. In either case, it would be more correct to determine the actual ionic activities of particular ions for which the solubility calculations are made, as these activities are influenced by all the constituents in the solution. The solubility product, and therefore solubility, generally increases with temperature, but not in all cases. For substances that exhibit phase changes at a certain temperature, such as inversion of gypsum to anhydrite, the solubility product decreases after the inversion temperature (e.g., 60°C for the gypsum-anhydrite inversion). The solubility of solid inorganic substances is independent of pressure in common groundwater systems. Figure 5.16 shows the dependence of solubility on temperature for some common solid inorganic substances. The solubility product also increases with the rising concentration of the solution. This effect is especially noticeable in highly concentrated mineralized water and NaCl brines, in which the concentration can reach very high values (Matthess, 1982).

When an ionic substance (AB), such as halite (HCl; A = H, B = Cl), dissociates in an ideal solution to form positively and negatively charged ions, the following relationship is true at equilibrium:

$$K = \frac{[A^+] \times [B^-]}{[AB]}$$
(5.11)

where AB = solid substance
  A^+ = cation part in dissolved phase
  B^- = anion part in the dissolved phase
  K = general constant of the reaction

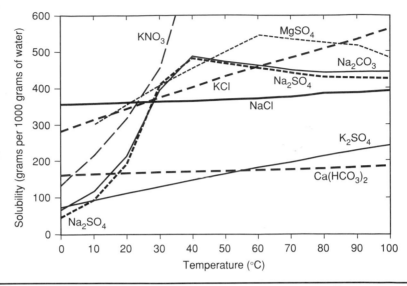

FIGURE 5.16   Solubilities of some solid inorganic substances in water at different temperatures. (Data from Matthess, 1982.)

At saturation, this general constant is exactly equal to the solubility product (K_{sp}) and the solution is in equilibrium between the solid and dissolved phases of the substance (no additional dissolution or precipitation of the substance from solution should occur). When $K < K_{sp}$, the solution is undersaturated with respect to AB and can dissolve (hold) more of it. When $K > K_{sp}$, the solution is supersaturated and precipitation of AB should occur.

Because of the complexity of dissolution and other chemical reactions in true groundwater systems (as opposed to chemical relationships derived from laboratory experiments), the use of geochemical models is arguably the only satisfactory approach for estimating dissolution and precipitation of inorganic solutes. The most widely used geochemical model for natural and contaminated waters is PHREEQC, developed at the USGS and in the public domain (Parkhurst and Appelo, 1999). The user-friendly version with a graphical user interface (PHREEQCI; USGS, 2002) includes the following modeling capabilities:

- Aqueous, mineral, gas, surface, ion-exchange, and solid-solution equilibria
- Kinetic reactions
- One-dimensional diffusion or advection and dispersion with dual-porosity medium
- A powerful inverse modeling capability, which allows identification of reactions that account for the chemical evolution in observed water compositions
- Extensive geochemical databases

Speciation modeling available in PHREEQC uses a chemical analysis of a water to calculate the distribution of aqueous species based on an ion-association aqueous model. The most important results of speciation calculations are saturation indices for minerals, which indicate whether a mineral should dissolve or precipitate. Speciation modeling is useful in situations where the possibility of mineral dissolution or precipitation needs to be known, as in water treatment, aquifer storage and recovery, artificial recharge, and well injection. Inverse modeling capability of PHREEQC can be used to deduce geochemical reactions and mixing in local and regional aquifer systems, and in aquifer storage and recovery studies. It calculates geochemical reactions that account for the change in the chemical composition of water along a flowpath. For inverse modeling, at least two chemical analyses of water are needed at different points along the flow path, as well as a set of minerals and gases that are potentially reactive. Mole transfers of phases are calculated, which account for the change in water composition along the flowpath. The numerical method applied accounts for uncertainties in analytical data. A book by Appelo and Postma (2006) includes numerous examples of geochemical reactions that can be solved with PHREEQC.

Solubility of Organic Substances

Organic liquids and solids are generally much less soluble in water when compared to inorganic salts, bases, and acids. However, there are organic liquids, such as some alcohols and solvent stabilizers (1,4-dioxane for example), that are highly soluble or even completely miscible with water due to their polar nature and hydrogen bonds. As a general rule, the most soluble organic species are those with polar molecular structure or those containing oxygen or nitrogen in a simple, short molecular structure. Examples include alcohols or carboxylic acids, which form hydrogen bonds with water molecules and fit easily into the structure of water. Without hydrogen bonding, the solubility of organic substances diminishes, since forcing a nonpolar organic molecule into the tetrahedral structure of water requires considerable energy. The importance of hydrogen bonding decreases with increasing molecule length and size. The larger the organic molecule is, the larger the space that is required in the water structure and the less soluble the compound is. For example, smaller alcohol molecules such as methanol and ethanol are infinitely soluble, whereas octanol is only slightly soluble. This effect is also clear when examining the solubility of aromatic compounds such as benzene, toluene, naphthalene, and biphenyl. The solubility is inversely proportional to the molecular mass or size of the molecules: it is the highest for benzene, which has the smallest molecular mass (78.0 g/mol) of the four, and the lowest for biphenyl, which has the largest molecular mass (154.0 g/mol). The respective reported aqueous solubilities of benzene and biphenyl are 1780 and 7.48 g/m^3 (Domenico and Schwartz, 1990).

Organic liquids that are hydrophobic (immiscible) because of their nonpolar nature will always dissolve in water to some extent, even though the bulk of their volume may remain in a separate ("free") liquid phase for considerable periods of time. As discussed earlier, such liquids are called NAPLs. Free NAPL phase may be distributed in a number of ways in the subsurface, such as relatively extensive contiguous volumes of the liquid called NAPL *pools*, or small globules and ganglia partially or completely surrounded by water (such globules and ganglia are often called *residual phase*). Dissolution of different NAPL phases in actual field conditions will depend on a number of factors such as effective and total porosities of the porous media, groundwater flow rates, surface contact area between the NAPL and groundwater (i.e., geometry of different NAPL phases), and fate and transport characteristics including adsorption and diffusion. Assuming that a NAPL would dissolve in the flowing groundwater, following some published solubility value for pure water would therefore be erroneous. The actual dissolution of NAPL in the subsurface is often referred to as rate limited because of various limiting factors, all of which act to decrease pure-water solubility. Even when assuming that some of these factors can be estimated fairly accurately at a certain site, it is almost impossible to reasonably accurately determine the distribution of free-phase and residual NAPL and the actual geometric shapes of various NAPL bodies, which are needed to determine their contact areas with water. When comparing two NAPL bodies with the same volume of product, all other factors being equal, the time needed for complete source depletion will be longer for the NAPL body with the smaller contact area. For example, an NAPL pool resting on an impermeable layer is exposed to groundwater flux only at the top, while irregular, suspended, and disconnected NAPL bodies have proportionally much greater contact areas and are more quickly dissolved (depleted).

Another important consideration when estimating site-specific solubilities of organic contaminants in the subsurface is that they are often commingled (there are more than one in a mixture), which decreases their individual aqueous solubilities. The exception is the presence of cosolvents, such as alcohols, which would act to increase solubilities of some NAPLs. Effective aqueous solubility (S_i^e) of an individual liquid compound i, in mg/L, is given as follows:

$$S_i^e = X_i S_i \tag{5.12}$$

where X_i = mole fraction of the individual compound and S_i = pure-phase solubility of liquid phase of the individual compound. This relationship is based on Raoult's law initially developed for gaseous compounds. Laboratory analyses suggest that Eq. (5.12) is a reasonable approximation for mixtures of sparingly soluble hydrophobic organic liquids that are structurally similar and that effective solubilities calculated for complex mixtures (e.g., gasoline and other petroleum products) are unlikely to be in error by more than a factor of 2 (Cohen and Mercer, 1993).

The true dissolution rate for NAPL is highly site specific and it changes in time. Published solubilities for different hydrophobic organic substances should, therefore, be used with care and only as a starting point for the related analyses. Cohen and Mercer (1993) provide a detailed discussion on different approaches to determining aqueous and field solubilities of NAPLs. Table 5.5 lists aqueous solubilities of common organic contaminants compiled by the USGS (Lawrence, 2006).

5.5.2 Volatilization

Volatilization refers to mass transfer from liquid and solid to the gaseous phase. Chemicals in the vadose zone gas may be derived from the presence of either NAPL-dissolved chemicals or adsorbed chemicals. Chemical properties affecting volatilization include vapor pressure, Henry's constant, and aqueous solubility. Other factors influencing volatilization rate are concentration of contaminant in soil, soil moisture content, soil air movement, sorptive and diffusive characteristics of the soil, soil temperature, and bulk properties of the soil such as organic-carbon content, porosity, density, and clay content (Lyman et al., 1982; from Cohen and Mercer, 1993). VOCs in soil gas can (1) migrate and ultimately condense, (2) sorb onto soil particles, (3) dissolve in groundwater, (4) degrade, and/or (5) escape to the atmosphere. Volatilization of flammable organic chemicals in soil can create a fire or explosion hazard if vapors accumulate in combustible concentrations in the presence of an ignition source (Fussell et al., 1981).

IUPAC Name	Common or Alternative Name	Water Solubility (mg/L) at 25°C	Henry's Law Constant (H)
2-methoxy-2-methylpropane	methyl tert-butyl ether, MTBE	36,200	0.070
1,2-dichloroethane	1,2-ethylidene dichloride, glycol dichloride	8,600	0.140
chloromethane	methyl chloride	5,320	0.920
chloroethane	ethyl chloride, monochloroethane	6,710	1.11
cis-1,2-dichloroethene	cis-1,2-dichloroethylene, cis-1,2-DCE	6,400	0.460
1,1-dichloroethane	1,1-ethylidene dichloride	5,000	0.630
1,1,2-trichloroethane	methyl chloroform	4,590	0.092
trans-1,2-dichloroethene	trans-1,2-dichloroethylene, trans-1,2-DCE	4,500	0.960
chloroethene	vinyl chloride, chloroethylene	2,700	2.68
1,1-dichloroethene	1,1-dichloroethylene, DCE	2,420	2.62
benzene	—	1,780	0.557
1,1,1-trichloroethane	methyl chloroform	1,290	1.76
1,1,2-trichloroethene	1,1,2-trichloroethylene, TCE	1,280	1.03
tetrachloromethane	carbon tetrachloride	1,200	2.99
methylbenzene	toluene	531	0.660
chlorobenzene	monochlorobenzene	495	0.320
stryrene	vinyl benzene	321	0.286
tetrachloroethene	perchloroethylene, tetrachloroethylene, PCE	210	1.73
1,2-dimethylbenzene	o-xylene	207	0.551
1,4-dimethylbenzene	p-xylene	181	0.690
1,3-dimethylbenzene	m-xylene	161	0.730
ethylbenzene	—	161	0.843
1,2-dichlorobenzene	o-dichlorobenzene	147	0.195
1,2,4-trimethylbenzene	pseudocumene	57	0.524
1,2,4-trichlorobenzene	1,2,4-trichlorobenzol	37.9	0.277
naphthalene	naphthene	31.0	0.043
1,2,3-trichlorobenzene	1,2,6-trichlorobenzene	30.9	0.242

IUPAC, International Union of Pure and Applied Chemistry.
From Lawrence, 2006.

TABLE 5.5 Aqueous Solubility (in mg/L at 25°C) and Henry's Law Constant (in kPa m³/mol at 25°C) for Common Organic Contaminants

Volatilization of gasses from liquids and their dissolution into liquids follow Henry's law, which states that the amount of the dissolved gas in a liquid is directly proportional to its pressure above the interface with liquid. Henry's law can be expanded to state that the dissolved concentration of a gas is also inversely proportional to its temperature: at a constant pressure, the solubility of gasses increases with the decreasing temperature and vice versa. At the boiling temperature of the liquid, all gasses escape the solution. When there is more than one gas in the mixture, Henry's law applies to the partial pressure of each individual gas. Henry's law relationship is expressed as follows:

$$C_g = \frac{P_g}{K_H} \tag{5.13}$$

where C_g = concentration of the gas in liquid (solution), in moles per volume (m^3 or L)
P_g = (partial) pressure of the gas in the gaseous phase above the interface with the liquid (the unit is in atmospheres, atm, or pascals, Pa)
K_H = Henry's constant, in atm/(mol/dm^3), or Pa/(mol/m^3)

Published values of Henry's constant for gasses, both common inorganic and various organic, vary greatly and should be used with caution when attempting to calculate gas concentrations in groundwater. Table 5.5 lists Henry's constant for some common organic contaminants compiled by the USGS (Lawrence, 2006).

The tendency of a dissolved organic chemical to volatilize from the aqueous solution increases with an increase in Henry's constant. The same is true for the free liquid phase of a chemical that has a high Henry's constant. Organic chemicals with a high Henry's constant are called volatile and include a very important group of common organic groundwater contaminants such as aromatic and halogenated hydrocarbons (e.g., benzene and TCE). Transport of these chemicals in the subsurface may occur in both the dissolved and the gaseous phases and may involve multiple exchanges between the two phases as the site-specific conditions change in time and space. According to Henry's law, a decrease in vapor (gas) pressure above the solution would cause volatilization (escape) of the dissolved gas into the gaseous phase above the interface. This phenomenon has been widely exploited in remediation of groundwater contaminated with VOCs through the use of various techniques that increase pressure gradients between the dissolved and gaseous phases. For example, application of a vacuum in the unsaturated zone above the water table would volatilize an aromatic organic compound from both the dissolved phase and the free phase (if present), which is the principle of a remediation technology called soil vapor extraction.

Raoult's law has been used to quantify the volatilization of individual constituents from a mixture of NAPLs. This law relates the ideal vapor pressure and relative concentration of a chemical in solution to its vapor pressure over the NAPL solution (Cohen and Mercer, 1993):

$$P_A = X_A \cdot P_A^o \tag{5.14}$$

where P_A = vapor pressure of chemical A over the NAPL solution
X_A = mole fraction of chemical A in the NAPL solution
P_A^o = vapor pressure of the pure chemical A

Volatilization losses from the subsurface NAPL are expected where NAPL is close to the ground surface or in dry pervious sandy soils, or where NAPL has a very high vapor pressure (Feenstra and Cherry, 1988). Estimating volatilization from soil involves (1) estimating the organic partitioning between water and air and between NAPL and air and (2) estimating the vapor transport from the soil. Henry's and Raoult's laws are used to determine the partitioning between water and air and between NAPL and air, respectively. Vapor transport in the soil is usually described by the diffusion equation, and several models have been developed where the main transport mechanism is macroscopic diffusion (e.g., Lyman et al., 1982). More complex models are also available (Falta et al., 1989; Sleep and Sykes, 1989; Brusseau, 1991).

5.5.3 Advection

Advection is movement of the dissolved contaminant with (in) groundwater, and it refers to the average linear flow velocity of the bulk of contaminant, i.e., to the center of mass of the moving contaminant. As explained in the next section, some dissolved contaminant particles, just like some water particles, will move faster than the others, resulting in the longitudinal, transverse, and vertical spreading of the dissolved contaminant mass. This spreading, or contaminant dispersion, results in mixing of the contaminant with the native groundwater. However, the bulk of the dissolved contaminant will move with the average linear groundwater velocity (v_L), which is given as follows:

$$v_L = \frac{K \times i}{n_{ef}} \qquad (5.15)$$

where K = hydraulic conductivity, which has units of length per time (m/d; ft/d)
   i = hydraulic gradient (dimensionless)
   n_{ef} = effective porosity of the porous aquifer material (dimensionless)

If no other fate and transport process were active (e.g., no dispersion, diffusion, sorption, or degradation), the contaminant mass would move as a sharp front. For a continuous source, the concentration behind the front would be the same as at the source, while for a discontinuous, one-time source (plug), the shape of the contaminant plume would stay the same as it travels away from the source (Fig. 5.17). However, because of ever-present dispersion, the contaminant breakthrough curve, the plume shape, and the concentrations inside the plume will change in both time and the space. Changes will also occur due to other site-specific processes such as sorption or biodegradation.

All other things in a porous medium being equal (hydraulic gradient and hydraulic conductivity), a change in effective porosity will change the groundwater velocity through that medium (e.g., higher effective porosity will decrease the groundwater velocity). However, if effective porosity changes, this means that the porous material also changes, which then means that the hydraulic conductivity changes as well. Therefore, it is very important to understand that one cannot arbitrarily assume a change in effective porosity without considering the related change of the hydraulic conductivity. Because the hydraulic conductivity can change a couple of orders of magnitude for the seemingly same material, it is a much more "sensitive" parameter (i.e., it has a greater impact on the calculated velocity) than the effective porosity, which can only change within a limited range. However, changing the effective porosity by a factor of 2 will simply double or

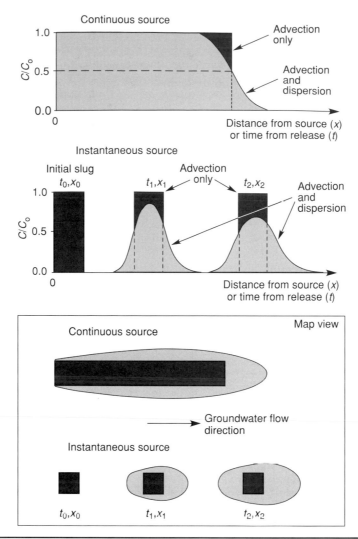

FIGURE 5.17   Breakthrough curves (top) and contaminant plumes (bottom) emanating from a continuous and a discontinuous source.

cut in half the time of travel. Finally, using the same effective porosity of, say 25 percent for "all" porous media, silt, gravel, or clay, would be completely erroneous, regardless of the intended level of effort (e.g., "it is just for screening purposes").

The advection term, which translates to contaminant residence time between the source zone and the receptor (or the point of discharge from the system), is particularly important when considering fate and transport of contaminants that (bio)degrade. The longer groundwater resident times would result in a more significant decrease in the contaminant concentration, since more time will be available for its degradation before it reaches the receptor.

Characterization and quantification of the dissolved phase of groundwater contamination can be performed with the following two approaches, which make the most sense when applied together: (1) measurement (prediction) of the contaminant concentration and (2) measurement (prediction) of the contaminant flux. Advective contaminant flux (Q_c) is simply the amount of contaminant, dissolved in groundwater and expressed with its concentration (C_c), that flows through a certain cross section of the aquifer (A) driven by the linear (effective) groundwater velocity (v_L):

$$Q_c = C_c \times v_L \times A \quad [\text{kg/d}] \tag{5.16}$$

Whatever the investigative approach, it is very important to collect three-dimensional, site-specific information on most (if not all) physical and chemical parameters needed to quantify the contaminant fate and transport. As emphasized throughout this book, groundwater flow takes place in a three-dimensional space, which is heterogeneous and anisotropic by default. Fate and transport of dissolved contaminants take place in the same three-dimensional heterogeneous space, and it is even more important not to represent it as a "sand box": a contaminant may move through a preferential, narrow, and convoluted flow zone at some high concentration, and it may even remain undetected, causing various negative impacts at distances far from the source. For this and other reasons, costs and efforts associated with contaminant fate and transport characterization, and subsequent groundwater remediation at an average "contaminated site," far outweigh most groundwater supply projects.

5.5.4 Dispersion and Diffusion

Dispersion is three-dimensional spreading of fluid particles as they flow through porous media. On the microscopic scale, dispersion is caused by deviation in velocity of the fluid particles. Particles within an individual pore will have different velocities; the ones in the center of the pore will travel the fastest, whereas those next to the pore walls will hardly move. Flow directions and velocities will also change as the particles navigate through tortuous travel paths around the individual grains of the porous material.

As discussed by Franke et al. (1990), on a larger (macroscopic) scale, local heterogeneity in the aquifer causes both the magnitude and the direction of velocity to vary as the flow concentrates along zones of greater permeability or diverges around pockets of lesser permeability. The term "macroscopic heterogeneity" is used to suggest variations in features large enough to be readily discernible in surface exposures or test wells, but too small to map (or to represent in a mathematical model) at a working scale. For example, in a typical problem involving transport away from a landfill or waste lagoon, macroscopic heterogeneities might range from the size of a baseball to the size of a building.

Although the phenomenon of dispersivity has a physical explanation that is relatively easy to understand, the process itself cannot be feasibly measured in the field. There are no widely accepted or routinely applied methods of quantifying field-scale dispersivity, and there are still very few credible large-scale field experiments that can help in better understanding dispersivity in heterogeneous porous media. It has been argued that defining the actual field distribution of hydraulic conductivity and its anisotropy to a satisfactory level of detail would eliminate the need for quantifying yet another uncertain parameter such as dispersivity. However, it is apparent that in many cases it would also not be feasible to determine distribution of the hydraulic conductivity and effective porosity (and therefore the velocity field) with very fine resolution, particularly in cases of large travel distances. It is for this reason that the dispersivity is explicitly included, through surrogate parameters, in common equations of contaminant fate and transport, in an attempt to somehow account for deviations from the average linear (advective) flow velocity.

The key assumption in deriving a term to represent dispersion is that dispersion can be represented by an expression analogous to Fick's second law of diffusion (Anderson, 1983):

$$\text{Mass flux due to dispersion} = \frac{\partial}{\partial x_i} \left(D_{ij}^* \frac{\partial c}{\partial x_j} \right) \tag{5.17}$$

where c = concentration and D_{ij}^* = the coefficient of dispersion (the i, j indices refer to cartesian coordinates). The coefficient of dispersion is assumed to be a second-rank tensor, where

$$D_{ij}^* = D_{ij} + D_d \tag{5.18}$$

D_{ij} = coefficient of mechanical dispersion and D_d = coefficient of molecular diffusion (a scalar). Molecular diffusion is a microscale process (it happens at the molecular level), which causes movement of a solute in water from the area of its higher concentration to the area of its lower concentration. An effective diffusion coefficient is generally assumed to be equal to the diffusion coefficient of the introduced liquid (or ion) in water times a tortuosity (convoluted travel paths between solids) factor, which accounts for the obstructing effects of solids and tortuous paths of the fluid particles. Effective diffusion coefficients are generally around 10^{-6} cm^2/s, which means that, except for systems in which groundwater velocities are very low, the coefficient of mechanical dispersion will be one or more orders of magnitude larger than D_d. Therefore, in many practical applications, the effects of molecular diffusion may be neglected and the coefficient of dispersion assumed to be equal to the coefficient of mechanical dispersion (Anderson, 1979, 1983). An additional parameter called dispersivity (α), which has units of length, relates the coefficient of mechanical dispersion (or just dispersion) to the average linear velocity in the main direction of groundwater flow (v):

$$\alpha = \frac{D_{ij}}{v_L} \tag{5.19}$$

The coefficient of mechanical dispersion and dispersivity are commonly expressed with three components: longitudinal (in the main direction of groundwater flow), transverse (perpendicular to the main direction in the horizontal plane), and vertical (perpendicular to the main direction in the vertical plane): D_x, D_y, and D_z, respectively, and α_x, α_y, and α_z, respectively.

A number of researchers have questioned the validity of the quantitative parameters of dispersion described above and their inclusion in equations of contaminant fate and transport. For example, the basic assumption of Fick's law is that the driving force for the diffusion is the concentration gradient. It is then assumed that Fick's law is applicable to hydrodynamic dispersion as shown with Eq. (5.17) and that coefficient of dispersion includes a coefficient of molecular diffusion—Eq. (5.18). However, the molecular diffusion is then ignored for all practical purposes. At the end, the pore channel veloci-ties and their variations are "left" to be driven by concentration gradients even though the molecular diffusion is excluded altogether (Knox et al., 1993).

Studies by various researchers have shown non-fickian behavior of dispersion in porous media flow, arguing that use of the coefficient of mechanical dispersion and dispersivity in fate and transport equations has to be reevaluated, including finding a better mathematical formulation for all time and space scales, possibly in favor of stochastic (probabilistic) approaches (e.g., Matheron and de Marsily, 1980; Smith and Schwartz, 1980; Pickens and Grisak, 1981a, 1981b; Gelhar and Axness, 1983; Dagan, 1982, 1984, 1986, 1988). The main issues when assuming a fickian behavior of dispersion and selecting a value for D or a (in any direction) are as follows (Anderson, 1983; Knox et al., 1993):

- The approach to fickian flow is asymptotic in many cases, resulting in significant non-fickian transport early in the process (dispersivity steadily increases with distance before it reaches an asymptotic value after a long time).

- During development of the dispersion process, there are significant departures from the classical normal concentration distribution associated with fickian pro-cess. Concentration-time curves are typically skewed on the right, except for long times or large distances from the source.

- There may be hydrogeologic settings where macroscopic dispersion never be-comes a fickian process.

Unfortunately, dispersion remains one of the more uncertain and unquantifiable fate and transport processes, posing a significant challenge to both the practitioners and the legal community (courts), in trying to find the most appropriate ways of incorporating it into predictive fate and transport models. The rule of thumb, suggested by the USEPA, is that longitudinal dispersivity in most cases could be initially estimated from the plume length as being 10 times smaller (Wiedemeier, 1998; Aziz et al., 2000). This means that, for example, if the plume length is 300 ft, the initial estimate of the longitudinal dispersivity is about 30 ft. However, recognizing the limitations of the available and reliable field-scale data on dispersivity, the agency also suggests that the final values of dispersivity used in fate and transport calculations should be based on calibration to the site-specific (field) concentration data.

The main reason why very few (if any) projects for practical groundwater remedi-ation consider field determinations of dispersivity is that they would require a large number of monitoring wells and the application of large-scale tracer tests. Such studies are expensive and are usually not feasible due to the generally slow movement of tracers in intergranular porous media over long distances. Two of the most widely analyzed and reported large-scale, controlled tracer field tests of dispersion, the Borden Landfill test (Sudicky et al., 1983; Mackay et al., 1986; Freyberg, 1986) and the Cape Cod test (Garabedian et al., 1991; LeBlanc et al., 1991), show similarly small macrodispersivity values compared to the overall travel distance of the injected tracers. At the Borden land-fill site, the dispersivities showed possible scale- and time-dependent behavior, with the approximated asymptotic value of 0.43 m for the longitudinal dispersivity and the travel distance of 65 m after 647 days. The transverse dispersivity was about 11 times smaller than the longitudinal. At the Cape Cod site, the plume did not show time dependence:

a constant longitudinal dispersivity of 0.96 m developed shortly after the tracer injection and a short period of nonlinear growth. The transverse and vertical dispersivities were 0.018 and 0.0015 m, respectively, or 53 and 640 times smaller than the longitudinal dispersivity. The travel distance of the plume was about 230 m after 461 days.

Xu and Eckstein (1995) authored one of the most widely cited practical studies relating the scale effects of dispersion and the selection of appropriate values of dispersivity. The authors conclude that "... the rate of increase of dispersivity declines as the scale increases. Theoretically, the rate will asymptotically approach zero as the scale approaches infinity. However, our analysis shows that the rate of increase is very small, and the curve of dispersivity versus scale is almost horizontal (with a slope angle of 0.24°) when the scale of flow exceeds 1 km. The increase in longitudinal dispersivity at that scale is so small that it can be practically ignored without causing significant error."

In their analysis, Xu and Eckstein used field data and model-calibration data reported by Gelhar et al. (1992) and shown here in Fig. 5.18. Based on the weighted least-squares data-fitting technique, which addresses the reliability of data and the nonlinear

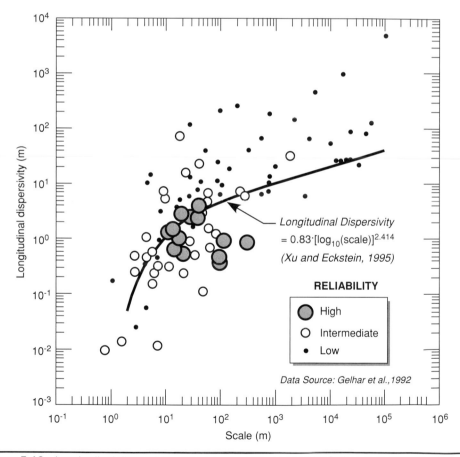

FIGURE 5.18 Longitudinal dispersivity versus scale data reported by Gelhar et al. (1992). Size of circle represents general reliability of dispersivity estimates. High reliability data are considered to be accurate within a factor of about 2 or 3. (Graph modified from Aziz et al., 2000.)

characteristics of the correlation between longitudinal dispersivity and the scale of observations, they proposed the following two equations for estimating the longitudinal dispersivity:

$$\alpha_L = 0.94(\log_{10} L)^{2.693} \quad \text{for 1:1.5:2 scheme} \tag{5.20}$$

$$\alpha_L = 0.83(\log_{10} L)^{2.414} \quad \text{for 1:2:3 scheme} \tag{5.21}$$

where α_L = longitudinal dispersivity (in units of length), L = scale (length of observation), and the first, second, and thirds numbers in the scheme correspond to the weighting factors for low, medium, and high reliability data, respectively.

Gelhar et al. (1992) include 106 data classified as of high, low, or intermediate reliability and state that the high reliability data are considered to be accurate within a factor of about 2 or 3. Their dataset includes model-calibrated longitudinal dispersivities, some of which are with low reliability and are at the same time extremely high (e.g., one data point has longitudinal dispersivity of 20 km for a "modeled" migration scale of 100 km?!). Neuman (1990) excludes available model-calibrated values of α_L in a similar analysis and proposes the following equation for longitudinal dispersivity when $L \geq 100$ m:

$$\alpha_L = 0.32L^{0.83} \tag{5.22}$$

Selecting a "representative" value of longitudinal dispersivity and then applying it to predictive models of contaminant fate and transport is a rather subjective process. Modeling should, therefore, include a thorough sensitivity analysis of this parameter, with the understanding that "more emphasis should be placed on field study and the accurate determination of hydraulic conductivity variations and other non-homogeneities and less on incorporating somewhat "arbitrary" dispersion coefficients into complex mathematical models" (Molz et al., 1983).

When the groundwater velocity becomes very low, due to small pore sizes and very convoluted pore-scale pathways, diffusion may become an important fate and transport process. Porosity that does not readily allow advective groundwater flow (flow under the influence of gravity), but does allow movement of the contaminant due to diffusion, is sometimes called diffusive porosity. Dual-porosity medium has one type of porosity that allows preferable advective transport through it; it also has another type of porosity where free gravity flow is significantly smaller than the flow taking place through the higher effective (advective) porosity. Examples of dual-porosity media include fractured rock, where advective flow takes place preferably through fractures, while the advective flow rate through the rest of the rock mass, or rock matrix, is comparably lower, is much lower, or does not exist for all practical purposes. This gradation depends on the nature of matrix porosity; in some rocks such as sandstones and young limestones, matrix porosity may be fairly high and it may allow a very significant rate of advective flow, often as high as or higher than through the fractures. In most hard rocks, matrix porosity is usually low, less than 5 to 10 percent, and it does not provide for significant advective flow. Other examples of dual-porosity media include fractured clay and residuum sediments. In some cases, various discontinuities and fractures in such media may serve as pathways for advective contaminant transport, while the bulk of the sediments may have a high overall matrix porosity and low effective porosity where advective transport is slow.

Flow of solutes with high concentration through the fractures may result in the solute diffusion into the surrounding matrix.

Diffusion is movement of a contaminant from higher concentration toward lower concentration solely due to concentration gradients; it does not involve "bulk," free-gravity movement of water particles (as in case of advection and dispersion). The contaminant will move as long as there is a concentration gradient, including when this gradient reverses, such as when fractures are flushed out by clean groundwater and the contaminant starts to move back from the invaded rock matrix into the fractures (the so-called back-diffusion).

The rate of diffusion for different chemicals (solutes) in water depends on the concentration gradient and the coefficient of diffusion, which is solute specific (different solutes have different coefficients of diffusion). The diffusion coefficients for electrolytes, such as major ions in groundwater (Na^+, K^+, Mg^{2+}, Ca^{2+}, Cl^-, HCO_3^-, and SO_4^{2-}) range between 1×10^{-9} and 2×10^{-9} m^2/s at 25°C (Robinson and Stokes, 1965; from Freeze and Cherry, 1979). Coefficient of diffusion is temperature dependent and decreases with the decreasing temperature (e.g., at 5°C these coefficients are about 50 percent smaller than at 25°C).

Flux (F) of a contaminant moving due to diffusion in a porous medium is described by Fick's first law:

$$F = -D_e \frac{\partial C}{\partial x} \tag{5.23}$$

The second Fick's law describes the change in concentration of a nonsorbing contaminant due to diffusion:

$$\frac{\partial C}{\partial t} = D_e \frac{\partial^2 C}{dx^2} \tag{5.24}$$

and if the contaminant is also subject to sorption as it moves through the porous media:

$$\frac{\partial C}{\partial t} = \frac{D_e}{R} \cdot \frac{\partial^2 C}{dx^2} \tag{5.25}$$

where D_e = effective coefficient of diffusion in the porous medium
R = coefficient of retardation (see next section on sorption and retardation)
C = contaminant concentration in groundwater

Because of tortuosity, effective diffusion coefficients in the subsurface are smaller than in free water. The effective diffusion coefficient (D_e) can be determined by using the known (experimentally determined) tortuosity of the porous media or by multiplying the aqueous diffusion coefficient (D_0) with an empirical coefficient, called apparent tortuosity factor (τ), which can range between 0 and 1. This empirical coefficient is related to the aqueous (D_0) and effective (D_e) diffusions, and the rock matrix porosity (θ_m) through the following expression (Parker et al.; from Pankow and Cherry, 1996):

$$\frac{D_e}{D_0} = \tau \cong \theta_m^p \tag{5.26}$$

where the exponent p varies between 1.3 and 5.4, depending on the type of porous geologic medium. Low porosity values result in small τ values and low D_e values. Laboratory studies of nonadsorbing solutes show that apparent tortuosity usually has values between 0.5 and 0.01. For example, for generic clay τ is estimated at 0.33, for shale/sandstone it is 0.10, and for granite it is quite small: 0.06 (Parker et al.; from Pankow and Cherry, 1996).

Concentration profile of a nonsorbing solute in subsurface, moving only due to diffusion and in one direction (x) from the high- into the zero-concentration layer, can be analytically calculated for various times (t) based on Fick's second law, using Crank's equation (Freeze and Cherry, 1979):

$$C_i(x, t) = C_0 \text{erfc} \cdot \left(\frac{x}{2} \cdot \sqrt{(D_e t)} \right) \tag{5.27}$$

where C_0 = initial concentration on the high-concentration side of the contact between two layers and erfc = complimentary error function.

5.5.5 Sorption and Retardation

Some critical processes that affect contaminant movement, without changing its chemical nature, are the result of various interactions between the three media: contaminant, water, and aquifer solids. Sorption is the general term that describes immobilization of the contaminant particles by the porous media, regardless of the actual mechanism. It may be the result of various more specific processes caused by geochemical interactions ("forces") between the solids and the dissolved contaminant. Cation exchange would be one example of sorption where the contaminant is immobilized by the mineral (usually clay) surfaces. This immobilization may not be permanent, and the contaminant may be released back into the water solution by the reverse process when geochemical conditions in the aquifer change (e.g., change of pH or inflow of another chemical species with the greater affinity for cation exchange with the mineral surfaces). Adsorption is a term often used to describe a process of contaminant particles (molecules) "sticking" to aquifer materials simply because of the affinity for each other. For example, many hydrophobic organic contaminants are adsorbed onto particles of organic carbon present in the aquifer and can be desorbed if conditions change. Adsorption is commonly used interchangeably with sorption, a more generic term, which sometimes may cause confusion. Absorption, a rather vague term, usually refers to contaminant incorporation "deep" into the solid particle structure, and it has chemical connotation. The term, however, is seldom used since its net effect would be equal to a complete destruction of the contaminant, i.e., its permanent removal from the flow system.

Sorption results in distribution of a solute between the solution (groundwater where it is dissolved) and the solid phase (where it is held by the solids of the aquifer). This distribution is called partitioning, and it is quantitatively described with the term distribution coefficient (or adsorption coefficient or partition coefficient), and denoted with K_d. Because of sorption, contaminant movement in groundwater is slower than the average groundwater velocity. This effect of sorption is called retardation, and the affinity of different solutes (chemicals dissolved in groundwater) to be retarded is quantified with a parameter called retardation factor, denoted by R. The overall effect of sorption is a decrease in dissolved contaminant concentration.

Adsorption/desorption will likely be the key process controlling contaminant migration in areas where chemical equilibrium exists, such as in areas far from the source. K_d is a generic term devoid of any particular mechanism and used to describe the general partitioning of aqueous-phase constituents to a solid phase due to sorption. In contrast, dissolution/precipitation is more likely to be the key process where there is chemical nonequilibrium, such as at a source, or in an area with high contaminant concentrations or with steep pH and redox gradients.

The distribution coefficient, K_d, is defined as the ratio of the contaminant concentration associated with the solid (C_s) to the contaminant concentration in the surrounding aqueous solution (C_w) when the system is at equilibrium:

$$K_d = \frac{C_s}{C_w} \tag{5.28}$$

Retardation due to sorption R is defined as follows:

$$R = \frac{v_w}{v_c} \tag{5.29}$$

where v_w = linear velocity of groundwater through a control volume and v_c = velocity of contaminant through a control volume. If the contaminant is sorbed (retarded), R will be greater than 1.

For the saturated groundwater flow in intergranular porous media, the retardation coefficient R is defined as follows (USEPA, 1999a):

$$R = 1 + \frac{\rho_b \cdot K_d}{n} \tag{5.30}$$

where ρ_b = bulk density of aquifer porous media (mass/length^3)
K_d = distribution coefficient (length^3/mass)
n = porosity of the media at saturation (dimensionless)

This equation can also be written as follows:

$$R = 1 + \left(\frac{1-n}{n}\right)\rho_s \times K_d \tag{5.31}$$

where ρ_s = particle density of porous media (mass/length^3), which is often assumed to be 2.65 g/cm^3 for most mineral soils (in the absence of actual site-specific information).

The relationship between the concentration of chemical sorbed onto solid surfaces (C_s) and the concentration remaining in aqueous solution (C_w) at equilibrium is referred to as the sorption isotherm because laboratory experiments for determining distribution coefficient values are performed at constant temperature. Sorption isotherms generally exhibit one of three characteristic shapes, depending on the sorption mechanism. These isotherms are referred to as the Langmuir isotherm, the Freundlich isotherm, and the linear isotherm, which is a special case of the Freundlich isotherm (Fig. 5.19).

The Langmuir isotherm model describes sorption in solute transport systems in which the sorbed concentration increases linearly with increasing solute concentration at low

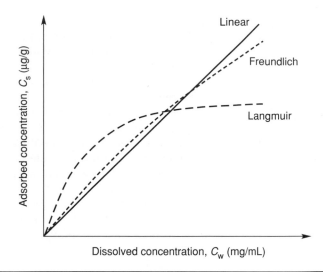

FIGURE 5.19 Sorption isotherms.

concentrations and approaches a constant value at high concentrations (experimental line flattens out). The sorbed concentration approaches a constant value because there are a limited number of sites on the aquifer matrix available for contaminant sorption. The Langmuir equation is described mathematically as follows (Devinny et al., 1990; from Wiedemeier et al., 1998):

$$C_s = \frac{KC_w b}{1 + KC_w} \qquad (5.32)$$

where C_s = sorbed contaminant concentration (mass contaminant/mass soil)
K = equilibrium constant for the sorption reaction ($\mu g/g$)
C_w = dissolved contaminant concentration ($\mu g/mL$)
b = maximum sorptive capacity of the solid surface

The Langmuir isotherm model is appropriate for highly specific sorption mechanisms where there are a limited number of sorption sites. This model predicts a rapid increase in the amount of sorbed contaminant as contaminant concentrations increase in a previously pristine area. As sorption sites become filled, the amount of sorbed contaminant reaches a maximum level equal to the number of sorption sites b.

The Freundlich isotherm is a modification of the Langmuir isotherm model in cases when the number of sorption sites is large (assumed infinite) relative to the number of contaminant molecules. This is generally a valid assumption for dilute solutions (e.g., downgradient from a petroleum hydrocarbon spill in the dissolved BTEX plume), where the number of unoccupied sorption sites is large relative to contaminant concentrations. The Freundlich isotherm is expressed mathematically as follows (Devinny et al., 1990; modified from Wiedemeier et al., 1998):

$$C_s = K_d C_w^n \qquad (5.33)$$

where K_d = distribution coefficient
$\quad\quad C_s$ = sorbed contaminant concentration (mass contaminant/mass soil, mg/g)
$\quad\quad C_w$ = dissolved concentration (mass contaminant/volume solution (mg/mL))
$\quad\quad n$ = chemical-specific coefficient

The value of n in this equation is a chemical-specific quantity that is determined experimentally. Values of n typically range from 0.7 to 1.1 but may be as low as 0.3 and as high as 1.7 (Lyman et al., 1992; from Wiedemeier et al., 1998).

The simplest expression of equilibrium sorption is the linear sorption isotherm, a special form of the Freundlich isotherm that occurs when the value of n is 1. The linear isotherm is valid for a dissolved species that is present at a concentration less than one-half of its solubility (Lyman et al., 1992). This is a valid assumption for BTEX compounds partitioning from fuel mixtures into groundwater. Dissolved BTEX concentrations resulting from this type of partitioning are significantly less than the pure compound's solubility in pure water. The linear sorption isotherm is expressed as follows (Jury et al., 1991; from Wiedemeier et al., 1998):

$$C_s = K_d C_w \tag{5.34}$$

where the notification is the same as in Eq. (5.33). Distribution coefficient K_d is the slope of the linear isotherm plotted using experimental laboratory data.

As discussed by the USEPA (1999a), soil and geochemists knowledgeable of sorption processes in natural environments have long known that generic or default K_d values can result in significant error when used to predict the absolute impacts of contaminant migration or site-remediation options. Therefore, for site-specific calculations, K_d values measured at site-specific conditions are absolutely essential. The general methods used to measure K_d values include the laboratory batch method, in situ batch method, laboratory flow-through (or column) method, field modeling method, and K_{oc} method. The advantages, disadvantages, and, perhaps more importantly, the underlying assumptions of each method are summarized in a two-volume reference text (USEPA, 1999a, 1999b), which also includes a conceptual overview of geochemical modeling calculations and computer codes as they pertain to evaluating K_d values and modeling of adsorption processes.

Sorption of Organic Solutes

Organic solutes tend to preferably adsorb onto organic carbon present in the aquifer porous media. This organic carbon in the soil is of various forms—as discrete solids, as films on individual soil grains, or as stringers of organic material in soil grains. Distribution coefficient (K_d) can be calculated for different organic solutes based on fraction soil organic carbon contents (f_{oc}) and partition coefficient with respect to the soil organic carbon (K_{oc}), using the following equation:

$$K_d = f_{oc} \cdot K_{oc} \tag{5.35}$$

For various reasons (including time needed and the expense), practitioners often forgo site-specific measurement of K_d and will instead calculate K_d using published K_{oc} values. However, as discussed by Gurr (2008), researchers and practitioners alike are far from united around standard K_{oc} values for nonpolar organics, even for well-studied

Source	log K_{oc} (mL/g)	K_{oc} (mg/L)	Retardation Coefficient	Time to Migrate 100 m (years)
USEPA, 2008	1.94	87	10.2	28
USEPA, 2008	2.18	150	16.9	46
Baker et al., 1997	1.81	65	7.8	21
SRC, 2007	2.02	104	12	33
Baker et al., 1997	2.16	145	16.4	45
Sabljic et al., 1995	1.95	90	10.6	29
Seth et al., 1999	1.46	29	4	11
Seth et al., 1999	2.06	116	13.3	36

The following conditions were assumed: linear groundwater velocity = 10 cm/day; effective porosity = 25%; soil bulk density = 2.5 g/cm^3; soil fraction organic carbon = 0.01.
From Gurr (2008); courtesy of Malcolm Pirnie, Inc.

TABLE 5.6   Example Variation in Subsurface Migration Period Due to Varying K_{oc} Values for Trichloroethylene

contaminants. It is easy to find multiple K_{oc} values for a single contaminant that vary by orders of magnitude. And when one considers that these values are often reported as log K_{oc}, small variations in the reported log value can translate into decades when the retardation factor is used to determine contaminant migration time periods. Table 5.6 illustrates how the variation of published log K_{oc} values for TCE translates into differing estimates of time to migrate 100 m in flowing groundwater.

The range in measured K_{oc} values is understandable since sorption to natural organic matter is an inherently complicated process. In addition to inevitable laboratory errors and use of nonstandard laboratory procedures, the differences in the measured K_{oc} values can be explained by natural organic carbon chemistry, which can vary substantially from one soil to the next. Nonpolar organics can also sorb to organic carbon still dissolved in the aqueous phase, and also to colloids suspended in the water. If these factors are not accounted for in the experimental method, the measured K_{oc} value may be skewed (Schwarzenbach et al., 2003).

While measuring K_{oc} values can be difficult and error prone, the simplicity of calculating partitioning from K_{oc} is attractive and easy to understand and, thus, is not likely to soon be replaced with another method. Given such diversity in quality and result, attempting to systematically choose reliable values from either published databases or calculated values is a much more responsible approach than choosing a number at random from published sources (Gurr, 2008).

The USEPA scientist Robert Boethling and his collaborators have provided guidance in a 2004 article titled *Finding and estimating chemical property data for environmental assessment* (Boethling et al., 2004). This paper provides a listing of online databases of physicochemical parameters, and a list of papers and resources for estimating soil/sediment sorption coefficients. Readers are cautioned, however, that many resources do not document the methods used to measure K_{oc} and that a reader must do his or her own due diligence on data quality. Boethling et al. recognize that each individual conducting his or her own review of data quality would be a waste of resources. The reader is instructed, instead, to use only the online databases or published compendiums where data evaluation has already been conducted.

The USEPA web page contains "technical fact sheets" for some—but not all—common groundwater contaminants that discuss a contaminant's environmental fate, including K_{oc} values (USEPA, 2008). Although the USEPA is the primary regulator for contaminated soil and groundwater, it does not appear that these values were published on their website after rigorous analysis (such as values for MCLs in drinking water). The presentation of values from one contaminant to another is uneven. A text description of the origins of the values is sometimes presented, other contaminants have a range of values, and others are called "estimated." The USEPA provides no citations of sources of the data, nor does it give indications that the data have been validated. Nonetheless, values published by the USEPA cannot be discounted outright and have to be given due consideration (Gurr, 2008).

Three online databases listed by Boethling et al. (2004) include two hosted by federal agencies, the Department of Agriculture and the National Institutes of Health (USDA, 2007; NIH, 2007), and one hosted by the Syracuse Research Corporation (SRC, 2007), a nonprofit research and development (R&D) organization that companies and government agencies use for outsourced R&D. The Department of Agriculture focuses on pesticides and only lists a range of K_{oc} values for a given contaminant. The primary focus of the National Institutes of Health database is toxicology, and environmental fate is included secondarily. It likewise only lists a range of values. It is not immediately apparent that the values in these databases have been validated. The database hosted by the Syracuse Research Corporation, CHEMFATE, is a quality-controlled database of physicochemical parameters. The description of the database on the Web site includes discussion on the data validation process.

As Boethling et al. mentioned, many papers do not adequately state the origin of their K_{oc} datasets. A search of literature by Gurr (2008) found one important departure from this trend—a paper published by Baker et al. (1997). These authors only accepted K_{oc} values if the investigators followed the American Society for Testing and Materials (ASTM) method for determining a sorption constant (ASTM, 2001). ASTM standards are widely used by environmental engineering and science professionals, making the Baker dataset particularly attractive.

Table 5.7 compiled by Gurr (2008) lists the Baker dataset and the CHEMFATE dataset side by side, with the USEPA values published on the agency's web page within technical fact sheets (with quotes from the text description of the USEPA values if available). Since the source and quality of the USEPA values are not listed (and the compilation cannot be discarded), the values must be scrutinized against the more reliable datasets of Baker et al. (preferred due to ASTM standard application) and CHEMFATE. Using the following set of criteria, Gurr selects the preferred K_{oc} values, listed in Table 5.8:

1. If the three datasets agree, use the agreed value.

2. If values vary to some extent, but not substantially, default to the USEPA value.

3. If USEPA presents a range or set of values, choose the value that is corroborated by the other sources (with preference to the Baker value).

4. If no USEPA value is listed, or is listed as "calculated" or "estimated", use the measured value from Baker or CHEMFATE.

5. If all the values vary substantially, do not recommend a value.

Analyte Name	EPA Range		EPA Text Description	Baker et al. log K_{oc}	CHEMFATE log K_{oc}
	Low Value log K_{oc}	High Value log K_{oc}			
Alachlor	2.08	2.28	log K_{oc} values for alachlor have largely been in the range 2.08–2.28	N/A	2.28
Atrazine	2.09	—	average K_{oc} value for 4 soils was determined to be 122	2.33	N/A
Endrin	4.53	—	estimated	N/A	4.06
Heptachlor	4.48	—	estimated	N/A	3.54
Lindane	3.03	—	a mean K_{oc} of 1080.9 was obtained from K_{oc} determinations on three soils	N/A	3.03
Chlordane	4.19	4.39	estimated	N/A	N/A
1,1,1-Trichloroethane	1.91	1.95	based upon experimental measurement, the mean K_{oc} range of 1,1,1-trichloroethane in a silty clay soil and sandy loam soil is 81–89	N/A	2.25
1,1,2-Trichloroethane	1.92	2.32	experimentally determined K_{oc} values of 83–209	N/A	1.90
1,1-Dichloroethylene	2.18	—	no experimental data is available on the adsorption of 1,1-dichloroethylene. A low K_{oc} of 150 is calculated from a regression equation	N/A	2.54
1,2,4-Trichlorobenzene	3.00	3.70	no details on origin of EPA values	4.02	3.16
1,2-Dibromo-3-chloropropane (DBCP)	1.66	2.11	observed	N/A	2.01
1,2-Dichloroethane	1.52		experimental K_{oc} of 33 for silt loam which in agreement with values calculated from the water solubility	N/A	1.51
Chlorobenzene	—	—	no values listed	N/A	2.44

TABLE 5.7 Log K_{oc} values from the USEPA (2008), Baker et al. (1997) and CHEMFATE Data Sets (*Continued*)

Analyte Name	EPA Range		EPA Text Description	Baker et al. log K_{oc}	CHEMFATE log K_{oc}
	Low Value log K_{oc}	High Value log K_{oc}			
Tetrachloroethylene (PCE)	2.32	2.38	reported and estimated K_{oc} (209–1685); also "K_{oc} = 210 (exp.) to 238 (est.)"	2.42	2.56
Trichloroethylene (TCE)	1.94	2.18	two silty clay loams (K_{oc} = 87 and 150)	1.81	2.02
cis-1,2-Dichloroethylene	1.56	1.69	estimated	N/A	1.70
trans-1,2-Dichloroethylene	1.56	1.69	estimated	N/A	1.54
Vinyl chloride	1.75	—	based on a reported water solubility of 2700 mg/L, a K_{oc} of 56 was estimated	N/A	1.47
Benzene	1.99	—	estimated	1.92	1.69
Toluene	1.57	2.25	reported K_{oc} values: Wendover silty loam, 37, Grimsby silt loam, 160, Vaudreil sandy loam, 46; sandy soil, 178; 100 and 151	2.06	1.98
Ethylbenzene	2.21	—	the measured K_{oc} for silt loam was 164	N/A	2.40
m-xylene	1.68	1.83	no details on origin of EPA values	2.22	2.28
o-Xylene	1.68	1.83	no details on origin of EPA values	2.11	2.11
p-xylene	1.68	1.83	no details on origin of EPA values	2.31	2.41

N/A, value was not present in the dataset; —, USEPA listed only a single value.
From Gurr (2008); courtesy of Malcolm Pirnie, Inc.

TABLE 5.7 Log K_{oc} values from the USEPA (2008), Baker et al. (1997) and CHEMFATE Data Sets (*Continued*)

Analyte Name	Recommended K_{oc}	Justification
Alachlor	2.28	High EPA is identical to CHEMFATE value
Atrazine	2.33	EPA value was average of a small dataset; see reliable Baker data
Endrin	No recommended value	Reliable values are not available
Heptachlor	No recommended value	Reliable values are not available
Lindane	3.03	EPA value is identical to CHEMFATE value
Chlordane	No recommended value	Reliable values are not available
1,1,1-Trichloroethane	No recommended value	Single reliable value is not available
1,1,2-Trichloroethane	1.92	EPA value is very close to CHEMFATE value
1,1-Dichloroethylene	No recommended value	Reliable values are not available
1,2,4-Trichlorobenzene	3.7	High EPA value is between Baker and CHEMFATE values
1,2-Dibromo-3-chloropropane (DBCP)	2.01	CHEMFATE value is within EPA range
1,2-Dichloroethane	1.52	EPA value is very close to CHEMFATE value
Chlorobenzene	2.44	No EPA value listed; use CHEMFATE value
Tetrachloroethylene	2.38	High EPA is close to Baker value
Trichloroethylene	1.94	Low EPA value is between Baker and CHEMFATE values
cis-1,2-Dichloroethylene	1.69	High EPA value is very close to CHEMFATE value
trans-1,2-Dichloroethylene	1.56	Low EPA value is near CHEMFATE value
Vinyl chloride	1.47	EPA value is estimated; use CHEMFATE value.
Benzene	1.99	EPA value is very close to Baker value
Toluene	2.06	Baker value is in the middle of the EPA range
Ethylbenzene	2.4	EPA value is based on a single measurement; use CHEMFATE value.
m-xylene	2.22	No details on EPA values; Baker and CHEMFATE are similar; use more reliable Baker values
o-Xylene	2.11	No details on EPA values; Baker and CHEMFATE are similar; use more reliable Baker values
p-xylene	2.31	No details on EPA values; Baker and CHEMFATE are similar; use more reliable Baker values

From Gurr, 2008; courtesy of Malcolm Pirnie, Inc.

TABLE 5.8 Recommended log K_{oc} Values (in mL/g)

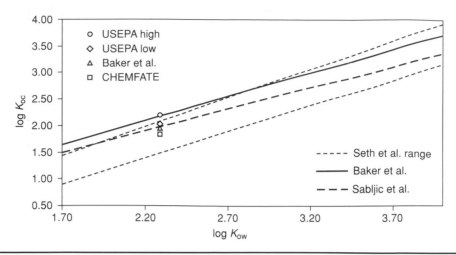

Figure 5.20   Calculated relationships between log K_{oc} and log K_{ow}, together with several reported values of K_{oc} for TCE. (From Gurr, 2008; courtesy of Malcolm Pirnie, Inc.)

The octanol-water partition coefficient (K_{ow}) is a much easier value to measure experimentally compared to K_{oc}. The simple process of allowing a chemical to reach equilibrium concentrations in a mixture of pure octanol and pure water was developed by the pharmaceutical industry to estimate the lipophilicity of pharmaceuticals. Environmental chemists have adopted this value to estimate hydrophobicity of organic contaminants (Schwarzenbach et al., 2003). And since hydrophobicity is also an important element in soil/water partitioning (K_d), many investigators have searched for a mathematical relationship between K_{ow} and K_{oc}.

To date, no investigator has proven that there is an exact relationship between K_{ow} and K_{oc}. However, several investigators have used experimentally derived data for log K_{ow} and log K_{oc} to construct linear empirical relationships that, while not exact for all compounds, do provide general estimates for nonpolar organic compounds (Sabljić et al., 1995; Baker et al., 1997; Seth et al., 1999).

Figure 5.20 is a plot of the relationships derived by Sabljić, Baker, and Seth for the log K_{ow} range of 1.7–4. Included in the figure are TCE data points from the USEPA, Baker, and CHEMFATE datasets. Note that even though K_{ow} is easier to measure than K_{oc}, different values still exist in the literature for the log K_{ow} of TCE. A value of 2.29 was assumed for plotting the relationships.

An accepted and scientifically rigorous relationship between a value that can be easily measured in the laboratory, K_{ow}, and the value that is key to estimation of soil/water partitioning at contaminated sites, K_{oc}, would be very useful. However, Fig. 5.20 demonstrates that such a precise relationship has not been established. For preliminary or comparative modeling, it is reasonable to use a recommended *measured* K_{oc} value, since the variability in K_{ow} values and K_{oc}/K_{ow} relationships adds a degree of uncertainty. It must be concluded that the primary value of K_{oc}/K_{ow} relationships is for estimating parameters for emerging contaminants that are not well known. Regulators and chemical manufacturers, especially, can use these relationships to evaluate environmental risk (Gurr, 2008).

5.5.6 Biodegradation

Soils and porous media in groundwater systems contain a large variety of microorganisms, ranging from simple prokaryotic bacteria and cyanobacteria to more complex eukaryotic algae, fungi, and protozoa. Over the past several decades, numerous laboratory and field studies have shown that microorganisms indigenous to the subsurface environment can degrade a variety of organic compounds, including components of gasoline, kerosene, diesel, jet fuel, chlorinated ethenes, chlorinated ethanes, the chlorobenzenes, and many other compounds (Wiedemeier et al., 1998). To obtain energy for growth and activity, under aerobic conditions (in the presence of molecular oxygen) many bacteria couple the oxidation of organic compounds (food) to the reduction of oxygen in the surrounding porous media. In the absence of oxygen (anaerobic conditions), microorganisms may use compounds other than oxygen as electron acceptors.

As discussed by Wiedemeier et al. (1998), biodegradation of organic compounds in groundwater occurs via three mechanisms:

1. Use of the organic compound as the primary growth substrate

2. Use of the organic compound as an electron acceptor

3. Cometabolism

The first two biodegradation mechanisms involve the microbial transfer of electrons from electron donors (primary growth substrate) to electron acceptors. This process can occur under aerobic or anaerobic conditions. Electron donors include natural organic material, fuel hydrocarbons, chlorobenzenes, and the less oxidized chlorinated ethenes and ethanes. Electron acceptors are elements or compounds that occur in relatively oxidized states. The most common naturally occurring electron acceptors in groundwater include dissolved oxygen, nitrate, manganese (IV), iron (III), sulfate, and carbon dioxide. In addition, the more oxidized chlorinated solvents such as PCE, TCE, DCE, TCA, DCA, and polychlorinated benzenes can act as electron acceptors under favorable conditions. Under aerobic conditions, dissolved oxygen is used as the terminal electron acceptor during aerobic respiration. Under anaerobic conditions, the electron acceptors listed above are used during denitrification, manganese (IV) reduction, iron (III) reduction, sulfate reduction, methanogenesis, or reductive dechlorination. Chapelle (1993) and Atlas (1984) discuss terminal electron-accepting processes in detail.

The third biodegradation mechanism is cometabolism. During cometabolism the compound being degraded does not benefit the organism. Instead, degradation is brought about by a fortuitous reaction wherein an enzyme produced during an unrelated reaction degrades the organic compound (Wiedemeier et al., 1998).

Fuel hydrocarbons are rapidly biodegraded when they are utilized as the primary electron donor for microbial metabolism under aerobic conditions. Biodegradation of fuel hydrocarbons occurs naturally when sufficient oxygen (or other electron acceptors) and nutrients are available in the groundwater. The rate of natural biodegradation is generally limited by the lack of oxygen or other electron acceptors rather than by the lack of nutrients such as nitrogen or phosphorus. The rate of natural aerobic biodegradation in unsaturated soil and shallow aquifers is largely dependent on the rate at which oxygen enters the contaminated media. Biodegradation of fuel hydrocarbons is discussed in detail by ASTM (1998), and Wiedemeier et al. (1999).

Bouwer et al. (1981) were the first to show that halogenated aliphatic hydrocarbons (such as chlorinated solvents PCE and TCE) could be biologically transformed under anaerobic conditions in the subsurface environment. Since that time, numerous investigators have shown that chlorinated compounds can degrade via reductive dechlorination under anaerobic conditions. Anaerobically, biodegradation of chlorinated solvents most often proceed through a process called reductive dechlorination. Highly chlorinated compounds such as PCE, TCE, or TCA are more oxidized and thus are less susceptible to oxidation. Therefore, they are more likely to undergo reductive reactions than oxidative reactions. The reductive dechlorination of PCE to TCE, and then TCE to DCE, and finally DCE to vinyl chloride is one of such reactions during which the halogenated hydrocarbon is used as an electron acceptor, not as a source of carbon, and a halogen atom is removed and replaced with a hydrogen atom (Fig. 5.21). Each step requires a lower redox potential than the previous one. PCE degradation occurs in a wide range of reducing conditions, whereas VC is reduced to ethene only under sulfate-reducing and methanogenic conditions. During each of these transformations, the parent compound (R-Cl) releases one chloride ion and gains one hydrogen. Two electrons are transferred during the process, which may provide a source of energy for the microorganism. This reductive anaerobic reduction of chlorinated solvents is expressed with the following general equation:

$$R - Cl + H^+ + 2e \rightarrow R - H + Cl^- \tag{5.36}$$

where R–Cl = chlorinated solvent structure.

Because chlorinated compounds are used as electron acceptors during reductive dechlorination, there must be an appropriate source of carbon for microbial growth in order for reductive dehalogenation to occur. Potential carbon sources can include low-molecular-weight organic compounds (lactate, acetate, methanol, and glucose), fuel hydrocarbons, by-products of fuel degradation (e.g., volatile fatty acids), or naturally occurring organic matter (Wiedemeier et al., 1998). Bioremediation technologies often involve injection of carbon sources ("food"), which stimulate the native microorganisms and accelerate biodegradation.

Reductive dechlorination processes result in the formation of intermediates that are more reduced than the parent compound. These intermediates are often more susceptible to oxidative bacterial metabolism than to further reductive anaerobic processes. For example, because of the relatively low oxidation state of VC, this compound more commonly undergoes aerobic biodegradation as a primary substrate than reductive dechlorination. For this reason, there may be accumulation of VC or DCE as a result of reductive dechlorination in some cases, as these cannot be further degraded in the absence of oxygen and specific microorganisms capable of a complete dechlorination. Bioaugmentation, which is a remedial technology of adding exogenous bacteria (i.e., bacteria not native to the contaminated aquifer), helps in preventing accumulation of the reductive dechlorination intermediates such as VC. Vinyl chloride is more toxic and more mobile in groundwater than the parent compounds and its accumulation is an undesired result of biodegradation.

As discussed by Wiedemeier et al. (1998), biodegradation causes measurable changes in groundwater chemistry. Table 5.9 summarizes these trends. During aerobic respiration, oxygen is reduced to water, and dissolved oxygen concentrations decrease. In anaerobic systems where nitrate is the electron acceptor, the nitrate is reduced to NO_2^-, N_2O, NO, NH_4^+, or N_2 via denitrification or dissimilatory nitrate reduction, and nitrate concentrations decrease. In anaerobic systems where iron (III) is the electron acceptor, it is reduced to iron (II) via iron (III) reduction, and iron (II) concentrations increase. In anaerobic

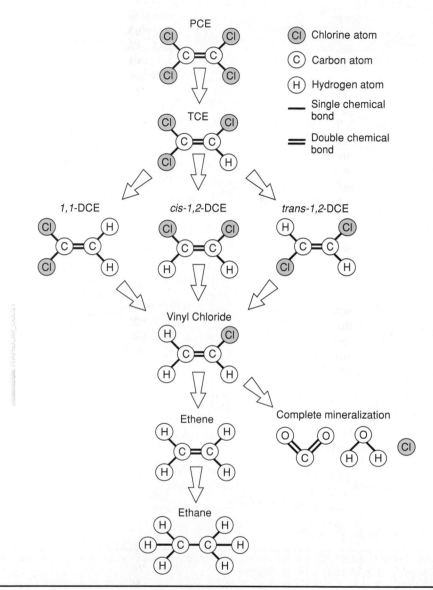

Figure 5.21 Reductive dehalogenation of chlorinated ethenes. (From Wiedemeier et al., 1998.)

systems where sulfate is the electron acceptor, it is reduced to H_2S via sulfate reduction, and sulfate concentrations decrease. During aerobic respiration, denitrification, iron (III) reduction, and sulfate reduction, total alkalinity will increase. In anaerobic systems where CO_2 is used as an electron acceptor, it is reduced by methanogenic bacteria during methanogenesis, and CH_4 is produced. In anaerobic systems where contaminants are being used as electron acceptors, they are reduced to less chlorinated daughter products; in such a system, parent compound concentrations will decrease and daughter product concentrations will increase at first and then decrease as the daughter product is used as

Analyte	Terminal Electron Accepting Process	Trend in Analyte Concentration During Biodegradation
Fuel hydrocarbons	Aerobic respiration, denitrification, manganese (IV) reduction, iron (III) reduction, methanogenesis	Decreases
Highly chlorinated solvents and daughter products	Reductive dechlorination	Parent compound concentration decreases, daughter products increase initially and then may decrease
Lightly chlorinated solvents	Aerobic respiration, denitrification, manganese (IV) reduction, iron (III) reduction (direct oxidation)	Compound concentration decreases
Dissolved oxygen	Aerobic respiration	Decreases
Nitrate	Denitrification	Decreases
Manganese (II)	Manganese (IV) reduction	Increases
Iron (II)	Iron (III) reduction	Increases
Sulfate	Sulfate reduction	Decreases
Methane	Methanogenesis	Increases
Chloride	Reductive dechlorination or direct oxidation of chlorinated compound	Increases
Oxidation-reduction potential	Aerobic respiration, denitrification, manganese (IV) reduction, iron (III) reduction, methanogenesis	Decreases
Alkalinity	Aerobic respiration, denitrification, iron (III) reduction, and sulfate reduction	Increases

From Wiedemeier et al., 1998.

Table 5.9 Trends in Contaminant, Electron Acceptor, Metabolic By-product and Total Alkalinity Concentrations During Biodegradation

an electron acceptor or is oxidized. As each subsequent electron acceptor is utilized, the groundwater becomes more reducing and the redox potential of the water decreases.

Lawrence (2006) provides a detailed literature review and discussion on various degradation mechanisms of VOCs commonly found in groundwater.

5.5.7 Analytical Equations of Contaminant Fate and Transport

General equation of contaminant fate and transport in one dimension (e.g., along horizontal X axis), known as advection-dispersion equation, is as follows:

$$\frac{\partial C}{\partial t} = \frac{D_x}{R}\frac{\partial^2 C}{\partial x^2} - \frac{v_x}{R}\frac{\partial C}{\partial x} \pm Q_s \tag{5.37}$$

where C = dissolved contaminant concentration (kg/m^3, or mg/L)
t = time (day)
D_x = hydrodynamic dispersion in x direction (m^2/d)
R = retardation coefficient (dimensionless)
x = distance from the source along X axis (m)
v_x = linear groundwater velocity in X direction (m/d)
Q_s = general term for source or sink of contaminant, such as due to biodegradation (kg/m^3/d)

This term can also be expressed using the first rate degradation constant, λ (1/d), which gives the following:

$$\frac{\partial C}{\partial t} = \frac{D_x}{R}\frac{\partial^2 C}{\partial x^2} - \frac{v_x}{R}\frac{\partial C}{\partial x} - \lambda C \tag{5.38}$$

Eq. (5.37) does not have an explicit solution, and approximate solutions, based on simplifying assumptions, have been proposed by various authors.

One of the most popular analytical solutions of the advection-dispersion equation is the Domenico (1987) solution. This is an approximate three-dimensional solution that describes the fate and transport of a decaying contaminant plume evolving from a finite planar source. This solution was based on an approach previously published by Domenico and Robbins (1985) for modeling a nondecaying contaminant plume. Prior to this work, several authors presented exact solutions to the same or similar problems (Cleary and Ungs, 1978; Sagar, 1982; Wexler, 1992). However, these solutions are not closed form expressions since they involve numerical evaluation of a definite integral. This numerical integration step can be computationally demanding and can also introduce numerical errors (Srinivasan et al., 2007). The key advantage of the Domenico and Robbins (1985) approach is that it provides a closed-form solution without involving numerical integration procedures. Due to this computational advantage, the Domenico solution has been widely used in several public domain design tools, including the USEPA tools BIOCHLOR and BIOSCREEN (Newell et al. 1996; Aziz et al., 2000).

The analytical Domenico and Robbins (1985) solution for concentration of a semi-infinite contaminated parcel, which moves in a homogeneous aquifer with a one-dimensional velocity in the positive x direction away from the continuous finite source (Fig. 5.22), including three-dimensional dispersion, and no degradation, has the following form:

$$c(x, y, z, t) = \frac{c_0}{8}\text{erfc}\left[\frac{x - vt}{2(D_x t)^{1/2}}\right] \times \left\{\text{erf}\left[\frac{y + \frac{Y}{2}}{2(D_y x/v)^{1/2}}\right] - \text{erf}\left[\frac{y - \frac{Y}{2}}{2(D_y x/v)^{1/2}}\right]\right\}$$

$$\times \left\{\text{erf}\left[\frac{z + \frac{Z}{2}}{2(D_z x/v)^{1/2}}\right] - \text{ert}\left[\frac{z - \frac{Z}{2}}{2(D_z x/v)^{1/2}}\right]\right\} \tag{5.39}$$

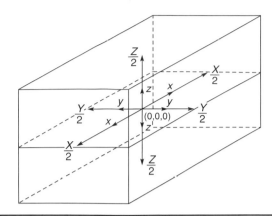

FIGURE 5.22 Parallelepiped source in the Domenico-Robbins analytical solution of the one-dimensional contaminant transport with three-dimensional dispersion. (From Domenico and Robbins, 1985; copyright Groundwater Journal; printed with permission.)

where

c = the concentration in time t at the location with coordinates x, y, z

c_o = initial concentration at the source

erf and erfc = error function and complimentary error function, respectively

v = groundwater (advection) velocity in the direction of flow (x direction)

D_x, D_y, and D_z = dispersion coefficients in x, y, and z directions (longitudinal, transverse, and vertical), respectively

X, Y, and Z = source dimensions (see Fig. 5.22)

In the 1987 solution, Domenico included a first-order decay term (k) leading to the following approximate equation for concentration of a decaying (degrading) contaminant:

$$c(x, y, z, t) = \frac{c_o}{8} f_x(x, t) f_y(y, x) f_z(z, x)$$

where

$$f_x(x, t) = \exp\left\{\frac{x}{2\alpha_x}\left[1 - \left(1 + \frac{4k\alpha_x}{v}\right)^{1/2}\right]\right\} \times \operatorname{erfc}\left\{\frac{x - vt\left(1 + \frac{4k\alpha_x}{v}\right)^{1/2}}{2(\alpha_x vt)^{1/2}}\right\}$$

$$f_y(y, x) = \left\{\operatorname{erf}\left[\frac{y + \frac{Y}{2}}{2(\alpha_y x)^{1/2}}\right] - \operatorname{erf}\left[\frac{y - \frac{Y}{2}}{2(\alpha_y x)^{1/2}}\right]\right\}$$

$$f_z(z, x) = \left\{\operatorname{erf}\left[\frac{z + \frac{Z}{2}}{2(\alpha_z x)^{1/2}}\right] - \operatorname{erf}\left[\frac{z - \frac{Z}{2}}{2(\alpha_z x)^{1/2}}\right]\right\} \tag{5.40}$$

where $\alpha_x = D_x/v$, $\alpha_y = D_y/v$, and $\alpha_z = D_z/v$ are the dispersivities in the x, y, and z directions, respectively.

Although the Domenico solution has been extensively used in the industry and several widely used analytical groundwater transport models are based on it, its approximate nature continues to be a subject of scientific debate. For example, West and Kueper (2004) compared the BIOCHLOR model against a more rigorous analytical solution and concluded that the Domenico solution can produce errors up to 50 percent. Guyonnet and Neville (2004) compared the Domenico solution against the Sagar (1982) solution and presented the results in a nondimensional form. They concluded that for groundwater flow regimes dominated by advection and mechanical dispersion, discrepancies between the two solutions can be considered negligible along the plume centerline. However, these errors may increase significantly outside the plume centerline. Based on a rigorous mathematical analysis, Srinivasan et al. (2007) conclude that the approximate Domenico solution can be expected to produce reasonable estimates for advection-dominated problems; however, it can introduce significant errors for longitudinal dispersion-dominated problems. Within the advective front, the longitudinal dispersivity plays a very important role in determining the accuracy of the solution. The key assumption used to derive the Domenico solution is the time reinterpretation step, where the time t in the transverse dispersion terms is replaced with x/v. This substitution process is valid only when the longitudinal dispersivity is 0. For all nonzero longitudinal dispersivity values, the solution will have a finite error. The spatial distribution of this error is highly sensitive to the value of α_x and the position of the advective front (vt) and is relatively less sensitive to other transport parameters. The authors conclude that the error in the Domenico solution will be low when solving transport problems that have low longitudinal dispersivity values, high advection velocities, and large simulation times.

After their analysis of the approximate Domenico solution, West et al. (2007) conclude that its accuracy is highly variable and dependent on the selection of input parameters. For solute transport in a medium-grained sand aquifer, the Domenico (1987) solution underpredicts solute concentrations along the centerline of the plume by as much as 80 percent, depending on the case of interest. Increasing the dispersivity, time, or dimensionality of the system leads to increased error. Because more accurate exact analytical solutions exist, the authors suggest that the Domenico (1987) solution and its predecessor and successor approximate solutions need not be employed as the basis for screening tools at contaminated sites (West et al., 2007).

Karanovic et al. (2007) present an enhanced version of BIOSCREEN that supplements the Domenico (1987) solution with an exact analytical solution for the contaminant concentration. The exact solution is derived for the same conceptual model as Domenico (1987) but without invoking approximations in its evaluation that introduce errors of unknown magnitude in the analysis. The exact analytical solution is integrated seamlessly within a modified interface BIOSCREEN-AT. The Excel user interface for BIOSCREEN-AT is nearly identical to that for BIOSCREEN, and a user familiar with BIOSCREEN will have no difficulty using BIOSCREEN-AT. BIOSCREEN-AT provides a simple and direct way to calculate an exact solution to the transport equation and, if desired, to assess the significance of the errors introduced by the Domenico (1987) solution for site-specific applications.

The analytical models of fate and transport can be used for simple screening-level analyses, since they assume simple planar geometry of the source zones and homogeneous isotropic aquifers. Numeric models are irreplaceable for actual field problems where all parameters of groundwater flow, and contaminant fate and transport change in all three dimensions.

References

Aiken, G.R., 2002. Organic matter in ground water. In: Aiken, G.R., and Kuniansky, E.L., editors. U.S. Geological Survey Open-File Report 02-89, pp. 21–23.

Alekin, O.A., 1953. *Osnovi gidrohemii* [Principles of hydrochemistry, in Russian]. Gidrometeoizdat, Leningrad.

Alekin, O.A., 1962. *Grundlages der wasserchemie. Eine einführung in die chemie natürlicher wasser.* VEB Deutsch. Verl., Leipzig, 260 p. (originally published in Russian in 1953).

Anderson, M.P., 1979. Using models to simulate the movements of contaminants through groundwater flow systems. *Critical Reviews in Environmental Controls*, vol. 9. no. 2, pp. 97–156.

Anderson, M.P., 1983. Movement of contaminants in groundwater transport: Groundwater transport—advection and dispersion. In: *Groundwater contamination*. Studies in Geophysics, National Research Council, National Academy Press, Washington, DC, pp. 37–45.

Anderson, M.E., and Sobsey, M.D., 2006. Detection and occurrence of antimicrobially resistant *E. coli* in groundwater on or near swine farms in eastern North Carolina. *Water Science and Technology*, vol. 54, pp. 211–218.

Appelo, C.A.J., and Postma, D., 2005. *Geochemistry, Groundwater and Pollution*, 2nd ed. Taylor & Francis/Balkema, Leiden, the Netherlands, 649 p.

Aurelius, L.A., 1989. *Testing for Pesticide Residues in Texas Well Water*. Texas Department of Agriculture, Austin, TX.

ASTM (American Society for Testing and Materials), 1998. *Standard Guide for Remediation of Ground Water by Natural Attenuation at Petroleum Release Sites*. E 1943-98, West Conshohocken, PA.

ASTM, 2001. *ASTM E 1195-01, Standard Test Method for Determining a Sorption Constant (K_{oc}) for an Organic Chemical in Soil and Sediments*. American Society for Testing and Materials, West Conshohocken, PA.

Atlas, R.M., 1984. *Microbiology—fundamentals and applications.* Macmillan, New York, 880 p.

Aziz, C.E., Newell, C.J., Gonzales, J.R., Haas, P., Clement, T.P., and Sun, Y., 2000. *BIOCHLOR: Natural Attenuation Decision Support System v. 1.0; User's Manual.* EPA/600/R-00/008. U.S. Environmental Protection Agency, Cincinnati, OH.

Baker, J.R., Mihelcic, J.R., Luehrs, D.C., and Hickey, J.P., 1997. Evaluation of estimation methods for organic carbon normalized sorption coefficients. *Water Environment Research*, vol. 69, pp. 1703–1715.

Banks, W.S.L., and Battigelli, D.A., 2002. Occurrence and distribution of microbiological contamination and enteric viruses in shallow ground water in Baltimore and Harford Counties, Maryland. U.S. Geological Survey Water-Resources Investigations Report 01-4216, Baltimore, MD, 39 p.

Barrett, M.H., Hiscock, K.M., Pedley, S., Lerner, D.N., Tellam, J.H., and French, M.J., 1999. Marker species for identifying urban groundwater recharge sources: a review and case study in Nottingham, UK. *Water Resources Research*, vol. 33, pp. 3083–3097.

Barlow, P.M., 2003. Ground water in freshwater-saltwater environments of the Atlantic coast. U.S. Geological Survey Circular 1262, Reston, VA, 113 p.

Benson, N.R., 1976. Retardation of apple tree growth by soil arsenic residues from old insecticideal treatments. *Journal of the American Society of Horticultural Science*, vol. 101, no. 3, pp. 251–253.

BGS (British Geological Survey), 1999. Denitrification in the unsaturated zones of the British Chalk and Sherwood Sandstone aquifers. Technical Report WD/99/2, British Geological Survey, Keyworth.

Boethling, R.S., Howard, P.H., and Meylan, W.M, 2004. Finding and estimating chemical property data for environmental assessment. *Environmental Toxicology And Chemistry*, vol. 23, pp. 2290–2308.

Bourg, A.C.M., and Loch, J.P.G., 1995. Mobilization of heavy metals as affected by pH and redox conditions. In: *Biogeodynamics of Pollutants in Soils and Sediments; Risk Assessment of Delayed and Non-linear Responses*, Chapter 4. Salomons, W., and Stigliani, W.M., editors. Springer, Berlin, pp. 815–822.

Bouwer, E.J., Rittman, B.E., and McCarty, P.L., 1981. Anaerobic degradation of halogenated 1- and 2-carbon organic compounds. *Environmental Science & Technology*, vol. 15, no. 5, pp. 596–599.

Bölke, J.K., and Denver, J.M., 1995. Combined use of groundwater dating, chemical, and isotopic analyses to resolve the history and fate of nitrate contamination in two agricultural watersheds, Atlantic coastal plain, Maryland. *Water Resources Research*, vol. 31, pp. 2319–2339.

Brown, C.J., Colabufo, S., and Coates, D., 2002. Aquifer geochemistry and effects of pumping on ground-water quality at the Green Belt Parkway Well Field, Holbrook, Long Island, New York. U.S. Geological Survey Water-Resources Investigations Report 01-4025, Coram, New York, 21 p.

Brusseau, M., 1991. Transport of organic chemicals by gas advection in structured or heterogeneous porous media: Development of a model and application of column experiments. *Water Resources Research*, vol. 27, no. 12, pp. 3189–3199.

Buss, S.R., Rivett, M.O., Morgan, P., and Bemment, C.D., 2005. Attenuation of nitrate in the sub-surface environment. Science Report SC030155/SR2, Environment Agency, Bristol, England, 100 p.

Carson, R., 2002. *Silent Spring*. Mariner Books/Houghton Mifflin Company, New York, 378 p.

Chapelle, F.H., 1993. *Ground-Water Microbiology and Geochemistry*. John Wiley & Sons, New York, 424 p.

Cleary, R., and Ungs, M.J., 1978. Analytical models for groundwater pollution and hydrology. Report No. 78-WR-15. Water Resources Program, Princeton University, Princeton, NJ.

Cohen, R.M., and Mercer, J.W., 1993. *DNAPL Site Evaluation*. C.K. Smoley, Boca Raton, FL.

Colborn, T., Dumanoski, D., and Meyers, J.-P., 1997. *Our Stolen Future; Are We Threatening Our Fertility, Intelligence and Survival? A Scientific Detective Story*. Plume, Penguin Group, New York, 336 p.

Craun, G.F., 1986. *Waterborne Diseases in the United States*. CRC Press, Boca Raton, FL, 192 p.

Dagan, G., 1982. Stochastic modeling of ground water flow by unconditional and conditional probabilities, 2, The solute transport. *Water Resources Research*, vol. 18, no. 4, pp. 835–848.

Dagan, G., 1984. Solute transport in heterogeneous porous formations. *Journal of Fluid Mechanics*, vol. 145, pp. 151–177.

Dagan, G., 1986. Statistical theory of ground water flow and transport: Pore to laboratory, laboratory to formation, and formation to regional scale. *Water Resources Research*, vol. 22, no. 9, pp. 120S–134S.

Dagan, G., 1988. Time-dependent macrodispersion for solute transport in anisotropic heterogeneous aquifers. *Water Resources Research*, vol. 24, no. 9, pp. 1491–1500.

Daughton, Pharmaceuticals and Personal Care Products (PPCPs) as environmental pollutants. National Exposure Research Laboratory, Office of Research and Development, Environmental Protection Agency, Las Vegas, Nevada. Presentation available at: http://www.epa.gov/nerlesd1/chemistry/pharma/index.htm. Accessed May 2007.

Davenport, J.R., and Peryea, F.J., 1991. Phosphate fertilizers influence leaching of lead and arsenic in a soil contaminated with lead and arsenic in a soil contaminated with lead arsenate. *Water, Air and Soil Pollution*, vol. 57–58, pp. 101–110.

Davis, S.N., and J.M. DeWiest, 1991. *Hydrogeology*. Krieger Publishing Company, Malabar, FL, 463 p.

Devinny, J.S., et al., 1990. *Subsurface Migration of Hazardous Wastes*. Van Nostrand Reinhold, New York, 387 p.

Domenico, P.A. 1987. An analytical model for multidimensional transport of a decaying contaminant species. *Journal of Hydrology*, vol. 91, no. 1–2, pp. 49–58.

Domenico, P.A., and Robbins, G.A., 1985. A new method of contaminant plume analysis. *Ground Water*, vol. 23, no. 4, pp. 476–485.

Domenico, P.A., and Schwartz, F.W., 1990. *Physical and Chemical Hydrogeology*. John Willey & Sons, New York, 824 p.

Dowdle, P.R., Laverman, A.M., and Oremland, R.S., 1996. Bacterial dissimilatory reduction of arsenic (V) to arsenic (III) in anoxic sediments. *Applied Environmental Microbiology*, vol. 62, no. 5, pp. 1664–1669.

Embrey, S.S., and Runkle, D.L., 2006. Microbial quality of the Nation's ground-water resources, 1993–2004. U.S. Geological Survey Scientific-Investigations Report 2006-5290, Reston, VA, 34 p.

Falta, R.W., et al., 1989. Density-drive flow of gas in the unsaturated zone due to evaporation of volatile organic chemicals, *Water Resources Research*, vol. 25, no. 10, pp. 2159–2169.

Feenstra, S., and Cherry, J.A., 1988. Subsurface contamination by dense non-aqueous phase liquid (DNAPL) chemicals. In: Proceedings: International Groundwater Symposium, International Association of Hydrogeologists, May 1–4, 1988, Halifax, Nova Scotia, pp. 62–69.

Feth, J.H., 1966. Nitrogen compounds in natural waters—a review. *Water Resources Research*, vol. 2, pp. 41–58.

Feth, J.H., Rogers, S.M., and Roberson, C.E., 1961. Aqua de Ney, California, a spring of unique chemical character. *Geochimica et Cosmochimica Acta*, vol. 22, pp. 75–86.

Florida Department of Environmental Protection, 2003a. Average annual daily flow by county-2000. http://www.dep.state.fl.us/water/uic/docs/2002v_ADF.pdf. Accessed February 10, 2006.

Florida Department of Environmental Protection, 2003b. Class I injection well status. http://www.dep.state.fl.us/water/uic/docs/Class_I_Table11_2003.pdf. Accessed February 10, 2006.

Focazio, M.J., 2001. Occurrence of selected radionuclides in ground water used for drinking water in the United States: A targeted reconnaissance survey, 1998. U.S. Water-Resources Investigations Report 00-4273, Reston, VA, 40 p.

Foster, S.S.D., 2000. Assessing and controlling the impacts of agriculture on groundwater – from barley barons to beef bans. *Quarterly Journal of Engineering Geology and Hydrogeology*, vol. 33, pp. 263–280.

Foster, S., Garduño, H., Kemper, K., Tiunhof, A., Nanni, M., and Dumars, C., 2002–2005. Groundwater quality protection; defining strategy and setting priorities. Sustainable Groundwater Management; Concepts & Tools, Briefing Note Series Note 8, The Global Water Partnership, The World Bank, Washington, DC, 6 p. Available at: www.worldbank.org/gwmate.

Franke, O.L., Reilly, T.E., Haefner, R.J., and Simmons, D.L., 1990. Study guide for a beginning course in ground-water hydrology: Part 1–course participants. U.S. Geological Survey Open File Report 90-183, Reston, VA, 184 p.

Freeze, R.A., and Cherry, J.A., 1979. *Groundwater*. Prentice-Hall, Englewood Cliffs, NJ, 604 p.

Freyberg, D.L., 1986. A natural gradient experiment on solute transport in a sand aquifer, (2) spatial moments and the advection and dispersion of nonreactive tracers. *Water Resources Research*, vol. 22, no. 13, pp. 2031–2046.

Fukada, T., Hiscock, K.M., and Dennis, P.F., 2004. A dual-isotope approach to the nitrogen hydrochemistry of an urban aquifer. *Applied Geochemistry*, vol. 19, pp. 709–719.

Fussell, D.R., et al., 1981. Revised inland oil spill clean-up manual. CONCAWE Report No. 7/81, Management of manufactured gas plant sites, GRI-87/0260, Gas Research Institute, Den Haag, 150 p.

Garabedian, S.P., LeBlanc, D.R., Gelhar, L.W., and Celia, M.A., 1991. Large-scale natural gradient tracer test in sand and gravel, Cape Cod, Massachusetts, 2, analysis of spatial moments for a nonreactive tracer. *Water Resources Research*, vol. 27, no. 5, pp. 911–924.

Gelhar, L. W., and Axness, C.L., 1983. Three-dimensional stochastic analysis of macrodispersion in aquifers. *Water Resources Research*, vol. 19, no. 1, pp. 161–180.

Gelhar, L.W., Welty, C., and Rehfeldt, K.R., 1992. A critical review of data on field-scale dispersion in aquifers. *Water Resources Research*, vol. 28, no. 7, pp. 1955–1974.

Gerba, C.P., 1999. Virus survival and transport in groundwater. *Journal of Industrial Microbiology and Biotechnology*, vol. 22, no. 4, pp. 247–251.

Gilliom, R.J., et al., 2006. Pesticides in the Nation's streams and ground water, 1992–2001. U.S. Geological Survey Circular 1291, Reston, VA, 172 p.

Gurr, C., 2008. Recommended K_{oc} values for nonpolar organic contaminants in groundwater. Groundwater Modeling and Geostatistics Knowledge Team White Paper, Malcolm Pirnie, Inc., Arlington, VA, 12 p.

Guyonnet, D., and Neville, C., 2004. Dimensionless analysis of two analytical solutions for 3-D solute transport in groundwater. *Journal of Contaminant Hydrology*, vol. 75, no. 1, pp. 141–153.

Harden, S.L., Fine, J.M., and Spruill, T.B., 2003. Hydrogeology and ground-water quality of Brunswick County, North Carolina. U.S. Geological Survey Water-Resources Investigations Report 03-4051, Raleigh, NC, 92 p.

Harper, C., and Liccione, J.J., 1995, Toxicological profile for gasoline. U.S. Department of Health and Human Services, Public Health Service, Agency for Toxic Substances and Disease Registry, p. 196 + appendices. Available at: http://www.atsdr.cdc.gov/toxprofiles/tp72.pdf.

Harrington, J.M., Fendorf, S.B., and Rosenzweig, R.F., 1998. Biotic generation of arsenic (III) in metal(loid)-contaminated sediments. *Environmental Science and Technology*, vol. 32, no. 16, pp. 2425–2430.

Heathwaite, A.L., Johnes, P.J., and Peters, N.E., 1996. Trends in nutrients. *Hydrological Processes*, vol. 10, pp. 263–293.

Heaton, T.H.E., 1986. Isotopic studies of nitrogen pollution in the hydrosphere and atmosphere – A review. *Chemical Geology*, vol. 59, pp. 87–102.

Hem, J.D., 1989. Study and interpretation of the chemical characteristics of natural water; 3rd ed. U.S. Geological Survey Water-Supply Paper 2254, Washington, DC, 263 p.

Hinrichsen, D., Robey, B., and Upadhyay, U.D., 1997. Solutions for a water-short world. Population reports. Series M, No. 14., Johns Hopkins School of Public Health, Population Information Program, Baltimore, MD. Available at: http://www.infoforhealth.org/pr/m14edsum.shtml.

Jeyanayagam, S., 2008. *Microconstituents in Wastewater Treatment – The Current State of Knowledge*. Professional Engineer Continuing Education Series. Malcolm Pirnie, Columbus, OH.

Jury, W.A., Gardner, W.R., and Gardner, W.H. 1991. *Soil Physics*. John Wiley & Sons, New York, 328 p.

Karanovic, M., Neville, C.J., and Andrews, C.B., 2007. BIOSCREEN-AT: BIOSCREEN with an exact analytical solution. *Ground Water*, vol. 45, no. 2, pp. 242–245.

Kauffman, W.J., and Orlob, G.T., 1956. Measuring ground water movement with radioactive and chemical tracers. *American Water Works Association Journal*, vol. 48, pp. 559–572.

Keeney, D., 1990. Sources of nitrate to ground water. *Critical Reviews in Environmental Control*, vol. 16, pp. 257–304.

Knox, R.C., Sabatini, D.A., and Canter, L.W., 1993. *Subsurface Transport and Fate Processes*. Lewis Publishers. Boca Raton, FL. 430 p.

Kreitler, C.W., and Jones, D.C., 1975. Natural soil nitrate – The cause of nitrate contamination of ground water in Runnels County, Texas. *Ground Water*, vol. 13, no. 1, pp. 53–61.

Kresic, N., 2007. *Hydrogeology and Groundwater modeling*. CRC Press, Taylor & Francis Group, Boca Raton, FL, 807 p.

Krumholz, L.R., 2000. Microbial communities in the deep subsurface. *Hydrogeology Journal*, vol. 8, no. 1, pp. 4–10.

Lawrence, S.J., 2006. Description, properties, and degradation of selected volatile organic compounds detected in ground water—a review of selected literature. U. S. Geological Survey, Open-File Report 2006-1338, Reston, VA, 62 p. A web-only publication at: http://pubs.usgs.gov/ofr/2006/1338/.

LeBlanc, D.R., Garabedian, S.P., Hess, K.M., Gelhar, L.W., Quadri, R.D., Stollenwerk, K.G., and Wood, W.W., 1991. Large-scale natural gradient tracer test in sand and gravel, Cape Cod, Massachusetts, 1, experimental design and observed tracer movement. *Water Resources Research*, vol. 27, no. 5, pp. 895–910.

Lee, F.F., and Jones-Lee, A., 1993. Public health significance of waterborne pathogens. Report to California Environmental Protection Agency Comparative Risk Project, 22 p.

Lopes, T.J., and Furlong, E.T., 2001. Occurrence and potential adverse effects of semivolatile organic compounds in streambed sediment, United States, 1992–1995. *Environmental Toxicology and Chemistry*, vol. 20, no. 4, pp. 727–737.

Lyman, W.J., Reehl, W.F., and Rosenblatt, D.H., 1982. *Handbook of Chemical Property Estimation Methods, Environmental Behavior of Organic Compounds*. McGraw-Hill, New York.

Lyman, W.J., Reidy, P.J., and Levy, B., 1992. *Mobility and Degradation of Organic Contaminants in Subsurface Environments*. C.K. Smoley, Chelsea, MI, 395 p.

Mackay, D.M., Freyberg, D.L., Roberts, P.V., and Cherry, J.A., 1986. A natural gradient experiment on solute transport in a sand aquifer, (1) approach and overview of plume movement. *Water Resources Research*, vol. 22, no. 13, pp. 2017–2029.

Maliva, R.G., Guo, W., and Missimer, T., 2007. Vertical migration of municipal wastewater in deep injection well systems, South Florida, USA. *Hydrogeology Journal*, vol. 15, pp. 1387–1396.

Masters, R.W., Verstraeten, I.M., and Heberer, T., 2004. Fate and transport of pharmaceuticals and endocrine disrupting compounds during ground water recharge. *Ground Water Monitoring & Remediation*, Special Issue: Fate and transport of pharmaceuticals and endocrine disrupting compounds during ground water recharge, pp. 54–57.

Matheron, G., and de Marsily, G., 1980. Is transport in porous media always diffusive? A counterexample. *Water Resources Research*, vol. 16, pp. 901–907.

Matthess, G., 1982. *The Properties of Groundwater*. John Willey & Sons, New York, 406 p.

Mehlman, M.A., 1990. Dangerous properties of petroleum-refining products: Carcinogenicity of motor fuels (gasoline). *Teratogenesis, Carcinogenesis, and Mutagenesis*, vol. 10, pp. 399–408.

Metz, P.A., and Brendle, D.L., 1996, Potential for water quality degradation of interconnected aquifers in westcentral Florida. U.S. Geological Survey Water-Resources Investigations Report 96–4030, 54 p.

Molz, F.J., Güven, O., and Melville, J.G., 1983. An examination of scale-dependent dispersion coefficients. *Ground Water*, vol. 21, no. 6, pp. 715–725.

Mueller, D.K., et al., 1995. Nutrients in ground water and surface water of the United States–An analysis of data through 1992. U.S. Geological Survey Water Resources Investigations Report 95-4031, 1995.

National Institutes of Health (NIH), 2007. TOXNET Toxicology Data Network. Available at: http://toxnet.nlm.nih.gov/. Accessed December 2007.

National Research Council, 1978. *Nitrates: An Environmental Assessment*. National Academy of Sciences, Washington, DC, 723 p.

Neuman, S.P., 1990. Universal scaling of hydraulic conductivities and dispersivities in geologic media. *Water Resources Research*, vol. 26, no. 8, pp. 1749–1578.

Neumann, I., Brown, S., Smedley, P., Besien, T., Lawrence, A.R., Hargreaves, R., Milodowski, A.E., and Barron, M., 2003. Baseline Report Series: 7. The Great Inferior Oolite of the Cotswolds District. British Geological Survey and Environment Agency, Keyworth, Nottingham, 62 p.

Newell, C.J., et al., 1995. *Light Nonaqueous Phase Liquids. Ground Water Issue*, EPA/540/S-95/500. Robert S. Kerr Environmental Research Laboratory, Ada, OK, 28 p.

Newell, C.J., McLeod, R.K., and Gonzales, J.R., 1996. *BIOSCREEN: Natural Attenuation Decision Support System User's Manual*, EPA/600/R-96/087. Robert S. Kerr Environmental Research Center, Ada, OK.

Nowell, L., and Capel, P., 2003. Semivolatile organic compounds (SVOC) in bed sediment from United States rivers and streams: Summary statistics; preliminary results from Cycle I of the National Water Quality Assessment Program (NAWQA),

1992–2001. Provisional data – subject to revision. Available at: http://ca.water.usgs.gov/pnsp/svoc/SVOC-SED_2001_Text.html.

Palmer, C.D., and Johnson, R.L., 1989. Physical processes controlling the transport of non-aqueous phase liquids in the subsurface. In: USEPA, Seminar Publication; Transport and fate of contaminant in the subsurface, EPA/625/4-89/019, U.S. Environmental Protection Agency, pp. 23–27.

Pankow, J.F., and Cherry, J.A., 1996. *Dense Chlorinated Solvents and Other DNAPLs in Groundwater*. Waterloo Press, Guelph, ON, Canada, 522 p.

Parkhurst, D.L., and Appelo, C.A.J., 1999. User's guide to PHREEQC (Version 2)—a computer program for speciation, batch-reaction, one-dimensional transport, and inverse geochemical calculations. U.S. Geological Survey Water-Resources Investigations Report 99–4259, 310 p.

Peryea, F.J., 1991. Phosphate-induced release of arsenic from soils contaminated with lead arsenate. *Soil Science Society of America Journal*, vol. 55, pp. 1301–1306.

Peryea, F.J., and Kammereck, R., 1997. Phosphate-enhanced movement of arsenic out of lead arsenate-contaminated topsoil and through uncontaminated subsoil. *Water, Air and Soil Pollution*, vol. 93, no. 1–4, pp. 243–254.

Petrusevski, B., Sharma, S., Schippers, J.C., and Shordt, K., 2007. Arsenic in drinking water. IRC International Water and Sanitation Centre, Thematic Overview Paper 17, Delft, the Netherlands, 57 p.

Pickens, J.F., and Grisak, G.E., 1981a. Scale-dependent dispersion in a stratified granular aquifer. *Water Resources Research*, vol. 17, no. 4, pp. 1191–1211.

Pickens, J.F., and Grisak, G.E., 1981b. Modeling of scale-dependent dispersion in hydrogeologic systems. *Water Resources Research*, vol. 17, no. 4, pp. 1701–1711.

Piper, A.M., 1944. A graphic procedure in the geochemical interpretation of water analyses. *American Geophysical Union Transactions*, vol. 25, pp. 914–923.

Porter, WP, Jaeger, J.W., and Carlson, I.H., 1999. Endocrine, immune and behavioral effects of aldicarb (carbamate), atrazine (triazine) and nitrate (fertilizer) mixtures at groundwater concentrations. *Toxicology and Industrial Health*, vol. 15, pp. 133–150.

Puckett, L.J., 1994. Nonpoint and point sources of nitrogen in major watersheds of the United States. U.S. Geological Survey Water-Resources Investigations Report 94-4001, 9 p.

Rees, T.F., et al., 1995. Geohydrology, water quality, and nitrogen geochemistry in the saturated and unsaturated zones beneath various land uses, Riverside and San Bernardino Counties, California, 1991–93. U.S. Geological Survey Water-Resources Investigations Report 94-4127, Sacramento, CA, 267 p.

Reneau, R., Hagedorn, C., and Degen, M., Jr., 1989. Fate and transport of biological and inorganic contaminants from on-site disposal of domestic wastewater. *Journal of Environmental Quality*, vol. 18, pp. 135–144.

RGS (Royal Geographic Society), 2007. Arsenic in drinking water a global threat to health. Media release, Embargoed, 12:50pm, Wednesday, 29 August 2007.

Rice, K.C., Monti, M.M., and Etting, M.R., 2005. Water-quality data from ground- and surface-water sites near concentrated animal feeding operations (CAFOs) and non-CAFOs in the Shenandoah Valley and Eastern Shore of Virginia, January–February, 2004. U.S. Geological Survey Open-File Report 2005–1388, Reston, VA, 78 p.

Robinson, R.A., and Stokes, R.H., 1965. *Electrolyte Solutions*, 2nd ed. Butterworth, London.

Rosenfeld, J.K., and Plumb, R.H., Jr., 1991. Groundwater contamination at wood treatment facilities. *Ground Water Monitoring Review*, vol. 11, no. 1, pp. 133–140.

Sabljić, A., Gusten, H., Verhaar, H., and Hermens, J., 1995. QSAR modeling of soil sorption, improvements, and systematics of log K_{oc} vs. log K_{ow} correlations. *Chemosphere*, vol. 31, pp. 4489–4514.

Sagar, B., 1982. Dispersion in three dimensions: Approximate analytical solutions. *ASCE Journal of Hydraulic Division*, vol. 108, no. HY1, pp. 47–62.

Sapkota, A.R., Curriero, F.C., Gibson, K.E., and Schwab, K.J., 2007. Antibiotic-resistant enterococci and fecal indicators in Surface water and groundwater impacted by a concentrated swine feeding operation. *Environmental Health Perspectives*, ehponline.org, National Institute of Environmental Health Sciences, 33 p. doi:10.1289/ehp.9770. Available at: http://dx.doi.org/. Accessed April 2007.

Sayah R.S., Kaneene, J.B., Johnson, Y., and Miller, R. 2005. Patterns of antimicrobial resistance observed in *Escherichia coli* isolates obtained from domestic- and wild-animal fecal samples, human septage, and surface water. *Applied and Environmental Microbiology*, vol. 71, pp. 1394–1404.

Schwarzenbach, R.P., Gschwend, P.M., and Imboden, D.M., 2003. *Environmental Organic Chemistry*, 2nd ed. Wiley Interscience, Hoboken, NJ.

Seiler, R.L., 1996. Methods for identifying sources of nitrogen contamination of ground water in valleys in Washoe County, Nevada. U.S. Geological Survey Open-File Report 96-461, Carson City, NV, 20 p.

Seth, R., Mackay, D., and Muncke, J., 1999. Estimating the organic carbon partition coefficient and its variability for hydrophobic chemicals. *Environmental Science and Technology*, vol. 33, pp. 2390–2394.

Sleep, B.E., and Sykes, J.F., 1989. Modeling the transport of volatile organics in variably saturated media. *Water Resources Research*, vol. 25, no. 1, pp. 81–92.

Sloto, R.A., 1994. Geology, hydrology, and ground-water quality of Chester County, Pennsylvania. Chester County Water Resources Authority Water-Resource Report 2, 127 p.

Smith, L., and Schwartz, F.W., 1980. Mass transport, 1, A stochastic analysis of macroscopic dispersion. *Water Resources Research*, vol. 16, pp. 303–313.

Smith, A.H., Lingas, E.O., and Rahman, M., 2000. Contamination of drinking-water by arsenic in Bangladesh: A public health emergency. *Bulletin of the World Health Organization*, vol. 78, no. 9, pp. 1093–1103.

Southwest Florida Water Management District, 2002. Artesian well plugging annual work plan 2002: Brooksville, Southwest Florida Water Management District Quality of Water Improvement Program, 86 p.

Spalding, R.F., and Exner, M.E., 1993. Occurrence of nitrate in groundwater – a review. *Journal of Environmental Quality*, vol. 22, pp. 392–402.

Srinivasan, V., Clement, T.P., and Lee, K.K., 2007. Domenico solution–is it valid? *Ground Water*, vol. 45, no. 2, pp. 136–146.

Sudicky, E.A., Cherry, J.A., and Frind, E.O., 1983. Migration of contaminants in ground water at a landfill: A case study, 4, a natural gradient dispersion test. *Journal of Hydrology*, vol. 63, no. 1/2, pp. 81–108.

Swoboda-Colberg, N.G., 1995. Chemical contamination of the environment—sources, types, and fate of synthetic organic chemicals. In: *Microbial Transformations and Degradation of Toxic Organic Chemicals*. Young, L.Y., and Cerniglia, C.E., editors. John Wiley & Sons, New York, pp. 27–74.

Syracuse Research Corporation, 2007. CHEMFATE. Available at database: http://www.syrres.com/esc. Accessed December 2007.

Tiemann, M., 1996. 91041: Safe Drinking Water Act: Implementation and Reauthorization. National Council for Science and the Environment, Congressional Research Service Reports. http://ncseonline.org/nle/crsreports. Accessed January 21, 2006.

Toccalino, P.L., 2007. Development and application of health-based screening levels for use in water-quality assessments. U.S. Geological Survey Scientific Investigations Report 2007–5106, Reston, VA, 12 p.

Toccalino, P.L., Rowe, B.L., and Norman, J.E., 2006. Volatile organic compounds in the Nation's drinking-water supply wells—what findings may mean to human health. U.S. Geological Survey Fact Sheet 2006-3043, 4 p.

Toccalino, P.L., et al., 2003. Development of health-based screening levels for use in state- or local-scale water-quality assessments. U.S. Geological Survey Water-Resources Investigations Report 03-4054, 22 p.

UNESCO, 1998. Soil and groundwater pollution from agricultural activities. Learning material. International Hydrological Programme, IHP-V Technical Documents in Hydrology No. 19, Project 3.5, Paris, France.

UNICEF, 2006. Arsenic migration in Bangladesh Fact Sheet. Available at: http://www.unicef.org/Bangladesh/Arsenic.pdf. Accessed February 2007.

USDA (U.S. Department of Agriculture), 2006a. Farms, land in farms, and livestock operations: 2005 summary. Sp Sy 4 (06). Agricultural Statistics Board, NASS, USDA, Washington, DC. Available at: http://usda.mannlib.cornell.edu/MannUsda/viewDocumentInfo.do?documentID=1259. Accessed September 13, 2006.

USDA (U.S. Department of Agriculture), 2006b. Meat animals production, disposition, and income: 2005 summary. Mt An 1-1 (06). Agricultural Statistics Board, NASS, USDA, Washington, DC. Available at: http://usda.mannlib.cornell.edu/MannUsda/viewDocumentInfo.do?documentID=1101. Accessed September 13, 2006.

USDA, 2007. Pesticide properties database. Available at: http://www.ars.usda.gov/services/docs.htm?docid=14199. Accessed December 2007.

USEPA (United States Environmental Protection Agency), 1981. Radioactivity in drinking water. Glossary. EPA570/9-81-002, U. S. Environmental Protection Agency, Health Effects Branch, Criteria and Standards Division, Office of Drinking Water, Washington, DC.

USEPA, 1983. Surface impoundment assessment national report. EPA-570/9-84-002, Office of Drinking Water, Washington, DC.

USEPA, 1990. Handbook: Ground Water, Volume I: Ground Water and Contamination. EPA/625/6-90/016a.

USEPA, 1992. Estimating potential for occurrence of DNAPL at Superfund sites. Publication 9355.4-D7FS, R.S. Kerr Environmental Research Laboratory, Office of Solid Waste and Emergency Response, U.S. Environmental Protection Agency, 9 p.

USEPA, 1997. Treatment technology performance and cost data for remediation of wood preserving sites. Washington, DC, USA, US EPA Office of Research and Development. Available at: http://www.epa.gov/nrmrl/pubs/625r97009/625r97009.pdf.

USEPA, 1999a. Understanding variation in partition coefficient, Kd, values; Volume I: The Kd model, methods of measurements, and application of chemical reaction codes. EPA 402-R-99-004A, U.S. Environmental Protection Agency, Office of Air and Radiation, Washington, DC.

USEPA, 1999b. Understanding variation in partition coefficient, Kd, values; Volume II: Review of geochemistry and available Kd values for cadmium, cesium, chromium, lead, plutonium, radon, strontium, thorium, tritium (^{3}H), and uranium. EPA 402-R-99-004B, U.S. Environmental Protection Agency, Office of Air and Radiation, Washington, DC.

USEPA, 2000a. National Water Quality Inventory; 1998 Report to Congress; Ground water and drinking water chapters. EPA 816-R-00-013, Office of Water, Washington, DC.

USEPA, 2000b. Radionuclides notice on data availability: Technical support document. Targeting and Analysis Branch, Standards and Risk Management Division, Office of Ground Water and Drinking Water. Available at: www.epa.gov/safewater/rads/tsd.pdf.

USEPA, 2002. Technical program overview: underground injection control regulations, revised July 2001. EPA 816-R-02-025, Office of Water, Washington, DC.

USEPA, 2003a. Overview of the Clean Water Act and the Safe Drinking Water Act. Available at: http://www.epa.gov/OGWDW/dwa/electronic/ematerials/. Accessed September 2005.

USEPA, 2003b. An overview of the Safe Water Drinking Act. Available at: http://www.epa.gov/OGWDW/dwa/electronic/ematerials/. Accessed September 2005.

USEPA, 2003c. Regulating microbial contamination; unique challenge, unique approach. Available at: http://www.epa.gov/OGWDW/dwa/electronic/ematerials/. Accessed September 2005.

USEPA, 2005a. Guidelines for carcinogen risk assessment. EPA/630/P-03/001B, Risk Assessment Forum, Washington, DC. Available at: http://cfpub.epa.gov/ncea/raf/recordisplay.cfm?deid=116283.

USEPA, 2005b. Available at: http://www.clu-in.org/contaminantfocus/default.focus/sec/arsenic/. Accessed May 2006.

USEPA, 2005c. Fact sheet: The drinking water contaminant candidate list – the source of priority contaminants for the Drinking Water Program. Office of Water, 6 p. Available at: http://www.epa.gov/safewater/ccl/ccl2_list.html.

USEPA, 2006. Setting standards for safe drinking water. Office of Water, Office of Ground Water and Drinking Water, Updated November 28, 2006. Available at: http://www.epa.gov/safewater/standard/setting.html. Accessed February 2007.

USEPA, 2008. Technical fact sheets for drinking water contaminants. Available at: http://www.epa.gov/OGWDW/hfacts.html. Accessed January 2008.

USGS (United States Geological Survey), 2004. Natural remediation of arsenic contaminated ground water associated with landfill leachate. U.S. Geological Survey Fact Sheet 2004-3057, 4 p.

Van der Perk, M., 2006. *Soil and Water Contamination from Molecular to Catchment Scale.* Taylor & Francis/Balkema, Leiden, the Netherlands, 389 p.

Welch, A.H., et al., 2000. Arsenic in ground water of the United States–occurrence and geochemistry. *Ground Water*, vol. 38 no. 4, pp. 589–604.

West, M., and Kueper, B.H., 2004. Natural attenuation of solute plumes in bedded fractured rock. In: Proceedings of USEPA/NGWA Fractured Rock Conference, National Ground Water Association, Portland, Maine, pp. 388–401.

West, M.R., Kueper, B.H., and Ungs, M.J., 2007. On the use and error of approximation in the Domenico (1987) solution. *Ground Water*, vol. 45, no. 2, pp. 126–135.

Wexler, E.J., 1992. *Analytical Solutions for One-, Two-, and Three-Dimensional Solute Transport in Ground-Water Systems with Uniform Flow,* U.S. Geological Survey TWRI Book 3, Chapter B7. Reston, VA.

Widory, D., Kloppmann, W., Chery, L., Bonninn, J., Rochdi, H., and Guinamant, J.-L., 2004. Nitrate in groundwater: An isotopic multi-tracer approach. *Journal of Contaminant Hydrology*, vol. 72, pp. 165–188.

Wiedemeier, T.H., et al., 1998. Technical Protocol for Evaluating Natural Attenuation of Chlorinated Solvents in Ground Water. EPA/600/R-98/128, U.S. Environmental Protection Agency, Office of Research and Development, Washington, DC.

Wiedemeier, T.H., et al., 1999. *Technical Protocol for Implementing Intrinsic Remediation with Long-Term Monitoring for Natural Attenuation of Fuel Contamination Dissolved in Ground-water*, Vol. I (Revision 0). Air Force Center for Environmental Excellence (AFCEE), Technology Transfer Division, Brooks Air Force Base, San Antonio, TX.

Wilhelm, S.R., Schiff, S.L., and Cherry, J.A., 1994. Biogeochemical evolution of domestic waste water in septic systems, Pt. 1, Conceptual model. *Ground Water*, vol. 32, no. 6, pp. 905–916.

Wikipedia, 2007. Endocrine disruptor. Available at: http://en.wikipedia.org/wiki/Endocrine_disruptor. Accessed December 2007.

Winter, T.C., Harvey, J.W., Franke, O.L., and Alley, W.M., 1998. Ground water and surface water: A single resource. U.S. Geological Survey Circular 1139, Denver, Colorado, 79 p.

Woolsen, E.A., Axley, J.H., and Kearney, P.C., 1971. The chemistry and phytotoxicity of arsenic in soils: 1. Contaminated field soils. *Soil Science Society America Proceedings*, vol. 35, no. 6, pp. 938–943.

Woolsen, E.A., Axley, J.H., and Kearney, P.C., 1973. The chemistry and phytotoxicity of arsenic in soils: II. Effects of time and phosphorous. *Soil Science Society America Proceedings*, vol. 37, no. 2, pp. 254–259.

Xu, M., and Eckstein, Y., 1995. Use of weighted least-squares method in evaluation of the relationship between dispersivity and field scale. *Ground Water*, vol. 33, no. 6, pp. 905–908.

Zogorski, J.S., Carter, J.M., Ivahnenko, T., Lapham, W.W., Moran, M.J., Rowe, B.L., Squillace, P.J., and Toccalino, P.L., 2006. The quality of our Nation's waters—Volatile organic compounds in the Nation's ground water and drinking-water supply wells. U.S. Geological Survey Circular 1292, Reston, VA, 101 p.

Zwirnmann, K.H., 1982. Nonpoint nitrate pollution of municipal water supply sources: Issues of analysis and control. IIASA Collaborative Proceedings Series CP-82-S4, International Institute for Applied Systems Analysis, Laxenburg, Austria, 303 p.

CHAPTER 6

Groundwater Treatment

Alessandro Franchi, PhD, PE *Metcalf & Eddy, Inc., Orange, California*

6.1 Introduction

Compared to surface water sources, many groundwater supplies are characterized by reduced seasonal variability and microbial counts, lower turbidity, and lower concentration of synthetic organic substances. Because of this, and its normally high overall quality, groundwater is often preferred to surface water as drinking water source. When it does not contain excessive mineral concentrations or contaminants, groundwater may be suitable for direct pumping to the distribution system and consumption, without prior treatment. However, this depends on specific regulations, typically on a country (federal) level. For example, in the United States, the Ground Water Rule, published in 2006 (US Federal Register, 2006), requires only those groundwater systems that are identified as being at risk of fecal contamination to disinfect water. "At-risk" supplies are identified by using sanitary surveys and periodic monitoring of fecal indicator organisms conducted, on a frequency determined by the size of the system, under the Total Coliform Rule (US Federal Register, 1989). Secondary disinfection (i.e., maintaining chlorine residual through the distribution system) is, instead, mandatory, regardless of the initial quality of raw water. Often the treatment of groundwater is limited to the removal of inorganic contaminants such as iron and manganese and disinfection (see Fig. 6.1 for a typical schematic for this type of system). This, by no means, implies that groundwater cannot be contaminated, as discussed in Chap. 5. When contaminants are found above the concentrations specified by the regulatory agencies or tolerated by the public, treatment of groundwater is required before consumption.

Groundwater treatment processes are designed to remove a variety of natural and anthropogenic contaminants. Efficacy, capital, and operation and maintenance (O&M) costs, and owner's preference determine the selection of a specific treatment technology. Individual process units are arranged in a "treatment train" to achieve the removal of multiple contaminants from the water. At the municipal scale, commonly used processes include some form of oxidation, coagulation/clarification process, filtration, and disinfection (a flow schematic for a typical conventional groundwater treatment process is presented in Fig. 6.2).

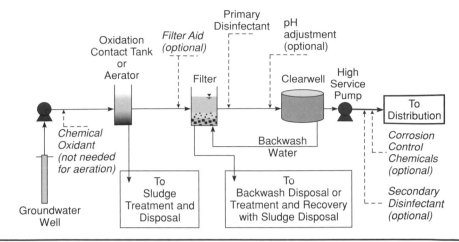

FIGURE 6.1 Basic groundwater treatment plant with iron and manganese oxidation and removal.

In cases where an individual household or a small number of connections is served by an individual well, it is often more economical to install a treatment system directly at the location where water is consumed. Point of entry (POE) and point of use (POU) devices are designed for this particular use (USEPA, 2006a). POE devices treat all water entering a single home, business, school, or facility, while POU devices treat only the water from a particular faucet. Typically, POE/POU devices rely on the same treatment technologies used by centralized treatment plants. However, it should be noted that POU/POE systems are designed to function at much higher specific flow rate than those

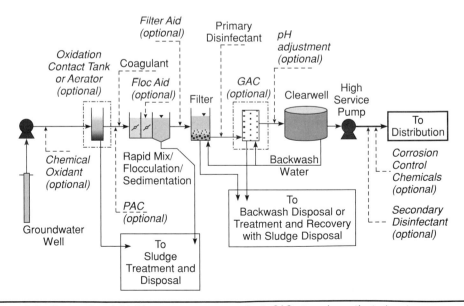

FIGURE 6.2 Conventional drinking-water treatment plant. GAC: granular activated carbon.

for municipal systems. For example, flow rates for ion-exchange resins up to 70 m^3/m^{2}h (approximately 30 gpm/ft^2) are used in POU/POE versus 2.5 to 12 m^3/m^{2}h (approximately 1 to 5 gpm/ft^2) in municipal systems. POU units are typically installed under the kitchen sink, providing purified water for drinking and cooking, while other water faucets deliver untreated water for washing and cleaning. This configuration provides treatments only where needed thus helping to contain the O&M costs of the treatment. POE units are instead typically installed to treat all the water that is entering a single home, building, or facility.

Municipal systems that supply water to larger communities, small systems treating water only for a few users, and individual well owners use a variety of technologies to remove contaminants from groundwater water. The most commonly used of these technologies are described in the following sections.

6.2 Oxidation

In groundwater treatment, oxidation is typically implemented at the plant headworks. The selection of oxidant type, dose, and reaction time depends on the type of contaminant to be removed and overall raw water quality conditions.

The simplest method of oxidation is aeration and is often used for oxidation of iron, manganese, and arsenic and the removal of hydrogen sulfide and volatile organic compound (VOC). Aeration is a physical process aimed at transferring oxygen from air to water, decreasing carbon dioxide dissolved in water and stripping volatile compounds from water. The efficiency of the exchange process is typically enhanced by providing a large surface area. Aeration systems may be classified into four general categories: waterfall aerators, diffusion or bubble aerators, mechanical aerators, and pressure aerators (AWWA and ASCE, 1998). For oxidation of groundwater, waterfall aerators are often used. The more common types include spray aerators, multiple-tray aerators, cascade aerators, cone aerators, and packed columns. The selection of the specific type of aerator system depends on several factors, such as required transfer efficiency, footprint, and cost.

Chemical oxidants used in groundwater treatment are chlorine, permanganate, chlorine dioxide, and ozone. Table 6.1 presents the standard potentials of common chemical oxidants used in water treatment.

Many of these oxidants are also used as disinfectant and can be added both at the plant headworks to provide oxidation and disinfection or/and downstream in the treatment process for disinfection purposes (further information on disinfection is provided later in this chapter).

Traditionally, chlorine has been the most common oxidant used by small groundwater systems because, in addition to its effectiveness, it requires limited equipment and capital investment both in the gaseous (Fig. 6.3) and in the liquid forms (Fig. 6.4), and relatively low O&M costs. However, stricter standards for and concerns over disinfection by-products (DBPs), which are formed by the reaction of chlorine with natural organic matter (NOM), have caused many plants to discontinue the practice of prechlorination, i.e., the addition of chlorine at the plant headwork (USEPA, 1999a). Many water treatment plants have switched to alternative oxidants. Permanganate is widely used, especially by systems with high concentrations of manganese, which is more difficult to oxidize than iron. One problem posed by permanganate is that the dosage has to be carefully targeted.

Oxidizing Compound	Standard Half-Cell Potential (V)
Ozone	2.08
Hydrogen peroxide	1.78
Hypochlorite[1]	1.64
Permanganate	1.68
Chlorine dioxide[2]	0.95
Hypochlorous acid[1]	1.48
Monochloramine	1.40
Chlorine gas[1]	1.36
Dichloramine	1.34

[1] Chlorine is available as chlorine gas and as dry and aqueous hypochlorite. Chlorine gas hydrolyzes rapidly in water to form hypochlorous acid (HOCl) (White, 1999).
[2] The oxidation potential of chlorine dioxide may be misguiding, as it is a very effective oxidant due to its selective (only reacts with certain compounds) characteristics (White, 1999).
Source: Singer and Reckhow, 1999; Stumm and Morgan, 1996.

TABLE 6.1   Oxidation Potential of Common Water Treatment Oxidants

An insufficient dosage will not provide adequate oxidation, while overfeeding can lead to pink water entering the distribution system. Pink water does not pose a health hazard but can cause customer concern and complaints. Furthermore, overfeeding can cause operational problems in conventional filters by contributing to the formation of mudballs (spherical conglomerate of sand and silt, which form in the top layer of granular filters due to inadequate backwashing).

Ozone is the strongest oxidant available in water treatment (see Table 6.1). It is very reactive and must be produced on site. It does not form chlorinated DBPs although, depending on the quality of the source water, may form other DBPs. Because of the relatively high capital and O&M costs, ozonation is more frequently used by larger systems (Singer and Reckhow, 1999; Kawamura, 2000).

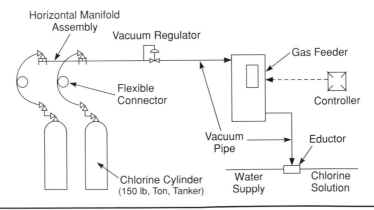

FIGURE 6.3   Gaseous chlorine feed system. (From USEPA, 1999a.)

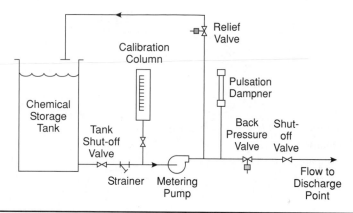

Figure 6.4 Hypochlorite feed system. (From USEPA, 1999a.)

Chlorine dioxide is another very strong oxidant that in recent years has gained more attention because it does not form chlorinated DBPs and capital and operational costs are lower than those for ozone. Like ozone, it is highly reactive and must be produced on site. Dosage of chlorine dioxide must be carefully calibrated because residual concentrations greater than 0.4 mg/L in the water delivered to consumers cause taste and odor problems (USEPA, 1999a; Singer and Reckhow, 1999).

Further details on the oxidation of specific contaminants are given later in this chapter.

6.3 Clarification

Following oxidation, a clarification step is commonly used to promote the aggregation and settling of suspended and dissolved constituents such as clay, silt, finely divided inorganic and organic matter, soluble organic compounds, algae, and microscopic organisms including microbes. Conventional clarification comprises three steps. First is the addition of coagulant or a pH-adjustment chemical in the flash mixer to destabilize (reduction of the electrostatic repulsion) particles and, therefore, promote their agglomeration. Second is the flocculation step where water is gently mixed to produce larger and heavier "flocs" (agglomerated particles). The third step is settling in which heavier particles and the flocs formed during the previous steps settle by gravity at the bottom of a tank (see Fig. 6.5 for an illustration of a circular radial flow clarifier). Extensive reviews of this process unit are found in Gregory et al. (1999) and Letterman et al. (1999).

Variations on the clarification process include (1) high rate clarification, which involves using smaller basins and higher surface loading rates than conventional clarifiers, and (2) dissolved air flotation, which uses rising air bubbles to float flocs to the surface of a tank where the material is skimmed off (LeChevallier and Au, 2004; Gregory et al., 1999).

The most commonly used coagulants are alum (aluminum hydroxide), iron salts, and polyaluminum chloride (PACl). Coagulant dosages vary, based on several factors including water pH, alkalinity, turbidity, quality of suspended solids (size and negative charge), concentration of TOC (a measure of the concentration of organic substances), and water temperature. Typical dosages for sources with low turbidity (less than 2 NTU), low TOC (less than 2.5 mg/L), low alkalinity (less than 30 mg/L), and average temperature around 15°C range between 5 and 15 mg/L for ferric salts; 10 to 25 mg/L for alum; and 2 to

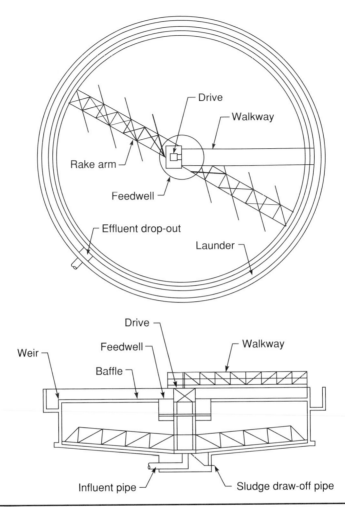

Figure 6.5 Circular radial flow clarifier. (AWWA and ASCE, 1998; McGraw-Hill, printed with permission.)

8 mg/L for PACl. Significantly larger dosages (>30 mg/L for ferric salts and >50 mg/L for alum) are normally required for groundwaters with higher TOC and lower temperatures. Various types of polymer additives with different molecular weights and charge may also be added to improve particle removal and filterability at the flash mix or flocculation step, or as sole coagulant in place of metallic coagulant. The selection of the appropriate polymer is based on water quality, point of addition, and treatment objectives.

As alternative to coagulation/flocculation, chemical precipitation through softening achieved through the addition of lime or caustic (sodium hydroxide) to increase pH can be used in the clarification step for removing particulate matter if hardness removal is one of the treatment objectives.

The performance of the clarification process is dependent on many factors, including physical design of the facility, concentration and type of particulate to be removed, type and dosage of coagulant, pH, ionic strength, temperature, and concentration of NOM. Turbidity is an expression of the optical property that causes light to be scattered and absorbed by particles and molecules rather than transmitted through water (USEPA, 1999a). Although not a true measurement of particle concentration, turbidity is used to measure the performance of the clarification process. High turbidity is an indication of potential for drinking water contamination with microbial agents because suspended particles can "shelter" microbes from disinfectants (USEPA, 1999b, 1999c). Typically, a turbidity of less than 2 NTU is often considered a minimal target for the clarification effluent.

Finally, the treatment of many groundwater supplies may not require the flocculation and sedimentation steps. For high-quality source waters, such as those characterized by low turbidity, in-line coagulation can be implemented by directly adding coagulant to the raw water pipeline before direct filtration. The addition of the coagulant improves the filterability of particles, thus improving filter performance (LeChevallier and Au, 2004).

6.4   Filtration

Following clarification, a filtration step is often used to remove remaining particulates. In most treatment plants, filtration is the final step to accomplish removal of suspended solids. Filtration consists of passing water through a bed of granular filter media such as sand, anthracite, or other filtering material. Most of the suspended matter in the influent water is retained onto the media. Filtered water should be clear with turbidity below 0.2 NTU (Binnie et al., 2002). Media size, filtration rates, effectiveness of the clarification system, and filter-aid polymer addition greatly affect filter performance. In the United States, the USEPA mandates (as of January 1, 2002) that the turbidity of the filter effluent combined for all filters in a drinking water plant may never exceed 1 NTU, and must not exceed 0.3 NTU in 95 percent of daily samples in any month (USEPA, 2006b).

6.4.1   Rapid Sand Filters

Rapid sand filters are the most common type of granular filter used in municipal water treatment. They consist of concrete boxes filled with one or more layers of porous media such as sand and anthracite (see Fig. 6.6). In some cases, a layer of granular activated carbon (GAC) may also be placed in filter to adsorb chemicals and DBP precursors dissolved in water. Typical loading rates for these filters are 5 to 12 m^3/m^2h (approximately 2 to 5 gpm/ft^2) of filter bed surface area. Water enters the filter from above the media and flows by gravity downward through the filter media. At the bottom of the filter unit water it is collected in the underdrain system, where it is removed from the filter. The filter is cleaned through the "backwash" process, which consists of reversing the flow of water through the filter, to remove the solids accumulated on the media surface and within the media bed. Air can be added to the backwash water to improve scouring of the solids. At the end of the backwash cycle, backwash water and the solids it contains are removed from the filter with a series of collection troughs. Rapid sand filtration requires advanced operator training and skills, particularly for starting and conducting backwash operations and bringing back filters on line without affecting the quality of finished water. Extensive monitoring is required during operation of media filtration units.

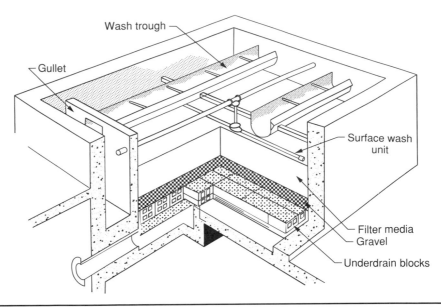

Wash trough

Gullet

Surface wash unit

Filter media
Gravel

Underdrain blocks

FIGURE 6.6 Rapid sand filter. (From AWWA and ASCE, 1998; McGraw-Hill; printed with permission.)

6.4.2 Slow Sand Filters

Slow sand filters are a form of filtration that has been in use for nearly two centuries (USEPA, 1999b). Typically, they are not preceded by coagulation or flocculation processes and can treat water with turbidity up to 50 NTU (Schultz and Okum, 1984). Hydraulic loading rates for this type of filter range from 0.1 to 0.4 m^3/m^2h, or approximately 0.2 to 1.0 gpm/ft^2 (Huisman and Wood, 1974; Schultz and Okum, 1984). The filter media is contained in a box and is composed of a bed of relatively uniform in size and fine-grained sand, which is supported by a layer of gravel (see Fig. 6.7). Similarly to rapid sand filters, water flows down the filter media drawn by gravity. Water is treated as a combination of physical straining and biological removal to remove suspended solids, and some microbes. Removal of solids by physical straining occurs within the upper 0.5 to 2 cm of the sand layer. During operation, a layer of dirt and biologically active organisms develops at the surface and within the uppermost part of the sand. This layer is known as the "schmutzdecke" and is essential to the effectiveness of the filter in removing suspended solids and reducing turbidity (Huisman and Wood, 1974; Schultz and Okum, 1984; USEPA, 1999b; Binnie et al., 2002). When the filter becomes clogged by the removed impurities, the top layer of the media is scraped off to start a new filter cycle. The cleaning can be done by unskilled labor (Schultz and Okum, 1984). Depending on water quality, filter runs between cleaning can last up to 2 to 6 months (Kawamura, 2000). Slow sand filters can effectively remove suspended solids; however, they have been found to have limited capability for removing clay particles and color (AWWA and ASCE, 1998). Their simplicity of operation, low cost of installation, simple design, and no power requirements make them a very effective "low-tech" method of treating water, especially for urban and rural communities of low-income areas (WHO, 2007). Prefabricated slow sand filter units are also on the market. These systems are particularly

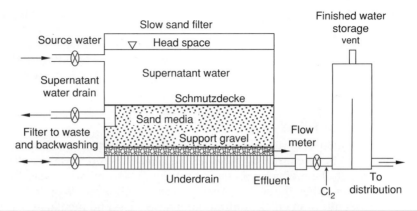

FIGURE 6.7 Typical covered slow sand filter installation. (From AWWA and ASCE, 1998; McGraw-Hill; printed with permission.)

suitable for installation in small communities where construction capabilities are limited and in emergency situations.

6.4.3 Pressure Filters

Pressure filters and rapid sand filters have many similar characteristics. They use the same types of media, and the removal of suspended solids is accomplished in the same way. Their efficacy to remove suspended solids and reduce turbidity is also similar (USEPA, 1999b). The main difference is that in pressure filters the media is contained within a pressurized vessel, usually made of steel (see Fig. 6.8). Pressure, not gravity, is the driving force to push water through the media. Some of the major advantages of pressure filters are that they have a compact design since they do not require several feet of water above the filter bed to provide a static pressure head and that water leaves

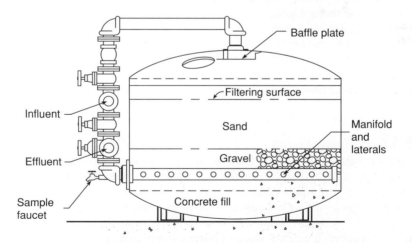

FIGURE 6.8 Typical cross section of a pressure filter. (From Cleasby and Logsdon, 1999; copyright McGraw-Hill; printed with permission.)

the filter under pressure. The latter occurrence eliminates the need for repumping water before delivery to the distribution systems and the potential for air binding, which is sometimes associated with rapid sand filters. The disadvantages of pressurized filters include the absence of visual observation of the filter during operation and the propensity to experience turbidity "breakthrough" because of the high pressure driving the filtration process (USEPA, 1999b).

6.4.4   Precoat Filters

This type of filter uses a thin layer of very fine media such as diatomaceous earth or perlite, which is supported by a permeable rigid structure or fabric element. Water is forced through the media by pressure or vacuum. The media physically strains solids from the water. When the surface cake of strained solids builds up to impart a headloss that does not allow an efficient filtration process, cake and media are washed out from the support and a new layer of media is deposited in a slurry to start a new filtration cycle (USEPA, 1999b; LeChevallier and Au, 2004).

Typically, loading rates are in the range of 1.2 to 5 m^3/m^2h (approximately 0.5 to 2 gpm/ft^2). Advantages of this process include relatively low capital cost and no need for the clarification. Disadvantages include the inability to handle water with high turbidity, the potential for particle breakthrough if the precoat phase is not properly done or cracks develop in the precoat layer during operation, and the poor capability to remove color- and taste- and odor-causing compounds (USEPA, 1999b).

6.4.5   Bag and Cartridge Filters

Bag and cartridge filters are used to remove microbes and turbidity. The filtration process for these units is based on physical screening—particles larger than the filter pore size are removed. The pore size of these filters is typically designed to be small enough to remove protozoa such as *Cryptosporidium* and *Giardia*. However, smaller particles, including viruses and most bacteria, can pass through the filters. Bag and cartridge filters can be used only for water with low-turbidity loadings, because high turbidity can quickly clog them. Upon fouling, bag filters can be backwashed, while cartridge filters must be replaced. A disinfectant is typically added to water before it enters cartridge filters to inhibit bacterial growth in the unit. In larger systems, these filters typically find their application as pretreatment to protect reverse osmosis (RO) membranes or other process units sensitive to suspended solids. They are easy to operate and maintain, which makes them suitable for small systems and for POE/POU applications (LeChevallier and Au, 2004).

6.4.6   Ceramic Filters

Ceramic filters can remove bacteria and parasites by passing water through a porous ceramic cartridge. The volume of water that can be treated with these filters is small, and their application is limited to kitchen sink units that produce drinking and cooking water. These filters are not able to remove all viruses if they are present in the water, and additional disinfection should be provided. Ceramic filters tend to plug up quickly if the water contains significant loads of particulate matter.

A type of "low-tech" ceramic filtration system (Filtron) developed by Dr. Fernando Mazariegos of the Central American Industrial Research Institute (ICAITI) is used in low-income areas and emergency situations to purify water (Potters for Peace, 2007). This filtration system is composed of a porous clay filter medium, a larger storage recipient canister or a bucket with a lid, and a spigot attached to the bottom. Water is poured in the clay filter at the top and percolates into the storage vessel. Users can draw water from the spigot. The clay filter is manufactured with a simple process that can be easily replicated on a local level, which includes a mix of local terra-cotta clay and a combustible material such as sawdust or rice husks. The three ingredients are mixed together. A hand-operated press and two-piece aluminum mold can be used to form the clay filter. The combustible material burns out in the firing process, leaving a network of fine pores (ranging between 0.6 and 3.0 .μm). After firing, the filter is coated with colloidal silver to inhibit bacterial growth (Potters for Peace, 2007). The *Asociación Guatemalteca para la Familia de las Americas* (AFA Guatemala), in association with other organizations, conducted a study from 1993 to 2005 on the effectiveness of this filter. The study found that the filters could reduce diarrhea by 50 percent (references in Johnson, 2006). Bench-scale testing has shown that the majority of the bacteria and protozoa are removed mechanically through the filter's fine pores. The colloidal silver inactivates 100 percent of the bacteria (Lantagne, 2001). The effectiveness of the filter in inactivating or removing viruses is unknown.

6.5 Membrane Filtration

Membranes are the new generation of water filters that are slowly replacing conventional filters due to their ease of operation and robust performance. Membrane filtration can also be considered, to a certain extent, as a form of disinfection because it can completely remove those pathogens whose size is larger than the pore size of a specific type of membrane (see Fig. 6.9). Capital and O&M costs, which used to be a deterrent for widespread implementation of membrane filtration, have considerably declined in recent years,

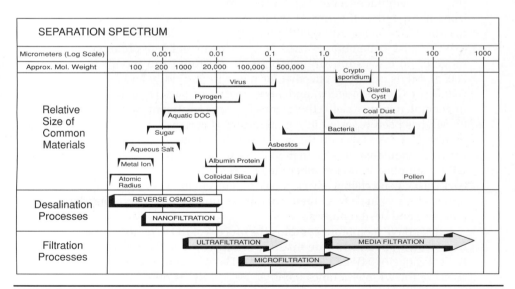

FIGURE 6.9 Pressure-driven membrane process classification (Westerhoff and Chowdhury, 1996; copyright McGraw-Hill; printed with permission.)

making membranes a very attractive treatment alternative. Small membrane units are also used at POE/POU for the purification of groundwater from individual wells.

The membrane filtration process is driven by pressure (or vacuum) to force water to the other side of a semipermeable membrane while retaining impurities and some of the feed water. The most commonly used membranes for drinking water treatment are microfiltration (MF), ultrafiltration (UF), nanofiltration (NF), and reverse osmosis (RO). Detailed descriptions of the fundamentals, design, and operation of these processes are available in Mallevialle et al. (1996), Taylor and Weisner (1999), and USEPA (2005a).

There are a variety of membranes of different materials and operational configurations (referred as modules) available on the market. Nearly all membranes used in drinking water are made of synthetic polymers. NF and RO are made of semipermeable cellulose acetate (CA) or polyamide materials, while UF and MF are membranes made of CA, polyvinylidene fluoride (PVDF), polyacrylonitrile (PAN), polypropylene (PP), polysulfone (PS), polyethersulfone (PES), or other polymers (USEPA, 2005a). Membranes are characterized by their capability to exclude (or reject) a solute (also referred as pore size) or by molecular weight cutoff (MWCO). The MWCO is expressed in terms of Daltons: a unit of mass designating 1/16 of the lighter and most abundant isotope of oxygen. Typically, manufacturers use the MWCO value—the nominal molecular weight of a known species that is always being rejected at a specific percentage under specific test conditions—to characterize individual types of membranes. However, as MWCO protocols vary among manufacturers, there is a certain level of ambiguity in defining the "true" cutoff of a membrane (Taylor and Weisner, 1999).

UF and MF are membranes made of polymeric porous material designed specifically to remove suspended solids via a sieving mechanism based on the size of the membrane pores. The distribution of pore sizes characteristic of each membrane varies with material type and manufacturing process. UF membranes are characterized by a pore size range of approximately 0.01 to 0.05 μm (nominally 0.01 μm) or less (USEPA, 2005a). UF is the primary membrane technology for the removal of viruses that, in general, range in size from about 0.01 to 0.1 μm. At the lower end of the UF spectrum some larger organic macromolecules, including DBP precursors, can be retained by the membrane. Typical MWCO for UF membranes used in water treatment is approximately 100,000 Da. The range of pore sizes for MF membranes is 0.1 to 0.2 μm (nominally 0.1 μm) (USEPA, 2005a). MF is primarily effective for removing turbidity and larger pathogens such as *Giardia* or *Cryptosporidium*, and some species of bacteria that are larger than 0.1 μm. In general, it is not an effective means for virus treatment, although some virus removal by MF has been reported in the literature. This removal is generally attributed to formation of a cake layer on the surface of the membrane. By the same process, MF can remove coagulated organic matter. In terms of module configuration, typically MF and UF are supplied in hollow-fiber membranes. MF and UF membranes must be regularly backwashed and chemically cleaned on recurrent intervals (USEPA, 2005a). The disposal options for residuals from these operations must be taken into consideration during the planning and design phases.

RO and NF membranes employ the process of RO to remove dissolved contaminants from water. NF and RO are made of semipermeable CA or polyamide materials. The typical range of MWCO is less than 100 Da for RO membranes, and between 200 and 1000 Da for NF membranes (USEPA, 2005a). RO is often used for the desalination of sea and brackish water, but also has high rejection capability for many synthetic organic compounds (SOCs) (Taylor and Weisner, 1999). NF is used, primarily, for softening or

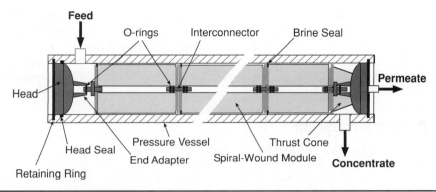

FIGURE 6.10 Typical spiral-wound (NF/RO) module pressure vessel. (From USEPA, 2005a.)

the removal of dissolved organic contaminants. NF and RO membranes are specifically designed for the removal of total dissolved solids (TDS) and not particulate matter. They are not intended to be sterilizing filters and some passage of particulate matter due to manufacturing imperfections may occur. Further, because these membranes cannot be backwashed, particulate matter can cause rapid irreversible fouling. Therefore, NF and RO are not typically used to directly treat raw water with significant concentrations of suspended solids. NF and RO use spiral-wound membranes (Fig. 6.10). A typical schematic for these systems is shown in Fig. 6.11. RO and NO filtration produce a continuous stream of concentrated brine, and the membranes must undergo chemical cleaning on a periodic basis. Both residuals must be disposed of. RO and NO do not require backwash (USEPA, 2005a).

An additional type of membrane filtration is electrodialysis (ED). In this process water does not pass through the membrane and only contaminant ionic species are transported across selectively permeable membranes driven by an electric potential (USEPA, 2005a).

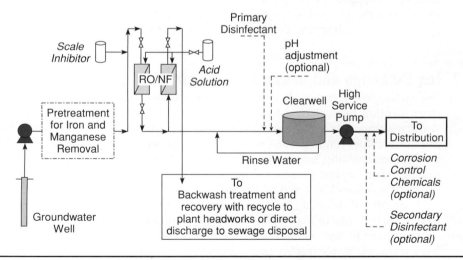

FIGURE 6.11 Schematic of a reverse osmosis/nanofiltration treatment system.

A variation of ED is electrodialysis reversal (EDR) in which the polarity of the electrodes is periodically reversed to change the direction of ion movement, in order to reduce scaling. As water is not physically filtered, ED and EDR do not provide a physical barrier to pathogens and suspended solids. The USEPA recognizes their effectiveness for removing dissolved ionic constituents but does not strictly consider them as filters (USEPA, 2005a).

6.6 Carbon Adsorption

The manufacturing of GAC for drinking water treatment involves a process of grinding, roasting, and activation of the source materials—such as bituminous coal, coconut shell, petroleum coke, wood, and peat—with high-temperature steam. The end product is a porous material with very high internal surface area and high adsorptive properties.

GAC filters can be placed after conventional filters as an additional process in conventional treatment (postcontactors) or in place of conventional filter media (Kawamura, 2000), since GAC media is able to remove suspended particles as efficiently as conventional media. But the main reason for using GAC is its capability for removing organic compounds. GAC can effectively remove SOCs such as aromatic solvents (benzene, toluene, and nitrobenzenes), chlorinated aromatics (PCBs, chlorobenzenes, and chloronaphthalene), phenol and chlorophenols, polynuclear aromatics (acenaphthene and benzopyrenes), pesticides and herbicides (DDT, aldrin, chlordane, and heptachlor), chlorinated aliphatics (carbon tetrachloride and chloroalkyl ethers), high-molecular-weight hydrocarbons (dyes, gasoline, and amines) (Tech Brief, 1997; Faust and Aly, 1999; Snoeyink and Summers, 1999), and methyl tertiary butyl ether (MTBE) (Stocking et al., 2000). GAC can be used to remove natural organic compounds such as humics (DBP precursors) and taste- and odor-causing compounds (Snoeyink and Summers, 1999).

During operation, when the adsorption capacity of a GAC filter is eventually exhausted, the media must be replaced and regenerated. Depending on the quality of the water to be treated and filtration rates, the interval between media replacement may be months or years. Pretreatment to reduce organic loading and remove suspended solids that may decrease the adsorptive capacity of GAC or clog the adsorption column is, in some cases, a valuable option. Powdered activated carbon (PAC) is rarely used in groundwater treatment with the exception of hydrogen sulfide removal.

6.7 Ion Exchange and Inorganic Adsorption

Ion exchange is a reversible chemical process wherein a charged molecule is removed from solution by exchange for a similarly charged ion attached to a solid matrix. In water treatment, the solid matrix is provided by resins composed of synthetic or natural high-molecular-weight polymeric or inorganic material. These resins have high porosity and, consequentially, large surface-area-to-weight ratio. This high ratio provides a large number of adsorption sites for the removal of pollutants. Typically, resins are in the form of small (less than 1 to 2 mm in diameter) beads that are packed in a filter column.

Examples of charged (anionic or cationic) constituents that can be removed through ion exchange include the calcium (hardness) and other metals, nitrates, heavy metals, arsenic, fluoride, and radionuclides. Ion exchange is most effective for water sources that have relatively stable quality (like most groundwater sources) and low loading of suspended solids and organic matter. In some cases, disposal of water used for regenerating

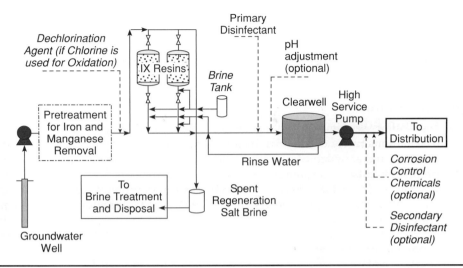

FIGURE 6.12 Schematic of an ion-exchange treatment system.

ion-exchange resins can be a problem as it contains high salt concentration and can negatively affect the performance of the wastewater plant. Otherwise, ion exchange is a relatively simple inexpensive process that requires low capital costs and labor. Ion-exchange media targeting specific compounds (e.g., nitrate and perchlorate) have been developed, improving the performance of this technology in situations where several ions may be competing for the same adsorption sites (Montgomery, 1985; Clifford, 1999). The schematic of a typical ion-exchange treatment system is shown in Fig. 6.12.

Activated alumina is a physical and chemical process for the removal of ions by adsorption. It is used in a similar way to ion-exchange resins. In this process, water flows through the alumina-packed bed typically contained in a fiberglass canister. The contaminant is adsorbed onto the surface of the alumina. Activated alumina is used in the removal of constituents such as arsenic, uranium, beryllium, selenium, silica, fluoride, and humics. Once the adsorptive capacity of activate alumina is exhausted, the media is replaced or regenerated (Montgomery, 1985; Clifford, 1999).

6.8 Biological Treatment

Biological water treatment processes are used in small and large municipal water treatment plants in Europe and are now gaining some consideration in North America. Biological treatment involves optimizing conditions to promote a permanent active biofilm for the biodegradation or conversion of unwanted constituents such as biodegradable organic matter, taste- and odor-causing substances, some SOCs, iron and manganese, and arsenic. Granulated active carbon is an excellent media to support the growth of bacteria (particularly when preceded by ozonation), which can be used to promote the biological removal of certain compounds. However, biological growth must be controlled to avoid operational problems such as sloughing-off of clumps of microbial growth from the filter and transport of microbes to the distribution on activated carbon fine particles that may be present in the plant effluent, which may cause high bacterial counts in the distribution system and taste and odor problems (AWWA and ASCE, 1998; Geldreich and LeChevallier, 1999). Disinfection of the GAC effluent is, therefore, recommended to ensure its acceptable microbiological quality before discharge into receiving surface water bodies.

6.9 Distillation

Distillation uses temperature change to evaporate and condense water. Metals and other mineral contaminants are generally retained in the boiling chamber, and purified water is collected in the condensate chamber. Some organic contaminants are not removed by this process. Because of the high energy cost and reject heat, this system is typically used only by small-scale installations and POE/POU devices for groundwater treatment. However, it is used in many large-scale systems in the Persian Gulf region for the desalination of sea water.

6.10 Disinfection

Typically, some form of disinfection is the last step in the treatment process to ensure that the water is microbiologically safe (free from bacteria, viruses, and protozoan parasites) before the water is consumed. Because water quality deteriorates as soon as it enters the distribution system, many utilities apply a secondary disinfectant to maintain the microbiological quality of water. Water in the distribution system can be contaminated by a variety of pollution sources, such as backflow, pipe leaks and intrusion, and bacterial regrowth in the distribution pipes. In the United States, the USEPA mandates that treated water contains sufficient excess disinfection chemical to maintain a residual in the distribution system and to ensure that no microbial regrowth and recontamination occur in the water as it is being distributed (USEPA, 1999a, 2006b). However, the idea of the need for a residual disinfectant is not universally accepted. For example, some European municipalities do not implement secondary disinfection.

The advantages and disadvantages of various disinfection methods are presented in Table 6.2, and the effectiveness of various disinfection methods for various pathogens are listed in Table 6.3.

6.10.1 Chlorine

Free chlorine can be introduced to water directly as a primary or secondary disinfectant. Chlorine is effective for disinfection of bacteria, viruses, and some protozoa (e.g., *Giardia*). However, the protozoan *Cryptosporidium* has demonstrated resistance to disinfection by chlorination. The major problem with chlorine is that it is known to react with organic substances and form variety of halogenated DBPs, including significant amounts of three halo methanes (THMs) and halo acetic acids (HAAs). In general, more THMs and HAAs are produced during chlorination than with all other disinfection methods. Further, excessive concentrations of chlorine can be harmful to human health. The USEPA has set a maximum residual disinfectant level (MRDL) for chlorine of 4.0 mg/L (USEPA, 1999a, 2006b; Singer and Reckhow, 1999). Besides disinfection, chlorine is used to achieve several other objectives during treatment (for example, oxidation and previously mentioned in this chapter). Table 6.4 summarizes some of the reasons for adding chlorine and other process considerations.

Consideration	Chlorine	Chloramines	Ozone	Chlorine Dioxide	UV
Equipment reliability	Good	Good	Good	Good	Medium
Relative complexity of technology	Less	Less	More	Medium	Medium
Safety concerns	Low to high[1]	Medium	Medium	High	Low
Bactericidal	Good	Good	Good	Good	Good
Virucidal	Good	Medium	Good	Good	Medium
Efficacy against protozoa	Medium	Poor	Good	Medium	Good
By-products of possible health concern	High	Medium	Medium	Medium	None
Persistent residual	High	High	None	Medium	None
pH dependency	High	Medium	Low	Low	None
Process control	Well developed	Well developed	Developing	Developing	Developing
Intensiveness of operations and maintenance	Low	Moderate	High	Moderate	Moderate

[1] Safety concern is high for gaseous chlorine, but it is low for hypochlorites.
Source: Earth Tech (Canada), 2005.

TABLE 6.2   Summary of Advantages and Disadvantages of Disinfection Techniques

Disinfectant	Microorganism Reduction Ability			
	E. Coli	Giardia	Cryptosporidium	Viruses
Chlorine	Very effective	Moderately effective	Not effective	Very effective
Ozone	Very effective	Very effective	Very effective	Very effective
Chloramines	Very effective	Moderately effective	Not effective	Moderately effective
Chlorine dioxide	Very effective	Moderately effective	Moderately effective	Very effective
Ultraviolet radiation	Very effective	Very effective	Very effective	Moderately effective

The reduction levels in the table are for normal dose and contact time conditions and they are only for general comparison purposes. The effectiveness of different disinfectants depends on the dose, contact time, and water characteristics.
Modified from USEPA, 1999a.

TABLE 6.3   Effectiveness of Disinfectants on Selected Pathogens

Application Typical	Dose	Optimal pH	Reaction Time	Effectiveness	Other Considerations
Disinfection	Min. 2 mg/L[1] in effluent[2]	7–8.5	"CT" requirements	Good for virus and bacteria Fair for protozoa	DBP formation possible Taste and odor problems
Iron oxidation	0.62 mg/ mg Fe	7.0	Less than 1 hour	Good	
Manganese oxidation	0.77 mg/ mg Mn	7–8 9.5	1–3 hours minutes	Slow kinetics	Reaction time increases at lower pH
Biological growth control	1–2 mg/L	6–8	NA	Good	DBP formation
Taste/odor control	Varies	6–8	Varies	Varies	Dosage and effectiveness depends on compound DBP formation
Color removal	Varies	4.0–6.8	Seconds to minutes	Good	DBP formation may result in taste and odor problems at high dosages
Zebra mussels	2–5 mg/L 0.2–0.5 mg/L[1]		Shock level Maintenance level	Good	DBP formation
Asiatic clams	0.3–0.5 mg/L[1]		Continuous	Good	DBP formation

[1] Residual, not dose.
[2] For systems implementing secondary disinfection.
Modified from USEPA, 1999a.

TABLE 6.4   Typical Chlorine Applications in Drinking Water Treatment and Doses

The forms of chlorine most commonly used in water disinfection are chlorine gas, liquid sodium hypochlorite, and (especially for smaller installations) calcium hypochlorite. Selection of the most appropriate addition method for a specific application should take into consideration several factors such as safety, cost and operation requirements, stability, availability, odor-control ability, corrosiveness, solubility, and ability to respond instantaneously to initiation and rate changes.

6.10.2   Chloramines

Chloramines are often used as a secondary disinfectant because of their capability to maintain disinfection residual for a long time in the distribution system and because they form less DBP compared to chlorine (USEPA, 1999a; Singer and Reckhow, 1999; Faust and Aly, 1999). However, in recent years, concern over the formation of N-nitrosodimethylamine (NDMA), iodoacetic acid (the most genotoxic DBP to mammalian cells ever identified), and leaching of lead from pipes have raised many doubts over the extensive use of chloramination to replace chlorination (Renner 2004a, 2004b; Edwards and Dudi, 2004). Similarly to chlorine, the USEPA has set an MRDL of 4.0 mg/L (as total chlorine) for chloramines (USEPA, 2006b).

6.10.3 Ozone

Ozone is very effective against a wide range of microorganisms, including bacteria, viruses, and protozoa. However, ozonation can form DBPs such as haloketones, aldehydes, ketoacids, carboxylic acids, and other forms of biodegradable organic matter, which must be adequately controlled, typically by a biologically active granular media filter, to avoid increased biofilm production within the transmission system. Further, in waters with sufficient bromide concentrations, ozonation can result in the formation of bromate and other brominated DBPs. Bromate is a regulated DBP (0.010 mg/L in the USA) and limits the use of ozone for many water supplies (USEPA, 1999a, 2006b; Singer and Reckhow, 1999).

Ozone is not used for secondary disinfection because, due to its highly reactive nature, ozone residual cannot be sustained for an extended duration. Generally, ozone is more costly than other commonly used disinfectants/oxidants. Its use is typically justified when disinfection objectives include inactivation of chlorine-resistant protozoa or where advanced oxidation is required for DBP, taste and odors, and color control.

6.10.4 Chlorine Dioxide

Chlorine dioxide is a powerful oxidant typically used as preoxidant (i.e., added at the plant headworks), because of its capability of enhancing the removal of iron and manganese. However, in recent years it has been increasingly used as a primary disinfectant due to its effectiveness for the inactivation of protozoa and bacteria (USEPA, 1999a).

Chemically, chlorine dioxide (ClO_2) is unstable and degrades to form by-products chlorite (ClO_2^-) and chlorate (ClO_3^-). The presence of all three species, chlorine dioxide, chlorite, and chlorate, is a health concern, and taste and odor complaints are also often associated with the use of chlorine dioxide (USEPA, 1999a; White, 1999). In the United States the first two are regulated; the MRDL for chlorine dioxide is 0.8 mg/L and the MCL for chlorite is 1 mg/L (USEPA, 2006b).

6.10.5 Ultraviolet Light Disinfection

When ultraviolet (UV) light is applied to a microorganism, DNA and RNA absorb the light energy and their structure is altered. These alterations inhibit DNA replication and diminish the capability of the microorganism to infect a host. UV light can be effectively used for primary disinfection. Generally, it has been observed that the more complex the microorganism, the more sensitive it is to UV inactivation. This means that UV is most effective for the disinfection of protozoa and least effective for the disinfection of viruses. Particularly, UV is effective for the disinfection of chlorine-resistant protozoa (i.e., *Giardia* and *Cryptosporidium*) (USEPA, 1999a, 2006c).

UV disinfection is a physical process rather than a chemical disinfectant. For this reason, there is no residual disinfectant associated with UV light, and an additional disinfectant (e.g., chloramines or chlorine) must be applied to achieve secondary disinfection. At the dosages appropriate for drinking water application, UV light is not believed to be an effective oxidant. However, when used in combination with another oxidant such as ozone, UV light can enhance the oxidation of contaminants. UV light can promote a free radical reaction pathway that increases the potency of ozone when used as an oxidant (USEPA, 1999a, 2006c).

To date, there is no evidence to suggest that UV irradiation results in the formation of any DBPs; however, little research has been performed in this area. Most of the research regarding application of UV light and DBP formation has focused on the impact on chlorinated DBP formation as a result of UV application prior to the addition of chlorine or chloramines.

One of the main drawbacks with UV is the possibility of microbes passing through at times the lamp is operating off specification. At low UV intensities, some microbes have shown the ability to repair damage done by UV light. Thus, it is important that drops in lamp intensity are minimized (USEPA, 2006c).

6.11 Corrosion Control

Corrosion is of concern for all drinking water suppliers because of its potential impact on aesthetics, economics, and human health. Discoloration and metallic taste caused by corrosion raise customer concern over the quality of drinking water. The reduced life of metallic and asbestos–cement pipe is a major economic loss due to corrosion. Leaching of toxic elements including lead, copper, cadmium, and asbestos into the finished water from the plumbing within a home and or well pump can present a health threat.

The presence of high concentrations of nitrate and sulfate ions may cause low pH in some groundwater. Low pH (typically, less than 7.0) inhibits the formation of a protective calcium carbonate scale on pipe and increases metal solubility. This requires that the pH of water is adjusted before being delivered to the distribution system. In many cases, pH is controlled by adding lime or caustic soda. In addition to raising the pH, lime increases the alkalinity and calcium content of water. Caustic soda converts excess CO_2 (if present) to alkalinity species. Some utilities also use chemical corrosion inhibitors in combination with pH control or alone to limit the effects of corrosion. Phosphate, zinc phosphate, and silicates are widely used corrosion control additives. The effectiveness of these chemicals for a specific system needs to be carefully evaluated and tested before full-scale implementation (DVGW/AWWARF/AWWA, 1996).

6.12 Removal of Specific Constituents from Groundwater

Removing specific contaminants in an affordable and effective manner is a challenge for groundwater providers and well owners. Specific treatment strategies employed to remove some of the most common contaminants found in groundwater are described below.

6.12.1   Iron and Manganese Removal

Until recently, it was thought that neither iron nor manganese causes adverse health effects and that high concentrations of these elements were only responsible for aesthetic effects. However, in recent study, Hafeman et al. (2007) have suggested that exposure to high manganese concentrations in drinking water may contribute to Bangladesh's extremely high infant mortality. In terms of aesthetic effects, water containing high concentrations of iron can discolor water, spot laundry, and stain plumbing fixtures. In addition, the growth of iron-oxidizing bacteria can result in abnormal taste and odor and can contribute to the growth of iron bacteria in distribution systems (Kawamura, 2000). Manganese causes similar reactions problems to form "black water" and can result in brownish-black stains on contact surfaces. Levels of 0.5 mg/L of iron and 0.05 mg/L of manganese are known to cause objectionable taste (Montgomery, 1985; Faust and Aly, 1999).

In general, iron and manganese problems arise in water containing low oxygen levels and high iron and manganese concentrations. The reason is that both iron and manganese are soluble under anoxic conditions, and problems occur when this type of water is pumped to the surface. When chemical equilibrium is changed upon exposure to the atmospheric pressure, the precipitation of iron and manganese will occur in plumbing, on fixtures, and on clothing, dishes, and utensils. These conditions are typical of waters that do not have regular contact with the atmosphere, such as groundwater from confined aquifers and deep wells. However, sometimes oxygen-poor conditions can also occur in relatively shallow wells that have stagnant water. In low-alkalinity groundwaters (less than 50 mg/L), iron concentrations can be up to 10 mg/L (or greater) and manganese concentrations up to 2 mg/L (Kawamura, 2000).

Oxidation

Iron and manganese removal is one of the most common objectives of groundwater treatment. Many systems remove iron and manganese with a combination of oxidation, coagulation/precipitation, and filtration. Oxidation is normally placed at the plant headworks to changes the form of iron and manganese from the bivalent form (Fe^{2+} or Mn^{2+}), which is soluble to the trivalent form (Fe^{3+} or Mn^{3+}), which is both insoluble and colored (Mongtomery, 1985). Oxidation methods used in drinking water treatment are aeration, chlorine, permanganate, ozone, and chlorine dioxide (Kawamura, 2000). Oxidant dosages required for the oxidation of iron and manganese are reported in Table 6.5.

Oxidant	Iron (II) (mg/mg Fe)	Manganese (II) (mg/mg Mn)
Chlorine, Cl_2	0.62	0.77
Chlorine dioxide, ClO_2	1.21	2.45
Ozone, O_3	0.43	0.88[1]
Oxygen, O_2	0.14	0.29
Potassium permanganate, $KMnO_4$	0.94	1.92

[1] Optimum pH for manganese oxidation using ozone is 8–8.5.
Source: Reckhow et al., 1991; Williams and Culp, 1986; Langlais et al., 1991.

TABLE 6.5   Oxidant Doses Required for Oxidation of Iron and Manganese

Cascading tray aerators are used for iron and manganese oxidation. In waters where iron and manganese form complexes with humic substances or other organic molecules, aeration is not effective because oxygen is a weak oxidant and cannot break the bonds between metal and organics. Further, the oxidation of manganese by oxygen is a slow process (on the order of hours) unless the pH is raised above 9.5 (USEPA, 1999a).

Chlorine, permanganate, ozone, and chlorine dioxide are very effective for converting iron and manganese to insoluble compounds. The presence of high concentrations of NOM can hinder the oxidation of iron and manganese and increase oxidant demand. This situation may pose a problem in terms of overfeeding of oxidant. Elevated concentrations of chlorine can result in high DBP formation. High permanganate concentrations can carry through the plant and result in pink water reaching the distribution system. High ozone dosages can turn the oxidation of manganese to permanganate, which may result in pink color to develop in the water. High chlorine dioxide residuals can cause taste and odor problems.

Coagulation/Precipitation

Following the oxidation step, a clarification is commonly used to remove small iron and manganese particulate. Manganese colloids are sometimes difficult to remove because of their small size and high specific surface area if the clarification process is not carefully optimized and may end up being removed by the downstream filters or carried through the plant and end up in the distribution system.

In alternative to coagulation/flocculation, chemical precipitation is through softening, and this process can also remove iron and manganese particles. Softening is, however, too expensive for removing these two metals alone. But when it is used for hardness removal it can also be used for removing iron and manganese. The high pH used during the softening process results in rapid oxidation and precipitation of iron and manganese. Also, the two metals can be removed by incorporation in calcium and magnesium precipitate (Montgomery, 1985; Faust and Aly, 1999).

Filtration

Filtration is usually the last step for removing iron and manganese either following sedimentation or, depending on water quality conditions, directly following oxidation. Generally, manganese oxidation determines the hydraulic detention time needed before filtration because it has a slower oxidation reaction rate than iron. Different types of filters are used for the direct removal of iron and manganese (Montgomery, 1985).

The most commonly used type of filter for removing iron and manganese for small- and medium-sized treatment plants is pressure manganese greensand filter. Greensand is manufactured using grains of the zeolite mineral glauconite—a green clay mineral. The glauconite is treated with various chemicals to produce a durable greenish-black coating that has ion-exchanging properties. This coating behaves as a catalyst, facilitating the chemical oxidation necessary for the removal of iron and manganese. As water is passed through the filter, soluble iron and manganese are removed from solution and oxidized to form insoluble iron and manganese. Insoluble iron and manganese will build up in the greensand filter and must be removed by backwashing. The greensand is regenerated by feeding continuously or intermittently a permanganate solution.

Anthracite/sand dual media can also be used for iron and manganese removal. In this case, the oxide coating is obtained after the installation in the filter boxes by feeding permanganate. Also, in this case, the coating acts as a catalyst for the oxidation of iron and manganese. Precoat filtration can also be used in potable water systems for removing low concentrations of iron and manganese precipitants after oxidation.

Ion Exchange

Ion exchange is usually considered for removing iron and manganese only if water hardness is also a problem. This method is suited to remove only low concentrations (less than 0.5 mg/L combined) of these metals because of the risk of precipitation and rapid clogging of the media. Removal of iron by ion exchange should be run in the absence of oxygen to prevent oxidation and precipitation of iron and manganese oxides. Iron and manganese precipitates can coat and foul the media requiring acid or sodium bisulfate washing to resume operations. Further, ion exchange is not effective if the iron has combined with organic material, or if the growth of iron bacteria, which secrete large amounts of iron oxide, is not adequately controlled (Montgomery, 1985; Faust and Aly, 1999; Kawamura, 2000).

Biological Treatment

Ion and manganese removal can be achieved by biological filtration. To optimize biological removal, the unit feed water should have low levels of oxygen and should not be chlorinated. After the biological step, aeration and disinfection are usually required to stabilize treated water (for more information of this topic see Sharma et al., 2005).

Sequestration

In systems with low concentrations of iron and manganese of, respectively, less than 1.0 and 3.0 mg/L, polyphosphates can be added as a sequestering agent before iron and manganese come in contact with air or an oxidant to prevent precipitation. Normally, a dosage of 2 to 4 ppm of a polyphosphate compound, such as sodium hexametaphosphate (SHMP), sodium tripolyphosphate (STP), or tetrasodium pyrophosphate (TSPP), forms colorless phosphate complexes. This method is effective in mitigating the negative effects of iron and manganese precipitation. However, consumers may still be affected as iron and manganese precipitation still takes place when water is heated—creating deposits in water heaters, boilers, and kitchen pots—and when water age—affecting areas at the margins of the distribution system or served by tanks where water is held for a long time (Kawamura, 2000).

6.12.2   Hardness

Excessive water hardness can result in soap deposits, scaly deposits in plumbing, appliances and cookware, and decreased cleaning action of soaps and detergents. It is caused by high concentration of dissolved calcium and magnesium. Other bivalent ions such as strontium, barium, zinc, and aluminum can also have some minor effects on hardness levels. Table 6.6 presents a classification of different types of water according to their hardness content.

Total hardness is calculated by summing up the concentrations of bivalent metals expressed as mg $CaCO_3$/L and is differentiated between carbonate and noncarbonated hardness. Carbonate hardness is the portion of hardness where ions are associated with carbonate and bicarbonate compounds. The most common source of this type of hardness is when groundwater flows through aquifers formed of limestone or chalk. Noncarbonate hardness is associated with chloride, sulfate, and other anions, and it is typical of

Description	Milligrams per Liter as $CaCO_3$
Soft	0–60
Moderately hard	61–120
Hard	121–180
Very hard	More than 180

From USGS, 2007.

TABLE 6.6 General Hardness Scale

groundwater passing through sedimentary formations containing high concentrations of sulfates.

Because of the tangible effects of high hardness levels, it is quite common to treat groundwater. However, because of concerns over the association between softened water and heart diseases, some systems prefer not to remove hardness.

For large potable systems, the most common method to soften water is precipitative softening by lime, lime–soda ash, or caustic soda. The selection of one of these processes depends on costs, concentration of TDS, sludge production, prevalent type of hardness, and chemical storage issues. In general, softening with caustic soda is more costly than lime and lime–soda ash. In most cases, lime and lime–soda ash softening lower TDS while caustic soda increases their concentration. More sludge is produced by lime and lime–soda ash than by caustic soda. Water mostly containing carbonate hardness can be softened by lime alone. However, if noncarbonate hardness is significant, lime and lime–soda ash may be used in combination to obtain the desired degree of hardness reduction. Chemical stability during storage and clogging of feeding systems are issues for softening plants. Hydrated lime absorbs water and carbon dioxide forming lumps, and quicklime may slake. Conversely, caustic soda does not deteriorate during storage and does not clog feeding lines. High concentrations of carbon dioxide should be reduced before precipitative softening because its removal through precipitation requires the use of more chemicals and more sludge is produced. An aeration step can be used for this purpose (Montgomery, 1985; AWWA and ASCE, 1998; Kawamura, 2000).

In precipitative softening plants, it is very important to carefully monitor pH. Typically, precipitative softening is conducted by raising the pH of water to approximately 10 for calcium carbonate precipitation. When the removal of magnesium is also required, the pH needs to be increased up to approximately 11. Operating the plant at an improper pH can lead to the precipitation and deposition of lime in the filter bed. This may lead to the "cementification" of media with consequent irreparable damages to the integrity of the media (AWWA and ASCE, 1998; Kawamura, 2000).

Water softened by chemical precipitation is characterized by high potential for carbonate scale formation. Recarbonation through the application of carbon dioxide and pH control is often used post precipitation to lower pH and mitigate scaling downstream of the softening process unit within the plant and in the distribution system. Scale formation can also be controlled by the application of low levels of polyphosphates (1 to 10 ppm) to inhibit scale formation (AWWA and ASCE, 1998; Kawamura, 2000).

Smaller potable systems are less likely to use precipitative softening because of high capital costs and operational complexity associated with this process. A popular alternative is ion exchange. Water is passed through a column filled with resins beads where bivalent ions are exchanged with sodium ions. For potable uses, only a portion of the total flow needs to undergo this process before being blended with untreated water to lower sodium concentration. Other less commonly methods that are used to lower hardness are EDR and RO/NF.

6.12.3 Nitrates

Nitrate contamination is a common problem for many groundwater supplies. In most cases, it is a result of the application of fertilizers containing nitrogen, livestock facilities, or sewage disposal areas. However, in some instances, high nitrate concentrations can originate from natural deposits of nitrates. The main concern with high concentrations of nitrate in drinking water is methemoglobinemia, which causes "blue baby" syndrome, especially in bottle-fed infants younger than 6 months of age, and the potential conversion of nitrate into carcinogenic nitrosamine. The current MCL set by the USEPA (USEPA, 2006b) for nitrate in drinking water is 10 mg/L as nitrate-nitrogen (NO_3-N). The WHO guideline (WHO, 2006) for nitrate is 50 mg/L as nitrate (NO_3); note that 10 mg/L NO_3-N = 44.3 mg/L NO_3.

The most common treatment processes used to reduce nitrates in drinking water are ion exchange and RO. In the ion-exchange process, a portion of the flow is passed through a special resin, which replaces nitrate anions with another ion, usually chloride. Because sulfate is preferentially exchanged over nitrate by chloride anions, special resins that preferentially remove nitrates are normally used. These special resins (IX process) that are used to remove nitrate/nitrite also reduce the possibility of nitrate "dumping" (release of a large amount of nitrates that may occur when the ion-exchange column is saturated).

An example of the application of the ion-exchange process is reported at the Village of White Oak Water System in North Carolina, USA (Mitchell and Campbell, 2003). This system supplies water to 42 homes from a single 95 L/minute (25 gpm) well. High nitrate concentrations above 10 mg/L (nitrogen) were detected in the raw water. After pilot testing, the water provider chose to install ion-exchange units to reduce nitrate levels. The design includes a 50–50 flow split, so that only 50 percent of the flow is treated. This allowed the reduction of costs and volume of reject water. After blending of the treated and untreated water, the residual nitrate levels are on average 3 mg/L (as nitrogen), well below the 10 mg/L MCL. The reject water from backwash of the ion-exchange units is collected in a storage tank and transported to a wastewater treatment plant for treatment.

Although nitrate is not very well rejected by most membrane materials, RO can be used for nitrate removal. Typical rejection rates for nitrate removal are around 90 percent, and they are sufficient to lower nitrate to acceptable concentrations. An example of the application of RO for nitrate removal is the system in the greater Milan area, Italy (Elyanow and Persechino, 2005). GE Italba installed 13 RO membrane plants in nine locations to control high nitrate concentrations from a series of wells providing drinking water. The nitrate concentrations in these wells range between 50 and 60 mg/L (NO_3), and the treatment goal was to lower nitrate levels to less than 40 mg/L (NO_3), while discharging in the sewer a waste stream with a nitrate concentration lower than 132 mg/L. To minimize treatment costs, permeate from the RO plants is blended with untreated raw water at a ratio of approximately 20 percent permeate to 80 percent blend water. The average water recovery for the RO plants averages around 60 percent, but

	Bermuda	**Delaware**	**Italy**
Production (m³/hour)	94.6	63.1	47.3
Recovery (%)	90	90	90
Desalting stages	3	3	3
Feed nitrate (mg/L NO_3)	66	61	120
Product nitrate (mg/L NO_3)	8.8	4.5	37
Feed TDS (mg/L)	278	11	474
NO_3 removal (%)	86.7	92.6	69.2
TDS removal (%)	81	88	53

From Elyanow and Persechino, 2005.

TABLE 6.7 Example of General Electric (GE) Electrodialysis Reversal for Nitrate Removal

considering blending actual recovery ranges between 77 and 88 percent. The system is run by a central control and data logging room in the GE Milan office.

ED reversal process is also used for nitrate removal, using the electrostatic properties of nitrate anions, and, especially where high recovery is required, EDR has been implemented as it has demonstrated recoveries greater than 90 percent (Elyanow and Persechino, 2005). This technology has been used by several plants across the world. For example, General Electric reports that plants in Bermuda, Delaware, and Italy have achieved reliable service (nearly 10 years) and nitrate removal rates from 69 to 93 percent. Performance data from these plants are summarized in Table 6.7. Recent upgrades to this technology have improved the performance of EDR systems (Elyanow and Persechino, 2005).

At the POE/POU scale, RO is the most frequently used point-of-use-sized treatment system for nitrate/nitrite removal. It is the most cost-effective method for producing only a few liters of treated water per day. Ion exchangers can also be used. However, because of the risk of "dumping" mentioned earlier in this section, only nitrate-specific resins should be used. Distillation is used at the POE/POU scale, since it is a very energy-intensive technique, not suited for larger-scale installations. It involves boiling and collecting water after condensation. Nearly all nitrates are removed in the process. However, it should be pointed out that just boiling the water does not remove nitrates; on the contrary, it increases their concentration.

6.12.4 Total Dissolved Solids

TDS refer to all dissolved solids that are present in water. Typically, these comprise inorganic salts such as calcium, magnesium, potassium, sodium, bicarbonates, chloride and sulfate, some organic materials, and soluble minerals (e.g., iron and manganese, arsenic, aluminum, copper, and lead). TDS originate from dissolution of minerals in geologic formations, sewage, urban runoff, and industrial wastewater.

In practice, TDS are measured by filtering a sample of water through a very fine filter (usually 0.45 μm), which removes the suspended solids, evaporating the filtrate and weighing the residue after evaporation (US Federal Register, 2007). It is important to notice that this type of test provides a qualitative measurement of the overall amount of

Water Source	Total Dissolved Solids (mg/L)
Potable water	<1000–1200
Mildly brackish water	1000–5000
Moderately brackish water	5000–15,000
Heavily brackish water	15,000–35,000
Average sea water	35,000

Sources: WHO, 2006; NRC, 2004.

TABLE 6.8   Classification of Source Water, According to Quantity of Dissolved Solids

dissolved solids, but it does not give any information on the specific nature of the solids. Thus, TDS should only be used as a general indicator of water quality.

High TDS levels do not per se pose a public health concern; however, they may create aesthetic problems such as limiting the effectiveness of detergents, corroding plumbing fixtures, resulting in scale formation, and causing salty or brackish taste. Driven by these considerations, the USEPA has set a secondary minimum contaminant level of 500 mg/L. The WHO (2006) guidelines for drinking water quality do not suggest a health-based value for TDS. However, it is noted that drinking water with high levels of TDS (greater than 1200 mg/L) may be objectionable to consumers, and extremely low concentrations of TDS may as well be unacceptable because of its flat, insipid taste (WHO, 2006). By comparison, seawater has an average TDS of about 35,000 mg/L. In Table 6.8, a classification of source water according to the concentration of dissolved solids is presented.

The selection of treatment options for elevated TDS depends on the type of dissolved species to be removed. If the high TDS concentration is due to a constituent such as calcium, magnesium, and iron, removal is possible through a softening or ion-exchange process (see hardness removal). If the problem is, instead, associated with an elevated concentration of sodium or potassium salts, removal techniques include RO, ED, or distillation.

In recent years, desalination with RO and ED due to advances in technologies and decreasing capital and operating costs has become an option for the development and management of water resources. This is of great importance in regions where fresh-water supplies are limited; especially in arid inland areas where the groundwater is characterized by high TDS and in coastal regions. Desalination of groundwater can be implemented for directly providing potable water to communities in areas where brackish groundwater is found underground or to limit the overexploitation and degradation of groundwater sources.

Numerous RO and ED desalination plants are in operation in the Western Region of the United States, Middle-East, Mediterranean Region, Central Asia, and North Africa for treating brackish groundwater for potable, irrigation, and industrial uses. Large RO desalination plants have been built to supply water for major urban centers (see Table 6.9), but smaller-sized plants have also have been installed to provide water for smaller communities.

For example, as one of several water management initiatives for limiting the overexploitation of the Hueco and Mesilla Bolson aquifers, El Paso Water Utilities, and Ft. Bliss,

Location	Capacity (m³/d)	Start of Operation
Malaga, Spain	165,000	2001
Al Wasia, Saudi Arabia	200,000	2004
Negev Arava, Israel	152,000	2006
Zara Maain, Jordan	145,000	2005
Undisclosed location, Iraq	130,000	2005
El Paso, USA	104,000	2007
Gwadar, Pakistan	95,000	2006
Bandar Imam, Iran	94,000	2002

Sources: Wagnick/GWI, 2005; El Paso Water Utilities, 2007.

TABLE 6.9   World's Largest Brackish Groundwater Desalination Plants

Texas, have jointly begun to operate a 104,000 m^3/day (27.5 mgd) desalination plant. This facility contributes to the stabilization of the levels of fresh groundwater by intercepting and treating brackish groundwater before it intrudes into historically freshwater areas and limiting the need for pumping of freshwater. For the two cities, brackish water represents an important alternative and plentiful water source in addition to surface water from the Rio Grande. The amount of brackish water in the Hueco and Mesilla Bolson aquifers exceeds the amount of potable water by approximately 600 percent (El Paso Water Utilities, 2007; Hutchison, 2007). In the desalination plant, the brackish water is filtered to reduce suspended solids and then is passed through RO membranes (approximately 83 percent recovery rate). The permeated (desalted) water from the RO process is blended with water from freshwater wells and, after pH adjustment and disinfection, is pumped into the distribution system. The concentrated brine from the desalination process is disposed of through injection into an underground rock formation (El Paso Water Utilities, 2007).

Another example is the Al Wasia Treatment Plant commissioned by the Saudi Arabia Ministry of Agriculture to supply to the city of Riyadh. The plant is expected to start operation in 2007 and treat 200,000 m^3/day (approximately 53 mgd) of brackish groundwater through softening, sand filtration, and RO membranes. The reject flow from the membranes will be finally conveyed to an evaporation pond, where it is further concentrated by solar energy.

Typically, RO treatment decreases the pH of treated water due to the removal of part of the constituents of the dissolved alkalinity. The EDR system does not lower pH as much, and the treated water tends to have a pH closer to that of source. However, both processes decrease TDS to very low concentrations. Pumping water with low pH and TDS in a distribution system can have negative consequences on water quality, especially if the pipes are used to carry water with different chemical characteristics. Suppliers of desalinated water must evaluate the impact of changing water chemistry on corrosion and pipe scale stability and implement adequate preventive measures. These measures can include blending of desalinated water with other water sources before entering the distribution system and adding chemicals such as sodium hydroxide (caustic soda) or lime—to raise the pH to the 7.5 to 8.0 units range—and corrosion inhibitors.

6.12.5 Radionuclides

Radioactive minerals are commonly present in bedrock aquifers. They are soluble in water, and, in some wells, radionuclides can be found at concentrations that exceed drinking water standards. There are different processes for the removal of different radionuclides. A general problem linked with the removal of these contaminants is that most treatment processes concentrate the radioactivity. Thus, handling and shielding of the treatment devices and waste must then be provided.

The following methods can be used for the removal of radionuclides (Montgomery, 1985; Faust and Aly, 1999):

- Distillation is capable of removing all types of mineral radionuclide types in one treatment process with the exception of radon gas. Because of the high energy costs associated with providing the heat source for distillation, this method is only practical as a POU/POE device.

- RO and EDR remove uranium, radium-226 and radium-228, gross alpha, gross beta, and proton emitters. There is no concern about the accumulation of radioactive substances in the membrane, as rejected particles are carried away in the waste stream.

- Cation exchangers can remove radium-226 and radium-228 and some gross alpha contaminants and that portion of the gross alpha that is positively charged.

- Anion exchange can be used to remove uranium and the negatively charged fraction of gross alpha. Particularly, uranium is removed only above pH 6.0 when it becomes negatively charged (below pH 6.0, uranium may be either an anion or nonionic). Above pH 8.2, uranium may precipitate and, as a solid, is not removed by anion exchange. Thus, ion exchange for uranium has to occur within the pH 6 to 8 window.

- When an oxidation filter such as potassium permanganate greensand is used to treat iron and manganese, removal of radionuclides (in particular radium) can also be expected.

- Some removal of radionuclides (especially uranium, radium-226, and radium-228) may also occur during coagulation/sedimentation and lime softening. To determine if these processes provide sufficient removal, it is necessary to monitor the treated water to determine the effluent concentration of radionuclides.

A special case among groundwater radioactive contaminants is represented by radon. Radon is a radioactive gas produced by the natural radioactive decay of radium and uranium during the weathering of rocks. According to the United States National Research Council (NRC, 1999): "Of all the radioisotopes that contribute to natural background radiation, radon presents the largest risk to human health." Radon is particularly harmful to the lungs through breathing of the gas. Water can transport the gas from the soil into households, and exposure can take place when taking shower, doing laundry, or washing dishes. Radon represents a greater problem for individual well users and small systems than for large systems, because radon decays relatively quickly (3.825-day half-life) and storing water (as done by most large systems) considerably reduces public exposure. In the United States, radon is not yet regulated at the federal level; however, the USEPA has proposed (1999 Federal Register 64 FR 59246) new regulations to reduce the public health risks from radon. Water should be treated where it enters a house or building. POU devices, such as those installed on a tap or under the sink, are not effective and should not be used for radon removal because they treat only a small portion of the water used in a house or building and do not address radon vapors that are released during showers or laundry. Aeration, GAC, and POE GAC are the recommended technologies for radon removal (NRC, 1999; USEPA, 2003). Aeration is the most efficient method. Proper venting of the aeration systems to the atmosphere must be installed to avoid indoor contamination. Removal by activated carbon requires large amounts of carbon media and long contact times. When activated carbon is used, it should be placed in sealed canisters to avoid leaks of accumulated gas. Furthermore, special consideration should be given to the disposal of the spent media or cartridge disposal because of radon accumulation to high concentrations (USEPA, 2003).

6.12.6 Hydrogen Sulfide

Hydrogen sulfide, H_2S, is found in groundwater as a result of the bacterial decomposition of vegetation and other organic matter under anoxic conditions. It can cause taste and odor problem such as rotten-egg odor and metallic taste even at very low concentrations. Furthermore, it reacts with many metals, causing black stains or black deposits of iron sulfite.

Aeration can be used to remove hydrogen sulfite from water, and groundwater supplies are often aerated for this reason. Oxidation is also often used to reduce sulfur species and limit taste and odor problems. Chlorine, ozone, permanganate, and hydrogen peroxide have been successfully used for this purpose. Adsorption through GAC or PAC is capable of adsorbing H_2S. This is a particularly advantageous system to use for H_2S at the POU/POE (Montgomery, 1985; Faust and Aly, 1999).

6.12.7 Volatile Organic Compounds and Synthetic Organic Compounds

Organic chemicals may be present in groundwater because of contamination from various sources. Many of these compounds represent serious public health threats if consumed in drinking water. In general, the three treatment methods that have been shown to be effective in removing organics from drinking water are aeration, adsorption using activated carbon, and oxidation. Carbon adsorption with GAC is effective for removing both VOCs and SOCs. Treatment with PAC is effective for removing some of the SOCs.

VOC chemicals, including trichloroethane and tetrachloroethane, are easily removed by air stripping. The off-gases from the air-stripping process must be further treated to avoid pollution of the atmosphere. Air stripping can be accomplished through packed tower aeration, incline cascade aeration, or membrane air stripping. In all cases, GAC columns are typically used to remove VOCs from off-gas.

Some organic contaminants will chemically react with oxygen and oxygen-like compounds. After this treatment, the resultant compounds either may be fully neutralized or will have a lower level of hazard. Further treatment may still be necessary.

Oxidizing chemicals could include potassium permanganate, hydrogen peroxide, and hypochlorite. Ozone oxidation is effective for removing certain classes of VOCs and SOCs, and certain RO membranes and UV treatment can also be effective against VOCs and SOCs (Montgomery, 1985; AWWA and ASCE, 1998; Faust and Aly, 1999; Kawamura, 2000).

6.12.8 Total Organic Carbon

Total organic carbon (TOC) is a gross measurement that quantifies the amount of NOM in water. TOC per se is not a harmful chemical, but its combination with disinfectant, and particularly with chlorine, results to the formation of DBPs, which are regulated compounds and pose a risk to public health. Further, natural organic compounds must be removed to improve the aesthetic quality of water (color) and to reduce the growth of biofilms in the distribution system as they constitute a food source for the bacterial population. Effective treatment strategies to remove TOC form water include the following (Singer, 1999; Letterman et al., 1999):

- Enhanced coagulation (low pH)
- Modified lime-softening (pH greater than 10 and addition of small amounts of ferric- or aluminum-based coagulants)
- GAC columns
- Ozone/biofiltration
- Synthetic iron-based resins
- RO

6.12.9 Arsenic

Arsenic occurs naturally in rock and soil and is released to groundwater due to its solubility. In recent years, arsenic contamination of groundwater has received increasing attention. The WHO guideline and the US and EU standard for arsenic is 10 parts per billion (ppb). According to an assessment conducted by the National Academy of Sciences (NAS), consumption of water with this concentration of arsenic still poses a significant health hazard (NAS, 2001). An individual drinking 2 L of water per day with 10 ppb of arsenic has a lifetime risk of fatal bladder or lung cancer greater than 1 in 300. This is far greater than the risk that the USEPA has traditionally used for cancer-causing compounds, which is no larger than 1 in 10,000.

As part of the regulatory process for lowering arsenic standards, the USEPA identified seven best available technologies (BATs) at the municipal scale in the Final Arsenic Rule based on removal efficiency, history of full-scale operation, geographic applicability, costs based on large and metropolitan water systems, service life, compatibility with other water treatment processes, and ability to bring all of the water in a system into compliance (US Federal Register, 2001). The performance of most of these treatment technologies is contingent on the state of oxidation of arsenic. Arsenic is removed more efficiently as arsenic V than arsenic III. Arsenic V ions are negatively charged, and they can be removed efficiently by several processes through electrostatic interaction. On the contrary, arsenic III is uncharged and its removal from water is difficult. All the BATs identified by the USEPA are for arsenic V, and preoxidation may be required to convert arsenic III to arsenic V before treatment (USEPA, 2000; US Federal Register, 2001).

Ion (Anion) Exchange

Ion exchange removes arsenic anions by exchange with chloride or other anions. There are several designs of ion-exchange systems. Some are nonproprietary, while others are proprietary designs by various manufacturers. Ion exchange combined with an oxidation pretreatment step has been shown to reduce total arsenic effluent concentrations as low as 0.003 mg/L (Health Canada, 2006). Laboratory studies using ion-exchange columns treating water with an arsenic concentration of 0.021 mg/L have achieved effluent concentrations as low as 0.002 mg/L (Clifford et al., 1999). A variety of adsorption media have been developed for treating arsenic. Sulfate, TDS, selenium, fluoride, and nitrate compete with arsenic for adsorption sites and can shorten run length. Suspended solids and iron precipitation can clog the adsorption column. Systems containing high concentrations of these constituents may require pretreatment. A typical schematic of an ion-exchange system in presented in Fig. 6.12. The groundwater is first oxidized with chlorine (or an alternative oxidant) and then passed to a filter cartridge to remove silt and salt. If chlorine is used for oxidation, a dechlorination agent should be added before the resin to limit the rapid deterioration of the resin and the potential for NDMA formation in the resin. The dechlorination agent is expected to increase the concentration of TDS in the filtered water. The brine from regeneration usually contains high TDS and arsenic. Often it cannot be discharged or accepted by a wastewater treatment plant. In those cases when the residual cannot be discharged it is treated with iron salts. The resulting sludge containing high levels of arsenic is then disposed of in a landfill while the supernatant is treated in an evaporation pond.

Activated Alumina
Activated alumina is a granular media manufactured for the purpose of removing ions from water. It is prepared by dehydration of aluminum hydroxide at high temperature. It has high adsorptive capacity, and it is the most common arsenic removal process for municipal-scale treatment of arsenic. Several studies have reported high arsenic removal efficiencies. A pilot study has reported effluent arsenic levels of <0.01 mg/L (Health Canada, 2006; Simms and Azizian, 1997). Activated alumina preferentially adsorbs arsenate over sulfate and other major ions; therefore, it is able to achieve long run lengths. The performance is sensitive to pH, which must be kept in the 5.5 to 6 range for optimum performance. This means that acid must be added upstream of the activated alumina column. After treatment, the pH must be adjusted before releasing into the water distribution system to avoid corrosion problems. It is possible to regenerate the media, but regeneration is incomplete and requires the addition of a strong base followed by acid neutralization. Replacement with new media and landfilling of the exhausted media is the other alternative. Operationally, the safest configuration is to have two columns operating in a series, with the first column removing the bulk of the arsenic and the second column providing a polishing step. When the medium in the first column is exhausted, it is replaced, and the two columns are switched using the first column for the polishing step. In this way, the column with the fresher media is always the last step, making the event of a breakthrough less likely. The waste produced from activated alumina processes includes the water caustic solution used in the regeneration process and the arsenic removed from the media. Typically, the combined volume of these waste streams is less than 1 percent of the processed water.

RO Membrane Technologies
RO is effective for removing both positive and negative forms of arsenic. However, because RO is nonselective in its rejection of contaminants and fouling, it requires frequent backwashing and water rejection can be high (on the order of 20 to 25 percent). The waste of a large portion of water may be an issue in water-scarce regions, which may prompt the implementation of more efficient, although more costly, recovery processes.

Although RO is included among the BATs for its high removal efficiency, based on costs of treatment and brine disposal, it is unlikely that it would be installed exclusively for arsenic removal.

Modified Coagulation/Filtration

Arsenic can be removed by coagulation with both aluminum and ferric hydroxide coagulants. Adsorbed arsenic is removed by precipitation or/and filtration. The optimal pH for adsorption is around 7 for aluminum and up to 8 for ferric hydroxides. Silica is known to compete with arsenic for adsorption onto ferric hydroxides particularly when pH is above 7. High concentrations of TOC may also reduce arsenic removal capabilities. The disadvantage of this treatment process includes the production of large amounts of arsenic-contaminated sludge that, depending on the arsenic concentration, might need to be disposed of in a hazardous waste landfill and large settling tanks. Typically, groundwater systems are not designed for particulate removal and use small clarification facilities. Because of high costs, it is unlikely for this process to be installed solely for the removal of arsenic.

Modified Lime Softening

Arsenic is effectively removed by the lime-softening process as long as the pH is high enough to precipitate $Mg(OH)_2$ (normally 10.5 or higher). As per conventional treatments, the production of arsenic-contaminated sludge is a disadvantage and the selection of lime softening solely for arsenic removal is not cost effective.

Electrodialysis Reversal

EDR can, in some instances, produce effluent water quality comparable to that of RO. A major advantage of EDR is that the system is fully automated and it requires limited operator attention, and no chemical addition. However, EDR systems are, in general, more expensive than other membrane systems including RO. Furthermore, typical recovery rates are low: in the 70 to 80 percent range. This may be a factor preventing the use of this technology in regions where water is scarce. Other options are more cost effective and have much smaller waste streams.

Oxidation/Filtration

This removal process is based on the coprecipitation of arsenic with iron during iron removal. The presence of a sufficiently high concentration of iron is critical to obtain significant removal of arsenic. One study conducted by Subramanian and coworkers (1997) suggested that a 20:1 iron to arsenic ratio result in lower removal rates. For a lower 7:1 iron to arsenic ratio, a 50 percent arsenic removal rate was reported. Competition with other ions is not a major factor limiting the effectiveness of this technology. Although a preoxidation step is not needed, when the arsenic is present as arsenic III, sufficient contact in an aerobic environment needs to be provided for the conversion to arsenic V. This USEPA suggests that this inexpensive technology may be appropriate for systems treating raw water with high iron and low arsenic concentrations. Table 6.10 summarizes the maximum arsenic removal rates as reported by the USEPA in the final version of the Arsenic Rule in 2001 (US Federal Register, 2001).

Additional technologies not listed as BAT by the USEPA but reputed to be effective to some extent for arsenic treatment are described below.

Treatment Technology	Maximum Percent Removal
Ion exchange (sulfate ≤50 mg/L)	95
Activate alumina	95
Reverse osmosis	>95
Modified coagulation/filtration	95
Modified lime softening (pH > 10.5)	90
Electrodialysis reversal	85
Oxidation/filtration (20:1 iron:arsenic)	80

Source: US Federal Register, 2001.

TABLE 6.10   USEPA Best Available Technologies for Arsenic Removal and Removal Rates for Arsenic V (Preoxidation May Be Required)

Coagulation/Membrane Filtration

The use of small doses (less than 10 mg/L) of ferric coagulant (added inline) in combination with MF or UF membranes removes arsenic. Arsenic adsorbs to the small floc formed by the ferric coagulant and then is filtered with the membrane. Typically, ferric dosages range between 5 and 20 mg/L, depending on the quality of raw water. This method reduces the amount of sludge produced, as there is no need for a large settleable floc to be formed. Other advantages of this technology include low chemical dosages and a smaller footprint compared to conventional treatment. Similar to conventional treatment, a pH above 8 or the presence of high levels of silica or TOC can reduce the arsenic removal performance of this technology. Membranes require occasional cleaning with caustic soda and citric acid to remove dissolved organic matter that may clog pores and reduce the efficiency of filtration. The frequency of the cleaning depends on the quality of the water to be treated. Typically, membranes must be backwashed to remove the ferric–arsenic precipitate that collects on the surface on the membrane. Backwash is, typically, started every 30 to 60 minutes, depending on the type of membrane and ferric dose. This process has two waste streams: the spent chemical-cleaning solution, which does not contain significant levels of arsenic but high concentrations of sodium and organic carbon, and the spent backwash water, which contains high concentrations of arsenic. The first stream is commonly discharged directly to the sanitary sewer. The second can be discharged to the sewer (if TDS concentrations are low) or decanted and returned to the head of the plant. The concentrated solids from the decanted can be normally discharged in the sanitary sewer.

Pilot tests conducted at Albuquerque, New Mexico, on iron coagulation followed by a direct MF showed that this process can effectively remove arsenic V and yield arsenic concentrations consistently below 2 mg/L (Clifford et al., 1997). Although there is extensive full-scale experience of coagulation and MF as separate processes, there are no full-scale applications of the combined coagulation/MF process (USEPA, 2000).

A full-scale application of this technology is provided by the Pall Aria® MF system (Fig. 6.13). To optimize arsenic removal, ferric chloride is added to the water before filtration. Arsenic anions are adsorbed onto positively charged ferric hydroxide particles, which are then removed by MF. Backwash is used to remove the ferric hydroxide–arsenic cake from the membrane surface. One of such systems has been installed for the Fallon

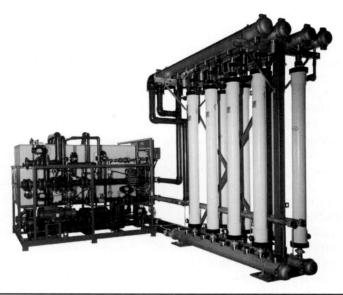

Figure 6.13 Pall Aria® microfiltration system. (Photograph courtesy of Pall Corporation.)

Paiute-Shoshone Tribe in Fallon, Nevada, to comply with the USEPA arsenic MCL. It is reported that at this location the Pall Aria® system is able to reduce arsenic from concentrations as high as 160 μg/L to undetectable levels, less than 2 ppb (Wachinski et al., 2006).

Granular Ferric Hydroxide

Granular ferric hydroxide (GFH) is a very promising technology based on arsenic adsorption onto iron-based granular media. It can effectively remove arsenic up to pH 8. Preoxidation is not required. High levels of silica and phosphate reduce arsenic removal through this process. The medium is contained in a filter vessel. Similarly to the operation of a conventional filter, water enters the filter from the top and is filtered through the media bed. As water is being filtered, the amount of arsenic being adsorbed onto the media increases. When the adsorptive capacity of the media is used up, the arsenic concentration in the treated water increases to a breakthrough value. At or before breakthrough, the operator must switch the operation to another filter with fresh media. It is possible to regenerate GFH with caustic; however, because of potential headloss issues, it may be more suitable for one-time use (US Federal Register, 2001). The simplicity of this process makes it very attractive for small installations and single-well applications.

Nanofiltration

NF membranes primarily reject arsenate by electrostatic repulsion. They do not foul as easily and can be operated at higher recovery than RO. Sato et al. (2002) compared the performance of rapid sand filters and NF for arsenic removal. When arsenic concentration in the raw water exceeded 50 μg/L, NF achieved 95 percent arsenic V removal and 75 percent arsenic III removal without chemical addition. Rapid sand filtration was ineffective for the removal of soluble arsenic III. Sato et al. (2002) suggested that NF membrane can be used with any type of water for arsenic removal.

Greensand Filtration

In manganese greensand filtration, the arsenic contained in the raw water is oxidized as it passes through the filter and is deposited onto the filter media. Similarly to oxidation/filtration, removal is dependent on water quality and particularly on the iron-to-arsenic ratio (Subramanian et al., 1997). In laboratory studies conducted by Subramanian et al. (1997) using tap water spiked with high concentrations (200 mg/L) of arsenic III, removal increased from 41 percent to more than 80 percent as the Fe/As ratio increased from 0 to 20. Sulfate and TDS did not seem to strongly affect arsenic removal. A full-scale study showed arsenic removal rates ranging between 90 and 98 percent (an average effluent arsenic concentration of 2.2 μg/L), with an influent arsenic concentration of 54 μg/L for the greensand system in Village Kelliher, Saskatchewan, Canada (Magyar, 1992). This water was characterized by a total iron concentration of 1.79 mg/L (a 33 Fe/As ratio) and an average pH between 7.2 and 7.3. Potassium permanganate was added before the filters at a feed rate of about 2.3 mg/L. The filters were operated at a continuous pressure between 4.0 and 5.5 psig and a flow rate of 76 gpm (4.8 L/s). The Guideline Technical Document on Arsenic prepared by Health Canada (2006) suggests that this technology is appropriate for systems that do not require high arsenic removal, because of the high iron to arsenic ratio that is required to obtain high arsenic removal rates. Furthermore, because of its simplicity of operation and relatively low cost, it may be attractive for installations in remote areas and developing countries.

Additional Technologies

Additional technologies that have received some attention for their arsenic removal capabilities include the following:

- Slow sand filters: removal without preoxidation of up to 96 percent from groundwater containing 14.5 to 27.2 μg/L arsenic (Pokhrel et al., 2005).
- Biological activated carbon filter: removal of up to 97 percent of arsenic without preoxidation and 99 percent after ozone addition (Pokhrel et al., 2005).
- Nano-sized polymer beads: 100 percent removal from Idaho groundwater at pH greater than 7 and in the presence of significant concentrations of silica in experiments conducted by the Idaho National Laboratory (Patel-Predd, 2006).
- Titanium-based media: the Dow Chemical Company manufactures the ADSORBSIA^{TM} GTO^{TM}, a granular titanium oxide media designed with strong affinity for arsenic, lead, and other heavy metals.
- Synthetic ceramic materials: Kinetico Inc. of Newbury, Ohio, has patented the Macrolite$^{®}$ pressure filtration process based on a ceramic media designed to remove arsenic.

POU and POE Treatment Devices for Arsenic Removal

A variety of technologies can be adapted to POE/POU devices to remove arsenic on the residential and commercial scale. These types of devices are affordable and can remove arsenic from drinking water to concentrations below 0.010 mg/L. Before selecting a POE/POU device, the groundwater should be tested to determine the concentrations of arsenic, and substances such as competing ions (e.g., fluoride, iron, sulfate, and silica) and organic matter. which could hinder arsenic removal. Furthermore, because most technology cannot effectively remove arsenic III, an oxidation step should be implemented to convert arsenic III to arsenic V (US Federal Register, 2001).

The most common POE/POU devices available on the market for arsenic removal are RO and steam distillation (Health Canada, 2006), but other their types of systems based on alternative technologies such as adsorption are also becoming more common. RO is easy to service and is affordable. The main disadvantage is the significant volume of reject water that is produced during treatment. Distillation can remove virtually all arsenic in drinking water. It is more complex than RO to operate and service. It is normally used in commercial-scale applications. It should be noted that both RO and distillation remove all minerals including those that are beneficial such as calcium and magnesium. Adsorption/filtration using media such as ferric hydroxide, aluminum, and titanium oxide is becoming increasingly popular for arsenic removal in small water treatment systems because of its relatively simple O&M.

Developing Countries' Systems and Applications

High arsenic levels in the alluvial aquifer underlying Bangladesh and India have been recognized as a major environmental and public health problem. Al-Muyeed and Afrin (2006) report that, in most of Bangladesh, arsenic in groundwater has been found to be higher than 0.05 mg/L. In many instances, high iron and arsenic levels have been found to coexist (Al-Muyeed and Afrin, 2006), and groundwater in approximately 65 percent of Bangladesh is characterized by high iron content above 2 mg/L. Al-Muyeed and Afrin (2006) also noted that, with the exception of a few cities and towns, centralized water treatment systems are rare, and a large part of the population is served by individual or community tube-wells. Therefore, the short-term solution of this environmental problem depends on the development of low-cost technologies that can be implemented at the community or household level.

Al-Muyeed and Afrin (2006) conducted an investigation on the efficiency of conventional iron removal plants for arsenic removal operating in small communities in Bangladesh. These systems are typically constructed, including aeration, sedimentation, and filtration steps in small units. The field survey of 60 community plants showed removal rates between 60 and 80 percent for 60 percent of the installations and less than 60 percent for 40 percent of the installations. The pH of water strongly affected arsenic removal, with increasing rates corresponding to increasing pH. For pHs above 7.0, arsenic removal was, in most cases, in excess of 70 percent. As previously mentioned, the removal of arsenic is strongly correlated with the removal or iron (Subramanian et al., 1997), and relatively high arsenic removal rates in excess of 70 percent were observed when Fe concentrations were in the 6 to 8 mg/L range.

A variety of simple POU technologies have been suggested and used for the treatment of contaminated water in rural communities in Bangladesh. Despite their limitations from a technological perspective, they are extremely valuable tools for protecting public health in less affluent communities. Among the most popular and effective technologies are the 3-Kolshi filter and the SONO® filter.

6.12.10   Trace Metals and Inorganic Compounds

A variety of trace metals and inorganic compounds can be found in groundwater. The most common treatment removal options for several of these contaminants are summarized in Table 6.11.

Compound	Treatment Technology
Antimony	RO
Asbestos	Chemical coagulation/filtration
Barium	Softening and ion exchange
Beryllium	Coagulation/filtration, lime softening, activated alumina, RO, and ion exchange
Cadmium	Chemical coagulation, lime softening, and RO
Chromium	GAC, PAC, coagulation at lower pH, oxidation, and lime softening
Copper	Coagulation at lower pH
Cyanide	Chemical and biological degradation
Lead	RO, distillation, customized GAC, chemical precipitation, and lime softening
Mercury	Coagulation at lower pH and coagulation/filtration
Nickel	Lime softening and RO
Selenium	Coagulation/filtration, lime softening, activated alumina, ion exchange, and RO
Silver	Coagulation/filtration and lime softening
Thallium	Ion exchange and activated alumina
Zinc	Conventional treatment
Fluoride	Activated alumina, coagulation, and lime softening

TABLE 6.11 Common Treatment Removal Technologies for Trace Metals and Inorganic Compounds

6.13 Drinking Water Treatment Costs

The cost of water treatment projects is highly variable and difficult to predict because the final cost of facilities depends on site-specific factors such as plant capacity, design criteria and selection of treatment processes, raw water quality, site characteristics, climate, land costs, regulation and permit requirements, status of national and local economic conditions, and cost of contractors.

In the United States, the cost of construction for a basic groundwater treatment plant (including well construction, chemical addition and filtration for iron and manganese removal, pumping, storage, and disinfection facility) and disinfection is in the US $250 to 500/m^3 ($1.00 to 2.00/gal) range, depending on plant size and local conditions. As discussed in previous sections, additional process units may be required to achieve other treatment objectives.

In terms of individual treatment technologies, costs for conventional clarification, lime softening, and conventional media filtration are, to a large extent, due to construction and land requirements; therefore, they vary from one location to another. Costs for other process units are more dependent on equipment cost. Indicative average capital costs for various technologies used in groundwater treatment and O&M for some of the individual process units that are often used in groundwater treatment are presented in Table 6.12. Costs are calculated for installation in the United States, and they include site work, electrical work and instrumentation, contractor overhead, and profit. They do not include land requirements, pretreatment, and sludge disposal options, as these factors are site specific and should be determined for each location.

Process Unit	Capital Cost (Including Construction Costs) ($/m³) ($/1000 gal)			O&M ($/m³) ($/1000 gal)		
	Small Plant <4000 m³/d (<1 mgd)	Medium Plant 4000–40,000 m³/d (1–10 mgd)	Large Plant >40,000 m³/d (>10 mgd)	Small Plant <4000 m³/d (<1 mgd)	Medium Plant 4000–40,000 m³/d (1–10 mgd)	Large Plant >40,000 m³/d (>10 mgd)
Oxidation						
Chlorine	14 (53)	5 (19)	1.3 (4.9)	0.021 (0.080)	0.006 (0.021)	0.001 (0.004)
Permanganate	15 (55)	4.3 (16.4)	1.3 (4.9)	0.024 (0.090)	0.013 (0.050)	0.001 (0.004)
Chlorine dioxide	95 (360)	15 (56)	6.3 (24)	0.021 (0.080)	0.005 (0.020)	0.003 (0.010)
Ozone (5 mg/L dose)	33 (125)	21 (78)	10 (36)	0.026 (0.100)	0.007 (0.026)	0.002 (0.007)
Aeration (packed tower)	84 (318)	40 (150)	12 (45)	0.011 (0.040)	0.005 (0.020)	0.004 (0.015)
Clarification						
Conventional	284 (1075)	73 (275)	30 (114)	0.001 (0.004)	0.013 (0.050)	0.003 (0.010)
Upflow	140 (528)	63 (237)	27 (101)	0.023 (0.090)	0.008 (0.030)	0.003 (0.010)
Media filtration						
Rapid sand	268 (1014)	146 (551)	50 (190)	0.050 (0.190)	0.018 (0.070)	0.018 (0.070)
Greensand	269 (1018)	147 (556)	52 (198)	0.050 (0.190)	0.018 (0.070)	0.018 (0.070)
Pressure	455 (1725)	89 (340)	83 (314)	0.016 (0.060)	0.034 (0.013)	0.018 (0.070)
Softening						
Lime softening	205 (777)	64 (241)	41 (155)	0.026 (0.100)	0.011 (0.040)	0.001 (0.005)
Ion exchanger softener	201 (761)	90 (339)	55 (208)	0.021 (0.080)	0.006 (0.022)	0.004 (0.015)

TABLE 6.12 Average Costs for Process Units Typically Used in Groundwater Treatment (*Continued*)

Table 6.12 Average Costs for Process Units Typically Used in Groundwater Treatment (Continued)

Process Unit	Capital Cost (Including Construction Costs) ($/m³) ($/1000 gal)			O&M ($/m³) ($/1000 gal)		
	Small Plant <4000 m³/d (<1 mgd)	Medium Plant 4000–40,000 m³/d (1–10 mgd)	Large Plant >40,000 m³/d (>10 mgd)	Small Plant <4000 m³/d (<1 mgd)	Medium Plant 4000–40,000 m³/d (1–10 mgd)	Large Plant >40,000 m³/d (>10 mgd)
Membrane filtration						
MF/UF	302 (1145)	183 (691)	167 (633)	0.40 (1.5)	0.066 (0.250)	0.028 (0.110)
RO/NF	957 (2623)	507 (1920)	320 (1210)	4 (15)	1.3 (4.8)	0.276 (0.015)
Electrodialysis reversal	172 (650)	121 (456)	102 (385)	0.150 (0.580)	0.120 (0.470)	0.250 (0.940)
GAC	819 (3101)	599 (2266)	553 (2093)	0.558 (2.113)	0.390 (1.490)	0.231 (0.873)
Ion exchange	14 (54)	9.8 (37)	6.7 (25)	0.05 (0.19)	0.036 (0.14)	0.025 (0.009)
Disinfection						
Chlorine	19 (71)	6.7 (25)	2.6 (9.9)	0.025 (0.090)	0.007 (0.026)	0.001 (0.004)
Ozone (5 mg/L)	44 (166)	27 (103)	17 (64)	0.026 (0.100)	0.009 (0.034)	0.002 (0.008)
Chloramines	26 (98)	9.2 (35)	3.2 (12)	0.0004 (0.0016)	0.0004 (0.0016)	0.0004 (0.0016)
Ultraviolet radiation	44 (165)	29 (111)	23 (87)	0.090 (0.034)	0.03 (0.007)	0.003 (0.012)
Corrosion control (lime or soda ash)	30 (113)	6.1 (23.2)	2.8 (11)	0.004 (0.015)	0.0024 (0.009)	0.001 (0.004)

Sources: USEPA, 2005b; Culp/Wesner/Culp, 2000; Cotton et al., 2001.

Chemical	Use	Cost in $/kg ($/lb)
Chlorine gas	Disinfection/oxidation	0.23 (0.5)
Sodium hypochlorite 12.5%	Disinfection/oxidation	0.36 (0.8) as free Cl_2
Ammonia, anhydrous	Chloramines formation	0.18 (0.4)
Ammonia, aqueous	Chloramines formation	0.23 (0.5)
Liquid oxygen (LOX)	Ozone production	0.06 (0.13)
Sodium chlorite ($NaClO_2$) 80%	Chlorine dioxide production	1.14 (2.5)
Potassium permanganate 80%	Oxidation	1.14 (2.5)
Aluminum sulfate (alum), dry	Coagulation	0.05–0.1 (0.1–0.2)
Ferric sulfate, dry	Coagulation	0.07–0.08 (0.15–0.18)
Ferric chloride, dry	Coagulation	0.07–0.08 (0.15–0.18)
Ferrous sulfate, dry	Coagulation	0.07–0.08 (0.15–0.18)
PACl, liquid	Coagulation	0.05 (0.11)
Anthracite coal	Filter media	0.07 (0.15)
Cationic polymer	Coagulation aid	0.3 (0.7)
Anionic polymer	Flocculation, filter aid	1.14 (2.5)
Sulfuric acid	pH control	0.03 (0.06)
Sodium chloride, salt	Hypochlorite production and ionic exchange regeneration	0.01 (0.03)
Lime, quick	Softening, lime precipitation, and pH control	0.02 (0.05)
Lime, hydrated	Softening, lime precipitation, and pH control	0.02 (0.06)
Caustic soda, dry	pH control	0.07–0.08 (0.15–0.18)
Soda ash, dry	pH control	0.07 (0.15)
Activated carbon, granular	Removal of dissolved contaminants, T&O compounds	0.45 (1.0)
Powdered activated carbon	Removal of dissolved contaminants, T&O compounds	0.3 (0.7)
Carbon dioxide, liquid	Recarbonation	0.08 (0.17)
Citric acid	Membrane cleaning	0.36 (0.8)
Hydrofluosilicic acid	Fluoridation	0.1 (0.2)
Phosphoric acid	Corrosion control	0.2 (0.45)
Zinc orthophosphate, liquid	Corrosion control	0.45 (1.0)

TABLE 6.13 Indicative Prices of Common Water Treatment Chemicals in the United States

Water treatment chemicals represent a major component of the cost of treatment. Chemicals used in drinking water should be high-grade food additive class or approved by the local regulatory agencies (Kawamura, 2000). Table 6.13 provides indicative prices (for the US market) of most chemicals commonly used in water treatment and their uses. In addition to price, other factors that should be considered in the selection of a specific chemical include appropriateness, supply reliability, sludge formational and disposal costs, possible effects on other process units, environmental impact, dosing and maintenance issues, and safety.

Finally, a variety of coagulant aids derived from natural products are available for the treatment of drinking water. They include adsorptive clays (e.g., bentonite and fuller's earth) and limestone to improve the coagulation of high-color or low-turbidity sources, and a variety of natural polyelectrolyte coagulant aids. The price of these chemicals tends to be low, although dosage requirements tend to be high. Because of their low cost and local availability, they have been mainly used in developing countries.

References

Al-Muyeed, A., and Afrin, R., 2006. Investigation of the efficiency of existing iron and arsenic removal plants in Bangladesh. *Journal of Water Supply Research and Technology - AQUA*, vol. 55, no., 4, pp. 293–299.

AWWA and ASCE (American Society of Civil Engineers), 1998. *Water Treatment Plant Design*, 3rd ed. AWWA and ASCE, McGraw-Hill, New York, 806 p.

Binnie, C., Kimber, K., and Smethurst, G., 2002. *Basic Water Treatment*. International Water Association Publishing, London, 291 p.

Cleasby, J.L., and Logsdon, G. S., 1999. Granular bed and precoat filtration. In: Letterman, R.D. (ed.), *Water Quality and Treatment*, 4th ed. American Water Works Association, McGraw-Hill, New York, pp. 8.1–8.92.

Clifford, D.A., 1999. Ion exchange and inorganic adsorption. In: Letterman, R.D. (ed.), *Water Quality and Treatment*, 4th ed. American Water Works Association, McGraw-Hill, New York, pp. 9.1–9.87.

Clifford, D.A., Ghurye, G., and Tripp, A.R., 1999. Development of anion exchange process for arsenic removal from drinking water. In: Chappell, W.R., Abernathy, C.O., and Calderon, R.L., editors. *Arsenic Exposure and Health Effects*. Elsevier, New York, pp. 379–388.

Cotton, C.A., Owen, D.M., and Brodeur T.P., 2001. UV disinfection costs for inactivating *Cryptosporidium*. *Journal AWWA*, vol. 93, no. 6., pp. 82–94.

Culp/Wesner/Culp, 2000. WATERCO$T Model – a computer program for estimating water and wastewater treatment costs, Version 3.0, CWC. Engineering Software, San Clemente, CA.

DVGW/AWWARF/AWWA, 1996. *Internal Corrosion of Water Distribution Systems*. AWWA Publishers, Denver, CO, 586 p.

Earth Tech (Canada), 2005. Chlorine and Alternative Disinfectants Guidance Manual. Prepared for: Province of Manitoba (Canada) Water Stewardship. Canada Office of Drinking Water. Winnipeg, Manitoba, various paging.

Edwards, M., and Dudi, A., 2004. Role of chlorine and chloramines in corrosion of lead-bearing plumbing materials. *Journal AWWA*, vol. 96, no. 10, pp. 69–81.

El Paso Water Utilities, 2007. Available at: http://www.epwu.org/water/desal_info. html/. Accessed August 1, 2007.

Elyanow, D., and Persechino, J., 2005. Advances in nitrate removal. Available at: http://www.gewater.com/pdf/Technical%20Papers_Cust/Americas/English/TP1033EN. pdf. GE Technical Papers. Accessed August 1, 2007.

Faust, S.D., and Aly, O. M., 1999. *Chemistry of Water Treatment*, 2nd ed. Lewis, Boca Raton, FL, 581 p.

Geldreich, E.E., and LeChevallier, M., 1999. Microbiological quality control in distribution systems. In: *Water Quality and Treatment*, 4th ed. Letterman, R.D., editor. American Water Works Association, McGraw-Hill, New York, pp. 18.1–18.38.

Gregory, R., Zabel, T.F., and Edzwald, J.K., 1999. Sedimentation and flotation. In: *Water Quality and Treatment*, 4th ed. Letterman, R.D., editor. American Water Works Association, McGraw-Hill, New York, pp. 7.1–7.82.

Gu, B., and Coates, J.D., editors, 2006. *Perchlorate: Environmental Occurrence Interaction and Treatment*. Springer-Verlag, New York, 412 p.

Hafeman, D., Factor-Litvak, P., Cheng, Z., van Geen, A., and Ahsan, H., 2007. Association between Manganese Exposure through drinking water and infant mortality in Bangladesh. *Environmental Health Perspectives*, vol. 115, no. 7, pp. 1107–1112.

Health Canada, 2006. Guidelines for Canadian Drinking Water Quality: Guideline Technical Document: Arsenic. Prepared by the Federal-Provincial-Territorial Committee on Drinking Water of the Federal-Provincial-Territorial Committee on Health and the Environment. Ottawa, ON, Canada.

Huisman, L., and Wood, W.E., 1974. *Slow Sand Filtration*. World Health Organization. Out of print. Available only in electronic form. Available at: http://www.who.int/water_sanitation_health/publications/ssfbegin.pdf. Accessed August 1, 2007.

Hutchison, W.R., 2007. Desalination of Brackish Ground Water in El Paso, Texas. Paper presented at the National Ground Water Association, Ground Water Summit San Miguel (Albuquerque Convention Center), May 1, 2007.

Johnson, S.M., 2006. Health and Water Quality Monitoring of Pure Home Water's Ceramic Filter Dissemination in the Northern Region of Ghana. Master in Civil and Environmental Engineering Thesis at the Massachusetts Institute of Technology, 146 p.

Kawamura, S., 2000. *Integrated Design and Operation of Water Treatment Facilities*, 2nd ed. John Wiley & Sons, New York, 691 p.

Langlais, B., Reckhow, D.A., and and Brink, D.R., 1991. *Ozone in Water Treatment: Applications and Engineering*. American Water Works Association Research Foundation and Compagnie Generale des Eaux. Lewis Publishers, New York, 569 p.

Lantagne, D.S., 2001. *Investigation of the Potters for Peace Colloidal Silver Impregnated Ceramic Filter: Report 1: Intrinsic Effectiveness*. Alethia Environmental, Boston, MA. Available at: www.alethia.cc. Accessed August 1, 2007.

LeChevallier, M.W., and Au, K., 2004. *Water Treatment and Pathogen Control: Process Efficiency in Achieving Safe Drinking Water*. WHO Drinking Water Quality Series. IWA Publishing. Available at: http://www.who.int/water_sanitation_health/dwq/en/watreatpath.pdf. Accessed August 1, 2007.

Letterman, R.D, Amirtharajah, A., and O'Melia, C.R., 1999. Coagulation and flocculation. In *Water Quality and Treatment*. 4th ed. Letterman, R.D., editor. American Water Works Association, McGraw-Hill, New York, pp. 6.1–6.61.

Magyar, J., 1992. Kelliher arsenic removal study. Report WQ-149 Saskatchewan Environment and Public Safety. Regina, Saskatchewan, pp. 1–24.

Mallevialle, J., Odendaal, P.E., and Wiesner, M.R., 1996. *Water Treatment Membrane Processes*. American Water Works Association Research Foundation, Lyonnaise des Eaux, Water Research Commission of South Africa, McGraw-Hill, New York.

Mitchell, L.W., and Campbell, R. A. 2003. Exploring options when there are nitrates in the well. *Opflow AWWA*, vol. April, pp. 8–12.

Montgomery, 1985. *Water Treatment Principles & Design*. James M. Montgomery Consulting Engineers, Wiley Inter-Science, New York, 696 p.

NAS, 2001. *Arsenic in Drinking Water: 2001 Update*. National Academy of Sciences, National Research Council. National Academy Press, 248 p.

NRC, 1999. *Risk Assessment of Radon in Drinking Water*. Committee on Risk Assessment of Exposure to Radon in Drinking Water, Board on Radiation Effects Research, Commission on Life Sciences. National Research Council of the National Academies. The National Academies Press, Washington, DC, 279 p.

NRC, 2004. *Review of the Desalination and Water Purification Technology Roadmap*. Water Science and Technology Board. National Research Council of the National Academies. The National Academies Press, Washington, DC, 84 p.

Patel-Predd, P., 2006. Nanoparticles Remove Arsenic from Drinking Water. ES&T Science News. June 21. Available at: http://pubs.acs.org/subscribe/journals/esthag-w/2006/jun/tech/pp_arsenic.html. Accessed October 9, 2007.

Pokhrel, D., Thiruvenkatachari, V., and Braul, L., 2005. Evaluation of treatment systems for the removal of arsenic from groundwater. *The Practice Periodical of Hazardous, Toxic, and Radioactive Waste Management*, vol. 9, ro. 3, pp. 152–157.

Potters for Peace, 2007. Filters. Available at: http://s189535770.onlinehome.us/pottersforpeace/?page_id=9. Accessed September 24, 2007.

Renner, R., 2004a. Plumbing the depths of DC's drinking water crisis. *Environmental Science & Technology*, vol. 38, no. 12, pp. 224A–227A.

Renner, R., 2004b. More chloramine complications. *Environmental Science & Technology*, vol. 38, no. 18, pp. 342A–343A.

Sato, Y., Kang, M., Kamei, T., and Magara, Y., 2002. Performance of nanofiltration for arsenic removal. *Water Research*, vol. 36, pp. 3371–3377.

Schroeder, D.M., 2006. Field Experience with SONO Filters. Report for SIM Bangladesh. Available at: http://www.dwc-water.com. Accessed August 1, 2007.

Schultz, C.R., and Okum, D.A., 1984. *Surface Water Treatment for Communities in Developing Countries*. John Wiley & Sons, New York, 300 p.

Sharma, S.K., Petrusevski, B., and Schippers, J.C., 2005. Biological iron removal from groundwater: A review. International Water Association. *Journal of Water Supply: Research and Technology - AQUA*, vol. 54, no. 4, pp. 239–247.

Simms, J., and Azizian, F., 1997. Pilot-plant trials on the removal of arsenic from potable water using activated alumina. Proceedings of the American Water Works Association Water Quality Technology Conference, Denver, CO.

Singer, P.C., editor, 1999. *Formation and Control of Disinfection By-Products in Drinking Water*. American Water Works Association, Denver, CO, 424 p.

Singer, P.C., and Reckhow, D.A., 1999. Chemical oxidation. In: *Water Quality and Treatment*, 4th ed. Letterman, R.D., editor. American Water Works Association, McGraw-Hill, New York, pp. 12.1–12.46.

Snoeyink, V.L., and Summers, R.S., 1999. Adsorption of organic compounds. In: *Water Quality and Treatment*, 4th ed. Letterman, R.D., editor. American Water Works Association, McGraw-Hill, New York, pp. 13.1–13.76.

Stumm, W., and Morgan, J.J., 1996. *Aquatic Chemistry*, 3rd ed. John Wiley & Sons, New York, 1022 p.

Subramanian, K.S., Viraraghavan, T., Phommavong, T., and Tanjore, S., 1997. Manganese greensand for removal of arsenic in drinking water. *Water Quality Research Journal Canada*, vol. 32, no. :3, pp. 551–561.

Taylor, J.S., and Wiesner, M., 1999. Membranes. In: *Water Quality and Treatment*, 4th ed. Letterman, R.D., editor. American Water Works Association, McGraw-Hill, New York, pp. 11.1–11.67.

Tech Brief, 1997. Organic Removal. National Drinking Water Clearing House. Available at: http://www.nesc.wvu.edu/ndwc/pdf/OT/TB/TB5_organic.pdf. Accessed August 1, 2007.

USGS, 2007. Ground Water Glossary. Available at: http://capp.water.usgs.gov/GIP/gw_gip/glossary.html. Accessed December 17, 2007.

USEPA, 1999a. Alternative Disinfectants and Oxidants Guidance Manual. Office of Water. EPA 815-R-99-014. Available at: www.epa.gov/safewater/mdbp/alternative_disinfectants_guidance.pdf. Accessed October 10, 2007.

USEPA, 1999b. Guidance Manual for Compliance with the Interim Enhanced Surface Water Treatment Rule: Turbidity Provisions. USEPA Office of Water. EPA 815-R-99-010. Available at: www.epa.gov/safewater/mdbp/mdbptg.html. Accessed October 10, 2007.

USEPA, 1999c. *Enhanced Coagulation and Enhanced Precipitative Softening Guidance Manual*. Office of Water. EPA 815-R-99-012. Available at: www.epa.gov/safewater/mdbp/coaguide.pdf. Accessed October 10, 2007.

USEPA, 2000. Technologies and Costs for Removal of Arsenic from Drinking Water. USEPA Office of Water. EPA 815-R-00-028. December 2000. Available at: www.epa.gov/safewater/arsenic/pdfs/treatments_and_costs.pdf. Accessed October 10, 2007.

USEPA, 2003. Small Systems Guide to Safe Drinking Water Act Regulations. Office of Water. EPA 816-R-03-017. Available at: www.epa.gov/safewater/smallsys/pdfs/guide_smallsystems_sdwa.pdf. Accessed August 1, 2007.

USEPA, 2005a. Membrane Filtration Guidance Manual. Office of Water. EPA 815-R-06-009. Available at: www.epa.gov/safewater/disinfection/lt2/pdfs/guide_lt2_membranefiltration_final.pdf. Accessed October 10, 2007.

USEPA, 2005b. Technologies and Costs Document for the Final Long Term 2 Enhanced Surface Water Treatment Rule and Final Stage 2 Disinfectants and Disinfection Byproducts Rule. Office of Water. EPA 815-R-05-013. Available at: www.epa.gov/safewater/disinfection/lt2/regulations.html - 21k. Accessed October 10, 2007.

USEPA, 2006a. Point-of-Use or Point-of- Entry Treatment Options for Small Drinking Water Systems. Office of Water. EPA 815-R-06-010. Available at: http://www.epa.gov/safewater/smallsys/ssinfo.htm. Accessed October 10, 2007.

USEPA, 2006b. Drinking Water Contaminants. Office of Water. November 2006. Available at: http://www.epa.gov/safewater/contaminants/index.html. Accessed August 1, 2007.

USEPA, 2006c. Ultraviolet Disinfection Guidance Manual for the Final Long Term 2 Enhanced Surface Water Treatment Rule. Office of Water. EPA 815-R-06-007. Available at: www.epa.gov/safewater/disinfection/lt2/compliance.html. Accessed October 10, 2007.

US Federal Register, 1989. Title 40 of the Code of Federal Regulations (40 CFR) Parts 141 and 142. June 29, 1989. Drinking Water; National Primary drinking Water Regulations; Total Coliforms (Including Fecal Coliforms and *E. coli*); Final Rule.

US Federal Register, 2001. 40 CFR Parts 9, 141, and 142 National Primary Drinking Water Regulations; Arsenic and Clarifications to Compliance and New Source Contaminants Monitoring; Final Rule.

US Federal Register, 2006. Title 40 of the Code of Federal Regulations (40 CFR) Parts 9, 141, and 142. November 8, 2007. National Primary drinking Water Regulations: Ground Water Rule; Final Rule.

US Federal Register, 2007. Title 40 of the Code of Federal Regulations (40 CFR) Part 136. February 27, 2007 update.

Wagnick/GWI, 2005. *2004 Worldwide Desalting Plant Inventory*. Global Water Intelligence, Oxford, England.

Wachinski, A., Scharf, M., and Sellerberg, W., 2006. New technologies for the effective removal of arsenic from drinking water. *Asian Water*, vol. April, pp. 17–19.

Westerhoff, G., and Chowdhury, Z.K., 1996. Water treatment systems. In: *Water Resources Handbook*. Mays, L.M., editor. McGraw-Hill, New York.

White, G.C., 1999. *Handbook of Chlorination and Alternative Disinfectants*, 4th ed. Wiley Inter-science, New York, 1569 p.

WHO, 2006. Guidelines for Drinking Water Quality Incorporating First Addendum. World Health Organization. Available at: www.who.int/water_sanitation_health/dwq/gdwq3rev/en/index.html. Accessed August 1, 2007.

WHO, 2007. Slow Sand Filtration. World Health Organization. Available at: www.who.int/water_sanitation_health/publications/ssf/en/index.html. Accessed August 1, 2007.

Williams, R.B., and Culp, G.L., editors, 1986. *Handbook of Public Water Systems*. Van Nostrand Reinhold, New York 1113 p.

CHAPTER 7

Groundwater Development

7.1 Introduction

The history of groundwater development in the United States is very similar to many other countries, and it reflects patterns of socioeconomic development. Groundwater as a resource is less understood than surface water, as it is "hidden" and inaccessible for direct observation. The first phase of groundwater development is generally characterized by unrestricted land use, growth of agriculture, and use of shallow aquifers, which allow human settlement of large areas including semiarid regions with scarce surface water resources (Fig. 7.1). The general result of uncontrolled early groundwater development is the lowering of the water table and widespread shallow groundwater contamination. The next phase reflects advances in well drilling technology (Fig. 7.2), and irrigation and energy availability, which enable development of deeper aquifers on a large regional scale, more intensive agriculture, and faster population growth. At the same time, rapid groundwater development is followed by advancements in various groundwater-related scientific and engineering fields, and a growing public awareness of the importance of groundwater as an irreplaceable resource. Finally, as large-scale groundwater development brings more obvious consequences such as land subsidence, depletion of aquifer storage, and diminished flows of springs and surface streams, the challenge for science and engineering changes from supporting the development of groundwater resources to understanding its sustainability and impact on the environment.

As in many other countries, the surface waters of the United States are largely developed, with little opportunity available to increase storage along main rivers because few suitable sites remain for dams, and there is general concern about the environmental effects of impoundments. The surface waters of the nation also receive and assimilate, to a large degree, significant quantities of point- and nonpoint-source contaminants (Anderson and Woosley, 2005). In contrast, approximately 800,000 boreholes are drilled for water in the United States each year. Although not all boreholes are completed as wells, this represents significant and continuing groundwater development. There are an estimated 16 million water wells in the United States, of which 283,000 are public supply wells with distribution systems (National Ground Water Association, 2003).

Rivers are an important component of the natural environment as well as the economic infrastructure. They are crucial water supplies for municipal, industrial, and agricultural uses and are sources for recreation, power generation, and transportation of goods. The history of water policy in the United States is dominated by the construction of structures such as dams, canals, dikes, and reservoirs (Gleick, 2000). As pointed out by Anderson and Woosley (2005), many of the 77,000 dams in the United States were

Figure 7.1 Many early immigrant families like this one settled in large areas of the American Midwest, in part thanks to readily available shallow groundwater, and helped turn it into one of the most important agricultural regions in the world. (Photograph courtesy of the Natural Resources Conservation Service.)

constructed largely without considering the environmental consequences. The dams on the Columbia River, for example, were constructed without the full understanding of the long-term consequences to fish populations. When Glen Canyon dam was proposed for the Colorado River, few concerns were expressed about the downstream ecosystem. The environmental debate focused on the submergence of Glen Canyon's sculptured canyon walls, beneath what is now Lake Powell in Utah. Today, the management of Glen Canyon dam is influenced strongly by numerous factors downstream from the dam, such as preservation of the limited sand supply that maintains beaches for recreation and wildlife habitat and protection of the Kanab amber snail and the humpback chub (Bureau of Reclamation, from Anderson and Woosley, 2005).

The storage, diversion, use, and reuse of limited surface water resources to support agriculture, industry, and the expanding human population in the American West have had an adverse effect on the health and sustainability of aquatic biological communities and associated riparian and wetland habitats (Postel and Richter, 2003). Some examples of adverse effects include the following (Anderson and Woosley, 2005):

- Loss of ecologically significant wetlands and other riparian habitat and invasion of less desirable native and nonnative, exotic vegetation owing to reduced flooding and increased groundwater pumpage. Riparian habitat, such as that along the Pecos River in New Mexico and Texas, supports more than 75 percent of the animal species in arid regions during some stage of their life cycles and is the sole

FIGURE 7.2   California well rig after completing 12-in. well flowing 5,250,000 gallons per 24 hours. (From Slichter, 1905; photograph courtesy of U.S. Geological Survey.)

habitat for amphibians and invertebrates that require moist conditions (Patten, 1997).

- Changes in the saltwater-freshwater interface in coastal estuaries, such as those in the San Francisco Bay in California, and the ecosystems dependent on them.

- Coastal land subsidence, such as that in the Houston-Galveston area of Texas, caused by subsurface fluid extraction.

- Contamination of fish and wildlife and their habitats as a result of irrigation return flows, urban runoff, point-source discharges, and numerous abandoned mine drainage areas in Colorado and South Dakota.

- Reduced populations of anadromous fish owing to in-stream restrictions to migration, such as those in the Columbia River in Oregon.

- Extinction or near extinction of native fish species, as in the Upper Klamath Basin in Oregon and California, resulting from habitat degradation.

- Increased erosion of stream banks associated with loss of bank-stabilizing riparian vegetation in numerous urbanized locations throughout the West.

The U.S. Fish and Wildlife Service and federal courts are using the authority under the Endangered Species Act (ESA) to divert water from past economic-based uses to support sensitive ecological communities and their habitats. The competition for water in the arid West, however, was intense well before considerations were being given to ecological water needs. As discussed in Chap. 4, surface waters of the Colorado River Basin managed by a series of reservoirs are overallocated for water supply and irrigation uses. An analysis of in-stream water needs showed that in-stream flows in the Rio Grande, the Upper Colorado, and the Lower Colorado water-resource regions are insufficient to meet current needs for wildlife and fish habitat, much less allow for any additional off-stream use (Guldin, 1989). As discussed in previous chapters, the competition for and overallocation of surface water resources are not limited to the American West. For example, the population growth and the recent droughts in the American Southeast are fueling the ongoing disputes over surface water rights between Georgia, Alabama, and Florida.

Sustainable groundwater development has become a focal point of the integrated water resources management in many countries, including where use of surface water resources had an undisputable priority. Following are some of the reasons for this trend in regions where both surface water and groundwater resources are available:

- Groundwater development requires lower capital investment and simpler distribution systems, since it can be executed in phases and closer to end users.

- Surface water intakes and storage (reservoirs) are more vulnerable to seasonal fluctuations in recharge, as well as periods of drought; they are also more vulnerable to the projected impacts of climate change, as discussed in Chap. 4.

- Evaporative loss from surface water reservoirs is large, especially in semiarid and arid regions, whereas such loss from groundwater systems (storage) is mostly negligible or nonexistent.

- The environmental impacts of surface water reservoirs are incomparably less acceptable to the general public than just a generation ago.

- The general quality of surface water bodies and their sediments has been impacted by point- and nonpoint sources of contamination to a much greater extent and for longer periods of time, requiring more expensive drinking water treatment and use of a variety of chemicals.

- Surface water supplies are more vulnerable to accidental or intentional contamination.

- The ability of surface water systems to balance between daily and seasonal periods of peak demand and periods of low demand is limited. In contrast, water wells can simply be turned off and on, and their pumping rates can be adjusted as needed.

Proper groundwater development can completely avoid some of the above-listed problems inherent to surface water supplies; it can also alleviate most of them as part of the integrated management of both surface water and groundwater. One such approach, described in more detail in Chap. 8, is artificial aquifer recharge with surface water and used water. In addition to aquifer recharge, which is usually considered more of a water management strategy, groundwater development is generally accomplished by the following: (1) installation of individual water wells or wellfields, (2) construction of underground dams (reservoirs), and (3) regulation of springs. In addition to classical vertical wells, groundwater extraction can be performed in a number of other ways, including with horizontal and slanted wells, collector wells, infiltration galleries, drainage galleries, trenches, and drains.

7.2 Water Wells

Probably the first thought that comes to mind for many people when discussing groundwater development and use, in general, is a well. For nonhydrogeologists and those who are not in a related water supply profession, a well usually means a nondescript hole in the ground that somehow produces water; this may include an image of a fenced well house or a picturesque country-side image of a dug well with a rotating wooden wheel and a bucket. In any case, a relatively small number of people fully understand the complexity, importance, and cost of a properly constructed well used for public water supply. The same is true in many developed countries where modern drilling technologies have been routinely used for a long time to construct wells for both the public and the domestic supply; the end users usually leave this "well business" to well drillers and do not care to learn much about their own hole in the ground. However, hydrogeologists and groundwater professionals think of wells in many different contexts, and some of them spend lifetimes trying to better understand and design them.

Selecting the "best" site for a well or wellfield is considered by some to be an art or a matter of luck, by some to be as simple as finding a recommended dowser, and by some as a natural part of collecting fees for drilling holes into the subsurface as deep as possible. Although a few groundwater professionals may agree with some of these statements (except, arguably, for the question of a dowser), siting and then designing wells for public supply or large irrigation projects is very complex and should result from thorough considerations of multiple design elements. In most cases, selecting the final well location(s) and well design is not a straightforward task but rather a compromise made after considering various factors such as

- Capital cost
- Vicinity to future users
- Existing groundwater users and groundwater permits
- Hydrogeologic characteristics and depth to different water-bearing zones (aquifers)
- Required flow rate of the water supply system and expected yield of individual wells
- Well drawdown and radius of well (wellfield) influence
- Interference between wells in the wellfield
- Water treatment requirements
- Energy cost for pumping and water treatment, and general operations and maintenance (O&M) costs
- Aquifer vulnerability and risks associated with the existing or potential sources of contamination
- Interactions with other parts of the groundwater system, and with surface water
- Options for artificial aquifer recharge, including storage and recovery

- Societal (political) requirements
- Existence or possibility of an open water market

The above factors are not all inclusive and are not listed in order of importance; sometimes just one or two factors are all that is needed for proceeding with the final design. However, as the development and use of groundwater resources is becoming increasingly regulated in the United States and in many other countries, it is likely that most of these factors will have to be addressed as part of a well-permitting process. Even in cases where permitting requirements are absent, it is prudent to consider most, if not all, of the listed factors, since they ultimately define the long-term sustainability of any new well or wellfield.

An overview of the technical design of water wells is discussed further in this chapter, while various aspects of groundwater management and protection, including optimization of groundwater extraction by wellfields, are presented in Chap. 8.

7.2.1 Vertical Wells

Vertical wells have been used for centuries for domestic and public water supply throughout the world. Their depth, diameter, and construction methods vary widely and there is no such thing as a "one size fits all" approach to well design. Answers to just about any question regarding well design can be found in the classic 1000-page book *Groundwater and Wells* by Driscoll (1986, published by Johnson Filtration Systems, now available from Johnson Screens/a Weatherford Company). Another exhaustive reference book on well design is *Water Well Technology* by Campbell and Lehr (1973). Various public-domain publications by U.S. government agencies provide useful information on the design and installation of water supply and monitoring wells (e.g., USEPA, 1975, 1991; U.S. Bureau of Reclamation, USBR, 1977; Aller et al., 1991; Lapham et al., 1997).

Well design, installation, and well construction materials should conform to applicable standards. In the United States, the most widely used water well standard is the ANSI/AWWA A100 standard, but the authority to regulate products for use in, or contact with, drinking water rests with individual states, which may have their own standard requirements. Local agencies may choose to impose requirements more stringent than those required by the state (AWWA, 1998).

The design elements of vertical water wells include the following: (1) drilling method, (2) boring (drilling) and casing diameter, (3) depth, (4) well screen, (5) gravel pack, (6) well development, (7) well testing, and (8) selection and installation of the permanent pump.

Whenever possible, a well design should be based on information obtained by a pilot boring drilled prior to the main well bore. Geophysical logging and coring (sample collection) of the pilot boring provide the following information: depth to and thickness of the water-bearing intervals in the aquifer, grain size, and permeability of the water-bearing intervals, and physical and chemical characteristics of the porous media and groundwater. Unknown geology and hydrogeology of the formation(s) to be drilled may result in the selection of an improper drilling technology, sometimes leading to a complete abandonment of the drilling location due to various unforeseen difficulties such as flowing sands, collapse of boring walls, or loss of drilling equipment in karst cavities.

If pilot boring is not feasible, some design parameters must be estimated, on the conservative side, and may significantly reduce well efficiency (e.g., selecting smaller screen openings or gravel-pack grain size to prevent entrance of fines).

The expected well yield, the well depth, and the geologic and hydrogeologic characteristics of the porous media (rock) all play an important role in selecting the drilling diameter and the drilling method. Deep wells, or thick stratification of permeable and low-permeable porous media, may require drilling with several diameters, and the installation of several casings of progressively smaller diameter, called telescoping casing. This is done to provide stable and plumb boreholes in deep wells and to bridge difficult or undesirable intervals (e.g., flowing sands, highly fractured and unstable walls prone to caving, and thick sequences of swelling clay). The cost of drilling increases progressively with the drilling diameter, and it is important to balance this cost with other design requirements, some of which may be desirable but not always necessary. For perspective, a public supply, high-capacity well that is several thousands feet deep (say, 1000 m) may easily cost well over US $1 million. Such wells are drilled with large drill rigs and may use special large-diameter drilling bits and include large-diameter casing as shown in Fig. 7.3.

Ultimately, the expected well capacity is the parameter that will define the last drilling diameter sufficient to accommodate the screen diameter, including thickness of any gravel pack, for that capacity. The relationship between the two diameters is not linear—doubling the screen diameter will not result in doubling the well yield as illustrated in Fig. 7.4. For example, for the same drawdown and radius of influence, an increase in diameter from a 6- to 12-in well will yield only 10 percent more water. In addition to the screen diameter, the diameter of the riser pipe (inner casing), which conducts pumped water to the surface, also plays a role in selecting the drilling diameter. The riser pipe may have the same diameter as the screen, or it may be larger in which case the screen and the casing are connected with a diameter reducer. In either case, the riser pipe diameter must satisfy two requirements: (1) the casing diameter must be large enough to accommodate the pump of required capacity and to provide for easy maintenance access, and (2) the diameter of the casing must be sufficient to assure that the uphole velocity is less than 5 ft/s (1.5 m/s) to avoid an excessive pipe loss (Driscoll, 1986).

All permanent well casings have to be continuous and watertight from top to bottom, except for the screen section, and have to be grouted (i.e., they cannot be left loose). Grouting prevents possible short-circuiting of groundwater along the boring walls and between various aquifer intervals or aquifers, and contamination from the land surface. In the United States, most states require that the upper casing be grouted a minimum depth from the land surface, usually 50 ft. Casing material must be compatible with the groundwater chemistry to prevent corrosion or other failures. The selection of materials for well casing is critical in locations where there is likelihood of its exposure to significant concentrations of contaminants comprising low-molecular-weight petroleum products or organic solvents and their vapors. Casing materials such as polyethylene, polybutylene, polyvinyl chloride (PVC), and elastomers, such as used in jointing gaskets and packing glands, may be subject to permeation by lower-molecular-weight organic solvents or petroleum products (AWWA, 1998). If the well casing extends through such a contaminated area or an area subject to contamination, the well casing material should be selected accordingly. Casing has to be strong and thick enough to provide for structural stability during its installation, well development, and use. This is especially important in deep wells where high formation pressures may cause casing to collapse if undersized

Figure 7.3 A 48-in.-diameter surface casing being placed in a 54-in.-diameter borehole. The target depth for the well is 1300 ft with an open interval from 1000 to 1300 ft in the Floridan aquifer. The final casing inside diameter is 20 in. The well is drilled for water supply of the city of Miramar, FL. (Photograph courtesy of Richard Crowles.)

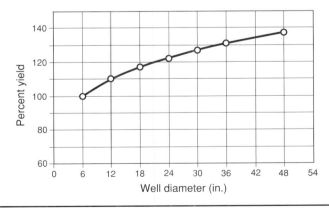

Figure 7.4 Graph showing well diameter versus percent yield increase from the basic 6-in.-screen-diameter well in an unconfined aquifer, pumping at 100 gpm and having a 400-ft radius of influence. (Data from Driscoll, 1986).

(underdesigned). In general, selecting an inferior casing material will initially reduce the capital cost but may result in irreparable damage to the casing and premature loss of a well. ANSI/AWWA A100 standard provides specifications for casing materials and diameters, and calculations of minimum acceptable casing strengths.

Wells in stable bedrock are in many cases completed as an open borehole intersecting as many fractures as possible to maximize well yield and minimize construction cost. Such wells should have appropriately grouted casing in the top portion to prevent caving-in of regolith material and pieces of weathered rock and contamination of the well from the land surface. It is highly recommended that the grouted casing extends through the entire thickness of the regolith and the highly weathered rock, and for some distance into the competent bedrock. This prevents deterioration of the well due to inflow of fine particles, and it extends the life of the well pump. Although the final boring diameter of unlined, open-borehole wells still has to accommodate the pump assembly and easy maintenance access, it is not limited by the well-screen diameter and gravel-pack thickness required for the screened wells.

Table 7.1 illustrates the selection of drilling methods based on geologic formation and well depth, and Table 7.2 lists the optimum and minimum size of well inner casing (riser pipe) for various pumping rates.

Figure 7.5 illustrates some of the more common well designs, which can vary widely based on project-specific requirements and can be combined in a single well. Continuing advances in drilling and well installation technology allow for elaborate designs such as under-reaming (widening of borehole below already installed and grouted casings), use of temporary casings for drilling in unstable conditions, telescopic screens, multiple screen intervals with or without continuous gravel packs, and slanted wells.

As discussed later in this chapter, iron bacteria can cause a variety of problems in water wells. Because it is difficult to eliminate iron bacteria once they exist in well systems, prevention is the best safeguard against accompanying problems. For well drillers, prevention means disinfecting everything that goes into the ground with a strong (250 parts per million (ppm)) chlorine solution. Iron bacteria are nourished by carbon and other organics, and it is essential that these are not introduced into any part of the well system during the drilling process. Tools, pumps, pipe, gravel-pack material, and even the water used in drilling should be disinfected. Use of a tank that circulates chlorinated water instead of digging a mud pit will help avoid contamination from soil. For owners of new wells in places where iron bacteria have been a problem, the best prevention is to be especially alert for signs of their occurrence. If the well driller and pump installer are scrupulous in keeping the new well "clean," iron bacteria even in such areas can be avoided (Wisconsin DNR, 2007).

Well Screen

The well screen is arguably the most important part of a well, since this is where groundwater enters the well and where the efficiency of an otherwise good design may be compromised, including loss of the entire well. In unconsolidated materials, and under certain conditions in consolidated materials, the well casing (lining) must be used to stabilize the formation materials and prevent their caving (entrance) into the well bore. To allow the water to enter the well, openings must be placed in the well casing opposite the water-bearing (targeted) aquifer intervals. These intervals with openings (perforations) are called well screens. Casing and screens both stabilize the formation materials, while screens, in addition to inflow of water, allow proper well development. It has

TABLE 7.1 Applicable Drilling Method for Different Types of Geologic Formations

Characteristics	Dug	Bored	Driven	Drilled				Jetted
				Percussion	Rotary			
					Hydraulic	Air		
General range of common depths	0–50 ft	0–100 ft	0–50 ft	0–1000 ft	0–1000 ft	0–750 ft		0–100 ft
Diameter	3–20 ft	2–30 in.	1¼ to 2 in.	4–18 in.	4–24 in.	4–10 in.		2–12 in.
Type of geologic formation								
Clay	Yes	Yes	Yes	Yes	Yes	No		Yes
Silt	Yes	Yes	Yes	Yes	Yes	No		Yes
Sand	Yes	Yes	Yes	Yes	Yes	No		Yes
Gravel	Yes	Yes	Finer size	Yes	Yes	No		1/4 in. pea gravel
Cemented gravel	Yes	No	No	Yes	Yes	No		No
Boulders	Yes	Yes if < than well diameter	No	Yes when in firm bedding	Difficult	No		No
Sandstone	Yes, if soft or fractured	Yes, if soft or fractured	Thin layers	Yes	Yes	Yes		No
Limestone			No	Yes	Yes	Yes		No
Dense igneous rock	No	No	No	Yes	Yes	Yes		No

From USEPA, 1991.

Anticipated Well Yield		Optimum Casing Size		Smallest Casing Size	
gpm	L/s	in	mm	in	mm
less than 100	less than 5	6 ID	152 ID	5 ID	127 ID
75–175	5–10	8 ID	203 ID	6 ID	152 ID
150–350	10–20	10 ID	254 ID	8 ID	203 ID
300–700	20–45	12 ID	305 ID	10 ID	254 ID
500–1,000	30–60	14 OD	356 OD	12 ID	305 ID
800–1,800	50–110	16 OD	406 OD	14 OD	356 OD
1,200–3,000	75–190	20 OD	508 OD	16 OD	406 OD
2,000–3,800	125–240	24 OD	610 OD	20 OD	508 OD
3,000–6,000	190–380	30 OD	762 OD	24 OD	610 OD

ID, inside diameter; OD, outside diameter.
Modified from Driscoll, 1986; reprinted by permission of Johnson Screens—A Weatherford Company.

TABLE 7.2 Recommended Well Diameters for Various Pumping Rates

been generally accepted that the screens of public water supply wells should be made of high-quality stainless steel (AWWA, 1998). To reduce the possibility of corrosion, the well screen and its fittings should be fabricated of the same material.

During the process of well development, the finer materials from the productive water-bearing zones, as well as any fines introduced by the drilling fluid, are removed so that only the coarser materials are in contact with the screen. In formations where the porous media grains surrounding the screen are more uniform in size (homogeneous) and are graded in such a way that the fine grains will not clog the screen, the developed aquifer materials will form a so-called natural pack consisting of grains coarser than further away from the well bore. Such wells are called naturally developed wells. In contrast, when the targeted aquifer (formation) intervals are heterogeneous and have predominantly finer grains, it may be necessary to place an artificial gravel pack around the screen intervals. This gravel pack (also called filter material) will allow proper well development and prevent the continuous entrance of fines and screen clogging by the fines during well operation.

The size of well-screen openings depends on the grain size distribution of the natural porous media. When natural well development is not possible, the size of screen openings is also dependent on the required gravel-pack characteristics (gravel-pack grain size and uniformity). The percentage of openings, the screen diameter, and the screen length should all be selected simultaneously to satisfy the following criteria: (1) maximize well yield, (2) maximize well efficiency by minimizing hydraulic loss at the screen, and (3) provide for structural strength of the screen, i.e., prevent its collapse due to formation pressure.

The following equation can be used to determine the optimum relationship between different screen parameters (AWWA, 1998):

$$L = \frac{Q}{A_e V_e (7.48)} \tag{7.1}$$

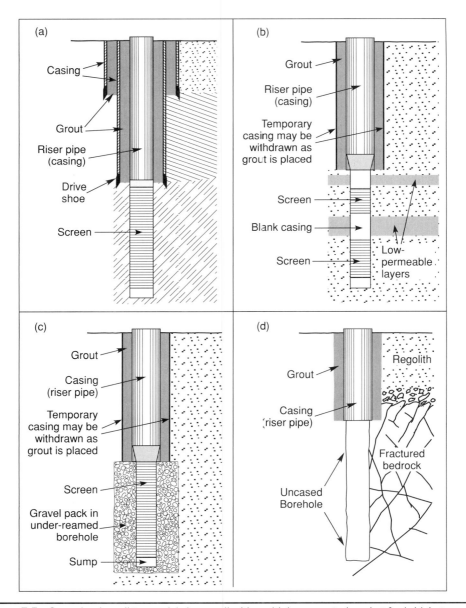

Figure 7.5 Some basic well types: (a) deep well with multiple cemented casing for bridging unstable and undesired formations; (b) naturally developed well in unconsolidated formation with intervals of low-permeability sediment that are not screened, and with telescoped screen; (c) well in unconsolidated sediments with telescoped screen and gravel pack in the under-reamed (widened) borehole below riser pipe cemented in place; and (d) well completed as open borehole in stable fractured bedrock and with casing cemented in place and extending through regolith and upper portion of the bedrock.

where L = length of screen (ft)

Q = discharge (gallons per minute)

A_e = effective aperture area per foot of screen, in square feet (the effective aperture area shall be taken as one-half of the total aperture area [ft^2/ft])

V_e = design entrance velocity (ft/min)

As a rule of thumb, the screen entrance velocity should be equal to or less than 0.1 ft/s (0.03 m/s), since it has been shown that higher velocities cause turbulent well loss, may accelerate various screen problems such as corrosion and incrustation, and can transport sand particles (Walton, 1962; Driscoll, 1986). A lower entrance velocity is recommended for water of significant incrusting potential (USEPA, 1975). ANSI/AWWA standard for the upper limit of entrance velocity is 1.5 ft/s (0.46 m/s); the users of the standard are cautioned to thoroughly examine the issue of well-screen entrance velocity and site-specific (aquifer) conditions before the final selection (AWWA, 1998).

In naturally developed wells, screen apertures should be sized according to the following criteria (AWWA, 1998):

1. Where the uniformity coefficient of the formation is greater than 6, the screen-aperture openings should retain from 30 to 40 percent of the aquifer sample.

2. Where the uniformity coefficient of the formation is less than 6, the screen-aperture openings should retain from 40 to 50 percent of the aquifer sample.

3. If the water in the formation is corrosive or the accuracy of the aquifer sample is in doubt, a size selected should retain 10 percent more than in items 1 and 2.

4. Where fine sand overlies coarse sand, use the fine-sand aperture size for top 2 ft (0.61 m) of the underlying coarse sand. The coarse-sand aperture size should not be larger than twice the fine-sand aperture size.

For gravel-packed wells, the screen-aperture openings should be sized to retain between 85 and 100 percent of gravel-pack material.

There are many different types of screens and screen opening configurations offered worldwide by a variety of vendors. Some types may have advantages in certain conditions, and the final selection should be made after careful considerations of site-specific design requirements. Continuous-slot screens (Fig. 7.6) provide maximum open area and access to the formation so that well development procedures are enhanced and through-screen head loss is reduced.

In general, the screen should be as long as possible and placed within a thick aquifer interval with the highest hydraulic conductivity. However, when the aquifer is stratified with less permeable interbeds, it is preferable to use multiple screen intervals separated by solid casing, including screens with varying slot size selected to match porous media in different water-bearing (production) intervals. This will prevent the continuous entrance of fines from the undesired intervals. It is recommended to select screen intervals such that the water level in the well during pumping always stays above the top of the screen. At the same time, the pump intake should not be placed within the screen interval but within the solid casing (also called a riser pipe or pump-housing pipe) above or below the screen. This prevents hydraulic stresses on the screen when the pump is turned on and off and problems associated with screen dewatering, which can accelerate screen corrosion and scaling (incrustation).

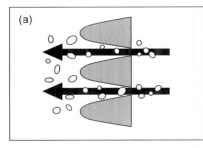 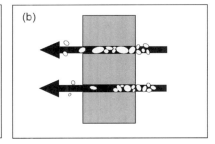

FIGURE 7.6 Top: Finishing of continuous-slot screen. Bottom: (a) slot openings in continuous-slot screen are V shaped and nonclogging because they widen inwardly; particles passing through the narrow outside opening can enter the screen and (b) elongated or slightly oversize particles can clog straight-cut, punched, or gauze-type openings. (Photograph courtesy of Johnson Screens—A Weatherford Company.)

Driscoll (1986) provides the following recommendations for the selection of screen intervals and their lengths for some common hydrogeologic situations:

- *Homogeneous unconfined aquifer.* Screening of the bottom one-third to one-half of an aquifer less than 150 ft (45 m) thick provides the optimum design for homogeneous unconfined aquifers. In some cases, however, particularly in thick, deep aquifers, as much as 80 percent of the aquifer may be screened to obtain higher specific capacity and greater efficiency, even though the total yield is less. A well in an unconfined aquifer is usually pumped so that, at maximum capacity, the pumping water level is maintained slightly above the top of the pump intake or screen. The well screen is positioned in the lower portion of the aquifer because the upper part is dewatered during pumping. Maximum drawdown should not exceed two-thirds of the saturated thickness because larger drawdown does not provide significant additional yield but increases well loss and energy cost for pumping.

- *Heterogeneous unconfined aquifer*. The basic principles of well design for homogeneous unconfined aquifers also apply to this type of aquifer. The only variation is that the screen or screen sections are positioned in the most permeable layers of the lower portion of the aquifer so that maximum drawdown of two-thirds of the aquifer saturated thickness is available.

- *Homogeneous confined aquifer*. In this type of aquifer, 80 to 90 percent of the thickness of the water-bearing sediment should be screened, assuming that the pumping water level is not expected to be below the top of aquifer. Maximum available drawdown for wells in confined conditions should be the distance from the hydraulic head (potentiometric) surface to the top of the aquifer. If the available drawdown is limited, it may be necessary to lower the hydraulic head below the aquifer top in which case the aquifer will respond like an unconfined aquifer during pumping.

- *Heterogeneous confined aquifer*. Most relatively thick confined aquifers are heterogeneous and screen sections should be placed in 80 to 90 percent of permeable layers, interspaced with blank casing in the less permeable (silt and clay) zones of the formation. Continuous screens of varying slot size (multiple-slot screens) can be successfully used in generally permeable, water-bearing aquifer sections, consisting of alternating layers of finer and coarser sediments. Two recommendations should be followed when selecting slot openings for such screens, to avoid entrance of the finer material into the well (Driscoll, 1986): (1) if fine material overlies coarse material, extend at least 3 ft (0.9 m) of the screen designed for the fine material into the coarse material below; and (2) the slot size for the screen section installed in the coarse layer 3 ft beneath the formation contact should not be more than double the slot size for the overlying finer material. Doubling of the slot size should be done over screen increments of 2 ft (0.6 m) or more.

Well Gravel Pack

Gravel pack is now routinely placed around well screens in wells completed in both uniform and heterogeneous (nonuniform) formations for the following reasons: (1) to stabilize the formation; (2) to minimize flow of fines and sand through the screen; (3) to enable larger screen openings, which can increase well efficiency and minimize rate of screen incrustation; and (4) to establish transitional velocity and pressure fields between the formation and the well screen, which also minimizes the incrustation rate.

The placement of a gravel pack makes the zone around the well screen more permeable and increases the effective hydraulic diameter of the well. The gravel pack allows the removal of finer formation material during well development, and it retains most of the aquifer fine material during the well exploitation. Gravel pack is particularly useful in fine-grained, uniformly graded formations, and in extensively laminated aquifers where there are alternating layers of silt, sand, and gravel. In addition to sand pumping, which can mechanically destroy the pump and the screen, another very important problem in well maintenance can be avoided or deferred by a properly installed gravel pack—chemical and biological incrustation of the well screen and the material adjacent to the well screen. Carbonate, iron, and manganese incrustation are the most common problems. They are partially related to the velocity-induced pressure changes that disturb the chemical equilibrium of the groundwater in the well-screen zone. Due to the permeability change, there is an abrupt increase of groundwater velocity at the contact between the aquifer material and the gravel pack before the well development. This increase corresponds to an equally abrupt decrease of pressure, which causes the precipitation of calcium carbonate, iron, and manganese. The precipitation may occur either in an improperly developed well with a gravel pack or in a well without a gravel pack (here the pressure drop occurs at the contact between the aquifer material and the screen). Well development results in a significant change in the velocity and pressure fields and the permeability in the screen zone. Most of the fine grains are removed from the aquifer material, resulting in an increased permeability. At the same time, the permeability of the gravel-pack portion is somewhat smaller due to the filling of the pore space between the pack grains with the aquifer material. The result is a gradually increasing groundwater velocity and a gradually decreasing pressure in the well-screen zone, which allows groundwater to carry its chemical load into the well rather than to precipitate it onto the screen and gravel-pack (aquifer) material.

In order to successfully retain the formation particles, the thickness of the gravel (filter) pack in ideal conditions does not have to be more than 0.5 in. according to laboratory tests made by Johnson Screens: "Filter-pack thickness does little to reduce the possibility of sand pumping, because the controlling factor is the ratio of the grain size of the pack material in relation to the formation material" (Driscoll, 1986). However, for practical purposes, the thickness of the gravel pack should be at least 3 in. to ensure its accurate placement and the complete surrounding of the screen. On the other hand, a filter pack that is more than 8 in. (203 mm) thick can make the final development of the well more difficult "because the energy created by the development procedure must be able to penetrate the pack to repair the damage done by drilling, break down any residual drilling fluid on the borehole wall, and remove fine particles (from the formation) near the borehole" (Driscoll, 1986).

As discussed by Campbell and Lehr (1973), probably the most common cause of well sanding is the use of a gravel pack that is too coarse for at least part of the formation. A relatively thin interval of fine sand, sandwiched between coarser sand and gravel, may continue to sift through the pack indefinitely. Generally, the problem of well sanding can be caused by three factors: (1) poor sampling of the formation (aquifer) materials, (2) lack of care in selecting the gravel-pack size, and (3) improper placement of gravel pack.

The choice of the gravel-pack size is based on the grain size distribution of the formation (aquifer) materials. If the aquifer material is uniform and well sorted, a uniform gravel pack is selected. The grading of the filter pack in this case is based on the grain size distribution curve of the finest aquifer material within the well-screen section. A graded gravel pack may be considered for aquifers with a wide range of particle sizes and large uniformity coefficients. In this case, the grading of the pack depends on both the coarsest and the finest aquifer materials.

Uniform gravel packs are designed for aquifers with uniformity coefficients (U) less than 2.5. The first criterion for choosing the gravel pack is as follows:

$$D_{70} = 4 \text{ to 6 times } d_{70} \qquad (7.2)$$

where D_{70} = sieve opening size which would retain 70 percent of the gravel-pack material and d_{70} = sieve opening size which would retain 70 percent of the finest formation (aquifer) material to be filtered. It is also common to use the fiftieth percentile retained of the gravel pack and the formation material in the above relationship.

The second criterion is

$$U_{\text{pack}} = \frac{d_{60}}{d_{10}} < 2.5 \tag{7.3}$$

where U_{pack} = uniformity coefficient of the gravel pack and d_{60} and d_{10} are the sieve opening sizes which would allow 60 and 10 percent of the aquifer material to pass, respectively. The gravel-pack grain size distribution should form a smooth and gradual curve, parallel to the formation curve. As described earlier, the screen aperture openings should retain 85 to 100 percent of the gravel-pack (filter) material. Fig. 7.7 is an example of the gravel-pack and the formation grain size characteristics for the above criteria.

Uniform gravel packs are generally preferred because well screens can be manufactured with varying sizes of openings to match the formation materials, and there is less possibility for separation during gravel-pack placement. To prevent separation, bridging, and voids, special equipment such as tremie pipe (ordinary 4-in. pipe) is needed. The tremie is lowered to the bottom of the annular space between the screen and the

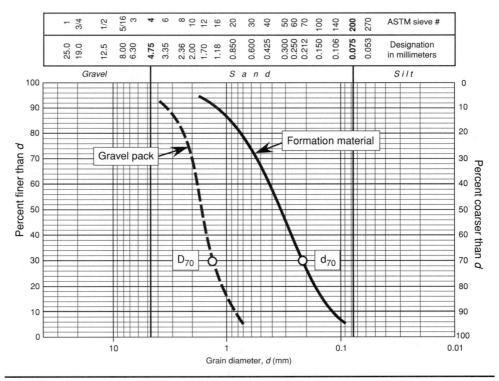

FIGURE 7.7 Example of a uniform gravel-pack and formation grains size distribution curves.

well bore wall and filled with pack material, which is allowed to settle 4 or 5 ft at each application as the tremie is slowly raised.

The placement of a gravel pack by a reverse-circulation method is generally accepted as being more effective than the placement with tremie pipes. This method is usable for wells of any depth, with certain modifications for relatively deep applications (Campbell and Lehr, 1973). If the velocity of the descending stream in the annular space is about the same as the velocity at which a particle of gravel falls in a fluid, no separation of sizes should occur.

The filter material should extend some distance above the top of the screen. Then it should be sealed with cement or a mixture of cement and bentonite, which is placed between the top of the gravel pack and the lower limit of the sanitary seal or outer casing.

It is very important that the gravel-pack material is made of well rounded silica grains with less than 5 percent soluble impurities and without organic materials. It should also not contain any iron, manganese, copper, lead, or any other heavy metals in a form or quantity that will adversely affect the quality of the well water.

Well Development

Proper well development will improve almost any well regardless of its type and size, whereas without development an otherwise excellent well may never be satisfactory. As discussed by USEPA (1975) and Driscoll (1986), in any well drilling technology the permeability around the borehole is reduced. Compaction, clay smearing, and driving fines into the wall of the borehole occur in the cable tool drilling method. Drilling fluid invasion into the aquifer and formation of a mud cake on the borehole walls are caused by direct rotary method. Silty and dirty water often clog the aquifer in the reverse rotary drilling method. In consolidated formations, compaction may occur in some poorly cemented rocks, where cuttings, fines, and mud are forced into fractures, bedding planes, and other openings, and a mud cake forms on the wall of the borehole.

Proper well development breaks down the compacted borehole wall, liquefies jelled mud, and draws it and other fines that have penetrated the formation (aquifer) into the well, from which they are removed by bailing or pumping. Well development also removes smaller grains initially present in the formation and creates a more permeable and stable zone adjacent to the well screen. Development of new wells is based on the mechanical action of water or air and should involve backwashing, i.e., movement of water in both directions through the well-screen openings. Chemicals should be used in well development only exceptionally, in small quantities, and with prior approval of both the well owner and the regulatory agency where applicable. Such chemicals include mud-dispersing agents such as crystalline and glassy polyphosphates and acids for wells completed in limestone. Various problems can arise from improper use of chemicals leading to an inferior well or sometimes to a complete screen clogging. For example, an overrich solution of sodium hexametaphosphate (SHMP) can precipitate glassy phosphates on contact with the cold groundwater. These glassy precipitates are gelatinous masses that are extremely difficult to remove because no effective solvents exist (Driscoll, 1986).

There are various methods of well development, and their selection depends primarily on the applied drilling technology and the formation characteristics. However, availability of the equipment and driller's preference in many cases play unjustifiably more important roles. It is often impossible to anticipate how a well will respond to certain types of development and how long it will take to achieve adequate development.

Since a lump-sum basis for well development may result in unsatisfactory work, it is better to provide for development on a unit price per hour basis and continue until the following conditions have been met (AWWA, 1998):

1. Sand content should average not more than 5 mg/L for a complete pumping cycle of 2-hour duration when pumping at the design discharge capacity.
2. No less than 10 measurements should be taken at equal intervals to permit plotting of sand content as a function of time and production rate and to determine the average sand content for each cycle.
3. There should be no significant increase in specific capacity during at least 24 hours of development.

The USEPA (1975) provides the following guidance as to the acceptable sand contents in clean well water:

- Wells supplying water for food-type irrigation and where the nature of the water-bearing formation and the overlying strata are such that pumping the following amount of sand will not seriously shorten the useful life of the well: 15 mg/L limit.
- Wells supplying water to sprinkler irrigation systems, industrial evaporative cooling systems, and other uses where a moderate amount of sand is not especially harmful: 10 mg/L limit.
- Wells supplying water to homes, institutions, municipalities, and industries other than those mentioned above: 5 mg/L limit.
- Wells supplying water to be used directly in contact with or in the processing of food and beverages: 1 mg/L limit.

General methods of well development are pumping, surging, fracturing, and washing, each of which has several variations (USEPA, 1975). It is recommended that at least two methods be applied for best results. One of the less effective—but commonly used—methods is overpumping the well. Here, water flows in one direction only—toward the well. The flow velocities are generally not fast enough to remove much of the fine material plugging the formation. During overpumping, a surging action can be created by periodically shutting off the pump and allowing the water in the pump column to flow back into the well. This is more effective than overpumping. However, water will reenter the most permeable parts of the formation or those that have been least damaged during well construction. Thus, the portion of the formation that requires the most active development is largely excluded (Johnson Screens, 2007). Injecting water into the well and then pumping it back from the well (backwashing) will result in movement of water through the well's screen and gravel pack in both directions, thus increasing the effectiveness of development.

Pumping with compressed air, or airlift (Fig. 7.8), is probably the most common method of well development. However, an improper use of the technique can cause various problems and may even lead to screen collapse. In certain formations, such as where stratified, coarse sand and gravel lenses are separated by thin impermeable clay layers, it should be avoided altogether because it may cause air locking of the formation.

Figure 7.8   Development of water well with airlift, St. Louis, MI. (Photograph courtesy of James Brode, Fishbeck, Thompson, Carr and Huber, Inc.)

In formations where air trapping is a problem, other techniques such as surging with air and high-velocity jetting should be used. Driscoll (1986) provides a very detailed discussion on various airlift techniques, including determinations of quantitative parameters required for proper airlift design. Unfortunately, many well drillers and contractors always apply the same airlift method (one they are familiar with) without regard for site-specific conditions.

The most efficient form of development is the use of high-pressure water or air jetting combined with simultaneous airlift pumping. This method employs a jetting tool that is lowered into the well and injects water at high pressure (1000 to 1500 kPa) through a series of nozzles. Jetted water dislodges clogging material from the well screen and gravel pack and puts it into suspension. This fine suspended material is removed from the well by the simultaneous airlift pumping. The pumping rate should be substantially greater than the rate at which water is injected into the well by the jetting tool. High-velocity jetting can be localized to an area of the formation that requires the development.

As with unconsolidated formations, all drilling methods including air drilling cause some plugging of fractures and other openings in consolidated sediments and hard rock formations. Therefore, any material that clogs openings in such formations should be removed during well development. In many cases, the best method is the water jetting combined with airlift pumping. Inflatable packers can isolate the productive zones (fractures) supplying water to the well and increase efficiency of their development.

In general, well yield (specific capacity) in consolidated and hard rock formations can be increased significantly if one or more well stimulation methods are applied. Well stimulation is considered as a second level of development, which can increase well performance beyond that obtained through traditional methods. As discussed by Driscoll (1986), sandstone aquifers require the most careful considerations regarding both well development and well stimulation. For example, open-borehole wells in friable sandstone may never be developed to the extent of reducing sanding to an acceptable level, even when expensive and long-lasting techniques such as blasting and bailing are applied. For this reason, an increasing number of wells in sandstone are screened at the expense of some loss of specific capacity.

Hydrofracturing is used to stimulate both new and old wells in consolidated rock formations. In hydrofracturing, water at extreme pressures can be injected into the entire well or into discrete intervals sealed by packers. The injected water removes sediment from the existing fractures and creates new fractures, resulting in an increased permeability of the formation adjacent to the well.

Blasting with explosive charges lowered in an uncased borehole in consolidated rock is sometimes used to increase well-specific capacity. Similarly to hydrofracturing, this method enlarges the existing fractures and creates new ones, resulting in an increased hydraulic conductivity of the formation. However, blasting with explosives should be applied with utmost care and only after considering many factors including legal requirements and environmental impacts.

Acid can be used for well stimulation and formation development in limestone and dolomite aquifers and in some semiconsolidated aquifers with calcium carbonate cement. Acid dissolves carbonate minerals and enlarges voids and small fissures in the formation adjacent to the borehole. Acid can also be forced into discontinuities away from the well, resulting in dissolution and removal of a larger volume of the native material. This increases the overall hydraulic conductivity of the aquifer around the well and may result in a significant increase of the specific capacity of the well.

Well Testing and Performance

Testing of well performance and aquifer characteristics is conducted after developing the well and allowing for recovery and stabilization of the water level in the well. Figure 7.9 shows hydrographs and the corresponding drawdown for a pumping test designed for that purpose. The first part of the test, which has three steps, is designed to determine the well characteristics such as well loss and need for possible redevelopment. The duration of each step should be the same, usually not more than 6 to 8 hours. Data recorded during the first step are used to initially assess the transmissivity and the storage coefficient of the aquifer. The size of the pump and the long-term pumping rate for the second part of the test are selected based on drawdown development during the three-step test. The second part of the test should be performed after a complete recovery of the hydraulic head in the well and with a maximum feasible pumping rate. Duration of this part of the test, which is designed to determine the overall aquifer transmissivity for an extensive radius of influence, depends on specific project requirements and may vary from 24 hours to several weeks in case of aquifer development for major water supply purpose. Long-term pumping with a maximum rate should uncover aquifer characteristics that may be less obvious after a short test: distant boundaries, leakage, presence of dual porosity, or changes in storage. Both the drawdown and the recovery data should be used to find

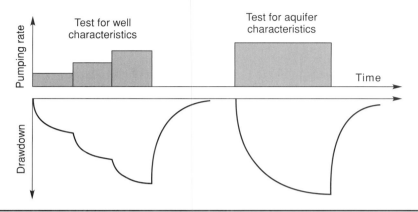

FIGURE 7.9 Pumping rate hydrographs and drawdown curves for a pumping test designed to determine well and aquifer characteristics.

the aquifer parameters. As explained in Chap. 2, at least one monitoring well near the pumping well should be available to analyze the test results.

Well loss is the difference between the actual measured drawdown in the pumping well and the theoretical drawdown due to groundwater flow through the aquifer's porous media. This theoretical drawdown is also called the formation loss. Equations of theoretical drawdown should be applicable to the actual aquifer (formation) conditions, such as confined, unconfined, leaky, with delayed gravity response, quasi steady state, or transient. Well loss is the result of various factors, such as an inevitable disturbance of the porous medium near the well during drilling, an improper well development (e.g., drilling fluid is left in the formation and mud cake along the borehole is not removed), a poorly designed gravel pack or well screen, and turbulent flow through the gravel pack and the well screen. Well loss is always present in pumping wells, and its evaluation is an important part in deciding if the well performance is satisfactory or not. All wells will experience a decrease in well efficiency sooner or later, as indicated by an increased well loss. Three-step pumping test is the only reliable means of quantifying the well loss, and it should be performed not only after well completion but also periodically during well exploitation to evaluate the well performance and needs for possible well rehabilitation.

The total measured drawdown (s_w) at a well is combination of the linear losses and turbulent losses:

$$s_w = AQ + BQ^2 \tag{7.4}$$

where A = coefficient of the linear losses, B = coefficient of turbulent losses, and
Q = pumping rate

The turbulent losses are usually assumed to be quadratic, but other powers may be used to describe it. The linear losses (A) include both the formation loss (A_0) and the linear loss (A_1) in the near-screen zone:

$$A = A_0 + A_1 \tag{7.5}$$

For practical purposes A_1 can usually be ignored. The formation loss or the theoretical drawdown in the well (s_0) is determined by using the appropriate equation for the specific flow condition. For example, in case of a quasi-steady-state flow in a confined aquifer, the equation is as follows:

$$s_0 = \frac{Q}{2\pi T} \ln \frac{R}{r_w} \tag{7.6}$$

where s_0 = drawdown due to groundwater flow through the aquifer
porous media
T = transmissivity
R = radius of well influence
r_w = well radius

The coefficient of linear formation loss (A_0) can be calculated as follows:

$$A_0 = \frac{1}{2\pi T} \ln \frac{R}{r_w} \tag{7.7}$$

A_0 can also be determined graphically if two or more monitoring well data are available, as shown in Fig. 7.10. The same graph shows that the s/Q ratio for the pumping well increases with the increasing pumping rate, whereas for the monitoring wells this ratio remains constant for all three steps.

The coefficients of the total linear loss (A) and the quadratic loss (B) can be determined from graph pumping rate (Q) versus drawdown-pumping rate ratio (s/Q), as shown in Fig. 7.11. The graph is a straight line of the following form:

$$\frac{s_w}{Q} = A + BQ \tag{7.8}$$

where A = intercept and B = slope of the best-fit straight line drawn through the experimental data from the three pumping steps. After substituting values of A and B

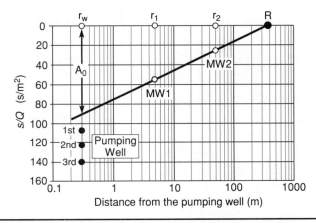

Figure 7.10 Graph distance versus s/Q for three-step pumping test in quasi-steady-state conditions, showing data for pumping well and two monitoring wells.

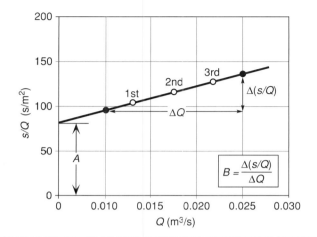

Figure 7.11 Graph pumping rate (Q) versus drawdown-pumping rate ratio (s/Q) for a three-step well pumping test.

determined from the graph into Eq. (7.4), it is possible to calculate the total (i.e., expected to be actually recorded) drawdown in the pumping well for any pumping rate. A graph similar to that in Fig. 7.12 can be used to plot the calculated drawdown expected to be actually measured in the well, versus the theoretical formation drawdown (which does not include well loss).

The coefficient of turbulent (quadratic) well loss, B, smaller than 2500 to 3000 s^2/m^5 is usually considered acceptable. Larger coefficients may indicate potential problems with the well such as inadequate well design and/or development, clogging of the well screen or other deterioration of the well. Theoretically, B is not time dependent and

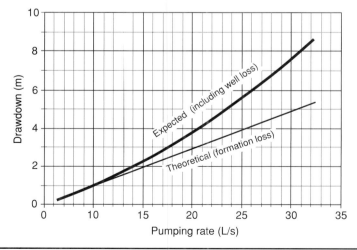

Figure 7.12 Pumping rate versus formation loss (theoretical drawdown) calculated with Eq. (7.6) and the expected drawdown in the well calculated with Eq. (7.4).

should remain the same for different pumping rates. A common exception is pumping from karst and fractured rock aquifers where turbulent well loss may increase with an increasing pumping rate. In such cases, the points on the graph Q versus s/Q would form a parabola rather than a straight line.

In case of transient conditions in a confined aquifer, the coefficient of formation loss is easily found by applying the Theis equation:

$$A_0 = \frac{1}{2\pi T} W(u) \tag{7.9}$$

Parameter u for the well is given as follows:

$$u = \frac{r_w^2 S}{4Tt} \tag{7.10}$$

where r_w = well radius
$\quad S$ = storage coefficient
$\quad T$ = aquifer transmissivity
$\quad t$ = time since the pumping started

As shown by Cooper and Jacob, for small values of parameter u ($u < 0.05$), i.e., sufficiently long pumping time, the well function $W(u)$ is

$$W(u) = \frac{2.25Tt}{r_w^2 S} \tag{7.11}$$

and the formation loss (i.e., theoretical drawdown s) can be written as

$$s = \frac{Q}{2\pi T} \frac{1}{2} \ln \frac{2.25Tt}{r_w^2 S} \tag{7.12}$$

$$s = \frac{Q}{2\pi T} \ln \sqrt{\frac{2.25Tt}{r_w^2 S}} \tag{7.13}$$

$$s = \frac{Q}{2\pi T} \ln \frac{\sqrt{\frac{2.25Tt}{S}}}{r_w} \tag{7.14}$$

$$s = \frac{Q}{2\pi T} \ln \frac{1.5\sqrt{Tt/S}}{r_w} \tag{7.15}$$

Notice that Eq. (7.15) looks similar to the steady-state equation describing groundwater flow toward a fully penetrating well in a confined homogeneous aquifer:

$$s = \frac{Q}{2\pi T} \ln \frac{R_D}{r_w} \tag{7.16}$$

where R_D = radius of well influence, which does not change in time (steady-state flow), also called Dupuit's radius of well influence. From the analogy between Eqs. (7.15) and

(7.16) it is apparent that, in transient conditions, Dupuit's radius of well influence is time dependent and is expressed as

$$R_D = 1.5 \times \sqrt{Tt/S} \tag{7.17}$$

Theoretically, for an infinite confined aquifer, the groundwater flow forms in infinity and reaches the well pumping rate at the well perimeter (r_w). The corresponding radius of well influence also approaches infinity for a long pumping period ($t \to \infty$), which means that Dupuit's radius of well influence does not have a real physical meaning. For most practical purposes, however, Dupuit's radius of well influence given with Eq. (7.17) will yield satisfactory results in various analytical calculations involving the Theis equation. Again, it should be noted that a definite real radius of well influence could not be formed in a homogeneous confined aquifer unless there is a source of recharge, such as from a boundary or from leakage. Using the expression for Dupuit's radius of well influence, the coefficient of the linear formation loss is

$$A_0 = \frac{1}{2\pi T} \ln \frac{1.5 \times \sqrt{Tt/S}}{r_w} \tag{7.18}$$

Similar to steady-state conditions, the coefficients of linear and turbulent well losses are found from a graph Q versus s/Q, as shown in Figs. 7.10 and 7.11. However, because the radius of well influence in transient conditions increases with time, the drawdown recorded at the end of each step may have to be corrected if not sufficiently stabilized. Figure 7.13 shows the components of drawdown recorded at the end of each step and the error made if the three drawdowns (s_1, s_2, and s_3) were used to draw a graph Q vs. s/Q without corrections. Kresic (2007) provides detailed explanation of this correction procedure.

Well Efficiency   Well efficiency is the ratio between the theoretical drawdown and the actual drawdown measured in the well. It is expressed in percent:

$$\text{Well Efficiency} = \frac{\text{Theoretical Drawdown } (s_0)}{\text{Measured Drawdown } (s_w)} \times 100\% \tag{7.19}$$

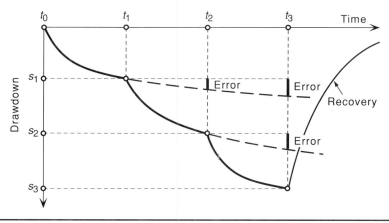

FIGURE 7.13   Components of drawdown recorded at the end of each step showing errors made if drawdowns s_1, s_2, and s_3 were used directly to draw graph Q versus s/Q.

As explained earlier, the theoretical drawdown is determined by applying an appropriate equation of groundwater flow toward a well (theoretical drawdown equals the formation loss). It can also be found graphoanalytically as explained earlier. In general, the difference between the theoretical drawdown and the measured drawdown increases with increasing pumping rate as shown in Fig. 7.12. Consequently, well efficiency decreases with an increasing pumping rate. Determining well efficiency and well loss is highly recommended because it provides valuable information about the well performance and can be used to make an informed decision regarding the well pumping rate, maintenance, and rehabilitation. A well efficiency of 70 percent or more is usually considered acceptable. If a newly developed well has less than 65 percent efficiency, it should not be approved without a thorough analysis of the possible underlying reasons. This may include well redevelopment followed by new performance testing.

Well efficiency has always been a major concern in the water well industry, but in the past this concern has focused primarily on pumping equipment. Pump efficiencies can be easily calculated, and most pump manufacturers make this information available to their customers. Efficiencies of different pumps may vary by only a few percentage points. Well efficiency, on the other hand, can vary greatly even between wells completed in very similar hydrogeologic conditions and close to one another. However, the determination of well efficiency is still largely ignored because it is difficult to quantify different factors that contribute to it. Many variables affect well efficiency including the drilling procedure, screen design, filter pack size, and development methods.

One aspect of well efficiency is operating costs over the life of the well. Efficiencies of two wells can be compared through these costs including cost savings from a more efficient well. Figure 7.14 compares the efficiency of two wells in the same aquifer. The direct cost for power can be calculated by the following formula (Johnson Screens, 2007):

$$Cph = \frac{m^3/hr \times tdh \times 0.746 \times kwh \; cost}{270 \times pump \; efficiency} \tag{7.20}$$

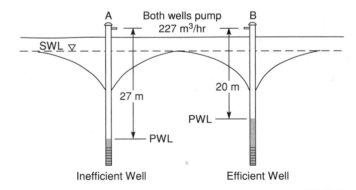

Figure 7.14 Two wells completed in the same aquifer can have nearly identical cones of depression and pump water at the same rate. The less efficient well (A), however, will have a lower pumping level. The extra drawdown will significantly increase the pumping costs of the well. PWL, pumping water level in the well; SWL, static water level before pumping. (Figure courtesy of Johnson Screens, 2007.)

where   Cph = cost of power per hour
           m³/hr = rate of pumping in cubic meters per hour
           tdh = total dynamic head, which includes the distance from the pump
                 discharge down to the pumping water level plus the elevation
                 pressure in meters beyond the pump discharge, and the head due to
                 friction and turbulence of flow
           0.746 = constant to convert brake horsepower to kilowatt-hours
           kwh cost = cost per kilowatt-hour for electricity in cents

Costs for wells A and B (Fig. 7.14) can be compared assuming the same pumping rate of 227 m³/hr for both wells, above-ground head of 50 m (i.e., the total dynamic head is 50 m + 27 m = 77 m for well A and 50 m + 20 m = 70 m, for well B), electricity cost of 10 cents per kilowatt-hour, and overall pump efficiency of 60 percent (0.60). Note that the cones of depression are the same. The inefficient well, however, requires more drawdown to produce the same volume of water, which significantly increases the pumping costs.

The cost for the inefficient well (A) with the above input parameters is $8.05 per hour of operation, whereas this cost for the efficient well (B) is $7.32. If these wells operated for 4000 hours during a year, the savings at well B would amount to $2920 in electricity cost per year. If these wells operated 20 years, that savings would amount to $58,400. While direct savings such as this is sufficient reason in itself for insisting upon an efficient well, the savings in indirect costs could exceed the savings in power costs. Indirect costs arise from maintenance expenses, short life span, and initial pumping costs. The principal causes of maintenance expenses and short life are corrosion, incrustation, and sand pumping. A correctly designed and constructed well will reduce these factors, contributing to well inefficiency to a minimum level (Johnson Screens, 2007).

The specific capacity of a well is given as

$$\text{Well Specific Capacity} = \frac{\text{Pumping Rate}\,(Q)}{\text{Measured Drawdown}\,(s_w)} \tag{7.21}$$

and expressed as the pumping rate per unit drawdown (e.g., liters per second per 1 m of drawdown). Similar to well efficiency, the well's specific capacity also decreases with the increasing pumping rate. It also decreases in time due to inevitable well deterioration. Choosing an optimum rate at which a well will be pumped is a decision based on numerous factors. For example, if the well will be used for a short-term construction dewatering, maintaining a desired drawdown may be the only relevant criterion. In some cases, where there are no alternatives, a certain pumping rate is all that matters. On the other hand, if the well is designed for a long-term water supply, in addition to the energy cost of pumping, the hydraulic criteria are the most important in deciding which pumping rate is the optimum one. This includes a comparative analysis of well losses and well efficiency for various pumping rates.

Well Completion

Well completion includes disinfection of the entire casing and the screen, installation of the permanent well pump, and construction of the well housing (wellhead) with all its sanitary requirements. It is highly recommended that a dedicated small-diameter perforated pipe ("sounding line") be installed in the well for water-level measurements

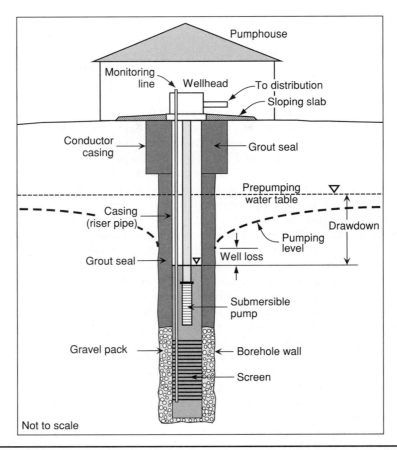

Figure 7.15 Schematic of a typical vertical well completed in unconfined aquifer. (Modified from Kresic, 2007; copyright Taylor & Francis Group; used with permission.)

and depth-discrete sampling. Figure 7.15 shows the main design elements of a well completed in a shallow, unconfined aquifer.

Probably the only design element of the entire well that cannot be guaranteed by any well driller, hydrogeologist, or engineer is the well yield and its long-term sustainability. It is not uncommon that, for various reasons, a well that costs several hundred thousand dollars to complete disappoints all stakeholders by actually producing just a fraction of its designed capacity. However, many such surprises can be avoided by following well-established hydrogeologic principles of aquifer evaluation and testing, and, of course, well design itself.

Well Maintenance and Rehabilitation

Continuous monitoring of well performance is the most important part of the overall O&M program. It enables early detection of well deterioration signs such as changes in the chemical and physical characteristics of the pumped water, or a notably increased drawdown for the same pumping rate. Well performance testing, as described earlier, is the easiest method for confirming the actual well deterioration. It also helps in excluding possible external reasons for apparent drawdown changes such as regional drawdown increases or the influence of other wells. In general, it is very important to have easy access for measuring water level in the well. This is best provided by a permanently installed small-diameter pipe ("sounding line"). Perforated sounding lines can also be used to sample well water at different depths and can host probes for continuous monitoring of water levels and various physical and chemical parameters such as temperature, pH, Eh, and electrical conductance. A permanent flow meter, preferably with the automated recording of the well pumping rate, is an essential tool for adequate O&M of any well. The metering system should allow measurements of the instantaneous pumping rate, the total pumped volume, and the indicator of actual hours of pumping and hours of standby. The permanent sounding line and the flow meter enable easy regular testing of well efficiency and specific capacity by well operators, eliminating the need for external consultants and minimizing interruptions of the water supply system.

The continuous monitoring of well performance does not eliminate the need for routine maintenance of the pump assembly, which includes pulling out the pump, cleaning it, and replacing various parts according to the manufacturer's schedule. Removal of the pump assembly also allows for optional inspection of the casing and the screen with a downhole video camera. Regular replacement of the pump discharge column every 3 to 4 years, or more frequently if needed, is also highly recommended, regardless of its apparent state. Although this entire process is dreaded by all well operators, it will assure an early detection of any well deterioration signs such as scaling and corrosion of the screen and the well pump.

When the three-step pumping test shows a decrease of well efficiency on the order of 20 percent or so, it will be necessary to pull out the pump assembly and inspect the entire casing and the screen using downhole video camera and geophysical tools. The inspection should include sampling of any screen-clogging (incrustation) materials to determine their type and select appropriate well rehabilitation methods. Well efficiency testing should be frequent enough to prevent a "sudden" discovery of various problems. When well efficiency decreases more than 40 to 50 percent, it is usually too late for any rehabilitation method to be effective. The well efficiency and specific capacity should be determined before and after any rehabilitation (redevelopment) to evaluate the effectiveness of the procedure.

Improper well operation such as excessive pumping rate, which results in excessive drawdown including partial dewatering of the well screen, can greatly accelerate well deterioration. High entrance velocities, as well as exposure of well screen to air, change chemical balance in the water entering the well and may cause or accelerate incrustation and corrosion. Similarly, frequent changes in pumping rates and cycling of pumping (turning the pump on and off) can damage the screen due to hydraulic stresses and promote incrustation due to turbulent flow and related water oxygenation (aeration). As long nonpumping periods also promote scaling, it is recommended that the pump operates continuously for 18 to 20 hours a day, followed by short period of rest during which the aquifer can recover to a certain degree and regain some form of equilibrium. Longer periods of operation can also allow for comparatively lower pumping rates and lower drawdown for the same daily demand.

Two general processes lead to well deterioration (aging) and may eventually result in a complete loss of the well: corrosion and incrustation (clogging). In general, however, proper well monitoring and O&M including timely rehabilitation can extend life of a well for decades. Chemical corrosion problems can be significantly delayed or avoided in the first place by using appropriate, corrosion-resistant, and high-quality casing and screen materials such as stainless steel. In situ cathodic protection of the already-installed well casing and screen may be effective in minimizing corrosion in some cases. This method, developed in the oil industry, is not widely applied in the water well industry mainly because of the cost. However, it can be very useful in extending the life of large-production municipal wells, which are expensive to replace or rehabilitate and whose operation cannot be interrupted for longer periods of time. Corroded casing of large-diameter wells can sometimes be lined with corrosion-resistant (inert) material.

In general, electrochemical corrosion is promoted by the presence of one or more constituents in the water including dissolved oxygen, carbon dioxide, hydrogen sulfide, water with low pH (acidic water), chlorides, and calcium sulfates such as gypsum. Mechanical corrosion is caused by continuous sanding, which enlarges screen openings and may result in irreparable screen damage.

Some form of incrustation is inevitable for any well, and, in extreme cases, no rehabilitation measures can save the well. Incrustation is defined as being any clogging, cementation, or stoppage of a well screen, gravel pack, and water-bearing formation, which is the result of a collection of material in and adjacent to the openings of the screen and pore spaces of the formation and gravel-pack materials (Driscoll, 1986). Incrustation may take the form of a hard, brittle, cement-like deposit, or it may be a soft and pasty sludge or jelly-like deposit. Incrustation can be triggered by a variety of chemical, biochemical, and hydraulic processes, but the main reason is installation of the well itself, which changes the natural equilibrium of the groundwater system by default. The most common causes of incrustation are the following:

- Chemical precipitation of materials carried to the screen in solution, such as carbonates of calcium and magnesium and hydroxides of iron and manganese
- Mechanical deposition of fine-grained materials such as clays and silts carried to the screen in suspension
- Biochemical precipitation caused by iron and manganese bacteria, including deposition of their organic matter
- Activity of slime-forming organisms other than iron bacteria (e.g., organisms that feed on ammonia and organic matter)

Mechanical well clogging with fine formation materials can be removed in the same manner that new wells are developed—using various pumping, surging, and jetting methods as described earlier. The mechanical action of water flushes fines from around the well screen and the filter pack. Flushing methods may also be effective for loosening and removing chemical and biological deposits unless they are particularly hard or thick. Flushing can be accompanied by screen scrubbing with mechanical brushes.

Methods such as acid washing can be used to remove scaling deposits from the well screen and gravel pack, but caution must be exercised in evaluating the applicability of any method that introduces a chemical to the subsurface. Considerations concerning the use of chemicals to rehabilitate a well include permitting requirements when approval of the regulatory agency is needed and possible chemical reactions between the introduced chemical and the well materials (screen and casing). For example, harsh or corrosive chemicals such as acids can damage certain well-screen materials such as PVC. In addition, acids can degrade the integrity of the adjacent formation material and actually cause a clogging problem to become worse. As discussed by Campbell and Lehr (1973), hydrochloric and other acids cause certain silicates to swell, expanding the individual particles to as much as five times the original size. Such a reaction could completely plug the formation and offset the increased permeability caused by the acid. In some cases, formation minerals containing iron are dissolved by the acid and form iron chloride. The iron chloride remains in solution until the acid spends to a pH of 3.5 or higher. At this time, the iron precipitates as iron hydroxide, a jelly-like material having remarkable plugging properties. These are just some of the reasons why the use of chemicals to rehabilitate wells must be evaluated very carefully and should include thorough analysis of possible interactions between the chemical, the well materials (screen and casing), and the natural formation materials.

Iron Bacteria   Iron-related bacteria grow naturally in environments with steep Eh gradients, such as groundwater seeps and wetlands. Similar conditions can develop in the vicinity of a well when oxygenated groundwater mixes with anoxic groundwater that contains ferrous iron. Favorable conditions for the development of iron bacteria are pH in the range of 5.4 to 7.2, ferrous iron concentration between 1.6 and 12 mg/L, presence of CO_2, redox potential (Eh) higher than -10 ± 20 mV, and high velocities of water entering the well (Detay, 1997). Biological clogging occurs regardless of the nature of the screen material (e.g., steel, copper, and synthetic materials), but metals are commonly susceptible to corrosion that results from clogging and bacterial activity.

Various terms such as screen clogging, screen encrustation, and screen (aquifer) biofouling have been used to describe the geochemical and microbial processes that contribute to declines in the specific capacity of a well caused by iron bacteria. Manganese-related well-screen encrustation, which involves the oxidation of manganese (II) and the precipitation of manganese oxyhydroxides, is less common but can occur in some hydrogeologic environments. Most iron bacteria can also form biofilms consisting of manganese oxyhydroxides and cellular material (Walter, 1997).

Iron bacteria can cause extensive physical clogging of well screens and promote growth of sulfur bacteria, which produce hydrogen sulfide (the "rotten egg" smell) that can corrode well casing and screen. Iron bacteria are oxidizing agents—they combine iron or manganese dissolved in groundwater with oxygen. A side effect of the process is a foul-smelling brown slime, which, although not a health hazard, can cause unpleasant odors and tastes and can change water color to yellow-red or orange. If conditions are right, the bacteria can grow at amazing rates and an entire well system may be rendered virtually useless in just a few months. Iron bacteria can also cause biochemical corrosion of the screen and pump beneath the slimy scale, which becomes rich in disaggregated metal.

Studies carried out in France show that biological clogging does not merely affect the screen or gravel pack but may extend much farther beyond into the aquifer formation. In most cases, such clogging is due to the presence of heterotrophic anaerobic and sulfate-reducing bacteria, which develop as a result of the nutrient flux brought in by pumping, thus forming a biomass several meters in size, which considerably reduces the permeability of the medium (Detay, 1997).

Treatment techniques that may be successful in removing or reducing iron bacteria include (1) physical removal, (2) pasteurization, and (3) chemical treatment. Treatment of heavily infected wells may be difficult, expensive, and only partially successful. Iron bacteria are widely spread in various hydrogeologic environments, and many state agencies in the United States offer guidance and technical help with iron-bacteria treatment.

The following discussion is from the Minnesota Department of Health, Division of Environmental Health (MDH, 2007).

Physical removal of iron bacteria is typically done as a first step in heavily infected wells. The pumping equipment in the well must be removed, cleaned, and disinfected, and the well casing is then scrubbed by use of brushes or other tools. Physical removal is usually followed by chemical treatment.

Pasteurization has been successfully used to control iron bacteria. Pasteurization involves a process of injecting steam or hot water into the well and maintaining a water temperature in the well of 60°C (140°F) for 30 minutes. Although pasteurization can be effective, its application may be cost prohibitive.

Chemical treatment is the most commonly used iron bacteria treatment technique. The three groups of chemicals typically used include (1) surfactants, (2) acids (and bases), and (3) disinfectants, biocides, and oxidizing agents. Surfactants are detergent-like chemicals such as phosphates. They are generally used in conjunction with other chemical treatment. It is important to use chlorine or another disinfectant if phosphates are used, since bacteria may use phosphates as a food source.

Acids have been used to treat iron bacteria because of their ability to dissolve iron deposits, destroy bacteria, and loosen bacterial slime. Acids are typically part of a series of treatments involving chlorine, and at times, bases. Extreme caution is required to use and properly dispose of these chemicals. Acid and chlorine should never be mixed together. Acid treatment should only be done by trained professionals.

Disinfectants are the most commonly used chemicals for treatment of iron bacteria, and the most common disinfectant is household laundry bleach, which contains chlorine. Chlorine is relatively inexpensive and easy to use but may have limited effectiveness and may require repeated treatments. Effective treatment requires sufficient chlorine strength and time in contact with the bacteria and is often improved with agitation. Continuous chlorine injection into the well has been used but is not normally recommended because of concerns that the chlorine will conceal other bacterial contamination and cause corrosion and maintenance problems (MDH, 2007).

Detailed discussions of various well rehabilitation methods and processes contributing to well deterioration is given by Detay (1997), McLaughlan (2002), and Houben and Treskatis (2007).

7.2.2 Collector Wells

This section—courtesy of Mr. Sam Stowe, Collector Wells International, Inc., a Layne Christensen Company.

For more than 70 years, horizontal collector wells (also referred to as Ranney™ wells and radial collector wells) have been used to develop groundwater primarily from alluvial aquifers formed around large rivers, but also from other unconsolidated and consolidated rock aquifers. They have also been used for groundwater recharge and aquifer storage and recovery (ASR) programs, as well as for extraction of filtered seawater in coastal areas. Collector wells, typically, are constructed in unconsolidated sand or sand and gravel deposits and generally consist of a central, reinforced concrete-caisson wet well that serves as the pump station, with horizontal well screens projected from within the caisson out into the aquifer (Fig. 7.16).

The horizontal collector well concept originated in the 1920s by petroleum engineer Leo Ranney, who developed a horizontal drilling approach for extracting oil in relatively

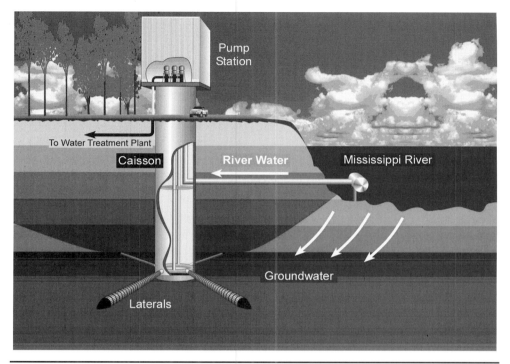

FIGURE 7.16 Typical collector well, showing possible additional direct intake of surface water. (Figure courtesy of Collector Wells International, Inc., a Layne Christensen Company.)

shallow oil-bearing rock formations. As the price of oil dropped in the 1930s, Mr. Ranney modified his horizontal drilling process to allow the installation of horizontal bores into unconsolidated deposits to develop water supplies.

The first horizontal collector well was constructed for the London Water Board, England, around 1933. Following that installation, Mr. Ranney introduced horizontal collector well technology to Europe, where the concept flourished, with utilities installing numerous collector wells using the original installation method bearing the inventor's name (The Ranney Method), whereby perforated pipe well screens were jacked horizontally into aquifer formations. This method was used exclusively until about 1946, when Swiss engineer Dr. Hans Fehlmann modified the jacking process to permit continuous wire-wound well screens to be installed in a collector well for the City of Bern in Switzerland. This technology involved projecting a solid pipe into the formation and then inserting a wire-wound well screen, designed to conform to the grain size of the formation, into the pipe. The projection pipe is then retracted, exposing the formation to the wire-wound well screen. This process allows fine slot screens to be used to match finer-grained formations with a more hydraulically efficient screen.

In 1953, German engineers modified this installation process to include the installation of an artificial gravel-pack filter around the well screens to accommodate finer-grained formations. This process, known as Preussag method, also involves a solid pipe that is projected full length into the formation. A special well screen is then inserted into the pipe, and gravel materials are pumped into the annulus between the projection pipe and the screen while the projection pipe is retracted. The use of an artificial gravel-pack filter provides a transition between fine-grained formation deposits and more efficient screen openings.

These two advances in collector well technology improved the hydraulic efficiency of collector wells and permitted collector well laterals to be installed in a wider range of geologic formations. Both the Fehlmann and the gravel-packing technologies were brought into the United States in the mid-1980s and have been used extensively since.

Collector wells can be used in virtually any unconsolidated geologic formation that contains sand, gravel, and cobbles, and in consolidated (rock) formations, if conditions are appropriate. Collector wells offer advantages in situations where aquifer formations are stratified or shallow, because the entire well screen can be installed in the most hydraulically efficient zone within the aquifer, thus minimizing head losses. The well screen is installed horizontally so that it can be placed toward the base of the aquifer, maximizing the amount of available drawdown and maximizing the possible yield. Since the length of screen is not restricted by the aquifer thickness, longer screens can be installed, thus minimizing head losses through the screen, and minimizing the rate of plugging.

Since the first collector well was installed in 1933 in England, hundreds of collector wells have been installed throughout the world, including over 300 in the United States. These have ranged from single-well to multiple-well installations for a single utility, such as 99 collector wells for the City of Belgrade in the Republic of Serbia. The individual well capacities (yields) have ranged from about 0.0044 to 1.75 m^3/s (70 to 27,700 gpm). The largest single collector well in the world for the Board of Public Utilities in Kansas City, Kansas, has been pumped at rates of up to 2.4 m^3/s (55 mgd). A second collector well in Kansas City; one in Prince George, British Columbia, Canada; and one in Sonoma County, California, have yields approaching this as well.

Collector Well Design and Construction

The expected yield and subsequent design of a collector well is determined much the same way vertical wells are, using data obtained from exploratory test drilling and aquifer testing. Specialized analytical equations for calculating the expected yields from horizontal collector wells are presented in Hantush (1964) for various hydrogeological settings. Numerical models can also be used to estimate yields and time of travel. A collector well is designed to infiltrate water from the adjacent surface water source, using the natural streambed and riverbank deposits to naturally filter out suspended materials from the source water. The first, obvious, requirement is that the well and its laterals be placed in close proximity to a source of recharge, such as a river. During the feasibility and siting stages of a project, a number of criteria must be considered, including

1. Availability of water from a surface water source that can recharge the aquifer
2. An efficient hydraulic interconnection between the river and the aquifer
3. Suitable aquifer, capable of developing and conveying infiltrated water to the well
4. Suitable water quality in the aquifer and source surface water
5. Sustainable flow in the river for the anticipated withdrawal rates

The key parameters to any riverbank filtration (RBF) evaluation are aquifer transmissivity and streambed permeability. A good understanding of these parameters along with the hydrogeological setting will result in the proper evaluation of expected yield and quality of a horizontal collector well and allow thorough evaluation of design options. Understanding the ability of the aquifer to provide sufficient RBF to recharge water pumped from a horizontal collector well is the key to ensuring that long-term capacities can be sustained and that target water quality can be maintained through a balance of infiltrated surface water and groundwater.

The caisson is constructed by forming and pouring concrete sections—or lifts—at grade, and then excavating soil materials from within the caisson and allowing the caisson walls to sink into the ground. One of the lower sections includes wall-port openings to be used for projecting the lateral well screens. As each lift section, usually 3–3.7 m (10 to 12 ft) high, sinks into the ground, the subsequent sections are tied together with reinforcing steel and water stops, formed, and poured. The sinking process continues until the lower portion of the caisson reaches the design depth for projection of the lateral well screens. Once the caisson has been placed to its design depth, a reinforced concrete bottom sealing plug is poured to enable the interior of the shaft to be dewatered for screen installation. The concrete caissons are typically constructed with an inside diameter ranging from 3 to 6 m (10 to 20 ft) or larger, if necessary. The caissons can be installed to depths of 46 m (150 ft) using normal construction methods and possibly deeper using special hydraulic-assist jacking equipment. The average depth of the caissons in the United States is 21 m (68 ft), and the average diameter is 4 m (13 ft).

Once the bottom seal has set, the water in the caisson is pumped down and the lateral well screens are jacked out into the aquifer formation hydraulically from inside the caisson (Fig. 7.17). Often, 150 to 300 m or more (500 to 1000 ft) of well screen is installed in a collector well at various lateral configurations, depending on the aquifer properties. The largest collector well to date has over 800 m (2600 ft) of well screen installed, divided into two tiers.

The original (Ranney) method to install lateral well screens involves projecting pipe sections that have been perforated by punching or sawing. The pipe sections are attached to a digging head that is used to direct the projection of the lateral pipe. In this approach, the pipe sections are projected into the aquifer and left in place. The openings on the pipe typically provide a maximum open area of 20 percent, which is limited since the pipe needs to have sufficient structural strength to accommodate the jacking forces used during projection. Because of the methods used for perforating the pipe, the minimum slot size that can be made is sometimes too large to sufficiently retain fine-grained formation materials for efficient well development, so it is used primarily for aquifers containing coarser-grained deposits with higher percentages of gravel.

The projection pipe method involves the use of a special heavy-duty pipe that is pushed out into the aquifer formation. During the projection process, formation samples are collected and analyzed for grain-size distribution. Once the pipe has been placed in the aquifer to the desired distance, a wire-wrapped continuous-slot well screen (with slot openings selected to conform to the aquifer deposits encountered) is inserted inside the projection pipe. The projection pipe is then withdrawn so that it may be used in projecting the next lateral. The lateral lengths range from about 30 to 75 m (100 to 250 ft) using this method, with 20- or 30-cm-diameter screens installed. This screen design can use a variety of slot size openings to accommodate almost any formation gradation, including fine- to medium-grained sands. These screens are developed by removing fine-grained deposits to create a natural gravel-pack filter around the well screens. Using this method, the well screen gains the following advantages:

Figure 7.17 Installing 12-in. diameter well screen in 13-ft-diameter collector well. (Photograph courtesy of Collector Wells International, Inc., A Layne Christensen Company.)

- The screen can have more open area (often 40 percent or more).
- The screen is more durable (usually composed of stainless steel).
- The screen has more flexibility in slot size, to accommodate a wider range of formation deposits.
- The method of installation allows the slot size of individual screen sections to be selected to conform to the specific gradation of the formation in which they are placed.
- This screening method allows the use of other screen materials (e.g., plastic) in selected applications.
- Screen can be installed in formations containing higher amounts of fine-grained (e.g., sand) deposits.

This method also provides the ability to use special well-screen materials that are applicable in saline and brackish environments. It is also possible to install laterals in formations containing large cobbles and boulders using this approach.

Examples of collector wells using wire-wrapped, continuous-slot well screens include the well referenced above in Kansas City; four municipal collector wells in Olathe, Kansas, that are about 23 m (77 ft) deep and each produces approximately 0.4 m^3/s (6300

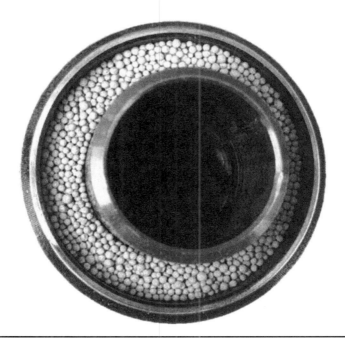

FIGURE 7.18 Johnson Muni-Pak™ prepacked well screen. Carbolite grains help minimize biofilm accumulation. (Photograph courtesy cf Johnson Screens—A Weatherford Company.)

gpm); a collector well for a municipality in Arizona that is about 32 m (104 ft) deep that can produce up to 1.1 m³/s (17,400 gpm); and a municipal collector well in Boardman, Oregon, that can produce 0.63 m³/s (10,000 gpm). Two municipal water utilities in central Iowa use collector wells (10 in total) that are in shallow alluvial aquifers about 9 to 12 m (30 to 40 ft) deep and can produce 0.09 to 0.13 m³/s (1400 to 2100 gpm) each.

Gravel-packed screens are installed in much the same manner as for the wire-wrapped design; however, an artificial gravel pack is installed around the well screen. For this method, formation samples are also collected as the pipe is projected. Once the projection pipe has been pushed to the full design length, specially designed well screens (usually stainless steel) are inserted and an artificial gravel-pack filter is placed around the well screens as the projection pipe is withdrawn, or a prepacked well-screen design can be installed that uses dual-screen sections and contains the artificial media between the two sets of screens, as shown in Fig. 7.18. This permits the installation of a gravel filter to act as a transition zone between a fine-grained aquifer formation and the openings in the well screen to prevent ongoing sand intrusion into the well. This method has been used for both seawater and freshwater (inland) applications.

Examples of several collector wells that have gravel-packed lateral well screens include a municipal utility in New Jersey that is 20 m (66 ft) deep and can produce 0.22 m³/s (3500 gpm) and a municipal utility in Missouri with a well about 42 m (137 ft) deep that can produce 0.53 m³/s (8400 gpm). At an industrial facility in Missouri, a collector well 27 m (90 ft) deep can produce 0.19 m³/s (3000 gpm). Three seawater collector wells with gravel-packed lateral screens in Mexico are about 30 m (98 ft) deep and each can produce approximately 0.19 m³/s (3000 gpm) for desalination.

Once installed, the well screens are developed to remove fine-grained formation materials from around the screens. This optimizes filter permeability and improves flow hydraulics within the filter as the water nears the screen. The development process is performed along the full length of each lateral well-screen section in an incremental manner to ensure that all sections of the well screen get uniformly developed to meet sand specifications.

Collector Well Maintenance

As with any water well, a collector well will eventually require maintenance to restore lost well efficiency and capacity. With the data collected and graphed as part of the monitoring program, declines in well performance can be observed and maintenance can be anticipated and scheduled for opportune times to minimize disruption to normal service. Maintenance is most effectively accomplished by inserting specialized equipment into the individual laterals while keeping the central caisson dewatered. This flushes out any loose scale, encrustation, and fine sand loosened during the cleaning process. This cleaning and redevelopment program removes debris from inside the well screen, accumulations within the well-screen slot openings, and deposits from outside the well screen in the aquifer and gravel-pack materials. Through the use of specialized procedures, well-screen cleaning and redevelopment can also be performed while the collector well remains in service, if the well is the only source of water supply for the utility. With one of these methods, redevelopment and cleaning can take place while the collector well continues to pump into the system, uninterrupted. This illustrates the unique flexibility of a collector well.

Maintenance can also include installation of new lateral well screens when older screens have corroded or deteriorated with time or become excessively plugged. Many of the older collector wells (constructed with mild steel screens) require new lateral screen after 40+ years of operation. However, most of the collector wells built in recent years have used newer technology that has allowed the use of stainless-steel well screens, which should last longer and be more resistant to normal corrosion processes. The new laterals can be installed into an existing facility by installing new port assemblies and projecting new lateral screens. This type of maintenance can be performed to restore well capacity where older screens need replacement, or to supplement existing well screens to develop additional capacity where available.

Riverbank Filtration

As the design and construction process for the horizontal collector well evolved, it became evident that wells installed adjacent to and, sometimes, underneath surface-water sources, were able to develop large quantities of water. As groundwater levels were lowered by pumping, the hydraulic gradients in the aquifer permitted water to be infiltrated from an adjacent river or lake, providing recharge into the aquifer to replenish water removed by pumping. This infiltration process prefilters river water as it percolates through riverbed sediments toward the aquifer (recharging it) and, ultimately, into the well screens, typically removing objectionable characteristics of the river water, such as turbidity and some microorganisms. Because the recharge water from the river is infiltrated over a large area, infiltration rates are low, providing a high degree of filtration in most cases. This process of recharging aquifers and supporting well yields through a natural filtration process is generally referred to as RBF.

As regulatory agencies began evaluating groundwater under the direct influence of surface water issues in the 1990s, siting and design philosophies for collector wells were revised to take full advantage of RBF. New installations are designed to

- Improve the filtration of surface water
- Site wells to minimize the potential for contamination from surface-water sources
- Improve caisson installation methods to minimize disturbance to the aquifer
- Improve surface-sealing techniques around the caisson

This involved the proper selection of the horizon (elevation) for projecting the lateral screens and sometimes locating the wells a sufficient distance back from the river to increase the degree of filtration and travel time for recharge water. The ability (or efficiency) of the streambed and aquifer materials to filter out objectionable microorganisms and to reduce the turbidity from surface-water sources will vary from region to region and from site to site. In most alluvial settings, it should be possible to achieve some degree of filtration to improve water quality. If adequate natural filtration occurs, RBF systems can qualify as an approved alternative treatment technology by regulatory agencies and receive filtration credits for removal of microorganisms.

Other Collector Well Applications

In addition to being used for municipal or industrial water supplies, collector wells have been used for other applications including construction dewatering, seawater collector wells, and for artificial recharge, such as in ASR programs. Collector wells have been installed at several coastal sites to develop a filtered seawater supply to be used for desalination. Installing the well screens beneath beaches minimizes environmental disturbance and impact on local aquatic life. Filtering the raw seawater through the native beach sands removes suspended particulates that would otherwise clog RO membrane equipment, serving as pretreatment for the main treatment process.

Collector wells have also been constructed in conjunction with a direct surface water intake to provide a system that can use either water source, depending on groundwater levels, river water quality, temperature needs, and system demands. Typical of this application is an installation constructed for an industry in Missouri that is about 26 m (85 ft) deep and can produce $0.13 \text{ m}^3/\text{s}$ or 2100 gpm (Fig. 7.16).

Case Study: Belgrade Waterworks

Courtesy of Urosevic, U., Vrvic, N., Dolinga, I., Teodorovic, M., and Miljevic, M., Belgrade Waterworks, Deligradska 28, Belgrade, Serbia; www.bvk.co.yu.

The first horizontal collector well for water supply of Belgrade was installed in 1953 using the original Ranney method. Over the next 40 years, 94 more collector wells were installed using a modified Ranney method (Fig 7.19). The last four collector wells were installed with the Preussag method. Currently, there are 98 operational collector wells and 44 vertical wells. Figure 7.20 shows the steady increase of the number of collector wells until 1986, at which point the Waterworks entered a phase of no growth due to economic and political problems in the country. The peak combined capacity of the collector wells was $6.2 \text{ m}^3/\text{s}$ in 1988, and it has been declining ever since due to minimal investment in well maintenance and rehabilitation. Most collector wells are aging rapidly and are

FIGURE 7.19   Ranney type collector well at the Ada Ciganlija wellfield. The well is adjacent to artificial Topcider Lake also shown in Figure 2.68 in Chap. 2. This lake, created by three dams-levees between the right bank of the Sava River and the Island of Ada Ciganlija, now serves as groundwater recharge basin receiving filtered river water. (Photograph courtesy of Belgrade Waterworks.)

creating excessive drawdowns, with operating water levels at 6 to 9 m below the bottom of the river channel. At the same time, the hydraulic heads across the wellfield have been rising steadily since about 1996 due to the decrease in the total pumping rate. The excessive drawdowns have locally created unconfined flow conditions from the initially confined to semiconfined aquifer system.

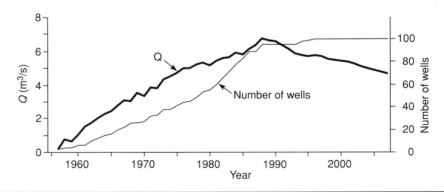

FIGURE 7.20   Number of collector wells and capacity of the Belgrade Waterworks wellfield, 1957–2007. (Figure courtesy of Belgrade Waterworks).

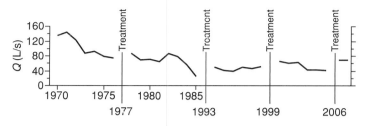

Figure 7.21 Changes in pumping rate of collector well RB-25 showing effects of well aging and rehabilitation. (Figure courtesy of Belgrade Waterworks.)

All drains of the collector wells are inserted into the lower portion of the alluvial aquifer, which rests on thick tertiary clays. The thickness of this highly permeable zone consisting of gravel and sandy gravel is between 5 and 15 m. The hydraulic conductivity ranges between 0.1 and 0.001 cm/s. This main productive zone is overlain with 0.5 to 10 m thick sands, silty sands, and silty clays, with the hydraulic conductivity range of 1×10^{-3} to 1×10^{-5} cm/s. The productive portion of the aquifer is overlain by low-permeable silts and clays with the hydraulic conductivity less than 1×10^{-6} cm/s, and thickness ranging between 1.5 and 10 m. These low-permeable sediments extend to the land surface and play an important role in aquifer protection from contamination. At the same time, when below the river channel, they often significantly reduce the rate of direct infiltration from the river.

During early years of operation, the pumping rate of collector wells in the system ranged between 150 and 250 L/s. Direct contribution of the Sava River water to the yield of collector wells ranged between 80 and 90 percent. Due to intensive groundwater extraction, large vertical gradients, and an uncontrolled deterioration (aging) of the wells, the pumping rates have declined and currently range between 30 and 100 L/s, with the average of 50 L/s for all wells (in comparison, the average pumping rate of vertical wells is 12 L/s). Well deterioration is caused by all of the following: chemical corrosion, biochemical incrustation including excessive iron bacteria development in most locations, and mechanical incrustation. In addition to the well scaling (incrustation), clogging of the aquifer porous media immediately below the river channel by fine river sediments has contributed significantly to the overall decline of the system's performance. The Waterworks have recently initiated a comprehensive assessment of the wellfield conditions, including a research and development program for rehabilitation of the collector wells and removal (dredging) of fine river channel deposits. Although sporadic well rehabilitation efforts in the past have helped in maintaining acceptable levels of water supply for the most part, they were not effective in significantly reversing the drop in specific capacity of individual wells (Fig. 7.21). The Waterworks is also looking at new drain technologies, since traditional Ranney punched-slot screen without gravel pack is installed in almost all collector wells.

7.3 Subsurface Dams

The benefits of storing water in the subsurface have been recognized in the United States since the late 1890s. Slichter (1902) gives this account of one of the first subsurface dams built in the United States:

Figure 7.22 Underground dam being constructed on the Pacoima Creek, Los Angeles County, CA, 1887–1890. (Photograph courtesy of the U.S. Geological Survey.)

Another method of recovering the underflow of a stream is by means of a subsurface dam. Such a dam is constructed by excavating a trench at right angles to the direction of the underflow and extending in depth to the impervious stratum, and then filling the trench with impervious material. If the underflow is confined within an impervious trough or canyon, it is obvious that such a construction must result in bringing it to the surface. An example of this is found on Pacoima Creek, Los Angeles County, Cal., where a subsurface dam was constructed in 1887–1890 (Fig. 7.22). It is claimed that by means of this dam the owners have been enabled to use the bedrock flow of water for the three dry years, 1898–1900, and thereby to successfully carry through the orange, lemon, and olive growing in Fernando Valley. This dam is described in the Eighteenth Annual Report of the United States Geological Survey, Part IV, pages 693 to 695; also in Reservoirs for Irrigation, Water Power, etc., by James D. Schuyler, 1901, page 205.

Compared with surface water reservoirs formed by conventional dams, use of groundwater impoundments behind underground dams has the following major advantages:

- Very limited or negligible evaporation loss.
- Land use above the groundwater reservoir can continue without change (there is no submergence of houses, infrastructure, and property in general).
- There is general improvement of water quality because of the porous media filtration of airborne and surface runoff contaminants and pathogens.

- Function of underground reservoir may be permanent since there is no accumulation of sediment, the main reason for a limited life span of surface reservoirs.

- There is no danger of dam failure and catastrophic loss of life and property.

- Overall impact on the environment and natural habitat of plants and animals is of much lower magnitude.

Reoccurring landslides and rock falls along reservoir banks present one of the more dramatic negative impacts of surface impoundments. Reservoir level changes due to water use, droughts, and floods, and the associated changes in groundwater levels cause activation of new landslides and rock falls and reactivation of old dormant landslides. For example, a recent study of 15 large artificial reservoirs for hydroelectric power generation in Serbia registered over 400 active landslides, which were classified based on the risk they pose to reservoirs and dams. Over 10 percent (47) were classified as high-risk landslides, requiring immediate stabilization measures, and another 40 percent (151) were classified as medium-risk, requiring continuous monitoring with permanently installed instrumentation. The remaining landslides, which were classified as having low risk for the direct operation of dams and reservoirs, nevertheless, all had significant risk levels with respect to regional and local infrastructure, buildings, and the environment (Abolmasov, 2007).

The main disadvantage of subsurface reservoirs is that virtually none of the feasible (cost-effective) construction methods can guarantee complete dam impermeability, although "practical" impermeability, i.e., a tolerable level of dam leakage, is often possible to achieve. In addition, storage capacity of the underground reservoir cannot be accurately determined and has to be estimated based on more or less limited field data on heterogeneous porous media. Another disadvantage of subsurface dams is that they intercept downstream groundwater flow. However, an appropriate design can allow controlled draining of the groundwater reservoir to reduce this impact if needed. Potential salinization in the reservoir area due to evaporation from the shallow water table can be avoided by an appropriate design, which keeps the water table from reaching critical depth of evaporation.

Subsurface dams can be defined as structures that intercept or obstruct the natural flow of groundwater and provide storage for water underground. Most commonly, they have been used in India, Africa, and Brazil, in areas where flows of groundwater vary considerably during the course of the year, from very high flows following rain to negligible flows during the dry season. Groundwater dams can be divided in two types: subsurface dams and sand storage dams. A subsurface dam is built entirely under the ground using low-permeable natural materials such as clay or impermeable materials such as concrete (Fig. 7.23). The underground dam causes water-table upgradient of the dam to raise, stores additional volumes of groundwater, and reduces water-table fluctuations (Fig. 7.24).

Another type of underground storage of water used in Africa's arid regions is the sand storage dam constructed above ground. Sand and soil particles transported during periods of high flow are allowed to deposit in front of the dam, and water is stored in these deposits. The sand storage dam is constructed in stages to allow sand to be deposited in thin coarser layers and finer material to be washed downstream. Figure 7.25 illustrates this principle. Whenever the dam basin has filled up with sand, the crest of the dam wall is topped by another barrier until the basin builds up a groundwater reservoir.

FIGURE 7.23 Construction of subsurface dam in Turkana District, Kenya, constructed by compacting clay in the narrow trench (dyke). (Photograph courtesy of VSF-Belgium, 2006.)

This process may take 4 to 10 years, depending on the frequency of flood occurrence. In order to improve storage capacity, in practical sediment engineering, the river floods are geared to flush the sands over a set of specially designed hydraulic weirs, which force the sand to meander. In the process, coarser sand grains tend to remain at the bottom of the channel, while small grains are lifted over the wedges (Diettrich, 2002). Figure 7.26 shows the initial stage of a sand dam after the first flood.

Similar to surface dams and reservoirs, underground dams and groundwater storage basins cannot be constructed "everywhere" and require certain favorable hydrogeologic conditions such as sufficient effective porosity of the aquifer materials, sufficient thickness of the unsaturated zone which will accommodate the rising water table, and natural lateral and vertical containment of the groundwater flow by low-permeable formations.

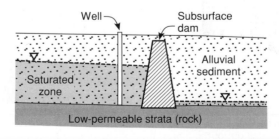

FIGURE 7.24 Schematic cross section of subsurface dam. (Modified from VSF-Belgium, 2006.)

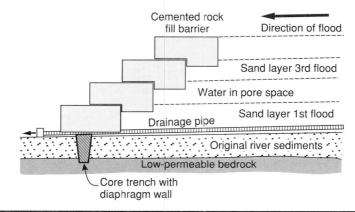

Figure 7.25   Schematic cross section of subsurface sand dam. (Figure courtesy of Thomas Diettrich.)

Figure 7.27 shows a subsurface dam built in a river valley filled with alluvial gravel and sand deposits, which are underlain by low-permeable bedrock.

There is considerable experience in building small-scale subsurface dams and sand dams in drier areas of sub-Saharan Africa where soil and water conservation is a high priority. For example, Vétérinaires Sans Frontières (VSF-Belgium, 2006) discuss results of

Figure 7.26   Initial stage of the subsurface sand dam located in Kwa Ngola, the Mwingi District, Kenya. Water leaking out at the foot of the dam is fetched by students of the primary school. (Photograph courtesy of Thomas Diettrich.)

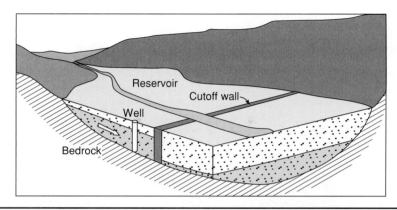

FIGURE 7.27 Conceptual design of subsurface dam with cutoff wall. (Modified from VSF-Belgium, 2006.)

the Turkana Livestock Development Program (TLDP) carried out in the Turkana District of Kenya where storage of water from the rainy season to the dry season, or even from wet years to dry years, is critical for the lives of pastoral people. The region is located in the dry savannah of northwest Kenya, an area characterized by desert conditions and a rainfall of 300 mm/yr or less. When sited and built properly, subsurface dams in Turkana store sufficient quantities of water for livestock and minor irrigation as well as for domestic use, thus providing an appropriate answer to the local water needs. Most rivers in the Turkana region only have water for a few days during the rainy season. These floodwaters are drained immediately except for water that remains stored in the subsurface in alluvial sediments deposited by the rivers. Subsurface dams increase the storage capacity of the sediments and, at the same time, minimize environmental degradation caused by the changes to the natural flow regime.

To estimate the effects of a proposed new subsurface dam on the Khumib River in Namibia for which little data are available, Diettrich (2002) used a model calibrated with long-term time series data for one of the first subsurface sand dams in Namibia, built in the channel of the Hoanib River in 1956 by the Department of Water Affairs. The proposed model is a discrete differential equation model ("dynamic model") for the operations research of the complex nonlinear hydrological time series. The model generates unit hydrographs, flood routing, and water balances of a river basin from topographical data, design criteria, rainfall records, and aquifer parameters. Based on the model simulations, the author concluded that subsurface dam operations improve bank storage and sustainable groundwater flow by approximately 60 percent. In view of Namibia's almost 5000 shallow farm dams suffering from 3000 to 8000 mm/yr of evaporation, their conversion into subsurface dams would add some 50 m^3/s of freshwater to agriculture and tourism at 10 US cents/m^3.

Detailed description of the design and construction of one of the largest subsurface earth-fill dams in Africa (Fig. 7.28), including discussion on extensive multidisciplinary investigations for the siting of the dam, is provided by the Ministry of the Environment, the Government of Japan (2004). The dam was built as part of a project financed by the Government of Japan to combat desertification in Nare Village, Burkina Faso. It was constructed by excavating the alluvial sediments down to the bedrock, and building the dam core from compacted clay. Parallel to the construction of the dam core, the excavated

Figure 7.28   Construction of one of the largest subsurface earth-fill dams in Africa located in Nare Village, Burkina Faso. (From Ministry of the Environment, 2004.)

space of both the upstream and the downstream sides was backfilled. Finally, the core was completely covered by the backfill.

7.4 Spring Development and Regulation

There are three basic types of spring capture: (1) with minimal or no artificial intervention when the spring is used "as is," (2) with some form of engineering intervention aimed at securing the source for reliable use and protecting it from surface contamination, and (3) with engineering aimed at artificially increasing the spring discharge rate. Figure 7.29 illustrates the first type, common for small springs issuing from fractured rocks and with a relatively uniform (steady) discharge rate. This form of spring capture is not recommended since the spring can be easily accessed by animals and contaminated by surface runoff.

When used for potable water supply, a spring should be completely enclosed, protected from contamination and equipped with fixtures for easy access, cleanup, and water distribution. One such design showing a typical capture of a contact gravity spring with a watertight basin ("spring box") constructed in place with reinforced concrete is illustrated in Fig. 7.30. The basin has one side open to the aquifer, allowing inflow of water. In case of ascending springs issuing from a horizontal surface without a clearly defined impermeable contact, the basin is open at the bottom. In either case, the side of the basin open to the inflow of groundwater should be stabilized with a gravel pack or rock

FIGURE 7.29 Simple capture of a low-yielding spring issuing from fractures in Paleozoic metamorphic rocks of the Piedmont physiographic province, the United States. This spring was tapped before the Civil War for water supply of a farmhouse in Virginia. Many other similar springs are still used as sources of both potable and nonpotable water, although drilled wells are now the main form of water supply in the region.

fragments (for fracture springs). The basin should be vented to the surface and have easy access for maintenance. Three pipes equipped with valves should allow for (1) water overflow, (2) complete basin drainage for the cleanout and maintenance, and (3) transfer of water to supply or storage. All pipes should have screens on either end. If the water is of such quality that disinfection is the only treatment required, a chlorination tank or UV equipment may be housed in the maintenance room adjacent to the basin. Sanitary protection of the spring at the surface is achieved by fencing, placement of an impermeable clay fill, and surface drainage ditches located uphill from the spring to intercept surface-water runoff and carry it away from the source. Depending on site-specific conditions, the basic configuration of spring capture shown in Fig. 7.30 may include additional features such as drainage pipes (or galleries in case of fractured rock aquifers) extending into the saturated zone behind the spring box for intercepting more flow.

Spring capture should be built at the location of primary discharge of water from the subsurface, since secondary springs may move in time. If the discharge is from colluvium and other types of rock debris, there is high probability that the spring is secondary and located away from the primary discharge location, which may not be visible. In such cases, every attempt should be made to clear the rock debris and locate the primary spring.

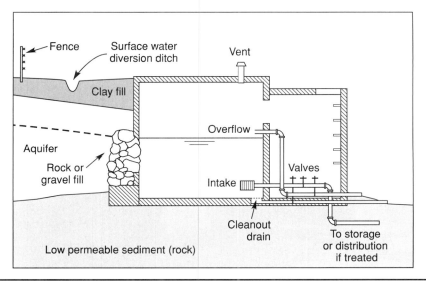

FIGURE 7.30 Typical capture of contact gravity spring using spring box. (Modified from Kresic, 2007; copyright Taylor & Francis Group; used with permission.)

Capture of seepage springs, which occur where groundwater seeps from the soil over large areas, is illustrated in Fig. 7.31. The development process for seepage springs consists of intercepting flowing groundwater over a wide area underground and channeling it to a collection point. The basic steps are the following (Jennings, 1996):

- Dig test holes uphill from the seep to find a point where the impervious layer below the water-bearing layer is about 3 ft (1 m) underground. Water flows on top of this layer in sand or gravel toward the surface seep.

- Dig a 2-ft-wide trench across the slope to a depth of 6 in. below the water-bearing layer and extending 4 to 6 ft beyond the seep area on each side. Install a 4-in. collector tile and completely surround the tile with gravel.

- Connect the collector tile to a 4-inch line leading to the spring box. The box inlet must be below the elevation of the collector tile.

Springs are often contaminated with bacteria during construction or maintenance. All new and repaired water systems should be disinfected using shock chlorination. If bacterial contamination occurs on a regular basis because of surface sources above the spring, continuous disinfection using chlorination or some other method may be necessary. Shock chlorination requires concentration of at least 200 ppm chlorine (Jennings, 1996).

The majority of all springs used for public water supply of more than several hundred users are karst springs, which in general have the highest average flow rate among different types of springs. Large permanent karst springs usually discharge at topographically low contacts with low-permeable formations, and along surface streams, which act as regional erosional bases of karst aquifers (Fig. 7.32). Because of the diversity of karst

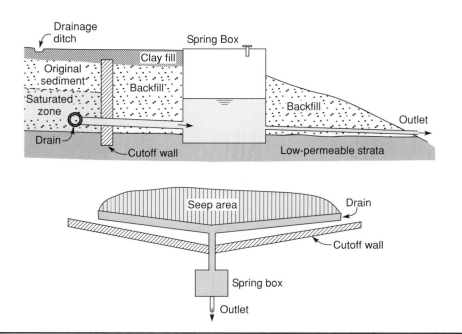

Figure 7.31 Capture of seepage spring. (Modified from Jennings, 1996.)

Figure 7.32 Photograph of the outflow from the spring box of Hughes Spring near Zack, Arkansas, used for public water supply. (Photograph courtesy of Joel Galloway, U.S. Geological Survey; Galloway, 2004.)

formations, the complexity of karst processes, and the role of geology and tectonics in directing groundwater flow in karst, karst springs can be of any type: ascending, descending, cold, thermal, with uniform discharge, or with discharge varying between 0 and >200 m^3/s. For these reasons, the capture of karst springs can vary between a simple spring box–type and artificial concrete basins and dams built to capture flow of large first-magnitude springs.

Advances in cave diving in the last few decades have made possible important revelations regarding major ascending springs in karst terrains (Touloumdjian, 2005). In most cases, such springs are issuing from deep vertical or subvertical conduits formed at the lateral contact of karstified carbonate rocks with noncarbonates or nonkarstified rock. They are also called vauclusian springs after the Fontaine de Vaucluse, the source of the river Sorgue in Provance, France:

> At the moment the shaft is explored to a depth of 315 m. This exploration was done using a small submarine robot called MODEXA 350. The camera of the robot showed a sandy floor at this depth, leads were not visible. The water table is most time of the year below the rim of the shaft. The Fontaine appears as a very deep and blue lake. Small caves below lead to several springs in the dry bed of the river, just 10 m below the lake. The source is fed by the rainfall on the Plateau de Vaucluse. In spring, or sometimes, after enormous rainfall, the water table rises higher than the rim. During these periods the Fontaine de Vaucluse really is a spring, producing more than 200 m^3/s of water (Showcaves, 2005).

The ascending karst spring can be tapped "as is," or, more invasively, by overpumping it with deep submersible pumps placed in the spring shaft or using vertical wells drilled into the deep karst channels connected to the main shaft.

The rate of natural discharge has often been a limiting factor when considering the use of a particular spring for public water supply. Springs with discharge hydrographs similar to the one shown in Fig. 7.33 may have potential for regulation, i.e., may be amenable to artificially increasing their minimum and average annual flows. The basic idea is to take advantage of the fact that the spring is capable of discharging large quantities of water during periods of nonpeak demand such as in spring or late fall when natural aquifer recharge is the highest. This volume of "surplus" water may be regulated in two basic ways: (1) by using it to naturally recharge the volume of the aquifer drained

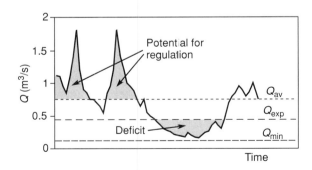

Figure 7.33 Hydrograph of a spring with potentially exploitable reserves higher than the minimum spring discharge. Q_{av}, average spring discharge; Q_{min}, minimum spring discharge; Q_{exp}, potentially secure exploitable reserves. (From Stevanovic et al., 2005.)

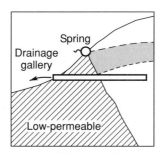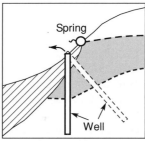

FIGURE 7.34 Potentially favorable conditions for spring regulation using drainage galleries or wells for overpumping. An additional volume of water that may be extracted from aquifer storage during peak demand, assuming recovery during main periods of natural aquifer recharge, is shaded.

by overpumping during periods of peak demand (e.g., summer-early fall) and (2) by storing it in the aquifer above the elevation of natural spring discharge, i.e., by creating a surface or underground dam and groundwater impoundment. These two concepts are illustrated in Figs. 7.34 and 7.35.

In either case, the main prerequisite is that the aquifer has adequate storage capacity, below or above the spring elevation, respectively. A key additional requirement for case 2 is that there should not be an uncontrollable water loss around or below the dam. This means that the spring must issue from a "V"-shaped land-surface contact between the aquifer and the impermeable barrier. The dam is keyed into the impermeable barrier and, in combination with a drainage gallery, is used to control the hydraulic head in the aquifer and water flow. Spring regulation in this way often enables a generation of hydroelectric power because of the elevated hydraulic head in the aquifer behind the dam.

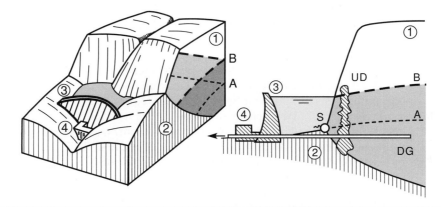

FIGURE 7.35 Possible spring regulation using surface or subsurface dam in the case of favorable geologic and geomorphologic conditions. (1) Aquifer, (2) impermeable base, (3) surface dam with impoundment, and (4) hydropower plant and water treatment. S, original spring; A, original natural water table before impoundment; B, water table after impoundment; DG, drainage gallery; UD, underground dam, such as grout curtain, instead of surface dam. (From Kresic, 1991.)

References

Abolmasov, B., 2007. Evaluation of geological parameters for landslide hazard assessment. Ph.D. Thesis, University of Belgrade, Belgrade, 258 p.

Aller, L.T., et al., 1991. *Handbook of Suggested Practices for the Design and Installation of Ground-Water Monitoring Wells.* EPA160014-891034. Environmental Monitoring Systems Laboratory, Office of Research and Development, Las Vegas, NV, 221 p.

Anderson, M.T., and Woosley, L.H., Jr., 2005. Water availability for the Western United States—Key scientific challenges. U.S. Geological Survey Circular 1261, Reston, VA, 85 p.

AWWA (American Water Works Association), 1998. AWWA standard for water wells; American National Standard. ANSI/AWWA A100-97, AWWA, Denver, CO.

Campbell, M.D., and Lehr, J.H., 1973. *Water Well Technology.* McGraw-Hill, New York, 681 p.

Detay, M., 1997. *Water Wells; Implementation, Maintenance and Restoration.* John Wiley & Sons, Chichester, England, 379 p.

Diettrich, T.E.K., 2002. Dynamic modelling of the ephemeral regimen of subsurface dams in Namibia. A Method also for Jordan and other arid countries. RCC Dam Construction Conference Middle East, April 7th–10th 2002, Irbid, Jordan, 11 p.

Driscoll, F.G., 1986. *Groundwater and Wells.* Johnson Filtration Systems Inc., St. Paul, MN, 1089 p.

Galloway, J.M., 2004. Hydrogeologic characteristics of four public drinking-water supply springs in northern Arkansas. U.S. Geological Survey Water-Resources Investigations Report 03-4307, Little Rock, AR, 68 p.

Gleick, P.H., 2000. *The World's Water 2000–2001. The Biennial Report on Freshwater Resources.* Island Press, Washington, DC, 335 p.

Guldin, R.W., 1989. An analysis of the water situation in the United States: 1989–2040. U.S. Forest Service General Technical Report RM–177, 178 p.

Hantush, M.S., 1964. Hydraulics of wells. In: *Advances in Hydroscience*, Vol. 1. Ven Te Chow, editor. Academic Press, New York, pp. 281–432.

Houben, G., and Treskatis, C., 2007. *Water Well Rehabilitation and Reconstruction.* McGraw-Hill Professional, New York, 391 p.

Jennings, G.D., 1996. Protecting water supply springs. North Carolina Cooperative Extension Service, Publication no. AG 473-15. Available at: http://www.bae.ncsu.edu/programs/extension/publicat/wqwm/ag473-15.html.

Johnson Screens, 2007. Well screens and well efficiency. Johnson Screens a Weatherford Company. Available at: www.weatherford.com/weatherford/groups/public/documents/general/wft029882.pdf. Accessed November 2007.

Kresic, N., 1991. *Kvantitativna hidrogeologija karsta sa elementima zaštite podzemnih voda* (in Serbo-Croatian; Quantitative karst hydrogeology with elements of groundwater protection). Naučna knjiga, Belgrade, 192 p.

Kresic, N., 2007. *Hydrogeology and Groundwater Modeling*, 2nd ed. CRC Press, Taylor & Francis Group, Boca Raton, FL, 807 p.

Lapham, W.W., Franceska, W.D., and Koterba, M.T., 1997. Guidelines and standard procedures for studies of ground-water quality: Selection and installation of wells, and supporting documentation. U.S. Geological Survey Water-Resources Investigations Report 96-4233, Reston, VA, 110 p.

McLaughlan, R.G., 2002. *Managing water well deterioration*. International Contribution to Hydrogeology, Vol. 22. International Association of Hydrogeologists, A.A. Balkema Publishers, Lisse, the Netherlands, 128 p.

MDH, 2007. Iron Bacteria in Well Water. Minnesota Department of Health, Division of Environmental Health, 4 p. Available at: http://www.seagrant.umn.edu/groundwater/pdfs/MDH-IBinWW.pdf. Accessed November 2007.

Ministry of the Environment, 2004. Model project to combat desertification in Nare village, Burkina Faso. Technical report of the subsurface dam. Overseas Environmental Cooperation Center, Government of Japan, 77 p. Available at: http://www.env.go.jp/en/earth/forest/sub_dam.html.

National Ground Water Association, 2003. The ground water supply and its use. Available at: http://www.wellowner.org/agroundwater/gwsupplyanduse.shtml/. Accessed August 4, 2003.

Patten, D.T., 1997. Sustainability of western riparian ecosystems. In: *Aquatic Ecosystems Symposium*, Denver, CO, Minckley, W.L., editor. Western Water Policy Review Advisory Commission, National Technical Information Service, pp. 17–31.

Postel, S., and Richter, B.D., 2003. *Rivers for Life—Managing Water for People and Nature*. Island Press, Washington, DC, 253 p.

Slichter, C.S., 1902. The motions of underground waters. U.S. Geological Survey Water-Supply and Irrigation Papers 67, Washington, DC, 106 p.

Slichter, C.S., 1905. Field measurements of the rate of movement of underground waters. U.S. Geological Survey Water-Supply and Irrigation Papers 140, Series 0, Underground Waters, Vol. 43. Washington, DC, 122 p.

Stevanovic, Z., et al., 2005. Management of karst aquifers in Serbia for water supply – achievements and perspectives. In: *Proceedings of Internationl Conference on Water Resources and Environmental Problems in Karst – Cvijić 2005*, Stevanovic, Z., and Milanovic, P., editors. Univ. of Belgrade, Institute of Hydrogeology, Belgrade, pp. 283–290.

Touloumdjian, C., 2005. The springs of Montenegro and Dinaric karst. In: *Proceedings of Internationl Conference on Water Resources and Environmental Problems in Karst – Cvijić 2005*, Stevanovic, Z., and Milanovic, P., editors. Univ. of Belgrade, Institute of Hydrogeology, Belgrade, pp. 443–450.

USBR, 1977. Ground Water Manual. U.S. Department of the Interior, Bureau of Reclamation, Washington, DC, 480 p.

USEPA, 1975. Manual of Water Well Construction Practices. EPA-570/9-75-001. Office of Water Supply, Washington, DC, 156 p.

USEPA, 1991. Manual of Small Public Water Supply Systems. EPA 570/9-91-003. Office of Water, Washington, DC, 211 p.

VSF-Belgium, 2006. Subsurface Dams: A Simple, Safe and Affordable Technology for Pastoralists. A Manual on Subsurface Dams Construction Based on an Experience of Vétérinaires sans Frontières in Turkana District (Kenya). VSF-Belgium, Brussels, Belgium, 51 p.

Walter, D.A., 1997. Geochemistry and microbiology of iron-related well-screen encrustation and aquifer biofouling in Suffolk County, Long Island, New York. U.S. Geological Survey Water-Resources Investigations Report 97-4032, Coram, New York, 37 p.

Walton, W.C., 1962. Selected analytical methods for well and aquifer evaluation. *Illinois State Water Survey Bulletin*, vol. 49, 81 p.

Wisconsin DNR (Department of Natural Resources), 2007. Iron bacteria in drinking water. Available at: http://www.dnr.state.wi.us/org/water/dwg/febact.htm. Accessed November 2007.

CHAPTER 8

Groundwater Management

8.1 Introduction

At the most basic level, water supply management can be defined as a process that secures enough water of suitable quality to meet demand at all times, if this demand is reasonable and that there is no waste of water. More broadly, water supply management is an integral part of water governance, which refers to the range of political, social, economic, and administrative systems that are in place to develop and manage water resources, and the delivery of water services, at different levels of society (Rogers and Hall, 2003). Water governance includes the ability to design public policies and institutional frameworks that are socially acceptable, equitable, and environmentally sustainable. Given the complexities of water use within society, effective water governance requires the involvement of all stakeholders and must ensure that disparate voices are heard and respected in decisions on development, allocation, and management of common waters, and in using financial and human resources. Governance aspects overlap with the technical and economic aspects of water, but include the ability to use political and administrative elements to solve a problem or exploit an opportunity (Rogers and Hall, 2003).

One recent example of groundwater governance on a grand political scale is the Groundwater Directive by the European Parliament, which refers to groundwater as "the most sensitive and the largest body of freshwater in the European Union and, in particular, also a main source of public drinking water supplies in many regions." (The European Parliament and the Council of the European Union, 2006). This directive establishes specific measures in order to prevent and control groundwater pollution, defined as the direct or indirect introduction of pollutants into groundwater as a result of human activity. These measures include (a) criteria for the assessment of good groundwater chemical status and (b) criteria for the identification and reversal of significant and sustained upward trends of contamination and for the definition of starting points for trend reversals. The directive also requires the "establishment by Member States of groundwater safeguard zones of such size as the competent national body deems necessary to protect drinking water supplies. Such safeguard zones may cover the whole territory of a Member State." Some of the more telling statements in the Directive are as follows:

> Groundwater is a valuable natural resource and as such should be protected from deterioration and chemical pollution. This is particularly important for groundwater-dependent ecosystems and for the use of groundwater in water supply for human consumption.
>
> The protection of groundwater may in some areas require a change in farming or forestry practices, which could entail a loss of income. The Common Agricultural Policy provides for funding mechanisms to implement measures to comply with Community standards.

European Environmental Bureau (EEB), Europe's biggest environmental nongovernmental organization, stated the following in response to the Groundwater Directive: "Members of the European Parliament have successfully fought off attempts by governments to re-nationalise groundwater protection. They ensured that preventing pollution and achieving quality standards is robust and legally binding. Without this kind of EU-wide approach, countries would have been left exposed to pressure from powerful, globalised businesses" (EEB, 2006).

Groundwater management is commonly divided into supply-side management and demand-side management. This division is more for technical and administrative purposes, however, since the two aspects are interdependent. At the same time, as discussed above, the overall water (groundwater) management is not "just" about making sound engineering, scientific, economic, and environmental decisions. Water governance is in many cases disproportionately influenced by policies that favor growth of one or more groups of water users—urban, industrial, or agricultural—without much regard for the sustainability of water use or the environmental impacts. The worst possible outcome of failed policies is an uncontrolled spiral of increasing demand causing increasing groundwater withdrawals, which in turn result in aquifer mining and overall environmental degradation. When groundwater is viewed as both an economic and a public good, and its use is overseen by most if not all stakeholders, it is less likely that this spiral would continue unchecked. In contrast, when selling water is viewed only as a source of profit for the water purveyor, possibly shared by others through tax revenues for example, it is more likely that an unsustainable use of groundwater resources would continue.

As emphasized throughout this book, groundwater is an essential element of the overall hydrologic cycle. It is inseparable from surface water resources because it provides baseflow and sustains aquatic life in surface water streams, lakes, and wetlands, as well as in the aquifer itself (such as in karst caves and conduits). Withdrawal of groundwater may affect surface water flows and quality and vice versa. Surface water may become groundwater at some point, and the same water may again emerge as surface water after flowing through a groundwater system for miles and centuries. Upstream users (in the case of surface water) or upgradient users (in the case of groundwater), water diversions, and wastewater discharges will affect downstream users, water availability, and water quality. Groundwater management should therefore be viewed as part of an integrated surface water and groundwater resources management on a watershed scale, or a regional groundwater system scale in the case of confined aquifers not connected with surface water.

The least desirable form of water supply management is the use of "tools" such as the one shown in Fig. 8.1. It is in situations of repeated water shortages, caused by either insufficient water resources or prolonged droughts, that water utilities and politicians alike are forced to reassess their water management practices and policies. Unfortunately, a seemingly abundant water resource is often taken for granted and little is done to evaluate or promote various aspects of its sustainable use. One such example is the approach to water management by the city of Chicago. The city switched its water supply from groundwater to Lake Michigan water in the early 1980s after realizing that extensive aquifer mining had created one of the largest and deepest cones of depression in the United States and was not sustainable. For many years, the Great Lakes were considered an inexhaustible source and Chicago kept charging a flat fee for water use, lower than in any other major city in the country. Notably, wastewater generated by the city is not discharged back into the lake but flows into the Mississippi drainage basin.

Figure 8.1 Water management tool utilized by the town of Warrenton, VA, during the drought of 2007. Incidentally, Warrenton is one of many fast growing communities near the metropolitan Washington, DC, area.

As this water management practice continues to the present day, the lakes are arguably trying to send signals, through their declining levels, that they are exhaustible sources. Although various factors are contributing to the water level decline of the Great Lakes, and no quantitative connection to city water withdrawals has been established thus far, it appears that Chicago did not learn from its not so distant past.

Managing the quality of "raw" groundwater, before it is extracted from an aquifer, is a much more complex task compared to managing its quantity. As discussed in Chap. 5, there are many sources, past and present, of potential or actual groundwater contamination and an infinite variety of natural and anthropogenic contaminants. It is often very difficult to pinpoint every single source of groundwater contamination, and even more difficult to quickly restore the resource to beneficial use. In most cases, the quality of the resource cannot be directly controlled by the end user due to legal, financial, and other constraints. A simple example is contamination that is (has been) occurring miles away, outside the jurisdiction of the utility that extracts groundwater and is affecting its well field for public water supply. Even when the sources of contamination are well defined and the legal authority for groundwater restoration is clearly established, it may take years before any measures to mitigate the situation are taken. One common reason is the high cost of groundwater remediation, which can prohibit small and large users alike from attempting to solve the problem on their own. This is the main reason why in some societies such as the United States, where the legal rights of both water users and alleged polluters are very strict and highly protected, exorbitant amount of money is spent each year on litigation over groundwater contamination rather than on its direct mitigation.

In some complex hydrogeologic environments, aquifer restoration to pristine or near-pristine conditions (which legally are defined as all contaminants present below their maximum allowed concentrations) may not be technically feasible. This fact is often not acceptable to some stakeholders including regulatory agencies, and there are many examples of groundwater users with false hopes waiting for "someone else" to pay for solving their groundwater contamination problem. In such cases "someone else" should be, whenever possible, considered as "something else" including at least two options: (1) groundwater treatment to drinking water standards after extraction from the aquifer and (2) innovative approaches to overall water management including water reuse and public outreach (education). Unfortunately, water supply often tends to be a local responsibility and its costs are seldom fully covered. Most utilities in both developed and developing countries rely heavily on subsidies and are far from being able to create their own capital. Since there is little political will to raise tariffs, and central governments are either unwilling or unable to provide financing, utilities are trapped in a vicious spiral of underspending on essential maintenance, resistance to fee increase, and inability to cope with groundwater contamination. In litigious societies such as the United States, this situation is a fertile ground for the legal profession (attorneys). In societies that are more pragmatic or not so rich, central governments and water-sector agencies play key roles in the various disputes that arise from groundwater contamination. They also help both groundwater users and polluters by creating a more flexible regulatory environment that emphasizes risk- and economy-based approaches to the allocation of resources for groundwater quality restoration and management.

As discussed by Winpenny (2003), serious defects in the governance of water resources in most undeveloped and many developing countries hamper the ability of water utilities to effectively manage them. Thus, the water sector is unable to generate or attract finance, and there is a tradition, especially in poorer countries, of reliance on foreign aid for new investments. Following are some of the issues important for understanding many difficulties in water resources management in such countries (modified from Winpenny, 2003):

- The apparent low priority that central governments give to water resources development. Since the mass of people, which is served water, is politically weak or disempowered, those in power find it tempting to postpone investments in the water sector and use available resources, including those saved from debt relief, for more important political gains.
- Confusion of social, environmental, and commercial aims.
- Political interference.
- Poor management structures and imprecise objectives for water development projects.
- Lack of transparency in the award of contracts.
- Nonexistent or weak and inexperienced regulators, as well as engineering, scientific, and technical resources.
- Resistance to cost-recovering tariffs.

- Ineffective and corruption-prone irrigation agencies and water utilities, over-staffed in a misguided attempt to create employment.

Effective groundwater management in such socioeconomic environments is poor or nonexistent and may be very difficult to improve or implement without radical changes at various political, institutional, and technical levels.

Regardless of socioeconomic and regulatory environments, calls for groundwater management do not usually arise until a decline in well yields or water quality affects one of the stakeholder groups. If further uncontrolled pumping and groundwater contamination are allowed, a "vicious circle" may develop (Fig. 8.2), seriously damaging the resource. This damage is evident in excessive groundwater level decline, groundwater contamination, and in some cases, saltwater intrusion or land subsidence (Tuinhof et al., 2002–2005a). To transform this "vicious circle" into a "virtuous circle" (Fig. 8.3), it is essential to recognize that managing groundwater is as much about managing people (water and land users) as it is about managing water (aquifer resources). In other words, the socioeconomic dimension (demand-side management) is as important as the hydrogeologic dimension (supply-side management) and integration of both is always required (Tuinhof et al., 2002–2005a).

In most situations, groundwater management will need to keep a balance between the costs and benefits of management activities and interventions. It should take into account the susceptibility to degradation of the hydrogeologic system involved and the legitimate interests of water users, including ecosystems and those dependent on downstream baseflow. Figure 8.4 illustrates a common evolution of groundwater resources development and the associated stages based on the impacts of the hydraulic stress (groundwater extraction) on the system.

The condition of excessive and unsustainable extraction (3A—Unstable Development) is also included in Fig. 8.4. For this case, the total abstraction rate (and usually the

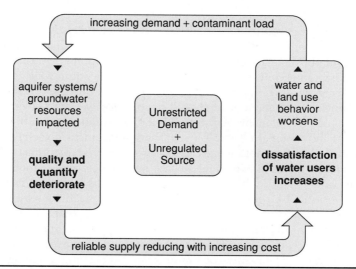

FIGURE 8.2 Supply-driven groundwater development—leading to a vicious circle. (From Tuinhof et al., 2002–2005a.)

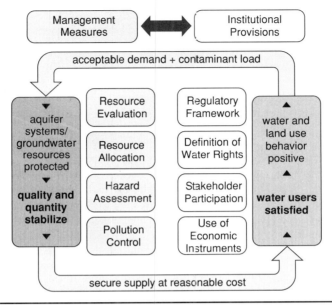

FIGURE 8.3 Integrated groundwater resource management—leading to a virtuous circle. (From Tuinhof et al., 2002–2005a.)

number of production wells) will eventually fall markedly as a result of near irreversible degradation of the aquifer system itself.

As suggested by Tuinhof et al. (2002–2005a), the framework provided in Table 8.1 can be used as a diagnostic instrument to assess the adequacy of existing groundwater management arrangements for a given level of resource development (both in terms of technical tools and institutional provisions). By working down the levels of development of each groundwater management tool or instrument, a diagnostic profile is generated and can be compared to the actual stage of resource development to indicate priority aspects for urgent attention. Such a diagnostic exercise can also be undertaken by each major group of stakeholders to promote communication and understanding. Necessary management interventions for a given hydrogeologic setting and resource development situation can be agreed upon through this type of approach.

Groundwater management should have a clearly stated objective. This is true for any level of management, starting with a local water agency or water purveyor and ending at the national (federal) level. The management objective should include the establishment of threshold values for readily measured quantities such as groundwater levels, groundwater quality, land surface subsidence, and changes in streamflow and surface water quality where they impact or are impacted by groundwater pumping. When a threshold level is reached, the rules and regulations require that groundwater extraction be adjusted or stopped to prevent exceeding that threshold.

Management objectives may range from entirely qualitative to strictly quantified. At a local level, each management objective would have a locally determined threshold value that can vary greatly. For example, in establishing a management objective for groundwater quality, one area may simply choose to establish an average value of total dissolved solids (TDS) as the indicator of whether a management objective is met, while

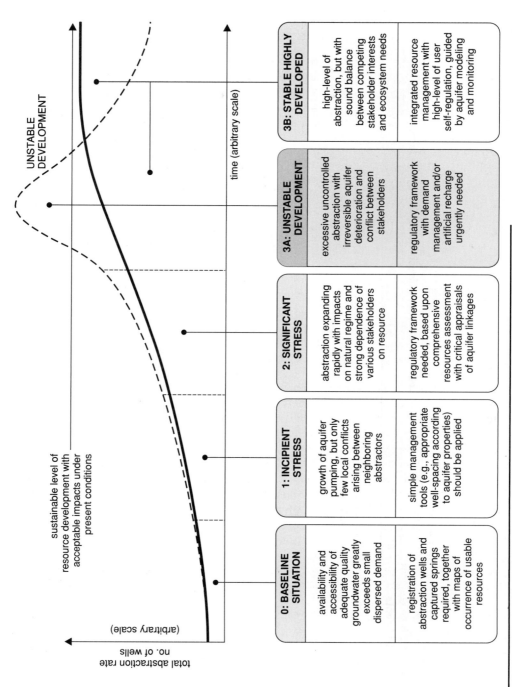

Figure 8.4 Stages of groundwater resource development in a major aquifer and their corresponding management needs. (From Tuinhof et al., 2002–2005a.)

Groundwater Management Tools & Instruments	Level of Development of Corresponding Tool or Instrument[1]			
	0	**1**	**2**	**3**
Technical Tools				
Resource assessment	basic knowledge of aquifer	conceptual model based on field data	numerical model(s) operational with simulation of different abstraction scenarios	models linked to decision-support and used for planning and management
Quality evaluation	no quality constraints experienced	quality variability is issue in allocation	water quality processes understood	quality integrated in allocation plans
Aquifer monitoring	no regular monitoring program	project monitoring, ad-hoc exchange of data	monitoring routines established	monitoring programs used for management decisions
Institutional Instruments				
Water rights	customary water rights	occasional local clarification of water rights (via court cases)	recognition that societal changes override customary water rights	dynamic rights based on management plans
Regulatory provisions	only social regulation	restricted regulation (e.g., licensing of new wells, restrictions on drilling)	active regulation and enforcement by dedicated agency	facilitation and control of stakeholder self-regulation
Water legislation	no water legislation	preparation of groundwater resource law discussed	legal provision for organization of groundwater users	full legal framework for aquifer management
Stakeholder participation	little interaction between regulator and water users	reactive participation and development of user organizations	stakeholder organizations co-opted into management structure	stakeholders and regulators share responsibility for aquifer management

Awareness and education	groundwater is considered an infinite and free resource	finite resource (campaigns for water conservation and protection)	economic good and part of an integrated system	effective interaction and communication between stakeholders
Economic instruments	economic externalities hardly recognized (exploitation is widely subsidized)	only symbolic charges for water abstraction	recognition of economic value (reduction and targeting of fuel subsidies)	economic value recognized (adequate charging and increased possibility of reallocation)
Management actions				
Prevention of side effects	little concern for side effects	recognition of (short- and long-term) side effects	preventive measures in recognition of *in-situ* value	mechanisms to balance extractive uses and *in-situ* values
Resource allocation	limited allocation constraints	competition between users	priorities defined for extractive use	equitable allocation of extractive uses and *in-situ* values
Pollution control	few controls over land use and waste disposal	land surface zoning but no proactive controls	control over new point source pollution and/or siting new wells in safe zones	control of all point and diffuse sources of pollution; mitigation of existing contamination

[1] According to hydraulic stress stage (see Fig. 8.4). From Tuinhof et al., 2002–2005a.

TABLE 8.1 Levels of Groundwater Management Tools, Instruments and Interventions Necessary for a Given Stage of Resource Development Shown in Fig. 8.4 (*Continued*)

another agency may choose to have no constituents exceeding the maximum contaminant level for public drinking water standards. While there is great latitude in establishing management objectives, local managers should remember that the objectives should serve to support the goal of a sustainable supply for the beneficial use of the water in their particular area (DWR, 2003).

This chapter covers, in detail, the key aspects of groundwater management. Ideally, groundwater management starts with an agreement of all stakeholders as to what constitutes a sustainable use of groundwater resources (Section 8.2). This agreement has to be binding and within a clearly defined regulatory framework (Section 8.3). Whenever applicable, groundwater management should be seamlessly integrated with the management of surface water, storm water, and used water (wastewater) thus constituting an integrated water resources management (IWRM; Section 8.4). In order for both to be effective or even possible, the groundwater management and the IWRM must rely on monitoring of water quantity, quality, and their spatial and temporal changes (Section 8.5). This monitoring should include the ambient water (before it is extracted), the extracted water, the storm water, and the wastewater, or, in other words, the entire cycle of water use. All monitoring data, as well as all data generated during water resource evaluation, development, and exploitation (operations and maintenance), should be stored and organized within an interactive GIS database (Section 8.6). One of the most critical aspects of successful water management is source protection, which includes delineation of source protection zones, assessment of contamination risks, and development of strategies for land use control (Section 8.7). Predictive computer modeling of quantity and quality of available groundwater, as well as various impacts of its extraction, is important for all phases of resource development and use (Section 8.8); it includes optimization of groundwater extraction (e.g., location and number of wells and their pumping rates) and artificial groundwater recharge where applicable. Finally, artificial aquifer recharge (Section 8.9) is becoming a focal point of groundwater management in many regions. The sustainable use of groundwater, surface water, storm water, and recycled (used) water relies increasingly upon storing water in the subsurface.

Demand management is not the topic of this book, but its importance for the effectiveness and sustainability of any IWRM cannot be overemphasized. The concept of water demand management generally refers to initiatives with objective of satisfying existing needs for water with a smaller amount of available resources, normally through increasing the efficiency of water use. Water demand management can be considered a part of water conservation policies, which describe initiatives with the aim of protecting the aquatic environment and ensuring a more rational use of water resources.

Unfortunately, water demand management is often given a low priority or practiced reluctantly, in part due to a false premise that it only involves raising fees. As discussed in Chap. 1, the water sector is heavily subsidized in both developed and developing countries and politicians are usually very hesitant to "wrestle" with the issue of water pricing, especially during elections. However, in addition to pricing, which is usually an effective means of demand management, there are many other measures that, when combined, can be as effective. A very detailed discussion on demand management tools, water use, and water conservation, with examples from European countries is presented in the report "Sustainable water use in Europe. Part 2: Demand management" published by the European Environment Agency (Lallana et al., 1999). Following are brief excerpts from this comprehensive report illustrating the importance of two demand management

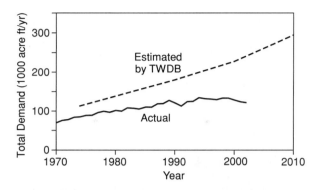

FIGURE 8.5   Total water demand, in thousands acre-feet per year, in service area of El Paso Water Utility. (Modified from Hutchinson, 2003.)

measures, other than water pricing, which can be effective in reducing pressure on water resources:

> Losses in water distribution networks can reach high percentages of the volume introduced. Thus, leakage reduction through preventive maintenance and network renewal is one of the main elements of any efficient water management policy. Leakage figures from different countries indicate the different states of the networks and the different components of leakage included in the calculations (e.g. Albania up to 75%, Croatia 30–60%, Czech Republic 20–30%, France 30%, and Spain 24–34%).

> In agriculture, the aim of the education programs is to help farmers optimize irrigation. This can be achieved through training (on irrigation techniques), and through regular information on climatic conditions, irrigation volume advice for different crops, and advice on when to start/stop adjusting irrigation volumes according to rainfall and type of soil.

Figure 8.5 shows the effect of demand management measures implemented by the city of El Paso, TX, after it became clear that the available groundwater and surface water resources were insufficient to sustain the population growth. The measures included block tariffs, public education, and incentives for water-saving devices in households.

Although the most obvious means of demand management is to limit population, industrial, or agricultural growth in regions where water resources are insufficient to support it, in many cases this is the last measure politicians are willing to contemplate. Notwithstanding various socioeconomic and geographic realities (e.g., poor nations and countries in desert environments), sometimes this option may be the only viable one left. One such example is from California where the state legislature has recognized the need to consider water supplies as part of the local land use planning process. Three bills (Senate Bill 2211, SB 6102, and AB 9013) were enacted in 2001 to improve the assessment of water supplies. The new laws require the verification of sufficient water supply as a condition for approving certain developments. They compel urban water suppliers to provide more information on the reliability of groundwater as an element of supply (DWR, 2003).

8.2 Concept of Groundwater Sustainability

The determination of the sustainable use of groundwater is not solely a scientific, engineering, or a managerial question. Rather, it is a complex interactive process that

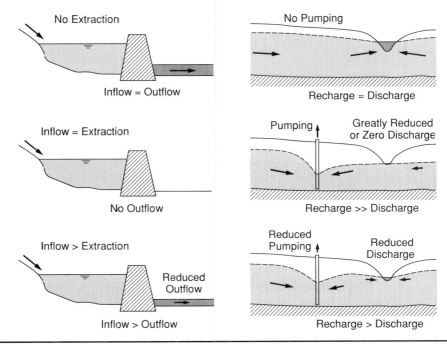

FIGURE 8.6 Illustration of pumping from an aquifer in comparison to surface water reservoir. Yield that is considered sustainable can be achieved only by accepting some consequences of groundwater pumping. Safe yield often is referred to as pumping equal to recharge, but as shown, can result in reduced or no discharge to surface stream and unacceptable consequences.

considers societal, economic, and environmental values, and the respective consequences of different decisions. One commonly held but inaccurate belief, when estimating water availability and developing sustainable water supply strategies, is that groundwater use can be sustained if the amount of water removed is equal to recharge—often referred to as "safe yield." However, there is no volume of groundwater use that can be truly free of any adverse consequence, especially when time is considered. The "safe yield" concept is therefore a myth because any water that is used must come from somewhere. It falsely assumes that there will be no effects on other elements of the overall water budget. Bredehoeft et al. (1982) and Bredehoeft (2002) provide illustrative discussions about the safe yield concept and the related "water budget myth."

In order to examine the "safe yield myth" more carefully, an analogy is made comparing an aquifer and a reservoir behind a dam on a river (Fig. 8.6). If withdrawals from a reservoir equal inflows, the river below the dam will be dry because there will be no outfall from the reservoir. The same principle can be applied to a groundwater reservoir. If pumping (withdrawal) equals inflows (recharge), the outflows (subsurface flow or discharge to springs, streams, or wetlands) from the aquifer will decrease and may eventually reach zero, resulting in some adverse consequence at some point in time. The direct hydrologic effects will be equal to the volume of water removed from the natural system, but those effects may require decades to centuries to be manifest. Because aquifer recharge and groundwater withdrawals can vary substantially over time, these changing rates can be critical information for developing groundwater management strategies (Anderson and Woosley, 2005).

With an increased demand for water and pressures on groundwater resources, the decades-long debate among water professionals about what constitutes "safe" withdrawal of groundwater has now changed into a debate about "sustainable use" of groundwater. The difference is not only semantic, and confusion has occasionally resulted. For example, there are attempts to distinguish between "safe yield" and "sustainable pumping" where the latter is defined as the pumping rate that can be sustained indefinitely without mining or dewatering the aquifer. Devlin and Sophocleous (2005), and Alley and Leake (2004) provide a detailed discussion of these and other related concepts.

What appears most difficult to understand is that the groundwater system is a dynamic one—any change in one portion of the system will ultimately affect its other parts as well. Even more important is the fact that most groundwater systems are dynamically connected with surface water. As groundwater moves from the recharge area toward the discharge area (e.g., a river), it constantly flows through the saturated zone, i.e., the groundwater storage (reservoir). If another discharge area (such as a well for water supply) is created, less water will flow toward the old discharge area (river). This fact seems to be paradoxically ignored by those who argue that groundwater withdrawals may actually increase aquifer recharge by inducing inflow from recharge boundaries (such as surface water bodies!) and therefore result in "sustainable" pumping rates. Although such groundwater management strategy may be "safe" or "sustainable" for the intended use, another question is whether it has any consequences for the sustainable use of the surface water system, which is now losing water to rather than gaining it from the groundwater system.

Another argument for sustainable pumping is based on managing groundwater storage. This management strategy adjusts withdrawal (pumping) rates to take advantage of natural recharge cycles. For example, during periods of high demand, some water may be withdrawn from the storage by greatly increasing pumping rates and lowering the hydraulic heads. During periods of low demand (low pumpage) and high natural recharge, this depleted storage would then be replenished (this is also the principle of spring regulation discussed in Chap. 7). However, the same question of the sustainability of this approach remains. Any portion of the natural recharge that does not contribute to the natural (nonanthropogenic) discharge will have some consequences for the water users and water uses which rely on it. Depending on the volumes and rates of the denied groundwater discharge, the affected users may or may not be able to adapt to the new reality.

In order to sustain valued ecosystems and endangered species, segments of societies worldwide expect water to be made available, in volumes not easily quantified, to meet key habitat requirements. This relatively recent trend is accompanied by actions of environmental groups, which include legal challenges and lawsuits against various government agencies in charge of water governance. Prior to 1970s similar involvement of nongovernmental groups or the public was virtually nonexistent.

A perfect example of an unchecked promotion of industrial growth, disregard for the environment and groundwater resources, and past regulatory decisions that would hardly be possible today is the story of historic Kissengen Spring in Florida (Fig. 8.7).

This historic second-magnitude spring, located approximately 4 mi south of the city of Bartow, in Polk County, central Florida, had an average flow of nearly 20 million gal/d

FIGURE 8.7 Photograph of historic Kissengen Spring in Florida taken in 1941. (Unknown photographer; possibly taken by Myra Haus; courtesy of Wayne Lewis). (Inset) Sink-spring formed near the main historic spring in 2000. (Photograph courtesy of Thomas Jackson.)

and served as a popular recreational area until continuous flow ceased in February 1950, primarily resulting from aquifer dewatering by nearby phosphate mining companies (Peek, 1951; see Fig. 8.8). After the mining operations were discontinued, the aquifer water levels started to slowly recover. Although the original spring vent has been physically plugged since 1962, flow resumed temporarily in January 2006 from a sink-spring that had recently formed nearby (inset photograph). The loss of continuous flow at

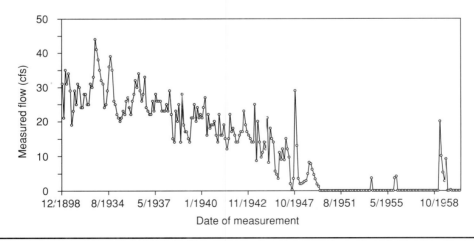

FIGURE 8.8 Hydrograph of the historic Kissengen Spring, Polk County, central Florida. (Modified from Jackson, in preparation.)

Kissengen Spring is an example of the potential fate of Florida springs when their drainage areas (called "springsheds" in Florida) are not adequately understood or properly managed. Recent flow from the adjacent sink-spring offers hope that with proper characterization and management, this natural treasure might be restored to the delight of the people of the region. It may also serve as a model for springshed protection and management (Jackson, in preparation).

As discussed by Anderson and Woosley (2005) and Alley et al. (1999), the implications of long-term droughts should also be considered in long-term water management strategies. Droughts, resulting in reduced groundwater recharge, may be viewed as a natural stress on a groundwater system that, in many ways, has effects similar to groundwater withdrawals—namely, reductions in groundwater storage and accompanying reductions in groundwater discharge to streams and other surface water bodies. At the same time, aquifers in many cases provide a much larger volume of water storage compared to constructed surface water storage (reservoirs). These large underground storage reservoirs can function effectively as a buffer against the annual to decadal variability of climate. Despite this buffering capacity against even annual declines in precipitation, the stress on some aquifers can be increased immediately if pumping increases. In semiarid climates with high evapotranspiration rates and deep water tables, effective recharge to aquifers may occur only as the result of infrequent climatic conditions (El Niño) or hydrologic events rather than a simple percentage of annual precipitation, and it is greatly delayed. Water availability from storage in an aquifer, therefore, depends upon long-term recharge (climatic) patterns rather than short-term climatic events such as seasonal droughts and floods.

When reductions in aquifer storage do occur, they can have serious adverse consequences—lowered groundwater levels may cause compaction of groundwater systems and land subsidence (see Chap. 2) and may induce upconing of poorer quality saline water from deeper portions of the system.

Dependence of communities or regions solely on groundwater in storage is a management strategy that is not sustainable for future generations. When it is obvious that the natural aquifer recharge cannot offset the reduction in groundwater storage in any meaningful way over a reasonable time, prudent groundwater management must consider strategies that rely on surface water and used water for aquifer recharge. In areas where surface water use predominates, groundwater will undoubtedly be an integral part of sustainable-use strategies because of the ability to buffer short-term fluctuations in supply. Large-scale withdrawals from groundwater storage can be used in times of crisis or episodic shortage to achieve sustainability, and likewise during periods of overabundance, water can be stored or banked in aquifers (Anderson and Woosley, 2005).

Groundwater of high quality, protected from surface contamination by competent thick aquitards, is an invaluable resource and should be given special consideration whenever possible. Using such a vital resource for watering golf courses and lawns or cleaning (washing) city streets would be characterized as beyond unsustainable in any hydrologic and climatic conditions. One example of a groundwater system from a temperate humid climate, fully recognized for its value, is the Albian-Neocomian aquifer system located in France, in the Parisian basin. The system is composed of two aquifers, the Albian and the Neocomian, which are hydraulically connected. According to information provided by the French Agency *Agence de l'Eau Seine-Normandie*, the system covers 84,000 km^2 and its total estimated reserves are around 655 billion m^3. The Albian aquifer has unique characteristics: full protection against surface pollution, high groundwater reserves, and a very low natural recharge. The Neocomian aquifer is still not well known, but likely has similar characteristics. The Albian aquifer has been developed since the middle of the nineteenth century. The result was a large drop of the hydraulic head in the aquifer. Public authorities reacted and in 1935 adopted a regulation imposing a licensing regime on all drillings of more than 80-m depth in the Parisian basin. The Neocomian aquifer has been tapped only recently, starting in 1982. Today, the Albian-Neocomian system is considered an important strategic resource, exploited primarily for drinking water purposes. In 2003, the *Schéma Directeur d'Aménagement et de gestion des eaux du bassin Seine-Normandie* was amended in order to emphasize the valuable function of the system, including its use in emergency situations. Indications concerning the total annual volume of water that can be extracted from the system in case of emergency are carefully provided (Foster and Loucks, 2006).

Another example of an aquifer protected by lawmakers because of its high intrinsic value is the Lloyd aquifer of Long Island, NY (Fig. 8.9). Nearly 3 million people on Long Island rely entirely on groundwater for their water supply needs and the island's groundwater system is classified as a sole-source aquifer. The Lloyd aquifer is the deepest in the system. It has been estimated to contain about 9 percent of Long Island's freshwater (Garber, 1985), but receives only 3.1 percent of the recharge that enters the Long Island aquifer system (Buxton et al., 1991; Buxton and Modica, 1992). The thick Raritan clay unit restricts the flow between the Lloyd and the Magothy aquifers and provides good protection of the Lloyd's high water quality. The recharge of the Magothy aquifer, the main source of water supply, and the Lloyd aquifer is predominantly at the groundwater divide near the center of the island. The Lloyd aquifer receives recharge through a corridor generally less than 0.5-mi wide. Travel time from the water table to the top of the Lloyd aquifer is approximately several hundred years (Buxton et al., 1991). The water in

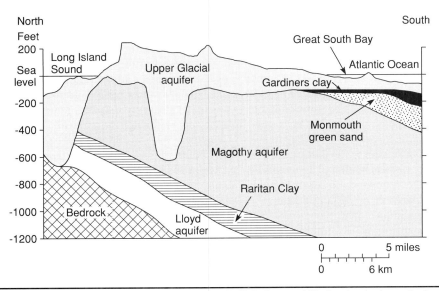

Figure 8.9 Vertical cross section through central Suffolk County, NY, showing hydrogeologic units of the Long Island groundwater system. (Modified from Smolensky et al., 1989, and Brown et al., 2002.)

the Lloyd aquifer is oldest at the southern coast of Long Island—more than 8000 years; it is more than 2000 years old at the northern coast (Buxton and Modica, 1992). This difference in age is attributed mainly to the proximity of the recharge area to the north shore.

Nearly all pumping from the Lloyd aquifer has been in the western part of Long Island (Chu et al., 1997). Excessive pumpage has led to saltwater intrusion in some coastal areas. Withdrawal of water from the Lloyd greatly diminished about 1996 when the Jamaica Water District in New York City was closed.

The following excerpts from an article published in *The New York Times* (October 28, 2007) illustrate the ongoing debate about the Long Island's water supply and especially about the use of the Lloyd aquifer:

> The State Department of Environmental Conservation has made the right decision to protect Long Island's most precious resource. Its commissioner, Pete Grannis, ruled this month that the Suffolk County Water Authority may not tap Long Island's oldest, deepest and purest water source, the Lloyd Aquifer, which has been protected since 1986 by a moratorium on drilling.
>
> Mr. Grannis, stepping into a dispute that has divided public officials, civic groups and scientists in both Nassau and Suffolk Counties, overruled an administrative law judge who had sided with the water authority. The authority had wanted the moratorium lifted so it could mix clean water from the Lloyd with polluted water from a closed well in Northport. The mixture, with the bad stuff diluted enough to meet quality standards, would have saved the agency the trouble and cost of removing the contamination in Northport or piping water in from somewhere else.
>
> The authority had argued that it and its customers would suffer "extreme hardship" if it were not able to exploit a pristine, protected resource. Mr. Grannis was right not to buy that, and to take a strong position in favor of conserving Long Island's endangered water supply.
>
> Long Island lives above the water it drinks. The water is buried deep underground in aquifers that have been relentlessly tapped with thousands of wells stuck into them like so many soda straws. There is no other significant source of water here, except maybe those trucks loaded with bottles of Poland Spring. But generations of unchecked growth and the limitless thirst of pools, lawns, hot tubs, showers and toilets have left some water sources badly compromised, contaminated with nitrates from fertilizer and sewage and tainted with saltwater that intrudes through sandy soil to replace the freshwater that was sucked out.
>
> Some scientists argue that the supply's days are numbered; others say there is plenty of water and nothing to worry about. The debate does not alter our bottom line. The optimists could be correct about the water supply, but still wrong about drilling in the Lloyd Aquifer. The answer to a growing thirst is not to drill ever deeper into an ancient, untapped source of freshwater.
>
> Other parts of the country are waking up to the realization that fresh, clean water is a scarce resource that should be conserved with wisdom and care. On Long Island, where our aquifers have generously supported egregiously wasteful and polluting habits for decades, that understanding has been slow to sink in. The region's supply of cheap, abundant, excellent water is under stress, from our own mistreatment of it. The answer is not adding one more straw.

Any use of groundwater to be sustainable must recognize the interests of various stakeholders, including the public and nongovernmental environmental groups, and require their participation in decisions. More importantly, regulatory and management frameworks that will ensure this participation and take into account often diverging interests should be put in place. One such attempt is the establishment of Texas Groundwater Protection Committee (TGPC, 1999), charged by the state legislature to *"develop and update a comprehensive groundwater protection strategy for the state that provides guidelines for the prevention of contamination and for the conservation of groundwater and that provides for the coordination of the groundwater protection activities of the agencies represented on the committee"*

(Texas Water Code Section 26.405(2)). The TGPC is composed of nine state agencies and the Texas Alliance of Groundwater Districts (TAGD). Members of the TGPC represent the primary state agencies and groundwater districts entrusted by the legislature with the conservation, protection, and where necessary the remediation of groundwater. Notably, the TAGD does not include representatives of nongovernmental environmental groups. Nevertheless, in the updated 2003 Texas Groundwater Protection Strategy, TGPC reminds all the participating stakeholders that

> One of Texas' most valued natural resources is its ground and surface water resources. In 1999, groundwater provided approximately 58 percent of the water used in the state, and it is a fundamental component of the state's water supply. In addition, groundwater provides a significant amount of the base flow for the state's rivers and streams, and is, therefore, important to the maintenance of the state's environment and economy.

The groundwater protection and conservation strategy developed by the TAGD includes the following:

- Details on the state's groundwater protection goal as established by the legislature
- Statewide groundwater classification system and the process used by the state to identify groundwater contamination
- Roles and responsibilities of the various state agencies involved in groundwater protection and explanation of the TGPC as a coordinating mechanism
- Examples of how the various state agencies carry out groundwater protection programs through regulatory and nonregulatory models
- Explanation how the local, state, and federal agencies coordinate management of groundwater data for the enhancement of groundwater protection
- Role that research plays in understanding groundwater's importance and the importance of coordinating research efforts
- Overview of the groundwater public education efforts in the state
- Public participation in establishing and implementing groundwater policy
- Planning process for updating the groundwater strategy
- Proposal for inclusion in the next *Strategy* an identification and ranking of significant threats to the state's groundwater resources, consideration of the vulnerability of groundwater resources, and a prioritization of actions to address those threats
- Recommendations and possible actions to protect and conserve groundwater

In conclusion, when thinking of groundwater sustainability, regardless of any actual context, it would be hard to add anything to the following 100-year plus old quote from Slichter (1902) that shows a great foresight:

> The fundamental disadvantage in the utilization of underground sources of water is the danger of overdrawing the natural supply. In regions in which the rainfall is light and catchment areas are small, as in parts of southern California, it is easy to extend development of underground sources

so as to greatly exceed the natural rate of annual replenishment. In this way underground reservoirs are depleted which have been ages in filling; principal as well as interest is drawn upon, and much disappointment must inevitably follow.

8.2.1 Nonrenewable Groundwater Resources

As discussed by Foster et al. (2002–2005a), groundwater resources are hardly ever strictly nonrenewable. However, in certain cases the period needed for replenishment (100s to 1000s of years) is very long in relation to the normal period of human activity in general and water resources planning in particular. For this reason, it is valid in such cases to talk of the utilization of nonrenewable groundwater or aquifer mining. Two general groups of groundwater systems fall into the category of nonrenewable:

1. Unconfined aquifers in areas where contemporary recharge is very infrequent and of small volume and the resource is essentially limited to static groundwater storage reserves

2. Confined portions of large aquifer systems, where groundwater development intercepts or induces little active recharge and the hydraulic head falls continuously with groundwater extraction

Both groups involve the extraction of groundwater that originated as recharge in a distant past, including during more humid climatic regimes. The volumes of such groundwater stored in some aquifers are enormous. For example, total recoverable volume of freshwater in the Nubian Sandstone Aquifer System of Africa is estimated at about 15,000 km^3 and the present rate of annual groundwater extraction is 2.17 km^3. For comparison, combined volume of water stored in the Great Lakes of the North America is 22,684 km^3.

The term groundwater sustainability in the case of nonrenewable systems has an entirely social rather than physical (engineering, scientific) context. It implies that full consideration must be given not only to the immediate benefits, but also to the negative socioeconomic impacts of development and to the "what comes after" question—and thus to time horizons longer than 100 years (Foster et al., 2002–2005a).

There are two general situations under which the utilization of nonrenewable groundwater occurs: planned and unplanned. In the planned scenario, the management goal is the orderly utilization of groundwater reserves stored in the system with little preexisting development. The expected benefits and predicted impacts over a specified time-frame must be specified. Appropriate exit strategies need to be identified, developed, and implemented by the time that the groundwater system is seriously depleted. This scenario must include balanced socioeconomic choices on the use of stored groundwater reserves and on the transition to a less water-dependent economy. A key consideration in defining the exit strategy will be identification of the replacement water resource, such as desalination of brackish groundwater (Foster et al., 2002–2005a). Saudi Arabia is a good example of two main stages in exploitation of nonrenewable groundwater: initial very rapid, large-scale and unrestricted development for all uses, subsequently supplemented by desalinated water and treated wastewater. Saudi Arabia has become the largest desalinated water producer in the world. The present production presents about 50 percent of the total current domestic and industrial demands, with the rest met from groundwater resources (Abderrahman, 2006). However, irrigated agriculture is still the largest user of

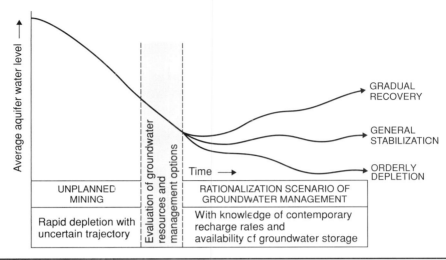

FIGURE 8.10 Targets for nonrenewable groundwater resource management in rationalization scenarios following indiscriminate and excessive exploitation. (From Foster et al., 2002–2005a.)

the nonrenewable groundwater in Saudi Arabia where food security concerns have the highest priority.

In an unplanned situation a rationalization scenario is needed in which the management goal is to achieve hydraulic stabilization (or recovery) of the aquifer or more orderly utilize groundwater reserves by minimizing quality deterioration, maximizing groundwater productivity, and promoting social transition to a less water-dependent economy (Fig. 8.10).

In both cases, the groundwater extraction rate will have to be reduced, and thus the introduction of demand management measures (including realistic water fees and incentives for real water saving) will be needed. In the longer run, potable water supply use will have to be given highest priority and some other lower productivity uses may have to be discouraged (Foster et al., 2002–2005a).

It is vital that the groundwater is used with maximum hydraulic efficiency and economic productivity, and this implies full reuse of urban, industrial, and mining water supplies and carefully controlled agricultural irrigation. An acceptable system of measuring or estimating the volumetric extraction will be required as the cornerstone for both realistic fees for water and enforcing regulations to discourage inefficient and unproductive uses.

Public awareness campaigns on the nature, uniqueness, and value of nonrenewable groundwater will be necessary to create social conditions conducive to aquifer management, including wherever possible full user participation. In this context, all groundwater data (reliably and independently synthesized) should be made regularly available to stakeholders and local communities. This transparency is absolutely essential because any depletion of nonrenewable groundwater resources is not sustainable in the long term by definition. Although it may sustain certain socioeconomic development and policies for one or several generations, this depletion will ultimately lead to an inevitable change of the socioeconomic and political environments it helped create in the first place. It is also important that the political entity engaged in mining of nonrenewable groundwater, as well as the society as a whole, clearly understand its negative environmental impacts. They include serious deterioration and ultimately a complete disappearance of springs, surface streams, marshes, and oases fed by the mined groundwater system as well as the disappearance of the associated flora and fauna. The only option for delaying or decreasing such impacts is artificial aquifer recharge with used water, accompanied by minimization of consumptive water use.

Foster et al. (2002–2005a) argue that nonrenewable groundwater in aquifer storage must be treated as a public property (or alternatively common property) resource. As such, it should be under jurisdiction of high levels of government where options on mining of groundwater reserves must be evaluated and the final decisions made. In countries with a water resources ministry, the decision could rest with the corresponding minister, but in other situations, it would be better taken by the president's, prime minister's, or provincial governor's office depending upon the territorial scale of the groundwater system (such systems are often transboundary, extending in more than one country). High-level political and transparent ownership of the rationalization plan for aquifers already been subject to mining on an unplanned basis is also highly desirable.

Various tools and instruments for nonrenewable groundwater management are identical as in the case of general water management (Table 8.1). A special emphasis, however, should be placed on predictive groundwater modeling, and on monitoring the effects of groundwater extraction over the entire extent of the groundwater system being mined. Three spatially distributed hydrogeologic parameters critical for estimating (modeling) fresh groundwater reserves available for extraction will have to be determined in the field in all three dimensions, and as accurately as possible throughout the system: (1) storage coefficients, (2) leakance rates between aquifers separated by aquitards where applicable, and (3) TDS or salinity of groundwater.

Predictive groundwater modeling of the available extraction rates, hydraulic head changes, and changes in groundwater quality will have varying degrees of uncertainty depending upon the scale of the system and the existing information. This uncertainty will be reduced as more observations become available during exploitation of the groundwater system and the model(s) are periodically updated. Based on the new modeling results, the management strategies may have to be modified with the agreement of all stakeholders.

A high priority should be establishing a system of groundwater extraction rights (i.e., permits, licenses or concessions) that is consistent with the hydrogeologic reality of continuously declining groundwater levels, potentially decreasing well yields, and possibly deteriorating groundwater quality due to saltwater intrusion and upconing. Thus the permits (for specified rates of extraction at given locations) will need to be time limited in the long term, but also subject to initial review and modification after 5 to 10 years, by which time more will be known about the aquifer response to extraction through operational monitoring. The existence of time-limited permits subject to periodic review will normally stimulate permit holders to provide regular data on well operations. It will be incumbent upon the water resources administration to make appropriate institutional arrangements—through some form of groundwater database or datacenter—for the archiving, processing, interpretation, and dissemination of this information (Foster et al., 2002–2005a).

As mentioned earlier, many major aquifers containing large reserves of nonrenewable groundwater are transboundary, either in a national sense or between autonomous provinces or states within a single nation. In such circumstances, the involved jurisdictions will all mutually gain through

- Operation of joint or coordinated groundwater monitoring programs
- Establishment of a common groundwater database or mechanism for information sharing
- Adoption of coordinated policies for groundwater resource planning, utilization, and management, as well as adoption of procedures for conflict resolution
- Harmonization of relevant groundwater legislation and regulations (Foster et al., 2002–2005a)

8.3 Regulatory Framework

In a publication prepared by the Global Water Partnership, a major international organization whose mission is to support countries in the sustainable management of their water resources, Rogers and Hall (2003) provide the following discussion on water governance and regulations:

> The theoretical bases of governance with regard to water are a subset of theories of collective behavior. Unfortunately, no one simple theory explains every situation. There is often a marked difference between the philosophical Continental European and Latin American approaches and the pragmatic US-Anglo Saxon schools of thought. A relatively clear original demarcation of property rights and experimentation with these rights over time has led the US to flexible approaches to water governance. This approach allows for adjustments when economic and social conditions change because it does not aspire to build institutions that cover all possible eventualities. There are also systems that are hybrids of the Civil law (philosophical, descended from Roman law) and Common law (pragmatic, from Britain) approaches, as well as systems with other ancient roots, such as those of the pre-Colombian Americas, India and Islamic countries.
>
> There are also systems of social rights and responsibilities that remain traditional and uncodified, and are not necessarily less strong because they are manifested in cultural expectations rather than written rules. A social perception of equitable sharing is important to governance. The notion of flexibility and equitable sharing is, however, alien to many countries whose governance systems are rigid and do not allow for 'reasonableness'. Adaptive capability is often not present and without enforceable sanctions, poor governance systems favor the strong. This makes it very difficult and even dangerous to translate practices based on flexibility and pragmatism into many developing country governance environments, unless the prevailing social system can provide adequate sanction against miscreants (Solanes, 2002).

8.3.1 Groundwater Quantity

In the United States, the management of water and the system of issuing water rights has historically been an un-denied responsibility of the states, rather than the federal government. However, many aspects of federal law intrude today into this state-based system of water management. The Endangered Species Act, the Clean Water Act, and the Wild and Scenic Rivers Act are just a few of the federal laws that impinge upon state authority. As an example, the legal concept of federal-reserved water rights was thought to apply only to Indian reservations until the mid-1970s when the U.S. Supreme Court issued its ruling in the case of *Cappaert* v. *United States*, 1976 (from Anderson and Woosley, 2005):

> This Court has long held that when the Federal Government withdraws its lands from the public domain and reserves it for a Federal purpose, the Government, by implication, reserves appurtenant water then unappropriated to the extent needed to accomplish the purpose of the reservation. In doing so the United States acquires a reserved water right in unappropriated water which vests on the date of the reservation and is superior to the rights of future appropriators.

The quantity of the water is limited to the quantity needed to accomplish the purpose(s) of the reservation. A significant challenge today is to determine the amount of water required to sustain native peoples, a riparian system, or an endangered species. Under this ruling, the federal government can claim a volume of water, required to sustain the lands set aside (reserved) from the public domain for a particular purpose, with an early priority date. National forests, national parks, national wildlife refuges, and wild and scenic rivers have a water right that goes along with the land.

As discussed by Anderson and Woosley (2005), the unique circumstances of American expansion into the vast lands of the West gave rise to a body of surface water law that is markedly different from the laws governing water use in the eastern United States. Under the riparian doctrine, which is used in the eastern states, the water-right holder must own land adjacent to a water body. In the west, a water right can be held by a property owner regardless of the proximity of his land to water, so long as the water is being put to a beneficial use. Western water law, or the prior appropriation doctrine, can trace its origins to the placer gold mines of California and the cultural differences and attitudes of the early settlers. The Mormons, Native Americans, and Spanish settlers all existed in the west, using an approach to water use very different from prior appropriation; but to encourage the westward expansion of the United States, prior appropriation served a useful purpose (Glennon, 2002).

The course of water-rights law changed in the late 1840s when thousands of fortune seekers flocked to California following the discovery of gold in the gravels of the American River. Water development proceeded on a scale never before witnessed in the United States as these "forty-niners" built extensive networks of flumes and waterways to work their claims. Often the water carried in these systems had to be transported far from the original river or stream. The self-governing, maverick miners applied the same "finders-keepers" rule to water that they did to their mining claims—it belonged to the first miner to assert ownership. It allowed others to divert available water from the same river or stream, but their rights existed within a hierarchy of priorities. This "first in time, first in right" principle became an important feature of modern water law in the West (Sax et al., 2000).

Western water law historically has placed a higher value on water being used off-stream. As an example, the Constitution of the State of Colorado states the following: "The right to divert the unappropriated waters of any natural stream to beneficial uses shall never be denied." Notice the requirement that the water be put to beneficial use. If a diverter fails to use his full allocation of water, after some period of time, he can be forced to forfeit some or all of his right. In this way, the prior appropriation doctrine discourages conservation, for there is a serious disincentive to conserve water on the part of the water-right holder. Such provisions, established in law and set by historical precedent, make it difficult to change the allocation of water to other uses, such as instream use for aquatic life and habitat maintenance and enhancement (Anderson and Woosley, 2005).

In contrast to surface water, groundwater was largely ignored in the early years of water use regulation in the United States. For example, at the beginning of the twentieth

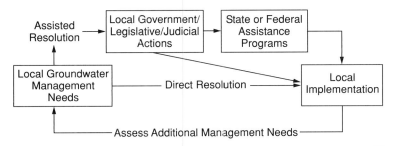

Figure 8.11   Process of addressing groundwater management needs in California. (From DWR, 2003.)

century, groundwater was seen largely as just a convenient resource that allowed for settlement in nearly any part of the American West, given its widespread occurrence, and was subject to common-law rule. The common-law rule regarding groundwater is the rule of capture or the English rule, which essentially provides that, absent malice or willful waste, landowners have the right to take all the water they can capture under their land and do with it what they please, and they will not be liable to neighboring landowners even if in so doing they deprive their neighbors of the water's use. The rule of capture is in contrast to "reasonable use" or the "American rule," which provides that the right of a landowner to withdraw groundwater is not absolute, but limited to the amount necessary for the reasonable use of his land, and that the rights of adjoining landowners are correlative and limited to reasonable use. Today, Texas stands alone as the only western state that continues to follow the rule of capture (Potter, 2004).

Except for Texas, all other states have some form of the American rule, usually left for interpretation to local regulators. For example, although the regulation of groundwater has been considered on several occasions, the California Legislature has repeatedly held that groundwater management should remain a local responsibility. Figure 8.11 depicts the general process by which groundwater management needs are addressed under existing law. They are identified at the local water agency level and may be directly resolved at the local level. If groundwater management needs cannot be directly resolved at the local agency level, additional actions such as enactment of ordinances by local governments, passage of laws by the legislature, or decisions by the courts may be necessary to resolve the issues. Upon implementation, local agencies evaluate program success and identify additional management needs. The state's role is to provide technical and financial assistance to local agencies for their groundwater management efforts, such as through the local groundwater assistance grant program (DWR, 2003).

8.3.2   Groundwater Quality

In the 1999 Ground Water Report to Congress, the United States Environmental Protection Agency (USEPA) emphasized the need for more effective coordination of groundwater protection programs at the federal, state, and local levels. Similar efforts in surface water programs have led many states to adopt watershed-based management approaches that coordinate the activities of agencies and programs that play a role in water quality protection. At the federal level, the 1998 Clean Water Action Plan is designed to promote similar coordination among federal agencies. While the Clean Water Action Plan and some state watershed protection approaches address groundwater, true coordination of groundwater management efforts has not been achieved in most states (USEPA, 1999). However, the agency also indicated that at the time 47 states have approved wellhead protection programs which mandate delineation of wellhead protection areas (WPAs) for public water supplies. WPA is a designated surface and subsurface area surrounding a well or well field for a public water supply and through which contaminants or pollutants are likely to pass and eventually reach the aquifer that supplies the well or well field. The purpose of designating the area is to provide protection from the potential of contamination of the water supply. These areas are designated in accordance with laws, regulations, and plans that protect public drinking water supplies.

In the United States, federal support is available for comprehensive groundwater protection planning, primarily through the Clean Water Act (CWA) and Safe Drinking Water Act (SDWA). However, the vast majority of federal resources allocated for groundwater have been devoted to groundwater remediation. Millions, and in many cases hundreds of millions of dollars from public and private funds have been spent in each state on cleanup activities, or government oversight of cleanup performed by private parties. The need for such spending will continue. In 1994, the National Academy of Sciences estimated that over a trillion dollars, or approximately $4000 per person in the United States, will be spent in the next 30 years on cleanup of contaminated soil and groundwater. However, comparatively few of those cleanup resources will be used to manage future threats to the resource in a comprehensive way that may prevent the need for future, costly cleanup efforts. According to USEPA, a comprehensive protection program would help determine the most significant threats to the resource, help establish the local priorities and direct funds to those programs that would deal with the most significant threats first (USEPA, 1999).

Title 40-Protection of the Environment of the Code of Federal Regulations (CFR) contains all the regulations governing USEPA's programs. The CFR is a collection of all federal regulations codified and enforced by all federal agencies. Pertinent to surface water and groundwater protection are subchapters D (Water Programs), I (Solid Wastes), J (Superfund), N (Effluent Guidelines and Standards), O (Sewage Sludge), and R (Toxic Substances Control Act). Electronic versions of regulations under Title 40 are available online at http://www.epa.gov/lawsregs/search/40cfr.html.

Clean Water Act

Growing public awareness and concern for controlling surface water pollution led to enactment of the Federal Water Pollution Control Act Amendments of 1972. As amended in 1977, this law became commonly known as the Clean Water Act. The Act established the basic structure for regulating discharges of pollutants into the surface waters of the United States. It gave USEPA the authority to implement pollution control programs such as setting wastewater standards for industry. The Clean Water Act also continued requirements to set water quality standards for all contaminants in surface waters. The Act made it unlawful for any person to discharge any pollutant from a point source into navigable waters, unless a permit was obtained under its provisions. It also funded the construction of sewage treatment plants under the construction grants program and recognized the need for planning to address the critical problems posed by nonpoint source pollution.

The key elements of the CWA are establishment of water quality standards (WQS), their monitoring and, if WQS are not met, developing strategies for meeting them. If all WQS are met, then antidegradation policies and programs are employed to keep the water quality at acceptable levels. Ambient monitoring is also implemented to ensure that this is the case. The most common strategy of meeting WQS is the development of a total maximum daily load (TMDL). TMDLs determine what level of pollutant load would be consistent with meeting WQS. TMDLs also allocate acceptable loads among sources of the relevant pollutants.

More details on the Clean Water Act can be found on the following USEPA Web site: http://www.epa.gov/r5water/cwa.htm.

Safe Drinking Water Act

The Safe Drinking Water Act (SDWA) was originally passed by the United States Congress in 1974 to protect public health by regulating the nation's public drinking water supply. The law was amended in 1986 and 1996 and requires many actions to protect drinking water and its sources: rivers, lakes, reservoirs, springs, and groundwater wells. SDWA applies to every public water system in the United States (the Act does not regulate private wells which serve fewer than 25 individuals). There are currently more than 160,000 public water systems providing water to almost all Americans at some time in their lives. Public water system is defined as water supply system that has at least 15 service connections or serves at least 25 people per day for 60 days of the year. There are two main types of public water supply systems (PWSs): community systems and noncommunity systems.

Community water systems (CWSs; there are approximately 54,000 in the country) serve the same people year-round. Most residences including homes, apartments, and condominiums in cities, small towns, and mobile home parks are served by CWSs.

Non-CWSs serve the public but do not serve the same people year-round. There are two types of non-CWSs: (1) Nontransient non-CWS (there are approximately 20,000) serves the same people more than 6 months/yr, but not year-round; for example, a school with its own water supply is considered a nontransient system and (2) transient non-CWS (there are approximately 89,000) serves the public but not the same individuals for more than 6 months; for example, a rest area or campground may be considered a transient water system.

Originally, SDWA focused primarily on treatment as the means of providing safe drinking water at the tap. The 1996 amendments greatly enhanced the existing law by recognizing source water protection, operator training, funding for water system improvements, and public information as important components of safe drinking water. This approach ensures the quality of drinking water by protecting it from source to tap.

Source Water Assessment Programs (SWAPs) require that states and water suppliers conduct an assessment of its sources of drinking water (rivers, lakes, reservoirs, springs, and groundwater wells) to identify significant potential sources of contamination and to determine how susceptible the sources are to these threats. Water systems may also voluntarily adopt programs to protect their watershed or wellhead and states can use legal authorities from other laws to prevent pollution. Section 1429 of SDWA authorizes the USEPA Administrator to make grants to the States for the development and implementation of these programs to ensure the coordinated and comprehensive protection of groundwater resources.

SDWA authorizes the USEPA to set national health-based standards for drinking water to protect against both naturally occurring and anthropogenic contaminants that may be found in drinking water. USEPA sets these standards based on sound science for protecting human health risks and considering available technology and costs. The

National Primary Drinking Water Regulations set enforceable maximum contaminant levels for particular contaminants in drinking water or required ways to treat water to remove contaminants (see also Chap. 5). Each standard includes requirements for water systems to test for contaminants in the water to make sure standards are achieved. Maximum contaminant levels are legally enforceable, which means that both USEPA and states can take enforcement actions against water systems not meeting safety standards. USEPA and states may issue administrative orders, take legal actions, or fine water utilities. In contrast, the National Secondary Drinking Water Standards are not enforceable; the Agency recommends them to water utilities but does not require systems to comply. However, states may choose to adopt them as enforceable standards or may relax them.

Sole Source Aquifer Protection Program

The Sole Source Aquifer (SSA) Protection Program is authorized by Section 1424(e) of the Safe Drinking Water Act of 1974 which states

> If the Administrator determines, on his own initiative or upon petition, that an area has an aquifer which is the sole or principal drinking water source for the area and which, if contaminated, would create a significant hazard to public health, he shall publish notice of that determination in the Federal Register. After the publication of any such notice, no commitment for federal financial assistance (through a grant, contract, loan guarantee, or otherwise) may be entered into for any project which the Administrator determines may contaminate such aquifer through a recharge zone so as to create a significant hazard to public health, but a commitment for federal assistance may, if authorized under another provision of law, be entered into to plan or design the project to assure that it will not so contaminate the aquifer.

The USEPA defines a sole or principal source aquifer as one which supplies at least 50 percent of the drinking water consumed in the area overlying the aquifer. Such area cannot have an alternative drinking water source(s) which could physically, legally, and economically supply all those who depend upon the aquifer for drinking water.

Although USEPA has statutory authority to initiate SSA designations, it has a long-standing policy of only responding to petitions. Any person may apply for SSA designation. A "person" is any individual, corporation, company, association, partnership, state, municipality, or federal agency. A petitioner is responsible for providing USEPA with hydrogeologic and drinking water usage data, and other technical and administrative information required for assessing designation criteria.

If an SSA designation is approved, proposed federal financially assisted projects which have the potential to contaminate the aquifer are subject to USEPA review. Proposed projects that are funded entirely by state, local, or private concerns are not subject to USEPA review. Examples of federally funded projects which have been reviewed by USEPA under the SSA protection program include the following:

- Highway improvements and new road construction
- Public water supply wells and transmission lines
- Wastewater treatment facilities
- Construction projects that involve disposal of storm water
- Agricultural projects that involve management of animal waste
- Projects funded through Community Development Block Grants

SSA designations help increase public awareness on the nature and value of local groundwater resources by demonstrating the link between an aquifer and a community's drinking water supply. Often, the realization that an area's drinking water originates from a vulnerable underground supply can lead to an increased willingness to protect it. The public also has an opportunity to participate in the SSA designation process by providing written comments to USEPA or by participating in an USEPA-sponsored public hearing prior to a designation decision.

Important information on the boundaries, hydrogeologic characteristics, and water use patterns of an area's aquifer must be documented by a petitioner seeking SSA designation. Following USEPA's technical review of a petition, this information is summarized by the Agency in a technical support document that is made available for public review. Following designation, a Federal Register notice is published to announce and summarize the basis for USEPA's decision.

Sole source aquifer designation provides only limited federal protection of groundwater resources which serve as drinking water supplies. It is not a comprehensive groundwater protection program. Protection of groundwater resources can best be achieved through an integrated and coordinated combination of federal, state, and local efforts such as called for under the Comprehensive State Ground Water Protection Program (CSGWPP) approach. For example, local wellhead protection programs designed to protect the recharge areas of public water supply wells should work in concert with contaminant source control and pollution prevention efforts being managed at various levels of government. This coordination ensures that all groundwater activities meet the same protection goal without duplication of time, effort, and resources.

Although designated aquifers have been determined to be the sole or principal source of drinking water for an area, this does not imply that they are more or less valuable or vulnerable to contamination than other aquifers which have not been designated by USEPA. Many valuable and sensitive aquifers have not been designated simply because nobody has petitioned USEPA for such status or because they did not qualify for designation due to drinking water consumption patterns over the entire aquifer area. Furthermore, groundwater value and vulnerability can vary considerably both between and within designated aquifers. As a result, USEPA does not endorse using SSA status as the sole or determining factor in making land use decisions that may impact groundwater quality. Rather, site-specific hydrogeologic assessments should be considered along with other factors such as project design, construction practices, and long-term management of the site. More information on the USEPA's sole source aquifer program can be found on the following Web page: http://yosemite.epa.gov/r10/water.nsf/Sole+Source+Aquifers/Program.

Various detailed information on the Safe Drinking Water Act can be found on the following dedicated USEPA Web page: http://www.epa.gov/safewater/sdwa/index.html.

CERCLA

The Comprehensive Environmental Response, Compensation, and Liability Act (CERCLA), commonly known as Superfund, was enacted by the United States Congress in 1980. This law created a tax on the chemical and petroleum industries and provided broad federal authority to respond directly to releases or threatened releases of hazardous substances that may endanger public health or the environment. The law authorizes two types of response actions:

- Short-term removals, where actions may be taken to address releases or threatened releases requiring prompt response.

- Long-term remedial response actions that permanently and significantly reduce the dangers associated with releases or threats of releases of hazardous substances that are serious, but not immediately life threatening. These actions can be conducted only at sites listed on USEPA's National Priorities List (NPL).

CERCLA also enabled the revision of the National Contingency Plan (NCP). The NCP provided the guidelines and procedures needed to respond to releases and threatened releases of hazardous substances, pollutants, or contaminants. The NCP also established the NPL. The NPL is the list of national priorities among the known releases or threatened releases of hazardous substances, pollutants, or contaminants throughout the United States and its territories. The NPL is intended primarily to guide the USEPA in determining which sites warrant further investigation.

The Superfund Amendments and Reauthorization Act (SARA) amended CERCLA in 1986. SARA reflected USEPA's experience in administering the complex Superfund program during its first 6 years and made several important changes and additions to the program.

The hazard ranking system (HRS) is the principal mechanism USEPA uses to place uncontrolled waste sites on the NPL. It is a numerically based screening system that uses information from initial, limited investigations—the preliminary assessment and the site inspection—to assess the relative potential of sites to pose a threat to human health or the environment. Any person or organization can petition USEPA to conduct a preliminary assessment using the Preliminary Assessment Petition.

HRS scores do not determine the priority in funding USEPA remedial response actions, because the information collected to develop HRS scores is not sufficient to determine either the extent of contamination or the appropriate response for a particular site. The sites with the highest scores do not necessarily come to the USEPA's attention first—this would require stopping work at sites where response actions were already underway. USEPA relies on more detailed studies in the remedial investigation/feasibility study, which typically follows listing.

After a site is listed on the NPL (becomes a "Superfund" site), a remedial investigation/feasibility study (RI/FS) is performed at the site (see Chap. 9) to determine if the site requires a cleanup (remediation), as well as the appropriate remedial technologies to achieve the cleanup goals if they are set.

One of USEPA's top priorities is to have those responsible for the contamination (the so-called potentially responsible parties or PRPs) remediate the site. If the PRPs cannot be found, are not viable, or refuse to cooperate, USEPA, the state, or tribe may cleanup the site using Superfund money. USEPA may seek to recover the cost of clean up from those parties that do not cooperate.

More detail on CERCLA regulation can be found on the following USEPA Web site: http://www.epa.gov/superfund/policy/cercla.htm.

The Resource Conservation and Recovery Act (RCRA)

The Resource Conservation and Recovery Act—commonly referred to as RCRA—is the primary law governing the disposal of solid and hazardous waste in the United States. Congress passed RCRA in 1976 to address the increasing problems the nation faced from the growing volume of municipal and industrial waste. RCRA, which amended the Solid Waste Disposal Act of 1965, set national goals for

- Protecting human health and the environment from the potential hazards of waste disposal
- Conserving energy and natural resources
- Reducing the amount of waste generated
- Ensuring that wastes are managed in an environmentally sound manner

To achieve these goals, RCRA established three distinct, yet interrelated, programs:

- The solid waste program, under RCRA Subtitle D, encourages states to develop comprehensive plans to manage nonhazardous industrial solid waste and municipal solid waste, sets criteria for municipal solid waste landfills and other solid waste disposal facilities, and prohibits the open dumping of solid waste.
- The hazardous waste program, under RCRA Subtitle C, establishes a system for controlling hazardous waste from the time it is generated until its ultimate disposal—in effect, from "cradle to grave."
- The underground storage tank (UST) program, under RCRA Subtitle I, regulates underground storage tanks containing hazardous substances and petroleum products.

RCRA banned all open dumping of waste, encouraged source reduction and recycling, and promoted the safe disposal of municipal waste. RCRA also mandated strict controls over the treatment, storage, and disposal of hazardous waste.

RCRA was amended and strengthened by Congress in November 1984 with the passing of the Federal Hazardous and Solid Waste Amendments (HSWA). These amendments to RCRA required the phasing out land disposal of hazardous waste. Some of the other mandates of this strict law include increased enforcement authority for USEPA, more stringent hazardous waste management standards, and a comprehensive underground storage tank program.

RCRA focuses only on active and future facilities and does not address abandoned or historical sites which are managed under the CERCLA (Superfund) regulations. A facility where contamination of soil, sediment, surface water or groundwater water has been documented is listed as RCRA site and required to take all necessary measures to remediate the contamination.

More details on RCRA regulations can be found on the dedicated USEPA Web site at: http://www.epa.gov/rcraonline/.

Groundwater Rule

The USEPA promulgated the final Groundwater Rule (GWR) in October 2006 to reduce the risk of exposure to fecal contamination that may be present in public water systems that use groundwater sources. The rule establishes a risk-targeted strategy to identify groundwater systems that are at high risk for fecal contamination. The GWR also specifies when corrective action (which may include disinfection) is required to protect consumers who receive water from groundwater systems from bacteria and viruses (USEPA, 2006).

In the explanation of its ruling, the agency points out that groundwater occurrence studies and recent outbreak data show that pathogenic viruses and bacteria can occur in public water systems that use groundwater and that people may become ill due to exposure to contaminated groundwater. Most cases of waterborne disease are characterized by gastrointestinal symptoms (e.g., diarrhea, vomiting, etc.) that are frequently self-limiting in healthy individuals and rarely require medical treatment. However, these same symptoms are much more serious and can be fatal for persons in sensitive subpopulations (such as young children, the elderly, and persons with compromised immune systems). Viral and bacterial pathogens are present in human and animal feces, which can, in turn, contaminate drinking water. Fecal contamination can reach groundwater sources, including drinking water wells, from failed septic systems, leaking sewer lines, and by passing through the soil and fractures in the subsurface.

The GWR applies to more than 147,000 public water systems that use groundwater (as of 2003). It also applies to any system that mixes surface and groundwater if the groundwater is added directly to the distribution system and provided to consumers without treatment equivalent to surface water treatment. In total, these systems provide drinking water to more than 100 million consumers.

The GWR addresses risks through a risk-targeting approach that relies on four major components:

1. Periodic sanitary surveys of groundwater systems that require the evaluation of eight critical elements and the identification of significant deficiencies (e.g., a well located near a leaking septic system). States must complete the initial survey by December 31, 2012, for most CWSs and by December 31, 2014, for CWSs with outstanding performance and for all non-CWSs.

2. Source water monitoring to test for the presence of *Escherichia coli*, enterococci, or coliphage in the sample. There are two monitoring provisions: (a) *Triggered monitoring* for systems that do not already provide treatment that achieves at least 99.99 percent (4-log) inactivation or removal of viruses and that have a total coliform-positive routine sample under Total Coliform Rule sampling in the distribution system. (b) *Assessment monitoring* as a complement to triggered monitoring—a state has the option to require systems, at any time, to conduct source water assessment monitoring to help identify high risk systems.

3. Corrective actions required for any system with a significant deficiency or source water fecal contamination. The system must implement one or more of the following correction action options: (a) correct all significant deficiencies; (b) eliminate the source of contamination; (c) provide an alternate source of water; or (d) provide treatment which reliably achieves 99.99 percent (4-log) inactivation or removal of viruses.

4. Compliance monitoring to ensure that treatment technology installed to treat drinking water reliably achieves at least 99.99 percent (4-log) inactivation or removal of viruses.

The Agency estimates that the GWR will reduce the average number of waterborne viral (rotovirus and echovirus) illnesses by nearly 42,000 illnesses each year from the current baseline estimate of approximately 185,000 (a 23 percent reduction in total illnesses). In addition, nonquantified benefits from the rule resulting in illness reduction from other viruses and bacteria are expected to be significant.

More details on the GWR can be found on the following USEPA Web page: www.epa.gov/safewater.

Bank Filtration

As explained by the USEPA in its 2006 Enhanced Treatment for Cryptosporidium, Pre-filtration Treatment Rule (40CFR141.717), commonly referred to as the Bank Filtration Rule, bank filtration is a water treatment process that uses one or more pumping wells to induce or enhance natural surface water infiltration and to recover that surface water from the subsurface after passage through a river bed or bank(s). Under the rule, bank filtration that serves as pretreatment to a filtration plant is eligible for cryptosporidium treatment credit if it meets the following criteria:

- Wells with a groundwater flow path of at least 25 ft receive 0.5-log treatment credit; wells with a groundwater flow path of at least 50 ft receive 1.0-log treatment credit. The groundwater flow path must be determined as specified in this section.

- Only wells in granular aquifers are eligible for treatment credit. Granular aquifers are those comprising sand, clay, silt, rock fragments, pebbles or larger particles, and minor cement. A system must characterize the aquifer at the well site to determine aquifer properties. Systems must extract a core from the aquifer and demonstrate that in at least 90 percent of the core length, grains less than 1.0 mm in diameter constitute at least 10 percent of the core material.

- Only horizontal and vertical wells are eligible for treatment credit.

- For vertical wells, the groundwater flow path is the measured distance from the edge of the surface water body under high flow conditions (determined by the 100 year floodplain elevation boundary or by the floodway, as defined in Federal Emergency Management Agency flood hazard maps) to the well screen. For horizontal wells, the groundwater flow path is the measured distance from the bed of the river under normal flow conditions to the closest horizontal well lateral screen.

- Systems must monitor each wellhead for turbidity at least once every 4 hours while the bank filtration process is in operation. If monthly average turbidity levels, based on daily maximum values in the well, exceed 1 NTU, the system must report this result to the state and conduct an assessment within 30 days to determine the cause of the high turbidity levels in the well. If the state determines that microbial removal has been compromised, it may revoke treatment credit until the system implements corrective actions approved by the state itself to remediate the problem.

- Springs and infiltration galleries are not eligible for treatment credit under the above rule sections, but are eligible for credit under the demonstration of performance provisions.

PWSs may apply to the state for cryptosporidium treatment credit for bank filtration using a demonstration of performance. States may award greater than 1.0-log cryptosporidium treatment credit for bank filtration based on a site-specific demonstration. For a bank filtration demonstration of performance study, the rule establishes the following criteria:

- The study must follow a state-approved protocol and must involve the collection of data on the removal of cryptosporidium or a surrogate for cryptosporidium and related hydrogeologic and water quality parameters during the full range of operating conditions.

- The study must include sampling both from the production well(s) and from monitoring wells that are screened and located along the shortest flow path between the surface water source and the production well(s). The Toolbox Guidance Manual provides guidance on conducting site-specific bank filtration studies, including analytical methods for measuring aerobic and anaerobic spores, which may serve as surrogates for cryptosporidium removal.

PWSs using existing bank filtration as pretreatment to a filtration plant at the time of the enactment of the rule must begin source water cryptosporidium monitoring under the rule and must sample the well for the purpose of determining bin classification. These PWSs are not eligible to receive additional treatment credit for bank filtration. In these cases, the performance of the bank filtration process in reducing cryptosporidium levels will be reflected in the monitoring results and bin classification.

As explained by the USEPA, directly measuring the removal of cryptosporidium through bank filtration is difficult due to the relatively low oocyst concentrations typically present in surface water and groundwater. During development of the rule, USEPA reviewed bank filtration field studies that measured the removal of cryptosporidium surrogates, specifically aerobic and anaerobic bacterial endospores. These microorganisms are suitable surrogates because they are resistant to inactivation in the subsurface, similar in size and shape to cryptosporidium, and present in both surface water and groundwater at concentrations that allow calculation of log removal across the surface water–groundwater flow system.

8.3.3   Transboundary (International) Aquifers

This section is courtesy of Dr. Neno Kukuric, International Groundwater Resources Assessment Centre, Utrecht, The Netherlands.

Many aquifers cross political borders. To groundwater specialists, this fact looks too obvious to be mentioned; however, it is a necessary starting point in any political negotiations on joint management of internationally shared water resources. The issue of sustainable and equitable use of transboundary waters is as old as the boundaries, and the related conflicts can be easily traced through history until the present day. During the last century, a significant improvement has been made in the regulation of common management of surface watercourses; many international river-, lake- or basin commissions have been set up and the legal treaties signed. Major comparable activities related to invisible groundwaters have started just a several years ago.

During one of the meetings dedicated to the preparation of international law on transboundary groundwaters, the expert group was able to come up with just a few examples of successful joint management of transboundary aquifers. The examples were coming from the heart of Europe—Switzerland, Germany, and France. Although additional successful stories can probably be found, the examples of poor management or a complete lack of management are overwhelming.

Deterioration of international groundwater resources was the main incentive for the International Association of Hydrogeologists (IAH) to set up a Commission on Transboundary Aquifers (TARM). UNESCO embraced this initiative and realized it with the ISARM program (Internationally Shared Aquifer Resources Management). ISARM (www.isarm.net) has developed a framework document and launched several joint activities in Africa, South East Europe, Americas, and Asia. For example, cooperation with Organization of American States (OAS) has led to publication of the *Atlas of Transboundary Aquifers of Americas* in 2007. The main technical work on the atlas was carried out by the International Groundwater Resources Assessment Centre (IGRAC). As the UNESCO/WMO's groundwater knowledge center, IGRAC (www.igrac.nl) is involved in practically all ISARM activities.

In Europe, UNESCO is working closely with the United Nations Economic Commission for Europe (UN/ECE). The UN/ECE produced an inventory of Transboundary Groundwaters in Europe in 1999 (Almássy and Buzás, 1999), and just recently (2007) a first inventory of groundwaters in Caucasus and Central Asia and an improved inventory of groundwaters in South East Europe. Finally, in cooperation with UNESCO, a world map showing locations of major transboundary aquifers has been produced by the World-wide Hydrogeological Mapping and Assessment Programme (WHYMAP).

For several years (and in the framework of its International Waters focal area), the Global Environment Facility (GEF) has been endorsing projects with a significant groundwater component. Several large internationally shared aquifers in Africa (Limpopo, North-West Sahara, Iullemeden, Nubian) and in South America (Guarani) have been included in the GEF program. Three UN organizations (UNDP, IAEA, and UNEP) and the World Bank are responsible for implementation of these projects. At this moment, the projects are in various stages of execution, facing various challenges; one of the most interesting is certainly a socially sustainable use of nonrenewable groundwater resources.

Experience gained through activities such as those listed above is helping in understanding the main requirements for a proper management of internationally shared groundwaters. The ISARM program distinguishes five aspects of transboundary aquifers, namely, hydrogeologic, legal, socioeconomical, institutional, and environmental. The hydrogeologic aspect, as the most interesting for groundwater specialists, is addressed first. The legal aspect is highlighted as well, recognizing the significant progress made in preparation of the international law on transboundary groundwaters. The other aspects of transboundary aquifers still need elaboration, certainly including the political one; experience shows that without the political will, no significant progress is possible.

The progress toward a sufficient hydrogeologic knowledge on internationally shared aquifers includes the following steps:

- Inventory
- Delineation and description
- Classification and prioritization
- Data harmonization and information management

The inventories conducted so far provide valuable first information on transboundary aquifers in some parts of the world. Yet, the information obtained from various aquifer-sharing countries is often not consistent, sometimes even contradictory. This is attributable, in part, to the lack of precision in questionnaires. Nevertheless, discrepancies among countries are still substantial; it appears regularly that one country sees a particular aquifer as transboundary and the other country does not. This emphasizes the need for additional hydrogeologic information, including delineation and description of aquifers.

Delineation and description of aquifers should be carried out by teams of hydrogeologists from aquifer countries and facilitated by international organizations. The important task of international organizations is to work toward a set of consistent delineation and description criteria. For instance, in some countries, aquifer delineation is based primarily on the extent of the aquifer drainage area or geological formation rather than on multiple hydrogeologic criteria.

Water Framework Directive (WFD) of the European Community has introduced a notion of "groundwater bodies" as primary management units. Quite general criteria for delineation of groundwater bodies have led to different interpretation by different countries (e.g., 3 bodies defined by one country and 13 by the other—for the same transboundary region). In most of the cases this problem has been solved through the intensified cooperation among the countries. The WFD has also introduced a set of attributes (monitoring variables) that are necessary for classification and further prioritization of the transboundary aquifers.

The European Community is developing a Water Information System Europe (WISE). At this moment, WISE is primarily a reporting mechanism where relatively small sets of representative observations from member countries are stored. Harmonization (i.e., the same reference level, ranges, units, etc.) is required to allow further processing and analysis of these transboundary data.

Ideally, all the transboundary data should be made available online and in real-time. Figure 8.12 shows locations of groundwater wells in the border area between Germany and the Netherlands. This is an example of online synchronized access to distributed information services (where data and information remain at the source). Available groundwater level time-series data from two countries can be plotted together on one graph. The ultimate information management step is semantic (language) harmonization and translation of geological and hydrogeologic information, where each unit is automatically translated into the equivalent unit according to classification used in a neighboring country. For Germany and the Netherlands this was a complex task, because one country uses chronostratigraphic and the other lithostratigraphic classification. The task was completed through intensive and constructive cooperation.

Legal Context

As groundwater quickly emerges into the limelight and gains strategic importance as a source of often high-quality freshwater in the face of the impending water crisis worldwide, the need for rules of international law addressing groundwater management and protection becomes ever more compelling. However, international law has so far only rarely taken account of groundwater. While surface water treaties abound, groundwater is either nominally included in the scope of these instruments, mainly if it is "related" to surface waters, or it is not mentioned at all. Only a few legal instruments contain groundwater-specific provisions, and even fewer address groundwater exclusively. In 2005, FAO and UNESCO published "*Groundwater in international law. Compilation of treaties and other legal instruments*" (Burchi and Mechlem, 2005). This publication brings together a variety of binding and nonbinding international law instruments that, in varying degrees and from different angles, deal with groundwater. Its aim is to report developments in international law and to contribute to detecting law in-the-making in this important field.

Several bilateral agreements containing issues of groundwater management were signed between various European countries in the 1960s and 1970s. The first major

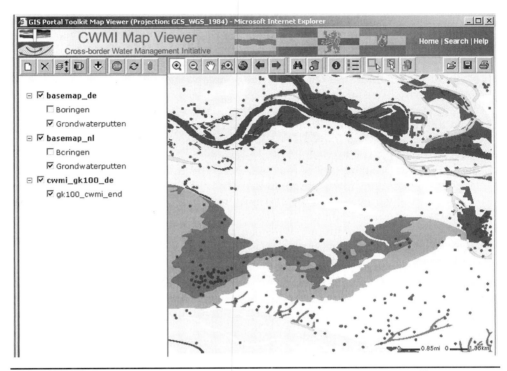

FIGURE 8.12   Screen shot of the CWMI Map Viewer, GIS tool for transboundary aquifer management developed at the International Groundwater Resources Assessment Centre (IGRAC). (Figure courtesy of Neno Kukuric.)

international agreement on transboundary waters was the 1992 "Convention on the Protection and Use of Transboundary Watercourses and International Lakes" in Helsinki, Finland (UN/ECE, 1992). The preparation of the Helsinki convention took about 17 years and, to date, it is ratified by only 34 ountries. Nevertheless, the convention has been used on various occasions, and also as a guideline by the countries that have not ratified it (e.g., in the Middle East and Southern Africa). Although the Helsinki convention addresses some aspects of groundwater management, it does it in a rather limited form.

In 2002, the UN International Law Commission (ILC) decided to include the topic "Shared natural resources" comprising groundwater, and oil and gas in its program of work, appointed a Special Rapporteur, and established a Working Group to assist the Special Rapporteur in preparing draft articles on law on international groundwaters. The Working Group completed its task and submitted a report including nineteen draft articles. These articles were adopted by the ILC at first reading with their related commentaries. The ILC also decided to transmit the draft articles to the States members of the UN, with a deadline for comments and observations by January 1, 2008. More information on the report can be found on the ILC and ISARM (www.isarm.net) Web sites.

It has not been decided yet on the final form the legal groundwater articles will take; eventually, it could be a set of guidelines, a convention, or a treaty. The final form will also have an impact on the presence and the practical importance of some articles. For example, the document suggests the possibility of the aquifer states to utilize an independent fact-finding body to make an impartial assessment of the effect of planned activities. In a treaty, one can expect the binding character of this article, whereas guidelines could offer only a suggestion. In any case, the proposed articles will provide a solid legal basis for international groundwater management. A comprehensive critical review of the articles is provided by Eckstein (2007) and can be downloaded from http://www.internationalwaterlaw.org/.

The fact that many aquifers cross political borders is in practice related to potential or actual groundwater problems, such as changes in groundwater flows, levels, volumes (quantity), and dissolved substances (quality). There are many specific obstacles for finding effective solutions to these problems (invisible groundwater, usually slow changes, various approaches to aquifer characterization, lack of information), but the main one is very common: the political will. When this obstacle is overcome, transboundary aquifer characterization (as briefly described earlier) is required to provide a solid basis for appropriate management of internationally shared groundwaters. The potential outcome of such management is obvious and very rewarding: elimination of potential sources of conflict and improvements from the beneficial use of groundwater.

8.4 Integrated Water Resources Management

As discussed by Rogers and Hall (2003), the Integrated Water Resources Management (IWRM) eschews politics and the traditional fragmented and sectoral approach to water and makes a clear distinction between resource management and the water service delivery functions. It should be kept in mind, however, that IWRM is itself a political process, because it deals with reallocating water, the allocation of financial resources, and the implementation of environmental goals. The political context affects political will and political feasibility. Much more work remains to be done to establish effective water governance regimes that will enable IWRM to be applied. This pertains to both the management of water resources and the delivery of water services. Nevertheless, there is a general agreement in the water community that IWRM provides the only viable way forward for sustainable water use and management, although there are no universal solutions or blueprints and there is much debate on how to put the process into practice.

One of the best examples of IWRM is Orange County Water District (OCWD) in California. The district was established in 1933 to manage Orange County's groundwater basin and protect the Santa Ana River rights of water users in north-central Orange County. The district manages the groundwater basin, which provides as much as 75 percent of the water supply for its service area. The district strives for a groundwater-based water supply with enough reserves to provide a water supply through drought conditions. An integrated set of water management practices helps achieve this, including the use of groundwater recharge, alternative sources, and conservation.

Recharge. The Santa Ana River provides the main natural recharge source for the county's groundwater basin. Increased groundwater use and lower-than-average rainfall during the late 1980s and early 1990s forced the district to rely on an aggressive program to enhance recharge of the groundwater basin. Following are the programs used currently to optimize water use and availability

- Construction of levees in the river channel to increase infiltration
- Construction of artificial recharge basins within the forebay

- Development of an underwater basin cleaning vehicle that removes a clogging layer at the bottom of the recharge basin and extends the time between draining the basin for cleaning by a bulldozer
- Use of storm water captured behind Prado Dam that would otherwise flow to the ocean
- Use of imported water from the State Water Project and Colorado River
- Injection of treated recycled water to form a seawater intrusion barrier

Alternative Water Use and Conservation. OCWD has successfully used nontraditional sources of water to help satisfy the growing need for water in Orange County. The following are projects that have added to the effective supply of groundwater:

- Use of treated recycled water for irrigation and industrial use
- In-lieu use to reduce groundwater pumping
- Change to low-flow toilets and showerheads
- Participation of 70 percent of Orange County hotels and motels in water conservation programs
- Change to more efficient computerized irrigation

Since 1975, Water Factory 21 has provided recycled water that meets all primary and secondary drinking water standards set by the California Department of Health Services. A larger, more efficient membrane purification project called the Groundwater Replenishment System (GWRS), begun operating at 70,000 acre-ft/yr in 2007. By 2020, the system will annually supply 121,000 acre-ft of high quality water for recharge, for injection into the seawater intrusion barrier, and for direct industrial uses.

The new facility uses reverse osmosis (RO), and ozonation with advanced oxidation treatment process, which is designed to eliminate any emerging and recalcitrant contaminants that may pass the RO treatment. This level of treatment creates water safe to drink under all existing regulations and goes several steps beyond (note that bottling water companies use RO to completely "purify" water from various sources, including municipal water supplies, and market it as such). The treated water has near-distilled quality, which helps reverse the trend of rising TDS in groundwater caused by the recharge of higher TDS-content Santa Ana River and Colorado River waters. The process uses about half the energy required to import an equivalent amount of water to Orange County from Northern California (DWR, 2003; Orange County Water District, personal communication).

The biggest challenge for IWRM is and will be coping with two seemingly incompatible imperatives: the needs of ecosystems and the needs of growing population. The shared dependence on water of both makes it natural that ecosystems must be given full attention within IWRM. At the same time, however, the Millennium Declaration 2000, agreed upon by world leaders at the United Nations, involves a set of human livelihood imperatives that are all closely water related: to halve by 2015 the population suffering from poverty, hunger, ill-health, and lack of safe drinking water and sanitation. A particularly crucial question will be the water-mediated implications for different ecosystems of the needs for an increasing population: growing food, biomass, employment, and shelter needs. (Falkenmark, 2003).

The most fundamental task of IWRM is the realization, by all stakeholders, that balancing and trade-offs are necessary in order to sustain both humanity's and the planet's life support systems. Therefore, a watershed-based approach should have a priority with the following goals:

- Satisfy societal needs while minimizing the pollution load and understanding the water consumption that is involved
- Meet an ecological minimum criteria in terms of fundamental ecosystem needs such as secured (uncommitted) environmental flow in the rivers, secured flood flow episodes, and acceptable river water quality
- Secure "hydro-solidarity" between upstream and downstream societal and ecosystem needs (Falkenmark, 2003)

On a more technical level, one of the most important roles of the hydrogeologists is to educate both the public and water professionals about the importance of groundwater and its "invisible role" in the watershed and the hydrologic cycle as a whole. Often, water resource managers and decision makers have little background in hydrogeology and thus a limited understanding of the processes induced by pumping groundwater from an aquifer. Both irrational underutilization of groundwater resources (compared to surface water) and excessive complacency about the sustainability of intensive groundwater use are thus still commonplace (Tuinhof et al., 2002–2005a). Table 8.2 includes some

Feature	Groundwater Resources & Aquifers	Surface Water Resources & Reservoirs
Hydrological characteristics		
Storage volumes	very large	small to moderate
Resource areas	relatively unrestricted	restricted to water bodies
Flow velocities	very low	moderate to high
Residence times	generally decades/centuries	mainly weeks/months
Drought propensity	generally low	generally high
Evaporation losses	low and localized	high for reservoirs
Resource evaluation	high cost and significant uncertainty	lower cost and often less uncertain
Abstraction impacts	delayed and dispersed	immediate
Natural quality	generally (but not always) high	variable
Pollution vulnerability	variable natural protection	largely unprotected
Pollution persistence	often extreme	mainly transitory
Socioeconomic factors		
Public perception	mythical, unpredictable	aesthetic, predictable
Development cost	generally modest	often high
Development risk	less than often perceived	more than often assumed
Style of development	mixed public and private	largely public

From Tuinhof et al., 2002–2005a.

TABLE 8.2 Comparative Features of Groundwater and Surface Water Resources

comparative features of groundwater and surface water resources that should be considered when planning for IWRM.

IWRM is sometimes referred to as integrated water cycle management (IWCM) such as in Australia, where water supply and management in general are two top priorities for all levels of government, all water-related agencies, water utilities, and everybody else. This is understandable since Australia is the driest continent, and in the midst of experiencing a continuous ≥ 1000-year drought. Table 8.3 is from a comprehensive publication by the New South Wales Department of Energy, Utilities and Sustainability (DEUS) entitled "*Integrated water cycle management guidelines for NSW local water utilities.*" This useful publication includes a review of the IWCM principles, steps how to achieve it, and the regulatory framework for the involvement of various stakeholders.

8.5 Monitoring

Regular and systematic monitoring of groundwater resources is the most important prerequisite for their effective management. Unfortunately, according to a worldwide inventory of groundwater monitoring compiled by the International Groundwater Resources Assessment Centre (IGRAC), in many countries systematic monitoring of groundwater quantity or quality, even at a regional scale, is minimal or nonexistent (Jousma and Roelofsen, 2004). This lack of monitoring may result in undiscovered degradation of water resources due to either overexploitation or contamination, leading to the following scenarios:

- Declining groundwater levels and depletion of groundwater reserves
- Reductions in stream/spring baseflows or flows to sensitive ecosystems such as wetlands
- Reduced access to groundwater for drinking water supply and irrigation
- Use restrictions due to deterioration of groundwater quality
- Increased costs for pumping and treatment
- Subsidence and foundation damage

A number of factors contribute to the lack of groundwater monitoring. Insufficient financial resources and lack of technical capacity to implement monitoring are perhaps the major factors. Other factors that may contribute are a lack of clear institutional responsibilities and legal requirements for monitoring. Even where monitoring programs are operating, they may fail to provide adequate information to support effective management because

- The objectives are not properly defined
- The program is established with insufficient knowledge of the groundwater system
- There is inadequate planning of sample collection, handling a groundwater monitoring guideline for countries with limited financial resources

In order to improve this situation, IGRAC assembled an international working group of groundwater professionals and assigned it the task of developing a groundwater

Feature	Older and Non Integrated System	In an Integrated System
Major infrastructure (e.g., sewage treatment plants, or STPs, and water treatment plants, or WTPs)	Aging facilities Inadequate capacity High running costs Limited adjustments Restricted reuse High environmental impact	Replaced or upgraded facilities Adequate present and future capacity Reduced running costs Greater flexibility in output quality High level of reuse Greater income potential
Collection and distribution systems (mains and pipes)	Reaching end of design life Frequent failures (blockages and leaks) Infiltration and exfiltration problems Very high replacement costs Unmonitored pumping stations Costly to maintain	Adequate capacity with smaller sizes Little to no leakage Greater use of local storage (OSD, sewer ouflow) Monitored pumping stations and reservoirs
Revenue collection	Set annual fee (tariff) Limited metering Little or no reuse income	2 or 3 part-tariffs Universal metering Reuse income Water trading
Environmental impact	High level of impact Limited control Potential for incurring fines	Very low impact High level of control Very low potential for incurring fines
Resource dependency (rivers and groundwater)	Highly dependent on river and groundwater allocations Increasing restrictions due to environmental requirements	Integrated supply from all three urban water sources (freshwater, recycled effluent, and stormwater) Increased independence from environmental restrictions
Future planning and growth	Uncertain resources for growth Simple planning process where resources must meet demand	Growth managed through demand management and use of other water sources Planning based on higher self-sufficiency in resources
Access to funds	Traditional state support with limitations Difficulty obtaining other funding	Higher self-funding Greater access to funds through diverified services and product delivery
Accountability	Decissions based on processes internal to utility	Open system with community input and sign-off
State and Commonwealth environmental reforms	Minimum compliance Difficult to achieve compliance	Strong focus to comply with wide reform agenda

TABLE 8.3 Comparison of Older and Nonintegrated Water Systems with Integrated Water Systems

Feature	Older and Non Integrated System	In an Integrated System
Rainwater tanks	Highly limited urban use Extensive remote area	Part of urban water supply and stormwater management systems
Links with CMAs	Limited linkages and ability to comply with catchment targets	Inclusion of committee objectives at planning stages Ongoing contribution to catchment planning
Responsibility for pollution	Strong focus on urban centers to comply	Accurate assessment of the balance between nonurban and urban pollution roles
Distribution of water system costs	Cross subsidy by community for some user and polluter costs	Strong focus on user and polluter paying
Roles and responsibilities	Unclear roles and responsibility for impacts of urban water systems	Clear definition of roles and responsibilities between key groups (councils, state agencies, and industry)
Stormwater management	Pipe collection and disposal High pollution potential Aging system with high replacement costs Large Gross Pollutant Traps (GPTs) and other costly structures Continuing high maintenance, cleaning and disposal costs Limited and low-efficiency on site detention systems 50–100 yr design focus	Considered as local urban water Collected with maximum reuse Reduced impervious areas and runoff Separating of pollutants from stormwater at source Significant potential contribution (up to 100%) to urban water needs Minimal long-term maintenance, cleaning and disposal costs Long planning time frame and shorter and more flexible structural design life (5–20 yr)

From DEUS, 2004.

TABLE 8.3 Continued

monitoring guideline for countries with limited financial resources. This document is the result of concerted action by the working group. It focuses on the first stage of groundwater monitoring for general reference—a prerequisite for sound groundwater management. The guideline includes a detailed discussion of monitoring principles and objectives, institutional requirements, design, methods, implementation, and data management (International Working Group I and Jousma, 2006).

Groundwater monitoring is a scientifically designed, continuing measurement, and observation of the groundwater characteristics (status). It also includes data evaluation and reporting procedures. Within a monitoring program, data should be collected at set locations and regular time intervals as much as possible. Although the regulatory basis, institutional framework and funding situation will impose their own objectives and constraints, still the underlying scientific or technical objective is to describe groundwater characteristics in space and time.

Groundwater regulation should make a provision for monitoring of groundwater use and status by assigning facets of this task to the water resources administration and to water users. To be effective, this legislation should set realistic requirements that take account of existing institutional capacity. A typical division of responsibilities is as follows (modified from Tuinhof et al., 2002–2005b):

- Central Government/national Water Authority—ambient (basic reference) monitoring network
- Regional/Basin/Aquifer Water Resource Agency—monitoring as a function of resource regulation and protection
- Water-well Contractors/Drilling Companies—obligations for well logs and aquifer testing data
- Large Groundwater Developers (utilities)—records of metered groundwater extraction and groundwater levels; early-warning groundwater quality monitoring at sentinel wells near water supply wells (well fields)
- Small Groundwater Developers (small utilities, water purveyors, farms)—general feedback on well characteristics and performance
- Potential Groundwater Polluters—compliance and early-warning groundwater quality monitoring at site level

There are three basic types of monitoring applicable to both groundwater and surface water resources: (1) ambient, 2) compliance, and (3) performance monitoring. All three types can monitor both water quantity and quality, and can be short term or long term, depending upon the project-specific goals.

8.5.1 Ambient Monitoring

Ambient water monitoring programs measure background or existing water quality, and water quantity such as streamflows and spring flows, lake levels, and water levels in monitoring wells. This type of monitoring is designed to collect long-term data on regional water resources, but can also be implemented for short periods in order to collect site-specific information required by a particular project. An example of short-term ambient monitoring is quarterly concentration of certain water quality parameters (constituents) in background monitoring wells assumed not to be impacted by anthropogenic contamination at a Superfund site. Another example is measurement of spring flows, streamflows, or water levels in monitoring wells for several years. This short-term, site-specific ambient information can then be correlated with long-term data series collected at closest monitoring sites that are part of a regional network.

Long-term, regional ambient monitoring of water quality and quantity is the most important foundation of water resources management at any scale. Because of this importance, as well as associated costs and complexity, such monitoring programs are implemented and managed by state and federal government agencies. The United States Geological Survey (USGS) maintains one of the most extensive and sophisticated water

FIGURE 8.13   USGS monitoring well located at the headquarters of the National Ground Water Association in Columbus, OH. Water level data are recorded in real-time and transmitted via satellite to the USGS processing center. The data are available online 15 minutes after recording.

resources monitoring networks in the world. Data on streamflows and spring flows, lake levels, groundwater levels, and physical and chemical water parameters are collected for all major watersheds and aquifers in the country at approximately 1.5 million sites. All historic and current data, including sites with real-time monitoring (Fig. 8.13), are in the public domain and available online (http://waterdata.usgs.gov/nwis/). Users can browse the USGS National Water Information System online database, download the data, or choose to visualize the data on screen by selecting various options for plotting time-series graphs of selected parameters.

Stuart et al. (2003) provide an overview of national strategies for groundwater monitoring in England and Wales, European Union countries, and the United States, and a review of current availability of field chemistry methods, measurements techniques, novel analytical tools, and the use of chemical indicators.

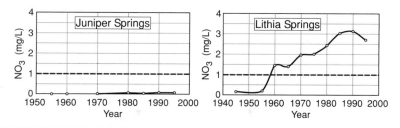

FIGURE 8.14 Nitrate concentration (in mg/L) in two Florida springs over 40 to 45 years. (From the Florida Springs Task Force, 2000.)

Figure 8.14 illustrates the importance of long-term ambient monitoring. The bold lines on the graph show nitrate concentrations (in milligrams per liter) in two Florida springs over 45 to 50 years. Juniper Springs is located within the Ocala National Forest, with a recharge basin that is mainly in conservation lands. Land uses within the Lithia Springs recharge basin are mainly agricultural. Field observations indicate that as nitrate levels approach 1 mg/L (doted line), spring biological communities become degraded by out-of control plant growth. The exact concentration that causes impairment is still not known.

8.5.2 Compliance Monitoring

Compliance monitoring programs of groundwater quality are required by federal or state regulations at and near facilities where groundwater contamination has occurred or where there is a potential for release of contaminants (such as at gas stations with underground storage tanks or at landfills). These programs measure concentrations of specific "constituents of concern" (COCs) in groundwater to ensure that there is no contaminant migration that would pose unacceptable risk to human health and the environment.

Groundwater monitoring networks at hydrogeologically complex sites with significant point source contamination are quite elaborate and may include multiple monitoring wells screened at different depths and in different hydrogeologic units. Figure 8.15 illustrates one such network with well clusters arranged in transects perpendicular to the main groundwater flow direction. This configuration enables determination of the contaminant concentrations and mass flux, and the effects of remedial actions undertaken in the source zone to reduce the contaminant mass and flux. In general, the monitoring costs associated with characterization, delineation, and remediation of contaminant plumes are very high, and often exasperated by long investigation and remediation times spanning multiple years.

When public supply wells (well fields) are located in a relative vicinity and downgradient of a known or suspected groundwater contamination site, early-warning ("sentinel") monitoring wells may have to be placed between the site and the water supply wells to provide for another layer of protection. As water supply wells often have long or multiple screens to maximize production, it may also be necessary to install cluster or multiport monitoring wells that enable the collection of discrete samples at different depths. When, for various reasons, a groundwater contamination site is poorly characterized (e.g., plumes are not delineated in all three dimensions), or there is likelihood that the contaminant plume has already expanded beyond the immediate area of

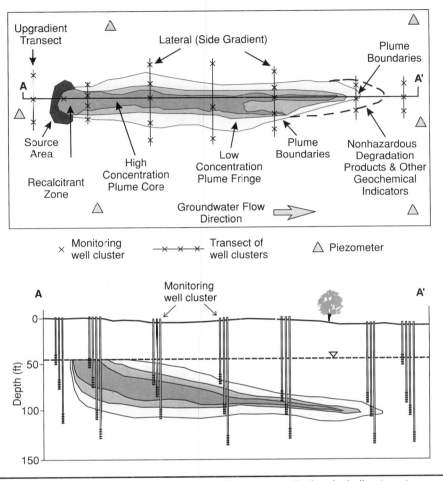

FIGURE 8.15   Example of a network design for performance monitoring, including target zones for monitoring effectiveness with respect to specific remedial objectives. (Top) Map view. (Bottom) Cross-sectional view AAʹ. (From Ford et al., 2007.)

investigation ("property boundary"), sentinel wells may sometimes be the only option left for ensuring that water supply wells are not affected.

Monitoring Optimization

The main purpose of monitoring programs at contaminated sites is to provide sufficient data to support site management decisions, including regulatory actions and possible or ongoing remediation measures. As indicated earlier, such programs can be very expensive and long-lasting and should therefore be frequently evaluated as more information on the site becomes available. The evaluation of the monitoring program focuses on the following objectives (USEPA, 2007):

- Evaluate well locations and screened intervals within the context of the hydrogeologic regime to determine if they meet site characterization and decision support objectives. Identify possible data gaps.

- Evaluate overall plume stability qualitatively and through trend and moment analysis.

- Evaluate individual well concentration trends over time for target constituents of concern (COCs) both qualitatively and statistically.

- Develop site-specific sampling location and frequency recommendations based on both qualitative and quantitative statistical analysis results.

The main outcome of the monitoring program evaluation is the adjustment in the sampling frequency and the number of sampling locations at the site which can sometimes lead to significant cost reduction. Typical factors considered in developing recommendations to retain a well in, or remove a well from, a long-term monitoring (LTM) program are summarized in Table 8.4. Once the decision has been made to retain a well in the network, data are reviewed to determine a sample frequency supportive of site monitoring objectives. Typical factors considered in developing recommendations for monitoring frequency are summarized in Table 8.5.

A very useful tool for evaluation and optimization of monitoring programs is the computer program MAROS developed for the United States Air Force Center for Environmental Excellence (AFCEE, 2003; Aziz et al., 2003). MAROS is a collection of tools in one software package that is used in an explanatory, nonlinear but linked fashion to evaluate individual well trend, plume stability, spatial statistics, and empirical relationships to assist the user in improving a groundwater monitoring network system. Results generated from the software tool are typically used to develop lines of evidence, which, in combination with results of the qualitative analysis, are used to recommend an optimized monitoring network (AFCEE, 1997). Program MAROS is in the public domain and available for download at http://www.gsi-net.com/software/maros/Maros.htm.

Reasons for Retaining a Well in Monitoring Network	Reasons for Removing a Well from Monitoring Network
Well is needed to further characterize the site or monitor changes in contaminant concentrations through time	Well provides spatially redundant information with a neighboring well (e.g., same constituents, and/or short distance between wells
Well is important for defining the lateral or vertical extent of contaminants	Well has been dry for more than 2 yr[1]
Well is needed to monitor water quality at a compliance or receptor exposure point (e.g., water supply well)	Contaminant concentrations are consistently below laboratory detection limits or cleanup goals
Well is important for defining background water quality	Well is completed in same water-bearing zone as nearby well(s)

[1] Periodic water-level monitoring should be performed in dry wells to confirm that the upper boundary of the saturated zone remains below the well screen. If the well becomes rewetted, then its inclusion in the monitoring program should be evaluated.
From USEPA, 2007.

Table 8.4 Typical Factors Considered in Developing Recommendations to Retain a well in, or Remove a Well from, a Long-Term Monitoring (LTM) Program

Reasons for Increasing Sampling Frequency	Reasons for Decreasing Sampling Frequency
Groundwater velocity is high	Groundwater velocity is low
Change in contaminant concentration would significantly alter a decision or course of action	Change in contaminant concentration would not significantly alter a decision or course of action
Well is necessary to monitor source area or operating remedial system	Well is distal from source area and remedial system
Cannot predict if concentrations will change significantly over time, or recent significant increasing trend in contaminant concentrations at a monitoring location resulting in concentrations approaching or exceeding a cleanup goal, possibly indicating plume expansion	Concentrations are not expected to change significantly over time, or contaminant levels have been below groundwater cleanup objectives for some prescribed period of time

From USEPA, 2007.

TABLE 8.5 Typical Factors Considered in Developing Recommendations for Monitoring Frequency

8.5.3 Performance Monitoring

Performance monitoring refers to the collection, on a regular basis, of quality and quantity data associated with water supply pumping or groundwater remediation. Monitoring at individual groundwater extraction locations is one of the key elements for successful groundwater management. It provides crucial information needed for the accurate determination of local and regional water budgets, interactions between individual pumping locations, optimization of pumping, assessment of impacts of groundwater extraction on groundwater levels and quality, and predictive groundwater modeling.

Performance monitoring of large-production wells is especially important. These wells should be equipped with a flow or volume metering device at a minimum. Ideally, monitoring should also include frequent (preferably continuous) recording of water levels and basic physical and chemical parameters such as water temperature, pH, and electrical conductance. As explained in Chap. 7, performance monitoring also includes periodic testing of well efficiency in order to detect early signs of well deterioration. At least once a year, or as required by the applicable regulations for drinking water quality, water analysis of the full suite of regulated chemicals should be performed. If the well has been impacted by certain natural or anthropogenic contaminants, constituent-specific analyses would have to be performed more frequently.

In the case of irrigation wells operated by individual farmers, or water supply wells for individual households, it is unlikely that any information regarding actual groundwater extraction rates or even quality would be available. In such cases, groundwater extraction (e.g., total volume per time or pumping rates) would have to be estimated based on energy consumption for pump operation, irrigated area, and water requirements of individual crops, or typical water use patterns per person in the given climatic and socioeconomic conditions.

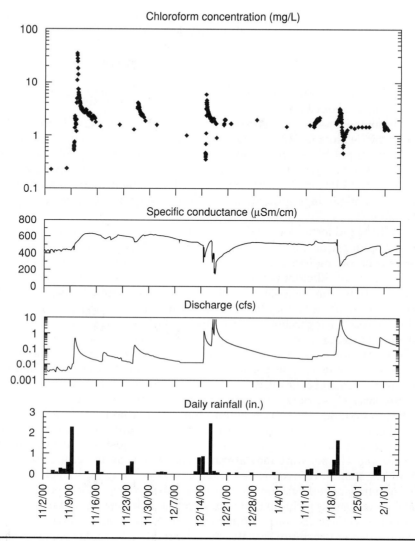

Figure 8.16 Results of high-frequency measurements at Wilson Spring in Central Basin karst region of Tennessee. (Modified from Williams and Farmer, 2003.)

In some cases, sampling frequency becomes the most critical parameter for the success of a monitoring program. Notwithstanding the limitations associated with the high-frequency (e.g., daily), continuous or real-time monitoring, such as relatively high cost and feasibility of implementation, a site-specific situation may deem any other sampling interval entirely inadequate. For example, Fig. 8.16 shows discharge, rainfall, temperature, pH, specific conductance, and dissolved oxygen measured at 10- or 15-minute intervals at the Wilson Spring in Central Basin karst region of Tennessee, the United States. Nonisokinetic dip-sampling methods were used to periodically (mostly during baseflow conditions) collect samples of volatile organic compounds (VOCs) from the spring. During selected storms, automatic samplers were used to collect samples, which were analyzed using a portable gas chromatograph (GC). Quality-control samples included trip blanks, equipment blanks, replicates, and field-matrix spike samples. Significant changes in water quality and discharge were detected with rapid changes observed during storms. Specific conductance ranged from 81 to 663 $\mu S/cm$, and chloroform concentrations ranged from 0.073 to about 34 mg/L. The greatest change was observed during the first storm during fall 2000, when chloroform concentrations increased from about 0.5 to about 34 mg/L.

From these results, it is apparent that a sampling interval of 1 week or 1 month would not provide any meaningful information as to the actual volumes of water discharging from the aquifer, response of the aquifer to precipitation events, and the actual range of concentrations (fluctuation) of a possible contaminant. It is precisely for this reason why the USEPA, in its GWR, emphasized vulnerability to contamination of aquifers with potentially high groundwater velocities, such as karst, fractured rock, and gravel aquifers.

With rapid technological development and commercialization, and the accompanying drop in prices, continuous monitoring and data recording of various water parameters is becoming increasingly affordable and should be always considered. Real-time monitoring of different parameters at multiple locations can be linked and integrated into a variety of management decisions which, in turn, can be automated, increasing efficiency. Many large water utilities and groundwater remediation sites are implementing this approach using technology commonly referred to as SCADA (or Supervisory Control And Data Acquisition). It is a computer-based facility used to provide a centralized operation of services, such as well pumpage, water treatment, and water distribution. With SCADA, various sites in the water supply (or groundwater remediation) system, can be monitored and controlled from a central operations facility. SCADA consists of three main parts: (1) sensors for continuous monitoring and programmable logic controllers (PLCs) at remote sites, (2) communications system, and (3) control center.

The control center computers run software that processes data and presents the status of the remote site to the operator, and allows the adjustment (control) of the remote site operation. The software may also provide some form of "cruise control" of the entire utility system to relieve the demands on the operating personnel. The communication system may operate via medium such as dedicated copper or fiber optic links, and dedicated radio links including satellite. In addition to sensors and monitoring equipment, remote sites may also be equipped with remote terminal units (RTUs) connected to the control center to capture its status and control it.

The basic control center system would consist of a computer, monitor, and a software package for human-machine interface (HMI), or sophisticated process control. In the basic HMI configuration, the operator can view graphic displays of the remote sites and send control functions to those sites. For a stand-alone remote system, the control functions are generated at the remote sites. In this application, the HMI software system is only monitoring the remote status and alarm conditions. Communications software can be added to allow for remote access to the computer at the control center. This enables an operator with a laptop or desktop computer to access the system, view the computer screens and various monitored parameters, and make control changes. Passwords should be used to prevent unauthorized people from accessing the system. In addition, an alarm reporting software package can be incorporated into SCADA and used to call out over a standard telephone systems or via Internet, alerting operator(s) that intervention is needed.

The city of Fresno Water Division, located in the central valley of California, is a municipal domestic water production and distribution provider. Fresno is approximately 110 mi^2 in area and its water supply is mainly based on groundwater. The city's 245 wells are controlled by one of the most sophisticated SCADA systems of its kind. The systems has 12,000 input/output points and 35,000 polled points which monitor 12 parameters at each well site including, among others, well pumping levels, pump motor energy consumption, pressure, flow, pump start/stop status, and flow totalizer. The host software polls all sites through a three-channel radio communications system, which uses the utility-owned radio towers and repeaters. One example of a very useful SCADA function is the energy and pressure optimization controller. It gathers the polled data, averages the pressure in a given geographical city zone (the city has 25 zones), and compares it against a high and low set-point for that zone. In addition, cost is calculated for each pump station in dollars per acre-foot pumped. If the zone average pressure falls below the zone low set-point, the most cost-effective pump which is off in the zone is turned on. If the zone average pressure rises above the zone high set-point, the least cost effective pump which is on in the zone is turned off. The Fresno Water Division has in the past saved approximately $500,000 per year in utility costs by using the energy and pressure optimization algorithm which is implemented through SCADA (Schneider Electric, 1999).

8.5.4   Detecting Contamination

Groundwater contamination may be detected in a number of ways. The most obvious is with human senses: "water that looks bad, smells bad and tastes bad is probably bad." It is also true that relying only on human senses to detect contamination does not make sense because water that looks good (clear), does not smell, and has no taste may still be contaminated and its ingestion may cause serious health problems. Unfortunately, very few individuals, even in most developed countries, can afford testing their drinking water for the full suite of chemicals included on the list of National Primary Drinking Water Standards (see Table 5.2 in Chap. 5), let alone doing it on a regular basis (one such analysis can cost well over 3000 US dollars). PWSs are required by law to regularly test their raw source water for drinking water standards thus protecting their customers. Individual well owners are not subject to such requirements and are therefore most at risk from drinking contaminated groundwater, especially if their wells are in shallow unconfined aquifers, in both rural and urban settings.

In the United States, the public and regulators alike are increasingly recognizing the vulnerability of private wells to groundwater contamination. For example, anyone thinking of buying one of the 20,000 homes in the New York's Westchester County, served by a private well, will enjoy safer drinking water thanks to a new law that took effect in November 2007. The law requires that a water test be conducted upon signing a contract of sale for any property served by a private drinking water well. The law also requires that new private wells and private wells that have not been used to supply drinking water for a period of 5 years be tested before use. Private wells on leased properties must also be tested regularly. The test ensures that the well water is safe for human consumption through analysis for the presence of coliform bacteria and chemical contaminants. Under the new law, only certified laboratories are authorized to collect and test the water samples. Test results must be submitted to the Westchester County Health Department, as well as to the person requesting the test. The law also makes clear the responsibilities of home sellers and buyers and other parties to ensure that drinking water quality problems are corrected. It also establishes penalties for noncompliance. Information on the law is available on the Westchester County Health Department's Web site at www.westchestergov.com/health.

Monitoring for and detecting groundwater contamination at sites that have confirmed releases of contaminants to the subsurface, or have potential for releases, is not an easy task. As illustrated by the following excerpts from the State of Texas regulations for landfills, it requires proper planning, execution and reporting, and a significant investment (Texas Administrative Code, 2006):

- A groundwater monitoring system must be installed and consist of a sufficient number of monitoring wells, installed at appropriate locations and depths, to yield representative groundwater samples from the uppermost aquifer as defined in §330.3 of this title (relating to Definitions).

- Background monitoring wells shall be installed to allow the determination of the quality of background groundwater that has not been affected by leakage from a unit.

- A groundwater monitoring system, including the number, spacing, and depths of monitoring wells or other sampling points, shall be designed and certified by a qualified groundwater scientist. Within 14 days of the certification, the owner or operator shall submit the certification to the executive director and place a copy of the certification in the operating record. The plan for the monitoring system and all supporting data must be submitted to the executive director for review and approval prior to construction.

- The design of a monitoring system shall be based on site-specific technical information that must include a thorough characterization of: aquifer thickness; groundwater flow rate; groundwater flow direction, including seasonal and temporal fluctuations in flow; effect of site construction and operations on groundwater flow direction and rates; and thickness, stratigraphy, lithology, and hydraulic characteristics of saturated and unsaturated geologic units and fill materials overlying the uppermost aquifer, materials of the uppermost aquifer, and materials of the lower confining unit of the uppermost aquifer. A geologic unit is any distinct or definable native rock or soil stratum.

- The owner or operator may use an applicable multi-dimensional fate and transport numerical flow model to supplement the determination of the spacing of monitoring wells or other sampling points and shall consider site-specific characteristics of groundwater flow as well as dispersion and diffusion of possible contaminants in the materials of the uppermost aquifer.

In addition to the requirement that the plan is certified by a qualified groundwater scientist, the following are several other important points applicable to any site (not just landfills):

- Determination of background concentrations of various physical and chemical parameters and potential constituents of concern (COCs)

- Determination of sampling frequency by taking into account seasonal and temporal fluctuations in groundwater flow such that contaminants are not missed

- Determination of the number and spacing of wells, and screen intervals (depths of monitoring), by taking into account fate and transport characteristics of the COCs, and hydrogeologic characteristics of the porous media including heterogeneity and anisotropy

Probably the most important factor when monitoring for possible groundwater contamination is understanding that groundwater systems are dynamic—groundwater is always flowing and hydraulic heads are fluctuating—and that contaminants can be introduced in a variety of ways. Sources can be intermittent or continuous, can be periodically flushed through the vadose zone due to varying recharge (precipitation) pattern, or may reside above the water table and be periodically stripped by the rising water levels. These and other factors should always be considered as they can greatly influence contaminant concentrations in monitoring wells and springs.

Decision as to the number of physical and chemical water parameters, and COCs to be monitored, will be site-specific depending upon the type of contaminant release or potential sources of contamination. Table 8.6 lists types of sites with potential groundwater contamination and related COCs, and Table 8.7 lists analytical laboratory methods for detection of various COCs.

Once the monitoring data is collected and its quality verified (see Section 8.6), it can be evaluated for contamination detection. Any detection of COCs that are not naturally occurring (e.g., anthropogenic, synthetic organic chemicals) would constitute groundwater contamination. When the COCs are not detected in the background monitoring wells (wells located upgradient from the suspected source), and there is a clearly defined and confirmed contaminant release from the suspected point source, this contamination may or may not (still) be linked to that particular source. Determination of contamination at this point depends on the skills of attorneys, consultants, and representatives if the case were to become the subject of a lawsuit. If the background wells are contaminated with the same COCs, which indicates the presence of another potential source, groundwater professionals involved in the project may have a hard time deciphering contaminant contribution from different source(s) without spending more money on investigation.

Many RCRA and Superfund sites are examples of multiple point sources of groundwater contamination. These sources may form individual plumes of individual contaminants, individual plumes of mixed contaminants from identifiable sources or, in the most complicated cases, commingled (merged) plumes of various contaminants from multiple sources, some of which are not easily, or not at all identifiable. Sites on military installations, large industrial complexes, and chemical manufacturing plants, are likely to have groundwater contaminated by multiple constituents, which may be distributed at various depths in the underlying aquifer(s) and form plumes with complicated shapes. "Attaching" contaminant sources (landowners) to their own plumes, or showing no such attachment, is the favorite goal of attorneys working for various PRPs. Heavy involvement of attorneys in groundwater contamination and remediation issues in the United States is understandable since costs associated with groundwater remediation may be astronomical, and the question of who is responsible for the plume(s) becomes of an utmost importance.

When COCs are detected in both the background (upgradient) and the source wells, but are also known to be naturally occurring, or are widely spread due to various nonpoint sources (such as nitrates, arsenic, and perchlorate), various statistical analyses will have to be performed in order to determine that a particular point source site is actually contributing the COCs. This is done by statistically comparing data from the two groups of wells—background and source or downgradient wells, and determining if there is a statistically significant difference between their concentrations. *ASTM D6312–98(2005) Standard Guide for Developing Appropriate Statistical Approaches for Ground-Water Detection Monitoring Program* explains in detail various applicable methods. Another excellent source is *Statistical Methods in Water Resources* published by the USGS (Helsel and Hirsch, 2002).

Type of Site	Potential Contaminants of Concern (COCs) in Groundwater
Small gas station (with ASTs and USTs)	TPH, PAH, and metals
Heating oil tank (AST or UST)	PAHs
Dry cleaners	VOC
Landfill (class C and D)	TDS, TPH, PAH, PCB, VOC, SVOC, metals, chloride
Airport	TPH, PAH
Rail switching yard	TPH, PAH, VOC
Livestock farm (i.e., dairy farm)	Ammonia, nitrate/nitrite, CFU
Crop farm	OP, OC, and herbicides
Orchards	OP, OC, metals (As), and herbicides
Open pit or strip mine	Metals, sulfate, sulfide, pH
Quarry	VOC, HMX, RDX, TNT, Perchlorate
Nuclear power plant	Tritium, strontium, cesium
Power plants (thermal)	PAH, TPH, metals
Small arms firing range	Metals (lead and tungsten) PAH (clay targets)
Millitary ranges	HMX, RDX, TNT, perchlorate, metals
Machine shop (i.e., plating facilities, airline parts manufacturers, auto mechanic)	VOC, SVOC, metals
Solid propellant manufacturer (i.e., fireworks, rocket motors)	Perchlorate, metals
Wood treatment plants	PAH, VOC, SVOC
Paper mill	PAH, metals, dioxin
Municipal wastewater treatment plant	Nitrate/nitrite, CFU, TOC, PBP, pharmaceuticals, surfactant (detergent)
Leach fields and septic systems	Ammonia, nitrate/nitrite, CFU, chloride, surfactant (detergent)
Industrial wastewater treatment plant	Metals, VOC
Automated car washes	TPH, PAH, VOC
Chemical manufacturing plant	Chemical specific + VOC, SVOC
Manufactured gas plant (MGP)	PAH, metals, TPH

TPH, total petroleum hydrocarbons; PAH, polynuclear aromatic hydrocarbons; OP, organophosphorous pesticides; OC, organochlorine pesticides; PCB, polychlorinated biphenyls; VOC, volatile organic compounds; SVOC, semivolatile organic compounds; CFU, fecal coloform; AST - above ground storage tank; UST, underground storage tank; PBP, disinfection biproducts.

Table 8.6 Types of Sites with Potential Groundwater Contamination and Related Contaminants of Concern

Contaminant of Concern (COC)	Laboratory Analytical Method
General Chemistry	
Alkalinity	USEPA 310
Bromide	USEPA 300/320
Chloride	USEPA 300/325
Conductivity	USEPA 120.1
Cyanide	USEPA 335
Flouride	USEPA 300/340
Ammonia	USEPA 350
Nitrate	USEPA 300/352/353
Nitrite	USEPA 300/354
Nitrate + nitrite	USEPA 353
Ortho phosphate	USEPA 365
Perchlorate	USEPA 314, USEPA 332, Draft USEPA 6850
Total phosphorous	USEPA 365
Total dissolved solids	USEPA 160.1
Total suspended solids	USEPA 160.2
Sulfate	USEPA 300/375
Sulfide	USEPA 300/376
Total organic carbon	USEPA 415.1
Turbidity	USEPA 180.1
Microbiology	
Total coliform (most probable number, MPN)	SM 9221B
Fecal coliform (colony forming units, CFU)	SM 9222D
Metals	
Metals (total or dissolved)—all but mercury	USEPA 6010/6020
Mercury (total or dissolved)	USEPA 7470
Toxicity characteristic leaching procedure (TCLP) metals	USEPA 1311
Organics	
PAH	SW 8310
VOC	SW 8260
SVOC	SW 8270
OP	SW 8141
OC	SW 8081
TPH	USEPA 418.1
Explosives	USEPA 8330
Herbicides	SW 8151
PCBs	SW 8082
Detergents—anionic surfactants	EPA 425
Pharmaceuticals	no single standard; varies

TABLE 8.7 Commonly Used Analytical Methods for Detection of Potential Contaminants in Groundwater

As discussed by Daughton (2007), since the 1970s, the impact of chemical pollution has focused almost exclusively on conventional "priority pollutants," especially on those collectively referred to as "persistent, bioaccumulative, toxic" (PBT) pollutants, "persistent organic pollutants" (POPs), or "bioaccumulative chemicals of concern" (BCCs). However, it is important to recognize that the current lists of priority pollutants were established in large part for expediency—that is, they could be measured with off-the-shelf chemical analysis technology. Priority pollutants were not necessarily selected solely on the basis of risk and, in any case, are only one piece of the larger risk puzzle.

Environmental regulators have traditionally approached chemical pollution by devoting resources solely to managing established, well-characterized risks. This is referred to as "list-based," target analyte monitoring and represents a reactive approach—only those compounds targeted for monitoring have the potential for being identified and quantified. A more proactive approach, on the other hand, can prevent the establishment of new risks so that their management is not needed. New and unanticipated chemicals (together with their transformation products) that have not previously occurred in the environment need to be identified as early as possible—well before they become pervasive in the environment. Such chemicals present unknown risks, some of which cannot be anticipated. Detecting and monitoring their presence in the environment is referred to as "nontarget" analyte approach (Daughton, 2007). More polar compounds, such as many pharmaceuticals and personal care products (PPCPs), represent a particular challenge for detection and emphasize the need for development of new analytical methodologies, as well as the establishment of a nationwide monitoring network to detect newly present (emerging) chemicals in surface water and groundwater.

Two characteristics of PPCPs distinguish them from other anthropogenic chemicals and give them unique capabilities for use as environmental monitoring tools: (1) Individual PPCPs or their metabolites in the environment result solely from human (or domestic animal) ingestion or use. (2) Each PPCP has a well-known date of introduction to commerce (generally a function of Federal Drug Administration (FDA) approval). Some also have dates of withdrawal from commerce. Use of PPCPs in natural waters as "tracers," "markers," "indicators," "sentinels," or "early warning" of environmental contamination by human sewage or animal excrement is an idea that has garnered significant attention. This idea capitalizes on the relative ease and speed of chemically analyzing water samples to determine the occurrence of PPCPs with a spectrum of half-lives, as opposed to cumbersome measurement of pathogens (Daughton, 2007).

PPCPs could possibly be used as a means of providing maximum ages for groundwater or sediments with respect to the intrusion of contaminated surface waters. Dates of intrusion could be bracketed by using drugs with different commercialization dates; they can also be established for particular geographic locales on the basis of regional prescription use. This approach could work with any parent compound or transformation product with sufficient persistence in the groundwater environment.

8.6 Data Management and GIS

This section is courtesy of Jeff Maruszak and Marla Miller, Malcolm Pirnie, Inc.

From groundwater purveyors managing entire groundwater systems to local environmental firms sampling a few monitoring wells, understanding and interpreting data collected from the subsurface is a key requirement to successfully accomplishing project

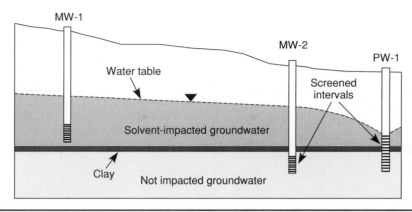

FIGURE 8.17 Illustration showing how data was mismanaged at a facility with a leaking above ground storage tank.

goals. If data is not sufficiently organized, the simple tasks can become overwhelming and often lead to a mismanagement of resources as illustrated with the following scenario.

A facility realized that it was responsible for releasing solvents into the subsurface due to a leaking above ground storage tank. With approval and oversight from the local regulatory agency, an initial monitoring well MW-1 (see Fig. 8.17) was installed to determine if released solvent affected the aquifer underlying the facility. Concentrations of the solvent were detected in the groundwater at MW-1 above applicable cleanup criteria, so a second well MW-2 was proposed in a downgradient location along the presumed direction of groundwater flow. The well was constructed exactly the same as MW-1 with the screened interval installed at the same depth below ground surface. MW-2 was tested for solvents and the results were nondetect. The facility and regulatory agency were satisfied with the characterization and began planning a strategy to remediate the "localized plume." Two months later, a down gradient drinking well, PW-1 on Fig. 8.17, was tested for a variety of parameters, including solvents. Solvents were detected in groundwater above applicable standards! After an exhaustive review of the existing dataset for the site, which included boring logs and previous sample results from the nearby production well, it was determined that the second monitoring well was not screened in the proper location to observe the contamination. The borehole logs from the second monitoring well, as well as the production well, indicated that a clay layer was acting as an aquitard. The aquitard was sealing off the shallower groundwater system from a lower unit. Monitoring well MW-2 should have been designed with a screened interval at a shallower depth above the aquitard, not just based on the "same depth" from ground surface. All the information was available to make the correct decisions; however, it was not readily accessible, and therefore utilized, by the project team.

This example is a simplification of a real world problem, but it highlights the fact that an embarrassing and costly mistake could have been avoided if the project data was properly maintained and readily accessible. In today's market, where sustainability of entire watersheds is becoming increasingly required by various stakeholders, quick access to the right information is vital to the appropriate allocation of efforts and resources. Specific database structures and geographic information systems (GISs) have been developed for managing, interpreting, and displaying surface and subsurface information for such purposes (e.g., Radu et al., 2001).

This section introduces the concept of data management and includes guidelines to organizing information collected in the course of groundwater resource evaluation, utilization, and restoration. The goal of this section is to provide an enterprise (or project) manager an understanding of the general concepts, utility, and available tools to assist them with the successful management of data. Instruction on how to use these tools is beyond the scope of this book.

8.6.1 Data Management

Data management can be defined as the "development and execution of architectures, policies, practices, and procedures that properly manage the full data lifecycle needs of an enterprise." This definition is from the Data Management Association (DAMA) which recognizes that information (data) is a key project asset (DAMA, 2007). A data management system facilitates data management and implements the standard processes and procedures that control the storage, retrieval, and processing of project information. The concepts of data management and data management systems are used throughout this book.

Data management can be as simple as organizing laboratory analytical data collected from one monitoring well, or as complex as an aquifer information system that integrates web-based technology with GIS and relational databases. With such a wide range of complexity, deciding what scale of data management to use is crucial to accomplishing project objectives in an efficient manner. Answering the following questions, in the order presented below, helps to make the right decision:

1. What are the project goals? What will be accomplished with the data?
2. What is the duration of the project?
3. What data is needed to accomplish the project goal(s)?
4. Is there a potential that data needs will change?
5. What tools are needed to collect the data?
6. How is the data going to be used?
7. What tools are needed to manage the data?

Table 8.8 provides an illustration of the increasing levels of complexity in data management related to the seven primary questions that any manager should ask before initiating a project that involves data collected from aquifer systems.

1. *What are the project goals? What will be accomplished with the data?*
This question needs to be answered before starting to collect, organize, and interpret data (Microsoft, 1994; Groff and Winberg, 1999). Identifying overall objectives will help set guidelines for establishing the scale and requirements of the data management system. The concept of scale is a key factor in maximizing return on investment and is a common theme throughout this chapter. The scale has relevance not only to the type of tools used to manage data, but also to how the tools are utilized. Typical examples of project goals for aquifer management include the following:

- Determine the local impacts to the shallow groundwater aquifer from a localized spill at an underground storage tank.

Project Goal (Project Requirements)	Project Duration	Data Requirements	Potential for Data Needs to Change	Data Use	Data Management Tools
Determine the impacts to shallow groundwater from a leaking 10,000 gal petroleum underground storage tank	2 yr	3–4 wells, analytical samples for hydrocarbons	Low	Tables that show analytical results; general site map	Flat file database
Determine the maximum pumping rate within a well field that will minimize yield loss over 20 years	20 yr	Aquifer testing, well properties, groundwater measurements at 10s of wells	Low to moderate	Generation of data from data, modeling of groundwater flow	Relational database, GIS, groundwater model
Characterize the environmental conditions at a Superfund site and develop a remedial strategy	10 yr	Soil, sediment, and ground water analytical data at 100s of locations. Aquifer testing	High	Generation of data from data, spatial analysis, fate and transport modeling	Relational database, GIS, and groundwater model

TABLE 8.8 Examples of Relationship Between Project Goals, Data Management, and Data Management Tools

- Determine the optimal pumping rate for groundwater wells in a production field that will minimize yield loss over the next 20 years.

Additional examples of project goals are provided in Table 8.8, as well as their relation to the remaining questions.

2. What is the duration of the project?

Is the project lifespan a few months, years, or decades? This question should not only apply to when the project is "active," but also to how long it is required to maintain the data after the active phase of the project is complete. For example, do regulatory requirements outline how long the data must be maintained? Does the consulting company (utility) have a data retention policy that needs to be considered? As a general rule, the longer the project duration, the more time one should spend upfront developing the data management system architecture. Examples of common responses to the duration of a project are provided in Table 8.8.

3. What data is needed to accomplish the project goal(s)?

This question addresses the type, quality, and amount of data needed to successfully conduct and complete the project. Having a defined project goal, as described in question 1, will focus data requirements and needs. In general, it helps to write down a list of the data needs, and how they will contribute to accomplishing the project goal. This often helps "weed-out" extraneous datasets. Special consideration should be given to regulatory requirements and frequency of data collection.

In real world situations, there is often a variety of "legacy data" associated with a project. It is important to remember that legacy data may have been obtained to accomplish goals that are inconsistent with the current project's objectives. Although this existing data may be quite useful, it should not determine the type and quality of data to be collected for the new project goals. Additionally, to avoid unnecessary expenditure of resources, it is advisable to determine what legacy data meets new project goals and adopt only that portion of the information into the project database. Examples of data requirements as they relate to project goals are provided in Table 8.8.

Consideration should also be given to data that is generated by processing information stored in the database. For example, fate and transport information generated from modeling programs can be stored and accessed in the data management system.

4. Is there a potential that data needs will change?

Consider the full lifecycle of the enterprise that is being undertaken. Is there a potential that the regulatory requirements will change? Will later phases of the project include different datasets? Functionality may need to be built into the data management system to accommodate foreseeable changes.

For example, a project may have the immediate goal of defining the hydrogeologic characteristics of a local aquifer system and how they relate to the screened interval of several hundred production wells. Later phases of the project will collect groundwater quality information from the production wells to determine the most suitable wells for irrigation versus potable use. It would be advisable to build functionality into the data management system at the project initiation to accept water quality data so that the integration of the future dataset is seamless and efficient. Table 8.8 provides additional examples, and how changes in data needs relate to the complexity of data management.

5. What tools are needed to collect the data?

This chapter does not go into detail concerning the multitude of data collection techniques or standard operating procedures used to collect data. However, relative to data management, consideration should be given to this topic. In ideal situations, one should predetermine how to collect, validate, and report data to maximize its efficient integration into the architecture of the data management system. This architecture can be modified based on the output of the tool collecting project data.

Best management practices, such as standardizing data collection forms (both digital and hard copy), are often effective solutions to maintaining an efficient transfer of data from the collection phase into a data management system. For example, standardized borehole log forms can be used to emphasize the key data elements required to characterize subsurface geology. Permutations that are more complex include automated SCADA systems that are extremely efficient at collecting and uploading information into data management systems. For example, data from a transducer, which records groundwater level fluctuations in a well at a specified time interval, can be automated to upload water level fluctuations into a database system automatically. Examples of other information that are commonly automated during their collection process include flow rates, water quality parameters, remote sensing information, and climate data. When using automated data collection, consideration should be given to the frequency of data collection.

6. How is the data going to be used?
Will data be plotted on graphs to show the decay of contamination over time at a specific monitoring well? Will the cost of water versus the productivity of the well be compared to conduct a cost-benefit analysis of rehabilitating the well? Will the project data be shown on maps or cross sections to relay a technical concept? Will the data be shared with others? Will there be just a few data users or several hundred?

An understanding of the project data requirements and their relevance to the project goals is critical to formulating the architecture of the data management system. The use of each aspect of various datasets should be related back to the project goal(s), and the data should be organized in the most efficient way to accomplish these goals. For example, if the project will have tens of users in different cities accessing the same dataset on a daily basis, a Web-based interface to the database is almost mandatory for project success.

7. What tools are needed to manage the data (i.e., what are the components of the data management system?)
After answering the first six questions, the architecture of the data management system can be outlined. This outline is most efficiently expressed in the early stages of a project as a flow diagram. The diagram should show how the different project elements relate to each other to accomplish the project goals. The diagram also establishes where specific tools will be required to manage the different project elements. Once the general framework of the data management system is established, additional effort can be focused on the development of detailed standard policies, practices, and procedures throughout the lifecycle of the project.

An example of a flow diagram produced for a real-world water utility is shown in Fig. 8.18. The project specifics included the following: immediate project objective of locating additional production well locations; future project objective of developing quarterly groundwater quality reports for all the production wells operated by the utility; project duration of at least 5 years; management of multiple datasets including aquifer properties, groundwater modeling data, existing GIS and subsurface data, and well construction information; generation of maps showing the best locations for potential production wells based on existing water quality and hydrogeology data; generation of tables in the future to display water quality information; and use of the data by multiple regulators and consultants on a daily basis.

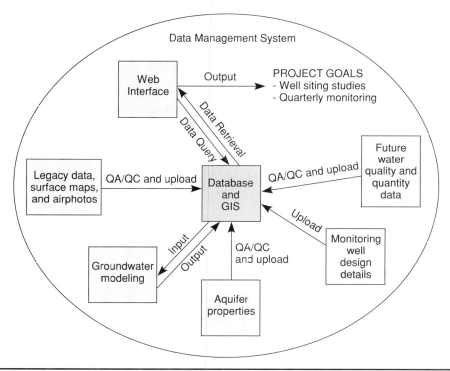

Figure 8.18 Example of a flow diagram showing the different elements for a larger scale project involving data collection, use of legacy data, groundwater modeling, database development, implementation of a project GIS, and a Web-based interface. Note how all the data elements relate back to accomplishing the project goal.

Using the flow diagram shown in Fig. 8.18, the following data management system requirements were identified for the project:

- A database tool to store project data (both spatial and tabular)
- A Web interface tool to allow access to the project data by multiple users
- Modeling software and GIS to meet project objectives
- Standard procedures developed to assure the quality of both new and existing data
- Standard practices for uploading new and existing data into the database

Now that a general understanding of how to manage information for aquifer systems has been established, a more detailed description of the data types, data management tools, and data quality requirements can be developed. These topics are developed further to provide project managers a better understanding of the typical features of a complete groundwater management system at various scales of complexity.

8.6.2   Types of Data

Table 8.9 provides a list of common datasets required for the proper management of groundwater systems. This list is by no means exhaustive, but is intended to provide the

Aquifer Properties (hydrology)	conductivity (k) permeability (k_i) porosity (n) hydraulic gradient (i) transmissivity (T) potentiometric surface recharge rate/location physical caracteristics (thickness/extent) aquifer type (e.g., confined, unconfined)
Well Properties	well type location (x, y, z) screened interval(s) injection rate pumping rate pump type well efficiency, specific capacity
Geology	borehole geophysics bedrock geology/stratigraphy seismic surveys soil type geomorphology structure (e.g., folds and faults)
Groundwater Chemistry	general chemistry (e.g., pH, hardness, TDS) contaminants (e.g., perchlorate, VOCs) temperature
Buried Infrastructure	underground storage tanks piping (e.g., gas lines and sewer pipes) modeling data (fate and transport) ownership
Other	aquifer compaction (subsidence) surface features (e.g., topography, rivers) climatic data (e.g., precipitation, air temp.) end use

TABLE 8.9 Some Common Datasets for Groundwater Management

reader with an understanding of the diverse set of data related to groundwater systems and their management.

Groundwater system data, and its related elements, can be divided into two primary components: (1) spatial (the extent of a feature in space) and (2) attribute (information that can be related to a location). These components are often maintained separately, but should be integrated together whenever possible using a GIS platform as explained further. The photograph in Fig. 8.19 illustrates this concept. Data is collected from a monitoring well, with known space coordinates and screen depth so that the sampling location is accurately defined in all three dimensions (spatial data). Attributes that could be assigned to this location include VOC concentrations (determined later in the laboratory), as well as other measured parameters such as depth to water level, water temperature, pH, and electrical conductance, all coded by the sampling date.

Figure 8.19 Collection of groundwater samples at a monitoring well screened in shallow aquifer in the panhandle region of Florida. (Left) Samples are being collected with a Teflon® bailer.

Spatial Data

The spatial component of data provides information about the location and geometry of a feature that is usually stored in some type of coordinate system. This information is used to provide an abstract representation of the feature on a planar surface (i.e., a map) or three-dimensional image (e.g., a model or block diagram). Spatial data is often displayed with the aid of GIS software packages.

Features are usually expressed as points, lines, areas (e.g., polygons), rasters (e.g., images and grids), and triangular irregular networks (TINS). A point generally represents a single location on a map that defines an object too small to be represented as a line or area (ESRI, 1994). Typical examples of point type data for subsurface datasets include well locations, sample locations, and injection points. A line is a set of connected, ordered coordinates representing a linear shape (ESRI, 1994). Roads, contour lines, geologic contacts, and streams are examples of linear features. A polygon represents a boundary that encloses a given area (ESRI, 2007). Features such as lakes, building footprints, and surface geology are commonly expressed as polygons. A raster is a cellular data structure composed of rows and columns (i.e., matrix) for storing images and/or grids. Groups of cells with the same value represent features (ESRI, 2007). Imagery such as aerial photography is an example of a raster dataset. TINS are used to represent surfaces as a set of linked triangles. TINS provide an efficient way of representing surfaces with irregular topography (Booth and Mitchell, 2001). Figure 8.20 provides an example of point, line, area, and raster feature types.

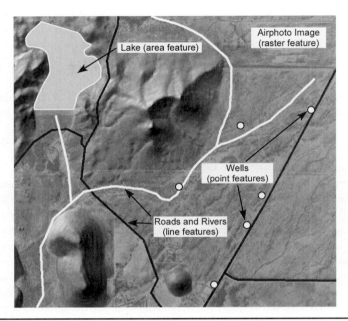

FIGURE 8.20 Illustration of point, line, area, and raster feature types.

Spatial data should be referenced to some type of coordinate system if the intention is to represent data geographically. A coordinate system has three components including a unit of measure (e.g., meter or feet), datum (e.g., NAD83 or WGS 84), and projection (e.g., UTM or GCS). These three basic components can be used to represent a feature from a curved surface (i.e., the earth's surface) onto a two or three-dimensional map. Defining a coordinate system at the initial stages of project lifecycle is necessary for proper management of spatial information.

Unreferenced maps, such as engineering design plans (CAD drawings), do not always have a coordinate system associated with them. Features on the drawings are represented as a distance from an arbitrary benchmark location. A drawing of this nature represents a "scaled drawing" and is not a geo-referenced map. Overlaying geographic information collected during later stages of a project can often become time consuming and therefore a costly process. Geospatial and consulting firms can often assist in the cost effective integration of legacy datasets into a real-world spatial system.

Attributes

The attribute component of data provides characteristic and/or reference information relative to a feature. For example, a water well (point feature) may have the following information associated with it:

- Date of installation
- Well construction information
- Pumping rate
- Analytical data results

- Regulatory compliance levels associated with the analytical data
- Photographs
- Monitoring and maintenance schedule

The attribute information is often organized in some type of tabular format (flat or relational) as discussed further in the text. Ideally, the attribute information and the features can be linked in one tool, such as a GIS.

8.6.3   Data Management Tools

In general, data should be organized into some type of data management system that allows quick and accurate access to information (Microsoft, 1994). As discussed previously, a data management system should be based on a series of standardized policies, practices, and procedures throughout the lifecycle of the data. The scale of data management system can help to dictate the type of tools used to organize and store information.

A sampling of tools (computer programs) used to manage subsurface datasets are provided in Table 8.10. This list is not exhaustive, but should provide the reader with a set of tools that are commonly used in the profession. Generally, databases are used to store information, while other tools, such as GIS, are used to manipulate, visualize, and interpret data to accomplish the project goals. A description of these two components is provided further in the text.

Databases

Databases are effective tools for organizing tabular (or attribute) type information. There are two basic types of databases: flat file and relational.

Function	Tool
Data Storage	Various text editors (e.g., Microsoft ® Notepad) Microsoft ® Excel Microsoft ® Access IBM ® Lotus Notes DB2 MySQL ESRI ® Geodatabase and SDE FileMaker Oracle
Data Visualization and Manipulation	Golden Software ® Surfer ESRI ® ArcView ESRI ® ArcGIS Mapinfo Visual MODFLOW GMS Processing Modflow Groundwater Vistas

TABLE 8.10   List of Programs That Are Commonly Used to Store, Manipulate, and Manage Data Related to Groundwater Systems

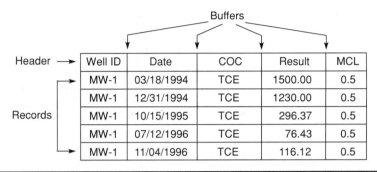

FIGURE 8.21 Example of a flat file type database generated using Microsoft Excel. The different components of a flat file database are illustrated on the figure.

1. *Flat File.* As defined by Databasedev.co.uk (2007), a flat file database is designed around a single table which stores information in a series of records. The record represents individual lines that contain fields separated by some type of buffer (e.g., comma, tab, or cell buffer). A header row is often placed at the start of the file to denote the significance of the series of records. Flat files are useful at storing information, manipulating and formatting fields, printing or displaying formatted information, and quickly exchange information with others (e.g., over the internet or through email). There are no structural relationships within or between multiple flat files databases. For example, updating an analytical result in one flat file will not populate (or update) a second flat file that compares recent data to a regulatory standard (Databasedev.co.uk, 2007).

   Designing flat file databases is simple, requiring little design knowledge. Flat file databases can be developed using a variety of software tools including, text editors, Microsoft Excel, IBM Lotus Notes, and a variety of text editing software that can manipulate ASCII file formats. Figure 8.21 is an example of a flat file type database that was generated using Microsoft Excel. Flat files can also be generated within relational databases, which are discussed below.

2. *Relational Database.* Relational databases are one of the most important foundation technologies in the computer industry today (Groff and Weinberg, 1999). A relational database stores multiple tables and provides methods for the tables to work together. The information held within the different tables can be merged, collated, queried, and displayed based on the relationships built into the database. The methods connecting the tables within the relational database include a variety of program languages, such as Structured Query Language (SQL), visual basic, and MySQL.

The connections, or links, between the various tables are based on Key Fields or Indexed Values. The links are the bases to performing queries that select, collate, organize, and display information stored within the tables. The relationships generated between the different tables can result in three basic scenarios: one-to-one correlation, one-to-many, and many-to-many. Figure 8.22 displays the basic concept of each of the relationship types. The grey boxes represent individual tables, and the dark black lines represent links. One-to-one and many-to-many correlations are atypical of relational

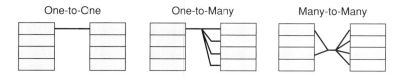

FIGURE 8.22　Illustration of the three possible relationship types in relational databases: one-to-one, one-to-many, and many-to-many.

database structures for aquifer system management projects and should be avoided unless necessary. One-to-one relationships usually indicated inefficiency in data structure. Many-to-many relationships usually indicate that a "link table" is required to clarify data relationships.

As recognized by Databasedev.co.uk (2007),

> Designing a relational database takes more planning than flat file databases. With relational databases, you must be careful to store data in tables such that the relationships make sense. Building a relational database is dependant upon your ability to establish a relational model. The model must fully describe how the data is organized, in terms of data structure, integrity, querying, manipulation and storage.

An example of a simple relational database model for subsurface data is shown in Fig. 8.23. This example was built in Microsoft Access 2000. The primary components of this model include a table that records geologic data collected from a series of well logs, monitoring well information, water level data, analytical data collected from the monitoring wells, and regulatory compliance levels. The lines drawn from each table show the relationships (links) between the various data sets. The links are based on

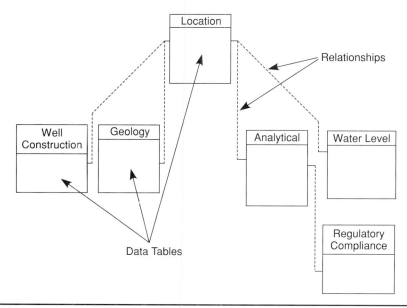

FIGURE 8.23　Example of relational database developed for a small-scale environmental investigation involving the collection of groundwater analytical data, water level measurements, and geologic information.

key fields and each represents a one-to-many relationship. The purpose of this database was to determine on a quarterly basis, which wells exceed regulatory action levels for a variety contaminants of concern. Additionally, the table provides which portions of the aquifer were most impacted. A query was set up to automate this process. Data is uploaded into the database on a quarterly basis, and a form is instantaneously produced showing which wells exceed the established criteria, and for which part of the aquifer.

Software commonly used to build relational database for aquifer management systems include DB2, MySQL, FileMaker, Microsoft Access, Microsoft SQL Server, and Oracle. Embedded, relational databases are packaged as part of other software packages, such as ESRI's ArcGIS software. Relational databases often form the backbone of many of today's Web-based data management systems. For example, analytical laboratories routinely provide access to groundwater analytical data via Web-based interfaces with tools to organize, report, plot, and export your project data (Pace, 2007).

8.6.4   Geographic Information Systems (GIS)

GISs include a variety of tools that are used to display, integrate, manipulate, store, and/or interpret spatially referenced data (ESRI, 2007). GIS is one of the most powerful tools at the disposal of individuals visualizing and managing subsurface information (ESRI, 2007; Strassberg, 2005; Gogu et. al., 2001; Goodchild, 1996). As recognized by Gogu et al., 2001, GIS provides utility in

- The management of subsurface data
- Elaboration of subsurface elements (such as hydrogeology)
- Vulnerability assessments
- Database support for numerical modeling

As summarized in Table 8.10, software commonly used to build a GIS for subsurface investigations and groundwater modeling, directly or indirectly, include ESRI's ArcGIS products, Mapinfo, Surfer, Groundwater VISTAS, Visual Modflow, and Groundwater Modeling System (GMS).

A GIS can be used at its most basic level to display and overlay various GIS datasets on a map. For example, in Fig. 8.20, an air photo (dataset 1) is overlain with rivers (dataset 2), and roads (dataset 3). At a more complex level, it can be used to merge both tabular data and spatial data to analyze information and create new datasets. For example, in Fig. 8.24, a groundwater modeling program, Visual Modflow, was used to generate a groundwater contour map.

The scale of a GIS should be directly related to the purpose defined in the initial stages of project development. For example, will laboratory results be displayed on a map for one sampling event, or does the project require fate and transport modeling through a known set of hydrogeologic conditions over a 20-year period? The first example could be managed using a simple GIS that displays X, Y, Z data; while the second example would require the integration of GIS with a groundwater modeling program. A defined project goal will help determine the scale of the GIS as well as how it is integrated into the data management system.

GIS data can be managed as individual files, akin to a flat file dataset (e.g., shape file), or within a relational database system (e.g., Access Geodatabase). GIS data can

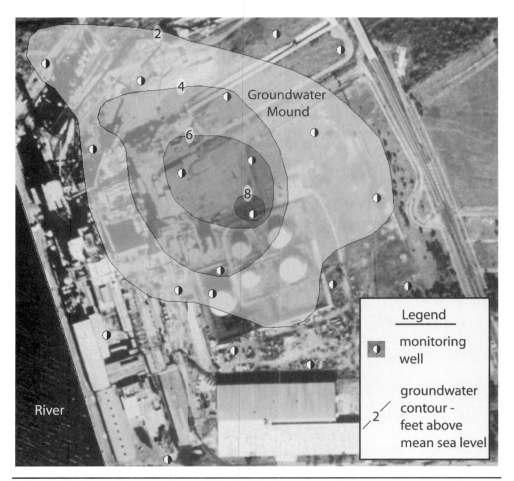

FIGURE 8.24 Groundwater contour map produced using modeling software. The groundwater contours (grey shaded areas) were produced from water level data collected at several monitoring wells. Statistics can be produced for the calculated surfaces to estimate its accuracy and precision.

effectively be managed as individual files for small scale, short duration projects. The data can be organized into standardized data directory structures (e.g., file folders) with best management practices to facilitate data integrity. A variety of information, such as state boundaries, counties, roads, soil type, and surface geology, is available online from various sources as individual GIS files.

GIS data can effectively be managed in relational database systems for both small scale and large-scale GIS projects. The advantage of using a geodatabase to store project GIS data is that data integrity rules can be set up in the initial stages of the project to control items such as data entry errors, standardized symbology (e.g., wells, piezometers, soil type), and enforce project coordinate systems (Booth and Mitchell, 2001). Additional functionality can be added to GIS systems utilizing relational database systems to allow for multiple users, versioning, and metadata management.

Geospatial and tabular data can be merged together to provide an efficient analysis of data to accomplish the project goals. The link between the two elements can be established with programming language such as SQL. Depending on the functionality of GIS, several programs, such as ESRI's ArcGIS have embedded relational databases (called geodatabases) that are packaged as part of the program. These geodatabases often allow the integration with, or connectivity to, other relational databases.

Specific database structures have been developed for the exclusive purpose of managing, interpreting, and displaying hydrogeologic information (Gogu et al., 2001).

8.6.5 Data Quality

The reliability and validity of decisions based on data evaluation is directly related to the data quality. Successful data management employs procedures to ensure high quality data (e.g., using standardized collection and analytical methods) as well as mechanisms for assessing the quality of the data. Data verification or validation is used to assess if the data collected meets project objectives.

The level of data assessment will depend on the project objectives and regulatory requirements. This can range from having a second person check data entry for groundwater elevations (data review or verification) to recalculating analytical results from raw laboratory analytical data (data validation). Data assessment identifies the level of confidence in a dataset. After all, if one starts out with inaccurate or biased data, it does not matter how the data is managed or organized, the overall evaluation will be flawed.

Data Verification and Validation

Data verification or validation is the process for evaluating quality control (QC) samples and parameters associated with a dataset. The appropriate level of data verification/validation will depend on the type of project and requirements of the end data user. For example, if the data will be used in support of risk assessment or litigation, the level of data review required may be more extensive than for a routine groundwater monitoring program. Examples of data verification and validation guidelines for analytical samples include the USEPA Contract Laboratory Program National Functional Guidelines for Organic and Inorganic Data Review, Test Methods for Evaluating Solid Waste (SW-846), and laboratory standard operating procedures.

For analytical data collected as part of groundwater studies, QC samples that are evaluated during verification/validation process can include blanks, spiked samples, and duplicates. Table 8.11 list some of the more frequently used QC samples and parameters for analytical data. The QC samples are compared to acceptance criteria that are identified during the project planning stages.

At a minimum, data review or verification for groundwater management systems should include evaluation of the following parameters: (1) chain-of-custody, (2) sample holding times, (3) blanks, (4) surrogate recoveries (for organic parameters), (5) Matrix spike/matrix spike duplicate (MS/MSD) recoveries and relative percent differences (RPDs), (6) laboratory duplicate RPDs, (7) laboratory control sample (LCS) recovery, and (8) field duplicate RPDs.

The more extensive data validation would require the raw laboratory data, as opposed to just QC summary, and could include (1) initial and continuing calibration results, (2) internal standard recoveries (for applicable parameters), (3) interference check samples (for inorganics), (4) serial dilution and postdigestion spike recoveries (for

Parameter	Description	Criteria
Examples of Field QC Samples		
Field duplicate	Assess sample heterogeneity and sample collection techniques	RPD
Trip blank	Measures potential contamination during sample transport to/from lab (for volatile organics)	Detection above reporting limit
Equipment rinsate blank	Measures potential cross-contamination from sampling equipment	Detection above reporting limit
Examples of Lab QC Samples/Parameters		
Method blank	Measures potential lab contamination	Detection above reporting limit
Laboratory Control Sample	Assess extraction/analysis efficiency	Percent recovery
Matrix spike/matrix spike duplicate	Assess sample matrix interference	Percent recovery
Laboratory duplicate	Assess sample heterogeneity and lab techniques	RPD
Surrogates	Assess extraction/analysis efficiency and matrix interference	Percent Recovery

TABLE 8.11   A List of Frequently Used QC Samples and Parameters for Analytical Data

inorganics), (5) result calculations and documentation procedures, (6) review of dilutions and reanalysis of samples, (7) sample preparation (extraction/digestion logs), and (8) other laboratory QC checks

8.6.6   Metadata

Metadata is data about data. It provides information about the source, content, purpose, limitations, and other characteristics of a given dataset. Metadata should be developed for both tabular and spatial information in groundwater management systems.

According to ESRI (2007), "Metadata for spatial data may describe and document its subject matter; how, when, where, and by whom the data was collected; availability and distribution information; its projection, scale, resolution, and accuracy; and its reliability with regard to some standard." For tabular datasets, metadata is structured information that describes, explains, locates, or otherwise makes it easier to retrieve, use, or manage an information resource (NISO, 2005).

8.7   Protection of Groundwater Resources

As discussed by Hötzl (1996), it is practical to distinguish between resource and source protection, although both concepts are closely related to each other—it is impossible to protect a source without protecting the resource. In European countries, for example, groundwater is considered a valuable resource that must be protected. Activities endangering its quality are forbidden by law (e.g., German regulation in WHG 1996). The European Water Framework Directive (The European Parliament and the Council of the European Union, 2000) emphasizes that water is not a commercial product like any other but a heritage which must be protected, defended, and treated as such. Thus, the directive demands for the protection of groundwater and surface water resources. The highest priority is to protect the groundwater used for drinking water supply. The source may be a captured spring, a pumping well, or any other groundwater extraction point. The European Groundwater Directive of 2006 extends the concept of overall resource protection in detail (The European Parliament and the Council of the European Union, 2006).

In the United States, at the federal level, groundwater as a resource is addressed either together with surface water (Safe Drinking Water Act, SDWA) or as a part of the overall environment. One exception is the Sole Source Aquifer Program within SDWA which is aimed more at promoting the importance of such aquifers than the actual protection (see section "Safe Drinking Water Act"). The Ground Water Task Force (GWTF), established in the fall of 2002 as part of the USEPA's *One Cleanup Program* to improve the planning and quality of USEPA cleanup programs dealing with brownfields, federal facilities, leaking underground storage tanks, and RCRA Corrective Action and Superfund, provides the following discussion on the importance of groundwater resources and their vulnerability (GWTF, 2007):

> **Ground water use** typically refers to the current use(s) and functions of groundwater as well as future reasonably expected use(s). Groundwater use can generally be divided into drinking water, ecological, agricultural, industrial/commercial uses or functions, and recreational. Drinking water use includes both public supply and individual (household or domestic) water systems. Ecological use commonly refers to groundwater functions, such as providing base flow to surface water to support habitat; groundwater (most notably in karst settings) may also serve as an ecologic habitat in and of itself. Agricultural use generally refers to crop irrigation and live-stock watering. Industrial/commercial use refers to in any industrial process, such as for cooling water in manufacturing, or commercial uses, such as car wash facilities. Recreational use generally pertains to impacts on surface water caused by groundwater; however, groundwater in karst settings can be used for recreational purposes, such as cave diving. All of these uses and functions are considered "beneficial uses" of groundwater. Furthermore, within a range of reasonably expected uses and functions, the maximum (or highest) beneficial groundwater use refers to the use or function that warrants the most stringent groundwater cleanup levels.
>
> **Groundwater value** is typically considered in three ways: for its current uses; for its future or reasonably expected uses; and for its intrinsic value. Current use value depends to a large part on need. Groundwater is more valuable where it is the only source of water, where it is less costly than treating and distributing surface water, or where it supports ecological habitat. Current use value can also consider the "costs" associated with impacts from contaminated groundwater on surrounding media (e.g., underlying drinking water aquifers, overlying air—particularly indoor air, and adjacent surface water). Future or reasonably expected values refer to the value people place on groundwater they expect to use in the future; the value will depend on the particular expected use or uses (e.g., drinking water, industrial). Society places an intrinsic value on groundwater, which is distinct from economic value. Intrinsic value refers to the value people place on just knowing clean groundwater exists and will be available for future generations, irrespective of current or expected uses. While the value of groundwater is often difficult to quantify, it will certainly increase as the expense of treating surface water increases, and as existing surface water and groundwater supplies reach capacity with continuing development.
>
> **Groundwater vulnerability** refers to the relative ease with which a contaminant introduced into the environment can negatively impact groundwater quality and/or quantity. Vulnerability depends to a large extent upon local conditions including, for example, hydrogeology, contaminant properties, size or volume of a release, and location of the source of contamination. Shallow groundwater is generally more vulnerable than deep groundwater. Private (domestic) water supplies can be particularly vulnerable because (1) they are generally shallower than public water supplies, (2) regulatory agencies generally require little or no monitoring or testing for these wells, and (3) homeowners may be unaware of contamination unless there is a taste or odor problem. Furthermore, vulnerability can

change over time. For example, anthropogenic activities, such as mining or construction, can remove or alter protective overburden thus making underlying aquifers more vulnerable.

Protection of groundwater resources is achieved by prevention of possible contamination, by remediation of already contaminated groundwater, and by detection and prevention of unsustainable extraction. The prevention aspect includes pollution prevention programs or control measures at potential contaminant sources, land use control, and public education. Following are some examples of prevention measures:

- Mandatory installation of devices for early detection of contaminant releases such as leaks from underground storage tanks at gas stations, and landfill leachate migration
- Banning of pesticide use in sensitive aquifer recharge areas
- Land use controls that prevent an obvious introduction of contaminants into the subsurface, such as from industrial, agricultural, and urban untreated wastewater lagoons
- Land use controls that minimize the interruption of natural aquifer recharge, such as the paving of large urban areas ("urban sprawl")
- Management of urban runoff that can contaminate both surface and groundwater resources (e.g., USEPA, 2005)

Figure 8.25 shows a prevention measure implemented in the city of Austin, TX, where the underlying karstic Edwards aquifer is directly exposed at the land surface or covered by a thin layer of residuum. Surface runoff from state highways is directed toward

FIGURE 8.25 Concrete catch basin by a state highway in Austin, TX, designed to strip oil and fines from surface runoff before water is allowed to infiltrate into the underlying Edwards aquifer.

specially designed concrete catch basins where oil and fine particulate matter are removed before water is allowed to infiltrate into the subsurface.

Probably the single most important aspect of groundwater protection is public education. Unfortunately, it is also often the most underfunded or completely disregarded. There are simple means of educating the public that can pay off many times more than the investment made. Some examples include public outreach with programs describing septic tank maintenance and proper use (e.g., see Riordan, 2007), disposal of toxic wastes generated in households (e.g., paints, solvents, garden pesticides), and proper disposal of unused pharmaceuticals. In terms of protecting the availability (quantity) of groundwater in the areas where it is used for water supply, public outreach programs on water conservation are irreplaceable. Perhaps the most receptive audience to groundwater education programs are the numerous visitors of state parks which have been established because of the groundwater (Fig. 8.26). Such parks should be used as "role models" for the importance of groundwater protection, and featured in media and school programs as much as possible.

Remediation of already contaminated groundwater is the second key aspect of resource protection. In its publication entitled *Protecting The Nation's Ground Water: EPA's Strategy for the 1990's. The Final Report Of The EPA Ground-Water Task Force*, the USEPA stated that groundwater remediation activities must be prioritized to limit the risk of adverse effects to human health first, and then to restore currently used and reasonably expected sources of drinking water and groundwater whenever such restorations are practicable and attainable (USEPA, 1991).

Figure 8.26 Recreational swimming pool built directly over discharge location of San Solomon Spring, one of the Balmorhea Springs in Balmorhea, TX; between 22 and 28 million gal of groundwater flow through the pool each day.

The Agency also stated that

"given the costs and technical limitations associated with groundwater cleanup, a framework should be established that ensures the environmental and public health benefit of each dollar spent is maximized. Thus, in making remediation decisions, EPA must take a realistic approach to restoration based upon actual and reasonably expected uses of the resource as well as social and economic values.

Finally, given the expense and technical difficulties associated with groundwater remediation, the Agency emphasizes early detection and monitoring so that it can address the appropriate steps to control and remediate the risk of adverse effects of groundwater contamination to human health and the environment.

In the 16 years since this publication, "the expense and technical difficulties associated with groundwater remediation" have become even more apparent. The author of this book has been involved, and has knowledge of, various hydrogeologically complex "megasites" where characterization of groundwater contamination has continued for more than a decade, costing sometimes over 100 million US dollars at individual sites. At the same time and for various reasons, the actual full-scale groundwater remediation (aquifer restoration to beneficial uses) at many of these sites has not yet begun. Often, this is caused by many remaining uncertainties as to the distribution of contaminants in the subsurface and their migration pathways (such as at many sites in karst and fractured rock environments). These uncertainties have often resulted in the failure of remediation pilot tests. At the same time, however, the regulators and the public alike are in many cases not ready to accept that restoration at a particular site may not be "practicable and attainable." It therefore appears that much more should be done to promote dialogue between various stakeholders, the public, and the regulators at all levels (federal, state, and local), as to the "realistic approach to restoration based upon actual and reasonably expected uses of the resource as well as social and economic values."

Two complementary approaches to the protection of quality of groundwater resources are protection of the existing water supplies ("sources," such as water wells and springs), and protection of the resource as a whole, for all present and future users. In different countries, and in different parts of the same country (such as in the United States), the emphasis may be given to one or both approaches, depending upon the resource development situation, the prevailing hydrogeological conditions, and the prevailing political environment. Source-oriented approaches are based on delineation of source protection zones ("groundwater zoning"). Such zones serve for setting priorities for groundwater quality monitoring, environmental audit of industrial premises, determining priorities for the clean-up of historically contaminated land, and determining which land use activities within which zones have acceptable risk. This approach is best suited to relatively uniform, unconsolidated aquifers exploited only by a small number of high-yielding municipal wells with stable pumping regimes. It is not so readily applied where there are a large and growing number of localized groundwater extraction centers, including individual households. (Foster et al., 2002–2005b).

Resource-oriented strategies are more universally applicable, since they are aimed at achieving protection for the entire groundwater resource and for all groundwater users. They involve aquifer pollution vulnerability mapping over extensive areas, and may include one or more important aquifers. Such mapping would normally be followed by an inventory of potential or existing sources of contamination, and possibly hazards and risks associated with various land use activities, at least in the more vulnerable areas.

8.7.1 Groundwater Vulnerability Maps

The fundamental concept of groundwater vulnerability is that some areas are more vulnerable to contamination than others. The goal of a vulnerability map is the subdivision of an area into several areas that have the different degrees of vulnerability. Vrba and Zaporozec (1994) emphasize that the vulnerability of groundwater is a relative, nonmeasurable, dimensionless property, and they make the distinction between intrinsic (natural) and specific vulnerability. The intrinsic vulnerability depends only upon the natural properties of an area, such as characteristics of the porous media, and recharge. It is independent of any particular contaminant. Specific vulnerability takes into account fate and transport properties of a contaminant. Simplified, this means that, for example, an aquifer may be vulnerable to an improper disposal or spill of chlorinated solvents at the land surface even though the groundwater flow directions and the presence of a low-permeable overlying aquitard may be protective enough in case of a nonpoint source contamination with nitrates. Another example is a thick unsaturated zone (e.g., ≥ 300 ft) in arid climates that may be highly protective of the underlying unconfined aquifer simply because of the insignificant present-day aquifer recharge that cannot facilitate migration of a contaminant through such thick vadose zone, all the way down to the water table. However, if there were some land use practices, such as waste disposal in ponds, which can facilitate contaminant migration, these aquifers would be considered vulnerable.

Although most definitions and methods for mapping groundwater vulnerability consider only contamination aspects, there are also quantitative aspects of groundwater protection and vulnerability, such as overexploration and aquifer mining (Vrba and Zaporozec, 1994). Maps depicting time-dependent quantities of groundwater available for extraction, and the associated development of drawdown (decrease in water levels), are a very useful tool for groundwater management of nonrenewable groundwater resources, as well as other stressed aquifer systems.

It is important to distinguish between the protection of groundwater resources in general, which is supported by vulnerability maps, and the protection of a drinking water source, which is supported by mapping the source zone protection area (or, as commonly referred to in the United States, delineation of the wellhead protection area). Although these two concepts are closely related, they usually involve different scales and the associated maps have different objectives. A wellhead protection area (described in more detail in the next section) is an area within which there is a complete pathway between any given location at the water table (top of the unconfined aquifer) and the groundwater extraction point such as water well. This zone is usually defined by the length of time required for all water particles inside the zone to be extracted, i.e., "captured" by the well (e.g., 5- or 10-year capture zone). In contrast, groundwater vulnerability, which is a qualitative concept, involves a vaguely expressed probability that a theoretical contamination at the land surface is more or less likely to reach the water table. In some cases, the objectives of resource and source protections are merged by creating a map, or several map overlays, that can be used to present both the general vulnerability of an area and the wellhead protection zones for the existing groundwater supplies.

When vulnerability maps include potential sources of contamination and their associated contaminants, possibly ranked by the hazard they pose on groundwater resources (supplies), they are referred to as groundwater contamination hazard or risk maps. Hazards can be grouped into different categories such as from infrastructure development, industrial and agricultural activities. They are commonly evaluated using a hazard index, which takes into account the "harmfulness" of a hazard to groundwater, the quantity of relevant substances that can be released in case of an accident, and the probability for a contaminant release to occur. The highest risk is present in situations where a dangerous hazard (high probability of large quantities of harmful substances to be released) is located in a highly vulnerable zone. The risk map shows areas where engineering, legislative (political), or management response is required (Drew and Hötzl, 1999).

Notably, however, none of the above-described concepts of vulnerability, hazard, and risk of groundwater contamination, fully addresses a three-dimensional likely pathway of any particular contaminant, and its time of travel, between the point of release at the land surface and the receptor (a well or a spring). Note that this pathway must include the vadose zone and in some cases a contaminant released to the subsurface may never reach the water table due to various attenuation mechanisms in the vadose zone. For these reasons, some authors criticize that a purely descriptive (qualitative) definition of groundwater vulnerability, which is not quantitatively defined in terms of physics, may be the source of misinterpretations. Andersen and Gosk (1989) point out that it is impossible to create a "general" vulnerability map which expresses in a comparable way permanent protective properties of an area. They emphasize that a vulnerability map cannot at the same time be applicable for both conservative and reactive contaminants, for both instantaneous and long-term contaminant releases, and for both point and diffuse contamination scenarios. They consequently conclude that vulnerability maps should be prepared for well-defined specific situations only.

As discussed by Goldscheider (2002), it is debatable if a descriptive, general definition and the lack of physical precision in vulnerability concepts should be considered as an advantage or as a disadvantage. The advantage of a descriptive definition is that the term vulnerability is often intuitively understood, particularly by decision makers in the planning process (Hötzl et al., 1995). A vulnerability map showing areas of different color symbolizing different degrees of vulnerability (or natural protection) is easy to interpret and can be a practical and applicable tool for land use planning, protection zoning, and qualitative risk assessment. There are also disadvantages to a purely descriptive definition. A property, which is not precisely defined in terms of physics, cannot be derived unambiguously from measurable physical quantities. Therefore, every method of vulnerability mapping is based on the individual point of view and experience of the person who developed it and is thus subjective (Goldscheider, 2002). It is difficult to compare different vulnerability methods or maps and to decide which one is the best. If different methods are tested in one area, the resulting maps are always different and sometimes contradictory (Gogu, 2000). Another important consequence of the lack of a physical definition is that it is difficult to validate (verify or negate) a vulnerability assessment or mapping (Broyère et al., 2001). The same authors suggest that the three practical questions to which a vulnerability assessment has to answer are the following: If pollution occurs, (1) when will it reach the target, (2) at which concentration level, and (3) for how long will the target be polluted? However, these three questions are more related to protection or vulnerability of a particular water supply source ("target"), such as a water supply well or a spring, than to groundwater resource protection in general.

Among various methods of qualitative groundwater vulnerability mapping, the so-called index (or parametric) methods are the most common (Fig. 8.27). For example, Magiera (2000) counted 34 different methods and the author of this book could add a few more, including one with an amazing acronym: GOD (G is for groundwater hydraulic confinement, O is for overlying strata, and D is for depth to groundwater table or strike;

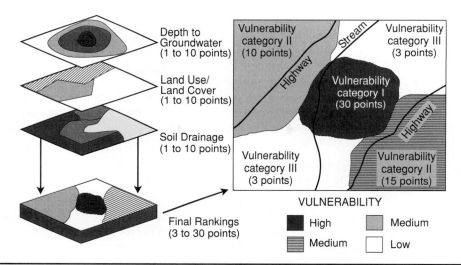

FIGURE 8.27   Schematic of index mapping of groundwater vulnerability. Some indices are based on quantitative data, but the individual and the final rankings are subjective. (Modified from Focazio et al., 2002.)

see Foster et al., 2002–2005b). However, the overall procedure for the various index methods is the same. The first step is the selection of factors (parameters) assumed to be significant for vulnerability. Each factor has a natural range which is subdivided into discrete intervals and each interval is assigned a value reflecting the relative degree of sensitivity to contamination. The vulnerability of an area is determined by combining the values for the different factors using a rating (index) system.

The most widely used index method is DRASTIC, named for the seven factors considered: **d**epth to water, **r**echarge, **a**quifer media, **s**oil media, **t**opography, **i**mpact of the vadose zone, and the hydraulic **c**onductivity of the aquifer (Aller et al., 1985). These seven factors are incorporated into a relative ranking scheme that uses a combination of ratings and weights to produce a numerical value, called the DRASTIC index. Each of the factors is ranked between 1 and 10 and the rank is multiplied by an assigned weighting, which ranges between 1 and 5. The weighted ranks are summed to give a score for the particular hydrogeologic unit with the higher scores indicating greater vulnerability to contamination. The DRASTIC method has been used to develop groundwater vulnerability maps in many parts of the United States as well as worldwide. However, the effectiveness of the method has met with mixed success due to its subjectivity since the maps are not calibrated to any measured contaminant concentrations or specific contaminants (USEPA, 1993a; Rupert, 1999, 2001). Basic DRASTIC maps may be improved by calibration of the point ratings based on the results of statistical correlations between groundwater quality and hydrogeologic and anthropogenic factors. For example, one of the significant weaknesses of the relative vulnerability maps developed for agricultural nitrate contamination in an area of Idaho is that soil permeability was not the primary soil factor. As discussed by Rupert (1999), there was no correlation between the nitrite, nitrate, and nitrogen concentrations in groundwater and the soil permeability, but there was a strong correlation with soil drainage types, presumably because soil drainage is a better indicator of nitrate leaching conditions. Calibration of the aquifer vulnerability maps with groundwater quality information is the most effective way to determine which hydrogeologic and anthropogenic factors are influencing vulnerability to a chemical compound of interest.

DRASTIC and most similar index methods do not take into account unique features of fractured rock and karst groundwater systems. The following characteristics of karst systems are significant with respect to groundwater vulnerability and should consequently be taken into account when attempting to create vulnerability maps (compiled from Hötzl, 1996; Leibundgut, 1998; Trimmel 1998; Drew and Hötzl, 1999; Goldscheider, 2002; Kresic, 2007a,2007b):

- Each karst system has its individual characteristics and any generalization is problematic; a detailed hydrogeologic investigation of a karst system is irreplaceable for any method of vulnerability mapping.

- Karst systems are highly heterogeneous and anisotropic. Interpolation and extrapolation of field data is problematic and the reliability of a vulnerability map is significantly lower for karst than for other areas.

- Karst aquifers are recharged both by diffuse infiltration and by concentrated point recharge via sinkholes and sinks ("swallow holes" in surface streams). The first case is considered less vulnerable than the second one.

- The overlying layers above the karst aquifer, such as top soil and residuum (regolith) sediments, may provide a limited form of protection depending on their thickness; however, surface runoff over these areas intersected by sinkholes or sinks may enhance aquifer vulnerability.

- The presence of an epikarst zone has to be expected. The main functions of the epikarst are water storage and concentration of flow. The first process increases the natural protection of the system while the second process increases the vulnerability. The structure and the hydrologic function of the epikarst are difficult to assess. A large portion of the epikarst is not visible at the land surface.

- Karst aquifers are characterized by a dual or triple porosity due to presence of intergranular pores and micro-fissures in the rock matrix, and fractures and dissolutional voids (conduits). Groundwater storage takes place in the pores and fractures, while conduits act as drains. Consequently, there are both extremely fast and slow flow components within a karst system. Contaminants can be transported very fast and/or stored and slowly transported for a very long time.

- Karst systems are often characterized by a fast and strong hydraulic reaction to hydrologic events. The temporal variations of the hydraulic head (water table) often reach several tens of feet and sometimes more. In many karst systems, the water table is discontinuous and difficult to determine.

- Karst drainage areas are often extremely large and hydraulically connected over long distances. Watersheds are often difficult to determine and variable in time, depending on the seasonal hydrologic conditions. Drainage areas of karst springs often overlap and the flow paths proved by tracer tests often cross each other.

- There are possible transitions between mildly fractured and intensely karstified portions in the same carbonate aquifer, and there may be several types of hydraulically connected aquifers in one area, such as a granular aquifer overlying a karst aquifer.

Among index methods proposed for karst environments, two have gained wider acceptance: the EPIK method (Doerfliger 1996; Doerfliger and Zwahlen, 1998) and the PI method, which adds a level of complexity to the EPIK method (Goldscheider et al. 2000a, 2000b; Goldscheider, 2002).

In conclusion, the qualitative vulnerability maps (source or resource, intrinsic or specific) show the sensitivity of different areas to groundwater contamination generated by human activities. They can be a useful tool for land use planning and groundwater management if the limitations of the concepts are clarified. They should be made for a well-defined purpose and should not be a stand-alone element but an integrated part of an overall groundwater protection plan.

8.7.2 Delineation of Source Water Protection Zones

The first wellhead protection zone in the United States was established on May 24, 1610, when Sir Thomas Gates, Lieutenant Governor of Virginia, instituted "laws divine moral and marshal," a harsh civil code for Jamestown. The wellhead protection law reads

> There shall be no man or woman dare to wash any unclean linen, wash clothes, . . . nor rinse or make clean any kettle, pot or pan, or any suchlike vessel within twenty feet of the old well or new pump. Nor shall anyone aforesaid within less than a quarter mile of the fort, dare to do the necessities of nature, since these unmanly, slothful, and loathsome immodesties, the whole fort may be choked and poisoned.

Almost 400 years later, in a memorandum entitled *"Source Water Assessment and Protection"* addressed to Regional Administrators, the Assistant Administrator of the USEPA, Tracy Mehan III reminds that

> The deadline is fast approaching by which States must complete their source water assessments. This year (2003) all 160000 Public Water Systems (PWS's) must have:
>
> 1. Delineated and mapped their sources of drinking water;
> 2. Inventoried the potential contamination activities and contaminants involved;
> 3. Assessed the susceptibility of the drinking water resources to those contaminants; and,
> 4. Made the assessment information public.
>
> The PWSs are expected to then develop (a) management measures to protect their sources of drinking water, and (b) contingency plans for man-made or catastrophic events. (Mehan, 2003a).

We can only imagine what the reaction of Sir Thomas Gates would have been if he were to find out that only 17 percent of the 160,000 systems in the country have completed the assessments, in the year when all of them were required by law to complete such an assessment. In all probability, Sir Gates would have been very pleased to learn that Virginia ranked rather high with an 80 percent completion rate (only two states ranked higher). He, of course, would also have been astonished to learn that there were 50 states on the continent, and probably would not have been able to focus on the fact that 19 of them did not complete a single assessment of their CWSs (many of which are based on groundwater).

This anecdote emphasizes the importance of the involvement of all stakeholders in groundwater management and protection, and the irreplaceable role of public education in achieving that goal. Both the public education about the importance of water resources

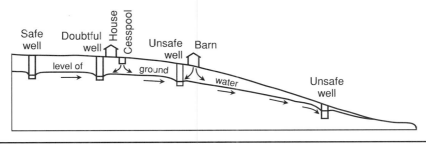

Figure 8.28 Diagram showing ordinary location of farm wells. (From Fuller, 1910.)

protection in the United States, and the pollution control programs that started in the early 1900s, were created in response to the large number of typhoid and other disease outbreaks that had occurred. States and local governments began establishing public health programs to protect surface water supplies by identifying and limiting sources of contamination. Early water pollution control programs concentrated on keeping raw sewage out of surface waters used for drinking water. Efforts were also made to site intakes used to collect drinking water upstream from sewage discharges (USEPA, 2003b). At the same time, public agencies such as the USGS were educating citizens, rural home-owners, and farmers on groundwater contamination issues as illustrated in Figs. 8.28 and 8.29. These efforts can be viewed as early examples of the concept of source protection areas.

Two excerpts from Fuller (1910), illustrate the early public education efforts by the USGS:

Farms, which are generally remote from towns, cities, or other areas of congested population, seem to be almost ideally situated for obtaining pure and wholesome water. In reality, however, polluted water is exceedingly common on them [Fig. 8.28. in this book] and typhoid-fever rates are usually greater in country districts than in cities. Typhoid fever is now almost universally believed to be transmitted solely through drink or food taken into the stomach, and is especially liable to be communicated by polluted waters obtained from shallow wells near spots where the discharges of typhoid patients have been thrown upon the ground and subsequently carried down through the soil and into the wells, and it is doubtless principally this fact that makes the disease so common in farming regions.

An example of danger from refuse of a more disgusting type is shown in Plate X, A [Fig. 8.29 in this book]. Located in the middle of a well-traveled street, only a few inches above a gutter filled with paper and refuse, a part of which is sure to enter whenever a heavy rain occurs; open to the rain which washes into it from the steps leading down to it such dirt from the street as is brought in by the feet of the users; subject to the dipping of all sorts of more or less dirty buckets and utensils; receiving the underground drainage and presumably more or less sewage from the buildings on the slopes above; and containing in its bottom several inches of decaying paper and other refuse, this spring is on the whole one of the worst and most dangerously located sources of drinking water in the United States.

As discussed by Kraemer et al. (2005) of the USEPA,

The main steps of the source water protection process involve the assessment of the area contributing water to the well or wellfield, a survey of potential contaminant sources within this area, and an evaluation of the susceptibility of the well to these contaminants. This includes the possibility of contaminant release and the likelihood of transport through the soil and aquifer to the well screen. The designation of the wellhead protection area is then a commitment by the community to source area management. Delineation of the wellhead protection area is often a compromise between scientific and

FIGURE 8.29 Spring in center of city street. (From Fuller, 1910.)

technical understanding of geohydrology and contaminant transport, and practical implementation for public safety. The EPA Office of Ground Water and Drinking Water established guidance on the criteria and methods for delineating protection areas (USEPA, 1993a; USEPA, 1993b).

The Agency maintains a Web page dedicated to source water protection, with downloadable publications and various links, at http://www.epa.gov/safewater/protect.html.

In its efforts to facilitate capture zone delineation and protection area mapping in support of the State's Wellhead Protection Programs (WHPP) and Source Water Assessment Planning (SWAP) for public water supply wells in the United States, the USEPA has recently released an updated version of WhAEM2000, a public domain and open source general-purpose groundwater flow modeling program. WhAEM2000 is analytic element model limited to steady-state conditions and two-dimensional flow conditions, and includes options for simulating the influence of flow boundaries, such as rivers, recharge, and no-flow contacts (Kraemer et al., 2005). Figure 8.30 illustrates just some of the complexities associated with the delineation of a wellhead protection area using several approaches outlined by the USEPA. As can be seen, the shapes of the four zones are quite different reflecting critically different assumptions, all of which would be valid based on USEPA guidance. In all four cases the same (single) well near a relatively large permanent river operates at a constant (steady state) pumping rate, pumping water from an unconfined aquifer in which the groundwater flow is assumed to be horizontal and

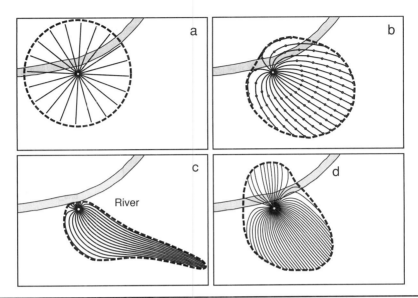

FIGURE 8.30 Five-year capture zone of a single well near a permanent river delineated using different options in the WhAEM2000 computer program. General groundwater flow direction is from the southeast toward the river. (*a*) Fixed radius volumetric method. (*b*) Uniform flow field method; small circles correspond to 1-yr intervals. (*c*) Zone delineated taking into consideration hydrologic boundaries (not shown). (*d*) Zone delineated taking into consideration possible resistance of the riverbed sediments. (Modified from Kraemer et al., 2005.)

with a "regional" hydraulic gradient from the southeast to the northwest (toward the river). It is obvious that the "correct" wellhead protection zone would have to be selected in a constructive discussion between professional hydrogeologists and all other stakeholders (nonhydrogeologists), along with the inevitable considerations of time and budgetary constraints. Unfortunately, most states in the United States were confronted with the task of developing wellhead protection programs for thousands of public drinking water supply systems in only a few years time. According to Kraemer et al. (2005), conducting an "extensive" groundwater modeling campaign for each individual drinking water well (or well field) was out of the question, both in view of the time involved and the cost. "The USEPA recognized this reality from the start and proposed a series of simplified capture zone delineation methods to facilitate a timely implementation of the States wellhead protection programs."

Consequently, it is very likely that many PWSs have delineated wellhead protection areas that are not based on any hydrogeologic reality, and may be overprotecting their water supply by unnecessarily restricting various land uses, or may not be protecting them at all thus creating a false sense of security. In the United States, the majority of community water supply systems serving fewer than 3300 people use groundwater, while many larger systems also depend on groundwater. Small systems use groundwater as a source because groundwater usually requires less treatment than surface water and is therefore more affordable. This is an important consideration since many small systems without a large, rate-paying base cannot afford extensive hydrogeologic studies for wellhead (source water) protection purposes. At the same time, wellhead protection efforts are often among the most cost-effective ways to ensure safe drinking water. These efforts prevent contamination from occurring rather than treating contamination after it has occurred.

Most methods used to delineate wellhead protection zones address the residence-time criterion. This criterion is based on the assumption that

- Nonconservative contaminants, subject to various fate and transport processes (e.g., sorption, diffusion, degradation) may be attenuated after a given time in the subsurface

- Detection of conservative contaminants (not subject to attenuation) entering the wellhead protection area will give enough lead time for the public water supply entity to take necessary action, including groundwater remediation and/or development of a new (alternative) water supply

- Detection of any contaminants already in the wellhead protection area would require an immediate remedial action

The most critical decision regarding an appropriate residence time is made by the stakeholders in each individual case, although 5-year, 10-year, and 20-year wellhead capture zones have been most widely used to delineate certain subzones of various land use restrictions within the main wellhead capture zone. This concept for a single well in a homogeneous isotropic aquifer, pumping with the same pumping rate, and assuming strictly horizontal flow with the same hydraulic gradient that does not change in time, is illustrated in Fig. 8.31.

Commonly, there are three wellhead protection zones, although the number may vary based on local hydrogeologic conditions and regulatory requirements:

- Zone 1 (sanitary zone or zone of strict protection) is essentially an administrative zone of physical protection of the water source, such as fencing and restricted access measures. Its purpose is to prevent accidental or deliberate damage and/or

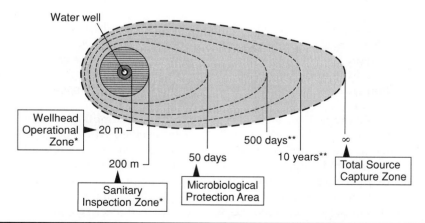

FIGURE 8.31 Idealized scheme of surface sanitary zones and groundwater flow perimeters for the protection of a water well in an unconfined aquifer. $*$, empirical fixed radius area; $**$, intermediate flow-time perimeters sometimes used. (From Foster et al., 2002–2005b.)

contamination of the source itself including contamination of the aquifer through the source. The size of zone 1 may vary between several tens and several hundreds of feet (meters) for simple systems consisting of one well for example, or it may be larger in case of several closely spaced springs or multiple wells.

- Zone 2, sometimes called the inner protection zone, is based on time-of-travel analysis. It reflects the site-specific understanding of the minimum time of potential contaminant travel between its introduction into the subsurface and the groundwater withdrawal locations during which it would be possible to initiate remediation activities and execute contingency plans for water supply if necessary. In some states, zone 2 is considered an attenuation zone in which pathogens entering the zone from septic systems or surface water bodies would be attenuated before reaching the well (e.g., see Wyoming DEQ, 2007). It is obvious that zone 2 in general may vary in shape and size widely, depending on aquifer type and hydraulic conductivity of the most permeable zones. Zone 2 is entirely inside the groundwater divide and corresponds to the aquifer volume from which all groundwater would discharge through the well(s) or spring(s) within the given time. Delineation of zone 2 may not be necessary or even meaningful in the case of deep confined aquifers protected by competent aquitards, distant aquifer recharge zones, and groundwater resident times measured by hundreds or thousands of years. Glacial gravel deposits and karst aquifers are examples of porous media where contaminant travel times may be extremely short (e.g., several days) over distances of miles. Consequently, zone 2 may cover a rather large area and pose a serious challenge for development of an adequate source water management and protection plan.

- Zone 3, also called the contributing area, includes the entire aquifer volume within the groundwater divide from which all groundwater will eventually discharge through the well(s) or spring(s), regardless of time of travel. This zone may be subdivided based on various criteria, including time of travel. Its main importance is for long-term planning of groundwater resources management, and the related land use and aquifer protection regulations.

Wellhead protection zone delineation methods can be divided into the following three categories: (1) nonhydrogeologic, (2) quasi-hydrogeologic, and (3) hydrogeologic. The nonhydrogeologic method is a selection of an arbitrary fixed radius or fixed-shape area around the well(s) in which authorized personnel implement some type of strict protection such as limited access. This method does not consider the residence-time criterion.

Quasi-hydrogeologic methods use very simple assumptions, which, in many cases, do not have much in common with the site-specific hydrogeologic conditions. Since they include the application of certain equations, it may appear, to nonhydrogeologists, that such methods must have some credibility. When involved in the application of quasi-hydrogeologic methods, hydrogeologists should clearly explain the limitations of various unrealistic assumptions and their implications on the final wellhead zone delineation. For example, these methods do not consider aquifer heterogeneity and anisotropy, stratification or the presence of confining layers, vertical flow components, interference between multiple wells screened at different or the same depths in the same well field, or depth to screen of individual wells. Three common quasi-hydrogeologic methods include

(1) calculated fixed radius, (2) well in a uniform flow field, and (3) groundwater modeling not based on hydrogeologic mapping (Kresic, 2007a). None of these methods should be applied in any of the situations listed above.

Hydrogeologic Mapping and Groundwater Modeling

Hydrogeologic mapping first investigates and then presents, in the form of maps and accompanying graphics and documentation, geologic, hydrologic, and hydraulic features that control groundwater flow within an area of interest. The capture zone of a well, well field, or a spring used for water supply can be quite complex and may include multiple interconnected aquifers and surface water features. Mapping the three-dimensional physical and hydraulic boundaries of such a flow system is therefore the key for a successful hydrogeologic map of the wellhead capture zone. It is important to understand that a hydrogeologic map is not simply a geologic map with different colors; a geologic or litostratigraphic unit (or formation) is not necessarily directly translatable into an aquifer or aquitard. Several geologic units may act as one aquifer, and there may be several aquifers separated by aquitards within the same geologic formation. In hydrogeologic studies, it is common to use the term hydrostratigraphic unit, which describes one or more geologic units that have the same porous media characteristics and act as one hydrodynamic entity (aquifer or aquitard). Tectonic fabric of the mapped area, including faults and fault systems, may also play an important role in directing groundwater flow within its boundaries. In short, the hydrogeologic map must show where the water is coming from into the flow system captured by the well(s), and how is it flowing toward the well(s), which act as the local discharge area within the aquifer. This is possible only if there is sufficient three-dimensional field information on the actual hydraulic heads within the flow system, including their seasonal variations. Once the geometry of the flow system is understood, it is mapped by showing three-dimensional equipotential lines and flowlines. The final step is to estimate the velocities of groundwater flow within the capture zone and to delineate aquifer volumes with the same residence time, i.e., show them in the three-dimensional space. It is obvious that a thorough hydrogeologic mapping may require substantial resources for more complex hydrogeologic conditions, and therefore may not be feasible for some water supply systems. Whatever the case may be, hydrogeologic mapping for the delineation of wellhead capture zones is most resourceful when used for a concurrent mapping of the aquifer vulnerability in its recharge area.

Hydrogeologic maps and vulnerability maps should be developed in a GIS environment to combine various data layers. With comprehensive hydrogeologic maps, data layers include geology, hydrostratigraphy, tectonics, topography, hydrologic features (hydrography), climate factors influencing aquifer recharge, land cover, land use, soil types, depth to saturated zone, aquifer parameters (transmissivity, hydraulic conductivity, porosity, effective porosity, storage properties), hydraulic head contour lines for different water-bearing zones, aquifer recharge and discharge areas and locations (both natural and artificial), and known or potential sources of soil and groundwater contamination.

Numeric models based on the results of hydrogeologic mapping are the best tool for the delineation of wellhead capture areas, and groundwater resources management in general. Regardless of the effort and resources invested into hydrogeologic mapping for any purpose, there will inevitably remain a certain level of uncertainty as to the true representative hydrodynamic characteristics of the groundwater flow system analyzed. Numeric models provide for quantitative analysis of this uncertainty and enable

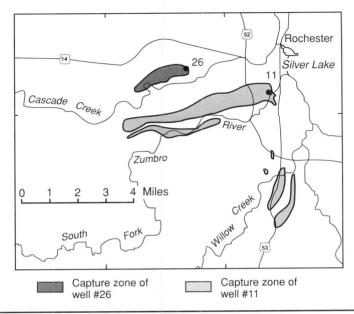

Figure 8.32 Long-term (steady-state) model-calculated contributing recharge areas for wells 11 and 26 near Rochester, MN, which are screened in the St. Peter-Prairie du Chien-Jordan aquifer. Contributing areas shown for the case when no other high-capacity wells are pumping. (Modified from Franke et al., 1998, and Delin and Almendinger, 1993.)

decision makers to analyze different "what if" scenarios of their source management and protection. Numeric models can take into account all important aspects of the three-dimensional hydrogeologic mapping and convert them into a quantitative description of the flow system. Consequently, they can delineate complex shapes and aquifer volumes contributing water to well fields as illustrated in Fig. 8.32. For example, any pumping wells in addition to the ones shown in Fig. 8.32 would capture their own subsurface flowpaths and have their own contributing recharge areas at the water table, thereby changing local flow patterns in the surrounding groundwater flow system. More on groundwater models, their selection, and proper uses is given in Section 8.8.

Karst Aquifers

As mentioned repeatedly, karst aquifers are particularly vulnerable to contamination. Contaminants can easily enter underground and may be transported rapidly over large distances in the aquifer. Processes of contaminant retardation and attenuation often do not work effectively in karst systems. Therefore, karst aquifers need special protection and attention. A detailed knowledge of karst hydrogeology is a precondition for the delineation of source protection areas in karst. However, karst drainage areas contributing water to a single spring or a large supply well can be very large and/or extensive and convoluted due to anisotropy (e.g., presence of karst conduits); it is therefore often unrealistic to designate maximum protection for the entire system, as the resulting land use restrictions would not be acceptable to some stakeholders. In addition, many PWSs do not have financial and other resources to "do it right" and embark on an adequate characterization of karst hydrogeology in their "little" service area. Unfortunately, this gamble may sometimes prove fatal, as illustrated with the following example provided by Worthington et al. (2003; also see Goldscheider et al., 2007).

Walkerton is a rural town in Ontario, Canada, with a population of some 5000 people. In May 2000, about 2300 of them became ill and 7 died from bacterial contamination of the municipal water supply. The principal pathogens were *Escherichia coli O157:H7* (a pathogenic strain of *E. coli*) and *Campylobacter jejuni*. Subsequent epidemiological investigations indicated that most of the contamination of the water supply has occurred within hours or days at most, after a heavy rain. Three municipal wells had been in use at the time of the outbreak. Soon after the outbreak, a hydrogeological investigation was carried out that included the drilling of 38 boreholes, surface and downhole geophysics, pumping tests, and testing of numerous water samples for both bacteriological and physical parameters. The aquifer at Walkerton consists of 70 m of thick flat-bedded Paleozoic limestones and dolostones, which are overlain by 3 to 30 m of till. A numerical model of groundwater flow (using MODFLOW) indicated that the 30-day time of travel capture zones extended 290 m from Well 5, 150 from Well 6, and 200 m from Well 7. These results suggested that if a groundwater pathway was implicated in the contamination, then the source must have been very close to one of the wells.

A public inquiry (the Walkerton Inquiry) was held to investigate the causes of what came to be known as the Walkerton Tragedy. During the inquiry, the question was raised as to whether the aquifer might be karstic and thus have rapid groundwater flow. The original hydrogeologic investigation, carried out after the outbreak, had not mentioned the possibility of karstic groundwater flow. Subsequent investigations by karst experts found that there were many indications that the aquifer is karstic (Worthington et al., 2003). These included a correlation between bacterial contamination in wells and antecedent rain, demonstrating rapid recharge and flow to wells; localized inflows to wells which video images showed to be dissolutionally enlarged elliptical openings on bedding planes; the presence of springs with discharges up to 40 L/s; rapid changes in discharge and chemistry at these springs following rain; and rapid, localized changes to electrical conductance in a well during a pumping test.

All these tests strongly suggested that the aquifer is karstic, but the most persuasive evidence were the results of aquifer tracing tests. Earlier numerical modeling had suggested that groundwater velocities were typically in the range of a few meters per day, but tracer tests demonstrated that actual velocities were some one hundred times faster (Fig. 8.33). In conclusion, the investigations by the karst experts demonstrated that the source for the pathogenic bacteria could have been much further from the wells than the earlier investigations and modeling had indicated (Goldscheider et al., 2007).

8.7.3 Management Strategies

Approaches to groundwater resources protection differ at various levels of government and may mean different things to different stakeholders, thus emphasizing the need for public education and dialogue. For an individual household that has just discovered (or was told) that the well water they were drinking was contaminated with a dangerous carcinogenic substance for years, it would be impossible not to state that their government has failed them. Unfortunately, similar cases are occurring daily and worldwide. On the other hand, in many developed countries there are quite a few governmental programs and regulations aimed at groundwater protection. For example, below is a list included in the 1998 report to Congress by the USEPA (2000) in which the Agency explains that

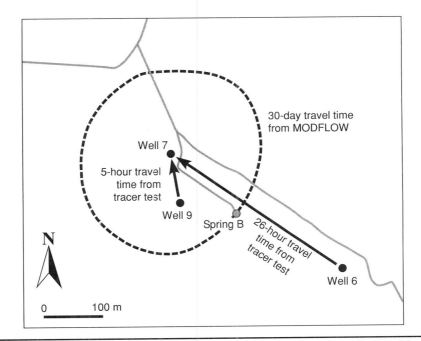

FIGURE 8.33 Trajectories and travel times for tracers injected in monitoring wells 6 and 9 and recovered in pumping Well 7 showing velocities ≥300 m/day, compared to a 30-day capture zone for Well 7 predicted using MODFLOW. (Modified from Worthington et al., 2003).

States develop prevention programs to prevent and reduce contamination of groundwater. They serve to:

- Analyze existing and potential threats to the quality of public drinking water.
- Focus resources and programs on drinking water source protection.
- Prevent pollution at the source whenever feasible.
- Manage potential sources of contamination.
- Tailor preventive measures to local ground water vulnerability.

Examples of programs that fully or in part address pollution prevention include: Source Water Assessment Program (SWAP), Pollution Prevention Program, Wellhead Protection Program (WHPP), aquifer vulnerability assessments, vulnerability assessments of drinking water/wellhead protection, Pesticide State Management Plan, Underground Injection Control (UIC) Program, and SARA Title III Program.

At the state level, this long list often translates into development and implementation of a Source Water Assessment Program (SWAP), which is focused on the delineation of wellhead (or "springhead") protection areas for the existing sources of water supply. Some states may include an inventory of potential contaminant sources within the delineated wellhead protection areas and release this information to the public. However, the overall protection of the groundwater resource still remains a rather vague concept and is certainly not subject to any legally enforced, overall regulation by the states or the federal government. As noted by the USEPA (1999),

> Ground water management in this country is highly fragmented, with responsibilities distributed among a large number of federal, state, and local programs. At each level of government, unique legal authorities allow for the control of one or more of the ground water threats described in Section 3.0. These authorities need to complement one another and allow for comprehensive management of the ground water resource.

The main reason for this lack of comprehensive, overall groundwater protection regulation, is that any such regulation would require a lot of political will because it would have to include strict land use controls and significant resources to monitor and enforce land use practices. This, by definition, includes any agricultural, industrial, or other activity that uses land. As this is not feasible in the near future, in most cases the real protection of both the existing sources and the resource is, in the end, left to the local communities and public water systems (PWSs). They have to develop management plans, involving all local stakeholders and the public, which would minimize risks to their water supplies and their resource. Simply put, if the community (all stakeholders included) believes that a certain industry or land use activity will not threaten their sources and the resource as a whole (both quantity and quality), everyone should be satisfied. If this is not the case, there are usually four options: (1) public education and outreach that may resolve the issue by voluntarily changing questionable land uses; (2) enactment of local regulations and ordinances that may leave some stakeholders still unhappy; (3) land acquisitions and land conservation; and (4) a lawsuit.

Protection of recharge areas, whether natural or artificial, through regulated land use practices that cannot adversely impact either the quantity nor quality of groundwater, is the foundation of any successful groundwater resource protection and management. Given many diverse human activities that can potentially lead to contamination of the subsurface, a lack of protection of recharge areas will eventually decrease the availability of usable groundwater and require expensive treatment or the substitution of a more expensive water supply, as demonstrated in many urban and rural areas worldwide.

An example of land use change that is increasingly affecting many groundwater basins in California is urban development. In addition to quality impacts, urban development, (pavement and buildings on former agricultural land, lining of flood control channels, and other land use changes) have reduced the capacity of recharge areas to replenish groundwater, effectively reducing the sustainable yield of some basins. As advised by the California Department of Water Resources (DWR, 2003), to ensure that recharge areas continue to replenish high quality groundwater, water managers and land use planners should work together to

- Identify recharge areas so the public and local zoning agencies are aware of the areas that need protection from paving and from contamination
- Include recharge areas in zoning categories that eliminate the possibility of contaminants entering the subsurface
- Standardize guidelines for pretreatment of the recharge water, including recycled water

- Install monitoring wells to collect data on changes in groundwater quality that may be caused by recharge
- Consider the functions of recharge areas in land use and development decisions

Groundwater protection and management in rural areas, including those areas where irrigation for agriculture is significant or predominant, presents a special challenge. Groundwater use in such areas is a decentralized activity with many private users normally involved. They have drilled their own wells, installed their own equipment, and follow their own pumping schedules. In the case of major aquifers, with thousands or hundreds of thousands of users, enforcement of any kind including, for example, well discharge metering, is impossible if users have no incentive to comply. The same is true with the use of pesticides, fertilizers, or more efficient irrigation practices (Figs. 8.34 and 8.35). It is therefore essential that incentives are created for users to participate actively in groundwater protection and management. This can be achieved by providing data on the status of groundwater resources (both quantity, such as trends in groundwater levels, and quality, such as concentrations of nitrates and pesticides for example), promoting aquifer management and protection associations (through which users exert peer pressure to achieve management and protection goals), and making increased use of innovative technologies (Kemper et al., 2002–2005). One such technology is remote sensing. Satellite images are now affordable and various organizations have developed interpretation tools to map crop distributions and to estimate actual evapotranspiration at high resolution. Groundwater management and protection associations, as well as individual users, can now be provided with such data. By so doing, the control of groundwater use becomes more transparent and enforceable.

FIGURE 8.34 Center pivot spray irrigation uses less water than furrow irrigation and decreases leaching of salt. (Photograph courtesy USDA National Resources Conservation Service.)

FIGURE 8.35 Center pivot irrigation on wheat growing in Yuma County, CO, using LESA (Low Elevation Spray Application) system. This type of application uses least water and reduces evaporation. (Photograph courtesy of Gene Alexander, USDA National Resources Conservation Service.)

In the United States, certain states rely on self-reporting by agricultural users of the groundwater volumes pumped on a quarterly or annual basis. This system developed because it is prohibitively expensive for water resource agencies to visit every well individually. Self-reporting, however, does not work in every setting. Another technological option is to link groundwater use with electrical energy bills. Since irrigated agriculture is heavily subsidized worldwide and by various means, one of the incentives for groundwater protection and conservation in large irrigated areas may be another temporary subsidy for switching to more efficient irrigation practices and less water-intensive crops. The resulting overall reduction in groundwater pumpage, which translates into energy and other cost savings, may significantly reduce pressure on governments to continue long-term subsidies.

In any case, the introduction of economic instruments into groundwater protection strategies in agriculture will depend upon hydrologic, economic, social, and political conditions. The feasibility analysis should include an assessment of costs and benefits of each instrument and their possible combination. It should also take into account long-term recurrent costs and institutional capacity (for administration, monitoring, and enforcement) and the transaction costs involved in setting up systems. The expected costs and benefits would also influence the trade-off between the use of economic instruments and other groundwater management tools (Kemper et al., 2002–2005).

Finally, agricultural and food policy is usually made at the highest political level and will typically be analyzed within the macro-socioeconomic context of the country or region (state) concerned. Here, the critical step will be for groundwater resource managers to establish a dialogue with macro policymakers in order to clarify the impacts of current policies. Making this link should lead to more effective groundwater management by placing this vital resource more centrally in the context of national socioeconomic development policy (Kemper et al., 2002–2005).

Figure 8.36 Bunker Hill dike in the San Bernardino area, California, circa 1904 (top) and 2004 (bottom). (Top photograph from Mendenhall, 1905, USGS Water Supply Paper 142, plate XI; bottom photograph courtesy of Wesley Danskin, USGS.)

Figure 8.36 illustrates how population growth and migration into large urban areas can replace agricultural water uses and the associated issues in an unplanned fashion, thus creating a different set of issues facing water managers. Land use planning is therefore the key component in any groundwater management strategy.

The National Research Council (NRC), as requested by the National Science Foundation, synthesized broad expertise from across the many disciplines of environmental science to offer its judgment as to the most significant environmental research challenges of the next generation. This analysis was based on the "potential to provide a scientific breakthrough of practical importance to humankind if given major new funding." Of the eight "grand challenges" identified in the NRC's report entitled *Grand Challenges in Environmental Sciences*," the author of this book finds the following "challenges" directly related to the overall theme of groundwater protection and sustainability (NRC, 2000):

- *Climate Variability.* The challenge is to increase our ability to predict climate variations, from extreme events to decadal time scales; to understand how this variability may change in the future; and to assess realistically the resulting impacts. Important research areas include improving observational capability, extending the record of observations back into the earth's history, improving diagnostic process studies, developing increasingly comprehensive models, and conducting integrated impact assessments that take human responses and impacts into account.

- *Hydrologic Forecasting.* The challenge is to develop an improved understanding of and ability to predict changes in freshwater resources and the environment caused by floods, droughts, sedimentation, and contamination. Important research areas include improving understanding of hydrologic responses to precipitation, surface water generation and transport, environmental stresses on aquatic ecosystems, the relationships between landscape changes and sediment fluxes, and subsurface transport, as well as mapping groundwater recharge and discharge vulnerability.

- *Institutions and Resource Use.* The challenge is to understand how human use of natural resources is shaped by institutions such as markets, governments, international treaties, and formal and informal sets of rules that are established to govern resource extraction, waste disposal, and other environmentally important activities. Important research areas include documenting the institutions governing critical lands, resources, and environments; identifying the performance attributes of the full range of institutions governing resources and environments worldwide, from local to global levels; improving understanding of change in resource institutions; and conceptualizing and assessing the effects of institutions for managing global commons.

- *Land Use Dynamics.* The challenge is to develop a systematic understanding of changes in land uses and land covers that are critical to ecosystem functioning and services and human welfare. Important areas for research include developing long-term, regional databases for land uses, land covers, and related social information; developing spatially explicit and multisectoral land-change theory; linking land-change theory to space-based imagery; and developing innovative applications of dynamic spatial simulation techniques.

- *Reinventing the Use of Materials.* The challenge is to develop a quantitative understanding of the global budgets and cycles of materials widely used by humanity and of how the life cycles of these materials (their history from the raw-material stage through recycling or disposal) may be modified. Important research areas include developing spatially explicit budgets for selected key materials; developing methods for more complete cycling of technological materials; determining how best to utilize materials that have uniquely useful industrial applications but are potentially hazardous to the environment; developing an understanding of the patterns and driving forces of human consumption of resources; and developing models for possible global scenarios of future industrial development and associated environmental implications.

8.8 Modeling and Optimization

Modeling is the corner stone of decision support systems (DSS) for water resources management. As water resources related projects and their management vary in scope and complexity, so do the models. In the most complex cases, an integrated management of water resources in large watersheds may require several hierarchic models connected with scenario-decision-feedback loops. High-level models serve to optimize water supply for multisectoral, and often competing, demands in the watershed (basin). They perform operational simulations of existing water systems as well as conceptual alternatives ("what if" scenarios), include any number of decision variables, and have automated sensitivity analyses and optimization algorithms. Such models also operate with a variety of management constraints set by regulations or agreed upon by stakeholders. One example is maintenance of guaranteed minimal environmental flows in the basin, or flows for certain downstream users. Figure 8.37 illustrates various water sector linkages within large river basins showing the importance of modeling in decision making.

The basin-scale management models are periodically updated by the output from physical models developed for smaller watersheds and groundwater systems, which, at the same time, are also periodically updated using input parameters from various monitoring locations. Some examples of output from local surface water models include predicted temporal changes in water quality due to application of fertilizers, turbidity loads after storm events, and changes in flows caused by controlled releases of water from

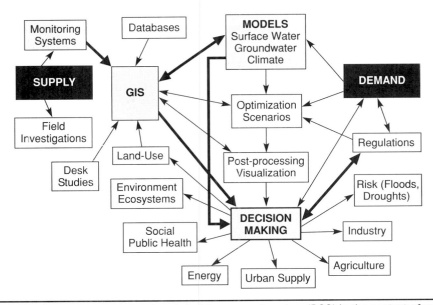

FIGURE 8.37   Resource management and decision support system (DSS) in the context of water sector. Decisions can be made based on monitoring, modeling, and regulations (thick arrow lines).

reservoirs. Groundwater models may provide predictions of baseflow, and the expected irrigation pumpage for seasonal weather or soil moisture conditions. In short, the level of complexity of integrated modeling systems will depend on the management goals set by end users and stakeholders.

As both water resources and their users are spatially distributed, all models are naturally best integrated in a GIS which links databases, models, and model outputs in a user friendly, graphics-based visual environment. Monitoring systems, model(s), and GIS are the three main pillars of decision support systems (DSS). DSS are quickly becoming a standard in water resources management as visual and multimedia tools are now irreplaceable in transferring technical knowledge to decision makers, policymakers, and the public. They are also the most efficient means of identifying the economic, environmental, and social impacts of different management alternatives. The results of such analyses are also displayed in understandable formats.

8.8.1 Numeric Groundwater Models

In general, a model simulates the aerial and temporal properties of a system, or one of its parts, in either a physical ("real") or mathematical ("abstract") way. An example of a physical model in hydrogeology would be a tank filled with sand and saturated with water—the so-called "sandbox," the equivalent of a miniature aquifer of limited extent. The application of real physical models has been limited to educational purposes. Models that use mathematical equations to describe the elements of groundwater flow are called mathematical. Depending upon the nature of equations involved, these models can be empirical (experimental), probabilistic, and deterministic. Empirical models are derived from experimental data that are fitted to some mathematical function. A good example is Darcy's law. (Note that Darcy's law was later found to be theoretically based and actually became a physical or deterministic law). Although empirical models are limited in scope, they can be an important part of a more complex numeric modeling effort. For example, the behavior of a certain contaminant in porous media can be studied in the laboratory or in controlled field experiments, and the derived experimental parameters can then be used for developing numeric models of groundwater transport.

Probabilistic models are based on laws of probability and statistics. They can have various forms and complexity starting with a simple probability distribution of a hydrogeological property of interest, and ending with complicated stochastic time-series models. The main limitations for a wider use of time-series (stochastic) models in hydrogeology are that (1) they require large datasets needed for parameter identification and (2) they cannot be used to answer (predict) many of the most common questions from hydrogeologic practice such as effects of a future pumping for example.

Deterministic models assume that the stage or future reactions of the system (aquifer) are predetermined by physical laws governing groundwater flow. An example is the flow of groundwater toward a fully penetrating well in a confined aquifer as described with the Theis equation (Chap. 2). Most problems in traditional hydrogeology are solved using deterministic models, which can be as simple as the Theis equation or as complicated as a multiphase flow through a multilayered, heterogeneous, anisotropic aquifer system. There are two large groups of deterministic models depending upon the type of mathematical equations involved: analytical and numeric. Simply stated, analytical models solve one equation of groundwater flow at a time and the result can be applied to one point or "line of points" in the analyzed flow field (aquifer). For example, if we want to find (i.e., to "model") what the drawdown at 50 m from the pumping well would be after 24 hours of pumping, we would apply one of the equations describing flow toward a well depending upon the aquifer and well characteristics (confined, unconfined, leaky aquifer; fully or partially penetrating well). To find the drawdown at 1000 m from the well, we would have to solve the same equation (say, the Theis equation) for this new distance. If the aquifer were not homogeneous, these solutions would be of limited value. Obviously, if our aquifer is quite heterogeneous, and we want to know drawdown at "many" points, we might spend a rather long period of time solving the same equation (with slightly changed variables) again and again. If the situation gets really complicated, such as when there are several boundaries, more pumping wells, and several hydraulically connected aquifers, the feasible application of analytical models terminates.

Numeric models describe the entire flow field of interest at the same time, providing solutions for as many data points as specified by the user. The area of interest is subdivided into many small areas (referred to as cells or elements; see Fig. 8.38) and the basic groundwater flow equation is solved for each cell considering its water balance (water inputs and outputs) based on the boundary and initial conditions of the whole system. The solution of a numeric model is the distribution of hydraulic heads at points representing individual cells. These points can be placed in the center of the cell, at intersections between adjacent cells, or elsewhere. The basic differential flow equation for each cell is replaced (approximated) by an algebraic equation so that the entire flow field is represented by x equations with x unknowns where x is the number of cells. This system of algebraic equations is solved numerically, through an iterative process, thus the name numeric models.

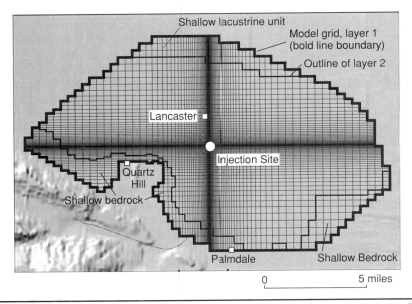

FIGURE 8.38 Schematic presentation of a groundwater model setup showing model grid. Cell size increases from about 100 ft^2 at the injection site, to maximum dimension of about 1000 ft^2. (Modified from Phillips et al., 2003.)

Based on various methods of approximating differential flow equations and methods used for numerically solving the resulting system of new algebraic equations, numeric models are divided into several groups. The two most widely applied groups are (1) finite differences and (2) finite elements. Both types of models have their advantages and disadvantages and for certain problems one may be more appropriate than the other. However, because they are easier to design and understand, and require less mathematical involvement, finite-difference models have prevailed in hydrogeologic practice. In addition, several excellent finite-difference modeling programs have been developed by the Unites States Geological Survey (USGS) and are in the public domain, which ensures their widest possible use. One of these is MODFLOW (McDonald and Harbaugh, 1988; Harbaugh and McDonald, 1996; Harbaugh et al., 2000), probably the most widely used, tested, and verified modeling program today. MODFLOW has become the industry standard thanks to its versatility and open structure: independent subroutines called "modules" are grouped into "packages," which simulate specific hydrologic features. New modules and packages can be easily added to the program without modifying the existing packages or the main code. The USGS has in recent years made public a significantly upgraded version of the finite element model SUTRA, now capable of simulating three-dimensional flow (Voss and Provost, 2002). This computer program can simulate both unsaturated and saturated flow, heat and contaminant transport, as well as variable density flow, which makes it a powerful tool for modeling just about any imaginable condition. Unfortunately, unlike Modflow, SUTRA3D is not yet part of any of the most widely used user-friendly commercial programs for processing model input and output data in a user-friendly graphic interface. This severely limits its greater application. For a thorough explanation of finite element and finite difference models, and their various applications, the reader should consult the excellent work by Anderson and Woessner (1992).

Numeric groundwater modeling in some form is now a major part of most projects dealing with groundwater development, protection, and remediation. As computer hardware and software continue to be improved and become more affordable, the role of models in groundwater management will continue to increase accordingly. It is essential, however, that for any groundwater model to be interpreted and used properly, its limitations should be clearly understood. In addition to strictly "technical" limitations, such as accuracy of computations (hardware/software), the following is true for any model:

- It is based on various assumptions regarding the real natural system being modeled.
- Hydrogeologic and hydrologic parameters used by the model are always just an approximation of their actual field distribution which can never be determined with 100 percent accuracy.
- Theoretical differential equations describing groundwater flow are replaced with systems of algebraic equations that are more or less accurate.

It is therefore obvious that a model will have a varying degree of reliability, and that it cannot be "misused" as long as all the limitations involved are clearly stated, the modeling process follows industry-established procedures and standards, and the modeling documentation and any generated reports are transparent, also following the industry standards.

Groundwater models can be used for three general purposes:

- To predict or forecast expected artificial or natural changes in the system (aquifer) studied. The term predict is more appropriately applied to deterministic (numeric) models since it carries a higher degree of certainty, while forecasting is the term used with stochastic time-series models. Predictive models are by far the largest group of models built in hydrogeologic practice.

- To describe the system in order to analyze various assumptions about its nature and dynamics. Descriptive models help to better understand the system and plan future investigations. Although not originally planned as a predictive tool, they often grow to be full predictive models.

- To generate a hypothetical system that will be used to study principles of groundwater flow associated with various general or more specific problems. Generic models are used for training and are often created as part of a new computer code development.

Predictive numeric models are divided into two main groups: (1) models of groundwater flow and (2) models of contaminant fate and transport. The later ones cannot be developed without first solving the groundwater flow field of the system studied; they use the solution of the groundwater flow model as the base for fate and transport calculations. Following are some of the more common questions that fully developed and calibrated groundwater flow, and fate and transport models may help answer (Kresic, 2007a):

- What is the sustainable yield of the aquifer portion targeted for groundwater development?

- At what locations and how many wells are needed to provide a desired flow rate?

- How will current or planned groundwater extraction affect the environment (e.g., on surface streamflows, wetlands)?

- Is there a potential for saltwater intrusion from an increased groundwater pumpage?

- Where is the contaminant flowing to, and/or where is it coming from?

- How long would it take the contaminant to reach potential receptors?

- What would the contaminant concentration be once it reaches a receptor?

- How long would it take to remediate (restore) the contaminated aquifer to its beneficial uses?

Once these questions are addressed by the model(s), many new questions may arise, which is exactly the purpose of a well-documented and calibrated groundwater model: to answer "all kinds" of possible questions related to groundwater flow, and fate and transport of contaminants. Here are just two of the common "big" questions that often come with a multimillion dollar price tag and the possibility of a protracted lawsuit: Who

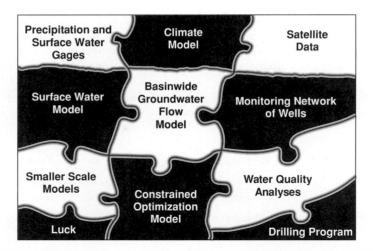

FIGURE 8.39 Example of modeling tools and information used to solve water management puzzle. (Modified from Danskin et al., 2006.)

is responsible for the groundwater contamination? What is the most feasible groundwater remediation option?

Models that are capable of answering most or all of the questions posed by a groundwater management puzzle would be rather complex and require significant investment of funds, resources, and time (Fig. 8.39).

Model Setup

A numeric groundwater model setup consists of the following stages:

- Development of the Conceptual Site Model (CSM), which is the most important part of modeling and the basis for all further, related activities.

- Selection of a computer code that can most effectively simulate the concept and meet the purpose of modeling.

- Definition of the model geometry: lateral and vertical extent of the area of interest, defined by model boundaries, grid layout, and position and number of layers.

- Input of hydrogeologic parameters, and fate and transport parameters when required, for each model cell such as horizontal and vertical hydraulic conductivities including possible anisotropy, storage properties, effective porosity, dispersivity, distribution coefficient, and others.

- Definition of initial conditions, such as an estimated distribution of the hydraulic head in the model domain, and a distribution of contaminant concentration when required (Chap. 2).

- Definition of external model boundary conditions that influence the flux of water, or are directly causing it; this flux of water (and contaminant when required) both enters and leaves the model domain and has to be provided for in the model design (e.g., boundaries with known hydraulic head, known flux, or head-dependent flux; see Chap. 2).

- Definition of external and internal hydraulic stresses acting upon the system, in addition to those assigned along model boundaries, such as aerial recharge, evapotranspiration, well pumpage, outflow through springs, drains, inflow of water from other sources (recharge wells, recharge basins, adjacent aquifers).

After the model has been set up, it is run and then adjusted (calibrated) to match the hydraulic (and chemical/contaminant where required) information collected in the field:

- The hydraulic head measured in monitoring wells and extraction wells
- The flux along model boundaries measured or calculated externally to the model; all such fluxes comprise the water budget of the model
- The contaminant concentration measured at monitoring wells and at model boundaries, in case of fate and transport models

During model development and the analysis of model results, it often becomes apparent that the CSM needs to be revised which includes the collection of more information and possibly additional field investigations. Developing a CSM is the most important part of every modeling effort. It requires a thorough understanding of hydrogeology, hydrology, and the dynamics of groundwater flow in and around the area of interest. Following is an example demonstrating how new investigation techniques and new information can result in a very significant hydrogeologic reinterpretation of an existing conceptual model. As discussed by Phillips et al. (2003), two key distinctions were found between the old and new conceptualization of a thick unconfined alluvial aquifer system in Antelope Valley, CA, leading to improvements of a numeric model. The modeling was part of a comprehensive study of the area where groundwater levels declined more than 200 ft during the twentieth century, resulting in reduced water supplies and more than 6 ft of subsidence.

The first distinction between two conceptual models is the recognition that the age of aquifer materials is a key factor controlling their hydraulic properties. Older materials have longer stress histories (compaction is more likely to have occurred) and are more likely to have undergone chemical cementation. Under the old conceptual model, surficial materials were considered part of the deep aquifer as were materials that started at a depth of about 900 ft below land surface in Lancaster. This configuration did not account for the differences in the age and depth of burial of the aquifer materials or the effects of these differences on hydraulic properties. The second distinction between the conceptual models is a recognition that aquifer properties above the regional aquitard in the study area vary significantly with depth. This includes a significant decrease in the storage, and the percent of total flow entering production wells with depth. Figure 8.40 shows the velocity log generated in 1998 during extraction using a dye-based method (Izbicki et al., 1999), which indicates that most of the water produced from well 7N/12W-27 P2 comes from the upper aquifer. At the water table and within the aquifer volume affected by the drawdown, water entering the well is obtained by gravity drainage of the materials above the new water table. Water entering the well at depth cannot be immediately drawn from the water table because the vertical flow path is partly blocked by numerous overlapping bodies of fine-grained deposits. Instead, the immediate demand for water is met primarily through compaction of these deposits in response to

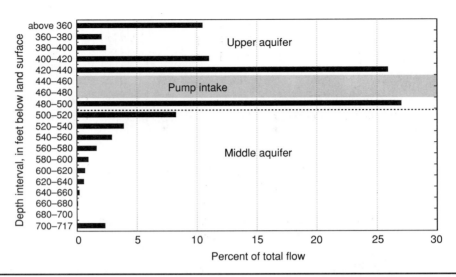

Figure 8.40 Velocity log, during extraction, for Los Angeles County Department of Public Works (LACDPW) well 4–32 (7 N/12 W-27 P2), Antelope Valley, CA. (From Phillips et al., 2003.)

decreased water pressure in pore spaces between grains of sediment. The storage values at depth, therefore, are much smaller than those at the water table. This is confirmed by different responses of the hydraulic head at cluster monitoring wells to water injection and withdrawal tests (Phillips et al., 2003). What all this means is that the thick portion of the alluvial system of the Antelope Valley above the regional aquitard, which was traditionally regarded as an unconfined aquifer, has to be modeled with multiple layers representing confined conditions at depth.

All the information collected during field investigations, monitoring, and desk studies for development of a CSM, should be incorporated into a computerized database, along with the simplified electronic maps and cross sections that will be used in the numeric model design. The most efficient way to organize all the information required for model development is to utilize a GIS environment. This enables all interested stakeholders, including nonmodelers, to provide invaluable input as to the validity of certain hydrogeologic assumptions, spatial information related to contaminant fate and transport, and other aspects of the model (Kresic and Rumbaugh, 2000). Most commercial GIS programs offer free software for viewing and sharing electronic files and maps which is arguably the most efficient way (other than meeting "face-to-face") to quickly exchange visual information. The modern development of groundwater models is a highly visual process, greatly enhanced by various graphical programs which facilitate quick and accurate input of data into models and visualization of model results (such programs are called graphic user interface or GUI programs).

Probably the single most important assurance that the model will be developed in a technically sound manner and efficiently, is the involvement of the "computer modeler" from the very beginning of concept development. Ideally, the leading modeler and the leading hydrogeologist on the project should be the same person since there is no valid excuse why any practicing hydrogeologist would not be intimately knowledgeable in groundwater modeling. Unfortunately, in many cases nonhydrogeologists (or even worse, nongeologists) may end up developing a groundwater model and calibrating it,

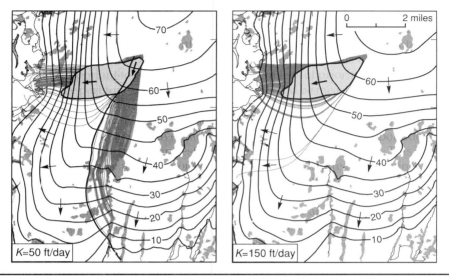

FIGURE 8.41 Simulated effects of increasing horizontal hydraulic conductivity of moraine sediments on the water table configuration and groundwater flowpaths near the Landfill-1 contaminant plume, Western Cape Cod, MA. (Modified from Franke et al., 1998, and Masterson et al., 1997.)

without realizing that some (or many) parts of such model simply do not make hydrogeologic sense. It cannot be emphasized enough that every numeric groundwater model is a nonunique solution of the underlying flow field. In other words, various combinations of various model parameters may produce very similar or identical results. The opposite is also true; what may seem a "slight parameter change" to some, can result in a dramatically different model output. Figure 8.41 shows particle tracking results for two different hydraulic conductivities for the groundwater flow model of the Landfill-1 contaminant plume, Western Cape Cod, MA. Hydraulic conductivity of the moraine sediments is 50 ft/d (left) and 150 ft/d (right). In the 50 ft/day simulation, flowpaths split in two directions, west and south, but predominantly to the south. In the 150 ft/day simulation, although the configuration of the water table changed very little at this scale, virtually all the flowpaths moved to the west and followed the known configuration of the contaminant plume (Franke et al., 1998).

Although some modelers still consider modeling as a process during which a few hydraulic heads measured in the field are matched by the model results, this is the least important part of modeling. It is infinitely more important that the model makes hydrogeologic sense, and that all its uncertainties and (inevitable) errors be fully documented. Then and only then, the model will be hard to misuse and it may be useful to most (if not all) stakeholders that have to make some decisions based on the modeling results.

Model Calibration, Sensitivity Analysis, and Error

The first model run is the fear (or joy) of every model designer. When dealing with a "real-life" model, it is almost certain that the first result will not be a satisfactory match between the calculated and measured hydraulic heads. The number of model runs during calibration will depend on the quantity and quality of available data, desirable accuracy of the model results, and the patience of the user.

Although often explained separately in modeling reports, calibration and sensitivity analysis are inseparable and are part of the same process. While performing calibration, which is composed of numerous single and multiple changes of model parameters, every user determines quickly which parameters are more sensitive to changes with regard to the final model result. By carefully recording all the changes made during calibration and commenting on their results, the model designer is engaged in the sensitivity analysis and can effortlessly finalize this part of the modeling effort later. Calibration is the process of finding a set of boundary conditions, stresses, and hydrogeologic parameters, which produce the result that most closely matches field measurements of hydraulic heads and flows. Calibration of every model should have the target of an acceptable error set beforehand. Its range will depend mainly on the model purpose. For example, a groundwater flow model for evaluation of a regional aquifer system can sometimes "tolerate" a difference between calculated and measured heads of up to several feet. This, however, would be an unacceptable error in the case of a model for the design of containment and cleanup of a contaminant plume spread over, say, 50 acres.

In many instances, the quality of calibration will depend on the amount and reliability of available field data. It is therefore crucial to assess the field data (calibration dataset) for their consistency, homogeneity and measurement error. Such assessment is the basis for setting the calibration target.

Model calibration can be performed for steady-state conditions, transient conditions, or both. Although steady-state calibration has prevailed in modeling practice, every attempt should be made to have a transient calibration as well for the following reasons:

- Groundwater flow is transient by its nature, and is often subject to artificial (man-made) changes.
- The usual purpose of the model is prediction which is by definition time-related.
- Steady-state calibration does not involve aquifer storage properties which are critical for a viable (transient) prediction.

A limited field dataset predetermines the steady-state calibration. In such a case, an appropriate approach would be to define boundary conditions and stresses that are representative for the period in which the field data are collected.

When a transient field dataset of considerable length is available, some meaningful average measure should be derived from it for a steady-state calibration. For example, this can be the mean annual water table elevation or the mean water table for the dry season, the average annual groundwater withdrawal, the mean annual precipitation (recharge), the average baseflow in a surface stream, and so on.

Transient calibration typically involves water levels recorded in wells during pumping tests or long-term aquifer exploitation. An ideal set that incorporates all common relevant boundary conditions and stresses would be

- Monthly water table (hydraulic head) elevations
- Monthly precipitation (recharge)
- Average monthly river stage
- Average monthly groundwater withdrawal

Transient calibration based on monthly values is preferred over daily or weekly data since groundwater systems usually react with a certain delay to surface stresses. In addition, monthly data enable accurate analysis of seasonal influences, which is very important for long-term predictions. Short-term pumping tests may often be the only accurate transient datasets available, which is the main reason why such tests are popular calibration targets.

There are two methods of calibration: (1) trial-and-error ("manual") and (2) automated calibration. Trial-and-error calibration was the first technique applied in groundwater modeling and is still preferred by most users. Although it is heavily influenced by the user's experience, it is always recommended to perform this type of calibration, at least in part. By changing parameter values and analyzing the corresponding effects, the modeler develops a better feeling for the model and the assumptions on which its design is based. During manual calibration boundary conditions, parameter values and stresses are adjusted for each consecutive model run until calculated heads match the preset calibration targets. The first phase of calibration typically ends when there is a good visual match between calculated and measured hydraulic heads at observation wells, as seen in (Fig. 8.42).

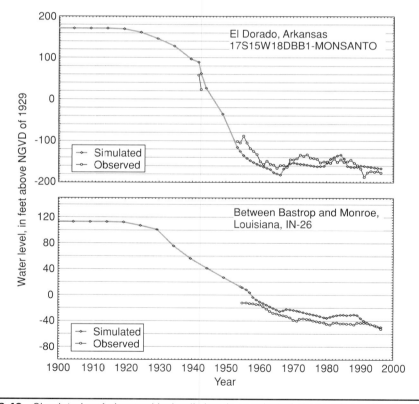

Figure 8.42 Simulated and observed hydraulic heads at selected observation wells; groundwater flow model for the Sparta aquifer of Southeastern Arkansas and North-Central Louisiana. (Modified from McKee and Clark, 2003.)

The next step involves quantification of the model error with various statistical parameters such as standard deviation and distribution of model residuals, i.e., differences between calculated and measured values. Once this error is minimized (through a lengthy process of calibration), and satisfies a preset criterion, the model is ready for predictive use. It will sometimes be necessary to change input values and run the model tens of times before reaching the target. The worst case scenario involves a complete redesign of the model with new geometry, boundaries, and boundary conditions.

During calibration, the user should focus on parameters that are determined with less accuracy or assumed, and change only slightly those parameters that are more certain. For example, hydraulic conductivity determined by several pumping tests should be the last parameter to change "freely" because it is usually the most sensitive. Most other parameters are less sensitive and can be changed only within a certain realistic range: it is obviously not possible to increase precipitation infiltration rate ten times from 10 to 100 percent. In general, hydraulic conductivity and recharge are two parameters with equivalent quality: an increase in hydraulic conductivity creates the same effect as a decrease in recharge. Since different combinations of parameters can yield similar, or even the same results, trial-and-error calibration is not unique. During calibration, it is recommended to plot residuals (measured values minus calculated values) on the model map using different symbols (or colors) for negative and positive values. This allows for a more accurate determination of the parameter value that produces the best overall fit. Another recommended procedure is to plot a graph of model error change versus parameter change as illustrated in Fig. 8.43. This describes parameter sensitivity—more sensitive parameters have steeper slopes than less sensitive parameters. In this example, model error is more sensitive to changes in natural recharge (x_1) than to changes in the

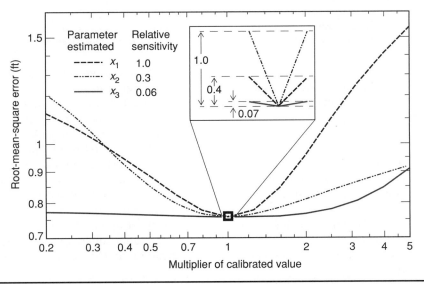

FIGURE 8.43 Comparison of the relative sensitivity of natural recharge (x_1), vertical anisotropy (x_2), and underflow recharge (x_3) to root-mean-square error change as these same parameters are varied independently. (From Halford, 2006.)

vertical anisotropy (x_2) or the underflow recharge rate (x_3). A detailed explanation of the calibration parameters, and a computer program for groundwater flow model calibration and parameter optimization, MODOPTIM, is given by Halford of the USGS (Halford, 2006).

An effective measure of calibration is the analysis of the water budget calculated by the model. The budget provides flows across boundaries, flows to and from all sources and sinks, and flows derived from storage. These calculated values should be compared with measurements and/or estimates made during concept development. Unrealistic components of the water balance should be analyzed in order to calibrate the parameter(s) and condition(s) that are causing them.

It is more difficult to calibrate a contaminant fate and transport model (F&T model) because aquifer heterogeneities and biochemical reactions usually have a much greater effect on contaminant flow pathways and concentrations, compared to "bulk" flow rate of groundwater. However, the process of calibration of F&T models is exactly the same as for groundwater flow models: various F&T parameters are being changed, within reasonable bounds, until a satisfactory match between the field-measured and model-predicted contaminant concentrations is achieved. The most critical parameter that should be "freely" adjusted the last is the rate of biodegradation. Following is the citation from a USEPA publication that speaks to that fact (Azdapor-Keeley et al., 1999):

> Many times during calibration, if a model does not fit observed concentrations, it is assumed that the biodegradation coefficient is the proper parameter to be adjusted. Using biodegradation to adjust a model without supporting field data should not be done until all abiotic mechanisms for reduction are explored. When using a model which incorporates a biodegradation term, care should be taken to verify that assumptions made about degradation rates and the amount and activity of biomass are valid for the site in question. Degradation rates are sensitive to a wide array of field conditions which have been discussed previously. Extrapolation of laboratory derived rates to a site can also lead to significant errors. Likewise, using models to derive degradation rates from limited field data where abiotic variables are not well defined can be misleading. ... Kinetic constants derived from laboratory microcosms or other sites are generally not useful on a wide scale to predict overall removal rates. Site specific degradation rates should be developed and incorporated into a model.

Automated calibration is gaining in popularity since several powerful computer programs are now widely available and are incorporated in most GUIs by default. It is a technique developed in order to minimize uncertainties associated with the user's subjectivity. As with any relatively new approach, it has been criticized, particularly because of its nonuniqueness (see Anderson and Woessner, 1992). However, this is the case with any calibration, including manual, and it is up to the modeler to use it wisely, as an aid, rather than some final solution that has to be accepted because of its "objectivity." Most computer codes for automated calibration search an optimal parameter set for which the sum of squared deviations between calculated and measured values is reduced to a minimum. Two well-known codes for parameter estimation used in groundwater modeling are PEST by Doherty, et al. (1994) and UCODE by Poeter and Hill (1998). Groundwater Vistas (Rumbaugh and Rumbaugh, 2004) also includes an easy-to-use and streamlined parameter optimization code developed specifically for this GUI.

In conclusion, the efficiency of automated calibration codes, coupled with the trial-and-error input from the user, is arguably the most appropriate calibration method. Readers interested in learning more about automated calibration should consult (Hill

1998; "Methods and guidelines for effective model calibration") and (Hill et al. 2000; "User guide to the observation, sensitivity, and parameter-estimation processes and three postprocessing programs").

Quantitative techniques for determining model error compare model results (simulations) to site-specific information, and include calculations of residuals, assessing correlation among the residuals, and plotting residual on maps and graphs (ASTM, 1999). Individual residuals are calculated by subtracting the model-calculated values from the targets (values recorded in the field, *not* extrapolated or otherwise assumed). They are calculated in the same way for hydraulic heads, drawdowns, concentrations, or flows; for example, the hydraulic head residuals are differences between the computed heads and the heads *actually measured in the field*:

$$r_i = h_i - H_i \qquad (8.1)$$

where r_i = residual
H_i = the measured hydraulic head at point i
h_i = computed hydraulic head at the approximate location, where
H_i was measured

If the residual is positive, the computed value was too high; if negative, the computed value was too low (ASTM, 1999).

Residual mean is the arithmetic mean of the residuals computed from a given simulation:

$$R = \frac{\sum_{i=1}^{n} r_i}{n} \qquad (8.2)$$

where R = residual mean and n = number of residuals. Of two simulations, the one with the residual mean closest to zero has a better degree of correspondence, with regard to this criterion, and assuming there is no correlation among residuals (ASTM, 1999). It is possible that large positive and negative residuals could cancel each other, resulting in a small residual mean. For this reason, the residual mean should never be considered alone, but rather always in conjunction with the other quantitative and qualitative comparisons (ASTM, 1999).

The weighted residual mean can be used to account for differing degrees of confidence in the measured heads:

$$R = \frac{\sum_{i=1}^{n} r_i}{n \cdot \sum_{i=1}^{n} w_i} \qquad (8.3)$$

where w_i = weighting factor for the residual at point i. The weighting factors can be based on the modeler's judgment or statistical measures of the variability in the water level measurements. A higher weighting factor should be used for a measurement with a high degree of confidence than for one with a low degree of confidence.

Second-order statistics give measures of the amount of spread of the residuals about the residual mean. The most common second-order statistics is the standard deviation of residuals:

$$s = \left\{ \frac{\sum_{i=1}^{n} (r_i - R)^2}{n - 1} \right\}^{\frac{1}{2}} \tag{8.4}$$

where s = standard deviation of residuals and R is given with Eq. (8.2). Smaller values of the standard deviation indicate better degrees of correspondence than larger values (ASTM, 1999).

Correlation among residuals—Spatial or temporal correlation among residuals can indicate systematic trends or bias in the model. Correlation among residuals can be identified through listings, scattergrams, and spatial and temporal plots. Of two simulations, the one with less correlation among residuals has a better degree of correspondence, with regard to this criterion (ASTM, 1999). *Spatial correlation* is evaluated by plotting residuals, with their sign (negative or positive) on a site map or cross sections. If applicable, the residuals can also be contoured. Apparent trends or spatial correlations in the residuals may indicate a need to refine aquifer parameters or boundary conditions, or even to reevaluate the conceptual site model. For example, if all the residuals in the vicinity of a no-flow boundary are positive, then the recharge may need to be reduced or the hydraulic conductivity increased (ASTM, 1999). For transient simulations, a plot of residuals at a single point versus time may identify temporal trends. Temporal correlations in residuals can indicate the need to refine input aquifer storage properties or initial conditions (ASTM, 1999).

Figure 8.44 shows a mandatory graph of calculated versus actually measured heads at monitoring wells. Only field-measured hydraulic heads, not those estimated (interpolated) by the user for the purposes of creating the initial CSM hydraulic head, can be used to plot such graph. If there were no calculation error for any of the control monitoring wells, all data would fall on the straight 1:1 ratio line, which can never happen (even if one were to engage in creating a nice-looking southwestern rug of hydraulic conductivity). Deviations of points or clusters of points, such as several monitoring wells in the same portion of the aquifer, from this line can reveal certain patterns and point toward the need for additional calibration and/or adjustment of the conceptual site model (CSM).

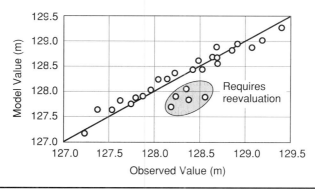

FIGURE 8.44 Model-calculated versus measured hydraulic heads at calibration targets (monitoring wells).

Model Documentation and Modeling Standards

Preparing model documentation and a report is the final phase of the modeling effort and arguably the most important from the client's standpoint. A poorly documented and confusing report can ruin days of work and an otherwise excellent model. Every effort should be made to produce an attractive and user-friendly document that will convey clearly all previous phases of the model design. Special attention should be paid to clearly state the model's limitations and uncertainties associated with calibrated parameters. Electronic modeling files (model input and output) and GIS files, if applicable, will have to be made available to the client in most cases. All modeling documentation should strictly follow widely accepted industry practices, guidelines, and standards for groundwater modeling as detailed in the widely accepted modeling standards.

The following industry standards, created by leading industry experts for the groundwater modeling community under the auspices of ASTM (American Society for Testing and Materials) cover all major aspects of groundwater modeling and should be followed when attempting to create a defensible groundwater model that can be used for predictive purposes:

- Guide for application of groundwater flow model to a site-specific problem (D 5447-93)
- Guide for comparing groundwater flow model simulations to site-specific information (D 5490-93)
- Guide for defining boundary conditions in groundwater flow modeling (D 5609-94)
- Guide for defining initial conditions in groundwater flow modeling (D 5610-94)
- Guide for conducting a sensitivity analysis for a groundwater flow model application (D 5611-94)
- Guide for documenting a groundwater flow model application (D 5718-95)
- Guide for subsurface flow and transport modeling (D 5880-95)
- Guide for calibrating a groundwater flow model application (D 5981-96)
- Practice for evaluating mathematical models for the environmental fate of chemicals (E 978–92)
- Guide for developing conceptual site models for contaminated sites (E 1689–95)

The following language accompanies the USEPA OSWER Directive #9029.00 entitled *"Assessment framework for ground-water model applications"* (USEPA, 1994): "The purpose of this guidance is to promote the appropriate use of ground-water models in EPA's waste management programs." More specifically, the objectives of the framework are to

- Support the use of ground-water models as tools for aiding decision making under conditions of uncertainty
- Guide current or future modeling
- Assess modeling activities and thought processes
- Identify model application documentation needs

Following is the introduction to *"Guidelines for Evaluating Ground-Water Flow Models"* published by the U.S. Geological Survey (Reilly and Harbaugh, 2004):

Ground-water flow modeling is an important tool frequently used in studies of ground-water systems. Reviewers and users of these studies have a need to evaluate the accuracy or reasonableness of the ground-water flow model. This report provides some guidelines and discussion on how to evaluate complex ground-water flow models used in the investigation of ground-water systems. A consistent thread throughout these guidelines is that the objectives of the study must be specified to allow the adequacy of the model to be evaluated.

Modeling Saltwater Intrusion

Saltwater intrusion (or seawater intrusion) is the encroachment of saline waters into zones previously occupied by fresh groundwater. Under stable natural conditions, hydraulic gradients of fresh groundwater in coastal aquifers are toward the sea, forming an interface between the discharging groundwater and seawater as illustrated in Figure 8.45. Persistent disturbances, however, such as groundwater extraction, can produce movements in the position of the seawater-freshwater interface, which can lead to the degradation of freshwater resources. The effects of seawater intrusion are widespread, and have led to significant losses in potable water supplies and in agricultural production (e.g., Barlow, 2003; Johnson and Whitaker, 2003; FAO, 1997). Superimposed on these local anthropogenic influences are the effects of a continuing global rise in sea water level, which may accelerate as projected by various climate change scenarios. The combined effects of local and global sea water intrusion will have serious negative consequences for water supplies in many coastal areas worldwide.

Seawater intrusion is typically a complex three-dimensional phenomenon influenced by the heterogeneous nature of coastal sediments, the spatial variability of coastal aquifer geometry and the distribution of extraction wells. It is therefore not possible to apply simple analytical equations to solving real-world problems, such as prediction of the effects of sea-level rise on underlying fresh groundwater (thus a question mark in Fig. 8.45). The effective management of coastal groundwater systems requires an understanding of the specific seawater intrusion mechanisms leading to salinity changes, including landward movements of seawater, vertical freshwater-seawater interface rise, or "upconing," and

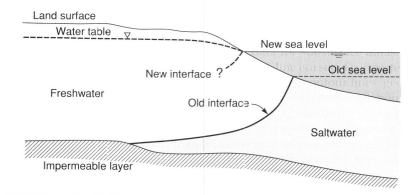

FIGURE 8.45   Schematic sketch of saltwater intrusion, sharp interface model. (Modified from Larabi, 2007)

the transfer of seawater across aquitards of multiaquifer systems (Simmons et al., 2007). Other processes, such as relic seawater mobilization, salt spray, atmospheric deposition, irrigation return flows, and water-rock interactions may also contribute to coastal aquifer salinity behavior, and need to be accounted for in water resource planning and operation studies (e.g., Werner and Gallagher, 2006; 2003; FAO, 1997). Variable density groundwater flow and solute transport modeling is arguably the only feasible tool for assessing the effects of the above listed processes.

As discussed by Simmons et al. (2007), when embarking on groundwater modeling in coastal areas, it is essential to realize that the textbook conception of fresh and saline groundwater distribution, which is classically conceived of as a freshwater lens overlying a wedge of saline groundwater, is seldom encountered in real field settings due to the dynamic nature of shorelines. The most conspicuous manifestations of transient effects are offshore occurrences of fresh groundwater and onshore occurrences of saltwater. In many instances, these waters occur too far offshore to be explained by active subsea outflow of freshwater due to topographic drive. Moreover, lowest salinities often occur at substantial depths beneath the seafloor and are overlain by more saline pore waters, suggesting absence of discharge pathways. These waters therefore are considered paleogroundwaters that were emplaced during glacial periods with low sea level. During subsequent periods of sea-level rise, salinization was apparently slow enough to allow relics of these freshwaters to be retained.

For proper predictive modeling of the effects of sea-level rise on groundwater resources, it is very important to understand that in many flat coastal and delta areas, the coastline during the recent geologic past was further inland than it is today. As a result, vast quantities of saline water were retained in the subsurface after the sea level retreated. Such occurrences of saline groundwater are sometimes erroneously attributed to seawater intrusion, i.e., the inland movement of seawater due to aquifer overexploitation. Effective water resource management requires proper understanding of the various forcing functions on groundwater salinity distribution on a geological timescale (Kooi and Groen, 2003). High salinities are maintained for centuries to millennia, or sometimes even longer, when the presence of low-permeability deposits prevents flushing by meteoric water (e.g., Groen et al., 2000; Yechieli et al., 2001). Rapid salinization due to convective sinking of seawater plumes occurs when the transgression is over a high-permeability substrate, as illustrated in Fig. 8.46. This process is responsible for the occurrence of saline groundwater up to depths of 400 m in the coastal area of the Netherlands (Post and Kooi, 2003).

The large spatial and time scales involved in modeling the effects of seawater intrusion pose special challenges. In particular, the high resolution model grid required to capture convective flow features imposes a severe computational burden that limits the size of the model domain. Other complications include the lack of information on boundary conditions, insufficient data for proper parameterization, especially for the offshore domain, and unresolved numerical issues with variable-density codes. Resolving these issues represents a continuing challenge for groundwater professionals.

The USGS finite-element, public domain model SUTRA (Voss and Provost, 2002), is one of the most widely applied simulators of seawater intrusion and other density-dependent groundwater flow and transport problems. Other popular computer programs capable of simulating seawater intrusion include FEFLOW (Diersch, 2005), FEMWATER (Lin et al., 1997) and MODHMS (HydroGeoLogic Inc., 2003). SUTRA has been applied to a wide range of seawater intrusion problems, ranging from regional-scale

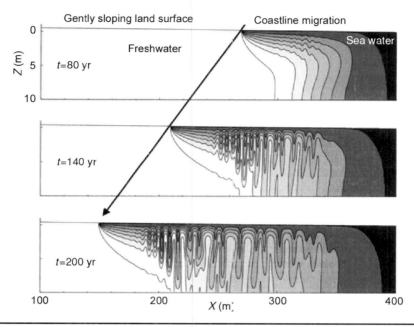

FIGURE 8.46 Variable density flow simulation showing salinization lags behind coastline migration during transgression on a gently sloping surface and development of offshore brackish groundwater. Highly unstable convective fingering is seen as dominant vertical salinization mechanism. (After Kooi et al., 2000; copyright American Geophysical Union).

assessments of submarine groundwater discharge (e.g., Shibuo et al., 2006) to riparian-scale studies of estuarine seawater intrusion under tidal forcing effects (e.g., Werner and Lockington, 2006). Gingerich and Voss (2005) demonstrate the application of SUTRA to the Pearl Harbour aquifer, southern Oahu, Hawaii, in analyzing the historical behavior of the seawater front during 100 years of pumping history.

Optimization

Optimization in general is defined as a process of finding a result that meets desired objectives subject to specified constraints which have to be strictly (completely) satisfied. Groundwater models provide the best tool for performing optimization because they can run a large number of scenarios with different input parameters enabling efficient comparison of results. An example of manual modeling optimization using one objective (desired pumping rate) and two constraints (chloride concentration and percent of aquifer recharge) is illustrated in Fig. 8.47. A groundwater model was used to study the sustainability of groundwater extraction on the island of Weizhou, off the South China coast, where an increase of pumping for the extension of tourism is planned. Modeling has determined that about one-third of the available recharge is required for the additional water supply. Because of the unfavorable hydraulic conductivity of the shallow strata, the freshwater lens on Weizhou has to be tapped in a deeper aquifer layer. If the pumping is concentrated in one well, within a few years, the salinity of the pumped water becomes unacceptable due to saltwater upconing. If the pumping is distributed over two wells at a 4-km distance, an acceptable final salinity is reached. If the pumping

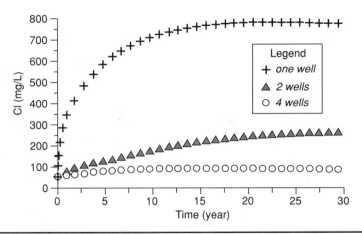

FIGURE 8.47 Development of chloride concentration in the pumping wells for different configurations of wells with the same total extraction rate. (From Kinzelbach et al., 2007.)

rate is distributed over four wells sitting on a square with a side length of 1 km, the chloride concentration stays below 200 mg/L all the time (Fig. 8.47). The model in this case helps to find a technical solution to the problem, as the total pumping rate stays well below the total recharge. However, as advised by the authors of the modeling study, "If the pumping rate is higher than total recharge the best model in the world cannot find a pumping strategy that avoids salt water intrusion" (Kinzelbach et al., 2007).

When approached in a qualitative way, the general definition of optimization leaves room for argument as to the degree of fulfillment of the objectives: "Is this the best we can get?" or "This is good enough!" This argument is illustrated by a common situation from groundwater remediation projects involving containment of plumes with pump-and-treat systems. The modeler on such a project can use a groundwater flow model to decide how many wells should be installed for the purpose, at what pumping rates, and at which locations. He/she may run the model two or twenty times before reaching the conclusion of which modeled scenario is the best one. Depending on the experience and the "intuition" of the modeler, the final selection may indeed be an optimum one for that particular situation. However, no one on the project team, or the client, would have an independent, objective measure of the final selection and how it compares to other scenarios that were not selected. There may also be other important factors for designing an optimum remediation system that were not considered by the modeler because he or she was not asked to consider them, such as energy cost for pumping, and water treatment cost.

As opposed to manual optimization, water management objectives and constraints are expressed in the constrained optimization model as mathematical equations. The objective function (equation) defines a specific objective that is to be maximized or min-imized subject to a set of constraint equations. To answer a slightly different water man-agement question, a slightly different objective function can be combined with the same or nearly the same set of constraint equations. Commonly, reformulation of the opti-mization model to answer a related water management question requires only that a specific constraint equation be used as the objective function, and the former objective function be included as a constraint. This capability of slight, but powerful, modifications makes optimization techniques an efficient way to investigate related water management questions. A slightly different formulation of the optimization model not only provides additional insight about overall water management, but also can be used to represent the specific viewpoint of a different water management entity (Danskin et al., 2006).

An example of a mathematical formulation of objectives and constraints used in the conjunctive-use optimization model for a portion of the Mississippi River Valley alluvial aquifer is given below (Czarnecki, 2007). "Conjunctive use" pertains to use of both groundwater and surface water. The model was used to calculate the maximum sustainable yield from wells and rivers where "sustainable yield" is defined as the maximum rate at which water can be withdrawn indefinitely from groundwater and surface water sources without violating specified constraints.

The optimization model was formulated as a linear programming problem with the objective of maximizing water production from wells and from streams subject to (1) maintaining groundwater levels at or above specified levels; (2) maintaining streamflow at or above minimum specified rates; and (3) limiting groundwater withdrawals to a maximum rate of the amount withdrawn in 1997. Steady-state conditions were selected (rather than transient conditions) because the maximized withdrawals are intended to represent sustainable yield of the system (a rate that can be maintained indefinitely without violating constraints). In this model, the decision variables are the withdrawal rates at 9979 model cells corresponding to well locations and at 1165 model cells which correspond to river locations.

The objective of the optimization model is to maximize water production from groundwater and surface water sources. The objective function of the optimization model has the form:

$$\text{maximize } z = \sum q_{\text{well}} + \sum q_{\text{river}} \tag{8.5}$$

where z = total managed water withdrawal (L^3/T, where L and T are dimensions of length and time, respectively)

$\sum q_{\text{well}}$ = sum of groundwater withdrawal rates from all managed wells (L^3/T)

$\sum q_{\text{river}}$ = sum of surface water withdrawal rates from all managed river reaches (L^3/T)

Equation (8.6) is computed such that the following constraints are maintained:

$$h_c \geq h_{\text{min}} \tag{8.6}$$

where h_c = hydraulic head (water-level altitude) at constraint location c (L) and h_{min} = water-level altitude at half the thickness of the aquifer (L). This constraint accommodates the critical regulatory criteria that water levels within the alluvial aquifer should remain above half the original saturated thickness of the aquifer.

Streamflow constraints for several rivers specified in the optimization model are based on 7-day, 10-year-recurrence low-flow data (7Q10). Streamflow constraints are specified as the minimum amount of flow required at individual river cells. The equation governing the relation between streamflow constraints and flow into and out of a stream is

$$q_{\text{head}}^R + \sum q_{\text{overland}}^R + \sum q_{\text{groundwater}}^R - \sum q_{\text{diversions}}^R - \sum q_{\text{river}}^R \geq q_{\text{min}}^R \qquad (8.7)$$

where
q_{head}^R = flow rate into the head of stream reach $R\,(L^3/T)$
$\Sigma q_{\text{overland}}^R$ = sum of all overland and tributary flow to stream reach R
$\Sigma q_{\text{groundwater}}^R$ = sum of all groundwater flow to or from stream reach R
$\Sigma q_{\text{diversions}}^R$ = sum of all surface water diversions from stream reach R
$\Sigma q_{\text{river}}^R$ = sum of all potential withdrawals, not including diversions, from stream reach R
q_{min}^R = minimum permissible surface water flow rate for stream reach R

Sustainable yield from wells was compared for four different management scenarios involving different groundwater level constraints and river-withdrawal specifications in 11 rivers, including the Mississippi River. A systematic relaxation of groundwater level constraints and removal of optimized river-withdrawal specifications resulted in up to 25.3 percent larger sustainable yield from groundwater. However, in all tested scenarios, sustainable yield from wells was less than 1997 withdrawal rates. Withdrawals from rivers represent a potential source of water that could be used to offset the unmet demand for groundwater (Czarnecki, 2007).

An example of a very complex constrained optimization model is the San Bernardino Basin, CA, area model developed in support of integrated water management (Danskin et al, 2006). The mathematically optimal value of the decision variables in the model is constrained by various equations representing water supply and water distribution constraints. These constraints assure that adequate water is supplied through the present distribution system and that the quantities determined by the optimization model are physically possible. For example, artificial recharge in each basin must be less than or equal to the maximum recharge capacity of that basin, and the sum of artificial recharge in all basins must be less than or equal to the total quantity of water that is available from the State Water Project. Water supply to each artificial-recharge basin also must be less than the capacity of the conveyance structures connecting the California Aqueduct to the basin.

Maximum pumpage from each individual site is restricted by well, pump, and aquifer characteristics, and total pumpage from all sites is restricted by a maximum value derived from legal adjudication or from an evaluation by local water managers based on distribution capabilities or on anticipated demand. Management of recharge and pumpage in the San Bernardino area is also constrained by requirements on groundwater levels. In the vicinity of the former marshland, groundwater levels need to be sufficiently low to prevent possible liquefaction and sufficiently high to prevent additional land subsidence. In the alluvial fan areas, groundwater levels need to be maintained sufficiently high to assure a continuous supply of groundwater to nearby wells.

The total number of constraint equations increases rapidly with various parameters involved. For example, a problem with 5 recharge sites, 15 well sites, 50 observation locations, and 32 modeling time periods can require more than 5000 constraint equations. Although optimization techniques are designed to address large problems, an optimization model with more than several thousand constraint equations can be cumbersome to

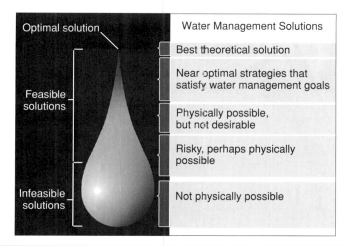

FIGURE 8.48 Water management solutions. (From Danskin et al., 2006.)

work with, and the results can be time-consuming to interpret. However, as in any modeling process, even initial formulation of the optimization model can be insightful. The formulation step requires choosing which components are most important. Objectives and constraints must be defined precisely and quantitatively—a process that commonly is more difficult and time-consuming than it first appears. Initial use of the optimization model can provide immediate insights by defining the feasibility space of potential solutions and by determining whether specific proposed operational plans are even feasible. Finally, if the optimization process results in greater hydrologic insight for the water managers and an improved solution to a water management problem (Fig. 8.48), then the optimization model will have been useful (Danskin et al., 2006).

The USGS has developed a public domain, Ground-Water Management Process (GWM) computer program for MODFLOW-2000 (Ahlfeld et al., 2005). GWM uses a response-matrix approach to solve several types of linear, nonlinear, and mixed-binary linear groundwater management formulations. Each management formulation consists of a set of decision variables, an objective function, and a set of constraints. Three types of decision variables are supported by GWM: flow-rate decision variables, which are withdrawal or injection rates at well sites; external decision variables, which are sources or sinks of water that are external to the flow model and do not directly affect the state variables of the simulated groundwater system (heads, streamflows, and so forth); and binary variables, which have values of 0 or 1 and are used to define the status of flow-rate or external decision variables. Flow-rate decision variables can represent wells that extend over one or more model cells and be active during one or more model stress periods; external variables also can be active during one or more stress periods. A single objective function is supported by GWM, which can be specified to either minimize or maximize the weighted sum of the three types of decision variables. Four types of constraints can be specified in a GWM formulation: upper and lower bounds on the flow-rate and external decision variables; linear summations of the three types of decision variables; hydraulic-head based constraints, including drawdowns, head differences, and head gradients; and streamflow and streamflow-depletion constraints.

8.8.2 Time Series Models

This section is courtesy of Ivana Gabric, School of Civil Engineering, University of Split, Split, Croatia, and Neven Kresic.

A time series, a common term in hydrologic studies, is a series of a time-dependent hydrologic variable such as the flow rate in a surface stream or at a spring. When analyzing a time series, one deals with a limited amount of recorded data—a sample. This sample, regardless of its size, consists of a limited number of realizations of the same hydrologic process. All possible realizations of that process constitute a population. The goal of most hydrologic and hydrogeologic studies is to understand and quantitatively describe the population, as well as the process that generates it, based on a limited number of samples (actual field measurements of limited duration).

A time series can be continuous (such as the flow rate in a perennial stream) or discrete (such as daily precipitation). For practical and computational purposes, most continuous time series are converted into discrete time series by introducing the recording (or modeling) time interval such as 1 day, 1 week, 1 month. When a time series is described with statistical and probabilistic parameters, it represents a probability of occurrence (realization) of one of its possible stages. A good example is a time series of monthly precipitation at a certain location in a moderate climate. Our long-term experience can tell us that, for example, April through June is the wet period, and July through September is the dry period of the year. Accordingly, it can be expected, with a high probability, that in the near future (say, next year), these two periods will again last about the same time. However, no one can state with 100 percent accuracy that this will indeed happen (for example, June may be an unusually dry month next year) because it is impossible to accurately predict the annual or monthly amount of precipitation using some physical laws of nature. One can only apply tools of statistics and make predictions about the future using probabilistic models based on past data. A time series studied in this way is called a stochastic time series. In contrast, the stage of a deterministic process at time t is defined with certainty knowing its stage at some earlier time t_0. In other words, a deterministic process is described with physical laws rather than laws of probability. An example is the flow of groundwater from point A to point B when described with equations such as the Dupuit equation, the Laplace equation or the Theis equation, to name just a few. Quantitative hydrogeology is based on the physical laws of groundwater flow as are the traditional numeric models presented earlier.

Strictly speaking, most time series in hydrologic and hydrogeologic studies are stochastic since they depend on at least one random variable, with precipitation often being the most important one. This also means that the result of a deterministic calculation, such as the drawdown in a well after 1 year of pumping, although given explicitly is actually just more or less probable.

In general, a time series has the following five components, all of which may or may not be present (adapted from McCuen and Snyder, 1986):

- *Trend*, which is a tendency to increase or decrease continuously for an extended period of time in a systematic manner. This component can often be described by fitting a functional form such as a line or polynomial. The coefficients of the equation are commonly evaluated using regression analysis. The trend is also referred to as a deterministic component even though its physical explanation may not always be clear.

- *Periodicity*, which is very common in hydrologic time series: annual and seasonal periodicity of precipitation, temperature, flows. The period(s) in time series can be identified using a moving-average analysis, an autocorrelation analysis, or a spectral analysis, after which it is described by one or more trigonometric functions.

- *Cycle*, which occurs with an irregular period and is hard to detect (for example, hydrometeorological time series are thought to be influenced by sunspot activity which has an irregular period).

- *Episodic variation*, which results from extremely rare or one-time events such as hurricanes. Identification of this component requires supplementary information.

- *Random fluctuations*, which are often a dominant source of the variation in time series and are the main target of a probabilistic identification.

Stochastic models describe time series formally and do not consider their physical nature. Simply stated, they statistically (mathematically) analyze the past of the time series, as system input, and then predict the present or the future as the system output. They can also analyze the past of one time series and use it to predict the present and future of some other time-dependent series proven to be correlated with first one. Stochastic models can also combine several inputs and give one or several outputs. Examples would be a model that predicts water table elevation based on its position in the past, a model that predicts water table based on its own past and the antecedent precipitation, or a model that includes past stages of a nearby river as well.

Two main applications of the time series models are the generation of synthetic samples, and forecasting of hydrologic events. Generated time series, which are statistically indistinguishable from historic time series, serve as input to the analysis of complex water resources systems. They can also be used to provide a probabilistic framework for analyses and design. Generated series show many possible hydrologic conditions that do not explicitly appear in the historic record. Consequently, using synthetic time series, different designs and operational schemes can be tested under many different conditions contained in these time series. Forecasted data from known historic observations can help in evaluating options for a real-time system operation.

Time series modeling originated from different scientific fields, but it has subsequently become very important in stochastic hydrology and the applications of generated time series are numerous. Development of stochastic modeling in hydrology began at the beginning of 1960s when time series analyses of hydrologic phenomena was extended to the synthetic generation of streamflow by using a table of normal random numbers. Thomas and Fiering (1962) were the first to propose a first-order Markov model to generate streamflow data. The classic book on time series analysis by Box and Jenkins (1976) presents the foundation of hydrologic stochastic modeling.

The general form of an input-output stochastic model of a discrete time series (or a continuous time series transformed into a discrete one) with the same recording time interval is

$$y_t = f(x_t, \ x_{t-1}, \ x_{t-2}, \ldots; y_{t-1}, y_{t-2}, \ldots; \theta_1, \theta_2, \ldots) + \varepsilon_t \tag{8.8}$$

where
f = selected mathematical function

y_t = predicted output at time t

$y_{t-1}, y_{t-2} \ldots$ = successive members of the output time series recorded at corresponding time intervals $t-1$ and $t-2$

$x_t, x_{t-1}, t_{t-2} \ldots$ = successive members of the input time series recorded at time intervals $t, t-1, t-2$

$\theta_1, \theta_2, \ldots$ = model parameters found by mathematically minimizing the differences between estimated (calculated) and observed y_t values

ε_t = model error (residual) given as the difference between the calculated and the recorded value of the output series at time t

Stochastic modeling generally follows the approach proposed by Box and Jenkins (1976), who introduced autoregressive moving average (ARMA) models. The mathematical formulation of ARMA models is

$$z_t = \sum_{j=1}^{p} \phi_j z_{t-j} + \sum_{j=0}^{q} \theta_j \varepsilon_{t-j} + \varepsilon_t$$

where
z_t = time-dependent series with mean zero and variance one

$\theta_1 \ldots \theta_p$ = time varying autoregressive coefficients

$\theta_0 \ldots \theta_q$ = time varying moving average coefficients

ε_t = an independent normal variable

Time series models used to generate synthetic time series can be classified into autoregressive models (AR(p)), moving average models (MA(q)), and their combination, autoregressive moving average (ARMA(p, q)) with variations such as autoregressive integrated moving average models ARIMA (p, d, q) and others, where p and q are the orders of autoregressive and moving average terms, respectively, and d is the differentiation order. An autoregressive model estimate values for the dependent variable Z_t as a regression function of previous values $Z_{t-1}, Z_{t-2}, \ldots, Z_{t-n}$. A moving average model is conceptually a linear regression of the current value of the series against the white noise or random shocks of one or more prior values of the series. Pure autoregressive (AR) model, commonly called Thomas-Fiering model, have been extensively applied in hydrology for modeling annual and periodic hydrologic time series. Because the parsimony ("the less the better") in the number of parameters is very desirable (since the parameters are estimated from data), the second order of these models is usually the highest lag necessary in representing hydrologic time series. A parsimonious model can be achieved using a mixed ARMA model as combination of a moving average process and an autoregressive process rather than a pure AR or MA model. Therefore, low-order ARIMA models have been widely used in hydrological practice (Salas et al. 1982; Padilla et al. 1996; Montanari et al. 2000).

An important aspect of stochastic modeling is the problem of nonstationarity in hydrologic time series. Stationarity is usually assumed when modeling annual time series. When dealing with monthly or weekly time-series, seasonal nonstationarity is present and it may be necessary to use a model that has seasonally varying properties. Significant contributions in developing periodic models were made by Hirsch (1979) and Salas et al. (1985). For modeling seasonal time series two approaches can be used. The first is a direct approach in which a model with periodic parameters is fitted directly to the seasonal flows. This method requires a considerable number of years of data. The number of coefficients involved can be very large. If available historical data are limited, the parameters are poorly estimated. Consequently, the main problem of all seasonal models with time varying coefficients is the lack of parsimony. The second approach is decomposition ("disaggregation") in which the seasonal flows are generated at two or more levels. For instance, the first level is modeling and generating annual flows and the second level is their decomposition into seasonal flows based on a linear model. However, if the autocorrelation structure of a historical time series shows a significant periodicity, then seasonal models that explicitly incorporate a periodic structure must be used. If the seasonality of time series under consideration is in the mean and the variance, then such seasonality can be removed by simple seasonal standardization, and a stationary model can be applied. Another peculiarity of hydrologic processes is the skewed distribution functions observed in most cases. Therefore, attempts have been made to adapt standard models to enable treatment of skewness (Bras and Rodriguez-Iturbe, 1994).

A generalized framework for a time series model development consist of three phases: (1) identification, (2) parameter estimation, and (3) verification/diagnostic checking. Model identification is not a standardized, automated procedure but it is rather heuristic. The usual approach is an iterative trial-and-error procedure. The first step is to investigate if the time series data is stationary and if there is any significant seasonality that needs to be modeled. A visual inspection of the time plot of the historic times series can help in deciding between seasonal and nonseasonal models, whether local differentiation is needed to produce stationarity, and to get a general feeling about the order of possible models. A further identification process is to examine the shapes of autocorrelation and the partial autocorrelation functions of the historic time series. To allow for the possible identification errors, a set of several models with a close structure are considered. It is advisable to always select the simplest acceptable model.

When the model order is selected, the estimation of parameters follows. The parameters are estimated from recorded data by either a method of moments or by methods of maximum likelihood. The final stage of modeling is verification as to what extent the selected historic statistics are reproduced by the model and to prove the adequacy of the model. The verification involves a check of possible overfitting (the confidence limits of the parameters), and a check of randomness of the residuals (the ACF function for the residuals resulting from a good ARIMA model should have statistically insignificant autocorrelation coefficients). Sometimes a sufficient objective for simulation purposes, and adequate for short term forecasting, is to preserve the first- and second-order moments of the time series. When comparing several possible models, the one with the best goodness of fit is selected based on the minimum Akaike Information Criterion (AIC) (Akaike, 1974).

When the selected mathematical function f in Eq. (8.8) ignores any physical laws that govern the transformation of input(s) into output(s), the model is a pure stochastic one. If, in any form, the mathematical function incorporates physical laws, the model is called a stochastic-conceptual model. The knowledge of various physical processes and relationships related to the system of interest is invaluable and provides the physical background for stochastic modeling (Klemes, 1978; Vecchia et al., 1983; Koch, 1985; Salas and Obeysekera, 1992; Knotteres and Bierkens, 2000; Lee and Lee, 2000). It is always preferable to conduct detailed structural and physical analyses of the hydrologic processes involved before performing stochastic modeling.

There are many possible uses of time series models. Typical applications of generated time series in surface water engineering are reservoir design, risk and reliability assessment, planning of hydropower production, and flood and drought hazard analysis. In groundwater studies, stochastic models can be used to analyze and forecast the hydraulic head fluctuations, fill-in data gaps, and detect and quantify trends. (Houston, 1983; Padilla et al., 1996; Ahn, 2000; Knotteres and Bierkens, 2000; Birkens et al., 2001; Kim et al., 2005). Particularly common are time series models of groundwater level fluctuation. They use groundwater level observations and incorporate factors that influence groundwater level, such as precipitation, evapotranspiration, and anthropogenic hydrologic disturbances.

A very important factor that limits wider use of stochastic models in hydrogeology is the lack of recorded data. Since these models are based on statistical and probabilistic calculations, very short time series do not allow for meaningful derivation of model parameters. Groundwater levels measured for a couple of years on a quarterly basis are obviously not good candidates for stochastic modeling. On the other hand, if an appropriate amount of data is available, every attempt should be made to develop one or several stochastic models of the input-output type. This is mainly because the process of building even the simplest stochastic model reveals a great deal of information on the possible structure(s) of the system, and connections between various hydrologic variables (Kresic, 1995, 1997).

Case Study: Jadro Spring, the City of Split, Croatia
The information presented in this case study is courtesy of Ivana Gabric, School of Civil Engineering, University of Split, Split, Croatia.

The following example illustrated with Figs. 8.49 through 8.51 shows the application of synthetic hydrologic time series in karst water resources management. Jadro karst

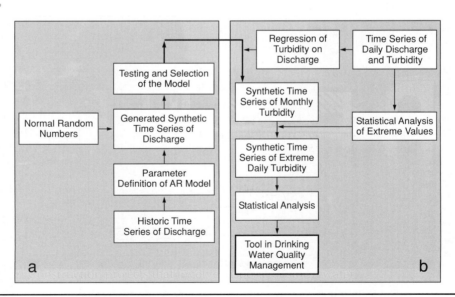

FIGURE 8.49 Building steps of the stochastic time series model of daily turbidity at Jadro karst spring. (Figure courtesy of Ivana Gabric, University of Split.)

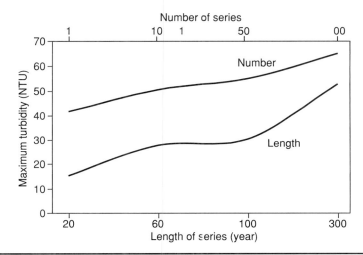

FIGURE 8.50 Maximum values of generated turbidity at Jadro spring versus number and length of generated time series. (Courtesy of Ivana Gabric, University of Split.)

spring (see Fig. 2.102 in Chap. 2), with an average discharge rate of 9.82 m³/s, provides water supply for the city of Split and its 270,000 inhabitants. The main characteristic of the spring discharge are its large fluctuations in response to precipitation. During high discharge rates, there is more intense washing of the soil and sediment accumulated in the subsurface, resulting in sudden, short-lived changes in water quality. Consequently, the spring water is characterized by the occasional occurrence of high turbidity exceeding allowable standards. Turbidity is the key problem in water quality management of this

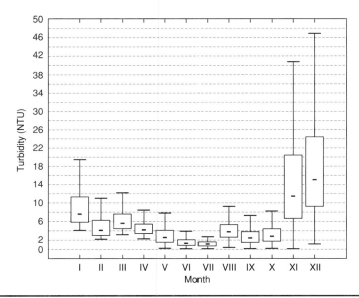

FIGURE 8.51 Box-whisker diagram of generated daily values of turbidity, without extremes and outliers. (Courtesy of Ivana Gabric, University of Split.)

spring as well as many other karst springs. For management purposes, it is important to know the nature of the turbidity and predict its occurrence as early as possible, because elevated turbidity is often associated with high bacteria counts indicating possible contamination. An accurate prediction of elevated turbidity can help optimize sampling strategies. In the case of Jadro spring, the turbidity monitoring was not systematic, leaving insufficient information for a reliable water supply system management. A stochastic time-series modeling can therefore provide for a more comprehensive understanding of the turbidity, using available short-term measurements of the turbidity and long-term recording of the spring discharge (Margeta and Fistanic, 2004; Rubinic and Fistanic, 2005).

Time series analyses of discharge and turbidity showed that the turbidity is higher during the first large rainfall following a dry period. Similar discharge rates following the first rainy period produce generally lower turbidities. Consequently, a reliable prognosis of turbidity can be carried out using different regression functions of discharge and turbidity for different parts of the year. A stochastic model was built based on 3 years of daily measurements of turbidity and discharge, and 28 years of daily discharge measurements, as illustrated in Figure 8.49.

Block A presents identification, estimation, and verification phases of the Jadro spring stochastic modeling. Synthetic time series of monthly spring discharge were generated using a Thomas-Fiering AR(2) model with constant coefficients, which provided the best preservation of historic statistical characteristics of the spring discharge—seasonal means, variances, and correlations of the processes. Block B describes the procedure of the turbidity time series generation. Based on the functional dependence between daily turbidity and discharge, the generated time series of the average monthly discharge (box A) are used to generate up to 100 time series of the average monthly turbidity. The time series of maximum daily turbidity are then generated using the time series of the average monthly turbidity.

Figure 8.50 shows the maximum daily values of the generated turbidity at Jadro spring as a function of the number and length of the generated time series. The analysis of the high turbidity occurrences indicates that their seasonal behavior is preserved in the model as shown in Fig. 8.51. The highest average, as well as maximum turbidity values, usually appear during late autumn months, when rainfall after long summer droughts mobilizes deposited sediment and brings it into the karst underground.

8.9 Artificial Aquifer Recharge

The following example illustrates why artificial aquifer recharge is gaining importance in integrated water resources management. The California Department of Water Resources (DWR) in its 2003 update on water supplies in the state and various related issues, stated that

> The extent to which climate will change and the impact of that change are both unknown. A reduced snowpack, coupled with increased seasonal rainfall and earlier snowmelt may require a change in the operating procedures for existing dams and conveyance facilities. Furthermore, these changes may require more active development of successful conjunctive management programs in which the aquifers are more effectively used as storage facilities. Water managers might want to evaluate their systems to better understand the existing snowpack-surface water-groundwater relationship, and identify opportunities that may exist to optimize groundwater and other storage capability under a new hydrologic regime that may result from climate change. If more water was stored in aquifers or

in new or re-tooled surface storage, the additional water could be used to meet water demands when the surface water supply was not adequate because of reduced snowmelt. (DWR, 2003).

The advantages of storing water underground as opposed to surface water reservoirs (see Chap. 7) are often made clear during long periods of drought. For example, since the 1987–1992 droughts in California, there has been an expansion in groundwater recharge and storage capacity in the state. Figure 8.52 shows some of the larger recharge projects in California. New projects, such as those operated by SWSD, Arvin-Edison Water Storage District, Kern Water Bank Authority, MWA, and Calleguas Municipal Water District, rely either wholly or in part on recharge supplies exported from the Sacramento River delta (the Delta). Operations of the projects are subject to the Delta export restrictions as well as to the availability of conveyance capacity (GADPP, 2000).

Although the main purpose of artificial aquifer recharge is to store water underground for its later use, another aspect of artificial recharge is protection of existing groundwater

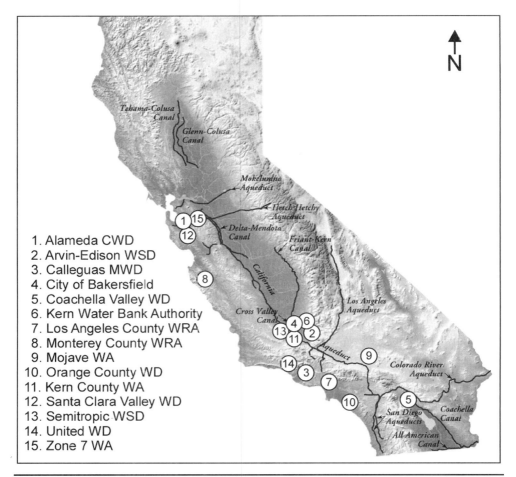

1. Alameda CWD
2. Arvin-Edison WSD
3. Calleguas MWD
4. City of Bakersfield
5. Coachella Valley WD
6. Kern Water Bank Authority
7. Los Angeles County WRA
8. Monterey County WRA
9. Mojave WA
10. Orange County WD
11. Kern County WA
12. Santa Clara Valley WD
13. Semitropic WSD
14. United WD
15. Zone 7 WA

Figure 8.52 Examples of larger California groundwater storage projects. (Modified from GADPP, 2000.)

from negative impacts such as the migration of contaminant plumes. This is achieved by injecting water into the saturated zone and creating areas of higher hydraulic heads thus changing the direction of the natural groundwater flow. Such areas are called hydraulic barriers. Depending on the scale of the contamination problem, aquifer recharge for this purpose may consist of one or two injection wells, or it may involve "many" wells and extensive pipelines for the transfer of water to be injected into the aquifer(s).

One example of large-scale aquifer recharge for groundwater protection purposes is the proposed Salinas Valley project. Located 100 mi south of San Francisco, the Salinas Valley supports a major portion of California's, and the nation's vegetable production (it is often called the "Salad Bowl of the World"). Virtually all the water used for municipal, industrial, and agricultural purposes is groundwater. Over the years, saltwater has intruded into valley aquifers because of excessive pumping and denied aquifer recharge. As of 2005, seawater had moved inland 6.5 mi in the Pressure 180-Foot Aquifer, and 3 mi in the Pressure 400-Foot Aquifer. This seawater intrusion has resulted in the degradation of groundwater supplies, reduction of crop yield, and abandonment of numerous urban and agricultural supply wells. The seriousness of these effects has generated the need for a combination of solutions, which fall into three general categories: (1) reduction in the amount of groundwater consumed, (2) increase of aquifer recharge, and (3) finding additional sources of water. One proposed solution is a 30,000 acre-ft/yr water recycling plant to recycle wastewater and a 45-mi long pipeline and well project to use recycled water on crops, reduce groundwater pumping, increase aquifer recharge, and stop seawater intrusion (USEPA, 1999; Federal Register, 2000; MCWRA, 2007).

The following discussion by Meinzer (1932) shows that early considerations and implementations of artificial aquifer recharge in the United States were made as early as at the beginning of the twentieth century:

> Artificial recharge can be accomplished in some places by draining surface water into wells, spreading it over tracts underlain by permeable material, temporarily storing it in leaky reservoirs from which it may percolate to the water table, or storing it in relatively tight reservoirs from which it is released as fast as it can seep into the stream bed below the reservoir. Artificial recharge by some of these methods has been practiced in the United States and other countries. It was suggested by Hilgard in 1902 for southern California, where it has since received considerable investigation and has been adopted as a conservation measure. Drainage into wells has been practiced in many parts of the United States, chiefly to reclaim swampy land or to dispose of sewage and other wastes. The drainage of sewage or other wastes into wells can not be approved because it may produce dangerous pollution of water supplies. Drainage of surface water into wells to increase the ground-water supply for rice irrigation in Arkansas is now under consideration. Water spreading has been practiced to a considerable extent in southern California partly to decrease the effects of flood but largely to increase the supply of ground water. Storage in ordinary reservoirs and subsequent release has frequently been considered and the unavoidable leakage of some reservoirs has been used to increase the ground-water supply. Artificial recharge by damming stream channels in the permeable lava rocks of the Hawaiian Islands has been considered. In ground-water investigations that involve the question of safe yield attention should as a matter of course be given to the possibilities of artificial recharge.

Two terms have gained popularity in describing the concept of artificial aquifer recharge: (1) managed aquifer recharge (MAR) as a more general description for a variety of engineering solutions and (2) aquifer storage and recovery (ASR), which describes injection and extraction of potable water with dual-purpose wells. Artificial aquifer recharge should not be confused with induced aquifer recharge, which is a response of the surface water system to groundwater withdrawal as discussed in Chap. 7 on collector wells. Sometimes the induced recharge is referred to as indirect artificial aquifer recharge. Many groundwater supply systems in alluvial aquifers near large streams are intentionally designed to induce aquifer recharge for two main reasons: (1) increased capacity and (2) filtration of the river water en route to the supply well, which improves water quality.

Two main types of systems take advantage of close hydraulic connection between surface streams and groundwater: (1) riverbank filtration and (2) dune filtration systems. The surface water can be infiltrated into the groundwater zone through the riverbank, percolation from spreading basins, canals, lakes, or percolation from drain fields of porous pipe. In all these cases, the river water may be diverted by gravity or pumped to the recharge site. The water then travels through an aquifer to extraction wells at some distance from the riverbank (or recharge site). In some cases, the residence time underground is only 20 to 30 days, and there is almost no dilution by natural groundwater (Sontheimer, 1980). In Germany, systems that do not meet a minimum residence time of 50 days are required to have posttreatment of the recovered water and similar guidelines are applied in the Netherlands. In the Netherlands, dune infiltration of treated Rhine River water has been used to restore the equilibrium between freshwater and saltwater in the dunes (Piet and Zoeteman, 1980; Olsthoorn and Mosch, 2002), while serving to improve water quality and provide storage for potable water systems. Dune infiltration also provides protection from accidental spills of toxic contaminants into the Rhine River. Some systems have been in place for over 100 years, and there is no evidence that the performance of the system has deteriorated or that contaminants have accumulated. The city of Berlin has greater than 25 percent reclaimed water in its drinking water supply, and no disinfection is practiced after bank filtration (USEPA, 2004).

The important factors that have to be considered for any scheme of artificial aquifer recharge are as follows:

- Regulatory requirements.
- The availability of an adequate source of recharge water of suitable chemical and physical quality.
- Geochemical compatibility between recharge water and the existing groundwater (e.g., possible carbonate precipitation, iron hydroxide formation, mobilization of trace elements).
- The hydrogeologic properties of the porous media (soil and aquifer) must facilitate desirable infiltration rates and allow direct aquifer recharge. For example, existence of extensive low permeable clays in the unsaturated (vadose) zone may exclude a potential recharge site from future consideration.
- The water-bearing deposits must be able to store the recharged water in a reasonable amount of time, and allow its lateral movement toward extraction locations at acceptable rates. In other words, the specific yield (storage) and the hydraulic conductivity of the aquifer porous media must be adequate.
- Presence of fine-grained sediments may have an advantage of improving the quality of recharged water due to their high filtration and sorption capacities. Other geochemical reactions in the vadose zone below recharge facilities may also influence water quality.

- Engineering solution should be designed to facilitate efficient recharge when there is an available surplus of water, and efficient recovery when the water is most needed.

- The proposed solution must be cost-efficient, environmentally sound, and competitive to other water resource development options.

Aquifers that can store large quantities of water and do not transmit them away quickly are best suited for artificial recharge. For example, karst aquifers may accept large quantities of recharged water but in some cases tend to transmit them very quickly away from the recharge area. This may still be beneficial for the overall balance of the system and the availability of groundwater downgradient from the immediate recharge sites, as illustrated with examples from the Edwards Aquifer in Texas (discussed later in this chapter). Alluvial aquifers are usually the most suited to storage because of the generally shallow water table and vicinity to source water (surface stream). Sandstone aquifers are in many cases very good candidates due to their high storage capacity and moderate hydraulic conductivity.

8.9.1 Methods of Artificial Aquifer Recharge

Three common methods of artificial aquifer recharge are (1) spreading water over the land surface, (2) delivering it to the unsaturated zone below the land surface, and (3) injecting water directly into the aquifer. A variety of engineering solutions and combinations of these three methods have been used to accomplish a simple goal—deliver more water to the aquifer (see Fig. 8.53).

Table 8.12 provides a comparison of major engineering factors that should be considered when installing a groundwater recharge system, including the availability and cost of land for recharge basins (Fox, 1999). If such costs are excessive, the ability to implement injection wells adjacent to the reclaimed water source tends to decrease the cost of conveyance systems for injection wells. Surface spreading basins require the lowest degree of pretreatment while direct injection systems require water quality comparable

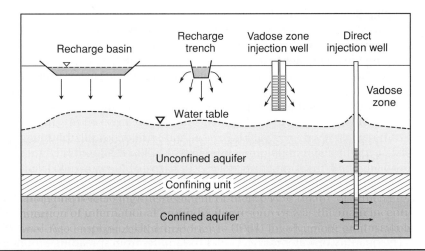

FIGURE 8.53 Engineering methods for artificial aquifer recharge. (Modified from USEPA, 2004.)

	Recharge Basins	Vadose Zone Injection Wells	Direct Injection Wells
Aquifer type	Unconfined	Unconfined	Unconfined or confined
Pretreatment requirements	Low technology	Removal of solids	High technology
Estimated major capital costs	Land and distribution system	$25,000–75,000 per well	$500,000–1,500,000 per well
Capacity	100–20,000 m³/hectare-day	1000–3,000 m³/d per well	2000–6,000 m³/d per well
Maintenance requirements	Drying and scraping	Drying and disinfection	Disinfection and flow reversal
Estimated life cycle	>100 yr	5–20 yr	25–50 yr
Soil aquifer treatment	Vadose and saturated zones	Vacose and saturated zones	Saturated zone

From USEPA, 2004.

Table 8.12   Comparison of Major Engineering Factors for Engineered Groundwater Recharge

to drinking water, if potable aquifers are affected. Low-technology treatment options for surface spreading basins include primary and secondary wastewater treatment with the possible use of lagoons and natural systems. RO is commonly used for direct injection systems to prevent clogging; however, some aquifer storage and recovery (ASR) systems have been operating successfully without membrane treatment when water was stored for irrigation. The cost of direct injection systems can be greatly reduced from the numbers presented in Table 8.12 if the aquifer is shallow and nonpotable (USEPA, 2004).

Spreading Structures and Trenches

Artificial recharge by spreading is the most common technique of artificial aquifer recharge. Recharge is accomplished by spreading water over the ground surface or by conveying the raw water to infiltration basins and ditches. The operational efficiency of the spreading depends on the following factors (modified from Pereira et al., 2002):

- Presence of sufficiently pervious layers between the ground surface and the water table (aquifer)
- Enough thickness and storage capacity of the unsaturated layers above the water table
- Appropriate transmissivity of the aquifer horizons
- Surface water without excessive particulate matter (low turbidity) to avoid clogging

The quantity of water that can enter the aquifer from spreading grounds depends on three basic factors:

FIGURE 8.54 Tonopah Recharge Site, Central Arizona Project. (Photograph courtesy of Philip Fortnam, Central Arizona Project).

- The infiltration rate at which the water enters the subsurface
- The percolation rate, i.e., the rate at which water can move downward through the unsaturated zone until it reaches the water table (saturated zone)
- The capacity for horizontal movement of water in the aquifer, which depends on the hydraulic conductivity and thickness of the saturated zone

The infiltration rate tends to reduce over time due to the clogging of soil pores by sediments carried in the raw water, growth of algae, colloidal swelling, soil dispersion, and microbial activity. A spreading basin is constructed with a flat bottom that is covered evenly by shallow water. This requires the availability of large surfaces of land for meaningful size recharge works. Several basins may be arranged in line so that excess water runs between the basins (Fig. 8.54). Retaining basins may be used for settlement of suspended sediments before water enters the spreading basins. The settling of sediments may also be assisted with the addition of coagulation agents. In addition to clogging, another major disadvantage of shallow infiltration basins is evaporation loss, which may be rather significant in the case of lower infiltration rates, caused either by low-permeable natural soils or clogging. Proper operations and regular maintenance of infiltration basins are therefore very important for the overall efficiency of spreading grounds and infiltration basins.

Though management techniques are site-specific and vary accordingly, some common principles of maintenance are practiced in most infiltration basins. A wetting and drying cycle with periodic cleaning of the bottom is used to prevent clogging. Drying cycles allow for desiccation of clogging layers and reaeration of the soil. This practice helps to maintain high infiltration rates, and microbial populations to consume organic matter, and helps reduce levels of microbiological constituents. Reaeration of the soil also promotes nitrification, which is a prerequisite for nitrogen removal by denitrification. Periodic maintenance by cleaning of the bottom may be done by deep ripping of the soils or by scraping the top layer of soil. Deep ripping sometimes causes fines to migrate to deeper levels where a deep clogging layer may develop (USEPA, 2004). The Orange County Water District in California has developed a device for continuous removal of clogging materials during a flooding cycle.

Spreading grounds can be managed to avoid nuisance conditions such as algae growth and insect breeding in the percolation ponds. Generally, a number of basins are rotated through filling, draining, and drying cycles. Cycle length is dependent on both soil conditions and the distance to the groundwater table. This is determined through field-testing on a case-by-case basis. Algae can clog the bottom of basins and reduce infiltration rates. Algae further aggravate soil clogging by removing carbon dioxide, which raises the pH, causing precipitation of calcium carbonate. Reducing the detention time of the recharge water within the basins minimizes algal growth, particularly during summer periods where solar intensity and temperature increase algal growth rates. The levels of nutrients necessary to stimulate algal growth are too low for practical consideration of nutrient removal as a method to control algae. Also, scarifying, rototilling, or discing the soil following the drying cycle can help alleviate clogging potential, although scraping or "shaving" the bottom to remove the clogging layer is more effective than discing it (USEPA, 2004).

Variations of the spreading ground technique consist of the use of trenches that are often easier to handle and for which clogging is a lesser problem since a major part of the sediment is carried out of the ditches by the slow-flowing water (Pereira et al., 2002). The main advantage of recharge trenches is that they are relatively inexpensive, and generally have higher infiltration rates per area than recharge basins because water also infiltrates through the trench sides. The disadvantage is that they eventually do clog up at their infiltrating surface because of accumulation of suspended soils and/or biomass. This, however, is the matter of maintenance where trenches may offer greater flexibility over basins in terms of taking them off line individually to minimize interruptions to the recharge operations.

Dams

Widely applied artificial recharge structures are floodwater retention dams, the purpose of which is to delay the runoff of water in surface streams and provide the time needed for recharge into the local aquifer. The structures usually consist of low dams, including earth walls and gabions built to be toppled by floods. For example, a number of such dams have been built in the Edwards Aquifer recharge zone (Fig. 8.55) and more are in various planning phases. The Edwards aquifer is the principal source of water supply for the city of San Antonio, as well as numerous other communities and agricultural interests throughout south-central Texas. The aquifer also supplies Leona, San Pedro, San Antonio, Comal, and San Marcos Springs, creating unique environments and recreational opportunities while providing baseflow to the Leona, San Antonio, Guadalupe, and San Marcos Rivers. Over the past several decades, increasing water demands on the Edwards Aquifer have raised concerns about the ability of the aquifer to meet these demands without causing social, economic, and environmental problems. Estimated benefits of the proposed four new recharge projects are additional pumpage of 21,440 acre-ft/yr (approximately 26.4 billion m^3/yr) from the aquifer, additional springflows of 15,240 acft/yr, and additional aquifer storage (in initial years), and then additional springflows

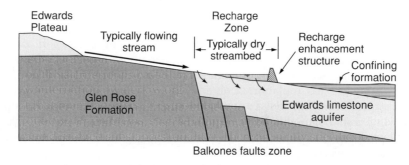

FIGURE 8.55 Principle of artificial recharge of Edwards Aquifer with recharge dams. (Figure Courtesy of Nueces River Authority, 2007, and HDR Engineering, Inc).

of 11,320 acre-ft/yr. In the Uvalde County, the additional artificial recharge would raise aquifer levels 11 ft on average, and result in 25 percent reductions in time the County is in drought management conditions (Nueces River Authority, 2007, from HDR Engineering, Inc.).

Wells

Vadose zone injection wells for groundwater recharge with reclaimed water were developed in the 1990s and have been used in several different locations in the Phoenix, AZ, metropolitan area. Typical vadose zone injection wells are 6 ft (2 m) in diameter and 100 to 150 ft (30 to 46 m) deep. They are backfilled with porous media and a riser pipe is used to allow for water to enter at the bottom of the injection well to prevent air entrainment. An advantage of vadose zone injection wells is the significant cost savings as compared to direct injection wells. The infiltration rates per well are often similar to direct injection wells. A significant disadvantage is that they cannot be backwashed and a severely clogged well can be permanently destroyed. Therefore, reliable pretreatment is considered essential to maintaining the performance of a vadose zone injection well. Because of the considerable cost savings associated with vadose zone injection wells as compared to direct injection wells, a life cycle of 5 years for a vadose injection well can still make the vadose zone injection well the economical choice. Since vadose zone injection wells allow for percolation of water through the vadose zone and flow in the saturated zone, one would expect water quality improvements commonly associated with soil aquifer treatment to be possible (USEPA, 2004; Bouwer, 2002).

The advantage of the saturated-zone injection wells is that they can be used to recharge any type of aquifer, at any depth, thus eliminating problems associated with low permeable surficial soils and low-permeable layers in general. Direct injection requires water of higher quality than for surface spreading because of the absence of vadose zone and/or shallow soil matrix treatment afforded by surface spreading and the need to maintain the hydraulic capacity of the injection wells, which are prone to physical, biological, and chemical clogging. Treatment processes beyond secondary treatment that are used prior to injection include disinfection, filtration, air stripping, ion exchange, granular activated carbon, and RO or other membrane separation processes. By using these processes or various subsets in appropriate combinations, it is possible to satisfy all present water quality requirements for water injection, including with reclaimed water. However, even when such high quality water is used for recharge, trouble-free operations cannot be guaranteed for longer periods of well operation due to various geochemical and mechanical processes, which tend to clog well screens and reduce permeability of the well gravel pack and adjacent aquifer material. A common practice is to inject water via dual-purpose wells, which are used to occasionally pump water back from the aquifer thus removing screen-clogging materials ("backwashing"). The most frequent causes of clogging are accumulation of organic and inorganic solids, biological and chemical contaminants, and dissolved air and gases from turbulence. Very low concentrations of suspended solids, on the order of 1 mg/L, can clog an injection well. Even low concentrations of organic contaminants can cause clogging due to bacteriological growth near the point of injection (USEPA, 2004).

In many cases, the wells used for injection and recovery are classified by the USEPA as Class V injection wells. Some states require that the injected water must meet drinking water standards prior to injection into a Class V well.

For both surface spreading and direct injection, locating the extraction wells as great a distance as possible from the recharge site increases the flow path length and residence time in the underground, as well as the mixing of the recharged water with the natural groundwater. Treatment of organic parameters does occur in the groundwater system with time, especially in aerobic or anoxic conditions (Gordon et al., 2002; Toze and Hanna, 2002).

Case Study: Central Arizona Project (Cap)

The information presented in this case study is courtesy of Crystal Thompson, Central Arizona Project, Phoenix, Arizona; http://www.cap-az.com/.

Today, more than 5 million people call Arizona home and this number is expected to exceed 8 million in less than 20 years. Arizonans use almost 8 million acre-ft of water every year and water resource leaders use a variety of tools to preserve and protect the state's water supplies. Arizona's water supplies are provided by three primary sources: surface water, groundwater, and used water. Use of groundwater is highly regulated by state law but continues to be relied upon for irrigation, municipal, and domestic uses.

Central Arizona Project (CAP) developed a recharge program more than a decade ago and it has been instrumental in helping protect groundwater supplies. CAP is a 336-mi-long system that brings more than 1.5 million acre-ft of Colorado River water to customers in the central and southern area of the state. CAP delivers water to cities for drinking, agricultural, and Native-American communities for farming, and recharge projects where it is stored underground for future use.

For more than 10 years, CAP has been building, maintaining, and operating numerous recharge projects so it can store Colorado River water underground for recovery during periods with water shortages. In addition to replenishing Arizona's depleted groundwater supplies, CAP's recharge program is helping to diminish the impacts of groundwater overdraft, including subsidence; improving water quality by natural filtration; and firming Arizona's water supply by providing a "reserve" of water to be recovered during prolonged droughts.

CAP currently maintains six recharge projects: Avra Valley, Pima Mine Road, Lower Santa Cruz, Agua Fria, Hieroglyphic Mountains, and Tonopah Desert. CAP's recharge projects are remotely monitored and controlled at the CAP Control Center located in Phoenix. Following are the main characteristics of the six recharge projects (note that 1000 acre-ft equals approximately 1.23 million m^3):

- The Avra Valley Recharge Project was built in 1996 and incorporates four recharge basins covering 11 acres. It is permitted to store 11,000 acre-ft of water per year.

- The Pima Mine Road Recharge Project is located on the Santa Cruz River flood plain. It was constructed in 1999 and provides a maximum permitted recharge capacity of 30,000 acre-ft/yr.

- In 2000, the Lower Santa Cruz Recharge Project was built in conjunction with a flood control levee along the Santa Cruz River. The facility has three basins covering more than 30 acres and has a permitted recharge capacity of 50,000 acre-ft/yr.

- The Agua Fria Recharge Project was completed in 2001 and has two operational components: a 4-mi river section used for in-channel recharge and conveyance of CAP water downstream to a constructed recharge facility incorporating 7 basins. Together the facilities have a combined permitted recharge capacity of 100,000 acre-ft of water per year.

- The Hieroglyphic Mountains Recharge Project began operating in January of 2003. The project is permitted to recharge 35,000 acre-ft/yr. It consists of seven basins that cover approximately 38 acres.

- Construction of the Tonopah Desert Recharge Project (Fig. 8.56; see also Fig. 8.54) began in August 2004 and was completed in January 2006. It has a total of 19 basins covering 207 acres and is designed to store 150,000 acre-ft/yr.

FIGURE 8.56 Delivery of recharge water to one of the basins at the Tonopah Desert Recharge Project. (Photograph courtesy of Philip Fortnam, Central Arizona Project.)

- An additional underground storage project is currently in the design and per-mitting phase: Superstition Mountains is designed to provide 56,500 acre-ft of storage capacity per year.

Central Arizona Project understands the vital role groundwater plays in maintaining the quality of life enjoyed by Arizonans. Through ongoing cooperative efforts with strategic partners including the Arizona Water Banking Authority, CAP is helping to ensure the state's water supply is protected against future shortages.

8.9.2 Aquifer Storage and Recovery

ASR has been widely applied in the coastal areas of the eastern United States, particularly in Florida, where population growth and demand for water resources are of great concern. Figure 8.57 illustrates the basic principle of ASR: potable water is injected (stored) into a portion of aquifer with brackish, nonpotable water during periods of low water demand and/or high water availability, and then extracted (recovered) during periods of increased demand. The success of such systems is measured by recovery efficiency, expressed as percentage of the potable water meeting a preset criterion (e.g., chloride concentration less than 250 mg/L) that can be recovered relative to the quantity injected, over one full cycle that includes injection, storage, and recovery. Recovery efficiency in most cases increase with the number of full cycles since more of the injected potable water remains in the aquifer after each cycle. Recovery of the stored water is dependent upon the effective placement of a relatively stable, thick lens or bubble of low-density recharge water during the injection phase. To form this lens, enough water must be injected to displace a large volume of saline water, the mixing of the injected and native waters must not be significant, and confinement must be sufficiently tight to prevent rapid vertical migration of the less dense recharged water (Rosenshein and Hickey, 1977, from Yobbi, 1997).

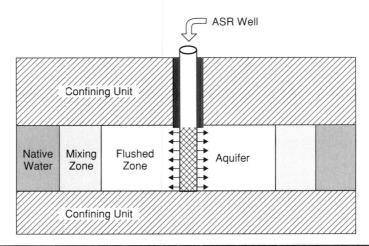

Figure 8.57 Aquifer storage and recovery well in a confined aquifer depicting idealized flushed and mixing (transition) zones created by recharge. Flushed zone contains mostly recharged water. (From Reese, 2002.)

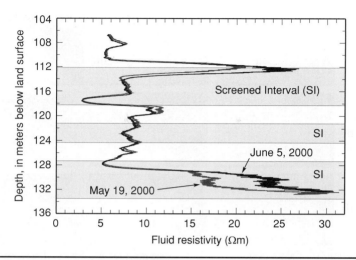

FIGURE 8.58 Electrical resistivity profiles collected in observation well CHN-809 during the injection phase of cycle 2 of an aquifer storage and recovery study in Charleston, SC. (Modified from Petkewich et al., 2004.)

Among various problems associated with ASR operations is the formation of lenses and bubbles of freshwater floating on, or embedded into, the denser native water as seen in Fig. 8.58. The lenses may have formed both because of a density contrast between the recharge water and the native groundwater, and because of the possible presence of different transmissive zones in the aquifer. Figures 8.59 and 8.60 show the results of variable-density groundwater modeling which illustrate the importance of aquifer characterization in planning ASR projects. This includes determining density differences between native (brackish to saline) groundwater and the recharge water, as well as aquifer heterogeneities, and then simulating various scenarios of injection and recovery to find an optimum one.

8.9.3   Source Water Quality and Treatment

In the United States, water used for well injection is usually treated to meet drinking water quality standards for two reasons. One is to minimize clogging of the well-aquifer interface, and the other is to protect the quality of the water in the aquifer, especially when it is pumped by other wells in the aquifer for potable uses. Direct injection into the saturated zone does not have the benefits of fine-textured unsaturated soils which often improve the quality of the recharged water in the case of recharge basins and vadose-zone infiltration systems. Thus, whereas secondary sewage effluent can in many cases be used in surface infiltration systems for soil-aquifer treatment and eventual potable reuse, effluent for direct well injection must at least receive tertiary treatment (sand filtration and chlorination). This treatment removes remaining suspended solids and some microorganisms like giardia (protozoa), and cryptosporidium and parasites, like helminth eggs, by filtration. Bacteria and viruses are removed by chlorination, ultra violet irradiation, or other disinfection (Bouwer, 2002).

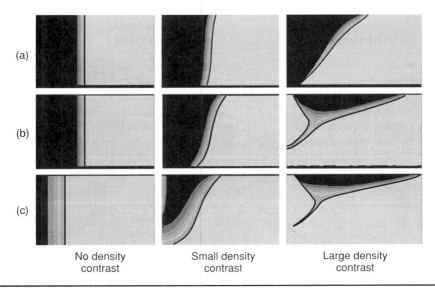

(a)

(b)

(c)

No density   Small density   Large density
contrast   contrast   contrast

FIGURE 8.59 Comparison between FEFLOW numerical model results for the cases of no density, small density, and large density contrasts. Results are given for (a) end of injection phase, (b) end of storage phase, and (c) end of recovery phase. Each image is symmetrical about its left-hand boundary, which represents the injection/extraction well where freshwater (in black) is injected into the initially more saline aquifer. The effects of variable density flow are clearly visible. Increasing the density contrast results in deviation from the standard cylindrical "bubble" most, as seen in the no density difference case. Injected fresh water is in black. (From Simmons et al., 2007.)

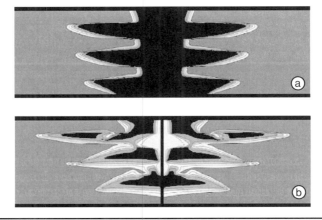

FIGURE 8.60 Modeling results showing distribution of injected water (in black) after 100 days of injection (a) and 250 days of storage (b), in an aquifer consisting of 6 even horizontal layers, with the hydraulic conductivity varying by a ratio of 10 between high and low K layers. Note the fingering and mixing processes. (From Simmons et al, 2007).

Where groundwater is not used for drinking, water of lower quality can be injected into the aquifer. For example, in Australia, stormwater runoff and treated municipal wastewater effluent are injected into brackish aquifers to produce water for irrigation by pumping from the same wells. Clogging is alleviated by a combination of low-cost water treatment and well redevelopment, and groundwater quality is protected for its declared beneficial uses (Dillon and Pavelic 1996; Dillon et al., 1997).

Turbidity of source water is often the main quality problem for all recharge facilities. In infiltration basins and trenches, fine suspended solids tend to clog contact surfaces and have to be regularly removed after infiltration basins and trenches are drained (infiltration drains with this problem cannot be efficiently rehabilitated). For this reason, large recharge facilities often include separate settling basins. Rehabilitation of injection well screens mechanically clogged with fine particulate matter is described in Chap. 7.

Fate of Contaminants in Recharge Systems

A very illustrative discussion on potential changes in quality of reclaimed water used for aquifer storage and recovery, as it interacts with aquifer porous media, is given by Clinton (2007). In this research effort, funded by Water Reuse Foundation, reclaimed water ASR projects located in Florida, Arizona, California, Texas, Hawaii, Australia, and Kuwait were surveyed to determine the state of the practice and screen for potential detailed study sites. By observing changes in the concentrations of over 90 compounds at four selected ASR sites with many variables, the study made a broad assessment of water quality changes in reclaimed water ASR storage. The study was designed to investigate the variables of aquifer characteristics, storage time, travel distance, recharge water quality, and operational history. The data support many aquifer process assumptions, such as enhanced chemical and biological activity near ASR wells, but do not statistically support conclusions regarding specific degradation rates for most of the constituents observed. A large emphasis was placed on the microcontaminant portion of the study. Several contaminants appeared in significantly higher concentrations in the recovered water than in the recharge water. Several possible causes for these increased concentrations were considered. The findings indicated that the most likely one were highly variable concentrations of endocrine-disrupting compounds, pharmaceuticals, and personal care products in reclaimed water. The concentrations measured in the recharge water were considered low and possibly not representative of typical conditions. Additional analyses indicated that the variability of input concentrations was the largest factor affecting measured concentrations in monitoring wells and in recovered water (Clinton, 2007).

Although the degree of raw water (surface water and wastewater) treatment prior to recharge may vary depending on the engineering solutions and applicable regulations, various stakeholders will likely have opposing opinions since the body of scientific evidence and accepted practices is still not definitive, particularly regarding disinfection requirements. The issue of emerging contaminants (see Chap. 5), which are not regulated, and often not removed by conventional drinking water or wastewater treatment technologies, is another major point of disagreement between advocates and opponents of artificial aquifer recharge. These contaminants include recalcitrant organic chemicals such as 1,4-dioxane, various pharmaceuticals, endocrine disruptors (hormones), and disinfection by-products formed during drinking water and injection water treatment such as trihalomethanes (THMs), chloroform, and total organic halide (TOX). Some of the disinfection by-product are carcinogenic and may be persistent for long periods in certain aquifer environments (Rostad, 2002). Anders and Schroeder (2003) describe in detail a methodology used to investigate: (1) the fate and transport of wastewater constituents as they travel from the point of recharge to points of withdrawal and (2) the long-term effects that artificial recharge using tertiary-treated municipal wastewater has on the quality of the groundwater in the Central Basin in Los Angeles County.

The most comprehensive health effects study of an existing groundwater recharge project was carried out in Los Angeles County, CA, in response to uncertainties about the health consequences of recharge for potable use raised by a California Consulting Panel in 1975–76. The primary goal of the Health Effects Study was to provide information for use by health and regulatory authorities to determine if the use of reclaimed water for the Montebello Forebay Project should be maintained at the present level, cut back, or expanded. Specific objectives were to determine if the historical level of reuse had adversely affected groundwater quality or human health, and to estimate the relative impact of the different replenishment sources on groundwater quality. During the course of the study, a technical advisory committee and a peer review committee reviewed findings and interpretations. The final project report was completed in March, 1984 as summarized by Nellor et al. (1985). The results of the study did not demonstrate any measurable adverse effects on either the area groundwater or health of the people ingesting the water.

The Rand Corporation has conducted additional health studies for the Montebello Forebay Project as part of an ongoing effort to monitor the health of those consuming reclaimed water in Los Angeles County (Sloss et al., 1996 and Sloss et al., 1999). These studies looked at health outcomes for 900,000 people in the Central Groundwater Basin who are receiving some reclaimed water in their household water supplies. These people account for more than 10 percent of the population of Los Angeles County. To compare health characteristics, a control area of 700,000 people that had similar demographic and socioeconomic characteristics was selected, but did not receive reclaimed water. The results from these studies have found that, after almost 30 years of groundwater recharge, there is no association between reclaimed water and higher rates of cancer, mortality, infectious disease, or adverse birth outcomes (USEPA, 2004).

One of the major concerns with any aquifer recharge project, using any type of recharge water, is the presence of bacteria and viruses. For example, enterovirus, *giardia* cysts, and *cryptosporidium* oocysts are often more than 10 percent of time tested positive in some treated and disinfected reclaimed water (USEPA, 2004). The survival or retention of pathogenic microorganisms in the subsurface depends on several factors including climate, soil composition, antagonism by soil microflora, flow rate, and type of microorganism. At low temperatures (below 4°C or 39°F) some microorganisms can survive for months or years. The die-off rate is approximately doubled with each 10°C (18°F) rise in temperature between 5°C and 30°C (41°F and 86°F) (Gerba and Goyal, 1985). Rainfall may mobilize bacteria and viruses that had been filtered or adsorbed, and thus, enhance their transport (USEPA, 2004).

The nature of the soil affects pathogen survival and retention. For example, rapid infiltration sites where viruses have been detected in groundwater were located on coarse sand and gravel types. Infiltration rates at these sites were high and the ability of the soil to adsorb the viruses was low. Generally, coarse soil does not inhibit virus migration. Other soil properties, such as pH, cation concentration, moisture holding capacity, and organic matter do have an effect on the survival of bacteria and viruses in the soil. Resistance of microorganisms to environmental factors depends on the species and strains present. Drying the soil will kill both bacteria and viruses. Bacteria survive longer in alkaline soils than in acid soils (pH 3 to 5) and when large amounts of organic matter are present. In

Factor	Comments
Soil type	Fine-textured retain viruses more effectively than light-textured soils. Iron oxides increase the adsorptive capacity of soils. Muck soils are generally poor adsorbents.
pH	Generally, adsorption increases when pH decreases. However, the reported trends are not clearcut due to complicating factors.
Cations	Adsorption increases in the presence of cations. Cations help reduce repulsive forces on both virus and soil particles. Rainwater may desorb viruses from soil due to its low conductivity.
Soluble organics	Generally compete with viruses for adsorption sites. No significant competition at concentrations found in wastewater effluents. Humic and fulvic acids reduce virus adsorption to soils.
Virus type	Adsorption to soils varies with virus type and strain. Viruses may have different isoelectric points.
Flow rate	The higher the flow rate, the lower virus adsorption to soils.
Saturated versus unsaturated flow	Virus movement is less under unsaturated flow conditions.

From USEPA, 2004; Gerba and Goyal, 1985.

TABLE 8.13 Factors That May Influence Virus Movement to Groundwater

general, increasing cation concentration and decreasing pH and soluble organics tend to promote virus adsorption. Bacteria and larger organisms associated with wastewater are effectively removed after percolation through a short distance of the soil mantle. Lysimeter studies showed a greater than 99 percent removal of bacteria and 95 to 99 percent removal of viruses (Cuyk et al., 1999). Factors that may influence virus movement in groundwater are given in Table 8.13. Proper treatment (including disinfection) prior to recharge, site selection, and management of the surface spreading recharge system can minimize or eliminate the presence of microorganisms in the groundwater. Once the microorganisms reach the groundwater system, the oxidation state of the water significantly affects the rate of removal (Medema and Stuyfzand, 2002; Gordon et al., 2002).

Several issues regarding quality of recovered water and geochemical reactions between injected and native aquifer water and porous media have emerged during recent years after more monitoring data became available at various ASR sites in Florida (Arthur et al., 2002). The current focus is primarily on mobilization of arsenic, uranium, and other trace elements at some locations within the Floridan aquifer due to introduction of imported water with higher dissolved oxygen content than native water. Once these waters are introduced into a reduced aquifer, selective leaching and/or mineral dissolution may release metals into the injected water. This amplifies the requirement that feasibility studies, design, construction, and operation of ASR facilities, including monitoring well placement and monitoring schedules, should take into account the possibility of water-rock interaction and mobilization of trace elements into recovered waters.

Constraints on groundwater recharge are conditioned by the use of the extracted water and include health concerns, economic feasibility, physical limitations, legal restrictions, water quality constraints, and recharge water availability. Of these constraints, health concerns are the most important as they pervade almost all recharge projects (Tsuchihashi et al., 2002). Where reclaimed water is used for recharge and will be extracted for drinking water (potable) use, health effects due to prolonged exposure to low levels of contaminants must be considered as well as the acute health effects from pathogens or toxic substances. One problem with recharge is that boundaries between potable and nonpotable aquifers are rarely well defined. Some risk of contaminating high quality potable groundwater supplies is often incurred by recharging (storing water in) "nonpotable" aquifers. The recognized lack of knowledge about the fate and long-term health effects of contaminants found in reclaimed water obliges conservative approach in setting water quality standards and monitoring requirements for groundwater recharge. Because of these uncertainties, some states in the United States have set stringent water quality requirements and require high levels of treatment—in some cases, organic removal processes—where groundwater recharge impacts potable aquifers (USEPA, 2004).

Large-scale artificial recharge projects are currently being implemented or considered by major water utilities and government agencies throughout the United States and the world. Although still challenging in many ways, the benefits they offer for sustainable management of water resources clearly makes MAR and ASR projects the trend of the future. More detailed information on artificial aquifer recharge can be found in National Research Council (1994), Pyne (1995), ASCE (2001), Bouwer (2002), Gale et al. (2002), IAH (2002), Gale (2005), Dillon and Molloy (2006), and UNESCO (2006). The International Association of Hydrogeologists (IAH) maintains a MAR-dedicated page on its Web site, including various useful links to related information.

References

Abderrahman, W.A., 2006. Saudi Arabia aquifers. In: *Non-Renewable Groundwater Resources; A Guidebook on Socially-Sustainable Management for Water-Policy Makers*. Foster, S., Loucks, D.P., editors. IHP-VI, Series on Groundwater No. 10, UNESCO, Paris, pp. 63–67.

AFCEE (Air Force Center for Environmental Excellence)., 1997. AFCEE Long-Term Monitoring Optimization Guide. Available at: http://www.afcee.brooks.af.mil.

AFCEE, 2003. Monitoring and Remediation Optimization System (MAROS) 2.1 Software Users Guide. Air Force Center for Environmental Excellence. Available at: http://www.gsi-net.com/software/maros/Maros.htm.

Ahlfeld, D.P., Barlow, P.M., and Mulligan, A.E., 2005. GWM—A ground-water management process for the U.S. Geological Survey modular ground-water model (MODFLOW-2000). U.S. Geological Survey Open-File Report 2005–1072, Reston, VA, 124 p.

Ahn, H., 2000. Modeling of groundwater heads based on second-order difference time series models. *Journal of Hydrology*, vol. 234, pp. 82–94.

Akaike, H., 1974. A new look at the statistical model identification. *IEEE Transactions on Automatic Control*, AC-vol. 19, no. (6), pp. 716–723.

Aller, L., Bennett, T., Lehr, J.H., and Petty, R.J., 1985. DRASTIC – A standardized system for evaluating ground water pollution potential using hydrogeologic settings. U.S. Environmental Protection Agency, Robert S. Kerr Environmental Research Laboratory, EPA/600/2–85/018, 163 p.

Alley, W.M., and Leake, S.A., 2004. The journey from safe yield to sustainability. *Ground Water*, vol. 42, no. 1, pp. 12–16.

Alley, W.M., Rielly, T.E., and Franke, O.L., 1999. Sustainability of ground-water resources. U.S. Geological Survey Circular 1186, Reston, VA, 86 p.

Almássy, E., and Buzás, ZS., 1999. Working Programme 1996/1999. Volume 1: Inventory of transboundary groundwaters. UN/ECE Task Force on Monitoring and Assessment, Lelystad, the Netherlands, 281 p.

Anders, R., and Schroeder, R.A., 2003. Use of water-quality indicators and environmental tracers to determine the fate and transport of recycled water in Los Angeles County, California. U.S. Geological Survey Water-Resources Investigations Report, 03–4279, Sacramento, CA, 104 p.

Andersen, L.J., and Gosk, E., 1989. Applicability of vulnerability maps. *Environmental Geology and Water Sciences*, vol. 13, no. 1, pp. 39–43.

Anderson, M.P., and Woessner, W.W., 1992. *Applied Ground Water Modeling; Simulation of Flow and Advective Transport*. Academic Press, San Diego, 381 p.

Anderson, M.T., and Woosley, L.H., Jr., 2005. Water availability for the Western United States—Key scientific challenges. U.S. Geological Survey Circular 1261, Reston, VA, 85 p.

Arthur, J.D., Dabous, A.A., and Cowart, J.B., 2002. Mobilization of arsenic and other trace elements during aquifer storage and recovery, southwest Florida. In: *U.S. Geological Survey Artificial Recharge Workshop Proceedings*, April 2–4, 2002, Sacramento, CA. Aiken, G.R., Kuniansky, E.L., editors. U.S. Geological Survey Open-File Report 02–89, pp. 47–50.

ASCE (American Society of Civil Engineers)., 2001. Standard guidelines for artificial recharge of groundwater. ASCE Standard No. 34–01, 120 p.

ASTM, 1999. *ASTM Standards on Determining Subsurface Hydraulic Properties and Ground Water Modeling*, 2nd ed. American Society for Testing and Materials, West Conshohocken, PA, 320 p.

Azdapor-Keeley, A., Russell, H.H., and Sewell, G.W., 1999. Microbial processes affecting monitored natural attenuation of contaminants in the subsurface. Ground Water Issue, U.S. Environmental Protection Agency, Office of Research and Development, EPA/540/S-99/001, 18 p.

Aziz, J.A., Newell, C.J., Ling, M., Rifai, H.S., and Gonzales, J., 2003. MAROS: a decision support system for optimizing monitoring plans. *Ground Water*, vol. 41, no. 3, pp. 355–367.

Barlow P.M., 2003. Ground water in freshwater-saltwater environments of the Atlantic Coast. U.S. Geological Survey Circular 1262, Reston, VA, 121 p.

Birkens, M.F.P., Knotters, M., and Hoogland, T., 2001. Space-time modeling of water table depth using a regionalized time series model and the Kalman filter. *Water Resources Research*, vol. 37, no. 5, pp. 1277–1290.

Booth, B., and Mitchell, A., 2001. *Getting started with ArcGIS, GIS by ESRI*. Environmental Systems Research Institute, Inc. (ESRI), Redlands, CA, 253 p.

Bouwer, H., 2002. Artificial recharge of groundwater: Hydrogeology and engineering. *Hydrogeology Journal*, vol. 10, pp. 121–142.

Box, G.E.P., and Jenkins, G.W., 1976. *Time Series Analysis: Forecasting and Control*. Holden-Day, San Francisco, 575 p.

Bras, R.L., and Rodriguez-Iturbe, I., 1994. *Random Functions and Hydrology*. Dover Publications, New York, 559 p.

Broyère, S., Jeannin, P.-Y., Dassargues, A., et al., 2001. Evaluation and validation of vulnerability concepts using a physically based approach. In: *7th Conference on Limestone Hydrology and Fissured Media*, Besançon 20–22 Sep. 2001. *Sci. Tech. Envir. Mém. H.S.*, vol. 13, pp. 67–72.

Brown, C.J., Colabufo, S., and Coates, D., 2002. Aquifer geochemistry and effects of pumping on ground-water quality at the Green Belt Parkway Well Field, Holbrook, Long Island, New York. U.S. Geological Survey Water-Resources Investigations Report 01–4025, Coram, New York, 21 p.

Burchi, S., and Mechlem, K., 2005. Groundwater in international law. Compilation of treaties and other legal instruments. FAO Legislative Study 86, Food and Agriculture Organization of the United States, Rome Italy, and UNESCO, Paris, France, 566 p.

Buxton, H.T., and Modica, E., 1992. Patterns and rates of ground-water flow on Long Island, New York. *Ground Water*, vol. 30, no. 6, pp. 857–866.

Buxton, H.T., Reilly, T.E., Pollock, D.W., and Smolensky, D.A., 1991. Particle tracking analysis of recharge areas on Long Island, New York. *Ground Water*, vol. 29, no. 1, pp. 63–71.

Bredehoeft, J.D., 2002. The water budget myth revisited: why hydrogeologists model. *Ground Water*, vol. 40, no. 4, pp. 340–345.

Bredehoeft, J.D., Papadopulos, S.S., Cooper, H.H., 1982. Groundwater—The water budget myth. In: *Studies in Geophysics, Scientific Basis of Water Resource Management*. National Academy Press, Washington, DC, 127 p.

Chu, A., Monti, J. Jr., and Bellitto, A.J., Jr., 1997. Public-supply pumpage in Kings, Queens, and Nassau Counties, New York, 1880–1995. U.S. Geological Survey Open-File Report 97–567, 61p.

Clinton, T., 2007. *Reclaimed Water Aquifer Storage and Recovery; Potential Changes in Water Quality*. Water Reuse Foundation, Alexandria, VA, 146 p.

Cuyk, S.V., Siegrist, R., Logan, A., Massen, S., Fischer, E., and Figueroa, L., 1999. Purification of wastewater in soil treatment systems as affected by infiltrative surface character and unsaturated soil depth. In: *WEFTEC99, Conference Proceedings of Water Environment Federation*.

Czarnecki, J.B., 2007. Ground-water models of the Mississippi River valley alluvial aquifer, USA: Management tools for a sustainable resource. In: *Procedings of Reg. IWA Conference on Groundwater Management in the Danube River Basin and Other Large Basins, Belgrade*, 7–9 June 2007. Dimkic, M.A., Brauch, H.J., and Kavanaugh, M.C., editors. Jaroslav Cerni Institute for the Development of Water Resources, Belgrade.

DAMA (Data Management Association), 2007. Available at: http://www.dama.org/public/pages/index.cfm?pageid=1. Accessed October 7, 2007.

Danskin, W.R., McPherson, K.R., and Woolfenden, L.R., 2006. Hydrology, description of computer models, and evaluation of selected water-management alternatives in the San Bernardino area, California. U.S. Geological Survey Open-File Report 2005–1278, 178 p. and 2 pl.

Daughton C.G., 2007. Pharmaceuticals and Personal Care Products (PPCPs) as environmental pollutants. National Exposure Research Laboratory, Office of Research and Development, Environmental Protection Agency, Las Vegas, NV. Presentation available at: http://www.epa.gov/nerlesd1/chemistry/pharma/index.htm. Accessed May 2007.

Databasedev.co.uk, 2007. Database solutions for Microsoft Access. Available at: http://www.databasedev.co.uk/flatfile-vs-rdbms.html. Accessed November 7, 2007.

DEUS (Department of Energy, Utilities and Sustainability)., 2004. Integrated water cycle management guidelines for NSW local water utilities. New South Wales Government, 81 p.

Delin, G.N., and Almendinger, J.E., 1993. Delineation of recharge areas for selected wells in the St. Peter- Prairie du Chien-Jordan Aquifer, Rochester, Minnesota. U.S. Geological Survey Water-Supply Paper 2397, 39 p.

Devlin, J.F., and Sophocleous, M., 2005. The persistence of the water budget myth and its relationship to sustainability. *Hydrogeology Journal*, vol. 13, pp. 549–554.

Diersch, H.J.G., 2005. FEFLOW Reference Manual. WASY GmbH, Institute for Water Resources Planning and Systems Research, Berlin.

Dillon P., and Pavelic, P., 1996. Guidelines on the quality of stormwater and treated wastewater for injection into aquifers for storage and reuse. Research Report No 109, Urban Water Research Association of Australia, Water Services Association of Australia, Melbourne.

Dillon, P., et al., 1997. Aquifer storage and recovery of stormwater runoff. *Water Journal of the Australian Water Association*, vol. 24, no. 4, pp. 7–11.

Dillon, P., and Molloy, R., 2006. Developing Aquifer Storage and Recovery (ASR) opportunities in Melbourne; Technical guidance for ASR. CSIRO Land and Water Science Report 4/06, 24 p.

Doerfliger, N., 1996. Advances in karst groundwater protection strategy using artificial tracer tests analysis and multiattribute vulnerability mapping (EPIK method). Ph.D. Thesis, University of Neuchâtel, Neuchâtel, France, 308 p.

Doerfliger, N., and Zwahlen, F., 1998. *Practical Guide, Groundwater Vulnerability Mapping in Karstic Regions (EPIK)*. Swiss Agency for the Environment, Forests and Landscape (SAEFL), Bern, Switzerland, 56 p.

Doherty J., Brebber L., and Whyte, P., 1994. *PEST–Model-independent Parameter Estimation*. User's manual. Watermark Computing. Australia.

Drew, D., and Hötzl, H., 1999. Conservation of karst terrains and karst waters: the future. In: *Karst Hydrogeology and Human Activities. Impacts, Consequences and Implications*. Drew, D., Hotzl, H., editors. International Contributions to Hydrogeology Vol. 20, International Association of Hydrogeologists, Balkema, Rotterdam/Brookfield, pp. 275–280.

DWR (Department of Water Resources)., 2003. California's groundwater. Bulletin 118, Update 2003. State of California, The Resources Agency, Department of Water Resources, 246 p.

Eckstein, G.E., 2007. Commentary on the U.N. International Law Commission's Draft Articles on the Law of Transboundary Aquifers. *Colorado Journal of International Environmental Law & Policy*, vol. 18, no. 3, pp. 537–610. Available at http://www.internationalwaterlaw.org/.

EEB (European Environmental Bureau), 2006. Protecting Europe's groundwater: Parliament achieves precautionary safeguards. Press Release, 18 October 2006 Brussels, Belgium. Accessed October 26, 2006 at: http://www.eeb.org/press/pr_groundwater_181006.htm

ESRI (Environmental Systems Research Institute, Inc.), 1994. Map projections, georeferencing spatial data. ESRI, Redlands, CA, 213 p.

ESRI, 2007. ESRI GIS Dictionary. Available at: http://support.esri.com/index.cfm?fa=knowledgebase.gisDictionary.gateway. Accessed November 1, 2007.

Falkenmark, M., 2003. *Water management and ecosystems: Living with change.* TEC Background Papers No. 9, Global Water Partnership Technical Committee (TEC), Global Water Partnership, Stockholm, Sweden, 50 p.

FAO., 1997. Seawater intrusion in coastal aquifers: Guidelines for study, monitoring and control. Water Reports 11, Food and Agriculture Organization of the United Nations, Rome, 152 p.

Federal Register., 2000. Notice of intent to prepare a draft Environmental Impact Statement for the proposed Salinas Valley Water Project, Monterey County, CA. March 3, 2000, vol. 65, no. 43, pp. 11561–11563.

Focazio, M.J., et al., 2002. Assessing ground-water vulnerability to contamination: providing scientifically defensible information for decision makers. U.S. Geological Survey Circular 1224, Reston, VA, 33 p.

Ford, R.G., Wilkin, R.T., and Puls, R.W., editors, 2007. Monitored natural attenuation of inorganic contaminants in ground water. Volume 1 – Technical basis for assessment. EPA/600/R-07/139, the U.S. Environmental Protection Agency, National Risk Management Research Laboratory, Ada, OK, 78 p.

Foster, S., Nanni, M., Kemper, K., Garduño, H., and Tuinhof, A., 2002–2005a. Utilization of non-renewable groundwater; a socially-sustainable approach to resource management. Sustainable Groundwater Management: Concepts and Tools, Briefing Note Series Note 11, GW MATE (Groundwater Management Advisory Team), the World Bank, Washington, DC, 6 p.

Foster, S., Garduño, H., Kemper, K., Tuinhof, A., Nanni, M., and Dumars, C., 2002–2005b. Groundwater quality protection defining strategy and setting priorities. Briefing Note Series Note 8, GW MATE (Groundwater Management Advisory Team), the World Bank, Washington, DC, 6 p.

Foster, S., and Loucks, D.P., editors, 2006. Non-renewable groundwater resources; A guidebook on socially-sustainable management for water-policy makers. IHP-VI, Series on Groundwater No. 10, UNESCO, Paris, 103 p.

Fox, P., 1999. Advantages of aquifer recharge for a sustainable water supply. United Nations Environmental Programme/International Environmental Technology Centre, International Symposium on Efficient Water Use in Urban Areas, Kobe, Japan, June 8–10, pp. 163–172.

Franke, O.L., Reilly, T.E., Pollock, D.W., and LaBaugh, J.W., 1998. Estimating areas contributing recharge to wells; lessons from previous studies. U.S. Geological Survey Circular 1174, 14 p.

Fuller, M.L., 1910. Underground waters for farm use. U.S. Geological Survey Water-Supply Paper 255, Washington, DC.

GADPP (Governor's Advisory Drought Planning Panel)., 2000. Critical water shortage contingency plan. December 29, 2000. Sacramento, CA.

Gale, I.N., Williams, A.T., Gaus, I., and Jones, H.K., 2002. ASR-UK: Elucidating the hydrogeological issues associated with Aquifer Storage and Recovery in the UK. British Geological Survey Report No. CR/02/156/N, UK Water Industry Research Limited, London, 45 p.

Gale, I., 2005. Strategies for Managed Aquifer Recharge (MAR) in semi-arid areas. UNESCO, Paris, 30 p.

Garber, M.S., 1986. Geohydrology of the Lloyd aquifer, Long Island, New York. U.S. Geological Survey Water-Resources Investigations Report 85–4159, 36 p.

Gerba, C.P., and Goyal, S.M., 1985. Pathogen removal from wastewater during groundwater recharge. In: *Artificial Recharge of Groundwater.* Asano, T., editor. Butterworth Publishers, Boston, MA, pp. 283–317

Gingerich, S.B., and Voss C.I., 2005. Three-dimensional variable-density flow simulation of a coastal aquifer in southern Oahu, Hawaii, USA. *Hydrogeology Journal*, vol. 13, pp. 436–450.

Glennon, R.J., 2002. *Water Follies—Groundwater Pumping and the Fate of America's Fresh Waters.* Island Press, Washington, DC, 304 p.

Gogu, R.C., 2000. Advances in groundwater protection strategy using vulnerability mapping and hydrogeological GIS databases. Ph.D. Thesis, University of Liège, Liège, France, 153 p.

Gogu, R., Carabin, G, Hallet, V., Peters, V., Dassargues, A., 2001. GIS-based hydrogeological databases and groundwater modeling. *Hydrogeology Journal*, vol. 9, pp. 555–569.

Goldscheider, N., 2002. Hydrogeology and vulnerability of karst systems – examples from the Northern Alps and the Swabian Alb. Ph.D Dissertation, University of Karlsruhe, Germany, 236 p.

Goldscheider, N., Klute, M., Sturm, S., and Hötzl, H., 2000a. Kartierung der Grundwasserverschmutzungsempfindlichkeit eines ausgesuchten Karstgebietes bei Engen, Baden-Württemberg. Final report, Karlsruhe (unpubl.), 83 p.

Goldscheider, N., Klute, M., Sturm, S., and Hötzl, H., 2000b. The PI method – a GIS based approach to mapping groundwater vulnerability with special consideration of karst aquifers. *Zeitschfirt fuer Angewandte Geologie,* vol. 46, no. 3, pp. 157–166.

Goldscheider, N., Drew, D., and Worthington, S., 2007. Introduction. In: *Methods in Karst Hydrogeology.* Goldscheider, N., Drew, D., editors. International Contributions to Hydrogeology 26. International Association of Hydrogeologists, Taylor & Francis, London, pp. 1–8.

Goodchild, M., 1996. The application of advanced information technology in assessing environmental impacts. In: *Applications of GIS to the Modeling of Non-Point Source of Pollutants in the Vadose Zone.* Corwin, D., and Loague, K., editors. American Society of Soil Sciences, Madison, WI, pp. 1–17.

Gordon, C., Wall, K., Toze, S., and O'Hara, G., 2002. Influence of conditions on the survival of enteric viruses and indicator organisms in groundwater. In: *Management of Aquifer Recharge for Sustainability.* Dillon, P.J., editor. A.A. Balkema Publishers, Lisse, the Netherlands, pp. 133–138.

Groen, J., Velstra, J., and Meesters, A.G.C.A., 2000. Salinization processes in paleowaters in coastal sediments of Suriname: evidence from $\delta^{37}Cl$ analysis and diffusion modelling, *Journal of Hydrology*, vol. 234, pp. 1–20.

Groff, J.R., and Weinberg, P.N., 1999. *SQL: The Complete Reference Guide.* Osborne/McGraw-Hill, Berkeley, CA, 994 p.

GWTF (Ground Water Task Force)., 2007. Recommendations from the EPA Ground Water Task Force; Attachment B: Ground water use, value, and vulnerability as factors in setting cleanup goals. EPA 500-R-07-001, Office of Solid Waste and Emergency Response, pp. B1–B14.

Halford, K.J., 2006. MODOPTIM: A general optimization program for ground-water flow model calibration and ground-water management with MODFLOW. U.S. Geological Survey Scientific Investigations Report 2006–5009, Carson City, NV, 62 p.

Harbaugh, A.W., and McDonald, M.G., 1996. User's documentation for Modflow-96, an update to the U.S. Geological Survey modular finite-difference ground-water flow model. U.S. Geological Survey Open-File Report 96–485, Reston, VA, 56 p.

Harbaugh, A.W., et al., 2000. Modflow-2000, the U.S. Geological Survey modular ground-water model – user guide to modularization concepts and the ground-water flow process. The U.S. Geological Survey Open-File Report 00–92, Reston, VA, 121 p.

Helsel, D.R., and Hirsch, R.M., 2002. Statistical methods in water resources. U.S. Geological Survey Techniques of Water-Resources Investigations, book 4, chap. A3, 524 p. Available at: http://water.usgs.gov/pubs/twri/twri4a3/

Hill, M.C., 1998. Methods and guidelines for effective model calibration. U.S. Geological Survey Water-Resources Investigations Report 98–4005.

Hill, M.C., et al., 2000. Modflow-2000, the U.S. Geological Survey modular ground-water model—User guide to the observation, sensitivity, and parameter-estimation processes and three post-processing programs. U.S. Geological Survey Open-File Report 00–184, 210 p.

Hirsch, R.M., 1979. Synthetic hydrology and water supply reliability. *Water Resources Research*, vol. 15, no. 6, pp. 1603–1615.

Houston, J.F.T., 1983. Ground-water systems simulation by time-series techniques. *Ground Water*, vol. 21, no. 3, pp. 301–310.

Hötzl, H., 1996. Grundwasserschutz in Karstgebieten. *Grundwasser*, vol. 1, no.1, pp. 5–11.

Hötzl, H., Adams, B., Aldwell, R., et al., 1995. Regulations. In: COST 65: Hydrogeological aspects of groundwater protection in karstic areas, Final report (COST action 65). European Comission, Directorat-General XII Science, Research and Development, Report EUR 16547 EN: 403–434; Brussels, Luxemburg.

Hutchinson, W.R., 2003. Hueco Bolson groundwater conditions and management in the El Paso area. El Paso Water Utility Hydrogeology Report 04–01. Available at: http://www.epwu.org/water/hueco_bolson.html

HydroGeoLogic Inc., 2003. MODHMS software (Version 2.0) documentation. Volume I: groundwater flow modules, Volume II: transport modules, Volume III: surface water flow modules. HydroGeoLogic Inc., Herndon, VA.

IAH (International Association of Hydrogeologists), 2002. Managing aquifer recharge. Commission on Management of Aquifer Recharge. IAH-MAR, 12 p.

International Working Group I, Jousma, G., editor, 2006. *Guideline on: Groundwater Monitoring for General Reference Purposes*. IGRAC (International Groundwater Resources Assessment Centre), Report GP 2006–1, Utrecht, the Netherlands.

Izbicki, J.A., Christensen, A.H., and Hanson, R.T., 1999. U.S. Geological Survey combined well-bore flow and depth-dependent water sampler. U.S. Geological Survey Fact Sheet 196–199, 1 p.

Jackson, T., in preparation. Management strategies for the potential restoration of Central Florida's historic Kissengen Spring. In: *Groundwater Hydrology of Springs: Engineering, Theory, Management and Sustainability*. Kresic, N., and Z. Stevanovic, Z., editors. Elsevier, New York.

Johnson, T.A., and Whitaker, R., 2003. Saltwater intrusion in the coastal aquifers of Los Angeles County, California. In: *Coastal Aquifer Management: Monitoring, Modeling, and Case Studies*. Cheng, A.H.D., Ouazar, D., editors. Lewis Publishers, Boca Raton, FL, pp. 29–48.

Jousma, G., and Roelofsen, F.J., 2004. World-wide inventory on groundwater monitoring. IGRAC, Report no. GP 2004-1. International Groundwater Resources Assessment Centre, Utrecht, the Netherlands.

Kemper, K., Foster, S., Garduño, H., Nanni, M., and Tuinhof, A., 2002–2005. Economic instruments for groundwater management: using incentives to improve sustainability. Sustainable Groundwater Management: Concepts and Tools, Briefing Note Series Note 7, GW MATE (Groundwater Management Advisory Team), The World Bank, Washington, DC, 8 p.

Kennedy, M., and Kopp, S., 2000. *Understanding Map Projections.* Environmental Systems Research Institute, Inc., Redlands, CA, 98 p.

Kim, S-J., Hyun, Y., and Lee, K-K., 2005. Time series modeling for evaluation of groundwater discharge rates into an urban subway system. *Geosciences Journal*, vol. 9, no. 1, pp. 15–22.

Kinzelbach, W., Bauer, P., Siegfried, T., and Brunner, P., 2007. Sustainable groundwater management—problems and scientific tools. In: G-WADI 2007, International Training Workshop on Groundwater Modeling in Arid and Semi-arid Areas, Course Material, 11–15 June 2007 Lanzhou, China, UNESCO, Asian G-WADI Network, Cold and Arid Regions Environmental Research Institute, Chinese Academy of Sciences, pp. 39–44.

Klemeš, V., 1978. Physically based stochastic hydrologic analysis. In: *Advances in Hydroscience*, vol 11. Chow, V.T., editor. Academic Press, New York, pp. 285–352.

Knotteres, M., and Bierkens, M.F.P., 2000. Physical basis of time series models for water table depths. *Water Resources Research*, vol. 36, no. 1, pp. 181–188.

Kooi, H., and Groen, J., 2003. Geological processes and the management of groundwater resources in coastal areas. *Netherlands Journal of Geosciences*, vol. 82, pp. 31–40.

Koch, R.W., 1985. A stochastic streamflow model based on physical principles. *Water Resources Research*, vol. 21, no. 4, pp. 545–553.

Kooi, H., Groen, J., and Leijnse, A., 2000. Modes of seawater intrusion during transgressions. *Water Resources Research*, vol. 36, no. 12, pp. 3581–3589.

Kraemer, S.R., Haitjema, H.M., and Kelson, V.A., 2005. Working with WhAEM2000; Capture zone delineation for a city wellfield in a valley fill glacial outwash aquifer supporting wellhead protection. EPA/600/R-00/022, United States Environmental Protection Agency, Office of Research and Development, Washington, DC, 77 p.

Kresic, N., 1995. Stochastic properties of spring discharge. In: *Toxic Substances and the Hydrologic Sciences.* Dutton, A.R., editor. American Institute of Hydrology, Minneapolis, Minnesota, pp. 582–590.

Kresic, N., 1997. *Quantitative Solutions in Hydrogeology and Groundwater Modeling.* Lewis Publishers/CRC Press, Boca Raton, FL, 461 p.

Kresic, N., 2007a. *Hydrogeology and Groundwater Modeling*, 2nd ed. CRC Press, Taylor & Francis Group, Boca Raton, FL, 807 p.

Kresic, N., 2007b. Hydraulic methods. In: *Methods in Karst Hydrogeology*, Goldscheider, N., Drew, D., editors. International Contributions to Hydrogeology 26, International Association of Hydrogeologists, Taylor & Francis, London, pp. 65–92.

Kresic, N., and Rumbaugh, J., 2000. GIS and data management for ground water modeling, National Ground Water Education Foundation Course, June 2001 San Francisco, National Ground Water Association, Westerville, OH.

Lallana, C., Krinner, W., Estrela, T., Nixon, S., Leonard, J., and Berland, J.M., 1999. Sustainable water use in Europe. Part 2: Demand management. Environmental Issue report No. 19, European Environment Agency, Copenhagen, Denmark, 94 p.

Larabi, A., 2007. Development and application of seawater intrusion models: theory, analytical, benchmarking and numerical developments, applications. In: G-WADI 2007. International Training Workshop on Groundwater Modeling in Arid and Semi-arid Areas, Course Material, 11–15 June 2007 Lanzhou, China, UNESCO, Asian G-WADI Network, Cold and Arid Regions Environmental Research Institute, Chinese Academy of Sciences, pp. 153–197.

Lee, J.-Y., and Lee, K.-K., 2000. Use of hydrologic time series date for identification of recharge mechanism in a fractured bedrock aquifer system. *Journal of Hydrology*, vol. 229, no. 3–4, pp. 190–201.

Leibundgut, C., 1998. Vulnerability of karst aquifers (keynote paper). *IAHS Publication*, vol 247, pp. 45–60.

Lin, H.C., Richards, D.R., Yeh, G.T., Cheng, J.R., Cheng, H.P., and Jones, N.L., 1997. FEMWATER: A three-dimensional finite element computer model for simulating density-dependent flow and transport in variably saturated media. Technical Report CHL-97–12. Waterways Experiment Station, U.S. Army Corps of Engineers, Vicksburg, MS, 39180–6199.

Magiera, P., 2000. Methoden zur Abschätzung der Verschmutzungsempfindlichkeit des Grundwassers. *Grundwasser*, vol. 3, pp. 103–114.

Margeta, J., and FistaniÊ, I., 2004. Water quality modelling of Jadro Spring. *Water Science and Technology*, vol. 50, no. 11, pp. 59–66.

McCuen, R.H., and Snyder, W.M., 1986. *Hydrologic Modeling. Statistical Methods and Applications*. Prentice-Hall, Englewood Cliffs, NJ, 568 p.

McDonald, M.G., and Harbaugh, A.W., 1988. A modular three-dimensional finite-difference ground-water flow model. U.S. Geological Survey Techniques of Water-Resources Investigations, book 6, chap. A1, 586 p.

McKee, P.W., and Clark, B.R., 2003. Development and calibration of a ground-water flow model for the Sparta aquifer of Southeastern Arkansas and North-Central Louisiana and simulated response to withdrawals, 1998–2027. U.S. Geological Survey Water-Resources Investigations Report 03–4132, Little Rock, AR, 71 p.

MCWRA (Monterey County Water Resources Agency), 2007. 2005 Ground water summary report. http://www.mcwra.co.monterey.ca.us/welcome_svwp_n.htm.

Medema, G.J., and Stuyzand, P.J., 2002. Removal of micro-organisms upon recharge, deep well injection and river bank infiltration in the Netherlands. In: *Management of Aquifer Recharge for Sustainability*. A.A. Balkema Publishers, Lisse, the Netherlands, pp. 125–131.

Mehan, G.T., 2003. Memorandum to regional administrators; source water assessment and protection. United States Environmental Protection Agency, Office of Water, Washington, DC, 6 p. Available at: http://www.epa.gov/safewater/sourcewater/pubs/fs_statusbyregion_mehan2003.pdf. Accessed February 14, 2008.

Meinzer, O.E., 1932 (reprint 1959). Outline of methods for estimating ground-water supplies. Contributions to the hydrology of the United States, 1931. U.S. Geological Survey Water-Supply Paper 638-C, Washington, DC, p. 99–144.

Mendenhall, W.C., 1905. Hydrology of San Bernardino Valley, California. U.S. Geological Survey Water-Supply Paper 142, 124 p.

Microsoft, 1994. *User's Guide to Microsoft Access, Relational Database Management System for Windows Version 2.0*. Microsoft Corporation, USA, 819 p.

Montanari, A., Rosso, R., and Taqqu, M.S., 2000. A seasonal fractional ARIMA model applied to the Nile River monthly flows at Aswan. *Water Resources Research*, vol. 36, no. 5, pp. 1249–1259.

National Research Council, 1994. *Ground Water Recharge Using Waters of Impaired Quality.* National Academy Press, Washington, DC, 382 p.

Nellor, M.H., Baird, R.B., and Smyth, J.R., 1985. Health effects of indirect potable reuse. *Water Journal of the Australian Water Association*, vol. 77, no. 7, pp. 88–96.

NISO (National Information Standards Organization), 2004. *Understanding Metadata.* NISO Press, Bethesda, MD, 16 p.

NRC (National Research Council), 2000. *Grand Challenges in Environmental Sciences.* Committee on Grand Challenges in Environmental Sciences, National Academy Press, Washington, DC, 96 p.

Nueces River Authority, 2007. Edwards Aquifer recharge dams. Available at http://www.nueces-ra.org/II/recharge/. Accessed date December 2007.

Olsthoorn, T.N., and Mosch, M.J.M., 2002. Fifty years artificial in the Amsterdam dune area. In: *Management of Aquifer Recharge for Sustainability.* A.A. Balkema Publishers, Lisse, the Netherlands, pp. 29–33.

Pace, 2007. PacePort, Pace Analytical Online Customer Data Management System. Available at: http://www.pacelabs.com/. Accessed November 17, 2007.

Padilla, A., Pulido-Bosch, A., Calvache, M.L., and Vallejos, A., 1996. The ARMA models applied to the flow of karstic springs. *Water Resources Research*, vol. 32, no. 5, pp. 917–928.

Peek, H.M., 1951. Cessation of flow of Kissengen Spring in Polk County, Florida. Florida Geological Survey Report of Investigation No. 7, part III.

Pereira, L.S., Cordery, I., and Iacovides, I., 2002. *Coping with Water Scarcity.* International Hydrological Programme VI, Technical Documents in Hydrology No. 58, UNESCO, Paris, 269 p.

Petkewich, M.D., et al., 2004. Hydrologic and geochemical evaluation of aquifer storage recovery in the Santee Limestone/Black Mingo aquifer, Charleston, South Carolina, 1998–2002. U.S. Geological Survey Scientific Investigations Report 2004–5046, Reston, VA, 81 p.

Phillips, S.P. et al., 2003. Analysis of tests of subsurface injection, storage, and recovery of freshwater in Lancaster, Antelope Valley, CA. U.S. Geological Survey Water-Resources Investigations Report 03–4061, Sacramento, California, 122 p.

Piet, G.J., and Zoeteman. B.C.J., 1980. Organic water quality changes during sand bank and dune filtration of surface waters in the Netherlands. *Water Journal of the Australian Water Association*, vol. 72, no. 7, pp. 400–414.

Poeter, E.P., and Hill, M.C., 1998. Documentation of UCODE, a computer code for universal inverse modeling. U.S. Geological Survey Water-Resources Investigations Report 98–4080.

Post, V.E.A., and Kooi, H., 2003. On rates of salinization by free convection in high-permeability sediments; insights from numerical modelling and application to the Dutch coastal area. *Hydrogeology Journal*, vol. 11, pp. 549–559.

Potter, H.G., 2004. History and evolution of the Rule of Capture. In: 100 *Years of Rule of Capture from East to Groundwater Management.* Mullican, W.F., Schwartz, S., editors. Texas Water Development Board Report 361, Austin, TX, pp. 1–9.

Pyne, R.D.G., 1995. *Groundwater Recharge and Wells: A Guide to Aquifer Storage Recovery.* Lewis Publishers, Boca Raton, FL, 375 p.

Radu, C.G., Carabin, G., Hallet, V., Peters, V., and Dassargues, A., 2001. GIS-based hydrogeological databases and groundwater modelling. *Hydrogeology Journal*, vol. 9, pp. 555–569.

Reese, R.S., 2002. Inventory and review of aquifer storage and recovery in southern Florida. U.S. Geological Survey Water-Resources Investigations Report 02–4036, Tallahassee, FL, 56 p.

Reilly, T.E., and Harbaugh, A.W., 2004. Guidelines for evaluating ground-water flow models. U.S. Geological Survey Scientific Investigations Report 2004–5038, 30 p.

Riordan, M.J., 2007. Septic system checkup: The Rhode Island Handbook for Inspection. Rode Island Department of Environmental Management. Available at: www.state. RI.us/dem.

Rogers, P., and Hall, A.W., 2003. Effective water governance. TEC Background Papers No. 7, Global Water Partnership Technical Committee (TEC), Global Water Partnership, Stockholm, Sweden, 44 p.

Rosenshein, J.S., and Hickey, J.J., 1977. Storage of treated sewage effluent and storm water in a saline aquifer, Pinellas Peninsula, Florida. *Ground Water*, vol. 15, no. 4, pp. 289–293.

Rostad, K., 2002. Fate of disinfection by-products in the subsurface. In: *U.S. Geological Survey Artificial Recharge Workshop Proceedings*, April 2–4, 2002, Sacramento, CA, Aiken, G.R., and Kuniansky, E.L., editors. U.S. Geological Survey Open-File Report 02–89, pp. 27–30.

Rubinic, J., and Fistanic I., 2005. Application of time series modeling in karst water management. In: *Water Resources and Environmental Problems in Karst, Proceeding of the International Conference and Field Seminar*, Stevanovic, Z., Milanovic, P., editors. Belgrade & Kotor, pp. 417–422.

Rumbaugh, J.O., and Rumbaugh, D.B., 2004. *Guide to Using Groundwater Vistas, Version 4*. Environmental Simulations, Inc., Reinholds, Pennsylvania, 366 p.

Rupert, M.G., 1999. Improvements to the DRASTIC ground-water vulnerability mapping method. U.S. Geological Survey Fact Sheet FS-066–99, 6 p.

Rupert, M.G., 2001. Calibration of the DRASTIC ground water vulnerability mapping method. *Ground Water*, vol. 39, no. 4, pp. 625–630.

Salas, J.D., Boes, D.C., and Smith, R.A., 1982. Estimation of ARMA models with seasonal parameters. *Water Resources Research*, vol. 18, no. 4, pp. 1006–1010.

Salas, J.D., Delleur, J.W., Yevjevich, V., and Lane, W.L., 1985. *Applied Modeling of Hydrologic Time Series*. Water Resources Research Publication, Littleton, CO.

Salas, J.D., and Obeysekera, J.T.B., 1992. Conceptual basis of seasonal streamflow time series models. *Journal of Hydraulic Engineering*, vol. 118, no. 8, pp. 1186–1194.

Sax, J.L., Thompson, B.H., Leshy, J.D., and Abrams, R.H., 2000. *Legal Control of Water Resources*, 3rd ed. American Casebook Series, West Group, St. Paul, MN, 956 p.

Schneider Electric, 1999. Automation and control of a ground water based municipal water system. Schneider Electric, North Andover, MA, 4 p.

Shibuo, Y., Jarsjo, J., and Destouni, G., 2006. Bathymetry-topography effects on saltwater-fresh groundwater interactions around the shrinking Aral Sea. *Water Resources Research*, vol. 42: Art. No. W11410.

Simmons, C.T., et al., 2007. Variable density groundwater flow: From modelling to applications. In: G-WADI 2007. International Training Workshop on Groundwater Modeling in Arid and Semi-arid Areas, Course Material, 11–15 June 2007 Lanzhou, China, UNESCO, Asian G-WADI Network, Cold and Arid Regions Environmental Research Institute, Chinese Academy of Sciences, pp. 85–152.

Slichter, C.S., 1902. The motions of underground waters. U.S. Geological Survey Water-Supply and Irrigation Papers 67, Washington, DC, 106 p.

Sloss, E, Geschwind, S.A., McCaffrey, D.F., and Ritz, B.R., 1996. Groundwater recharge with reclaimed water: An epidemiologic assessment in Los Angeles County, 1987–1991. RAND Corporation. Santa Monica, CA.

Sloss, E., McCaffrey, D.F., Fricker, R.D., Geschwind, S.A., and Ritz, B.R., 1999. Groundwater recharge with reclaimed water birth outcomes in Los Angeles County, 1982–1993. RAND Corporation. Santa Monica, CA.

Smolensky, D.A., Buxton, H.T., and Shernoff, P.K., 1989. Hydrologic framework of Long Island, New York. U.S. Geological Survey Hydrologic Investigations Atlas HA-709, 3 sheets, scale 1:250000.

Solanes, M., 2002. Water: rights, flexibility and governance: A balance that matters. ECLAC, paper for the Natural Resources Law Centre Conference, Boulder, CO.

Sontheimer, H. 1980. Experience with riverbank filtration along the Rhine River. *Water Journal of the Australian Water Association*, vol. 72, no. 7, pp. 386–390.

Strassberg, G., 2005. A geographic data model for groundwater systems. Doctor of Philosophy Dissertation, University of Texas at Austin, 232 p.

Stuart, M.E., Gaus, I., Chilton, P.J., and Milne, C.J., 2003. Development of a methodology for selection of determinand suites and sampling frequency for groundwater quality monitoring. National Groundwater and Contaminated Land Centre Project NC/00/35, Environment Agency, Olton, West Midlands, 89 p.

Texas Administrative Code, 2006. Title 30 (Environmental Quality), Part 1 (Texas Commission on Environmental Quality), Chapter 330 (Municipal Solid Waste), Subhapter J (Groundwater Monitoring and Corrective Action), Rule 330.403 (Groundwater Monitoring Systems), effective March 27, 2006, 31 TexReg 2502.

TGPC, 2003. Texas *groundwater protection strategy*. Texas Groundwater Protection Committee Publication AS-188, Texas Commission on Environmental Quality, the State of Texas, Austin, TX, 101 p.

The European Parliament and the Council of the European Union, 2000. Directive 2000/60/EC of the European Parliament and the Council of 23 October 2000 establishing a framework for Community action in the field of water policy (EU Water Framework). *Official Journal of the European Union*, 22 December 2000, pp. L 327/1-

The European Parliament and the Council of the European Union, 2006. Directive 2006/118/EC on the protection of groundwater against pollution and deterioration. *Official Journal of the European Union*, 27 December 2006, pp. L 372/19–31.

The New York Times, 2007. Long Island; Water Wisdom. Published October 28, 2007. http://www.nytimes.com/2007/10/28/opinion/nyregionopinions/LI-Aquifer.html?

Thomas, H.A., and Fiering, M.B., 1962. Mathematical synthesis of streamflow sequences for the analysis of river basin by simulation. In: *Design of Water Resources Systems*, Maas, A., et al., editors. Harvard University Press, Cambridge, MA, pp. 459–493.

Toze, S., and Hanna. J., 2002. The Survival Potential of Enteric Microbial Pathogens in a Reclaimed Water ASR Project. In: *Management of Aquifer Recharge for Sustainability*. A. A. Balkems Publishers, Lisse, the Netherlands, pp. 139–142.

Trimmel, H., editor, 1998. *Die Karstlandschaften der Österreichischen Alpen und der Schutz Ihres Lebensraumes Und Ihrer Natürlichen Ressourcen*. Fachausschuß Karst, CIPRA-Österreich, Vienna, 119 p.

Tsuchihashi, R., Asano, T., and Sakaji, R.H., 2002. Health aspects of groundwater recharge with reclaimed water. In: *Management of Aquifer Recharge for Sustainability*. A.A. Balkema Publishers, Lisse, the Netherlands, pp. 11–20.

Tuinhof, A., Dumars, C., Foster, S., Kemper, K., Garduño, H., and Nanni, M., 2002–2005a. *Groundwater resource management; an introduction to its scope and practice*. Sustainable Groundwater Management: Concepts and Tools, Briefing Note Series Note 1, GW MATE (Groundwater Management Advisory Team), The World Bank, Washington, DC, 6 p.

Tuinhof, A., Foster, S., Kemper, K., Garduño, H., and Nanni, M., 2002–2005b. *Groundwater monitoring requirements for managing aquifer response and quality threats*. Sustainable Groundwater Management: Concepts and Tools, Briefing Note Series Note 9, GW MATE (Groundwater Management Advisory Team), The World Bank, Washington, DC, 10 p.

UN/ECE (United Nations Economic Commission for Europe), 1992. Convention on the Protection and Use of Transboundary Watercourses and International Lakes, 17 March 1992 Helsinki, Finland.

UNESCO, 2006. Recharge systems for protecting and enhancing groundwater resources. Proceedings of the 5th International Symposium on Management of Aquifer Recharge ISMAR5, Berlin, Germany, 11–16 June 2005 United Nations Educational, Scientific and Cultural Organization, Paris, France, 913 p.

USEPA, 1991. Protecting the nation's ground water: EPA's strategy for the 1990's. The final report of the EPA Ground-Water Task Force. 21Z-1020, Office of the Administrator.

USEPA, 1993a. A review of methods for assessing aquifer sensitivity and ground water vulnerability to pesticide contamination: U.S. Environmental Protection Agency, EPA/813/R–93/002, 147 p.

USEPA, 1993b. Guidelines for delineation of wellhead protection areas. EPA/440/5–93-001, U.S. Environmental Protection Agency. Office of Water Office of Ground Water Protection, Washington, DC.

USEPA, 1994. Assessment framework for ground-water model applications. OSWER Directive 9029.00, U.S. Environmental Protection Agency, Office of Solid Waste and Emergency Response, Washington, DC.

USEPA, 1999. Safe Drinking Water Act, Section 1429, Ground Water Report to Congress. EPA-816-R-99–016, United States Environmental Protection Agency, Office of Water, Washington, DC.

USEPA, 2000. National Water Quality Inventory; 1998 Report to Congress; Ground water and drinking water chapters. EPA 816-R-00–013, Office of Water, Washington, DC.

USEPA, 2003b. Overview of the Clean Water Act and the Safe Drinking Water Act. http://www.epa.gov/OGWDW/dwa/electronic/ematerials/. Accessed September 12, 2005.

USEPA, 2004. Guidelines for water reuse. EPA/625/R-04/108, U.S. Environmental Protection Agency, Office of Wastewater Management, Office of Water, Washington, DC, 460 p.

USEPA, 2005. National management measures to control nonpoint source pollution from urban areas. EPA-841-B-05–004, United States Environmental Protection Agency, Office of Water, Washington, DC.

USEPA, 2006. Final Ground Water Rule. Fact Sheet, EPA 815-F-06–003, United States Environmental Agency, Office of Water, Washington, DC, 2 p.

USEPA, 2007. Long-term groundwater monitoring optimization Clare Water Supply Superfund Site, StageRight Area, Clare, Michigan. EPA 542-R-07–009, Solid Waste and Emergency Response.

Vecchia, A.V., Obeysekera, J.T.B., Salas, J.D., and Boes, D.C.,1983. Aggregation and estimation for low-order periodic ARMA models. *Water Resources Research*, vol. 19, no. 5, pp. 1297–1306.

Voss, C.I., and Provost, A.M., 2002. SUTRA; A model for saturated-unsaturated, variable-density ground-water flow with solute or energy transport. U.S. Geological Survey Water-Resources Investigations Report 02–4231, Reston, VA, 250 p.

Vrba, J., and Zaporozec, A., editors, 1994. *Guidebook on Mapping Groundwater Vulnerability*, International Contributions to Hydrogeology, Vol. 16. International Association of Hydrogeologists (IAH), Swets & Zeitlinger Lisse, Munich, 156 p.

WHG, 1996. Gesetz zur Ordnung des Wasserhaushalts (Wasserhaushaltsgesetz), BGBl, Bonn, Germany.

Werner, A.D., and Gallagher, M.R., 2006. Characterisation of sea-water intrusion in the Pioneer Valley, Australia using hydrochemistry and three-dimensional numerical modelling. *Hydrogeology Journal*, vol. 14, pp. 1452–1469.

Werner, A.D., and Lockington, D.A., 2006. Tidal impacts on riparian salinities near estuaries. *Journal of Hydrology*, vol. 328, pp. 511–522.

Williams, S.D., and Farmer, J.J., 2003. Volatile organic compound data from three karst springs in middle Tennessee, February 2000 to May 2001. U.S. Geological Survey Open-File Report 03–355, Nashville, TN, 69 p.

Winpenny, J., 2003. *Financing water for all*. Report of the World Panel on Financing Water Infrastructure. World Water Council, 3rd World Water Forum, Global Water Partnership, 54 p.

Worthington, S.R.H., Smart, C.C., and Ruland, W.W., 2003. Assessment of groundwater velocities to the municipal wells at Walkerton. In: *Proceeding of the 2002 Joint Annual Conference of the Canadian Geotechnical Society and the Canadian Chapter of the IAH*, Niagara Falls, ON, pp. 1081–1086.

Wyoming DEQ (Department of Environmental Quality), 2007. Wyoming's wellhead protection (WHP) program guidance document. Available at: http://www.wrds.uwyo.edu/wrds/deq/whp/whpcover.html. Accessed November 21, 2007.

Yechieli, Y., Kafri, U., Goldman, M., and Voss, C.I., 2001. Factors controlling the configuration of the fresh-saline water interface in the Dead Sea coastal aquifers: synthesis of TDEM surveys and numerical groundwater modeling. *Hydrogeology Journal*, vol. 9, no. 4, pp. 367–377.

Yobbi, D.K., 1997. Simulation of subsurface storage and recovery of effluent using multiple wells, St. Petersburg, Florida. U.S. Geological Survey Water-Resources Investigations Report 97–4024, Tallahassee, FL, 30 p.

CHAPTER 9

Groundwater Restoration

9.1 Introduction

The debate on the definition and objectives of groundwater restoration is an ongoing one. Its nature is often, and somewhat misleadingly, simplified to emphasize the diverging interests of "polluters" and "protectors" of groundwater, with the public caught somewhere in between. In reality, however, the complexity of groundwater contamination and the feasibility of groundwater restoration in many cases do not support this simplification. In addition, there are different opinions at various regulatory levels (local, state, and federal) as to the objectives and goals of groundwater restoration. The following excerpts are from a memorandum written by a regulator in the United States, referencing the United States Environmental Protection Agency (USEPA) discussion paper entitled *Ground Water Use, Value, and Vulnerability as Factors in Setting Cleanup Goals*:

> I think there is general agreement that it would be very difficult to set a national policy covering the many diverse issues raised in this discussion paper and by the commenters. There is abundant literature and discussion developed by the scientific community and commercial development interests that ground water has value; and its use, protection, and cleanup need to be addressed to varying degrees. Ground water Use, Value, and Vulnerability (UVV) has been, and most likely will remain, a state and local government issue because it is intrinsically bound to land use (e.g. zoning) and development. These have been traditionally viewed as being under the authority of state and local governments. Dealing with UVV issues is a site-specific process that depends on a large number of local factors, both scientific and political.
>
> The role of USEPA should be largely one of educator to the regulators, the regulated community, and the public. This education would be more technical than policy oriented. EPA could also act as discussion facilitator and, if asked, mediator on UVV issues affecting a site and, *as far as possible*, allow state and local regulators to continue to make decisions regarding setting cleanup goals protective of existing and potential drinking water supplies. These decisions would involve input from EPA and other federal interests as well as from an active public participation program. Informed decisions by local regulators are the most effective way to address UVV issues.
>
> Addressing these issues needs to be a dynamic interactive process and not just guidance documents and policy statements. There are certainly plenty of those. A comprehensive national UVV guidance would be hard to formulate and would be ignored by many. Any national policy should be general in nature recognizing that each State has its own UVV issues most of which require unique solutions (Pierce, 2004).

The main recommendation to the USEPA by this state regulator is to set an overall tone of good environmental stewardship and adopt a national policy statement regarding groundwater goals for the agency. This statement should reflect the view that (1) no aquifer will be degraded and (2) any degraded aquifer ultimately needs to be restored to its natural condition.

The above view of a state regulator is far from being a lonely one, and many stake-holders determined to protect the environment have commended this and similar views. What makes achieving restoration of some already-contaminated aquifers to their natural condition difficult is the nature of contamination and the "stubbornness" of certain complex hydrogeologic environments. Unfortunately, the nature of contamination can, only to a limited extent, be controlled or mitigated at a local "land use and development" level, whereas hydrogeologic conditions cannot be controlled at all. In some cases, in order to restore contaminated aquifers to their natural condition, the local socioeconomic structure, as well as the laws of the society, would have to be radically changed. Even then, it may take tens or hundreds of years for aquifers to return to their natural condition, assuming that the definition of natural conditions means the absence of any anthropogenic substances in groundwater.

Two examples of non-point-source groundwater contamination that require both regulatory and local land use changes, in order to restore aquifers to their natural condition, are the use of pesticides and of fertilizers. The United Kingdom Environment Agency reported that pesticides were found in over a quarter of groundwater-monitoring sites in England and Wales in 2004, and, in some cases, these exceeded the drinking water limit. Atrazine is a weed killer used mainly to protect maize (corn) crops, and it was used in the past to maintain roads and railways. It has been a major problem, but, since the nonagricultural uses were banned in 1993, concentrations in groundwater have gradually declined. A complete ban on all use of atrazine (and simazine, another pesticide) in the United Kingdom was planned to be phased in between 2005 and 2007 but has not been implemented as of February 2008. As noted by the agency, banned pesticides can remain a problem for many years after these were last used (Environment Agency, 2007). Some other European countries have banned the use of atrazine: France, Sweden, Norway, Denmark, Finland, Germany, Austria, Slovenia, and Italy. In contrast, the USEPA has concluded that the risks from atrazine for approximately 10,000 community drinking water systems using surface water are low and did not ban this pesticide, which continues to be the most widely used pesticide in the United States. Incidentally, as stated by the agency, 40,000 community drinking water systems using groundwater were not included in the related study, and private wells used for water supply were not mentioned in the agency's decision to allow continuous use of atrazine (USEPA, 2003).

The United Kingdom Environment Agency also reported that in 2004, almost 15 percent of monitoring sites in England (none in Wales) had an average nitrate concentration that exceeded 50 mg/L, the upper limit for nitrate in drinking water (for comparison, groundwater naturally contains only a few mg/L of nitrate). Water with high nitrate levels has to be treated or diluted with cleaner water to reduce concentrations. More than two-thirds of the nitrate in groundwater comes from past and present agriculture, mostly from chemical fertilizers and organic materials. It is estimated that over 10 million tons per year of organic material is spread on the land in the United Kingdom. More than 90 percent of this is animal manure; the rest is treated sewage sludge, green waste compost, paper sludge, and organic industrial wastes. Other major sources of nitrate are leaking sewers, septic tanks, water mains, and atmospheric deposition. Atmospheric deposition of nitrogen makes a significant contribution to nitrate inputs to groundwater. A study in the Midlands concluded that around 15 percent of the nitrogen leached from soils came from the atmosphere. The agency estimates that 60 percent of groundwater bodies in England and 11 percent in Wales are at risk of failing Water Framework Directive objectives because of high nitrate concentrations (Environment Agency, 2007).

An illustrative study of non-point-source contamination from agricultural activities was performed by the Kentucky Geological Survey in 1990s (Currens, 1999). The Pleasant Grove Spring Basin in southern Logan County, KY, was selected for the study because it is largely free of nonagricultural pollution sources. About 70 percent of the watershed is in crop production and 22 percent is pasture. The area is underlain by karst geology, and the groundwater flow in the basin is divided into a diffuse (slow) flow and a conduit (fast) flow regime. The diffuse- and conduit-flow regimes have a major influence on the timing of contaminant maxima and minima in the spring during and after major rainfall events. Nitrate is the most widespread, persistent contaminant in the basin, but concentrations average 5.2 mg/L basin-wide and generally do not exceed the drinking water maximum contaminant level (MCL) of 10 mg/L set by the USEPA. Atrazine has been consistently, and other pesticides occasionally, detected. Concentrations of triazines (including atrazine) and alachlor have exceeded drinking water MCLs during peak spring flows, as illustrated in Fig. 9.1. Maximum concentrations of triazines, carbofuran, metolachlor, and alachlor in samples from Pleasant Grove Spring were 44.0, 7.4, 9.6, and 6.1 μg/L, respectively (see Table 5.2 in Chap. 5 for the corresponding MCLs). Flow-weighted average concentrations for 1992 to 1993 were 4.91 μg/L for atrazine-equivalent triazines and 5.0 mg/L for nitrate nitrogen. In comparison, the maximum allowed concentration of any individual pesticide in drinking water in the European Union is 0.1 μg/L, and of all pesticides combined it is 0.5 μg/L.

The hydrogeology of the basin is a significant controlling influence on the temporal variation of contaminant concentrations. The fast-flow karst conduit region is characterized by intermediate concentrations of nitrate and pesticides during low flow but substantially higher concentrations of triazines and lower concentrations of nitrate during high flow. The diffuse (slow) flow regime, which is estimated to represent slightly less than half of the basin, drains into the area dominated by conduit flow. The diffuse-flow region has persistently higher concentrations of nitrate but lower, less variable

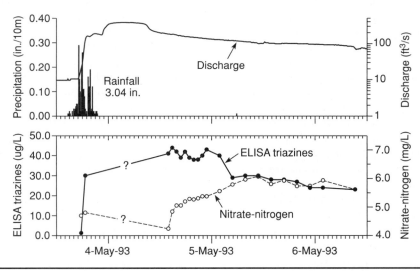

Figure 9.1 Chemographs and discharge hydrograph for the May 1993 high-flow event at Pleasant Grove Spring, illustrating highest triazine concentrations determined to date. (From Currens, 1999.)

concentrations of triazines. The diffuse, slow-flow area is acting as a reservoir of agricultural chemicals, maintaining a background level of triazines and nitrate during low flow in the conduit-flow regime. Triazine concentrations are significantly higher during high flow, while nitrate concentrations are diluted.

As concluded by Currens (1999), both municipal and domestic water supplies derived from groundwater can be adversely affected. Best management practices (BMP) implementation in the basin should focus on controlling animal waste, controlling crop field runoff with associated sediment and pesticide loss, and using more efficient methods of applying nutrients. A strong education program on groundwater protection is highly recommended.

The above examples illustrate that, even when the local community and local or possibly even state regulators are unified in their desire to restore an aquifer to its natural condition, it is not possible to do so without changing regulations at a higher (e.g., federal) level. Changing regulations, however, would be only the first necessary step. In order to restore a major aquifer contaminated from nonpoint sources to its natural condition in a meaningful period (e.g., several generations), the remediation measures would have to be extremely costly and decades long and would have to be paid and implemented by the wider society. Such efforts are, therefore, seldom, if ever, undertaken, and the restoration of aquifers to their "natural" condition is left to the natural attenuation processes, while the groundwater users are becoming accustomed to drinking treated water.

The following example shows that a widely spread, non-point-source groundwater contamination, is not the only one where restoration of groundwater (aquifer) to pristine conditions within several years or decades may not be a feasible approach.

Groundwater historically has been the sole source of water supply for the Town of Yucca Valley in the Warren sub-basin of the Morongo groundwater basin in California, the United States. From the late 1940s through 1994, water levels in the Warren sub-basin declined as much as 300 ft due to groundwater extraction. In response, the Hi-Desert Water District (HDWD) instituted an artificial recharge program in 1995 to replenish the groundwater basin using imported California State Water Project (SWP) water. The artificial recharge program resulted in water level recovery of about 250 ft between 1995 and 2001; however, NO_3 concentrations in some wells also increased from a background concentration of about 10 mg/L to more than the USEPA MCL of 44 mg/L (10 mg/L as nitrogen). In 1998, 3 years after the start of the artificial recharge program, the NO_3 concentrations increased to as high as 110 mg/L in hydrogeologic units where the recharge ponds were located. The highest NO_3 concentrations were in wells perforated in the upper and middle aquifers (Fig. 9.2).

Wastewater from all homes and businesses in Yucca Valley is disposed of using septic tanks that separate the floating and settleable solids from the wastewater and discharge the wastewater through leach lines. The wastewater percolates from the leach lines through the unsaturated zone and continues to travel toward the underlying saturated zone. The quantity of septic-tank wastewater ("septage") potentially recharging the underlying groundwater at any given time at a household can be estimated by assuming an average per capita septic-tank discharge of 70 gal/d (Eckenfelder, 1980). Bouwer (1978) reported that nitrogen concentrations in septage can range from 40 to 80 mg/L, mostly in the form of ammonium. If all of the nitrogen was converted to NO_3, then concentrations could range from 177 to 354 mg/L. A sample of septage collected from a residential septic tank in the Town of Yucca Valley had a NO_3 concentration of about 154 mg/L (Nishikawa et al., 2003). Samples of septage from five different septic tanks in

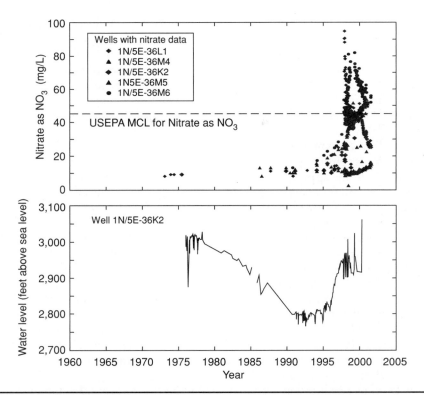

FIGURE 9.2 Nitrate concentrations in samples from selected Hi-Desert Water District production wells in the mid-west hydrogeologic unit (top), and water level hydrograph for the production well 1N/5E-36K2, Warren groundwater basin, San Bernardino County, CA. (Modified from Nishikawa et al., 2003.)

the nearby Victorville, California, had NO_3 concentrations ranging from 97 to 280 mg/L and averaged 208 mg/L (Umari et al., 1995).

Water-quality and stable-isotope data, collected after the start of the artificial recharge program in Yucca Valley, indicate that mixing had occurred between artificially recharged imported water and native groundwater. Nitrate-to-chloride and nitrogen-isotope data, as well as analyses for caffeine and selected human pharmaceutical products, indicated that septic tank leakage was the source of the measured increase in NO_3 concentrations. The rapid rise in water levels resulting from the artificial recharge program entrained the large volume of septage that was stored in the unsaturated zone, resulting in a rapid increase in NO_3 concentrations (Nishikawa et al., 2003). Presently, there are 18 public supply wells in the Warren groundwater basin, some of which have elevated nitrate levels and require water treatment.

Restoring the aquifer to its natural condition in the case of Yucca Valley would certainly be a very costly proposition, requiring various local ordinances and other measures, of which a centralized sewer system and a wastewater treatment plant would be necessary parts of the overall groundwater remediation efforts. Even if the ideal restoration measures were to be fully implemented in a short period, the residents of the Yucca Valley would still have to drink treated groundwater, likely for decades. In this and similar cases, the local community would have to make its own decision, based on cost-benefit and risk analyses, as to the aquifer restoration efforts (unless there is a legislation addressing the issue at higher levels of government). Since "everybody" is responsible for the contamination, and everybody is dependent on the contaminated groundwater for all uses including for drinking, it is very unlikely that the regulators would force individual households to remediate their own little plumes (point sources). It is even less likely that any lawsuits would be brought up by any of the stakeholders against individual households.

In contrast, when point-source groundwater contamination is caused by known "potentially responsible parties" (PRPs) whose pockets are deeper than those of individual households, the approaches to groundwater remediation would, in many cases, depend on the prevailing interpretation of the existing regulations at the local and state levels, and less on the cost-benefit and risk analyses. As illustrated earlier, Georgia and quite a few other states in the United States have adopted a zero tolerance for groundwater degradation by large polluters such as various industries and military installations; these PRPs are consequently required to restore "their" portions of contaminated aquifers to pristine natural conditions, often regardless of the underlying hydrogeologic characteristics, the risks, and the associated costs. Following the varying trends in groundwater restoration by individual states, different USEPA regions have also adopted varying strategies. For example, some regions have not yet approved a single technical impracticability (TI) waiver, whereas some regions have taken a more pragmatic approach to the feasibility of groundwater restoration in complex hydrogeologic environments, by approving such waivers at some groundwater contamination sites where the USEPA is the leading regulatory agency for cleanup.

Despite advances in technologies applicable to groundwater remediation (many termed "innovative" technologies), aquifer restoration for sites with complex geologic and contaminant characteristics has rarely been achieved. USEPA formally recognized the limitations on groundwater restoration with the publication of guidance on TI waivers in 1993 (USEPA, 1993). The primary goals of the USEPA for groundwater remediation are stated in the *National Oil and Hazardous Substances Pollution Contingency Plan* (NCP) as follows: "EPA expects to return usable groundwater to their beneficial uses wherever practicable, within a timeframe that is reasonable given the particular circumstances of the site. When restoration of groundwater to beneficial uses is **not practicable**, [emphasis added] USEPA expects to prevent further migration of the plume, prevent exposure to the contaminated groundwater, and evaluate further risk reduction." It should be noted that a "reasonable time frame" is sometimes generically applied to be 100 years. However, there is no accepted definition of "reasonable" as applied to groundwater restoration because it is dependent on the applicable technologies and site-specific conditions such as hydrogeology.

It has been estimated that there are more than 20,000 "mega-sites" in the European countries, requiring extensive remediation (cleanup) of contaminated soils and groundwater, with the projected costs of tens of billions of US dollars (Rügner and Bittens, 2006). At the vast majority of these sites, the cleanup efforts have not yet started as the regulatory agencies and various stakeholders are trying to develop restoration approaches that would balance risks to human health and the environment with the socioeconomic realities. One reason for this struggle is that many of the mega-sites, legacy of the cold war and industrial growth, are also owned by the governments, and the remediation costs would have to be paid by all citizens. A more recent factor in developing pragmatic and realistic groundwater restoration strategies in the European countries is the complex question of carbon footprint, energy consumption, and climate change. For example, there are many installed pump-and-treat (P&T) groundwater remediation systems in the United States that have very significant energy requirements, with the associated costs rising steadily. There are also examples where such systems are extracting tens of millions of gallons of water annually, with only tens of pounds or less of the actual groundwater contaminant. In addition, these large quantities of water, after treatment, are often being discharged to surface streams or sewers without being used or returned back to the aquifer. It is, therefore, likely that, even in the United States, the sustainability of certain groundwater remediation efforts will be increasingly questioned by various stakeholders, and not just by those who are paying the bill.

One possible roadmap to a more efficient and less confrontational approach to groundwater restoration is the establishment of a groundwater remediation trading system, similarly to the widely discussed carbon emissions trading. For example, large companies and parts of the government with "deep pockets" (such as Departments of Defense or Energy), responsible for groundwater contamination at multiple sites, would be able to trade their existing and future remediation costs based on priorities developed by risk and cost-benefit analyses. This, of course, would include providing secure drinking water supplies to the affected groundwater users, thus eliminating risks to human health.

At the same time, some endemic groundwater contamination problems caused by small businesses unable to pay the full cost of groundwater cleanup, such as dry-cleaner shops and gas stations, would be covered by common funds (this has already been practiced by some states in the United States). Such funds would also be used to support prevention of future groundwater contamination, including active protection of vulnerable groundwater recharge areas.

What this all may mean in an ideal world is that hundreds of millions of dollars would not be spent on a handful of sites where groundwater restoration to natural conditions may not be currently feasible within a reasonable time frame of, for example, 50 years, or may be too costly for the given benefits. Rather, the funds and the resources would be spent at other sites where the cost-benefit ratio would be much more favorable. Common funds could also be used to fully protect valuable uncontaminated sources. Although this approach may seem far-fetched, given the complex regulatory framework in the United States, it may work elsewhere to the benefit of both the society and the environment.

9.2 Risk Assessment

Risk analysis at any site contaminated with point sources includes four main steps: (1) data collection and evaluation, (2) exposure assessment, (3) toxicity assessment, and (4) risk characterization, for both human health and the environment. This analysis has to be performed for both present and future risks, considering current and reasonable future land use scenarios. In the case of groundwater, this includes contact with, and ingestion (drinking) of, contaminated water. Risk assessment also includes evaluation of risks that groundwater may become contaminated in the future from sources at the land surface or in the vadose zone (i.e., from contaminated soils). Baseline risk assessment is performed during the initial site investigation and helps determine whether additional response action is necessary at the site. For example, it may indicate an immediate, clear threat to human health and the environment requiring an immediate corrective action (e.g., removal of leaking drums with hazardous wastes or discontinued use of contaminated water wells). It may also support the "no-action" alternative if the risks are absent or minimal. When there are risks requiring further evaluation, baseline risk assessment sets preliminary remediation goals and identifies any data required for the full quantitative risk assessment (USEPA, 1991a).

Data collected during the remedial investigation (RI) and the remedial feasibility study (FS) phases are used to perform a full quantitative risk assessment and refine remedial goals for the site. The results of quantitative risk assessment play an important role in deciding which remedial measures at the site should be implemented. The final phase of risk assessment includes evaluation of remedial alternatives and the associated risks of implementing them. Figures 9.3 and 9.4 illustrate risk assessment activities during the RI/FS process and various elements of the baseline risk assessment. Note that acronym "ARARs" comes from the USEPA Superfund cleanup program and stands for legally "applicable or relevant and appropriate requirements," both federal and state, which have to be met by the site remediation (see Section 9.3).

As discussed earlier, regulations in some cases may require remediation of contaminated soil and groundwater regardless of the established risk, as part of a no-degradation policy.

Although human health risk assessment and environmental (ecological) risk assessment are different processes, these have much in common and generally can use some of the same chemical sampling and environmental setting data for a site. Planning for both assessments should begin during the scoping stage of the RI/FS, and site sampling and other data collection activities to support the two assessments should be coordinated. An example of this type of coordination is the sampling and analysis of fish or other aquatic

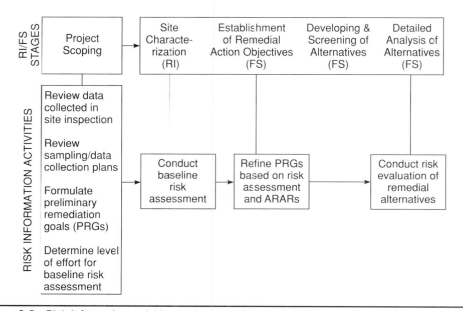

Figure 9.3 Risk information activities in the remedial investigation (RI) and feasibility study (FS) process. (From USEPA, 1989a.)

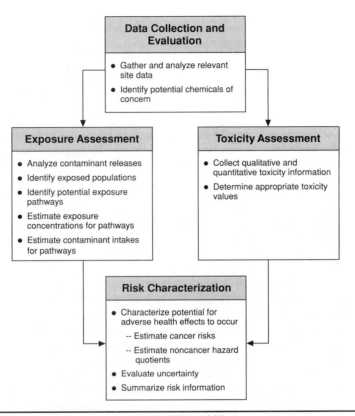

FIGURE 9.4 Baseline risk assessment. (From USEPA, 1989a.)

organisms; if done properly, data from such sampling can be used in the assessment of human health risks from ingestion and in the assessment of damages to and potential effects on the aquatic ecosystem (USEPA, 1989a). At a minimum, the baseline environmental (ecological) risk evaluation should include an assessment of any critical habitats and any endangered species or habitats of endangered species affected by contamination at the site. It should also provide the information necessary to adequately characterize the nature and extent of environmental risk or threat resulting from the site (e.g., see USEPA, 1991b).

9.2.1 Data Collection and Evaluation

The first step in the risk assessment and RI/FS process is data collection and evaluation. It involves collecting soil, air, and water samples, identifying locations where contaminants were (or may have been) disposed, and identifying contaminants or chemicals of potential concern (COPCs). The results of chemical analyses should include background (ambient) concentrations of COPCs, i.e., at locations not affected by the site activities. If the data show groundwater contamination at the site, the rest of the risk assessment will determine which ones are the contaminants of concern (COCs). In this case, a survey of any water supply wells and springs within a hydrogeologically reasonable radius from the site should be conducted, and water samples collected for chemical analyses of COCs whenever possible.

As discussed throughout the book, the characterization of hydrogeologic characteristics that influence groundwater flow, as well as various parameters of fate and transport (F&T) of contaminants, is often a complex and lengthy process. Data collection and evaluation and the risk assessment, therefore, continue during all phases of RI/FS.

9.2.2 Exposure Assessment

An exposure assessment is conducted to estimate the magnitude of actual and potential receptor (human and ecological) exposures, the frequency and duration of these exposures, and the pathways by which receptors are potentially exposed. In the case of groundwater exposures, all possible pathways that connect contaminant sources and receptors via groundwater should be evaluated. This, for example, includes contaminated surface water runoff or a contaminated stream that loses water to the underlying aquifer used as a drinking water source. Figure 9.5 shows some groundwater contamination pathways.

In the exposure assessment, reasonable maximum estimates of exposure are developed for both current and reasonable future land use assumptions. Current exposure estimates are used to determine whether a threat exists based on existing exposure conditions at the site and off the site. Future exposure estimates are used to provide decision makers with a qualitative estimate of the likelihood of such exposures occurring (USEPA, 1989a). Conducting an exposure assessment involves (1) analyzing contaminant releases, (2) identifying exposed populations, (3) identifying all potential pathways of exposure, (4) estimating COC concentrations at exposure point, based both on environmental monitoring data and predictive modeling results, and (5) estimating contaminant intakes for specific pathways. The results of this assessment are pathway-specific intakes for current and future exposures to individual COCs. For example, children might play in a surface

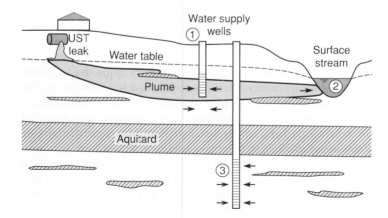

FIGURE 9.5 Examples of possible groundwater contamination pathways between contaminant sources and receptors. (1) Human receptor exposure at shallow water well, (2) exposure of ecological receptors in surface stream, and (3) absence of pathway between the source (contaminated groundwater) and the screen of deep water supply well.

stream contaminated by groundwater discharge and drink water with certain COC, or people might eat fish polluted with another COC that accumulates in fish tissue.

Intake dose estimates are based on five main factors:

1. Concentration of a chemical—at an exposure point, such as a drinking water well or contaminated fish fillet

2. Contact rate—amount of water, food, dust, or air that a person may take in over a specified time

3. Exposure frequency and duration—how often and for how long people could be exposed

4. Body weights for each age group that may be exposed

5. Exposure averaging time—is the time over which exposure is averaged in days

For a chemical that might cause cancer, USEPA prorates the total exposure over a lifetime to determine a lifetime average daily dose. This time is typically, but not necessarily, limited to 30 years. For a chemical that can cause noncancer effects, the averaging is over a year (365 days). These five factors are inserted into the following equation to calculate an intake dose for that pathway, called chronic daily intake (CDI):

$$\text{CDI (mg/kg day)} = \frac{C \times CR \times EF \times ED}{BW \times AT} \tag{9.1}$$

where C = contaminant concentration
CR = contact rate (also called intake rate)
EF = exposure frequency
ED = exposure duration
BW = body weight
AT = averaging time

A value for each factor is selected so that the combination of all factors results in a reasonable maximum exposure (RME) dose. To do this, risk assessors use statistics as well as professional judgment. For instance, because of uncertainty associated with estimates of contaminant concentration, risk assessors usually use the 95-percent upper confidence limit on the arithmetic average, which is a higher concentration than the straight average. To protect the majority of individuals in a population, they choose "high-end" values for contact rates and duration. However, to avoid unreasonable estimates, they use the average value for body weight over the exposure period. This combination of "high-end" contact rate and average body weight avoids the false assumption that a very small person would have the highest intake (USEPA, 2000a).

9.2.3 Toxicity Assessment

The toxicity assessment of each COC includes the types of adverse human health and environmental effects associated with chemical exposures (including potential carcinogenicity), the relationships between magnitude of exposures and adverse effects, and the related uncertainties of contaminant toxicity (e.g., the weight of evidence for a chemical's carcinogenicity). Information on human health effects of various contaminants is published in the USEPA database—the Integrated Risk Information System (IRIS). Risk assessors use IRIS to help evaluate cancer and noncancer effects for each COC.

A substance is toxic if it is hazardous or poisonous to living things. Toxicity refers to the inherent potential of a substance to cause damage to living things. A person must be exposed to a toxic substance before a damaging effect can occur. The term hazardous is more broadly defined than toxicity. Hazardous refers to the capability of a substance to cause harm due to its toxicity, flammability, corrosiveness, explosiveness, or other harmful property.

Frequency and length of exposure help determine how much, if any, harm will occur. Acute toxicity can occur after a single large exposure or limited number of exposures within a short time, generally less than 24 hours. Damaging effects, such as breathing difficulties, vomiting, rashes, and even death, can occur immediately or within hours of an acute exposure. Occasionally, acute exposure can produce delayed toxicity. Chronic toxicity, the main concern at most contaminated (Superfund) sites, can occur after repeated exposures over a long time—usually years—and damaging effects are seen months or years after exposure began.

The term "safe dose" in reference to chemical exposure levels usually refers to amounts that are too small to be a human health concern even though some level of risk remains. This is true for cancer risks as well as noncancer hazards and indicates that there are degrees of safety. As discussed by USEPA (2000), most chemicals cause cancer in different ways than these cause noncancer effects, such as damage to the liver or kidneys. Very small amounts of some substances are capable of starting the growth of cancers. For these substances, there is theoretically no level of exposure that is risk free. For other substances, however, scientists have discovered that exposure has to occur above a certain amount, called a threshold dose, before risks to humans become a concern. Most chemicals that cause noncancer effects as well as a few cancer-causing chemicals fall into the threshold category. Because of these differences, risk assessors report risks differently for cancer and for noncancer effects. When risk assessors estimate cancer risk, they try to predict a lifetime risk level for an exposed individual and how many additional cancer cases might occur in a population of exposed people. These are cancers that may or may not occur, but if these were to occur, these would be in addition to cancers from other causes, such as smoking tobacco. For noncancer toxicity, risk assessors estimate a daily exposure level that is likely to be of little risk to people.

9.2.4 Risk Characterization

The final step of the risk assessment, risk characterization, integrates the results of the exposure assessment and toxicity assessment. It reveals which chemicals are posing the risks and what the health risks are, in both quantitative expressions and qualitative statements. During risk characterization, chemical-specific toxicity information is compared against both measured contaminant exposure levels and those levels predicted through F&T modeling to determine whether current or future levels at or near the site are of potential concern.

Risk characterization estimates the potential health risks posed by the site if no remedial action is taken. It also explains the level of risk that may be left after different cleanup approaches are applied and describes the uncertainties associated with the data and risk estimates (USEPA, 1991c). Uncertainties may be associated with strengths and weaknesses of the data, the exposure assumptions, or the toxicity values. For these reasons, quantitative evaluation of risks has built in a large margin of safety to prevent underestimation of the risks.

Cancer risk from exposure to carcinogens is quantified using the CDI (see Eq. (9.1)) and a cancer risk slope factor (SF) that has units of risk per mg/kg day, using the following equation (USEPA, 1989a; Falta et al., 2007):

$$\text{Risk (dimensionless)} = 1 - \exp\left(-CDI \times SF\right) \tag{9.2}$$

which for small risks is equivalent to the following equation:

$$\text{Risk (dimensionless)} = CDI \times SF \tag{9.3}$$

The total carcinogenic risk, $Risk_T$, from exposure to multiple carcinogens (for example tetrachloroethylene (PCE), trichloroethylene (TCE), and vinyl chloride (VC)) is calculated as the sum of the individual risks:

$$\text{Risk}_T \text{ (dimensionless)} = \sum \text{Risk}_i \tag{9.4}$$

A major exposure route for contaminated groundwater is water from wells in the dissolved plume area. The contaminants contained in the water may be ingested directly, in drinking water, and if these are volatile, these may be inhaled as the contaminant partitions from the water into the air in the house (McKone, 1987). The cancer risk from inhalation is often as large as or larger than the risk from ingestion alone. Once the average tap water concentration is known, the ingestion and inhalation cancer risks can be calculated using standard methods (see, for example, Maxwell et al., 1998; McKone, 1987; Williams et al., 2004).

The lifetime excess cancer risk SFs vary widely among different chemicals, and these are often revised or withdrawn. Conflicting values of SF can be found in different sources in many instances (Falta et al., 2007). Table 9.1 lists relatively current (as of February 2006) recommended inhalation and oral SFs for PCE, TCE, and VC from the California Office of Environmental Health Hazard Assessment (Falta et al., 2007, from OEHHA, 2006).

Where applicable, the USEPA may set groundwater cleanup concentrations for individual COCs based on the information gathered in the risk assessment, such as location of chemical contamination, how people are exposed, and the concentrations that pose health risks. These risk-based cleanup standards, or maximum allowable concentrations of COCs, may be higher than their primary drinking water standards, expressed as MCLs. However, as discussed earlier, groundwater cleanup standards are always site specific

Chemical	Inhalation Slope Factor (mg/kg day)$^{-1}$	Oral Slope Factor (mg/kg day)$^{-1}$
Tetrachloroethylene (PCE)	0.021	0.540
Trichloroethylene (TCE)	0.007	0.013
Cis-1,2-Dichloroethylene (DCE)	Not a carcinogen	Not a carcinogen
Vinyl chloride (VC)	0.270	0.270

From Falta et al., 2007; OEHHA, 2006.

TABLE 9.1 California Cancer Risk Slope Factors for PCE and Its Degradation Products

and, in the United States, depend primarily on the state and local regulations and policies. For that reason, a cleanup standard for contaminated groundwater at a site may be risk based, may be MCL for specific COCs, or may be restoration to natural conditions, i.e., to nondetectable concentrations of COCs.

9.3 Remedial Investigation and Feasibility Study

The RI/FS process of the USEPA Superfund cleanup program has also been adopted, with minor modifications, by most states in the United States for their own cleanup programs. This process represents the methodology for characterizing the nature and extent of risks posed by uncontrolled hazardous waste sites and for evaluating potential remedial options. It is a site-specific process, which allows the project manager to determine how best to use the flexibility built into it and to conduct an efficient and effective RI/FS. A significant challenge project managers face in effectively managing an RI/FS is the inherent uncertainties associated with the remediation of uncontrolled hazardous waste sites. These uncertainties can be numerous, ranging from potential unknowns regarding site hydrogeology and the actual extent of contamination, to the performance of treatment and engineering controls being considered as part of the remedial strategy. As pointed out by USEPA (1988a), while these uncertainties foster a natural desire to want to know more, this desire competes with the Superfund program's mandate to perform cleanups efficiently and within designated schedules.

ARARs is the Superfund term used throughout RI/FS. It may be categorized as follows:

1. Chemical-specific requirements that may define acceptable exposure levels and therefore be used in establishing remediation goals

2. Location-specific requirements that may set restrictions on activities within specific locations such as floodplains or wetlands

3. Action-specific requirements, which may set controls or restrictions for particular treatment and disposal activities related to the management of hazardous wastes

The document, "CERCLA Compliance with Other Laws Manual" (USEPA, 1988b), contains detailed information on identifying and complying with ARARs.

Superfund cleanup standards are based on ARARs and have a strong statutory preference for remedies that are highly reliable and provide long-term protection. In addition to the requirement for remedies to be both protective of human health and the environment and cost effective, additional remedy selection considerations include the following (USEPA, 1988a):

- A preference for remedial actions that employ treatment that permanently and significantly reduces the volume, toxicity, or mobility of hazardous substances, pollutants, and contaminants as a principal element.

- Off-site transport and disposal without treatment is the least favored alternative where practicable treatment technologies are available.

- The need to assess the use of permanent solutions and alternative treatment technologies or resource recovery technologies and use them to the maximum extent practicable.

Standards also require a periodic review of remedial actions, at least every 5 years after initiation of such action, for as long as hazardous substances, pollutants, or contaminants that may pose a threat to human health or the environment remain at the site. If it is determined during a 5-year review that the action no longer protects human health and the environment, further remedial actions will need to be considered (USEPA, 1988a).

It is important to note that the RI and FS are conducted concurrently and that data collected in the RI influence the development of remedial alternatives in the FS, which in turn affects the data needs and scope of treatability studies and additional field investigations (Fig. 9.6). Two concepts are essential to the phased RI/FS approach. First, data should generally be collected in several stages, with initial data collection efforts usually limited to developing a general understanding of the site. As a basic understanding of site characteristics is achieved, subsequent data collection efforts focus on filling identified gaps in the understanding of site characteristics and gathering information necessary to evaluate remedial alternatives. Second, this phased sampling approach encourages identification of key data needs as early in the process as possible to ensure that data collection is always directed toward providing information relevant to selection of a remedial action. It is in this way that the overall site characterization effort can be continually scoped to minimize the collection of unnecessary data and maximize data quality (USEPA, 1988a).

The most important step at the beginning of the RI/FS process is the development of a conceptual site model (CSM). This model should include known and suspected sources of contamination, types of contaminants and affected media, known and potential routes of contaminant migration, and known or potential human and environmental receptors (Fig. 9.7). CSM, in addition to assisting in identifying locations where sampling is necessary, will also assist in the identification of potential remedial technologies. CSM is continuously updated and refined as the RI/FS process progresses.

Four main objectives of the remedial site investigation are to identify: (1) the site physical characteristics, (2) the characteristics of contaminant sources, (3) the nature and extent of groundwater contamination, and (4) the important contaminant F&T mechanisms. The site physical characteristics include geomorphology, soils, geology, hydrogeology, hydrology, meteorology, and ecology. This analysis should emphasize factors important in determining contaminant F&T for the identified exposure pathways of concern.

Sources of contamination are often hazardous substances contained in drums, tanks, surface impoundments, waste piles, and landfills. In a practical sense, heavily contaminated media (such as soils) may also be considered sources of contamination (secondary sources), especially if the original (primary) source, such as a leaking tank, is no longer present on the site or is no longer releasing contaminants. Source characterization involves the collection of data describing (1) facility characteristics that help to identify the source location, potential releases, and engineering characteristics that are important in the evaluation of remedial actions; (2) the waste characteristics, such as the type and quantity of contaminants that may be contained in or released to the environment; and (3) the physical or chemical characteristics of hazardous wastes present in the source (USEPA, 1988a).

The final objective of the field investigations is to characterize the nature and extent of contamination, and its possible future development, so that informed decisions can be made as to the level of risk presented by the site and the appropriate type(s) of remedial response. This means that, in addition to the sources (primary and secondary) of groundwater contamination, the extent of the dissolved-phase contamination as well as its rate

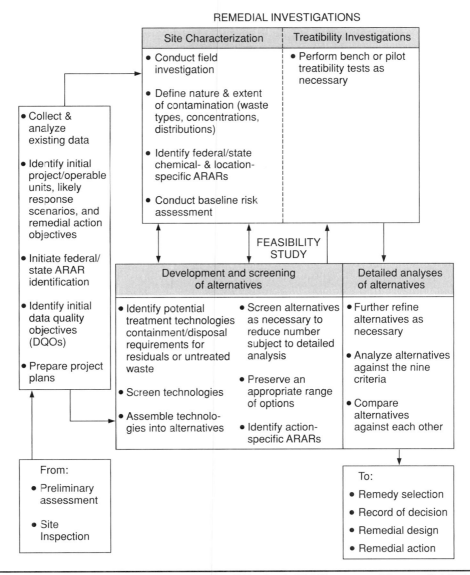

REMEDIAL INVESTIGATIONS

Site Characterization	Treatibility Investigations
• Conduct field investigation • Define nature & extent of contamination (waste types, concentrations, distributions) • Identify federal/state chemical- & location-specific ARARs • Conduct baseline risk assessment	• Perform bench or pilot treatibility tests as necessary

• Collect & analyze existing data

• Identify initial project/operable units, likely response scenarios, and remedial action objectives

• Initiate federal/state ARAR identification

• Identify initial data quality objectives (DQOs)

• Prepare project plans

FEASIBILITY STUDY

Development and screening of alternatives	Detailed analyses of alternatives
• Identify potential treatment technologies containment/disposal requirements for residuals or untreated waste • Screen technologies • Assemble technologies into alternatives	• Screen alternatives as necessary to reduce number subject to detailed analysis • Preserve an appropriate range of options • Identify action-specific ARARs	• Further refine alternatives as necessary • Analyze alternatives against the nine criteria • Compare alternatives against each other

From:
• Preliminary assessment
• Site Inspection

To:
• Remedy selection
• Record of decision
• Remedial design
• Remedial action

FIGURE 9.6   Phased Remedial Investigation/Feasibility Study (RI/FS) process. RD, remedial design; RA, remedial action; ROD, record of decision. (From USEPA, 1988a.)

of migration have to be defined and predicted in all three dimensions. While field data generally best define the extent of contamination, analytical or numeric modeling can be used as predictive tools (Figs. 9.8 and 9.9). More detailed numerical models provide greater accuracy and resolution because these are capable of representing spatial variations in site characteristics and irregular geometries commonly found at actual sites. Numeric models are useful during RIs as they provide insight into data gaps, help refine site conceptual model, and help organize all relevant site information in an interactive environment. Numeric models can also represent the actual configuration and effects of

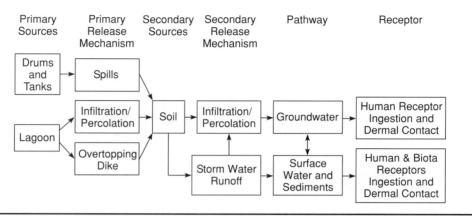

FIGURE 9.7 Example of conceptual site model (CSM). (Modified from USEPA, 1988a.)

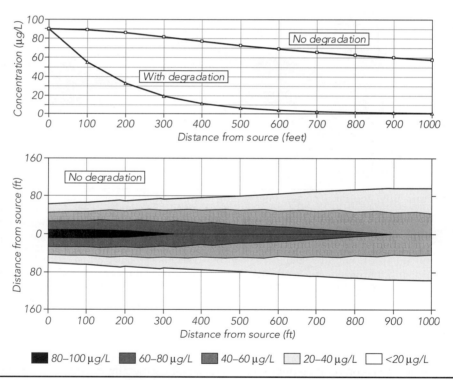

FIGURE 9.8 Concentrations of dissolved constituent after 10 years of migration from a continuous source as predicted with the analytical model BIOSCREEN AT (Karanovic, 2006; Karanovic et al., 2007). (Top) Concentration along the plume centerline without and with simulated biodegradation. (Bottom) Map view of the two-dimensional plume. Analytical models like this one assume homogeneous, isotropic aquifers and cannot simulate effects of pumping and various boundary conditions.

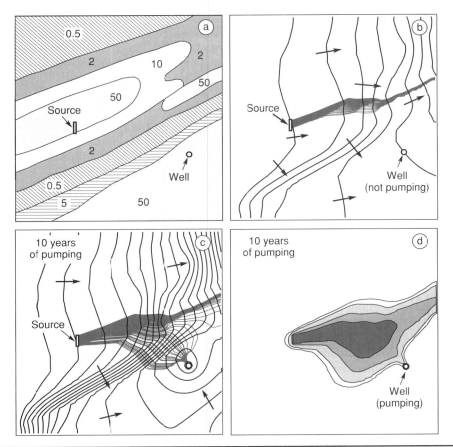

Figure 9.9 The influence of aquifer heterogeneity (varying hydraulic conductivity) can be modeled with numeric models. (a) Estimated distribution of the hydraulic conductivity (values are in ft/d). (b) Forward particle tracking from the source shows a very narrow steady-state flowpath between the source and the discharge zone (river). (c) Well pumpage causes wide spreading of particles leaving the source area, and traveling through several zones of varying hydraulic conductivity; refraction at the hydraulic conductivity boundaries causes additional spreading of the particles. (d) Dissolved contaminant concentrations are starting to be detected in the well between 9 and 10 years following the release. (Modified from Kresic, 2007; copyright Taylor & Francis Group, printed with permission.)

remedial actions and are therefore often used for screening remedial alternatives, as well as for a detailed analysis of alternatives during FS.

The FS may be viewed as occurring in three phases: the development of alternatives, the screening of the alternatives, and the detailed analysis of alternatives. However, in actual practice, the specific point at which the first phase ends and the second begins is not so distinct. Furthermore, in those instances in which circumstances limit the number of available options, and therefore the number of alternatives that are developed, it may not be necessary to screen alternatives prior to the detailed analysis.

9.3.1 Development and Screening of Remedial Alternatives

Alternatives for remediation are developed by assembling combinations of technologies, and the media to which these would be applied, into alternatives that address contamination on a site-wide basis or for an identified "operable unit" (OU) within the site (groundwater contamination is often addressed as a separate OU at Superfund sites). This process consists of six general steps (USEPA, 1988a):

1. Develop remedial action objectives specifying the contaminants and media of interest, exposure pathways, and preliminary remediation goals that permit a range of treatment and containment alternatives to be developed. The preliminary remediation goals are developed on the basis of chemical-specific ARARs when applicable, other available information, and site-specific risk-related factors.

2. Develop general response actions for each medium of interest defining containment, treatment, excavation, pumping, or other actions, singly or in combination, which may be taken to satisfy the remedial action objectives for the site.

3. Identify volumes or areas of media to which general response actions might be applied, taking into account the requirements for protectiveness as identified in the remedial action objectives and the chemical and physical characterization of the site.

4. Identify and screen the technologies applicable to each general response action to eliminate those that cannot be implemented technically at the site.

5. Identify and evaluate technology process options to select a representative process for each technology type retained for consideration.

6. Assemble the selected representative technologies into alternatives representing a range of treatment and containment combinations, as appropriate.

An example of initial screening of technologies and process options is shown in Fig. 9.10.

Information available at the time of screening should be used primarily to identify and distinguish any differences among the various alternatives and to evaluate each alternative with respect to its (1) effectiveness, (2) implementability, and (3) cost. Only the alternatives judged as the best or most promising on the basis of these evaluation factors should be retained for further consideration and analysis. Typically, those alternatives that are screened out will receive no further consideration unless additional information becomes available, which indicates further evaluation is warranted.

Alternatives should be developed, which will provide decision makers with an appropriate range of options and sufficient information to adequately compare alternatives against one another. In developing alternatives, the range of options will vary depending on site-specific conditions. For source control actions, the following types of alternatives should be developed to the extent practicable:

- A number of treatment alternatives ranging from one that would eliminate or minimize to the extent feasible the need for long-term management (including monitoring) at a site, to one that would use treatment as a primary component of an alternative to address the principal threats at the site. Alternatives within

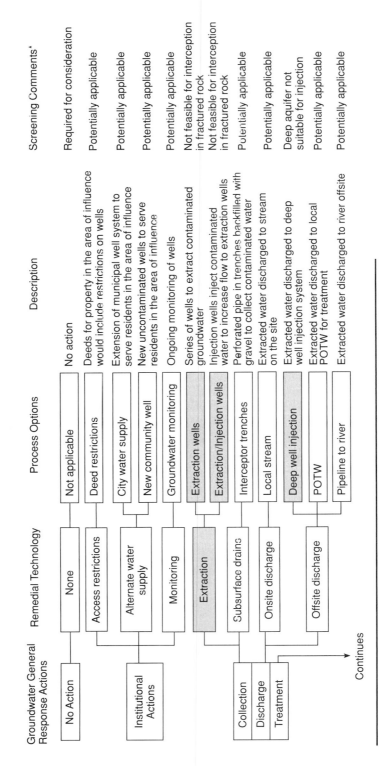

Groundwater General Response Actions	Remedial Technology	Process Options	Description	Screening Comments[*]
No Action	None	Not applicable	No action	Required for consideration
Institutional Actions	Access restrictions	Deed restrictions	Deeds for property in the area of influence would include restrictions on wells	Potentially applicable
	Alternate water supply	City water supply	Extension of municipal well system to serve residents in the area of influence	Potentially applicable
		New community well	New uncontaminated wells to serve residents in the area of influence	Potentially applicable
	Monitoring	Groundwater monitoring	Ongoing monitoring of wells	Potentially applicable
	Extraction	Extraction wells	Series of wells to extract contaminated groundwater	Not feasible for interception in fractured rock
		Extraction/Injection wells	Injection wells inject contaminated water to increase flow to extraction wells	Not feasible for interception in fractured rock
Collection	Subsurface drains	Interceptor trenches	Perforated pipe in trenches backfilled with gravel to collect contaminated water	Potentially applicable
Discharge	Onsite discharge	Local stream	Extracted water discharged to stream on the site	Potentially applicable
		Deep well injection	Extracted water discharged to deep well injection system	Deep aquifer not suitable for injection
Treatment	Offsite discharge	POTW	Extracted water discharged to local POTW for treatment	Potentially applicable
		Pipeline to river	Extracted water discharged to river offsite	Potentially applicable

Continues

FIGURE 9.10 An example of initial screening of technologies and process options. Shaded boxes are technologies that are screened out.
[*] Screening comments may or may not be applicable to actual sites. (Modified from USEPA, 1988a.)

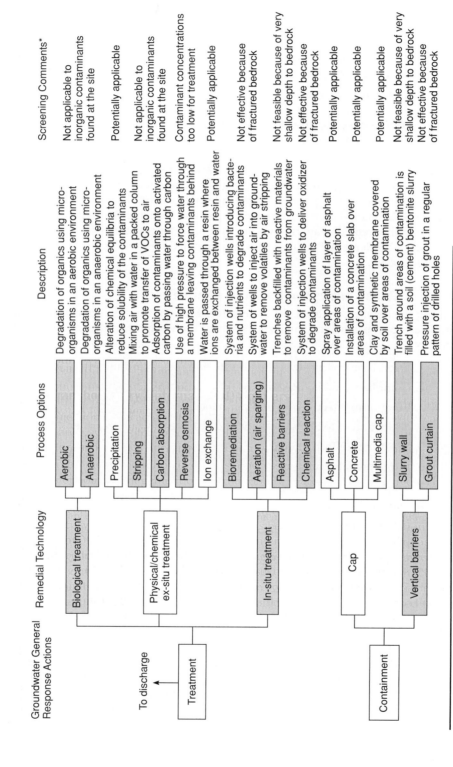

Groundwater General
Response Actions	Remedial Technology	Process Options	Description	Screening Comments*

Biological treatment	Aerobic	Degradation of organics using micro-organisms in an aerobic environment	Not applicable to inorganic contaminants found at the site
	Anaerobic	Degradation of organics using micro-organisms in an anaerobic environment	Potentially applicable
Physical/chemical ex-situ treatment	Precipitation	Alteration of chemical equilibria to reduce solubility of the contaminants	Not applicable to inorganic contaminants found at the site
	Stripping	Mixing air with water in a packed column to promote transfer of VOCs to air	Contaminant concentrations too low for treatment
	Carbon absorption	Adsorption of contaminants onto activated carbon by passing water through carbon	Potentially applicable
	Reverse osmosis	Use of high pressure to force water through a membrane leaving contaminants behind	
	Ion exchange	Water is passed through a resin where ions are exchanged between resin and water	
In-situ treatment	Bioremediation	System of injection wells introducing bacteria and nutrients to degrade contaminants	Not effective because of fractured bedrock
	Aeration (air sparging)	System of wells to inject air into groundwater to remove volatiles by air stripping	Not feasible because of very shallow depth to bedrock
	Reactive barriers	Trenches backfilled with reactive materials to remove contaminants from groundwater	Not effective because of fractured bedrock
	Chemical reaction	System of injection wells to deliver oxidizer to degrade contaminants	Potentially applicable
Cap	Asphalt	Spray application of layer of asphalt over areas of contamination	Potentially applicable
	Concrete	Installation of a concrete slab over areas of contamination	Potentially applicable
	Multimedia cap	Clay and synthetic membrane covered by soil over areas of contamination	Potentially applicable
Vertical barriers	Slurry wall	Trench around areas of contamination is filled with a soil (cement) bentonite slurry	Not feasible because of very shallow depth to bedrock
	Grout curtain	Pressure injection of grout in a regular pattern of drilled holes	Not effective because of fractured bedrock

FIGURE 9.10 (*Continued*)

this range typically will differ in the type and extent of treatment used, and the management requirements of treatment residuals or untreated wastes.

- One or more alternatives that involve containment of waste with little or no treatment but protect human health and the environment by preventing potential exposure and/or reducing the mobility of contaminants.
- A no-action alternative.

For groundwater response actions, alternatives should address not only cleanup levels but also the time frame within which the remedial goals might be achieved. Depending on specific site conditions and the aquifer characteristics, alternatives should be developed, which achieve ARARs or other health-based levels determined to be protective within varying time frames using different methodologies. For aquifers currently being used as a drinking water source, alternatives should be configured that would achieve ARARs or risk-based levels as rapidly as possible. More detailed information on developing remedial alternatives for groundwater response actions may be found in "Guidance on Remedial Actions for Contaminated Ground Water at Superfund Sites" (USEPA, 1988c).

9.3.2   Detailed Analysis of Remedial Alternatives

During the detailed analysis, the alternatives brought through screening are further refined, as appropriate, and analyzed in detail with respect to the following evaluation criteria:

1. Overall protection of human health and the environment
2. Compliance with ARARs
3. Long-term effectiveness and permanence
4. Reduction of toxicity, mobility, and volume through treatment
5. Short-term effectiveness
6. Implementability
7. Cost
8. State acceptance
9. Community acceptance

The first two criteria, which relate directly to the statutory requirements each remedial alternative must meet, are categorized as threshold criteria. The next five are the primary balancing criteria upon which the selection of the remedy is based. The final two, state acceptance and community acceptance, are called modifying criteria and are addressed in the record of decision (ROD) when comments are received on the RI/FS and the proposed remedial plan.

Alternatives may be further refined and/or modified based on additional site characterization or treatability studies conducted as part of the RI. The detailed analysis should be conducted so that decision makers are provided with sufficient information to compare alternatives with respect to the nine evaluation criteria and to select an appropriate remedy. Figure 9.11 illustrates a remedial alternative, which consists of one technology aimed at dense non-aqueous-phase liquid (DNAPL) removal in the source zone (in situ

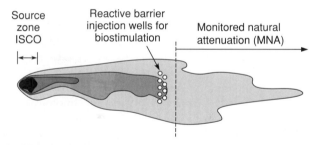

Figure 9.11 Remedial alternative consisting of three in situ technologies. ISCO: in situ chemical oxidation.

chemical oxidation—ISCO), and two technologies for in situ remediation of the dissolved plume: (1) permeable reactive barrier (PRB) formed by biostimulation of native microorganisms capable of degrading the contaminants and (2) monitored natural attenuation (MNA) downgradient of the reactive barrier.

9.3.3 Treatability Studies

Selection of remedial actions involves several risk management decisions, including uncertainties with respect to performance, reliability, and cost of treatment alternatives. These uncertainties underscore the need for well-planned, well-conducted, and well-documented treatability studies (pilot tests). In the absence of data in the available technical literature, treatability studies can provide the critical performance and cost information needed to evaluate and select treatment alternatives. Treatability studies can generally be divided into two groups: (1) pre-ROD and (2) remedial design/remedial action (RD/RA) treatability studies. The purpose of a pre-ROD treatability investigation is to provide the data needed for the detailed analysis of alternatives during the FS. Treatability studies conducted during RD/RA establish the design and operating parameters necessary for optimization of technology performance and implementation of a sound, cost-effective remedy. Although the purpose and scope of these two groups of studies differ, these complement one another because the information obtained in support of remedy selection may also be used to support the remedy design and implementation. Treatability studies can generally provide data to address the first seven of the nine criteria evaluated during FS (USEPA, 1992).

Pre-ROD treatability studies may be needed when potentially applicable treatment technologies are being considered for which limited (or no) performance or cost information is available in the literature with regard to the waste types and site conditions of concern. The need to conduct a treatability study on any part of a remedial alternative is a management decision. In addition to the technical considerations, certain nontechnical management decision factors must be considered. These factors include the expected level of state and community acceptance of a proposed alternative, time constraints on the completion of the RI/FS and the signing of the ROD, and the appearance of new site, waste, or technology data.

Although treatability studies of an innovative technology may be conducted during the RI/FS to support remedy selection, it may not be feasible to conduct sufficient testing to address all the significant uncertainties associated with the implementation of this option. This situation, however, should not cause the option to be screened out during the detailed analysis of alternatives in the FS. If the performance potential of an innovative technology indicates that this technology would provide the best balance of tradeoffs from among the options considered despite its uncertainties, CERCLA Section 121(b)(2) provides support for selecting such a technology in the ROD. Implementation of the technology, however, may be contingent upon the results of RD/RA treatability testing. When an innovative technology is selected and its performance is to be verified through additional treatability testing, a proven treatment technology may also be included in the ROD as a contingency remedy. In the event the RD/RA treatability study results indicate that the full-scale innovative remedy cannot achieve the cleanup goals at the site, the contingency remedy could then be implemented (USEPA, 1992).

To support the remedial action bid package, the lead agency or a PRP may choose to develop detailed design specifications. If technical data available from the RI/FS are insufficient for design of the remedy, an RD/RA treatability study may be necessary. Post-ROD treatability studies can provide the detailed cost and performance data required for optimization of the treatment processes and the design of a full-scale treatment system. Most RD/RA treatability studies are performed in the field with pilot- or full-scale equipment. Some prequalification treatability studies will be performed in the laboratory; however, the system should closely approximate the proposed full-scale operations (USEPA, 1992).

Post-ROD RD/RA treatability studies can also be performed to support the design of treatment trains. Although all parts of a treatment train may be effective for treating the wastes, matrices, and residuals of concern, issues such as unit sizing, materials handling, and systems integration must also be addressed. Treatability studies of one unit's operations can assist in identifying characteristics of the treated material that may need to be taken into consideration in the design of later units. A treatability study of the entire train can then provide data to confirm compliance with ARARs and the cleanup criteria outlined in the ROD. Because a treatment train will often involve several different technologies and vendors, the designer will coordinate treatability testing of the entire system and prepare the final treatability study report (USEPA, 1992).

Each level of treatability study requires appropriate performance goals. These goals should be specified before the test is conducted. The goals may need to be reassessed to determine appropriateness following testing performance as a result of new information (e.g., ARARs), treatment train considerations, or other factors. Pre-ROD treatability study goals will usually be based on the anticipated performance standards to be established in the ROD. This is because cleanup criteria are not finalized until the ROD is signed, due to continuing analyses and ARARs determinations. However, general expectation of the Superfund remediation program is that treatment technologies and/or treatment trains generally achieve a 90 percent or greater reduction in the concentration or mobility of individual COCs (USEPA, 1989b). This goal complements the site-specific risk-based goals. There will be situations where reductions outside this range that achieve health-based or other site-specific remediation goals may be appropriate (USEPA, 1989b).

9.4 Source-Zone Remediation

Although there are many different types of point sources of groundwater contamination, with varying spatial configurations and characteristics of contaminants, these can all be divided into two major groups from the practical remediation perspective: (1)

contamination with non-aqueous-phase liquids (NAPLs), and (2) contamination with chemicals readily or completely ("infinitely") soluble in groundwater. Remediation strategies and technologies for these two groups of groundwater contamination sources will differ for the most part. As discussed in Chap. 5, widespread production, transportation, utilization, and disposal practices of NAPLs (both lighter and denser than water) have resulted in groundwater contamination at numerous sites throughout the world. The potential for continuing long-term contamination at many such sites is high due to toxicity, limited solubility (but much higher than drinking water limits), and significant migration potential of NAPLs in vadose zone, soil gas, and groundwater, and as a separate phase. This migration can create secondary sources in the vadose and saturated zones, away from the locations of initial contaminant introduction into the subsurface. In addition, various NAPLs in source zones are often present in mixtures, which have different characteristics than pure chemicals, thus further complicating the analysis of their F&T mechanisms and the remediation options.

Many of the NAPL chemicals typically are proven or suspected carcinogens and have low MCLs in drinking water. Considering that source concentrations can be four or five orders of magnitude greater than MCLs, restoration of source zones to pristine conditions represents a great challenge and may be unlikely in many cases. However, reduction of dissolved plumes emanating from the NAPL source zones is a more realistic goal that can be achieved through combinations of source and plume remediation. Various in situ methods (e.g., soil vapor extraction (SVE), air sparging, thermal treatment, chemical oxidation, and surfactant and cosolvent flushing) are now available for removing or destroying NAPL mass contained in the source zone. Source containment methods (slurry walls, sheet-pile walls, and caps) can also be used to remove or reduce the contaminant loading to the plume (Falta et al., 2007).

Source remediation at individual sites can cost anywhere from several hundred thousand dollars to tens of millions of dollars (for example, see McDade et al., 2005), and it is rarely (if ever) possible to remove all the contaminant. The benefit of source remediation efforts is that by removing source mass, these tend to reduce the mass discharge to the plume (Rao et al., 2001; Falta et al., 2005a; Jawitz et al., 2005). The reduced plume loading following source remediation may or may not be sufficient to allow natural attenuation processes to keep the plume concentrations within acceptable limits (Falta et al., 2005a, 2005b).

Costs for plume remediation are usually considered to be smaller than those for source remediation because of the lower capital costs. At sites where the source is nearly depleted by dissolution or other processes, plume remediation would tend to be the most cost-effective strategy for site management. However, if substantial source mass is present, in the absence of source remediation, the plume remediation systems must be operated for a long period. In this case, the operating costs (in terms of present worth) can be comparable to the costs of source remediation. As emphasized by the USEPA, a reasonable strategy for many sites would be some combination of source and plume remediation. Selection of the optimal remedy for a site, in terms of the degree of remediation, must consider the inherent coupling of the source remediation to the plume remediation (Falta et al., 2007).

An informative study of the effectiveness of source-zone remediation was performed by McGuire et al. (2006). Performance and rebound of intensive source-depletion technologies were evaluated at 59 chlorinated solvent sites, where remediation targeted DNAPL source zones. The four technologies included in the study are chemical oxidation, enhanced bioremediation, thermal treatment, and surfactant/cosolvent flushing.

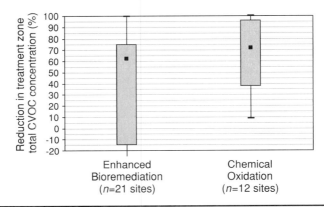

Figure 9.12 Reduction in total CVOC concentration within the treatment zone. Minimum reduction for enhanced bioremediation is –150 percent (value not shown; negative value indicates a concentration increase). Only sites implementing enhanced bioremediation and chemical oxidation had sufficient data records to evaluate total CVOCs. (From McGuire et al., 2006; copyright McGuire et al., 2006.)

All results reported in this study were calculated from actual concentration versus time data at monitoring wells, and concentration reduction values reported in the literature were not used. Data were available for 147 wells at 26 enhanced bioremediation sites, 23 chemical oxidation sites, six thermal sites, and four surfactant/cosolvent sites. Approximately 75 percent of the source-depletion projects were able to achieve a 70 percent reduction in parent compound concentrations. A median reduction in total chlorinated volatile organic compounds (CVOCs) concentrations (parent plus daughter compounds) of 72 percent was observed at 12 chemical oxidation sites and 62 percent at 21 enhanced bioremediation sites (Fig. 9.12). Note that enhanced bioremediation, which has primarily been used to treat dissolved-phase contamination, is increasingly being applied within and adjacent to DNAPL source zones to enhance dissolution rates (Parsons Corporation, 2004; U.S. DOE, 2002). A companion study by McDade et al. (2005) reports costs associated with the evaluated treatments at 59 chlorinated solvent sites.

In the study by McGuire et al. (2006), rebound was assessed at sites having at least 1 year of post-treatment data and included 43 wells at 20 sites (10 enhanced bioremediation sites, seven chemical oxidation sites, two surfactant/cosolvent sites, and one thermal site). On an individual well basis, rebound was observed in 20 percent of wells at enhanced bioremediation sites and in 81 percent of wells at chemical oxidation sites, and was not observed at surfactant/cosolvent and thermal sites. For example, concentrations in several wells at chemical oxidation sites rebounded by as much as one to two orders of magnitude throughout the post-treatment monitoring period. In fact, at 30 percent of the chemical oxidation rebound wells, rebound resulted in concentrations higher than pretreatment conditions. For rebound wells at enhanced bioremediation sites, the increased concentrations observed during the post-treatment period were still below pretreatment concentrations (McGuire et al., 2006).

Since the source-depletion technologies evaluated in this study were applied in DNAPL source zones that had high initial dissolved concentrations, common regulatory standards, such as MCLs, were not achieved in any of the cases. Although several sites achieved MCLs at some wells, none of the sites attained and sustained MCLs for all chlorinated compounds at all wells. Given the inability of most source-depletion technologies to achieve the primary remediation goal of returning groundwater to usable conditions, it is likely that some type of site management (e.g., institutional controls, long-term monitoring, MNA, or containment controls) will be necessary at many of these sites (McGuire et al., 2006).

The above-described study illustrates why there is considerable ongoing debate about the effectiveness and appropriateness of source depletion (i.e., the removal of contaminant mass from the source zone) at sites with groundwater plumes. For example, the USEPA commissioned an expert panel on DNAPL source remediation (Kavanaugh et al., 2003) and charged it with addressing several questions, including (1) what are the potential benefits and the potential adverse impacts of DNAPL source depletion as a remediation strategy? (2) what performance can be anticipated from source-zone mass-depletion technologies? and (3) what are currently available tools adequate to predict the performance of source-depletion options?

The Expert Panel concluded that quantitative predictions of potential benefits and adverse impacts of DNAPL source-depletion actions are highly uncertain and that these uncertainties represent significant barriers to more widespread use of source-depletion options (Kavanaugh et al., 2003). One of the panel's recommendations was the development of more user-friendly prediction tools that provide a basis for assessing the likely performance of source-zone depletion technologies. Partly in response to the panel's recommendations, and partly due to ongoing improvements of the existing software programs, there are now quite a few models that include various options for the simulation of NAPL source-zone depletion. These range from simple analytical screening tools such as REMChlor (Falta et al., 2007), and BIOBALANCE (Groundwater Services, Inc., 2007), to complex multiphase three-dimensional numeric models such as STOMP (White and Oostrom, 2000, 2004), TOUGH2 (Pruess et al., 1999; Pruess, 2004), RT3D (Clement et al., 2004), and SEAM3D (Waddill and Widdowson, 2000). Informative discussions on various models for simulation of multiphase F&T, and remediation of NAPLs in vadose and saturated zones, are given by Oostrom et al. (2005a, 2005b), White et al. (2004), and Zhu and Sykes (2004).

Contaminant loading to the saturated zone in most cases occurs after their passage through the vadose-zone soils. Depending on various F&T characteristics of the contaminant, release mechanisms, and characteristics of the porous media, the vadose zone may remain contaminated for varying periods, thus acting as a secondary source of groundwater contamination. Remediation of contaminated soils is, therefore, an integral part of groundwater remediation in source zones. Although the most obvious method is soil excavation and removal, in some cases it may be cost prohibitive or it may not be practicable (e.g., very thick and deep soil contamination, and presence of buildings and critical infrastructure). Many of the general and saturated-zone remediation methods discussed further in this chapter may also be applicable to the vadose-zone soils, depending on the site-specific conditions. These include SVE, soil flushing, bioremediation, phytoremediation (remediation by living plants), and ISCO, for example. When applicable from the regulatory standpoint, the most efficient and cost-effective method is soil capping with impermeable materials, which prevents water infiltration. This minimizes or eliminates further contaminant migration toward the water table.

General decisions on soil remediation alternatives are based on human health and environmental (ecological) risk analyses, whereas groundwater-related risks from soil contamination are evaluated based on contaminant leachability studies. These studies, which may include laboratory tests and modeling, determine likely mass loading (mass flux) of the contaminant to the saturated zone and the resulting dissolved concentrations in groundwater. If the study results show no adverse effects of contaminant leaching from the soils (e.g., groundwater concentrations are below regulatory threshold), or if the contaminant is immobile in the vadose zone, soil remediation would not be necessary from the groundwater contamination standpoint.

The term "soils" is somewhat loosely used in remediation engineering, and it may mean different things to hydrogeologists, engineers, and nonprofessionals. Remediation engineers often think of soils as all materials in the unsaturated zone (above the water table). Nonprofessionals may think of the first couple of feet of loose "dirt" below ground surface, whereas the definition of soils by geologists may be too lengthy to fit into one page of this book. In general, however, "soil remediation" refers to all relatively loose, unsaturated materials above the water table. This includes the first several feet or so of "real" soil formed on any parent geologic materials, unconsolidated sedimentary rocks (sand, gravel, clay, and silt), and residuum (regolith) sediments formed on the parent bedrock. In all cases, the unsaturated geologic (soil) materials are heterogeneous by default and often contain a significant portion of clay minerals due to weathering. The residuum heterogeneities are especially pronounced and are caused by selective weathering of different minerals and the presence of partially preserved parent-rock structures such as fractures and bedding planes. Characterization of the contaminant F&T in vadose zones, therefore, requires substantial effort and data collection on both micro- and macroscales. In addition to various physical and chemical parameters of the soils and residuum sediments, a successful remedial characterization must include analysis of contaminant distribution and properties in the field, and F&T modeling (Mikszewski and Kresic, 2006).

The following examples illustrate the importance of vadose-zone heterogeneities, contaminant F&T, and variably saturated modeling of both the vadose and the saturated zones when considering various groundwater remediation options. Figure 9.13 shows soil types and average position of the water table below a site where variably saturated model was used to predict dissolved concentrations as the contaminant leaches from the land surface. The model simulated 15 years of active contaminant loading, followed by additional 15 years after the loading was discontinued.

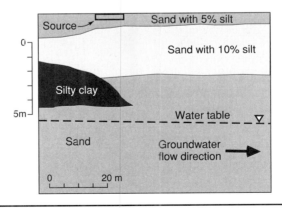

FIGURE 9.13 Conceptual cross-sectional model of the site with contaminant source at the land surface. (From Kresic and Mikszewski, 2006; copyright Taylor & Francis Group, printed with permission.)

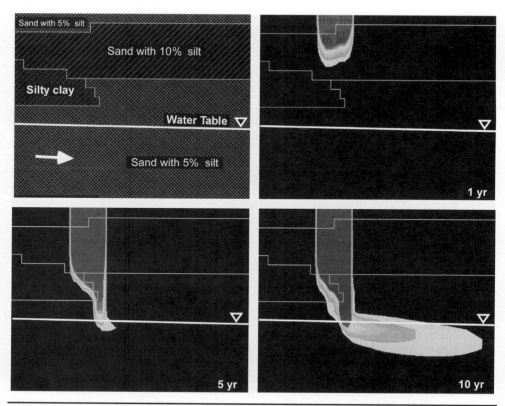

FIGURE 9.14 Model setup (top left) and model results for select years since the beginning of contaminant loading at the surface. (From Kresic and Mikszewski, 2006; copyright Taylor & Francis Group, printed with permission.)

Figures 9.14 and 9.15 show the results of model simulation for different time intervals. As the loading was discontinued at the end of year 15, the result for year 16 shows the effect of vadose-zone flushing by the newly percolating water with contaminant concentration of zero ("clean water"). The most interesting result is the effect of clay on contaminant F&T. Due to low hydraulic conductivity, and the contaminant sorption and diffusion, the clay lens acts as a long-term storage and a secondary source of dissolved contaminant in the saturated zone.

In case of a residual NAPL source area in the vadose zone, integrated saturated-unsaturated models can be used to assess the viability of remedial alternatives. Flow in variably saturated media adds a layer of complexity, which often causes industry practitioners to neglect vadose-zone modeling altogether. Unjustifiable and inaccurate assumptions regarding saturated-zone concentrations below the source area inevitably result from this exclusion. For example, modelers often assume a constant, uniform contaminant concentration to instantaneously result over the entire thickness of the aquifer in question. Such a methodology ignores the dependence of aquifer concentrations on groundwater recharge and contaminant flux from the (heterogeneous) vadose zone. Furthermore, key questions regarding the longevity of the vadose-zone source area usually remain unanswered (Kresic et al., 2007).

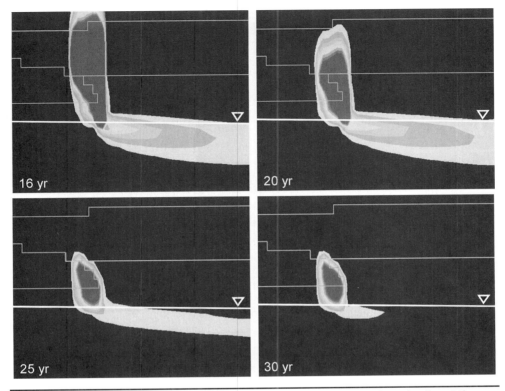

Figure 9.15 Model results for select years since the beginning of contaminant loading at the surface. The loading was discontinued at the end of year 15. (From Kresic and Mikszewski, 2006; copyright Taylor & Francis Group, printed with permission.)

9.4.1 NAPLs Problem

NAPLs are typically classified as either light non-aqueous-phase liquids (LNAPLs), which have densities less than that of water, or DNAPLs, which have densities greater than that of water (see Table 5.1). As discussed in Chap. 5, it is very important to make the distinction between the actual NAPL liquid in free phase and the chemical of the same name dissolved in water. For example, most common organic contaminants such as PCE, TCE, and benzene can enter the subsurface as both pure NAPL and dissolved in percolating water. However, the F&T of pure NAPL and the same chemical dissolved in groundwater are quite different.

NAPLs may be present in the subsurface as continuous (contiguous) bodies of a relatively significant extent (volume), which occupy all the pore space in the aquifer material. In such case, these are referred to as free-phase NAPLs. Detecting free-phase DNAPLs in the saturated zone is, in most cases, impractical, and, when it happens, it is often a matter of either luck or a significant lateral extent in the form of a pool resting on low-permeable porous media. LNAPL pools floating on the water table are often easier to detect; however, these too can surprise investigators assuming that all that is required to detect LNAPL is screening monitoring wells across the water table. Mobile LNAPL floating on the water table will move vertically as the groundwater elevation fluctuates.

As the water table rises or falls, LNAPL will be retained in the soil pores, and it may leave behind a residual LNAPL "smear zone," or the entire pool may migrate vertically. A decline in groundwater elevation may trap the LNAPL pool below the water table when groundwater elevations rise. For example, Oostrom et al. (2006) discuss the results of experiments and modeling, which show that viscous mobile LNAPL, subject to variable water-table conditions, does not necessarily float on the water table and may not appear in an observation well. A similar situation may develop during product recovery efforts. LNAPL will flow toward a recovery well or trench in response to the gradient induced by water-table depression. LNAPL may be retained below the water table as the water-table elevation returns to prepumping conditions.

At residual saturation, NAPL occurs as disconnected single- and multipore globules or ganglia (Fig. 9.16) within the larger pore spaces that have been cutoff and disconnected from the continuous NAPL body by the invading water (Cohen and Mercer, 1993; Pankow and Cherry, 1996). The actual saturation of pore space with NAPL can vary between 0 and 1, with the individual saturations of all fluids present (NAPLs and water) always summed to 1 in the saturated zone. Determining the percentage of NAPL saturation, even from the actual core samples, is difficult and requires application of complicated, laborious techniques, and care in obtaining, preserving, and transporting the samples. In any case, numeric saturation data from discrete sampling points would have to be extrapolated and interpolated ("contoured") in order to estimate the volume of DNAPL present in the corresponding volume of the aquifer. Cohen and Mercer (1993) present laboratory and field values of residual saturation for various NAPL fluids in intergranular porous media

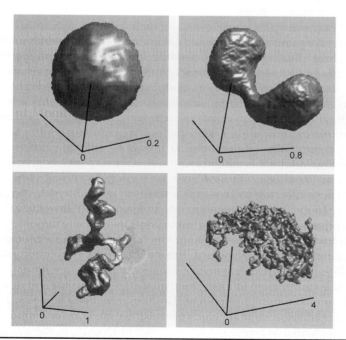

Figure 9.16 Electronic imaging of DNAPL blobs in porous media. Scale is in millimeters. (From Brusseau, 2005.)

Porous Medium	Entry Pressure (cm)
Clean sand ($K = 1 \times 10^{-2}$ cm/s)	45
Silty sand ($K = 1 \times 10^{-4}$ cm/s)	286
Clay ($K = 1 \times 10^{-7}$ cm/s)	4634
Fracture, 20 μ aperture	75
Fracture, 100 μ aperture	15
Fracture, 500 μ aperture	3

From Fountain (1998).

TABLE 9.2 Examples of DNAPL TCE Entry Pressure

(clays, silts, sands, gravels, and their mixtures). They conclude that values of residual saturation generally range between 0.10 and 0.50 in saturated porous media and tend to be higher in the preferential pathways of NAPL transport. Residual saturation also tends to increase with increasing pore aspect ratios and pore size heterogeneity, and with decreasing porosity, probably due to reduced pore connectivity and a decrease in mobile nonwetting fluid (NAPL) in smaller pore throats. Values of residual saturation in the vadose zone are generally smaller than in the saturated zone and range between 0.10 and 0.20. This is because NAPL drains more easily in the presence of air than in a water-saturated system. Residual saturation and retention capacity in the vadose zone increase with decreasing intrinsic permeability, effective porosity, and moisture content (Cohen and Mercer, 1993).

As discussed by Fountain (1998), below the water table, entry of DNAPL into water-filled pores requires overcoming an entry pressure resulting from interfacial tension (IFT) between DNAPL and water. The required entry pressure increases with decreasing grain size (Table 9.2). The downward flow of DNAPL may, therefore, be interrupted each time a layer with a smaller grain size is encountered. DNAPL tends to flow laterally above the fine-grained layer, accumulating until there is sufficient thickness of DNAPL to overcome the entry pressure. Even very subtle variations in grain size distribution may produce significant deflection of DNAPL flow. This results in a series of horizontal lenses of DNAPL connected by narrow vertical pathways. As in the vadose zone, a small amount of DNAPL is retained as residual saturation in every pore through which it flows. If the DNAPL encounters a layer that has a sufficiently high entry pressure, the DNAPL will accumulate on the top of this layer, forming a "pool." Thus, DNAPL is typically found as multiple horizontal lenses, connected by sparse vertical pathways, with one or more pools above fine-grained layers. Most of the horizontal lenses and vertical pathways will be at, or below, residual saturation; only pools will have higher saturations. The distinction between residual saturation and pools is important, since only the DNAPL in pools is expected to be mobile. However, changes in the IFT between water and DNAPL, which are produced either by heat (all thermal methods) or by chemicals (surfactants or cosolvents) may mobilize DNAPL at residual saturation by reducing capillary forces.

One potential indication of the presence of DNAPL in the saturated zone in a monitoring well is that the concentration of the dissolved contaminant is greater than 1 to 10 percent of the compound's effective solubility (Feenstra and Cherry, 1988; Pankow

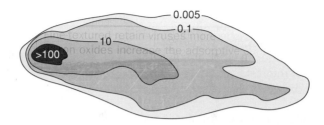

FIGURE 9.17 Delineation of potential aquifer zones with DNAPL trichloroethene (TCE) based on the 1–10% solubility rule of thumb. TCE has aqueous solubility approximately between 1100 and 1400 mg/L (1280 mg/L listed in Table 5.5). The aquifer area that may contain residual DNAPL is assumed to be within the 100 mg/L concentration contour, or approximately 8 percent of the pure-phase solubility. Note that in the case of a DNAPL mixture, the effective solubility of TCE would be less than the pure-phase solubility.

and Cherry, 1996). Figure 9.17 illustrates how concentrations in monitoring wells may be used to delineate aquifer zones with a potential presence of DNAPL. One reasoning behind this widely accepted "rule of thumb" is that, if DNAPL is present, it will generally either be present as a small lens in a small preferential pathway, as residual phase ganglia, or be diffused from a preferential pathway into a fine-grained matrix. If a 10-ft well screen is close to or intersects one of these areas, the area where the DNAPL is present will likely be thin when compared to the full length of the well screen. This is mainly because groundwater flow is generally laminar and will not mix quickly with the larger interval of the formation over short distances. As a consequence, the aqueous-phase contamination dissolving from the DNAPL into groundwater at a concentration close to its solubility limit will remain contained within a narrow (thin) interval some distance downgradient of the source zone. This contamination will be diluted in the monitoring well during sampling by the larger screened interval of the formation. Therefore, concentrations of a small percentage of solubility may indicate the presence of DNAPL in the vicinity of the monitoring well. If well screens are short, there will be less dilution and the contaminant concentration will be a higher percentage of solubility before it indicates DNAPL. This technique is subjective and must be applied very carefully because it may grossly over- or underestimate the presence and volumes of DNAPL in the aquifer if used alone. It should be considered only a part of the process used to determine if DNAPL is present, not a method that by itself will indicate the presence or absence of DNAPL (ITRC, 2003). An informative discussion on the behavior and dissolved concentrations expected to be found in DNAPL source zones is given by Anderson et al. (1987, 1992).

Another indirect method for detecting the potential presence of residual DNAPL is to calculate the hypothetical pore-water concentration from the measured total soil concentration by assuming equilibrium chemical partitioning between the solid phase, the pore water, and the soil gas, and assuming that no DNAPL is present in the collected sample. This pore-water concentration (C_w, in mg/L or $\mu g/cm^3$) can be expressed in terms of the total soil concentration (C_t, in $\mu g/g$ dry weight) as follows (Pankow and Cherry, 1996):

$$C_w = \frac{C_t \rho_b}{K_d \rho_b + \theta_w + H_c \theta_a} \qquad (9.5)$$

where ρ_b = dry bulk density of the soil sample (g/cm^3)

θ_w = water-filled porosity (volume fraction)

θ_a = air-filled porosity (volume fraction)

K_d = partition coefficient between pore water and solids for the compound of interest (cm^3/g)

H_c = dimensionless Henry's gas law constant for the compound of interest

If no DNAPL is present, there is a maximum amount of chemical, which can be contained in the soil sample at equilibrium with the soil pore water and air. In other words, for true aqueous solute (dissolved phase) in equilibrium, the calculated pore-water concentration (C_w) has to be equal to the solubility concentration ($C_w = S_w$). If the calculated pore-water concentration is higher than the solubility concentration, some DNAPL phase of the chemical has to be present in the sample. Note that, for a DNAPL mixture, effective solubilities of individual compounds will be lower than their pure-phase aqueous solubilities (see Chap. 5 and application of Raoult's law). Pankow and Cherry (1996) provide the following example of the application of Eq. (9.5):

- Measured TCE concentration (C_t) in the soil sample taken from the saturated zone (where θ_a equals 0) was 3100 mg/kg or 3100 μg/g.

- The partition coefficient (K_d) for TCE was calculated from the fraction of organic carbon in the soil sample ($f_{oc} = 0.001$) and the so-called organic-carbon partition coefficient for TCE ($K_{oc} = 126$) as follows: $K_d = f_{oc} \times K_{oc} = 0.001 \times 126 = 0.126$.

- The bulk density (ρ_b) was estimated to be 1.86 g/cm^3.

- The total porosity, equal to water-filled porosity (θ_w), was estimated to be 0.3.

Inserting the above values into Eq. (9.5) gives the calculated value for the pore-water concentration of 10,790 mg/L, which is much higher than the TCE pure aqueous solubility of approximately 1280 mg/L. The conclusion is that residual liquid TCE DNAPL is present in the sample.

Dissolution of different NAPL phases in the actual field conditions will depend on a number of factors, including effective and total porosities of the porous media, groundwater flow rates, surface contact area between the NAPL and groundwater, and F&T characteristics such as adsorption and diffusion (e.g., see Miler et al., 1990; Unger et al., 1998; Clement et al., 2004; Sale and McWhorter, 2001; Barth et al., 2003; Parker and Park, 2004). Assuming that NAPL would dissolve in the flowing groundwater following some published solubility value for pure water would, therefore, be erroneous. The actual dissolution of NAPL in the subsurface is often referred to as rate limited because of the various factors acting to decrease the pure-water solubility. The true dissolution rate for NAPL will be highly site specific and will change in time. Published solubilities for different hydrophobic organic substances should, therefore, be used with care and only as a starting point for the related analyses. Cohen and Mercer (1993) provide a detailed discussion on different approaches to determining aqueous and field solubilities of NAPLs. Literature values for aqueous solubility of common NAPL organic contaminants are given in Table 5.5 (Chap. 5).

One of the reasons why residual NAPL can persist in the aquifer for a long time is that blobs in dead end or otherwise restricted pores are surrounded by stagnant water and cannot dissolve as quickly as the blobs being constantly flushed by flowing groundwater

(groundwater advective flux). This flowing water may either be "clean" or have a much lower concentration of the dissolved constituent than water around the NAPL blob in the dead-end pore. As a result, the dissolution rate of the blobs constantly flushed by the flowing groundwater will be much higher compared to the diffusion-driven dissolution of the dead-end blobs.

In fractured rock aquifers contaminated with NAPLs, diffusion may transfer a significant mass of the contaminant from the fractures into the rock matrix. The magnitude of this transfer will depend on a variety of factors including the rock type (matrix porosity and tortuosity), matrix sorptive capacity, fracture spacing and aperture, and the advective flow rate in the fractures. In general, the contaminant removal from the advective flow component in the fractures into the surrounding rock matrix acts as an attenuating mechanism. However, once the rock matrix is invaded with the contaminant, it serves as a long-term storage that may play a significant role as a secondary source of groundwater contamination in some unfavorable hydrogeologic settings. This mass will diffuse back into the fractures as long as there is a concentration gradient, i.e., the concentration in the matrix is higher than in the fractures.

For example, Lipson et al. (2005) discuss the results of a model that simulates solute transport through a set of parallel, equally spaced fractures, using calibrated input parameters obtained from field investigation at a site contaminated with TCE DNAPL. The model was used to assess rates of diffusion and back-diffusion between the fractures and the matrix, and their impact on dissolved TCE concentrations in the fractures when the aquifer is subject to P&T remediation. Figure 9.18 presents a plot of TCE concentration versus distance into the matrix, between two parallel fractures 1.42 m apart. Time $t = 0$ corresponds to the start of P&T operations. The figure shows that TCE is diffusing into the rock matrix at this location along the flowpath in the fractures for the first 160 years of pumping and clean water injection (the location is 200 m downgradient from the point of injection of clean water at $t = 0$). After 160 years, concentrations in the first 20 cm of the rock matrix start to decline due to diffusion back out of the matrix from the zone adjacent to the fractures; however, concentrations within the interior of the rock matrix blocks continue to increase (e.g., compare $t = 160$ years with $t = 320$ years). At a time of 1280 years, all diffusion is directed out of the matrix blocks, with the peak

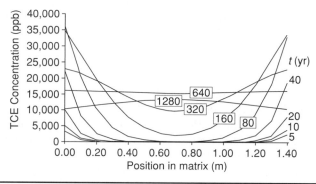

FIGURE 9.18 Simulated TCE concentration in sandstone matrix block at $z = 200$ m. Times shown are years since the DNAPL source has been completely removed. The block is in between two parallel fractures 1.42-m apart. (From Lipson et al., 2005; copyright National Ground Water Association, printed with permission.)

concentration (~13,000 ppb) occurring at the center of the blocks. Figure 9.18 shows that inward diffusion continues to occur for long periods after the start of pumping operations. It also shows that the back-diffusion process takes longer that the initial forward diffusion because of the fact that concentration gradients are directed both into and out of the matrix during back-diffusion stage (Lipson et al., 2005).

One important consequence of the residual NAPL persistency on groundwater remediation of the source zones is that, even after a significant mass of free-phase or residual product has been removed from the aquifer, the impact of the remaining mass in the stagnant water (dead-end pores) or in the rock matrix may still be great. The delivery of oxidizing fluids to stagnant zones and the rock matrix may be less successful or not feasible, so that the intact NAPL and the high dissolved concentrations in the stagnant zones and the rock matrix will continue to contaminate the faster advective flow of groundwater. Together with desorption, this mechanism is the main reason for the often-observed rebound of dissolved concentrations after ISCO applications.

A common simplified approach for quantifying the rate of mass transfer from the NAPL phase to the dissolved constituent phase is to lump various processes into one parameter called mass transfer coefficient. In general terms, this parameter is expressed as follows (Pankow and Cherry, 1996):

$$
\begin{aligned}
\text{Rate of Mass Transfer (M/T)} = {}& \text{Mass Transfer Coefficient (L/T)} \\
& \times \text{Concentration Difference (M/L}^3) \\
& \times \text{Contact Area (L}^2)
\end{aligned}
\tag{9.6}
$$

where M = mass
   L = length
   T = time

The mass transfer coefficient is usually first estimated by some of the proposed equations and then calibrated to match field data, if the related analysis is performed with the aid of numeric models. In practice, the majority of projects that include a more thorough quantitative analysis of contaminant F&T associated with NAPL source zones use numeric modeling techniques to account for the rate-limited dissolution processes. For example, MT3DMS and RT3D, two widely used F&T models based on the MODFLOW groundwater flow solution, include options for modeling a constituent transfer from the immobile phase (e.g., DNAPL) to the mobile phase (dissolved phase) by using the mass transfer coefficient as one of the model parameters.

Simple analytical screening models such as REMChlor (Falta et al., 2007), and BIOBALANCE (Groundwater Services, Inc., 2007) offer several options (equations) for simulating dissolution of NAPL in the saturated zone, as well as the effects of varying dissolution rates and NAPL mass removal on the downgradient dissolved concentrations. For example, Newell and Adamson (2005) discuss a source-depletion project that achieved 70 percent mass removal and reduced the remedial time frame (RTF) from 184 to 136 years (26 percent) when a first-order decay model of source mass depletion was applied, and a conservative concentration end point ($C_{goal}/C_0 = 0.01$) was selected. While a comprehensive cost analysis was not conducted for this scenario, the relatively minor improvement in the RTF led the authors to conclude that it might make it difficult to favor the selection of an aggressive source-depletion strategy based on reduction in RTF alone (Newell and Adamson, 2005).

The main requirement for any attempt of DNAPL source-zone remediation is the development of an adequate conceptual site-specific model of their F&T. The most important part of this effort is the realization that high density, low viscosity, and low IFT relative to water make chlorinated solvents mobile contaminants that are difficult to find or remove when released into the groundwater system. The USGS has developed five preliminary conceptual models, emphasizing accumulation sites for chlorinated DNAPL in karst aquifers (Wolfe and Haugh, 2001). Although the five models were developed for karst aquifers, which are hydrogeologically the most complex, one or more of them may be applicable for common situations found in other hydrogeologic settings. The five models of DNAPL accumulation in karst settings illustrated in Fig. 9.19 are (1) trapping in regolith, (2) pooling at the top of bedrock, (3) pooling in karst conduits, (4) pooling in bedrock diffuse-flow zones, and (5) pooling in isolation from active groundwater flow.

Trapping in regolith is most likely where the regolith is thick and relatively impermeable with few large cracks and fissures. Accumulation at the top of rock is favored by flat-lying strata with few fractures or karst features near the bedrock surface. Fractures or karst features near the bedrock surface encourage migration of chlorinated DNAPL into karst conduits or diffuse-flow zones in bedrock. DNAPL can migrate through one type of bedrock aquifer into an underlying aquifer of a different type or into openings that are isolated from significant groundwater flow (Wolfe and Haugh, 2001). The most problematic, from the aquifer restoration perspective, in addition to an extremely low

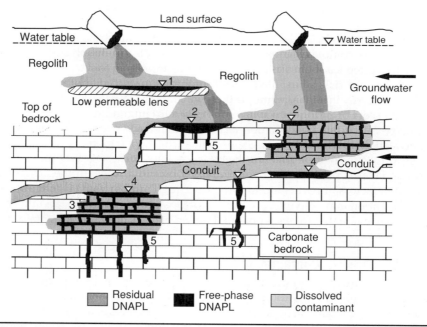

FIGURE 9.19 Distribution of potential DNAPL-accumulation sites in a hypothetical karst setting. (1) Pooling on low-permeability layer in regolith, (2) pooling on top of bedrock, (3) pooling in bedrock diffuse-flow zone, (4) pooling in conduit, and (5) pooling in fractures isolated from flow. (From Wolfe and Haugh, 2001.)

probability of actually finding DNAPL at any depth in fractured aquifers, is its migration into, and pooling in, bedrock diffuse-flow zones and fractures isolated from major zones of groundwater flow. DNAPL accumulated in such regions will be dissolved very slowly into the slow-flowing groundwater and will act as a long-term secondary source of groundwater contamination.

As emphasized by Wolfe and Haugh (2001), the five models shown in Fig. 9.19 are intended to be starting points for the analysis of chlorinated solvent contamination in karst settings and do not reduce the critical importance of careful characterization of the environmental settings and contaminant distributions at specific sites. These preliminary conceptual models are scale neutral. There is no minimum amount of DNAPL that could be stored in any of these environmental compartments, and the maximum amounts are a function of the size and nature of the release and the hydrogeologic character of specific sites. The models are mutually compatible in that more than one model may be applicable to a given site.

In conclusion, delineation of the DNAPL source zone in fractured rock and karst aquifers is generally much more difficult than in intergranular porous media. In relatively homogenous intergranular porous media, "clean" water in a sample provides reasonable evidence that there is no DNAPL further upgradient. In contrast "clean" water in one well in a fractured unit provides information only on those fractures that are both upgradient and in hydraulic contact with the well (Fountain, 1998).

The absence of an adequate conceptual model commonly leads to wasted effort and expense as data are collected, which do little to illuminate the problem at hand. Standard techniques of site characterization developed for aqueous-phase contaminants or for porous granular media may provide irrelevant or erroneous results at DNAPL sites in fractured rock and karst settings (Cohen and Mercer, 1993; Barner and Uhlman, 1995; Wolfe and Haugh, 2001). Photographs in Fig. 9.20 illustrate some of the features that cause a very complex nature of F&T of both DNAPL and dissolved-phase contamination in karst, often leading to infeasible aquifer restoration efforts.

There are risks associated with both the characterization and the remediation of DNAPLs. Any invasive technique that penetrates a DNAPL pool (e.g., drilling and push-in tools) could provide a pathway for downward migration of the DNAPL (Pankow and Cherry, 1996). Any remediation technique that alters the water table or decreases IFT between DNAPL and water may also enhance DNAPL mobility. The risks of mobilizing DNAPL are also dependent on the hydrogeology of the site. Whether a competent confining layer is present or if there are water resources beneath the DNAPL source zone are two factors of particular importance. Thus, the risk must be evaluated for each site. Remediation may also increase the risk of DNAPL mobilization, and this risk should be carefully evaluated for each potential remediation technology (Fountain, 1998).

9.4.2 Physical Containment

The enclosure of sources residing in the saturated zone with impermeable physical barriers is usually an option considered during FS for two main reasons: (1) given enough resources, physical barriers can be designed and constructed in virtually any geologic (hydrogeologic) environment, at almost any depth, and (2) these eliminate or substantially decrease the migration of dissolved contaminants from the source area and therefore the risks for human health and the environment. However, physical barriers are rarely selected as the final remedy for most common groundwater contamination problems

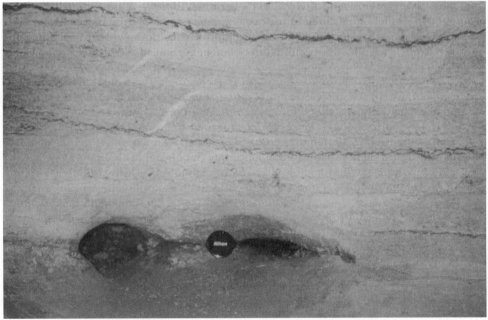

FIGURE 9.20 Some of the karst features that greatly complicate characterization and remediation of groundwater contamination. Top: Epikarst exposed in a road cut near Knoxville, TN. (Photograph courtesy of George Sowers; printed with kind permission of Francis Sowers). Bottom: Small karst conduits entering a cave passage (note camera lens cap for scale).

because the costs are often prohibitive compared to other technologies. In addition, there is no guarantee that any barrier would be completely impermeable and would maintain integrity indefinitely. For these reasons, all installed barriers must include monitoring of potential contaminant migration. Probably the main reason why physical barriers are a less desirable option even in the cases of relatively small containment volumes and low costs is the issue of a water-table rise inside the barrier due to infiltration from the land surface or a lateral inflow through the saturated zone. This buildup would have to be prevented by periodic or continuous pumping, depending on site-specific conditions. The fluid drainage system may include drains and wells in various configurations and will require a treatment system for the contaminated water. Any barrier technology that requires additional handling of fluids (groundwater) is referred to as "active." Capping of the source zone with an impermeable cover such as thick asphalt or concrete can eliminate the infiltration problem; however, the only theoretical option for preventing water table buildup is a complete enclosure of the source zone in the subsurface, including from the bottom, with a 100-percent impermeable barrier. Although rare, such "passive" systems have been installed in cases with some dangerous hazardous wastes.

In 1998, the USEPA published the results of a national survey of vertical barrier walls performance at Superfund and RCRA sites, and other hazardous waste management units at which such walls had been used as the containment method during a remedial or corrective action. Major differences were found in the monitoring of the containment systems. At some sites, very little monitoring of groundwater quality and levels was carried out, while at others monitoring well and piezometer networks downgradient of the site were used to measure trends in groundwater quality and to monitor groundwater levels. Essentially, no long-term monitoring of physical samples was performed to examine mechanisms of degradation affecting the barrier. Geophysical surveys along the wall alignment were used at several sites but were inconclusive because the available techniques cannot detect small changes in the permeability of the wall. Stress testing of the wall after construction was performed infrequently. However, monitoring data allowed the detection of leaks at four sites, and the leaks were repaired (USEPA, 1998a).

Of the 36 sites where detailed evaluation was performed, 22 had caps in addition to the barrier wall. In many cases, the caps were tied into the barrier wall. Cap design varied little among the sites, and most sites met the design requirements set forth under RCRA Subtitle C. Monitoring data for caps generally were not detailed enough to evaluate performance. Based on the findings of the survey, the Agency recommended the following measures for improvement of the performance and evaluation of subsurface engineered barriers:

- The design of subsurface barriers and caps should be based on more complete hydrogeological and geotechnical investigations than are usually conducted. In addition, designs should be more prescriptive (as appropriate) in terms of contaminant diffusion and compatibility with the barrier materials that could affect long-term performance.

- The construction quality assurance and quality control (QA/QC) effort for subsurface barriers requires further development and standardization, including nondestructive postconstruction sampling and testing.

- The importance of a systematic monitoring program in evaluating long-term performance of subsurface barriers cannot be overemphasized.

- Measures should be implemented to ensure the integrity of the barrier throughout its life, including comparative data reviews at 5-year intervals. Such reviews should address (1) hydraulic head data (specifically, the development and maintenance of a gradient inward to the containment), (2) trends in downgradient groundwater quality, and (3) data from monitoring points at the key-in horizon.

As emphasized by the USEPA, the type of containment system (active or passive) does not have a bearing on the complexity of the monitoring program. The complexity of the monitoring program is dictated by the hydrogeological characteristics of the site. It is recommended that, for passive or partially active containment sites, monitoring of groundwater quality be used to assess the performance of a barrier wall containment system. For active sites, groundwater head differentials should be the primary element monitored to assess performance of the containment system. It is recommended that the location of monitoring wells for the assessment of groundwater quality be based on a probabilistic approach to compliance monitoring. In addition, flow and transport mechanisms should be evaluated to assist in establishing the minimum necessary number and locations of monitoring points. Nests of monitoring wells, set at various depths in different strata, located close to the barrier system should also be used in identifying underflow or downward flow conditions that may allow the contaminants to migrate from the containment system (USEPA, 1998a).

The main concern in the application of engineered vertical barrier technologies is control of keying the barrier into the underlying aquitard in order to prevent the contaminant from passing under the barrier. The slurry trench excavation method, described below, is the only one that permits visual inspection of the key material and assurance of the key-in depth during construction.

Slurry Walls

Slurry walls are the most common type of subsurface wall and are considered baseline barrier technology. "It is the expert consensus" that, if properly designed and constructed, slurry walls can successfully contain waste at contaminated sites (Rumer and Mitchell, 1996). Slurry walls have been used for pollution control since 1970, and the technology is accepted and regarded as an effective method of isolating hazardous waste and preventing the migration of pollutants (Pearlman, 1999).

Barriers installed with the slurry trenching technology consist of a vertical trench excavated along the perimeter of the site, filled with bentonite slurry to support the trench and subsequently backfilled with a mixture of low-permeability material (1×10^{-6} cm/s or lower) (see Figs. 9.21 and 9.22). Such walls are keyed into an aquitard, a low-permeability soil or rock formation, or a few feet below the groundwater elevation when the objective is to contain LNAPL. Significant features of a vertical barrier are, at a minimum (USEPA, 1998a), as follows:

- Continuous wall of uniform low permeability
- Sufficient thickness to withstand earth stresses and hydraulic gradients and to provide long-term sorption capacity
- Wall backfill compatible with the groundwater quality and chemistry in the vicinity of the wall

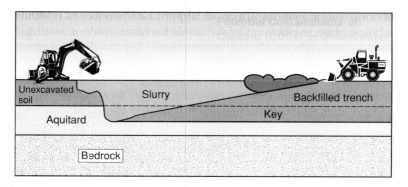

FIGURE 9.21   Soil-bentonite slurry wall installation. (Modified from USEPA, 1998a.)

- Continuous key, typically of 2 to 5 ft into the low-permeability soil or rock formation, if feasible (the quality of the key material can be verified continuously during excavation of the trench)

The most widely used technique for containment is the soil-bentonite slurry wall. It is typically the most economical, utilizes low-permeability backfill, and usually allows reuse of all or most of the material excavated during trenching. The low-permeability backfill is prepared by mixing a bentonite slurry with the excavated native soils. Additional borrow materials or dry bentonite can be added to the mixture to meet the design requirements. Recently, specialty additives have been used to increase the sorption

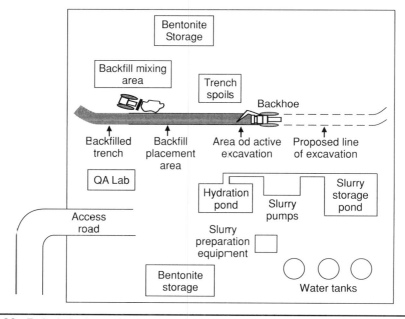

FIGURE 9.22   Typical site layout for slurry wall installation. (From USEPA, 1998a.)

capacity of the backfill. Geosynthetic liners can be used both in conjunction with slurry walls or as a stand-alone barrier.

Cement-bentonite and concrete slurry walls are used for containment when these are required by the site conditions. The techniques reduce the length of excavations held open under slurry at any given time and provide a backfill that exhibits strength. Typical applications would be trench excavation adjacent to an existing structure or through soft or unstable soil. However, the addition of the most common types of cement, such as Portland, increases the permeability of the backfill (USEPA, 1998a). Alternative self-hardening slurries incorporate ground-blast slag in with the cement to increase impermeabilities to 10^{-7} to 10^{-8} cm/s. Additions of slag can also increase the chemical resistance and strength of the barrier. Typically, the mixing ratio of Portland cement to slag is 3:1 or 4:1 (Mutch et al., 1997).

Deep Soil Mixing

Deep soil mixing technology was developed in Japan and consists of in situ mixing of soil and a slurry. The specially designed equipment typically consists of large-diameter counter-rotating augers mixing in situ soils with additives. A water-bentonite or cement slurry is injected into the soil as the augers are advanced, resulting in a column of thoroughly mixed soil. This technology is sometimes referred to as solidification stabilization. A continuous barrier is created by overlapping columns (Fig. 9.23).

Deep soil mixed barriers can achieve permeabilities of 10^{-7} cm/s. As with a vibrating beam barrier, the bottom of a deep soil mixed barrier cannot be inspected to confirm key penetration. However, deep soil mixed barriers are considerably wider than vibrating beam barriers and can achieve lower permeabilities. Because potentially contaminated materials are not excavated, the advantages of using deep soil mixing technology include reduction of health-and-safety risks and elimination of costs associated with handling and disposal of contaminated soils (USEPA, 1998a).

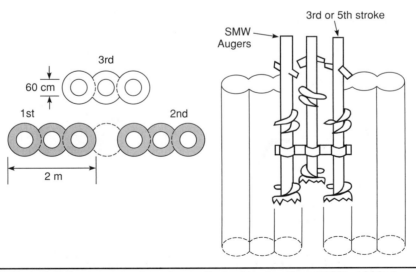

Figure 9.23 Installation sequence for a deep soil mixed barrier. (From USEPA, 1998a.)

Grouted Barriers

Grout barriers ("walls" or "curtains") have been used extensively for civil engineering projects (e.g., see USACE, 1984), but less frequently at hazardous waste sites. These are usually more expensive than other techniques, and the barriers have higher permeability. However, grout walls can have greater depths than any other type of subsurface barriers and are capable of extending the key through bedrock.

Construction of grouted barriers involves injection of a grout into the subsurface. Pressure grouting and jet grouting are both forms of injection grouting, in which a particulate or chemical grout mixture is injected into the pore spaces of the soil or rock. Particulate grouts include slurries of bentonite, cement, or both and water. Chemical grouts generally contain a chemical base, a catalyst, and water or another solvent. Common chemical grouts include sodium silicate, acrylate, and urethane. Particulate grouts have higher viscosities than chemical grouts and are therefore better suited for larger pore spaces, whereas chemical grouts are better suited for smaller pore spaces (USEPA, 1998a; USACE, 1995).

Sheet-Pile Walls

Sheet-pile cutoff walls are constructed by driving vertical strips of steel, precast concrete, aluminum, or wood into the soil, forming a subsurface barrier wall. The sheets are assembled before installation and driven or vibrated into the ground, a few feet at a time, to the desired depth. Sheet-pile walls traditionally have consisted of steel sheeting with some type of interlock joint. Recently, such sheeting includes an improved interlock design to accommodate sealing of joints; several innovative techniques have been developed recently to seal and test the joints between sheet piles. In addition, plastic has been substituted for steel in a number of applications.

Sheet-pile walls have long been used for a wide variety of civil engineering applications, but their use in environmental situations has been limited. Sheet-pile wall installation is limited to shallow depths and unconsolidated materials. Although steel sheet-pile walls are strong and steel will not hydrofracture, the interlocking joints present a leakage problem. The ability of steel sheet piling to meet a typical 10^{-7} cm/s design performance standard depends on the type of material used to seal the interlocking joints.

The Waterloo Barrier^{TM} is an adaptation of the sheet-pile wall that addresses the problem of leaky joints. The Waterloo Barrier^{TM} is specially designed to interlock sealable joints (Fig. 9.24). Installation involves driving sheet piles into the ground, flushing the interlocking joint cavity to remove soil and debris, and injecting sealant into the joints.

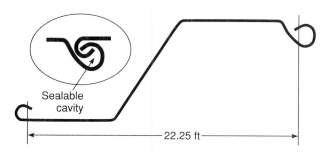

FIGURE 9.24   Waterloo Barrier^{TM} sealable joint steel sheet piling (WZ 75 profile). (Source: http://www.oceta.on.ca/profiles/wbi/barrier.html.)

Depending on site conditions, the cavity may be sealed with a variety of materials, including clay-based, cementitious, polymer, or mechanical sealants. Video inspection of the joint cavity prior to sealing ensures that the joint can be sealed. The Waterloo Barrier^{TM} can achieve bulk hydraulic conductivities of less than 10^{-8} cm/s (Mutch et al., 1997).

9.4.3 Fluid Removal Technologies

The main difficulty in applying NAPL fluid removal technologies in source zones is difficulty in finding pooled, free-phase NAPL precisely enough for feasible extraction. This difficulty is particularly pronounced in fractured rock and karst aquifers. While commonly applied for LNAPL sources, which are comparably easier to locate, as these tend to remain in the shallow saturated zone, fluid removal technologies have been less successfully demonstrated on DNAPL sources. In most cases, however, the removed fluids require some form of treatment and disposal, which is often a less desirable option compared to in situ technologies such as chemical oxidation and bioremediation.

Soil Vapor Extraction (SVE) and Bioventing

In situ SVE is the process of removing and treating volatile and some semivolatile organic compounds from the unsaturated zone. By applying a vacuum through a system of wells, contaminants are pulled to the surface as vapor or gas. Often, in addition to vacuum extraction wells, air injection wells are installed to increase the air flow and improve the removal rate of the contaminant (Fig. 9.25). An added benefit of introducing air into the soil is that it can stimulate bioremediation of some contaminants. Used alone, SVE cannot remove contaminants in the saturated zone. It is therefore often combined with groundwater extraction, which lowers the water table and increases the thickness of the unsaturated soil from which more residual NAPL and sorbed volatile organic compounds (VOCs) can be stripped (volatilized). SVE is also combined with air sparging, the technology described in more detail in the next section.

An example of combining groundwater removal with submersible pumps and SVE technology is illustrated in Fig. 9.26. Blowers attached to extraction wells alone or in

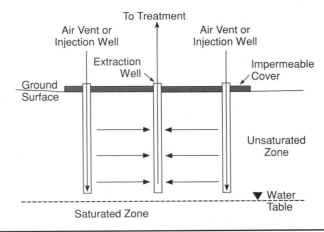

FIGURE 9.25 Typical soil vapor extraction system. (Modified from USEPA, 2006a.)

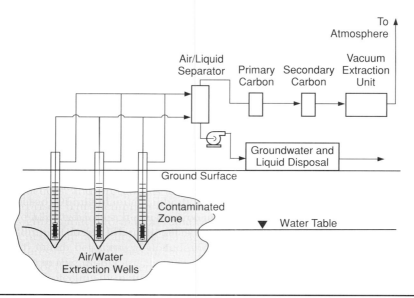

FIGURE 9.26   Illustration of combined groundwater removal and soil vapor extraction (SVE) technology implementation. (After BATTELLE, 1997.)

combination with air injection wells induce airflow through the soil matrix. The airflow strips the volatile compounds from the soil and carries them to extraction wells. The process is driven by the partitioning of volatile materials from the solid, dissolved, or nonaqueous liquid phases to the clean air being introduced by the vacuum extraction process. Air emissions from the systems typically are controlled aboveground by adsorption of the volatiles onto activated carbon, which is the most commonly used treatment for contaminated vapors and is adaptable to a wide range of VOCs. Vapor treatment technologies also include thermal destruction (incineration, catalytic oxidation, or internal combustion engine) and condensation by refrigeration. SVE is a mature technology that has been used routinely for remediation of soil contaminated with VOCs and, in combination with other technologies, for remediation of shallow groundwater.

Bioventing is a process in which a system consisting of injection and extraction wells is used to either push or pull air through the contaminated unsaturated zone. Airflow increases the availability of oxygen and promotes aerobic biodegradation of organics in the unsaturated zone. Adjustments to the soil moisture content, temperature, or other factors may be used to improve the biodegradation process. Bioventing relies on the ability to move air through the contaminated material. Adjustment chemicals may be applied as aqueous solutions, and modifications may be used to increase the soil temperature.

Relying on naturally occurring oxygen in the injected water is often not feasible, since large amounts of oxygen-saturated water are required for biotreatment and often cannot be delivered because the hydraulic conductivity of the soil is too low (Dupont et al., 1991). Bioventing is, therefore, applicable to sites where limited soil permeability makes the site unsuitable for biotreatment. A bioventing system moves air through the unsaturated zone with a system of vent wells and blowers. Air movement provides an oxygen source to speed metabolism of organic contaminants. Bioventing is an established technology for remediation of petroleum hydrocarbons in the unsaturated zone.

Air Sparging

In situ air sparging (IAS) is still an evolving technology applied to serve a variety of remedial purposes. While IAS has primarily been used to remove VOCs from the saturated subsurface through stripping, the technology can be effective in removing volatile and nonvolatile contaminants through other, primarily biological, processes enhanced during its implementation. The basic IAS system strips VOCs by injecting air into the saturated zone to promote contaminant partitioning from the liquid to the vapor phase. Off-gas may then be captured through an SVE system, if necessary, with vapor-phase treatment prior to its recirculation or discharge into the air (USACE, 2008). IAS appears to have first been utilized as a remediation technology in Germany in the mid-1980s, primarily to enhance cleanup of groundwater contaminated by chlorinated solvents (Gudemann and Hiller, 1988). Figure 9.27 depicts a typical combination of SVE and IAS systems.

Because injected air, oxygen, or an oxygenated gas can stimulate the activity of indigenous microbes, IAS can be effective in increasing the rate of natural aerobic biodegradation. This is particularly important when considering the use of IAS at sites with readily biodegradable hydrocarbons, particularly petroleum-contaminated sites. It has been speculated that, similarly, anaerobic conditions might be able to be created by injecting a nonoxygenated gaseous carbon source to remove the dissolved oxygen (DO) from the water. The resulting enhanced degradation of organic compounds, such as chlorinated VOCs, to daughter products would result in increased volatility, which could improve the effectiveness of stripping and phase transfer during IAS (USACE, 2008).

IAS is generally considered a mature technology. It is relatively easy to implement and well known to regulatory agencies, and the equipment necessary for IAS is generally inexpensive and easily obtained. Therefore, IAS is one of the most practiced engineered technologies for in situ groundwater remediation. Critical aspects governing the effectiveness of an IAS system, such as the presence and distribution of preferential airflow pathways, the degree of groundwater mixing, and potential precipitation and clogging of the soil formation by inorganic compounds, continue to be researched and reported in conference proceedings and technical journals. There are innovative field techniques that can aid the understanding of the effectiveness of IAS, such as neutron probes for measuring the effective zone of influence (ZOI) and distribution of the injected gas. As

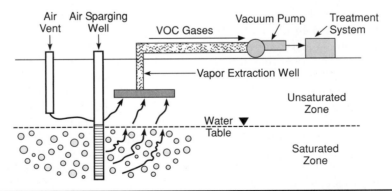

Figure 9.27 A combined soil vapor extraction (SVE) and in situ air sparging (IAS) system. (From USEPA, 1996a.)

IAS is often considered to be a straightforward technology, such techniques are not often implemented. However, when such data are collected, it is anticipated that the understanding of the mechanisms and processes induced by IAS will increase, as well as the ability to predict and measure its effectiveness (USACE, 2008).

There are fundamental physical limitations on the effectiveness of air sparging for treating NAPLs. LNAPLs tend to form pools above the water table or discontinuous ganglia throughout the capillary fringe and smear zone. These LNAPL pools and ganglia represent potentially large sources of VOCs with relatively limited surface areas. The small surface area of such NAPL bodies limits the rate of interphase mass transfer of VOCs from NAPL into sparge air, in much the same way as it limits the transfer of VOCs from NAPL into groundwater. However, over time, pooled volatile LNAPL, such as gasoline or jet fuel, and residual NAPL in the smear zone may be remediated by combined IAS/SVE approaches. Laboratory experiments performed with poorly graded coarse sand imbued with benzene NAPL "pools" demonstrated fairly rapid NAPL removal (Adams and Reddy, 2000). The potential for remediation of less volatile LNAPLs (e.g., diesel or fuel oils) is not as promising and relies more on biodegradation potential than enhanced volatilization of the LNAPL (USACE, 2008).

DNAPL sites are particularly difficult to remediate with IAS. In addition to the limitations of interphase mass transfer, the effect of capillary pressures on DNAPLs and sparged air operates to inhibit these two phases from contacting one another in the subsurface. In even moderately heterogeneous aquifers, DNAPLs tend to pool atop low-permeability lenses when these lack the entry pressure to penetrate the lower-permeability lens. Sparged air likewise often fails to enter lower-permeability lenses from below, because the capillary pressure resisting air flow through low-permeability units is even greater than that resisting the DNAPL. As a result, the sparged air tends to flow around the lower-permeability lens before continuing upward, never contacting the DNAPL resting atop the lens.

Steam can be injected in conjunction with, or instead of, air to incorporate a thermal treatment element to traditional air sparging technology. Steam injection has been employed successfully to remediate VOC-contaminated aquifers that would otherwise be difficult to remediate using traditional IAS and to remediate contaminants not amenable to traditional IAS (USEPA, 1997, 1998b; Davis, 1997). Steam injection design and operation are subject to many of the same constraints as air stripping. Considerations related to multiphase flow (i.e., preferential flow paths) are important in determining whether steam injection has the potential to succeed at a site. However, because steam incorporates an element of thermal treatment, the necessary vapor-water contact area can be substantially less than for traditional air sparging. Because the thermal conductivity rates are much higher than diffusive mass transfer between vapor-filled pores and the surrounding water-filled pores, steam injection can affect a larger volume of soil for a given vapor-phase saturation. The lateral distribution of heat is further enhanced by the horizontal flow of hot condensate from injection wells. As steam will condense in the cooler parts of the subsurface, the vapor phase will not initially reach the vadose zone and this condensation front will migrate from the steam sparging/injection well until breaking through to extraction wells or the water table. To enhance vapor-phase transfer of contaminants and to provide oxygen for destructive oxidation processes, the steam is sometimes amended with air (USACE, 2008).

A common application of IAS is for the treatment of dissolved-phase contamination in a plume, downgradient of source areas. Configurations used for aqueous-phase

Parameter	Limited Effectiveness	Likely Benefit	Well Suited
Contaminant type	Weathered fuels Lubricating oils Hydraulic fluids Dielectric fluids PCBs	Diesel fuel Jet fuel Acetone MTBE	MOGAS AVGAS Halogenated solvents[1] BTEX
Geology	Silt and clay (interbedded) Massive caly Lubricating oils Highly organic soils Fractured bedrock Stratified soil Confining layers	Weakly stratified soils Sandy silt Gravelly silt Highly fractured clay	Uniform coarse-grained soils (gravels and sands) Uniform silts
Contaminant phase	Free product	Sorbed	Dissolved
Contaminant location	Within confined aquifer; near bottom of unconfined aquifer	Within shallow aquifer	Near water table
Contaminant extent	Large plumes[2]	Modest-size plumes	Small plumes
Hydraulic conductivity (cm/s)	$<10^{-5}$	10^{-5} to 10^{-4}	$>10^{-4}$
Anisotropy	High degree	Moderate degree	Isotropic

[1] IAS is generally applicable to halogenated ethenes, ethanes, and methanes.
[2] Sparge curtains may be effective in managing migration within large plumes.
From USACE, 2008.

TABLE 9.3 Factor Affecting Applicability of In Situ Air Sparging (MOGAS: motor gasoline; AVGAS: aviation gasoline)

treatment include the installation of an array of air sparging points, spaced so that each individual ZOI overlaps. When the source is a release of LNAPL (e.g., gasoline or fuel oil), the dissolved plume is often situated near the water-table surface of an unconfined aquifer. In such cases, IAS points can be conveniently located just below the plume to obtain the desired coverage. In a survey of 32 IAS case studies, Bass and Brown (1996) concluded that performance of IAS systems was generally better in systems treating dissolved-phase plumes than in systems treating adsorbed contaminants. Table 9.3 includes various parameters influencing the effectiveness of IAS.

IAS does present some potential risks. One of the problems in applying air sparging is controlling the process. In bioventing, airflow is induced by air injection at low pressure or by air extraction. In groundwater extraction, the groundwater is collected and removed by pumping. In situ contaminant flow in these systems is under better control because the contaminants tend to migrate toward extraction points. By contrast, air sparging systems inject air at or above the local hydrostatic head. The injection of pressurized air can cause water and contaminants to move away from the air injection point. This migration can accelerate and aggravate the spread of contamination. A second problem with IAS is accelerated vapor travel. Air sparging can increase the partial pressure of

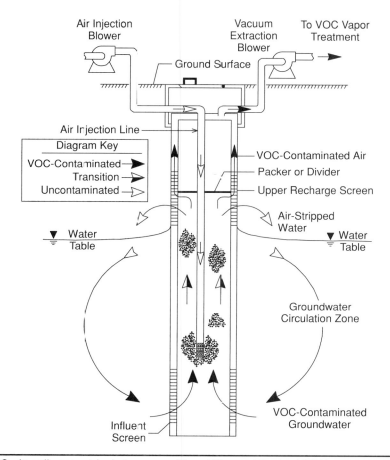

FIGURE 9.28 In-well vapor stripping process. (From Miller and Roote, 1997.)

volatile contaminants in the unsaturated-zone gas phase and induce gas migration out of the contaminated zone. The combination of increased contaminant concentration and increased gas migration can increase contaminant movement in the unsaturated zone (BATTELLE, 1997).

A variation of air sparging, called in-well vapor stripping, involves the creation of a groundwater circulation pattern and simultaneous aeration within the stripping well to volatilize VOCs from the circulating groundwater. Air-lift pumping is used to lift groundwater and strip it of contaminants. Contaminated vapors may be drawn off for aboveground treatment or released to the vadose zone for biodegradation. Partially treated groundwater is forced out of the well into the vadose zone where it reinfiltrates to the water table. Untreated groundwater enters the well at its base, replacing the water lifted through pumping (Fig. 9.28). Eventually, the partially treated water is cycled back through the well and through this process until contaminant concentration goals are met (Miller and Roote, 1997).

Modifications of the basic process involve combinations with SVE and aboveground treatment of extracted vapors and/or injection of nutrients and other amendments to enhance natural biodegradation of contaminants. Applications of in-well stripping have generally involved chlorinated organic solvents (e.g., TCE) and petroleum product contamination (e.g., benzene, toluene, ethylbenzene, and xylenes (BTEX) and total petroleum hydrocarbons (TPH)). Application of this technology, based on system modifications, may address nonhalogenated VOC, SVOC, pesticide, and inorganic contamination. In-well stripping has been used in a variety of soil types from silty clay to sandy gravel (Miller and Roote, 1997).

Reported advantages of in-well stripping include lower capital and operating costs due to use of a single well for extraction of vapors and remediation of groundwater and absence of the need to pump, handle, and treat groundwater at the surface. Additional advantages involve its easy integration with other remediation techniques such as bioremediation and SVE and its simple design with limited maintenance requirements. Limitations reported for this technology include limited effectiveness in shallow aquifers, possible clogging of the well due to precipitation, and the potential to spread the contaminant plume if the system is not properly designed or constructed (Miller and Roote, 1997).

A variety of resources are available to assist in assessing the feasibility of IAS and designing an effective system. Resources include models for system design and optimization and various manuals published by the United States Federal agencies. For example, TOUGH2/TMVOC is a multiphase, nonisothermal, saturated, and unsaturated numerical transport model that can be applied to IAS simulations. More information and the model are available from the Lawrence Berkeley Laboratory Web site at http://www.esd.lbl.gov. A limitation associated with IAS models is that the heterogeneities that control airflow paths are on a scale much finer than the available site characterization data. The processes that IAS models must incorporate include multiphase flow, buoyancy and capillary forces acting on air, and soil variability on a small and large scale (perhaps by stochastic methods).

USACE maintains a Web site that contains information on SVE, bioventing, and other air-based remediation technologies. This Web site lists useful documents and links to Federal bulletin boards and databases, located at http://www.environmental.usace.army.mil/info/technical/geotech/geotopical/sve/sve.html.

USEPA also maintains a web page cataloguing relevant IAS guidance documents, located at http://clu-in.org/techfocus/default.focus/sec/Air_Sparging/cat/Guidance/.

Free-Phase and Multiphase Extraction

A number of technologies are aimed at NAPL extraction from the subsurface using trenches and drains, and vertical and horizontal wells of sufficient diameter to accommodate pumping equipment. These are usually given names based on fluid phase targeted for removal and the principal extraction mechanism. Extraction mechanisms include vacuum pumps, skimmer pumps (for floating LNAPLs), downhole single- and dual-phase pumps, total fluid pumps, and their varying combinations applied in a single well.

Conventional LNAPL recovery uses an electric or pneumatic pump to remove LNAPL from the surface of the water table. This recovery can be accomplished with or without water-table drawdown. Skimming systems rely on passive movement of LNAPL into the well (trench) and use skimming pump for LNAPL-only recovery. Such systems have negligible drawdown and radius of influence (ROI) outside the well (trench). LNAPL recovery by drawdown can be performed using a single total fluids pump or separate groundwater and LNAPL recovery pumps. Single-pump systems are installed below the water table and extract groundwater and LNAPL in the same stream that is then

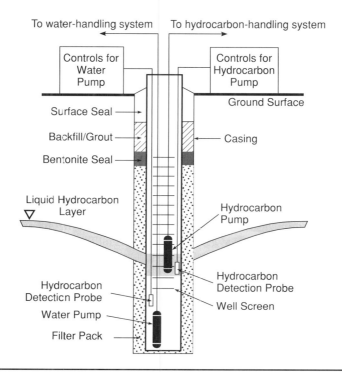

To water-handling system   To hydrocarbon-handling system

Controls for Water Pump

Controls for Hydrocarbon Pump

Ground Surface

Surface Seal

Backfill/Grout

Casing

Bentonite Seal

Liquid Hydrocarbon Layer

Hydrocarbon Pump

Hydrocarbon Detection Probe

Water Pump

Hydrocarbon Detection Probe

Well Screen

Filter Pack

FIGURE 9.29 Conventional LNAPL recovery using dual pump system. (From BATTELLE, 1997.)

separated aboveground. Dual-pump systems use a submersible water pump to lower the groundwater table and an LNAPL skimming pump to recover LNAPL that migrates into the well (Fig. 9.29). Drawdown systems for LNAPL increase recovery by lowering the water table, which increases the hydraulic gradient toward the well and accelerates the LNAPL flow into the collection system. Drawdown, however, can result in entrapment of LNAPL within the cone of depression, potentially deepening the smear zone of LNAPL in the soil, which can be difficult to remediate (Leeson and Hinchee, 1995). Conventional LNAPL recovery is best suited for sites with homogeneous, coarse-grained soils that will allow LNAPL to flow freely into a recovery well or trench.

Multiphase extraction (MPE) has evolved as a remediation method that applies the technology pioneered for construction vacuum dewatering to enhance the recovery of LNAPL. At many sites, LNAPL present in the capillary fringe cannot flow toward extraction wells due to capillary forces holding the LNAPL within soil pores (Baker and Bierschenk, 1995). This phenomenon is common in fine-textured soils such as fine sands, silts, and clays. By applying high vacuums at extraction wells, the capillary forces holding the LNAPL in the soil may be overcome to a degree, and LNAPL can flow toward the extraction well (USACE, 1999). Vacuum-enhanced recovery also improves recovery rates by increasing the hydraulic gradient and increasing the aquifer transmissivity. Vacuum-enhanced pumping systems use the same concept as conventional groundwater pumping, except that the cone of depression actually is a cone of reduced pressure around the well. Fluids then flow horizontally across the pressure-induced gradient, from higher pressure outside the well to lower pressure inside the well. Vacuum-enhanced pumping increases transmissivity by promoting flow along more permeable horizontal flow lines and by decreasing the local pressure above the aquifer to, in effect, increase the saturated thickness of the aquifer. The sum effect of the increase in hydraulic gradient and aquifer transmissivity is an enhanced liquid recovery rate (AFCEE, 1995).

MPE (with or without drawdown) will generate groundwater, air, and LNAPL to be managed and treated aboveground. MPE with drawdown will typically result in more groundwater extraction from a given well than MPE without drawdown. However, the most commonly perceived benefit of using this technique is to dewater the soil surrounding the MPE well to expose to air discontinuous ganglia of LNAPL trapped below the water table. As the water table is lowered, these ganglia may either drain toward the declining water table surface due to gravity and vacuum inducement or volatilize and be extracted in gas that flows to the MPE well.

MPE is accomplished using two distinct technologies. Dual-phase extraction (DPE) technology generally employs separate pumps to extract liquid and gas from a well (Fig. 9.30). Two-phase extraction (TPE) extracts liquid and gas from a well using a single-suction pipe or conduit. Liquid and gas flow from extraction wells can be measured and controlled more effectively in DPE systems than in TPE systems. Therefore, DPE provides more opportunity for developing a system in which flow rates from the MPE wells in a network can be balanced to accommodate differences in soil characteristics across the

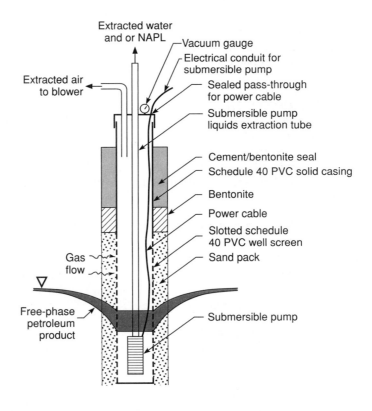

Figure 9.30 Typical dual-phase extraction (DPE) system. (From USACE, 1999.)

treatment area. A common problem with TPE systems is breaking suction at one or more of the wells in the network. If a single well is able to produce a high flow rate of air, then the vacuum in the entire system can be reduced to a level that is insufficient for liquid extraction at other wells (USACE, 1999).

At some sites, the physical/chemical properties of the DNAPL combined with the release history and geologic conditions result in the formation of zones of potentially mobile DNAPL (e.g., pools). Once the mobile DNAPL zone(s) are identified, the extraction system can be designed. The screen interval of DNAPL recovery wells should correspond to the subsurface zone containing the DNAPL. At sites where several zones are encountered at different elevations, it is advisable to begin extracting from the uppermost zone first and then extracting from progressively lower zones once the upper zone(s) have ceased DNAPL production. This will maximize recovery efficiency and minimize the potential for uncontrolled mobilization. Creating a shallow sump in a less permeable stratum at the bottom of the well for the collection of the DNAPL may also be advisable. The sump will provide a convenient and efficient location for placing the intake of the DNAPL pump (Michalski et al., 1995).

A total liquids approach can be used (i.e., water and DNAPL are removed from the well via one pump and then separated at the surface). This may minimize equipment costs; however, it is not the most efficient approach. As the DNAPL and water are extracted from the well, the DNAPL saturation is decreased in a zone around the well, the relative permeability of the formation with respect to DNAPL is decreased, and the DNAPL production rate decreases. Eventually, a zone of residual (nonmobile) DNAPL is created around the well and the well no longer produces DNAPL (USACE, 1999).

The ideal approach is to maintain or enhance DNAPL saturation around the well in order to increase removal efficiency. DNAPL extraction can be enhanced using a dual-pumping approach, where water is removed separately from the zone immediately above the mobile DNAPL (Sale and Applegate, 1997). This approach results in upwelling of DNAPL in the well and increased DNAPL saturations in the immediate vicinity of the well. A variation of this approach is to apply a vacuum to the upper of the two wells, to decrease the pressure head in the well. This has a similar effect as pumping water, in that it results in a decrease in the total head in the well (i.e., increased hydraulic gradients near the well) and increased DNAPL thicknesses, saturations, production rates, and removal efficiencies.

Under the most favorable conditions, direct recovery will remove between 50 and 70 percent of the DNAPL in the subsurface (Pankow and Cherry, 1996). The remaining residual DNAPL will still be sufficient to serve as a significant long-term source unless it is addressed through other means.

There are a number of enhancements to DPE aimed at increasing NAPL mobility and recovery rate. These include flushing (flooding) with surfactants and cosolvents, and heating (thermal) technologies.

Surfactant and Cosolvent Flushing

Surfactant-enhanced aquifer remediation (SEAR) is a source-zone remediation technology that may be used as an enhancement to conventional P&T systems, which are inefficient for recovering contaminants that are trapped as immiscible-phase liquid (NAPLs). The premise of this technology is that most organic NAPLs are only somewhat soluble in water and therefore will persist in the subsurface for a very long time. However, chemical amendments to groundwater can cause many types of NAPL to dissolve in groundwater much more readily. Surfactants such as detergents and cosolvents such as alcohols can, when added to the groundwater in high concentrations (e.g., 50 percent by volume in the case of cosolvents), enhance the rate of NAPL dissolution by orders of magnitude.

Surfactant stands for surface active agents, which are active ingredients in soaps and detergents and are common commercial chemicals. Two properties of surfactants are central to remediation technologies: the ability to lower interfacial tension (IFT) and the ability to increase solubility of hydrophobic organic compounds. Both properties arise from the fact that surfactant molecules have a hydrophobic and a hydrophilic portion. As a result, when water containing surfactant and NAPL come into contact, surfactant molecules will concentrate along the interface, with their polar ends in the water and their nonpolar ends in the NAPL; this lowers the IFT between the two immiscible fluids. When present in sufficient concentration (the critical micellar concentration), surfactant molecules form oriented aggregates, called micelles. In water, the molecules in a micelle are arranged with their polar ends outward and their nonpolar ends inward, forming a nonpolar interior to the micelle. Micelles can incorporate hydrophobic molecules in their interior, producing an apparent increase in solubility. The process of dissolving by incorporation into micelles is termed solubilization. Once solubilized, a compound is transported as if it were a typical dissolved phase (Fountain, 1998).

Over the last decade, the greatest demand for the SEAR technology has been for remediating chlorinated hydrocarbon DNAPLs such as TCE and PCE. Industrial solvents were often released to the environment as mixtures with oily contaminants. SEAR has been used to remove these variable-density contaminant mixtures and contaminants such as creosote, gasoline, jet fuels, and polychlorinated biphenyls (PCBs) (Lowe et al., 1999).

At suitable sites, implementation of surfactant or cosolvent flushing involves installation of injection wells to introduce the chemical amendment into the contaminated zone and extraction wells for fluid recovery (Fig. 9.31). Groundwater is typically recirculated through the contaminated zone in an effort to achieve the widest possible dispersion of the additive throughout the contaminated area. While this technology is promising, it is also relatively expensive. SEAR suffers from the same limitations as MPE in heterogeneous unsaturated soils, that is, the tendency of the surfactant/cosolvent laden water to preferentially flow through the highest permeability strata, which may not be where the bulk of the contaminant mass resides. At the same time, the risk of mobilizing contaminants requires a complete hydraulic control over the injected fluid and contaminants.

SEAR application uses the properties of surfactants to remove contaminants either primarily by solubilization or primarily by mobilization. Surfactant-induced mobilization can remove greater amounts of DNAPL in less time; however, there is greater risk of uncontrolled downward movement of DNAPL, as DNAPL is being physically displaced by the surfactant solution. Thus, to conduct a mobilization flood, it is necessary to have a competent aquitard as a barrier to prevent vertical DNAPL migration. Where there is no clay barrier underlying the contaminated zone, or a thin and/or discontinuous one, the surfactant flood must be designed only to solubilize contaminant. Solubilization increases the density of the contaminant-loaded surfactant solution by only several weight percent (wt percent) compared to mobilization, which involves a much denser, moving front of DNAPL. It is important to identify from the outset whether solubilization or mobilization of DNAPL is desired, because not all surfactants can accomplish the low IFT necessary to conduct a mobilization flood.

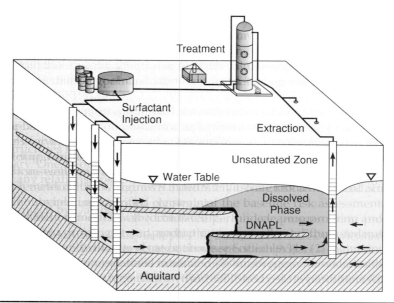

FIGURE 9.31 Conceptual design for a surfactant-enhanced aquifer remediation (SEAR) application. (From BATTELLE and Duke Engineering and Services, 2002.)

The surfactant solution formulated for a SEAR application typically consists of water and surfactant, plus additives such as an electrolyte (i.e., dissolved ionic salt) and a co-solvent. The surfactant must be able to efficiently remove the DNAPL and be compatible with the soils and groundwater. A cosolvent, such as isopropyl alcohol (IPA), often is added to improve the solubility of surfactant in water, so that the resulting surfactant-DNAPL solution (microemulsion) has an acceptable viscosity. The addition of a cosolvent also influences the surfactant-phase behavior, so the effects of cosolvent addition on the surfactant solution must be examined under a range of system salinities. Because the presence of a cosolvent complicates wastewater treatment, ongoing technology development has focused on surfactants that have no or minimal cosolvent requirements, as well as on cosurfactant substitutes to cosolvents (BATTELLE and Duke Engineering and Services, 2002).

Aquifer heterogeneities can cause significant channeling of the injected fluids and bypassing of contaminated zones, causing poor surfactant sweep of the area targeted for remediation. The success of SEAR depends on an accurate characterization of not only the aquifer lithology but also the DNAPL saturation and its spatial distribution in the aquifer. The most economical application of SEAR is in a relatively homogeneous and highly permeable subsurface ($K \geq 10^{-3}$ cm/s). As the permeability of soils decreases and/or the heterogeneity increases, remediation project costs will increase (BATTELLE and Duke Engineering and Services, 2002).

9.4.4 In Situ Chemical Oxidation

ISCO involves the introduction of a chemical oxidant into the subsurface for transforming groundwater or soil contaminants into less harmful chemical species. There are several different forms of oxidants that have been used for ISCO; however, the following four are the most commonly used: (1) permanganate (MnO_4^-), (2) hydrogen peroxide (H_2O_2) and iron (Fe) (Fenton-driven or H_2O_2-derived oxidation), (3) persulfate ($S_2O_8^{2-}$), and (4) ozone (O_3). The type and physical form of the oxidant indicate the general materials handling and injection requirements. As discussed by Huling and Pivetz (2006), the persistence of the oxidant in the subsurface is important, since this affects the contact time for advective and diffusive transport and ultimately the delivery of oxidant to targeted zones in the subsurface. For example, permanganate persists for long periods of time, and diffusion into low-permeability materials and greater transport distances through porous media are possible. H_2O_2 has been reported to persist in soil and aquifer material for minutes to hours, and the diffusive and advective transport distances will be relatively limited. Radical intermediates formed using hydrogen peroxide, persulfate, and ozone, which are largely responsible for various contaminant transformations, react very quickly and persist for very short periods of time (<1 seconds).

Some oxidants are stronger than others, and it is common to calculate a relative strength for all oxidants using chlorine as a reference. Table 9.4 lists the relative strengths of common oxidants. All the oxidants shown have enough oxidative power to remediate most organic contaminants. The standard potentials are a useful general reference of the strength of an oxidant, but these values do not indicate how these will perform under field conditions. Four major factors play a role in determining whether an oxidant will react with a certain contaminant in the field: (1) kinetics, (2) thermodynamics, (3) stoichiometry, and (4) delivery of oxidants. On a microscale, kinetics or reaction rates are, perhaps, most important. In fact, reactions that would be considered thermodynamically favorable based on E_0 values may be impractical under field conditions. The rates of oxidation reactions are dependent on many variables that must be considered simultaneously, including temperature, pH, concentration of the reactants, catalysts, reaction by-products, and system impurities such as natural organic matter (NOM) and oxidant scavengers (ITRC, 2005).

The remediation of groundwater contamination using ISCO involves injecting oxidants and potential amendments directly into the source zone and downgradient plume

Chemical Species	Standard Oxidation Potential (V)	Relative Strength (Chlorine = 1)
Hydroxyl radical (OH^{-o})[1]	2.8	2.0
Sulfate radical (SO_4^{-o})	2.5	1.8
Ozone	2.1	1.5
Sodium persulfate	2.0	1.5
Hydrogen peroxide	1.8	1.3
Permanganate (Na/K)	1.7	1.2
Chlorine	1.4	1.0
Oxygen	1.2	0.9
Superoxide ion (O^{-o})[1]	−2.4	−1.8

[1] These radicals can be formed when ozone and H_2O_2 decompose.
From ITRC, 2005.

TABLE 9.4 Oxidant Strengths

FIGURE 9.32   Direct-push drilling for oxidant injection into shallow aquifer contaminated with chlorinated solvents. Inset: Temporary injection point with the oxidant feed line. (Photographs courtesy of ECC)

(Fig. 9.32). The oxidant chemicals react with the contaminants, producing innocuous substances such as carbon dioxide, water, and—in the case of chlorinated compounds—inorganic chloride. However, there may be many chemical reaction steps required to reach those end points, and some reaction intermediates, as in the case of polyaromatic hydrocarbons (PAHs) and organic pesticides, are not fully identified at this time. Fortunately, in most cases if an adequate oxidant dose is applied, the reactions proceed to completion, and the end products are reached quickly (ITRC, 2005). Contaminants amenable to treatment by ISCO include

- BTEX
- Methyl *tert*-butyl ether (MTBE)
- TPH
- Chlorinated solvents (ethenes and ethanes)
- PAHs
- PCBs
- Chlorinated benzenes (CBs)
- Phenols
- Organic pesticides (insecticides and herbicides)
- Munitions constituents (RDX, TNT, HMX, etc.)

As discussed by Huling and Pivetz (2006), permanganate-based ISCO is more frequent and fully developed compared to other oxidants. Well-documented and widespread use of in situ permanganate oxidation involving a diversity of contaminants, in conjunction with long-term monitoring data and cost information, has contributed to the development of the infrastructure needed to support decisions to design and deploy permanganate ISCO systems. However, additional research and development is needed. Fenton-driven ISCO has been deployed at a large number of sites and involves a variety of approaches and methods of using hydrogen peroxide (H_2O_2) and iron (Fe). In general, Fenton chemistry and in situ Fenton oxidation is complex and involves numerous reactive intermediates and mechanisms, and technology development has been slower. Ozone (O_3) is a strong oxidant that has been used in the subsurface but in much more limited application than permanganate and Fenton-driven oxidation. Persulfate ($S_2O_8^{2-}$) is a relatively new form of oxidant that has mainly been investigated at bench scale. However, considerable research and applied use of this oxidant at an increasing number of field sites are resulting in rapid development (Huling and Pivetz, 2006).

There are two main advantages of using ISCO over other conventional treatment technologies: large volumes of waste material are not usually generated, and treatment is commonly implemented over a much shorter time frame. Both of these advantages often result in savings on material, monitoring, and maintenance. It should be noted, however, that chemical oxidation often requires multiple applications. In the special case of NAPLs, oxidants that are in a water-based solution will only be able to react with the dissolved phase of the contaminant, since the two will not mix. This property limits their activity to the oxidant solution/NAPL interface (USEPA, 2006b). Nevertheless, because all oxidants are nonselective, these also oxidize NOM present in the soil. Since organic contaminants sorb to NOM in the soil matrix, these can be released as the NOM is oxidized by the injected oxidant. After this initial contaminant release, the rate of continued desorption should be increased due to the shift in equilibrium partitioning that results as the aqueous-phase concentration of the target organic is depleted (ITRC, 2005).

As with any other fluid injection technology, ISCO application design should thoroughly address possible geochemical reactions with the aquifer (soil) porous media and all contaminants present. For example, naturally occurring or anthropogenic metals can be mobilized within the treatment zone due to a change in oxidation states and/or pH. This is of particular concern when using persulfate, as very low pH conditions (1.5 to 2.5) in water were observed due to persulfate decomposition. Natural soil buffering capacity can help alleviate this phenomenon but would have to be evaluated (tested) prior to ISCO application. Another observed problem at some sites is reduction of porous media permeability (hydraulic conductivity) due to chemical reactions that precipitate insoluble salts, such as MnO_2 in the case of ISCO with permanganate.

Like most technologies, ISCO has limitations that should be recognized. There are situations in which ISCO would be ineffective at degrading the contaminants present. It is also possible that, due to the total volume of oxidant required, it would not be cost effective to use ISCO for site remediation. Site-specific information—including the applicability of ISCO to the specific contaminants, the concentration range, and hydrogeologic conditions—must be gathered and reviewed when evaluating the appropriateness of using ISCO for a remediation strategy. Probably the most technically challenging factor in ISCO applications is delivery of the oxidant(s) to low-permeable porous media occupied by the contaminant(s). This includes clayey and silty lenses and layers in unconsolidated sedimentary rocks, and rock matrix in consolidated rocks. In order to degrade contaminants diffused into low-permeable media, oxidants must be persistent and must have long residence times. These two requirements in most cases result in infeasibility of ISCO applications in low-permeable porous media.

Cost estimates of ISCO depend on the heterogeneity of the site subsurface, soil oxidation demand, stability of the oxidant, and type and concentration of the contaminant. Care should be taken when comparing different technologies on a cubic yard basis without considering these site attributes. Cost data can be found in ITRC (2005) and Brown (2003). ISCO has been used at a number of sites and is available from a variety of vendors.

Case Study: BTEX Treatment with Fenton's Reagent

Fenton's reagent was selected as the oxidant of choice for in situ remediation of BTEX groundwater contamination at the former Pierce Service Station site in Loss Angeles, California (consultant: Gary Cronk, MECX, LLC; regulatory agency: Los Angeles Regional Water Quality Control Board (LARWQCB); from ITRC, 2005). An off-site gasoline plume extended approximately 150 ft to the southwest of the former Pierce Service Station site, across two high traffic streets (Fig. 9.33). Groundwater flows toward the southwest at a mild gradient of 0.008 ft/ft. The groundwater velocity is estimated to be 0.04 ft/d. The LARWQCB considers the shallow aquifer a potential drinking water source. A high-school campus (a sensitive receptor) is located directly downgradient of the site. Baseline iron levels in the groundwater ranged 6 to 338 mg/L, and total organic carbon 17 to 35 mg/L. The aquifer sediments comprise silty sands in the uppermost portion of the aquifer and low-permeability clayey silts in the lowermost. The COCs at this site included BTEX as well as TPH as gasoline (TPHg). No MTBE was identified. The plume was confined to a shallow alluvial aquifer at a depth of 30 to 45 ft bgs. The approximate areal extent of the targeted contamination was 7065 ft^2, and the aquifer volume was estimated at 5200 cubic yards. The highest pretreatment level of benzene (risk driver) was 2000 μg/L, and the highest TPHg was 65,000 μg/L.

Twenty-one injection wells (screened 31 to 46 ft bgs) were installed and used during a full-scale treatment of the site using Fenton's oxidation remediation technology. Based on prior experience with low-permeability soils, the injection wells were estimated to have a radius of influence (ROI) of about 15 ft. The ROI estimate was confirmed in the field by measuring changes in water quality parameters. The wells were spaced approximately 25 ft apart and staggered to provide overlapping treatment radii and cover the off-site plume. The groundwater was initially "conditioned" by injection of a small quantity (50 gallons/well) of a catalyst solution consisting of ferrous sulfate and hydrochloric acid. Hydrogen peroxide (17.5% solution) was then gravity-fed into the subsurface (not pumped or pressurized). Downhole temperatures were monitored during the injections, and the rate of injection of peroxide was controlled to ensure that the groundwater temperatures did not exceed 180°F. Over the course of 4 weeks, a total of 8600 gallons of hydrogen peroxide were injected in the groundwater. The average injection quantity was 430 gallons/well.

Overall, Fenton's oxidation remediation technology was successful at this site. Following treatment, the benzene level in the most contaminated well was reduced from 2000 to 240 μg/L (88 percent reduction), and TPHg was reduced from 62,000 to 4300 μg/L (93 percent reduction). Overall, the six monitoring wells showed an average 96 percent reduction in benzene and 93 percent reduction in TPHg. Figure 9.33 shows the

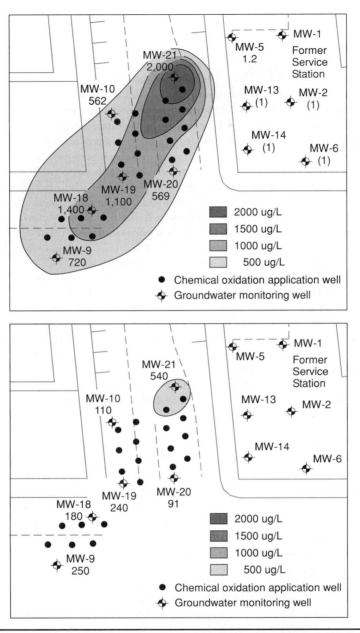

Figure 9.33 Top: Benzene isoconcentration map from May 2003 (baseline conditions). Bottom: Benzene isoconcentration map from January 2004 (6 months after Fenton's treatment), Pierce Service Station, Los Angeles, CA. (From ITRC, 2005.)

benzene plume map before the application of ISCO and 6 months after the full-scale remediation.

9.4.5 Enhanced Bioremediation

There has been an increasing interest in treatment of NAPL source zones with enhanced bioremediation. When conditions in a source zone are favorable, biostimulation of microorganisms by injection of nutrients may result in a rapid and large increase in their population. This increase results in an increase in production of natural biosurfactants and bioemulsifiers by the microorganisms. The result is desorption of the chlorinated contaminants adsorbed to the aquifer porous media and an increase in solubilization of DNAPL by partitioning into surfactant micelles. In addition, the fermentation by-products such as alcohols and ketones also increase the solubilization of DNAPL by cosolvency effects (Suthersan and Payne, 2005).

The natural biological surfactant and cosolvent effects are often observed as an increase in the dissolved constituent levels in the treatment zone and downgradient of the treatment zone (Fig. 9.34). The constituent concentrations in the treatment zone may also remain unchanged for some periods even when biodegradation end-product data support the conclusion that sufficient mass is being degraded by the reductive dechlorination. One potentially limiting factor in treating source zones with enhanced biodegradation is accumulation of *cis*-1,2-DCE and VC, as illustrated in Fig. 9.35. These components often cannot be degraded in the highly reductive environments of the source zones and may have to be treated downgradient with a reactive barrier or some other technology. As discussed by Suthersan and Payne (2005), the success of the NAPL source-zone treatment with biodegradation is site specific and depends on enhancing and maintaining degradation rates faster than the rate of mass transfer from NAPL to the dissolved phase.

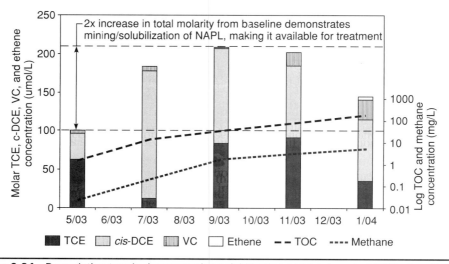

Figure 9.34 Degradation trends demonstrating solubilization and treatment of NAPL at a site in South Carolina. (From Suthersan and Payne, 2005; copyright CRC/Taylor & Francis Group, printed with permission.)

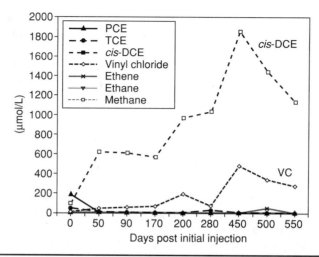

FIGURE 9.35 Performance data from a highly contaminated DNAPL site in Alabama. (From Suthersan and Payne, 2005; copyright CRC/Taylor & Francis Group, printed with permission.)

More detail on bioremediation technologies for dissolved-phase (plume) remediation is presented in Section 9.5.3.

9.4.6 Thermal Technologies

In situ thermal heating methods were first developed by the petroleum industry for enhanced oil recovery. These methods were adapted to the treatment of soil and groundwater. Initial variations included hot water injection, steam injection, hot air injection, and electric resistive heating (ERH). Thermal conductive heating was developed in the late 1980s and early 1990s. Currently, steam injection (or steam-enhanced extraction, SEE), ERH, and thermal conductive heating are used for remediation of soil and groundwater in source zones contaminated with chlorinated solvents. These in situ thermal treatment technologies have also been used for treating other volatile and semivolatile organic contaminants (SVOCs), such as PCBs, PAHs, pesticides, and various fuels, oils, and lubricants that are less amenable to other treatment methods. For example, hot water injection has been used to enhance the recovery of low-volatility and low-solubility oils. RF heating, a variety of ERH that uses radiofrequency energy, has been applied to remediation of various contaminants in the unsaturated zone, but its applicability in the saturated zone has been limited (USEPA, 2004; USACE, 2006).

All thermal technologies are used to lower the viscosity of NAPL and increase the vapor pressure and solubility of VOCs or SVOCs, thus enhancing their removal. Vapor extraction is an integral part of these remediation systems to ensure the removal and treatment of mobilized contaminants. Liquid extraction is also used during steam injection, and sometimes with other thermal technologies when groundwater flow rates are high and/or when the contaminant being recovered is semivolatile (USEPA, 2006b).

In situ vitrification is a unique thermal technology in that the temperatures used will vitrify soil. The stable glass that is formed by vitrification will immobilize any nonvolatile contaminants that are present, including metals and radioactive materials.

Davis (1997) provides a general discussion of the effects of heat on chemical and physical properties of organic contaminants. Vaporization is the main mechanism used in these technologies to enhance the recovery of VOCs. Vapor pressures of organic compounds increase exponentially with temperature, causing significant redistribution to the vapor phase as the subsurface is heated. When an NAPL is present, the combined vapor pressure of the NAPL and water determine the boiling temperature, and coboiling of the two liquids occurs at temperatures less than the boiling point of water. For example, an azeotropic mixture of PCE and water will boil at 88°C, more than 30°C less than the 121°C boiling point for pure PCE and significantly less than the boiling point of water (USEPA, 2004).

Typically, chlorinated solvents that boil at less than 100°C will have a five to seven times greater vapor pressure at 50°C than at 10°C (Fares et al., 1995). In addition, the liquid viscosity of a given chlorinated solvent generally decreases by 1 percent per °C of increased temperature up to its boiling point, enhancing its mobility in the subsurface. In the gas phase, a mass of chlorinated solvent occupies a larger volume than it does as a liquid, resulting in expansion and advective flow. For example, a mass of water occupies 1600 times more volume as a gas than it does as a liquid (Davis, 1997). As chlorinated solvents expand, the mass of a chlorinated solvent can be captured and removed from the subsurface. In addition, the viscosity and diffusivity rates (in air) allow for more efficient flow of chlorinated solvents as a gas than as a liquid. The viscosity of a chlorinated solvent as a gas is generally two orders of magnitude less than that of a liquid. Increasing the temperature from 10°C to 100°C will increase the diffusion in the vapor phase by approximately 50 percent (Davis, 1997).

As discussed by USEPA (2004), thermal effects also enhance the removal of chlorinated solvents dissolved in source-zone groundwater or pore water. Physical and chemical properties, such as solubility, Henry's law constant, octanol-water partition coefficient, and aqueous diffusivity rate, change in ways beneficial to remediation. For solubility, concentrations increase by a factor or two or more as an area is heated. The Henry's law constant for chlorinated solvents generally increases and the partitioning from the aqueous phase to soil (based on the octanol-water partition coefficient) generally decreases with elevated temperature. For example, the Henry's law constant for TCE increases by one order of magnitude, and its adsorption from the aqueous phase onto soils can be expected to decrease by a factor of approximately 2.2 when the temperature is increased from 20°C to 90°C (Heron et al., 1996). The aqueous diffusion rate will increase by approximately 30 percent when the temperature is increased from 10°C to 90°C (Treybal, 1980). In summary, increasing the temperature decreases viscosity, increases solubility, and decreases adsorption, all of which aid in the recovery of VOCs and SVOCs. For some SVOC NAPLs, such as creosote, viscosity reduction may be an important mechanism for increased contaminant recovery (Davis, 1997).

The elevated temperatures achieved during in situ thermal treatment can also enhance abiotic and biotic degradation or destruction of chlorinated solvents. Abiotic degradation pathways, such as hydrolysis, where the hydrogen ions in water replace the chlorine ions in the chlorinated solvent molecule, and hydrous pyrolysis oxidation (HPO), where chlorinated solvents under oxidizing and aqueous conditions may be oxidized (eventually to carbon dioxide), have been shown to increase substantially at elevated temperatures (USEPA, 2004). For example, the hydrolysis rates for chlorinated methanes and ethanes have been shown to result in relatively short half-lives for these contaminants at elevated temperatures (Jeffers et al., 1989). In addition, rates of HPO of chlorinated solvents have been shown to increase (up to a maximum rate) with temperature (Baker and Kuhlman, 2002).

Biological degradation pathways may also be enhanced at elevated temperatures. One commonly used rule of thumb (based on the Van't Hoff-Arrhenius relationship) states that, for every 10°C increase in temperature, there is roughly a twofold increase in biological activity resulting in an increase in degradation rate constants (USEPA, 2004). Extremely high temperatures may sterilize soils of some microbes. However, significant levels of thermophiles (microbes that thrive under high-temperature conditions) are present in many soils, and nearly all microbes benefit from elevated temperatures in the more moderately heated soil regions at the fringe of the treatment area. The overall effect of the elevated temperatures achieved during in situ thermal treatment on biological degradation pathways has not been fully determined and is dependent on site-specific conditions (USEPA, 2004).

Care should be taken in designing the systems to ensure that all plumbing, including monitoring wells, are capable of withstanding high heat. In the presence of clay, vadose-zone heating by resistivity, conductance, or radiofrequency may result in some settlement of the treatment area due to the drying of the clay (USEPA, 2006b).

Detailed description of design principles for in situ thermal remediation, including quantification of design parameters, is provided by USACE (2006).

Steam-Enhanced Extraction

SEE takes advantage of the relatively large heating capacity of steam, which provides a greater heat input to the subsurface than injecting hot air. In remedial applications, SEE typically involves the injection of steam into the subsurface to dissolve, vaporize, and mobilize contaminants that are then recovered. Mobilized contaminants are extracted from the subsurface using vapor and liquid extraction equipment. Extracted vapors and liquids are treated using conventional aboveground treatment technologies, such as condensation, air stripping, carbon adsorption, and thermal oxidation.

SEE is most effective when the steam is able to enter the pore space of the soils and best suited for zones of moderate-to-high permeability. In low-permeability soil, steam cannot penetrate the pore space as rapidly, resulting in higher heat losses and, in some cases, the inability to completely heat the area. In addition, smaller pore diameters create higher capillary pressures and, as a result, lower the rate of evaporation of contaminants (USEPA, 2004).

A SEE system typically consists of a series of injection and extraction wells. For small applications, a ring of injection wells typically surround a central extraction well located near the middle of the DNAPL area (Fig. 9.36). In this configuration, the injection wells are placed in clean areas around the source zone, if possible, to minimize the risk of contaminant spreading. In some cases, but less frequently, an inside-out configuration has been used, where the steam is injected centrally, and extraction wells on the perimeter provide hydraulic and pneumatic control, reducing the potential for contaminant spreading outward. For larger areas, multiple arrays of injection and extraction wells typically are used to heat the area and capture mobilized contaminants in the treatment area. The patterns and spacing of the injection/extraction wells depend on the geologic conditions (including whether the application is in unsaturated or saturated media), the permeability, and the depth of application. Typical spacing for SEE wells ranges from several to more than 10 m (Davis, 1998).

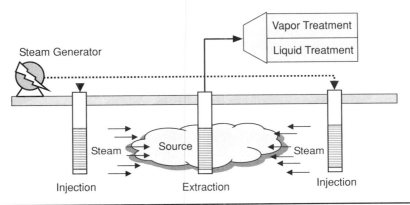

Figure 9.36   Schematic of steam-enhanced extraction (SEE) system. (From USEPA, 2004; Davis, 1998.)

The wells used for injection, extraction, or monitoring, and the steam distribution system need to be designed to handle the expected temperatures and changes in temperatures that are inherent to SEE. Steel is typically the preferred casing and screen material, because conventional polyvinyl chloride (PVC) or fiberglass wells can degrade or deform under high-temperature conditions. Well casing joints and grout must also be selected to handle pressures and thermal expansion. In some cases, grouts can be amended with quartz silica or silica flour for temperature stability and with sodium chloride for greater expansion capability. Temperature considerations are also relevant to the selection of groundwater extraction and monitoring wells and equipment, because some in situ groundwater extraction pumps do not function reliably under high-temperature conditions (USEPA, 2004).

Electrical Resistive Heating

ERH involves the application of electrical current through the subsurface, resulting in the generation of heat. ERH uses the natural electrical resistance within the subsurface where energy is dissipated through ohmic, or resistive, losses. This manner of in situ heating allows energy to be focused into a specific source zone. When the subsurface temperature is increased to the boiling point of the pore water or the saturated media in the treatment zone, steam is generated. The steam strips contaminants from the soils and enables them to be extracted from the subsurface. In addition, contaminants are directly volatilized from unsaturated soil (USEPA, 2004).

The necessary power input to the subsurface is inversely proportional to the soil resistivity and directly proportional to the square of the applied voltage, based on the following equation derived from Ohm's Law:

$$\text{Power} = (\text{voltage})^2 / \text{resistance} \tag{9.7}$$

The resistance of a subsurface matrix is largely determined by its water content, concentration of dissolved salts or ionic content in the water, and ion-exchange capacity of the soil itself. The organic carbon content of soils also affects resistivity but has a greater effect on the required treatment time as a result of the stronger partitioning of organic contaminants, such as chlorinated solvents, to the soils. In addition, the resistivity

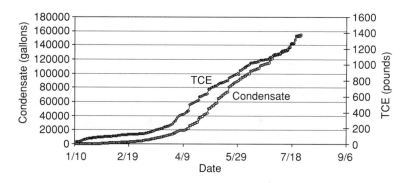

FIGURE 9.37 Condensate and TCE removed over time using ERH system. (From USEPA, 2004.)

is a function of temperature, and, as the water reaches its boiling point, the resistivity decreases with increased ion mobility. Soil resistance can be measured in the field or estimated from characterization data for soils and groundwater. The total resistance of an ERH system is determined based on the resistivity of the soil and the geometry of the electrode system (USEPA, 2004).

ERH is particularly suited to the treatment of lower-permeability strata and to DNAPLs that have become consolidated within lower-permeability zones with higher organic content. In some cases, ERH can be combined with SEE in aquifers interbedded with low-permeable lenses or in situations where a lower aquitard has been impregnated with DNAPLs. ERH is used to treat the lower-permeability zones, which the steam vapors cannot penetrate rapidly (Beyke, 1998). For example, Fig. 9.37 shows the results of TCE removal using ERH to treat TCE and DNAPL in soil and groundwater beneath Building 181 at Air Force Plant 4, in Fort Worth, TX. The geology at the site consisted of heterogeneous interbedded silt, clay, and gravel. The depth to groundwater was 27 ft (9 m) below ground surface. In April 2002, TRS, as a subcontractor to URS Corp., designed, installed, and operated a full-scale ERH system consisting of 60 electrodes and colocated vapor recovery wells covering an area of about half of an acre inside the building. The system was operated from April to December 2002, on a 24-hours-per-day, 7-days-a-week schedule. By the end, almost 1400 pounds of TCE were recovered. The average weekly power input ranged from about 450 to 675 kW between May and August, dropping to below 300 kW for the remainder of the system operation. A total of about 1,900,000 kWh of energy were input to the subsurface during ERH operations (USEPA, 2004).

Thermal Conductive Heating
Thermal conductive heating relies on conduction rather than convection to heat subsurface soils. It involves the simultaneous application of heat and vacuum to subsurface soils with an array of vertical heater/vacuum wells or, less commonly, with surface blanket heaters and a vacuum-insulated shroud. In both of these configurations, heat originates from a heating element and is transferred to the subsurface largely via thermal conduction and radiant heat transport. There is also a contribution through convective heat transfer that occurs during the formation of steam from pore water. Because this technology can achieve elevated soil temperatures (in excess of 500°C), a significant portion (reported up to 99 percent at some sites) of organic contaminants either oxidize (if sufficient air is

present) or pyrolize once high soil temperatures are achieved. Therefore, this technology is also considered to be an in situ destruction method (Baker and Kuhlman, 2002).

Soil heat conductivities are low but similar in magnitude. The movement of heat away from the heaters, whether vertically or radially outward, is therefore uniform. However, because the driving force for heat migration is the temperature gradient, soils initially are not heated to the same temperature within the treatment area, resulting in a temperature profile that decreases radially from the source. Over time, superposition of heat from adjacent heaters tends to even out these differences. Other factors, including advective heat transport, the anisotropic nature (variable thermal conductivity depending on flow direction) of the thermal conductivity of soils, or heat loss through groundwater flow, can also affect the uniformity of subsurface heating (USEPA, 2004).

Thermal conductive heating is suited to treating DNAPL source zones in most hydrogeologic conditions. Thermal conductive heating differs from other heating methods (SEE and ERH) in that it does not rely solely on steam as a heat source or water as a conductive path. It can heat soils to temperatures in excess of 500°C, making it particularly applicable to SVOCs such as PCBs, PAHs, pesticides, and herbicides (Vinegar and Stegemeier, 1998). However, these higher-boiling-point compounds typically require high temperatures (for example, 325°C) that typically can only be achieved in the unsaturated zone. Lower-boiling compounds such as chlorinated solvents can be treated with thermal conductive heating through achievement of steam distillation temperatures in the bulk of the interwell regions. Locations close to heaters may achieve temperatures well above the boiling point of water (USEPA, 2004).

9.5 Dissolved Phase (Plume) Remediation

As in the case of source-zone remediation, delineation of dissolved contaminant plumes in all three dimensions and understanding of the contaminant F&T mechanisms are the key for a successful groundwater remediation. Plume delineation is a complex task, especially in highly heterogeneous porous media such as fractured rock and karst aquifers. Failure or infeasibility to properly characterize the extent of contaminant plumes in such environments may lead to selection of an improper remedial technology and may even exasperate the problem by pulling contaminated water into previously clean portions of the groundwater system. As discussed in Chap. 5, many complex contaminated sites have multiple point sources of groundwater contamination. These sources may form individual plumes of individual contaminants, individual plumes of mixed contaminants from identifiable sources, or, in the most complicated cases, commingled (merged) plumes of various contaminants from multiple sources, some of which are not easily, or not at all, identifiable. Sites on military installations, large industrial complexes, and multiple chemical manufacturing plants are likely to have groundwater contaminated by multiple constituents, which may be distributed at various depths in the underlying aquifer(s) and form plumes with complicated shapes. Complex sites of groundwater contamination are often a nightmare for groundwater professionals trying to characterize possible contaminant sources and "attach" to them their own plumes. This, however, is the favorite topic of attorneys working for various PRPs. Heavy involvement of attorneys in groundwater contamination and remediation issues is understandable, since the costs associated with groundwater remediation may be astronomical, and the question of "plume ownership" is therefore the most important one.

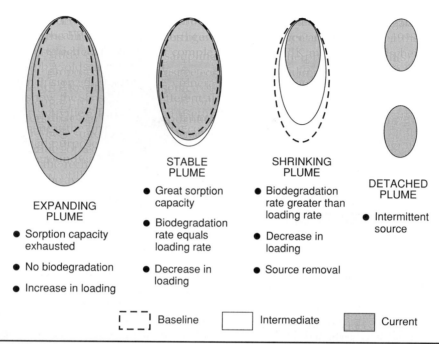

EXPANDING PLUME
- Sorption capacity exhausted
- No biodegradation
- Increase in loading

STABLE PLUME
- Great sorption capacity
- Biodegradation rate equals loading rate
- Decrease in loading

SHRINKING PLUME
- Biodegradation rate greater than loading rate
- Decrease in loading
- Source removal

DETACHED PLUME
- Intermittent source

Baseline   Intermediate   Current

FIGURE 9.38 Influence of various fate and transport (F&T) processes on plume development. While most F&T processes may be present in any given case, the bullets list only those with the possibly greatest net effect. (Modified from USEPA, 1977.)

In source zones, contaminants are usually subject to a variety of complex and rapid geochemical reactions. In contrast, the dissolved, mobile contaminant phase flowing away from the source is usually characterized by much slower reactions, and the flow system is often described as being in (quasi) equilibrium for practical purposes. Although this assumption helps to greatly simplify F&T calculations, it is not entirely correct in that whenever the front of a moving contaminant plume encounters uncontaminated groundwater, the system enters into nonequilibrium conditions. The exception would be a stable, nonexpanding plume. Figure 9.38 illustrates main types of contaminant plumes with respect to various possible F&T processes influencing their development. Each plume type, expanding, stable, or shrinking, is, to a varying degree, subject to most of the F&T processes discussed in Chap. 5. The bullets in Fig. 9.38 list only those that may have the greatest net effect on the particular plume type. An example of inadequate plume characterization and the resulting wasted effort and resources in plume remediation would be the installation of a P&T system at the leading edge of a naturally shrinking plume.

As explained in the following selection, P&T systems for plume remediation are increasingly "unpopular" with both the regulators and the PRPs because of their general inability to restore contaminated groundwater to drinking water standards. In addition, these extract groundwater, which, if not reinjected into the aquifer after treatment, is a wasted resource. These are the main reasons why in situ technologies for plume remediation, such as bioremediation, air sparging, ISCO, and MNA, should be given preference whenever possible. For example, as of 2005, P&T projects represented the largest

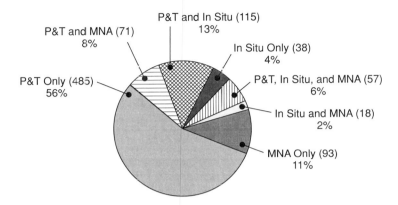

FIGURE 9.39 NPL sites with P&T, in situ treatment, or MNA selected as part of a groundwater remedy for the 1982–2005 period; total number of projects is 877. (From USEPA, 2007a.)

number of projects (725) of the total of 1915 projects at Superfund sites, including in situ and ex situ treatment for both source control and dissolved plumes. Only 10 percent of P&T projects have been completed or shut down for various reasons (including because of the change of remedy); the number of P&T projects actually achieving cleanup goals has not been reported. RODs that select P&T alone have decreased from about 80 percent before 1992 to an average of 20 percent during 2001-2005 (USEPA, 2007a). Figure 9.39 shows a comparison of different dissolved plume ("groundwater" in USEPA terms) remediation technologies implemented at 877 Superfund sites in the United States as of 2005. P&T is the sole groundwater (dissolved plume) treatment remedy at more than half of NPL sites, although many of these sites also have a source control remedy. Figure 9.40

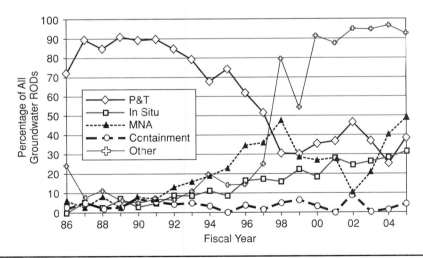

FIGURE 9.40 Trends in RODs selecting groundwater remedies for the 1986–2005 period; total number of groundwater RODs is 1458. No hierarchy is used in the figure; RODs may be counted in more than one category. "Other" includes institutional controls and other remedies not classified as treatment, MNA, or containment. (Modified from USEPA, 2007a.)

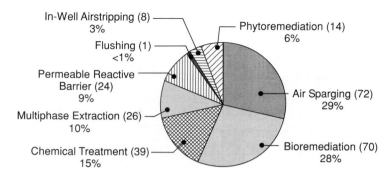

FIGURE 9.41   In situ groundwater treatment projects, 1982–2005; total number of projects is 254. (From USEPA, 2007a.)

shows that, since 1991, the percentage of groundwater RODs selecting conventional P&T remedies has steadily declined, while those selecting in situ or MNA remedies have increased.

As illustrated in Fig. 9.41, bioremediation and air sparging account for more than half of all in situ groundwater (dissolved plume) treatment projects, but in recent years (2002–2005) bioremediation and chemical treatment have become more commonly applied remediation technologies at Superfund sites (USEPA, 2007a).

9.5.1   Pump and Treat

P&T is the most widely used groundwater remediation technology today, although its effectiveness to restore contaminated aquifers to their natural condition was called into question almost two decades ago (e.g., USEPA, 1989c; Freeze and Cherry, 1989; Travis and Doty, 1990; NRC, 1994). After years of use, this technology remains an important component of groundwater remediation efforts (USEPA, 2007a).

As discussed by USEPA (1996b), the general failure of the P&T approach was identified as its inability to achieve groundwater restoration (i.e., reduction of contaminants to levels required by health-based standards) in 5 to 10 years, as anticipated in the design phase of early remediation projects. Although a variety of factors contribute to this shortcoming, the major barrier to achieving remediation goals is the slow processes of desorption and back-diffusion of contaminants trapped in stagnant groundwater zones and rock matrix. These processes result in the so-called "tailing" of contaminant concentrations observed in the extracted groundwater at many P&T sites and the rebound of contaminant concentrations after cessation of pumping. In addition, any dissolved contamination that continues to leave the source zone(s) would have similar negative effects on a P&T system performance.

The USEPA pointed out, more than a decade ago, that groundwater scientists and engineers generally agree that complete aquifer restoration is an unrealistic goal for many, if not most, contaminated sites and that expectations for the effectiveness of P&T technology may have been too high (USEPA, 1996b). Nonetheless, further experience with P&T systems has shown that at some sites with relatively simple characteristics the full restoration is possible and that combining P&T with in situ restoration technologies provides further opportunities for improving effectiveness of groundwater cleanup.

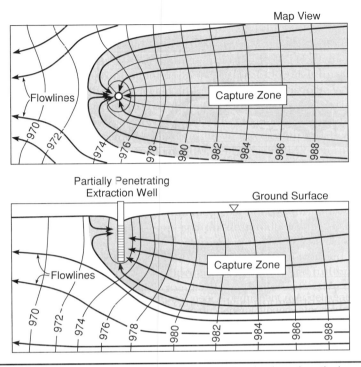

Figure 9.42 Illustration of horizontal capture zone in map view (top), and vertical capture zone in cross-sectional view (bottom) demonstrating the need for a three-dimensional approach to pump-and-treat system design and analysis. (Modified from USEPA, 2008.)

P&T systems are used primarily to accomplish the following (USEPA, 1996b):

- *Hydraulic containment.* To control the movement of contaminated groundwater and prevent the continued expansion of the contaminated zone. Figure 9.42 illustrates the main elements for evaluating both horizontal and vertical hydraulic containment.
- *Treatment.* To reduce the dissolved contaminant concentrations in groundwater sufficiently so that the aquifer complies with cleanup standards or the treated water withdrawn from the aquifer can be put to beneficial use.

As illustrated in Fig. 9.42, the zone of hydraulic containment ("capture zone") is the three-dimensional region that contributes the groundwater extracted by one or more wells or drains. This zone is not equivalent to, and should not be mistaken for, the zone of well(s) influence, also referred to as "radius of well influence." The main difference between these two concepts is that the capture zone encompasses the volume of porous media from which groundwater flows toward the well(s) and is eventually extracted by the well(s). In contrast, the ROI is the zone in which the hydraulic head is lowered because of the well(s) operation, but the groundwater flow direction is not necessarily reversed from the prepumping direction (Fig. 9.43). Consequently, the well(s) capture zones can be verified only if there are enough monitoring wells (piezometers)

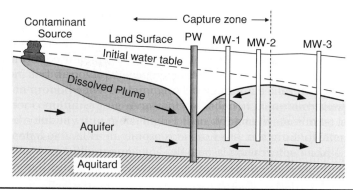

Figure 9.43 Concept of capture zone versus radius of well influence. Using data only from monitoring well MW-3 and pumping well PW would lead to an erroneous conclusion as to the extent of the real well capture zone.

that show the actual gradient reversal in all three dimensions. It is also very important to understand that the drawdown in the well, i.e., the pumping water level in the extraction well, cannot be used to demonstrate the capture zone if an inadequate number and locations of monitoring wells (piezometers) are selected. This is illustrated in Fig. 9.43, where data from the monitoring well MW-3 and the pumping well PW would not be sufficient to demonstrate the capture zone, even though the water elevation in MW-3 is higher than in PW. In addition, if the water elevation in PW is not corrected for the well loss or not available, three monitoring wells would be needed to confirm the well capture.

If a contaminant plume is hydraulically contained, contaminants moving with the groundwater will not spread beyond the capture zone. Failed capture, illustrated schematically on Fig. 9.44, can allow the plume to grow, which may cause harm to receptors and may increase the ultimate cost or duration of the groundwater remedy (USEPA, 2008). For this reason, regulatory agencies place special emphasis on the performance of P&T systems for hydraulic containment of contaminant plumes and require

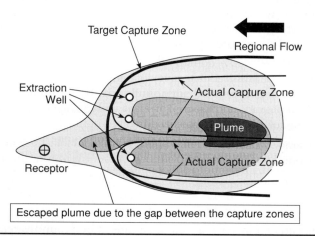

Figure 9.44 Failed capture of contaminant plume. (Modified from USEPA, 2008.)

that such systems be regularly monitored and periodically evaluated for their efficiency, and to confirm continuing containment (e.g., see USEPA, 1994, 1996b, 2002, 2008). Three-dimensional numeric models with particle tracking are the best tool for designing and evaluating performance of P&T systems. These can be used to incorporate any determined or suspected anisotropy and heterogeneity of the porous media and can be calibrated and recalibrated as more field data on the system performance become available.

An appropriate methodology for evaluating plume capture, including requisite monitoring locations, should be developed as part of the P&T system design. In addition, the implemented P&T system may differ substantially from the system that was originally designed, and the following issues should be assessed (USEPA, 1996b):

- Did the design account for system down time (i.e., when wells are not pumping)?
- Did the design consider time-varying influences such as seasons, tides, irrigation, or transient off-site pumping?
- Did the design account for declining well yields due to fouling or provide for proper well maintenance?
- Did the design address geologic heterogeneities?
- Did the design take into account other hydraulic boundary conditions such as a surface water boundary or a hard rock boundary?

Such issues may impact the effectiveness of capture relative to the designed system, highlighting the need to conduct capture zone evaluations for the operating P&T system.

Although hydraulic containment and groundwater cleanup can represent separate goals, more typically, remediation efforts are undertaken to achieve a combination of both. For example, if restoration is not feasible, the primary objective might be containment. In contrast, where a contaminated well is used for drinking water but the contaminant source has not been identified, treatment at the wellhead might allow continued use of the water even though the aquifer remains contaminated (USEPA, 1996b).

The USEPA (2007c) provides a detailed discussion on various options for discharging treated water from P&T systems including infiltration basins and galleries, injection wells, publicly owned treatment works (POTW), other on-site disposal, and treated water reuse. The return of treated water to the subsurface can play an important role in the performance of a groundwater remedy. The returned water can be designed to positively impact the groundwater remedy in the following ways (USEPA, 2007c):

- Contaminant flushing can be enhanced by returning treated water upgradient of the plume and extraction system, or in a zone where contamination is present in the unsaturated zone.
- Degradation of remaining contamination in the subsurface can be enhanced through the addition of oxygen and/or nutrients to the returned water.
- Hydraulic containment of impacted groundwater can be enhanced by returning treated water to the subsurface and creating a hydraulic divide, particularly downgradient of the extraction wells.
- Negative impacts that might be caused by groundwater extraction, such as reduced groundwater discharge to wetlands or dewatering of water supply well

screens, can potentially be mitigated by returning the treated water to the subsurface.

Return of treated water to the subsurface can also have a negative impact on the groundwater remedy in the following ways (USEPA, 2007c):

- Frequent fouling of the infiltration or injection structures can result in frequent system shutdowns.
- Returning treated water into the plume can result in spreading of the plume and could compromise plume capture.
- Returning treated water in close proximity to the extraction wells can result in the extraction of treated water rather than contaminated water, which can compromise plume capture and/or slow progress toward aquifer restoration.

Various options for returning treated water to the subsurface are best evaluated with the use of groundwater models.

Optimization

Optimization of a P&T system is accomplished during the design and implementation phases, as well as after the system has been operational for an extended period and more detailed information on its performance is available. The purpose of optimization is to identify potential changes that will improve the effectiveness of a system and reduce operating and monitoring costs without compromising the effectiveness of the remedy or the achievement of other cleanup objectives (USEPA, 2007a). More specifically, the goals of P&T optimization are to minimize life-cycle costs, annual costs, or cleanup time, while maximizing mass removal rates and minimizing pumping rate required for plume capture. Simultaneously, achieving most or all of these goals can (arguably) be accomplished only by applying simulation-optimization techniques that couple simulations of groundwater flow (e.g., MODFLOW) and/or contaminant transport (e.g., RT3D) with mathematical optimization algorithms, to determine an optimal solution when many possible solutions exist.

The simulation-optimization approach is more efficient than simulating a small number of pumping scenarios in a "trial and error" manner (the traditional modeling approach) and typically yields a much improved result. There are two general subclasses of simulation-optimization methods for groundwater P&T systems: (1) hydraulic optimization based on groundwater flow modeling, most appropriate when hydraulic containment is the primary concern, and (2) transport optimization based on groundwater flow and transport modeling, most appropriate when aquifer restoration is the primary concern.

An ideal outcome of P&T optimization is a reduction in the number of extraction wells, decrease of the combined pumping rate, or ultimately the shutdown of the system. All three options should be evaluated periodically against preset thresholds that can be developed based on the simulation-optimization modeling results. These thresholds do not have to be equal to the final groundwater cleanup goals. For example, a threshold can be a contaminant concentration that does not have to be contained by the P&T system anymore because natural attenuation will lower it to less than MCL by the time the plume reaches critical receptor(s).

An outline of the P&T optimization approach is given in USEPA (2007b), and general information on simulation-optimization techniques can be found on the Federal Remediation Technologies Roundtable (FRTR) Web site at http://www.frtr.gov/optimization/simulation.htm.

Case Study: 1,4-Dioxane Plume

Courtesy of Farsad Fotouhi and Saied Tousi, Pall Corporation, and James W. Brode and Jerry B. Lisiecki, FTC&H.

A full-scale treatment system has been installed at a site in Ann Arbor, Michigan, the United States, to remediate groundwater containing high concentrations of 1,4-dioxane. Treatment consists of pumping of contaminated groundwater to the on-site facility where ozone/hydrogen peroxide (O_3/H_2O_2) technologies are employed. Groundwater pumped from the aquifers is treated and discharged into a tributary under an NPDES permit. The company has treated over 4.2 billion gallons (15.9 million m^3) of groundwater and removed over 76,000 pounds (34,450 kg) of 1,4-dioxane from the contaminated aquifers since 1997.

The company used large quantities of 1,4-dioxane during 1966 to 1986. On-site wastewater disposal practices resulted in 1,4-dioxane release into the subsurface and ultimately into groundwater, where multiple plumes developed. When 1,4-dioxane contamination was identified in the mid-1980s, groundwater concentrations were as high as 210,000 $\mu g/L$ and several local drinking water wells were affected. The plumes, as defined by the State of Michigan Drinking Water Criterion of 85 $\mu g/L$, collectively encompass an area of approximately 0.6 mi^2.

Groundwater at the site is generally shallow, at a depth averaging approximately 15 ft below ground surface. Subsurface material consists of glacial deposits that are up to 300 ft thick and overly the Mississippian-Aged Coldwater (primarily shale) formation, which serves as the lower boundary of contamination. At least two primary sand/gravel aquifers with differing flow directions and groundwater flow rates exist between the area's clay-rich deposits. 1,4-Dioxane has migrated at least 8000 ft from the source areas in these aquifers.

Twenty-two strategically placed groundwater extraction wells and a 4479-ft (1365 m) long horizontal well (Fig. 9.45) are used to extract and divert groundwater at a rate of up to 1300 gal/min (82 L/s) to the facility for further treatment and disposal. During the treatment process, water from the extraction wells is mixed in a treatment pond. The water is transferred from the pond to the treatment facility equipped with a 1300 gal/min ozone/hydrogen peroxide technology. Once the water meets appropriate treatment criteria, it is discharged into a nearby stream.

Groundwater is monitored on a routine basis at approximately 150 locations. To date, treatment has resulted in only a slight reduction in the areal extent of the plumes. 1,4-Dioxane concentrations within the plume areas have decreased considerably as mass has been removed. In portions of the site, nearly a 100-fold reduction in 1,4-dioxane concentrations has been observed. Maximum 1,4-dioxane concentrations in the plume are now less than 10,000 $\mu g/L$.

Continued efforts are underway to evaluate alternative cleanup remedies capable of removing the contaminant mass from the aquifers while maintaining hydraulic control. In 2004, the company conducted an ISCO pilot test involving the injection of H_2O_2 and Fenton's reagent (iron catalyst) into one of the confined aquifers. The test results indicate, however, that a minimal reduction of 1,4-dioxane concentrations was achieved.

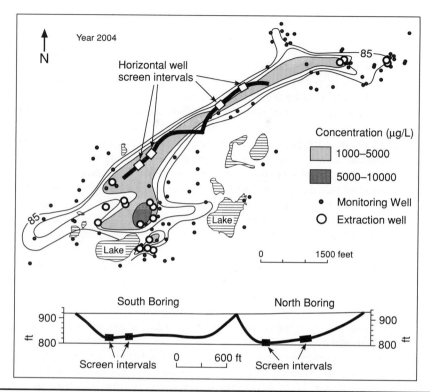

Figure 9.45 One of the longest and deepest horizontal wells in the world installed for pump-and-treat remediation of 1,4-dioxane plume in Ann Arbor, MI. Map view and cross section. (Figure courtesy of Pall Corporation.)

Additional field testing of ISCO using ozone resulted in a slightly higher rate of 1,4-dioxane removal, but bromate formation exceeded the 10-μg/L MCL set by USEPA. Use of the current P&T methods is anticipated to continue for many years until the 1,4-dioxane target cleanup criteria are reached. Figure 9.46 illustrates a typical tailing effect at one of the vertical extraction wells.

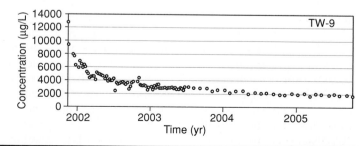

Figure 9.46 1,4-Dioxane concentration versus time at one of the extraction wells. (Figure courtesy of Pall Corporation.)

Fractured Rock and Karst Aquifers

Care should be applied when using numeric models to evaluate design options and estimate capture zone in fractured rock and karst aquifer. Equivalent porous media models should not be used for such purpose as illustrated earlier in Section 8.7.7 (Chap. 8). The USGS has released a computer program named Conduit Flow Process (CFP) for MODFLOW-2005 that addresses this problem (Shoemaker et al., 2008). The CFP has the ability to simulate turbulent groundwater flow conditions by (1) coupling the traditional groundwater flow equation with formulations for a discrete network of cylindrical pipes (Mode 1), (2) inserting a high-conductivity flow layer that can switch between laminar and turbulent flow (Mode 2), or (3) simultaneously coupling a discrete pipe network while inserting a high-conductivity flow layer that can switch between laminar and turbulent flow (Mode 3). Conduit flow pipes (Mode 1) may represent dissolution or biological burrowing features in carbonate aquifers, voids in fractured rock, and/or lava tubes in basaltic aquifers and can be fully or partially saturated under laminar or turbulent flow conditions. Preferential flow layers (Mode 2) may represent (1) a porous media where turbulent flow is suspected to occur under the observed hydraulic gradients; (2) a single secondary porosity subsurface feature, such as a well-defined laterally extensive underground cave; or (3) a horizontal preferential flow layer consisting of many interconnected voids.

The CFP was developed in response to a need for a computer program that accounts for the dual-porosity nature of many aquifers. There also was a desire to provide compatibility with recent advancements to the USGS modular groundwater model (MODFLOW). Many research computer programs are available for simulating dual-porosity aquifers but have not been fully documented for wider use (for example, Clemens et al., 1996; Kiraly, 1998; Bauer, 2002; Birk, 2002). Additionally, the structure of MODFLOW has changed with MODFLOW-2000 (Harbaugh et al., 2000) and MODFLOW-2005 (Harbaugh, 2005), making the groundwater flow computer code even more modular and allowing easier addition of new processes to the code.

The CFP was designed to be flexible enough for use in locations with limited or abundant field data. In some geologic environments, such as Mammoth Cave, KY, detailed information is available (or could be derived) on the location, diameter, tortuosity, and roughness of the subsurface caverns. CFP Mode 1 (CFPM1) was designed with these locations in mind. In other locations, such as the Biscayne aquifer of southern Florida, void connections and distributions are so complicated within preferential flow layers that a complete characterization is not possible. CFP Mode 2 (CFPM2) was designed with these locations in mind; specifically, laminar and turbulent flows through complicated void connections are represented with a limited number of "effective" or "bulk" layer parameters.

One of the powerful options in the CFP is that, in cases with abundant field data on the void architecture and hydraulic behavior, complex two- or three-dimensional networks of conduit flow pipes and nodes can be designed to represent interconnected or dead-end voids in the subsurface. Flow calculations assume that pipe nodes are located in the center of MODFLOW cells. An exception is in the vertical direction, for which there are two options. First, pipe nodes can be assigned elevations above a datum and, therefore, are not restricted to center elevations of MODFLOW cells. Second, pipe nodes can be assigned a distance above or below the center of the MODFLOW cell (Fig. 9.47). With this second option, if the distance is set to zero, pipe nodes are assumed to exist at the vertical center of the MODFLOW cell (Shoemaker et al., 2008).

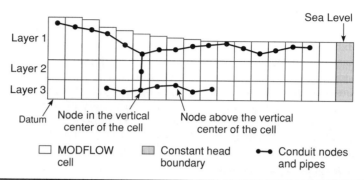

FIGURE 9.47   Possible variations in elevation of conduit nodes in MODFLOW cells. (From Shoemaker et al., 2008.)

9.5.2 Permeable Reactive Barriers

The concept of a PRB is relatively simple. Reactive material is placed in the subsurface where a plume of contaminated groundwater must flow through it, typically under its natural gradient (creating a passive treatment system), and treated water comes out the other side (Fig. 9.48). The PRB is not a barrier to the water, but it is a barrier to the contaminant. When properly designed and implemented, PRBs are capable of remediating a number of contaminants to regulatory concentration goals (USEPA, 1998c). PRBs can be installed in various configurations, using trenches or by injecting reactive materials into the subsurface via boreholes. If the injected materials serve to enhance biodegradation of the contaminant(s), such barriers are called biobarriers.

The groundwater moves passively through the treatment zone, where the contaminants are degraded, precipitated, or absorbed by the treatment media. The treatment zone may contain metal-based catalysts for degrading volatile organics, chelators for immobilizing metals, nutrients and oxygen for microorganisms to enhance biodegradation, or other agents. Degradation reactions break down the contaminants in the groundwater into benign by-products. A precipitation wall reacts with the contaminants to form insoluble products that remain in the wall as the water passes. A sorption wall adsorbs or chelates contaminants to the wall surface (USEPA, 1998b).

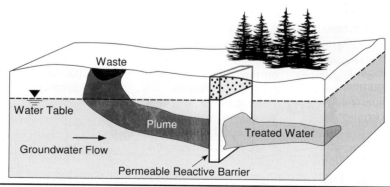

FIGURE 9.48   Example of a plume being treated by a permeable reactive barrier wall. (From USEPA, 1998c.)

Most treatment walls are designed to operate in situ for years with little or no mainte-nance. Some treatment walls are permanent, others are semipermanent, and still others are replaceable. The long-term stability of these walls has not been determined (USEPA, 1998a, 1998c).

To date, granular iron has been the most widely used reactive media in full-scale PRBs. The prevalent use of granular iron, or zero-valent iron (ZVI), stems mainly from its documented ability to abiotically degrade a variety of contaminant types, the most common of which are the chlorinated solvent compounds such as perchloroethylene (PCE) and trichloroethylene (TCE) (Wilkin and Puls, 2003). The abiotic process involves corrosion (oxidation) of ZVI and reduction of dissolved chlorinated hydrocarbons. The process induces highly reducing conditions that cause substitution of chlorine atoms by hydrogen in the structure of chlorinated hydrocarbons.

In the past few years, alternative PRB designs using non-iron-based reactive mate-rials to treat additional contaminants have also gained attention. For example, reactive materials such as compost, zeolites, activated carbon, apatite, and limestone are now being used to control pH, metals, and radionuclides (ITRC, 2005). Table 9.5 is a partial

Treatment Material Categories	Example Materials	Constituents Treated (Examples, Not Comprehensive)
Metal-enhanced reductive dechlorination for organic compounds	Zero-valent metals (Fe)	Chlorinated ethenes, ethanes, methanes, and propanes; chlorinated pesticides, Freons, and nitrobenzene
Metal-enhanced reduction for metal contaminants	Zero-valent metals (Fe), basic oxygen furnace slag, and ferric oxides	Cr, U, As, Tc, Pb, Cd, Mo, Hg, P, Se, and Ni
Sorption and ion exchange	Zero-valent iron, granular activated carbon, apatite (and related materials), bone char, zeolites, peat, and humate	Chlorinated solvents (some), BTEX, Sr-90, Tc-99, U, and Mo
pH control	Limestone and zero-valent iron	Cr, Mo, U, and acidic water
In situ redox manipulation	Sodium diothionite and calcium polysulfide	Cr and chlorinated ethenes
Enhancements for bioremediation (including carbon, oxygen, and hydrogen sources)	(Includes solid, liquid, and gaseous sources) Oxygen-release compounds, hydrogen-release compounds, carbohydrates, lactate, zero-valent iron, compost, peat, sawdust, acetate, and humate	Chlorinated ethenes and ethanes, nitrate, sulfate, perchlorate, Cr, MTBE, and polyaromatic hydrocarbons

From ITRC, 2005.

TABLE 9.5   Examples of Reactive Materials Used in PRBs

list of materials that have been used as components within PRB systems. Each of the materials, including iron, which is shown for reference, conditions the aqueous system to either directly reduce the presence or mobility of the target chemical or promote its destruction or immobilization by other chemical or biological changes to the aqueous system. The observation that most of the materials listed are natural materials (e.g., not manufactured or enhanced by human intervention) is encouraging, as PRBs can be promoted as remedies that take advantage of natural conditioning processes. The fact that most of the materials are well known to both the scientific community and the regulatory and public stakeholder community is also beneficial for receiving public approval for their use (ITRC, 2005).

Most PRBs are less than 10 years old, and it is not known whether these will remain effective over the lifetime of the contaminant plume, which could be on the order of decades or more. Therefore, much research has focused on changes in PRB reaction rates over time. Additionally, some PRBs have had problems with permeability and hydraulics, most of which seem to be an artifact of the construction techniques for PRB installation or inadequate predesign site characterization rather than chemical precipitation and clogging of the reactive media. As with any technology used to treat contaminants in the subsurface, successful implementation is contingent on effective site characterization, design, and construction. ITRC (2005) has produced a detail document highlighting many of the lessons learned over the 10-year history of iron-based PRB systems.

Trench emplacement of PRBs has a number of disadvantages. Trenches are limited to shallow treatment zones and require specialized trenching equipment, and the replacement and disposal costs of reactive material after contaminant breakthrough may be significant. The USGS (1999) has developed a tool to take advantage of the natural groundwater gradient to channel groundwater into highly permeable reactive material(s). This tool, named deep aquifer remediation tool (DART), is used in conjunction with nonpumping wells and offers a low-cost and virtually maintenance-free alternative to ex situ treatment methods. As the groundwater passes through the permeable reactive material, the contaminant is immobilized or transformed to a nontoxic form by a variety of chemical reactions.

The DARTs are deployed into an aquifer and corresponding contaminant plume through a series of nonpumping wells (Fig. 9.49). Wilson and Mackay (1997) have found that groundwater will converge to arrays of nonpumped wells in response to the difference in hydraulic conductivity between the well and aquifer. Numerical simulations conducted during DART development indicate that each well typically intercepts groundwater in the upgradient part of the aquifer that is approximately twice the inside diameter of the well.

Because DARTs are deployed through non-pumping wells, in-situ treatment of deeper contaminant plumes (greater than 100 ft below land surface) that could not be treated with currently available trenching technologies is possible. In addition, DARTs allow for easy retrieval, replacement, and disposal of reactive material after chemical breakthrough.

DARTs are designed to fit a variety of well dimensions and plume geometries. A DART is composed of three basic components (Fig. 9.50): (1) a rigid PVC shell with high-capacity flow channels to contain the permeable reactive material, (2) flexible wings to direct the flow of groundwater into the permeable reactive material, and (3) passive samplers to determine the quality of the treated water. Multiple DARTs can be joined together for the treatment of thicker contaminant plumes. DARTs also allow for vertical stacking

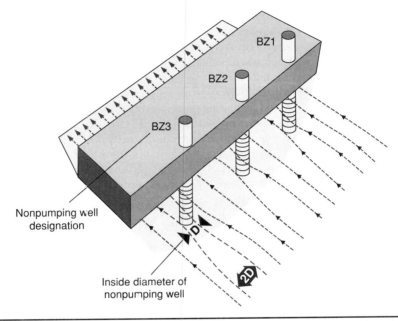

FIGURE 9.49 Schematic diagram showing nonpumping wells containing DARTs and modeled contaminant capture zones, Fry Canyon, UT. (From USGS, 1999.)

of different reactive materials for the treatment of chemically segregated contaminant plumes. The main disadvantage of this system is the very small ROI of individual wells, resulting in potentially high installation cost per plume area (width).

The ability of iron to degrade a variety of contaminant types is being increasingly utilized in nontraditional technologies, such as hydraulic injection of the nanoscale iron in an aqueous solution into the subsurface. When using granular iron, a biodegradable

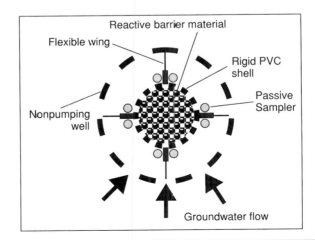

FIGURE 9.50 Schematic diagram of deep aquifer remediation tool (DART). (From USGS, 1999.)

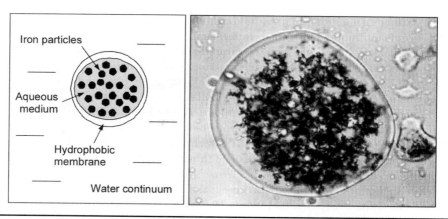

FIGURE **9.51** Left: Rendition of emulsified zero-valent iron droplets. Right: Micrograph of nanoscale iron particles (100–200 nm) contained within an emulsion droplet. (From ITRC, 2005.)

slurry containing iron and guar is injected through a borehole into the subsurface. Depending on the geologic material present, fractures are created or the material is intermixed with the more permeable soils. When using the biodegradable method, an enzyme is added after the injections to degrade the slurry over a short time period, leaving a lens of granular iron in the subsurface (ITRC, 2005). A discussion on use of nanoscale iron for environmental remediation is provided by Zhang (2003).

NASA has developed an emulsified, nanoscale-iron process that directly targets DNAPL. The emulsified system consists of a surfactant-stabilized, biodegradable water-in-oil emulsion with nanoscale-iron particles contained with the emulsion droplets (Fig. 9.51). The surfactant serves two functions: it increases the stability of the emulsion for injection into the DNAPL zone, and the surfactant micelles within the oil membrane of the emulsion droplet aid in the delivery of TCE to the iron. The DNAPL diffuses through the hydrophobic oil membrane of the emulsion droplet, whereupon it reaches the surface of the iron particle and dehalogenation takes place (Geiger et al., 2003). This technology has demonstrated that DNAPLs such as TCE diffuse through the oil membrane of the emulsion droplet, whereupon these reach the surface of the iron particle and dehalogenation takes place. Recent field work at NASA's Launch Complex 34 on Cape Canaveral Air Force Station demonstrated the effectiveness of this process in treating DNAPL (O'Hara et al., 2004).

Research and deployment of biobarrier systems are increasing in recent years, particularly for treatment of chlorinated solvents and petroleum hydrocarbon constituents such as BTEX and MTBE. Biobarriers are often described as in situ bioremediation deployed with the PRB design concept (i.e., a continuous, linear, flow-through zone where treatment occurs). These systems may use solid, liquid, or gaseous amendments, such as wood chips, compost, lactate, and molasses, to create an enhanced zone of biological activity where contaminant degradation occurs. In this way, the reactive treatment zone within a biobarrier is created indirectly through the addition of amendments. This addition is usually achieved through vertical temporary borings or permanent wells, although other technologies including trenching can be applied depending on the site-specific conditions and depth of the contaminated zone.

The economic benefits of PRBs drive the application of this technology. The passive functioning of a PRB means that relatively little energy or labor input (except for site monitoring) is necessary; therefore, the technology has a potential advantage over conventional groundwater treatment systems such as P&T. Regardless, a cost-benefit approach should be used to evaluate the economic feasibility of a PRB at a given site (ITRC, 2005).

PRB technology also has limitations and should not be considered as the only remedy for a site. For example, a PRB may be used in conjunction with one or more other remedies, such as MNA for the downgradient portion of a contaminant plume and/or source removal technologies for DNAPL or other contaminant residual upgradient from the PRB. Additionally, since most PRBs operate passively, site remediation may take several years or even decades, requiring the use of long-term institutional controls for site management. Therefore, a PRB should be considered within the context of overall and long-term site remediation goals (ITRC, 2005).

Biobarriers are considered a unique type of PRB. Some biobarrier designs, particularly those that require deep delivery and circulation of liquid amendments, can challenge the passive operation concept of PRBs. For example, although many biobarriers are designed to deliver amendments into the subsurface using relatively passive techniques (i.e., slow injection or diffusion of oxygen or air), some biobarriers require substantial energy input to deliver amendments to the proper aquifer depth and then to circulate and mix the amendments within the subsurface. Such designs function less passively than traditional PRBs and may incur greater operation and maintenance costs (ITRC, 2005).

9.5.3 Bioremediation

In situ enhanced bioremediation of halogenated VOCs, and in particular of chlorinated aliphatic hydrocarbons (CAHs), is the most rapidly growing groundwater remediation technology. This is not surprising since VOCs are the most frequently occurring type of contaminant in soil and groundwater at Superfund and other hazardous waste sites in the United States. The USEPA estimated that cleanup of these sites will cost more than $45 billion (in 1996 dollars) over the next several decades (USEPA, 1997). Innovative technologies, such as in situ bioremediation, are being developed and implemented in an effort to reduce the cost and time required to clean up those sites. In situ bioremediation is increasingly being selected to remediate hazardous waste sites because, when compared to other technologies, it is usually less expensive, does not require waste extraction or excavation, and is more publicly acceptable as it relies on natural processes to treat contaminants (USEPA, 2000b).

Engineered bioremediation is a technology of adjusting the concentration of various electron acceptors, electron donors, and nutrients in groundwater in order to stimulate biodegradation of contaminants by native (indigenous) microorganisms. This process is commonly referred to as biostimulation. Bioremediation also includes bioaugmentation, or the addition of microbes to the subsurface where organisms able to degrade specific contaminants are deficient. Such microbes may be "seeded" from populations already present at a site and grown in aboveground reactors, or these can be specially cultivated nonindigenous (exogenous) strains of bacteria having known capabilities to degrade specific contaminants.

An electron acceptor is a compound capable of accepting electrons during oxidation-reduction reactions. Microorganisms obtain energy by transferring electrons from electron donors, such as organic compounds (or sometimes reduced inorganic compounds, such as sulfide), to an electron acceptor. Electron acceptors are compounds that are reduced during the process and include oxygen, nitrate, iron (III), manganese (IV), sulfate, carbon dioxide, and, in some cases, chlorinated aliphatic hydrocarbons, such as carbon tetrachloride (CT), PCE, TCE, DCE, and VC. Electron donors are compounds that are oxidized during the process and include fuel hydrocarbons and native organic carbon (USEPA, 2000b).

Nutrients are elements required for microbial growth such as carbon, hydrogen, oxygen, nitrogen, and phosphorus. Substrate is a source of energy or molecular building block used by a microorganism to carry out biological processes and reproduce. Substrates include various forms of solid and liquid organic carbon such as carbohydrates.

The goal of engineered biodegradation is to promote growth and stimulate the activity of those groups of microorganisms best capable of degrading certain contaminants. For example, as illustrated earlier in Fig. 9.35, biostimulation may be successful in degrading PCE and TCE, but it may result in a buildup of cis-DCE and VC, which cannot be as successfully degraded by the present bacteria. Another type of biostimulation would then have to be used downgradient from the first treatment zone to create conditions that promote biodegradation of cis-DCE and VC.

Biodegradation involves the production of energy in a reduction-oxidation (redox) reaction within a bacterial system. This includes respiration and other biological functions needed for cell maintenance and reproduction. Bacteria generally are categorized by (1) the means by which these derive energy, (2) the type of electron donors these require, or (3) the source of carbon that these require. For example, bacteria that are involved in the biodegradation of CAHs in the subsurface are chemotrophs (bacteria that derive their energy from chemical redox reactions) and use organic compounds as electron donors and sources of organic carbon (organoheterotrophs). However, lithotrophs (bacteria that use inorganic electron donors) and autotrophs (bacteria that use carbon dioxide as a carbon source) may also be involved in degradation of CAHs (USEPA, 2000b).

Redox potentials measured in groundwater provide an indication of the relative dominance of the bacteria electron-acceptor classes. These classes determine the type of redox zone that will dominate in the subsurface (for example, an aerobic zone will dominate when aerobes are present). The typical electron-acceptor classes of bacteria are listed in Table 9.6, in the order of those causing the largest energy generation to those causing the smallest energy generation during the redox reaction. A bacteria electron-acceptor class causing a redox reaction that generates relatively more energy will dominate over a bacteria electron-acceptor class that generates relatively less energy during the redox reaction (USEPA, 2000b).

Figure 9.52 shows the redox zones of a typical petroleum plume in an aerobic aquifer, illustrating the progression from the source area to the edge of the plume. A plume moving with groundwater flow typically will develop distinct redox zones (bacteria will use the electron acceptor that causes the most energy to be generated during the redox reaction when compared with the energy generated from redox reactions using other available electron acceptors). As seen in Fig. 9.52, once an electron acceptor is depleted, a new redox reaction with the electron acceptor that will result in the next largest generation of energy during the redox reaction will dominate. The dominant redox reaction will determine the type of bacteria that typically will exist in a particular zone and determine the biodegradation mechanisms that may occur.

Dominance (as Determined by Relative Energy Generation)	Bacteria Electron Acceptor Class	Predominant CAH Biodegradation Mechanism	Approximate Redox Potential (volts)
Most dominant	Oxygen-reducing (aerobes)	Aerobic oxidation	+ 0.82
	Nitrate-reducing		+ 0.74
	Manganese (IV)-reducing		+ 0.52
	Iron (III)-reducing		− 0.05
	Sulfate-reducing		− 0.22
Least dominant	Carbon dioxide-reducing (methanotrops)	Reductive dechlorination	− 0.24

From USEPA, 2000.

TABLE 9.6   Redox Potential as Indicator of Bacteria Electron Acceptor Classes

There are many potential reactions that may degrade CAHs in the subsurface, under both aerobic and anaerobic conditions (Table 9.7). Not all CAHs are amenable to degradation by each of these processes. However, anaerobic biodegradation processes may potentially degrade all the common chloroethenes, chloroethanes, and chloromethanes. Enhanced in situ anaerobic bioremediation involves the delivery of an organic substrate into the subsurface for the purpose of stimulating microbial growth and development, creating an anaerobic groundwater treatment zone, and generating hydrogen through fermentation reactions. This creates conditions conducive to anaerobic biodegradation of chlorinated solvents dissolved in groundwater.

The most common chlorinated solvents released to the environment include tetrachloroethene (PCE, or perchloroethene), trichloroethene (TCE), trichloroethane (TCA), and CT. Because these chlorinated solvents exist in an oxidized state, these are generally not susceptible to aerobic oxidation processes (with the possible exception of cometabolism). However, oxidized compounds are susceptible to reduction under anaerobic conditions by either biotic (biological) or abiotic (chemical) processes. Enhanced anaerobic bioremediation is intended to exploit primarily biotic anaerobic processes

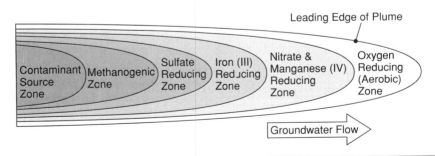

FIGURE 9.52   Areal view of redox zones in a typical petroleum plume in an aerobic aquifer. (Modified from USEPA, 2000b)

	Compound1											
	Chloroethenes				Chloroethanes				Chloromethanes			
Degradation process	PCE	TCE	DCE	VC	PCA	TCA	DCA	CA	CT	CF	MC	CM
Aerobic oxidation	N	N	P	Y	N	N	Y	Y	N	N	Y	P
Aerobic cometabolism	N	Y	Y	Y	P	Y	Y	Y	N	Y	Y	Y
Anaerobic oxidation	N	N	P	Y	N	N	Y	P	N	N	Y	P
Direct anaerobic reductive dechlorination	Y	Y	Y	Y	Y	Y	Y	Y	Y	Y	Y	Y
Cometabolic anaerobic reduction	Y	Y	Y	Y	P	Y	Y	P	Y	Y	Y	P
Abiotic transformation	Y	Y	Y	Y	Y	Y	Y	Y	Y	Y	Y	Y

PCE, tetrachloroethene; TCE, trichloroethene; DCE, dichloroethene; VC, vinyl chloride; PCA, tetrachloroethane; TCA, trichloroethane; DCA, dichloroethane; CA, chloroethane; CT, carbon tetrachloride; CF, chloroform; MC, methylene chloride; CM, chloromethane; N, not documented in the literature; Y, documented; P, potential for reaction to occur, but not well documented in the literature. From Parsons, 2004.

TABLE 9.7 Potential Degradation Processes for Chlorinated Aliphatic Hydrocarbons

to degrade CAHs in groundwater. Other common groundwater contaminants that are subject to reduction reactions are also susceptible to enhanced anaerobic bioremediation. These include chlorobenzenes, chlorinated pesticides (e.g., chlordane), PCBs and chlorinated cyclic hydrocarbons (e.g., pentachlorophenol), oxidizers such as perchlorate and chlorate, explosive and ordnance compounds, dissolved metals (e.g., hexavalent chromium), and nitrate and sulfate (Parsons, 2004).

Anaerobic reductive dechlorination is the degradation process targeted by enhanced anaerobic bioremediation. Through the addition of organic substrates to the subsurface, enhanced anaerobic bioremediation converts naturally aerobic or mildly anoxic aquifer zones to anaerobic and microbiologically diverse reactive zones, making them conducive to anaerobic degradation of CAHs. Examples of easily fermentable organic substrates typically used include alcohols, low-molecular-weight fatty acids (e.g., lactate), carbohydrates (e.g., sugars), vegetable oils, and plant debris (e.g., mulch). The substrates most commonly added for enhanced anaerobic bioremediation include lactate, molasses, Hydrogen Release Compound (HRC®), and vegetable oils. Substrates used less frequently include ethanol, methanol, benzoate, butyrate, high-fructose corn syrup (HFCS), whey, bark mulch and compost, chitin, and gaseous hydrogen.

Table 9.8 summarizes the attributes of several substrate types. These substrates are classified as soluble substrates, viscous fluids and low-viscosity fluids, solid substrates, and experimental substrates. The physical nature of the substrate dictates the frequency of addition, the addition technique, and potential system configurations. The selected organic substrate should be suitable for the biogeochemical and hydrodynamic character of the aquifer to be treated. A common goal is to minimize overall project cost by minimizing the number of required injection points, the number of injection events, and substrate cost (Harkness, 2000). The physical and chemical characteristics of the substrate (e.g., phase and solubility) may make certain substrates more suitable than others in

Substrate	Typical Delivery Techniques	Form of Application	Frequency of Injection
Soluble substrates			
Lactate and butyrate	Injection wells or circulation systems	Acid or salts diluted in water	Continuous to monthly
Methanol and ethanol	Injection wells or circulation systems	Diluted in water	Continuous to monthly
Sodium benzoate	Injection wells or circulation systems	Dissolved in water	Continuous to monthly
Molasses, high fructose corn syrup	Injection wells	Dissolved in water	Continuous to monthly
Viscous fluid substrates			
HRC® or HRC-X™	Direct injection	Straight injection	Annually to biannually for HRC® (typical); every 3–4 yr for HRC-X™; potential for one-time application
Vegetable oils	Direct injection or injection wells	Straight oil injection with water push, or high oil:water content (>20% oil) emulsions	One-time application (typical)
Low-viscosity fluid substrates			
Vegetable oil emulsions	Direct injection or injection wells	Low oil content (<10%) microemulsions suspended in water	Every 2–3 yr (typical); potential for one-time application
Solid substrates			
Mulch and compost	Trenching and excavation	Trenches, excavations, or surface amendments	One-time application (typical)
Experimental (few applications)			
Whey (soluble)	Direct injection or injection wells	Dissolved in water or slurry	Monthly to annually
Chitin (solid)	Trenching or injection of a chitin slurry	Solid or slurry	Annually to biannually; potential for one-time application
Hydrogen (gas)	Biosparging wells	Gas injection	Pulsed injection (daily to weekly)
Humic acids (electron shuttles)	Direct injection or injection wells	Dissolved in water	Unknown; potentially semiannually to annually

From Parsons, 2004.

TABLE 9.8   Substrates Used for Enhanced Anaerobic Bioremediation

particular applications. Furthermore, combinations of various substrates are becoming more common. For example, an easily distributed and rapidly degraded soluble substrate such as lactate may be combined with a slow-release substrate such as vegetable oil. HRC® is also available from the manufacturer as both a fast-acting primer and a longer-lasting HRC-X™ product (Parsons, 2004).

The most commonly used methods to deliver liquid substrates are via installed injection wells or direct-push well points, or by direct injection through temporary direct-push probes. Direct-push methods are commonly used for shallow groundwater applications in unconsolidated formations at depths less than approximately 50 ft. This technique is constrained by soil characteristics such as grain size or degree of cementation (i.e., gravel and cobbles, or caliché inhibits use of direct-push technology). Direct injection of liquid substrates can be made through direct-push probes (e.g., Geoprobe®). This technique does not leave well points in place and is only practical for long-lasting substrates such as HRC®, vegetable oil emulsions, or whey slurries. These substrates release carbon over periods of 6 months to several years and typically require injection on 7.5 to 15-ft centers to treat the target zone (Parsons, 2004).

Permanent wells are typically used for continuous or multiple injections or recirculation of soluble substrate. Use of permanent injection wells is also necessary where direct-push technology is impractical, such as greater treatment depths and difficult lithology. Existing monitoring or extraction wells from previous investigation or remediation activities may also be used where applicable.

Recirculation systems, consisting of closely spaced injection and extraction wells, are sometimes used to increase the retention time of contaminated groundwater in the treatment zone and to promote mixing of the substrate with the contaminant. The rate at which groundwater passes through the system depends on the rate of recirculation and the natural groundwater flux through the recirculation system. Therefore, design of recirculation systems must consider hydraulic conductivity, aquifer heterogeneity, and hydraulic gradient. Recirculation approaches may be the only effective method to achieve more uniform distribution of substrates and amendments at sites with difficult hydrogeological conditions (e.g., lack of a natural hydraulic gradient or pronounced heterogeneity). Recirculation may also be considered for shorter-term applications that cannot be achieved through less aggressive, more passive methods. For example, recirculation may be useful to circulate groundwater from the greater contaminant plume through an established bioaugmented treatment zone (Parsons, 2004). The most critical design element of recirculation systems is prevention of biofouling and clogging of well screens.

Effective mixing of substrate with the contaminant plume is one of the most difficult design challenges for enhanced anaerobic bioremediation. Injection of large volumes of substrate may cause significant displacement of the contaminant plume, and sometimes a decrease in permeability of the porous media adjacent to injection points due to biofouling.

Depending on the frequency of substrate delivery to the subsurface, bioremediation systems can be grouped into passive, semipassive, and active. Passive biobarriers typically use slow-release, long-lasting substrates (e.g., HRC©, vegetable oils, or mulch) that can be either injected or otherwise placed in a trench and that are designed to remain in place for long periods to maintain the reaction zone. Contaminant mass is delivered to the treatment zone via natural groundwater flow.

Semipassive or active biobarriers are similar to passive biobarriers except that a soluble substrate is typically injected periodically (semipassive) or via a recirculation system

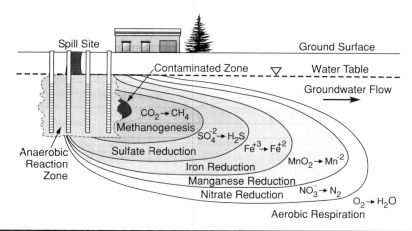

FIGURE 9.53 Reducing zones established downgradient of substrate injection. (From Parsons, 2004.)

(active). Soluble substrates migrate with groundwater flow, are depleted more rapidly, and require frequent addition. However, these systems offer the advantage of being able to adjust the rate or type of substrate loading over time, and soluble substrates may be easier to distribute throughout larger volumes of the contaminant plume.

Biodegradation of an organic substrate depletes the aquifer of DO and other terminal electron acceptors (e.g., nitrate or sulfate) and lowers the oxidation-reduction potential of groundwater, thereby stimulating conditions conducive to anaerobic degradation processes. After DO is consumed, anaerobic microorganisms typically use native electron acceptors (as available) in the following order of preference: nitrate, manganese and ferric iron oxyhydroxides, sulfate, and finally carbon dioxide. Figure 9.53 illustrates a CAH plume where substrate has been injected into the source area. An anaerobic treatment area is created with the development of progressively more anaerobic zones closer to the source of organic carbon as electron acceptors are depleted. Anaerobic dechlorination has been demonstrated under nitrate-, iron-, and sulfate-reducing conditions, but the most rapid biodegradation rates, affecting the widest range of CAHs, occur under methanogenic conditions (Bouwer, 1994).

Enhanced anaerobic bioremediation will not be effective in the following conditions:

- Sites with impacted receptors, or with short travel time or distance to potential receptors.
- The contaminant cannot be anaerobically degraded.
- Strongly reducing conditions cannot be generated.
- A microbial community capable of driving the process is not present or cannot be introduced to the subsurface.
- A fermentable carbon source cannot be successfully distributed throughout the subsurface treatment zone.
- There are unknown or inaccessible DNAPL sources.
- Difficult hydrogeologic characteristics that preclude cost-effective delivery of amendments, such as low permeability or a high degree of aquifer heterogeneity.

- Geochemical factors (e.g., unusually low or high pH) that inhibit the growth and development of dechlorinating bacteria.

Settings with the extremes of very high and very low rates of groundwater flow impose significant limits to applying bioremediation. It may be impractical to maintain reducing conditions in high flow settings, due to the magnitude of groundwater and native electron-acceptor flux. On the other hand, it may be difficult to inject substrates into tight formations, and, under low-flow settings, mixing of substrate with groundwater due to advection and dispersion may be limited (Parsons, 2004). Table 9.9 provides a summary of site characteristics related to suitability of enhanced bioremediation.

Engineered bioremediation is typically used to address the following remedial objectives: (1) remediation of source zones where good substrate/contaminant contact is possible, (2) reduction of mass flux from a source zone or across a specified boundary (e.g., plume containment), and (3) plume-wide treatment. Total treatment of an entire dissolved plume with active bioremediation may be feasible in some cases; ultimately, however, there will be an economic limit to the size of a plume that can be treated. For plume sizes greater than 10 to 20 acres, use of containment strategies combined with other remedial approaches may be more feasible (Parsons, 2004).

While anaerobic dechlorination may be effective in degrading chlorinated solvents, secondary degradation of groundwater quality may occur. Degradation reactions or excessive changes in groundwater pH and redox conditions may lead to solubilization of metals (e.g., iron, manganese, and potentially arsenic), formation of undesirable fermentation products (e.g., aldehydes and ketones), and other potential impacts to secondary water quality (e.g., total dissolved solids). Many of these changes are not easily reversed, and, in the case of a slow-release carbon source, it may take many years for the effects of the substrate addition to diminish. These issues should be considered during technology screening (Parsons, 2004).

In many cases, the sole use of an organic substrate (biostimulation) is sufficient to stimulate anaerobic reductive dechlorination. However, bioaugmentation may be considered at a site when an appropriate population of dechlorinating microorganisms is not present or sufficiently active to stimulate complete anaerobic reductive dechlorination of the CAH constituents present. As shown with examples presented by Suthersan and Payne (2005), at quite a few sites, biostimulation results in accumulation of *cis*-DCE and VC, which is not a desirable outcome and has to be addressed separately. To date, experience with bioaugmentation is limited, and there is some disagreement among practitioners as to its benefits. Bioaugmentation involves the injection of a microbial amendment comprising non-native organisms known to carry dechlorination of the targeted CAHs to completion. For example, the presence of *Dehalococcoides*-related microorganisms has been linked to complete dechlorination of PCE to ethene in the field (Major et al., 2001; Hendrickson et al., 2002). Commercial bioaugmentation products that contain these microorganisms are available.

Independent field studies demonstrating the effectiveness of in situ stimulation or augmentation with degrading bacterial consortia to remediate contaminant plumes are still relatively limited. Lendvay et al. (2003) performed a carefully controlled field experiment comparing the effects of biostimulation and bioaugmentation at the Bachman Road site in Michigan. The site had a PCE plume derived from former dry-cleaning operations along the Lake Huron shoreline. The plume showed evidence of naturally occurring reductive dechlorination, as indicated by the appearance of reductive

Site Characteristics	Suitable for Enhanced Bioremediation	Suitability Uncertain	Suitability Unclear—Possible Red Flag—Requires Further Evaluation
DNAPL presence	Residual DNAPL or sorbed sources	Poorly defined sources may require additional characterization	May not be appropriate for aggressive treatment of pools of DNAPL
Plume size	Small, a few acres or less	Medium to large, a few acres plus May require concurrent technology	Large plumes of many acres May require concurrent technology
On-site or near-site infrastructure	The risk of vapor intrusion from contaminants or biogenic gases is deemed acceptable	Target treatment zone in close proximity to sensitive infrastructure	Target treatment zone in an area where known vapor intrusion or high methane problem exists
Evidence of anaerobic dechlorination	Slow or stalled dechlorination	Limited evidence of anaerobic dechlorination	No evidence of any degradation
Depth	<50 ft to water	>100 ft to groundwater	Deep groundwater and deep contamination
Hydraulic conductivity	>1 ft/d (>3 × 10^{-4} cm/s)	0.01–1 ft/d (>3 × 10^{-6} to 3 × 1^{-4} cm/s)	<0.01 ft/d (<3×10^{-4} cm/s)
Groundwater velocity	30 ft/yr to 5 ft/d	10–30 ft/yr 5–10 ft/d	<10 ft/y >10 ft/d
pH	6.0–8.0	5.0–6.0 8.0–9.0	<5.0 >9.0
Solute concentration	<500 ppm	500–5000 mg/L (with caution)	>5000 mg/L or presence of mineral gypsum may not be suitable

ft/d, feet per day; ft/yr, feet per year; cm/s, centimeter per second; mg/L, milligrams per liter. From Parsons, 2004.

TABLE 9.9 Suitability of Site Characteristics for Enhanced Anaerobic Bioremediation

dechlorination products (TCE, *cis*-DCE, and VC) and the presence of culturable populations capable of complete dechlorination of PCE to ethene from the site provided impetus for a controlled comparison of side-by-side biostimulation and bioaugmentation strategies for plume control. The study showed that the bioaugmentation phase took 43 days, and the biostimulation phase took at least 121 days and probably longer, since the pumping during the control phase, which stimulated the indigenous microflora, is not included in this number. The shorter time and more complete detoxification suggest that significant savings are possible with the bioaugmentation approach, albeit at a higher cost, to produce the dechlorinating inoculum (Lendvay et al., 2003).

9.5.4   Monitored Natural Attenuation

MNA, sometimes referred to as passive bioremediation, is defined as use of natural F&T processes, such as contaminant dilution, volatilization, degradation, adsorption, diffusion, dispersion, immobilization, and chemical reactions with subsurface materials to reach site-specific remediation goals (Wiedemeier et al., 1998; USEPA, 1999a; Chapelle et al., 2007).

In current engineering practice, the effectiveness of MNA is evaluated on a site-by-site basis by considering three lines of evidence: (1) historical monitoring data showing decreasing concentrations and/or contaminant mass over time, (2) geochemical data showing that site conditions favor contaminant transformation or immobilization, or (3) site-specific laboratory studies documenting ongoing biodegradation processes. Various field and laboratory methods for assessing these three lines of evidence have been developed and are currently in use (Wiedemeier et al., 1998, 1999; Gilmore et al., 2006).

A major difficulty in confirming and measuring natural attenuation lies in its complexity and in the fact that most of the dominant and critical attenuation mechanisms cannot be determined directly. For example, in current practice, there is no direct measure that biodegradation is occurring. Instead, indicators that measure conditions suitable for biodegradation and breakdown products are collected to provide evidence that biodegradation is occurring. Therefore, multiple parameters are needed, plus an understanding of their relationships to accurately assess MNA. The stakeholder or decision makers determine the amount of evidence needed in order to make a defensible decision. The question then arises, how much evidence is enough? Direct field measurement of key processes and degradation rates is, therefore, one of the research priorities because it would eliminate the question of where biodegradation is actually occurring along the groundwater flow path (Gilmore et al., 2006).

It has been emphasized by regulatory agencies and practitioners alike that MNA is not a "do-nothing" approach because it involves (ITRC, 1999)

- Characterizing the F&T of the contaminants to evaluate the nature and extent of the natural attenuation processes
- Ensuring that these processes reduce the mass, toxicity, and/or mobility of subsurface contamination in a way that reduces risk to human health and the environment to acceptable regulatory levels
- Evaluating the factors that will affect the long-term performance of natural attenuation
- Monitoring of the natural processes to ensure their continued effectiveness

MNA has been the favorite first option for consideration at most (if not all) groundwater contamination sites for more than a decade now. Although it has been approved as the sole remedy at a relatively few complex sites, it is always very attractive as a supplemental remedy for three main reasons: it is noninvasive, does not require use of energy and working equipment, and has a much lower implementation cost compared to various engineered groundwater remediation systems. However, installation of monitoring wells, which is a necessary part of any MNA remedy, may involve a significant initial cost. One of the potentially most attractive aspects of MNA to general public is that it is "noninvasive": unlike many elaborate engineered site cleanup facilities, it is "quietly" working below ground so that the land surface aboveground may continue to be used. In its effort to educate the public on the benefits of natural attenuation and bioremediation in general and to alleviate concerns that MNA is not a "do-nothing" groundwater remedial alternative, USEPA has published various pamphlets (e.g., USEPA, 1996c, 1996d), which include general explanations such as the following:

> Bioremediation, is a process in which naturally occurring microorganisms (yeast, fungi, or bacteria) break down, or *degrade,* hazardous substances into less toxic or nontoxic substances. Microorganisms, like humans, eat and digest organic substances for nutrition and energy. (In chemical terms, "organic" compounds are those that contain carbon and hydrogen atoms.) Certain microorganisms can digest organic substances such as fuels or solvents that are hazardous to humans. Biodegradation can occur in the presence of oxygen (aerobic conditions) or without oxygen (anaerobic conditions). In most subsurface environments, both aerobic and anaerobic biodegradation of contaminants occur. The microorganisms break down the organic contaminants into harmless products—mainly carbon dioxide and water in the case of aerobic biodegradation (Figure 1). Once the contaminants are degraded, the microorganism populations decline because they have used their food sources. Dead microorganisms or small populations in the absence of food pose no contamination risk.
>
> Many organic contaminants, like petroleum, can be biodegraded by microorganisms in the underground environment. For example, biodegradation processes can effectively cleanse soil and ground water of hydrocarbon fuels such as gasoline and the BTEX compounds—benzene, toluene, ethylbenzene, and xylenes. Biodegradation also can break down chlorinated solvents, like trichloroethylene (TCE), in ground water but the processes involved are harder to predict and are effective at a smaller percentage of sites compared to petroleum-contaminated sites. Chlorinated solvents, widely used for degreasing aircraft engines, automobile parts, and electronic components, are among the most often-found organic ground-water contaminants. When chlorinated compounds are biodegraded, it is important that the degradation be complete, because some products of the breakdown process can be more toxic than the original compounds.

Many chemicals will undergo zero- or first-order degradation in the dissolved, solid, and/or gaseous phase. A first-order degradation is described by the following equation:

$$C = C_0 e^{-kt} \tag{9.8}$$

where C = concentration at time t
C_0 = initial concentration at time $t = 0$
k = first-order rate constant (units = time^{-1})
t = time

First-order degradation can also be stated in terms of a chemical's half-life. The half life is the time it takes for half the contaminant to degrade:

$$\frac{C}{C_0} = 0.5 = e^{-kt} \tag{9.9}$$

$$t_{1/2} = \frac{\ln 2}{k} = \frac{0.693}{k} \tag{9.10}$$

As can be seen, the contaminant half-life and the first-order degradation constant can both be used in quantitative analyses, but consistently, so that there is no confusion. For example, a 2-year half-life is equivalent to a first-order rate constant of 0.35 per year. Most practitioners, and most analytical equations, consistently use the first-order degradation constant in the same units of time as all other time-dependent parameters.

Some chemicals undergo zero-order degradation, which is described by the following equation:

$$C = C_0 - k_0 t \qquad (9.11)$$

where k = zero-order rate constant (units = mass/volume-time). Zero-order kinetics does not typically occur for most common organic compounds found in groundwater.

Obviously, the key parameter in all calculations involving contaminant degradation is the degradation constant. Unfortunately, it is also very difficult to be accurately determined in the field, and it may change in time as geochemical conditions at the site change. It is also the least likely parameter to be accepted by regulatory agencies in case all other abiotic parameters, such as advection, recharge (very important when considering effects of potential dilution), sorption, and dispersion, are not reasonably accurately established for the particular site. In the case of sequential decay reactions, such as those involving chlorinated solvents and munitions constituents, degradation constants are certain to vary for different daughter products and may not even be applicable for some of them. In other words, some daughter products may be less biodegradable or not biodegradable at all in certain conditions.

The USEPA published an informative work on different methods commonly used to determine the degradation rate constant including discussions on associated uncertainties and applicability of individual methods (Newell et al., 2002). The key point is that this constant may mean different things to different people and in different context. It is, therefore, very important to make a clear distinction between the general attenuation constant and the (bio)degradation constant. Although either constant can be used to quantitatively describe a general decrease in contaminant concentration in time (a first-order decay process), the degradation constant should be used only if the associated biological process is confirmed, and its rate quantified. The more general attenuation constant includes both abiotic and biotic processes, without making distinction between them. It is relatively easily determined from the measured contaminant concentrations at monitoring wells, at multiple times, by establishing a quantitative relationship between the concentration decrease and the time.

In summary, the biodegradation rate constant applies to both space and time, but only to one degradation mechanism. Quantification of this parameter is arguably the most critical part of contaminant F&T studies and remediation projects that consider biodegradation, in any form, as a potentially viable alternative. It can be performed in the laboratory, using controlled microcosm studies with soil and groundwater samples from the site, or in the field using extensive (and expensive) tracer studies. It can also be estimated during model calibration of all other abiotic F&T parameters. Whatever the case may be, selecting a literature value should be the last option, and only for the "screening" purposes. Since every site will have a very specific degradation rate for any particular contaminant, and this rate may change in time, the author deliberately did not include any table of "literature values" of degradation rate constants. This is, again, simply because every organic chemical may or may not degrade at any particular site. Every effort should be made to avoid using literature values because it may result in very misleading results, whatever any particular stakeholder were hoping for.

A detailed explanation of various biodegradation processes and their characterization in the field and in the laboratory are beyond the scope of this book (a brief discussion is provided in Chap. 5). As a result of the ever-increasing interest in the bioremediation of contaminated groundwater, there are many publications and resources available in the public domain and accessible via free download at various Web sites maintained by the United States Government agencies. A good starting point is works by Wiedemeier et al. (1998, 1999), Azdapor-Keeley et al. (1999), Lawrence (2006), and Gilmore et al. (2006).

Sustainability of MNA

The effectiveness of natural attenuation is typically assessed over relatively short periods of time, often only a few months or years. When MNA becomes part of a long-term remediation strategy, however, it must be assumed that processes observed during site assessment will remain intact over the system's operational lifetime. This operational lifetime depends on the length of time that contaminants are released from source areas to groundwater (NRC, 2000), a period of time that may encompass decades or even centuries. This, in turn, raises an important question. Will the natural attenuation processes observed during site assessment continue with the same efficiency in the future? In other words, will MNA be sustainable throughout the operational life of the remediation system? (Newell and Aziz, 2004; Chapelle et al., 2007).

The sustainability of MNA over time depends on (1) the presence of chemical and biochemical processes that transform contaminants to harmless by-products and (2) the availability of energy to drive these processes to completion. The presence or absence of contaminant-transforming chemical (biochemical) processes can be determined by observing contaminant mass loss over time and space (mass balance). The energy available to drive these processes to completion can be assessed by measuring the pool of metabolizable organic carbon available in a system, and by tracing the flow of this energy to available electron acceptors (energy balance). Natural attenuation is sustainable when the pool of bioavailable organic carbon is large, relative to the carbon flux needed to drive biodegradation to completion (Chapelle et al., 2007).

There are two approaches to quantify mass and energy balance of MNA in groundwater systems:

1. Empirical, using field monitoring data to directly calculate the mass loading of contaminants to a system and the system's natural attenuation capacity (NAC); similarly, the energy flow through a groundwater system can be determined in the field by evaluating the succession of electron-accepting processes along the plume.

2. Deterministic, by using numeric models (equations) that describe the physics and biochemistry of a system.

Given a well-characterized plume, the flux of contaminants across a given cross-sectional transect of an aquifer can be estimated from the following equation:

$$\text{Mass Flux}_{\text{transect}} = C_{\text{ave}} \times Q \qquad (9.12)$$

where C_{ave} = average contaminant concentration in units of mass per volume and Q = groundwater flux in units of volume per time. When this flux is calculated for a given

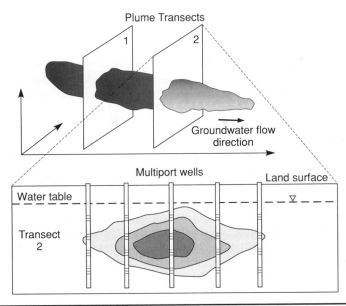

FIGURE 9.54 The empirical approach to describing the contaminant mass balance (mass flux) and natural attenuation capacity using site-specific monitoring data.

transect of the aquifer, it gives contaminant mass flux in units of mass per time (e.g., mg/d). The ambient NAC can be quantified as the difference in contaminant mass flux between two transects (Fig. 9.54):

$$NAC = \text{Mass Flux}_{\text{transect1}} - \text{Mass Flux}_{\text{transect2}} \qquad (9.13)$$

In this formulation, NAC is expressed in units of contaminant mass attenuated per unit time (e.g., kg/d or kg/yr).

As discussed by Chapelle et al. (2007), the main advantage of using the empirical mass-balance approach is that no a priori assumptions need to be made about the processes contributing to contaminant loading, contaminant attenuation, or the consumption of electron donors and acceptors. As a description of the present behavior of a system, therefore, the empirical approach is highly useful. Because the individual processes contributing to contaminant loading and natural attenuation are not explicitly considered, however, the empirical approach cannot predict how a system will respond if conditions change at some point in the future. Therefore, the utility of the empirical approach for assessing long-term questions, such as the sustainability of natural attenuation, is limited.

Chapelle et al. (2007) demonstrate the use of the three-dimensional sequential electron acceptor model (SEAM3D; Waddill and Widdowson, 2000) to quantify mass and energy balance of the sequential reduction of PCE, TCE, DCE, and VC at the Kings Bay Naval Submarine Base in southeastern Georgia, the United States. This model was selected because it explicitly calculates a mass balance for the amount of contaminant NAPL present in a system and because it tracks the flow of energy from electron donors (organic carbon) and the various competing electron-accepting processes. This, in turn, makes it possible to simultaneously track the dissolution of NAPL over time and to couple the biotransformation of dissolved contaminant mass to the organic carbon and electron acceptors available in an aquifer.

In order to use a deterministic model such as SEAM3D for quantifying mass and energy balance at any given site, a large number of parameters describing NAPL dissolution, advective and dispersive transport of contaminants, biotransformation of contaminants, and the delivery and utilization of electron donors and acceptors are required. The model-building process begins with conceptualizing the hydrologic system, followed by estimating values for the various parameters required, and then constraining these estimated values by using available hydrologic, geochemical, and microbiologic data. Once the model has been adequately constrained, it can be used to evaluate how the sustainability of natural attenuation in a particular hydrologic system responds to various environmental conditions over time (Chapelle et al., 2007).

An important capability of the SEAM3D code is that it includes an NAPL package designed to simulate the dissolution and mobilization of chlorinated ethenes from the NAPL present in the contaminant source area. This feature, when combined with the biodegradation and reductive dechlorination packages, allows simulation of a global mass balance that includes mineral and dissolved electron acceptors, dissolved organic carbon substrate, and the dissolution and natural attenuation of chlorinated ethenes. This capability, in turn, makes it possible to estimate times of remediation associated with NAPL dissolution and, thus, to assess the sustainability of natural attenuation.

9.6 Measuring Success of Remediation

The task of evaluating the efficiency and effectiveness of a remedial action in meeting the remediation and operational objectives established for the project is termed "performance assessment." System effectiveness is the ability of the system to achieve remediation goals at a given site, while "efficiency" refers to the optimization of time, energy, and cost toward the achievement of effectiveness. The USEPA defines performance monitoring as "the periodic measurement of physical and/or chemical parameters to evaluate whether a remedy is performing as expected." In terms of DNAPL source-zone treatment, performance assessment involves the collection and evaluation of conditions following treatment and the comparison of that information to pretreatment or baseline conditions (ITRC, 2004).

Goals for an NAPL source-zone cleanup usually differ from goals of cleaning a dissolved plume without a continuous source. An NAPL zone cleanup goal commonly falls into three categories: short-term, intermediate, and long-term performance goals. Short-term goals focus on controlling NAPL mobility and mitigating the potential for further contaminant migration. Long-term goals typically target the achievement of compliance with regulatory criteria applicable to contaminated media at the site, such as restoration of groundwater to drinking water standards. Intermediate performance goals are appropriate when guiding cleanup at an NAPL source zone, where complete removal of the source in one aggressive remedial effort is typically not feasible, yet the levels of contamination left behind are unacceptable. Examples of intermediate performance goals might include depleting the source sufficiently to allow for natural attenuation, preventing the migration of contaminated fluids beyond the treatment zone, reducing dissolved-phased concentrations outside the source zone, or reducing the mass discharge rate or flux emanating from the source (ITRC, 2004).

Assessing the performance of NAPL source-depletion technologies is necessary in order to determine whether such intensive, costly measures are capable of achieving remedial goals. However, assessing performance is complicated by the variation in remedial goals and metrics used to determine whether those goals are met (ITRC, 2004; Kavanaugh et al., 2003). At few sites, measurements of the change in NAPL mass and/or contaminant flux are used as a performance metric. However, the remedial goals at most sites with impacted groundwater are based on reducing groundwater concentrations to regulatory standards (e.g., MCLs or risk-based values). Since remedial goals are usually based on dissolved contaminant concentrations, most sites where source depletion has been applied rely on groundwater concentrations to track remediation performance. Unfortunately, as discussed earlier, there are very few, if any, sites with complex hydrogeologic conditions and NAPL sources that have achieved drinking water standards throughout groundwater contamination zones after attempts at NAPL removal have been made. It is for this reason that the scientific, engineering, and, more recently, regulatory communities are looking at contaminant mass flux reduction as a potential measure of the effectiveness and, ultimately, the success of groundwater remediation.

Methods for estimating contaminant flux include the use of hydraulic parameters and measured concentrations from traditional monitoring wells, the use of flux meters, total plume capture by a well, integrated pumping tests (steady- and unsteady-state using single and multiple wells), and recirculation zones. These methods have been described in detail by ITRC (2004).

The most common approach of calculating contaminant mass flux is to estimate the mass of solute at two cross-sectional areas perpendicular to the flow direction (Fig. 9.54). The basis of this approach is the law of continuity of groundwater flow (i.e., flow in equals flow out). The integrated mass flux of a solute across two separate plume transects intersecting the entire plume must be equal unless there is removal of solute through attenuation. For this approach to be accurate, however, the plume must be at steady state (i.e., the concentrations at the downgradient edge are not changing and the plume is not expanding or shrinking) and the aquifer geometry and groundwater flow conditions are known well enough to capture the entire plume (Gilmore et al., 2006).

The University of Florida's passive flux meter (PFM) is a self-contained permeable device inserted into a well or boring that acts as an integrating sampler. Groundwater flows through it and is not retained. The interior of the device is a matrix of hydrophobic and hydrophilic permeable sorbents that retain dissolved organic and inorganic contaminants present in fluid intercepted by the unit. The sorbent matrix is also impregnated with known amounts of water-soluble resident tracers (typically benzoate), which are leached from the sorbent at rates proportional to the fluid flux. Following exposure to groundwater flow for a period ranging from days to months, the passive flux indicator is removed from the monitoring well and the sorbent extracted to quantify the masses of contaminants intercepted and residual masses of resident tracers. The contaminant masses are used to calculate time-averaged contaminant mass fluxes; residual resident tracer masses are used to calculate time-averaged groundwater flux (Hatfield et al., 2002). The passive flux indicator is being validated through the Environmental Security Technology Verification Program (ESTCP) demonstrations at several Department of Defense sites and is being demonstrated as part of the Monitored Natural Attenuation and Enhanced Attenuation for Chlorinated Solvents Technology Alternative Project supported by Department of Energy (Gilmore et al., 2006).

The largest source of error in traditional transect and PFM methods of measuring flux is associated with aquifer heterogeneity and the subjective process of interpolating solute distribution. These flux measurement techniques rely on multiple point measurements from monitoring wells and subsequent interpolation, which potentially create error. For example, Fraser et al. (2005) evaluated mass flux versus sampling density for a naphthalene plume at the Borden research aquifer. When the sampling grid density was reduced from 1.7 points per meter squared (m^2) to 0.7 points per m^2, the range (as a standard deviation) in mass discharge increased to more than 50 percent. Guilbeault et al. (2005) showed that 75 percent of the mass flux occurred within 5 to 10 percent of the transect cross-sectional area for three plumes in Ontario, New Hampshire, and Florida and that a spacing no larger than 15 to 30 cm was needed at some locations to identify high-concentration zones. In any case, however, when the vertical distribution of the mass flux is needed, point measurement techniques are currently the best option.

The Mass Flux Toolkit developed for ESTCP by Farhat et al. (2006) is an easy-to-use, free software tool that allows comparison of different mass flux approaches, calculates mass flux from transect data, and applies mass flux to calculate downgradient dissolved concentrations. In this software, the term mass flux is used to describe the mass discharge rate in a groundwater plume in units of mass per time passing across a plume transect.

The Mass Flux Toolkit provides three methods for analyzing uncertainty in the total mass flux estimates derived from the transect method. One option utilizes the Monte Carlo-type approach to analyze uncertainty in the actual concentration, hydraulic conductivity, and gradient measurements. The second option provides a tool for estimating the contribution of each individual observation to the total mass flux. If a single monitoring point represents a high percentage of the total mass flux, then the uncertainty in the calculation is high and additional monitoring points should be added to reduce the uncertainty. The third method shows the uncertainty involved in the interpolation scheme that is used to calculate mass flux (Farhat et al., 2006).

The University of Tübingen, Germany, has developed a method to estimate mass flux using a transect of extraction wells or a single well capturing the entire plume. The integrated pumping test has two stages. In the first stage, each monitoring well in the transect is pumped at a constant rate and the time-varying concentration of the contaminant of interest is recorded. In the second stage, the concentration versus time data for all of the wells are combined using a numerical inversion technique to estimate the mass flux of contaminant across the transect. Unlike the PFM, it allows the maximum and average concentrations over the transect to be estimated. As an active method, the integrated pumping test overcomes localized issues (e.g., borehole damage) that interfere with passive methods and can obtain measurements over a length greater than the well diameter. On the other hand, water pumped from the wells must be disposed of, perhaps as hazardous waste, and multiple chemical analyses are required (Gilmore et al., 2006). This method has been tested at dozens of sites in Europe (e.g., Bayer-Raich et al., 2004, 2006; Jarsjö et al., 2005).

References

Adams, J.A., and Reddy, K.R., 2000. Removal of dissolved and NAPL-phase benzene pools from groundwater using in-situ air sparging. *Journal of Environmental Engineering, ASCE*, vol. 126, no. 8, pp. 697–707.

AFCEE, 1995. *Test Plan and Technical Protocol for a Field Treatability Test for POL Free Product Recovery–Evaluating the Feasibility of Traditional and Bioslurping Technologies*. Air Force Center for Environmental Excellence, Technology Transfer Division, Brooks AFB, TX, various pages.

Anderson, M.R., Johnson, R.L., and Pankow, J.F., 1987. The dissolution of residual dense non-aqueous phase liquid (DNAPL) from a saturated porous medium. In: *Proceedings of the NWWA/API Conference on Petroleum Hydrocarbons and Organic Chemicals in Ground Water: Prevention, Detection, and Restoration*, National Water Well Association-American Petroleum Institute, Houston, TX, pp. 409–428.

Anderson, M.R., Johnson, R.L., and Pankow, J.F., 1992. Dissolution of dense chlorinated solvents into groundwater. 3. Modeling of contaminant plumes from fingers and pools of solvent. *Environmental Science and Technology*, vol. 26, pp. 901–908.

Azdapor-Keeley, A., Russell, H.H., and Sewell, W.W., 1999. Microbial processes affecting monitored natural attenuation of contaminants in the subsurface. Ground Water Issue, U.S. Environmental Protection Agency, Office of Research and Development, EPA/540/S-99/001, 18 p.

Baker, R.S., and Bierschenk, J., 1995. Vacuum enhanced recovery of water and NAPL: Concept and field test. *Journal of Soil Contamination*, vol. 4, no. 1, pp. 57–76.

Baker, R.S., and Kuhlman, M., 2002. A description of the mechanisms of in-situ thermal destruction (ISTD) reactions. In: *Second International Conference on Oxidation and Reduction Technologies for Soil and Groundwater*, ORTs-2, Toronto, Canada, November 17–21, 2002.

Barner, W., and Kristine, U., 1995. Contaminant transport mechanisms in karst terrains and implications on remediation. In: *Karst Geohazards – Engineering and Environmental Problems in Karst Terrane*. Beck, B.F., editor. A.A. Balkema, pp. 207–212.

Barth, G., Illangasekare, T.H., Rajaram, H., and Hill, H., 2003. Demonstration of solute flux sensitivity to entrapped nonaqueous phase liquids: Intermediate-scale experiments in heterogeneous porous media. *Journal of Contaminant Hydrology*, vol. 67, pp. 247–268.

Bass, D.H., and Brown, R.A., 1996. Air sparging case study database update. Proceedings of the 1st International Symposium on In Situ Air Sparging for Site Remediation, October 24 and 25, 1996, Las Vegas, NV. INET, Potomac, MD.

BATTELLE, 1997. Engineering evaluation and cost analysis for bioslurper initiative (A005). Prepared for U.S. Air Force Center for Environmental Excellence, Brooks AFB, Texas. BATTELLE, Columbus, OH, 87 p.

BATTELLE and Duke Engineering and Services, 2002. Surfactant-enhanced aquifer remediation (SEAR) design manual. NFEC Technical Report TR-2206-ENV, Naval Facilities Engineering Command, Washington, DC, 86 p. + appendices.

Bauer, S., 2002. *Simulation of the Genesis of Karst Aquifers in Carbonate Rocks*. Vol. 62 of Tübinger Geowissenschaftliche Arbeiten: Tübingen, Germany, Reihe C. Institut und Museum für Geologie und Paläontologie der Universität Tübingen.

Bayer-Raich, M., Jarsjö, J., Liedl, R., Ptak, T., and Teutsch, G., 2004. Average contaminant concentration and mass low in aquifers from time dependent pumping well data: analytical framework. *Water Resources Research*, vol. 40, W083031 p.

Bayer-Raich, M., Jarsjo, J., Liedl, R., Ptak, T., and Teutsch, G., 2006. Integral pumping test analyses of linearly sorbed groundwater contaminants using multiple wells: Inferring mass flows and natural attenuation rates. *Water Resources Research*, vol. 42, W08411 p.

Beyke, G., 1998. 6-Phase electrical heating. Paper presented at the In Situ Thermal Treatment Conference, USEPA TIO, Atlanta, GA, December 1998.

Birk, S., 2002. *Characterization of Karst Systems by Simulating Aquifer Genesis and Spring Responses: Model Development and Application to Gypsum Karst*, Vol. 60 of Tübinger Geowissenschaftliche Arbeiten: Tübingen, Germany, Reihe C. Institut und Museum für Geologie und Paläontologie der Universität Tübingen.

Bouwer, H., 1978. *Groundwater Hydrology*. McGraw-Hill, New York, 480 p.

Bouwer, E.J. 1994. Bioremediation of chlorinated solvents using alternate electron acceptors. In: *Handbook of Bioremediation*. Norris, R.D., et al., editors. Lewis Publishers, Boca Raton, FL, pp. 149–175.

Brown, R., 2003. In situ chemical oxidation: performance, practice, and pitfalls. 2003 AFCEE Technology Transfer Workshop, February 25, 2003, San Antonio, TX. Available at: http://www.cluin.org/download/ techfocus/chemox/4_brown.pdf.

Brusseau, M.L., 2005. DNAPL source zones. In: DNAPLs – source zone behavior and mass flux measurement, Clue-In Seminar, August 10, 2005. U.S. Environmental Protection Agency, Technology Innovation Program. Available at: http://www.clu-in.org/conf/tio/dnapls1_081005/. Accessed September 2007.

Chapelle, F.H., Novak, J., Parker, J., Campbell, B.G., and Widdowson, M.A., 2007. A framework for assessing the sustainability of monitored natural attenuation. U.S. Geological Survey Circular 1303, Reston, VA, 35 p.

Clemens, T., Hückinghaus, D., Sauter, M., Liedl, R., and Teutsch, G., 1996. A combined continuum and discrete network reactive transport model for the simulation of karst development. In: *Calibration and Reliability in Groundwater Modelling—Proceedings of the ModelCARE 96 Conference*, September 1996. International Association of Hydrological Sciences Publication 237, Golden, CO, pp. 309–318.

Clement, T.P., Gautam, T.R., Lee, K.K., Truex, M.J., and Davis, G.B., 2004. Modeling of DNAPL-dissolution, rate-limited sorption and biodegradation reactions in groundwater systems. *Bioremediation Journal*, vol. 8, no. 1–2, pp. 47–64.

Cohen, R.M., and Mercer, J.W., 1993. *DNAPL Site Evaluation*. C.K. Smoley, Boca Raton, FL.

Currens, J.C., 1999. Mass flux of agricultural nonpoint-source pollutants in a conduit-flow-dominated karst aquifer, Logan County, Kentucky. Report of Investigations 1, Series XII, Kentucky Geological Survey, University of Kentucky, Lexington, KY, 151 p + 2 plates.

Davis, E. 1997. Ground water issue: How heat can enhance in-situ soil and aquifer remediation: Important chemical properties and guidance on choosing the appropriate technique, EPA 540/S-97/502, U.S. Environmental Protection Agency, Office of Research and Development, Ada, OK, 18 p.

Davis, E.L., 1998. Steam injection for soil and aquifer remediation. EPA/540/S-97/505, U.S. Environmental Protection Agency, Office of Research and Development, Ada OK, 16 p.

Dupont, R.R., Doucette, W.J., and Hinchee, R.E., 1991. Assessment of in situ bioremediation potential and the application of bioventing at a fuel-contaminated site. In: *In Situ Bioreclamation: Applications and Investigations for Hydrocarbon and Contaminated Site Remediation*. Hinchee, R.E., and Olfenbuttel, R.F., editors. Butterworth-Heinemann, Stoneham, MA, pp. 262–282.

Eckenfelder, W.W., Jr., 1980. *Principles of Water Quality Management*. CBI Publishing Company, Boston, 717 p.

Environment Agency, 2007. Underground, under threat. The state of groundwater in

England and Wales. Environment Agency, Almondsburry, Bristol, 23 p. Available at: http://www.environment-agency.gov.uk/.

Falta, R.W., Rao, P.S., and Basu, N., 2005a. Assessing the impacts of partial mass depletion in DNAPL source zones: I. Analytical modeling of source strength functions and plume response. *Journal of Contaminant Hydrology*, vol. 78, no. 4, pp. 259–280.

Falta, R.W., Basu, N., and Rao, P.S., 2005b. Assessing the impacts of partial mass depletion in DNAPL source zones: II. Coupling source strength functions to plume evolution. *Journal of Contaminant Hydrology*, vol. 79, no. 1, pp. 45–66.

Falta, R.W., Stacy, M.B., Ahsanuzzaman, A.N.M., Wang, M., and Earle, R.C., 2007. *REM-Chlor, Remediation Evaluation Model for Chlorinated Solvents; User's Manual*, Version 1.0. U.S. Environmental Protection Agency, Center for Subsurface Modeling Support, National Risk Management Research Laboratory, Ada, OK, 79 p.

Fares, A., Kindt, B., Lapuma, R., and Perram, G.P., 1995. Desorption kinetics of trichloroethylene from powdered soils. *Environmental Science & Technology*, vol. 29, no. 6, pp. 1564–1568.

Farhat, S.K., Newell, C.J., and Nichols, E.M., 2006. Mass Flux Toolkit to evaluate groundwater impacts, attenuation, and remediation alternatives. User's manual; Version 1.0. Environmental Security Technology Certification Program (ESTCP), 131 p.

Feenstra, S. and Cherry, J.A., 1988. Subsurface contamination by dense non-aqueous phase liquid (DNAPL) chemicals. In: *Proceedings of International Groundwater Symposium*, International Association of Hydrogeologists, May 1–4, 1988, Halifax, Nova Scotia, pp. 62-6-9.

Fountain, J.C., 1998. Technologies for dense nonaqueous phase liquid source zone remediation. Technology Evaluation Report TE-98-02, Ground-Water Remediation Technologies Analysis Center (GWRTAC), Pittsburgh, PA, 62 p.

Fraser, M., McLaren, R., and Barker, J., 2005. Multilevel monitoring wells to assess contaminant mass discharge: Magnitude of uncertainties based on Borden monitoring experience. The Abstract Book of the 2005 Ground Water Summit Program, National Ground Water Association, San Antonio, TX, April 17–20, 2005.

Freeze, R.A., and Cherry, J.A., 1989. What has gone wrong? *Ground Water*, vol. 27, no. 4, pp. 458–464.

Geiger, C.L., Clausen, C.A., Reinhart, D.R., Quinn, J., O'Hara, S., Krug, T., and Major, D., 2003. *Innovative Strategies for the Remediation of Chlorinated Solvents and DNAPL in the Subsurface*. ACS Books, Washington, DC.

Gilmore, T., et al., 2006. Characterization and monitoring of natural attenuation of chlorinated solvents in ground water: A systems approach. WRSC-STI-2006-00084, Rev 1, Savannah River National Laboratory, Savannah River Site, Aiken, SC, 53 p.

Groundwater Services, Inc., 2007. BIOBALANCE: A mass balance toolkit for evaluating source depletion, competition effects, long-term sustainability, and plume dynamics. Savannah River National Laboratory and U.S. Department of Energy, 55 p. + appendices.

Gudemann, H., and Hiller, D., 1988. In situ remediation of VOC contaminated soil and ground water by vapor extraction and ground water aeration. In: *Proceedings of the Third Annual Haztech International Conference*, Cleveland, OH, 1988.

Guilbeault, M.A., Parker, B.L., and Cherry, J.A., 2005. Mass and flux distributions from DNAPL zones in sandy aquifers. *Ground Water*, vol. 43, no. 1, pp. 70–86.

Harbaugh, A.W., 2005. MODFLOW-2005, the U.S. Geological Survey modular groundwater model—the ground-water flow process: U.S. Geological Survey Techniques and Methods 6-A16, variously pages.

Harbaugh, A.W., Banta, E.R., Hill, M.C., and McDonald, M.G., 2000. MODFLOW-2000, The U.S. Geological Survey modular ground-water model—User's guide to modularization concepts and the ground-water flow process: U.S. Geological Survey Open-File Report 00-92, 121 p.

Harkness, M.R., 2000. Economic considerations in enhanced anaerobic biodegradation. In: *Proceedings of the Second International Conference on Remediation of Chlorinated Recalcitrant Compounds, May 22–25, 2000*, Vol. 4, Wickramanayake, G.B., et al., editors. Battelle Press, Columbus, OH, pp. 9–14.

Hatfield, K., Annable, M.D., Kuhn, S., Rao, P.S.C., and Campbell, T., 2002. A new method for quantifying contaminant flux at hazardous waste sites. In: *Groundwater Quality: Natural and Enhanced Restoration of Groundwater Protection*, Thornton, S.F., and Oswald, S.E., editors. IAHS Publication No. 275. IAHS Press, Oxfordshire, United Kingdom, pp. 25–32.

Hendrickson, E.R., Payne, J.A., Young, R.M., Starr, M.G., Perry, M.P., Fahnestock, S., Ellis, D.E., and Ebersole, R.C., 2002. Molecular analysis of *Dehalococcoides* 16S ribosomal DNA from chloroethene-contaminated sites throughout North America and Europe. *Applied Environmental Microbiology*, vol. 68, no. 2, pp. 485–495.

Heron, G., Christensen, T.H., and Enfield, C.G., 1996. Temperature effects on the distribution of organics in soils and groundwater and implications for thermally enhanced in-situ remediation. *Environmental Science & Technology*.

Huling, S.G., and Pivetz, B.E., 2006. In-situ chemical oxidation. Engineering Issue, EPA/600/R-06/072, U.S. Environmental Protection Agency, Office of Research and Development, National Risk Management Research Laboratory, Cincinnati, OH, 58 p.

ITRC (The Interstate Technology & Regulatory Council), 1999. Natural Attenuation of Chlorinated Solvents in Groundwater: Principles and practices. Technical/Regulatory Guidelines, In Situ Bioremediation Work Team, Interstate Technology & Regulatory Council, Washington, DC, 25 p. + appendices. Available at: http://www.itrcweb.org.

ITRC, 2003. An introduction to characterizing sites contaminated with DNAPLs. Technology Overview, Dense Nonaqueous Phase Liquids Team, Washington, DC.

ITRC, 2004. Strategies for Monitoring the Performance of DNAPL Source Zone Remedies. Technical and Regulatory Guidelines, Interstate Technology & Regulatory Council, Washington, DC, 94 p. + appendices.

ITRC, 2005. Technical and Regulatory Guidance for In Situ Chemical Oxidation of Contaminated Soil and Groundwater, 2nd ed. In Situ Chemical Oxidation Team, Interstate Technology & Regulatory Council, Washington, DC. 71 p. + appendices.

Jarsjö, J., Bayer-Raich, M., and Ptak, T., 2005. Monitoring groundwater contamination and delineating source zones at industrial sites: Uncertainty analyses using integral pumping tests. *Journal of Contaminant Hydrology*, vol. 79, no. 3–4, pp. 107–134.

Jawitz, J.W., Fure, A.D., Demmy, G.G., Berglund, S., and Rao, P.S.C., 2005. Groundwater contaminant flux reduction resulting from nonaqueous phase liquid mass reduction. *Water Resources Research*, vol. 41, no. 10, W10408 p.

Jeffers, P.M., Ward, L.M., Woytowitch, L.M., and Wolfe, N.L., 1989. Homogeneous hydrolysis rate constant for selected chlorinated methanes, ethanes, ethenes, and propanes. *Environmental Science & Technology*, vol. 23, pp. 965–969.

Karanovic, M., 2006. *BIOSCREEN-AT Natural Attenuation Decision Support System*, Version 1.41. S.S. Papadopulos & Associates, Bethesda, MD.

Karanovic, M., Neville, C.J., and Andrews, C.B., 2007. BIOSCREEN-AT: BIOSCREEN with an exact analytical solution. *Ground Water*, vol. 45, no. 2, pp. 242–245.

Kavanaugh, M.C., and Rao, S.C. (Co-Chairs), 2003. The DNAPL remediation challenge: Is there a case for source depletion? EPA/600/R-03/143, National Risk Management Research Laboratory, Office of Research and Development, U.S. Environmental Protection Agency, Cincinnati, OH, 112 p.

Kiraly, L., 1998. Modelling karst aquifers by the combined discrete channel and continuum approach. *Bulletin d'Hydrogeologie*, vol. 16, pp. 77–98.

Kresic, N., 2007. *Hydrogeology and Groundwater Modeling*, 2nd ed. CRC Press, Taylor & Francis Group, Boca Raton, FL, 807 p.

Kresic, N., and Mikszewski, A., 2006. Modeling of leachate concentration in unsaturated and saturated zones. In: *Hydrogeology and Groundwater Modeling*. Taylor & Francis Group, Boca Raton, FL. Available at: http://www.crcpress.co.uk/e_products/downloads/.

Kresic, N., Mikszewski, A., Manuszak, J., and Kavanaugh, M.C., 2007. Impact of contaminant mass reduction in residuum sediments on dissolved concentrations in underlying aquifer. *Water Science & Technology: Water Supply*, vol. 7, no. 3, pp. 31–39, DOI: 10.2166/ws.2007.064.

Lawrence, S.J., 2006. Description, properties, and degradation of selected volatile organic compounds detected in ground water—A Review of Selected Literature. U.S. Geological Survey Open-File Report 2006-1338, Atlanta, GA, 62 p. A Web-only publication at: http://pubs.usgs.gov/ofr/2006/1338/.

Leeson, A., and Hinchee, R.E., 1995. *Principles and Practices of Bioventing*. Vols. I and II. Prepared by Battelle Memorial Institute, Columbus, OH, for Environics Directorate, Armstrong Laboratory, Tyndall AFB, FL; National Risk Management Research Laboratory, USEPA, Cincinnati, OH; and, US Air Force Center for Environmental Excellence, Technology Transfer Division, Brooks AFB, TX.

Lendvay, J.M., et al., 2003. Bioreactive barriers: A comparison of bioaugmentation and biostimulation for chlorinated solvent remediation. *Environmental Science & Technology*, vol. 37, no. 7, pp. 1422–1431.

Lipson, D.S., Kueper, B.H., and Gefell, M.J., 2005. Matrix diffusion-derived plume attenuation in fractured bedrock. *Ground Water*, vol. 43, no. 1, pp. 30–39.

Lowe, D.F., Oubre, C.L., and Ward, C.H., 1999. *Surfactants and Cosolvents for DNAPL Remediation – A Technology Practices Manual*. CRC Press, Boca Raton, FL.

Major, D.W., McMaster, M.L., Cox, E.E., Lee, B.J., Gentry, E.E., Hendrickson, E., Edwards, E., and Dworatzek, S., 2001. Successful field demonstration of bioaugmentation to degrade PCE and TCE to ethene. In: *Proceedings of the Sixth International In Situ and On-Site Bioremediation Symposium*, vol. 6, no. 8, San Diego, CA, pp. 27–34.

Maxwell, R.M., Pelmulder, S.D., Tompson, A.F.B., and Kastenberg, W.E., 1998. On the development of a new methodology for groundwater-driven health risk assessment. *Water Resources Research*, vol., 34, no. 4, pp. 833–847.

McDade, J.M., McGuire, T.M., and Newell, C.J., 2005. Analysis of DNAPL source-depletion costs at 36 field sites. *Remediation Journal*, vol. 15, no. 2, pp. 9–18.

McGuire, T.M., McDade, J.M., and Newell, C.J., 2006. Performance of DNAPL source depletion at 59 chlorinated solvent-impacted sites. *Groundwater Monitoring & Remediation*, vol. 26, no. 1, pp. 73–84.

McKone, T.E., 1987. Human exposure to volatile organic compounds in household tap water: The indoor inhalation pathway. *Environmental Science and Technology*, vol. 21, pp. 1194–1201.

Michalski, A., Metlitz, M.N., and Whitman, I.L., 1995. A field study of enhanced recovery of DNAPL pooled below the water table. *Ground Water Monitoring Review*, vol. Winter, pp. 90–100.

Mikszewski, A., and Kresic, N., 2006. Modeling the transport of munitions constituents to the saturated zone using estimated surface loading rates: A novel approach. *Hydrological Science and Technology*, vol. 22, no. 1–4, pp. 153–160.

Miller, R.R., and Roote, D.S., 1997. In-well vapor stripping. Technology Overview Report, TO-97-01. Ground-Water Remediation Technologies Analysis Center, Pittsburgh, PA, 17 p.

Mutch, R.D., Ash, R.E., and Caputi, J.R., 1997. Contain contaminated groundwater. *Chemical Engineering*, vol. 104, no. 5, pp. 114–119.

Newell, C.J., et al., 2002. Calculation and use of first-order rate constants for monitored natural attenuation studies. Ground Water Issue, EPA/540/S-02/500, U.S. Environmental Protection Agency, National Risk Management Research Laboratory, Cincinnati, OH, 27 p.

Newell, C.J., and Adamson, D.T., 2005. Planning-level source decay models to evaluate impact of source depletion on remediation time Frame. *Remediation*, pp. 27–47. Published online in Wiley Interscience (www.interscience.wiley.com). DOI: 10.1002/rem.20058.

Newell, C.J., and Aziz, C.E., 2004. Long-term sustainability of reductive dechlorination reactions at chlorinated solvent sites. *Biodegradation*, vol. 15, no. 6, pp. 387–394.

Nishikawa, T., Densmore, J.N., Martin, P., and Matti, J., 2003. Evaluation of the source and transport of high nitrate concentrations in ground water, Warren Subbasin, California. U.S. Geological Survey Water-Resources Investigations Report 03-4009, Sacramento, CA, 133 p.

NRC (National Research Council), 1994. *Alternatives for Ground Water Cleanup*. National Academy Press, Washington, DC, 315 p.

NRC, 2000. *Natural Attenuation for Groundwater Remediation*. National Academy Press, Washington, DC, 274 p.

OEHHA, 2006. Toxicity Criteria Database. California Office of Environmental Health Hazard Assessment. Available at: http://www.oehha.ca.gov/risk/ChemicalDB/index. asp. Accessed February, 2006.

O'Hara, S., Krug, T., Major, D., Quinn, J., Geiger, C., and Clausen, C.A., 2004. Performance evaluation of dehalogenation of DNAPLs using emulsified zero-valent iron. In: Proceedings, Fourth International Conference on Remediation of Chlorinated and Recalcitrant Compounds, Monterey, California. Battelle Press, Columbus, OH.

Oostrom, M., White, M.D., Lenhard, R.J., van Geel, P.J., and Wietsma, T.W., 2005a. A comparison of models describing residual NAPL formation in the vadose zone. *Vadose Zone Journal*, vol. 4, pp. 163–174.

Oostrom, M., Dane, J.H., and Wietsma, T.W., 2005b. Flow behavior and remediation of a multicomponent DNAPL in a heterogeneous, variable saturated porous medium. *Vadose Zone Journal*, vol. 4, pp. 1170–1182.

Oostrom, M., Hofstee, C., and Wietsma, T.W., 2006. Behavior of a viscous LNAPL under variable water table conditions. *Soil & Sediment Contamination*, vol. 15, no. 6, pp. 543–564.

Pankow, J.F., and Cherry, J.A., 1996. *Dense Chlorinated Solvents and Other DNAPLs in Groundwater*. Waterloo Press, Guelph, Ontario, Canada, 522 p.

Parker, J.C., and Park, E., 2004. Modeling field-scale dense non-aqueous phase liquid dissolution kinetics in heterogeneous aquifers. *Water Resources Research*, vol. 40, W05109 p.

Parsons (Parsons Corporation), 2004. Principles and practices of enhanced anaerobic bioremediation of chlorinated solvents. Air Force Center for Environmental Excellence (AFCEE), Brooks City-Base, Texas; Naval Facilities Engineering Service Center Port Hueneme, California; Environmental Security Technology Certification Program, Arlington, VA.

Pearlman, L., 1999. Subsurface containment and monitoring systems: Barriers and beyond (Overview Report). U.S. Environmental Protection Agency, Office of Solid Waste and Emergency Response, Technology Innovation Office, Washington, DC, 61 p.

Pierce, B., 2004. Comments & Recommendations from the Draft Ground Water Use and Vulnerability Discussion Paper. Memorandum to Guy Tomassoni, USEPA, Ken Lovelace, USEPA, September 30, 2004, Georgia Department of Natural Resources, Environmental Protection Division, Atlanta, Georgia, 4 p. Available at: http://gwtf.clu-in.org/docs/options/comments/45.pdf. Accessed December 12, 2007.

Pruess, K., Oldenburg, C., and Moridis, G., 1999. TOUGH2 User's guide, version 2.0. Report LBNL-43134, Lawrence Berkeley National Laboratory, Berkeley, CA.

Pruess, K., 2004. The TOUGH codes—a family of simulation tools for multiphase flow and transport processes in permeable Media. *Vadose Zone Journal*, vol. 3, pp. 738–746.

Rao, P.S.C., Jawitz, J.W., Enfield, C.G., Falta, R.W., Annable, M.D., and Wood, A.L., 2001. Technology integration for contaminated site remediation: Cleanup goals and performance criteria. In: *Groundwater Quality: Natural and Enhanced Restoration of Groundwater Pollution*. Publication No. 275. International Association of Hydrologic Sciences, Wallingford, United Kingdom, pp. 571–578.

Rumer, R.R., and Mitchell, J.K. (editors), 1996. Assessment of Barrier Containment Technologies; a Comprehensive Treatment for Environmental Remedial Application. Product of the International Containment Technology Workshop. National Technical Information Service, PB96-180583.

Rügner, H., and Bittens, M., 2006. Revitalization of contaminated land and groundwater at megasites; SAFIRA II Research Program 2006–2012, UFZ Center for Environmental Research Leipzig-Halle GmbH, Germany.

Sale, T., and Applegate, D., 1997. Mobile NAPL recovery: Conceptual, field, and mathematical considerations. *Ground Water*, vol. 35, no. 3, pp. 418–426.

Sale, T.C., and McWhorter, D.B., 2001. Steady state mass transfer from single-component dense non-aqueous phase liquids in uniform flow. *Water Resources Research*, vol. 37, pp. 393–404.

Shoemaker, W.B., Kuniansky, E.L., Birk, S., Bauer, S., and Swain, E.D., 2008. Documentation of a Conduit Flow Process (CFP) for MODFLOW-2005. U.S. Geological Survey Techniques and Methods 6-A24.

Suthersan, S.S., and Payne, F.C., 2005. *In Situ Remediation Engineering*. CRC Press, Boca Raton, FL, 511 p.

Travis, C.C., and Doty, C.B., 1990. Can contaminated aquifers at Superfund site be remediated? *Environmental Science & Technology*, vol. 24, no. 1, pp. 1464–1466.

Treybal, R.E., 1980. *Mass-Transfer Operations*, 3rd ed. McGraw-Hill, New York.

Umari, A.M.J., Martin, P., Schroeder, R.A., Duell, L.F.W., Jr., and Fay, R.G., 1995. Potential for ground-water contamination from movement of wastewater through the unsaturated zone, Upper Mojave River Basin, California. U.S. Geological Survey Water-Resources Investigations Report 93-4137, 83 p.

Unger, A.J.A., Forsyth, P.A., and Sudicky, E.A., 1998. Influence of alternative dissolution models and subsurface heterogeneity on DNAPL disappearance times. *Journal of Contaminant Hydrology*, vol. 30, no. 3–4, pp. 217–242.

USACE (U.S. Army Corps of Engineers), 1984. Grouting technology. Engineer Manual 110-2-3506, Washington, DC, various pages.

USACE, 1995. Engineering and design: Chemical grouting. Engineer Manual 110-2-3500, Washington, DC, various pages.

USACE, 1999. Engineering and design: Multi-phase extraction. Engineering Manual 1110-1-4010, Washington, DC, various pages.

USACE, 2006. Design: In situ thermal remediation. UFC 3-280-05. Unified Facilities Criteria (UFC). U.S. Army Corps of Engineers, Naval Facilities Engineering Command (NAVFAC), Air Force Civil Engineer Support Agency (AFCESA).

USACE, 2008. Engineering and design; In-situ air sparging. Manual No. 1110-1-4005, Washington, DC, various pages.

U.S. DOE (U.S. Department of Energy), 2002. DNAPL bioremediation-RTDF. Innovative Technology Summary Report, DOE/EM-0625. Office of Environmental Management. Dover, DE.

USEPA, 1977. The report to Congress: Waste disposal practices and their effects on ground-water. EPA 570977001, 531 p.

USEPA, 1988a. Guidance for conducting remedial investigations and feasibility studies under CERCLA; Interim Final. EPA/540/G-89/004, OSWER Directive 9355.3-01, Office of Emergency and Remedial Response, U.S. Environmental Protection Agency Washington, DC.

USEPA, 1988b. CERCLA compliance with other laws manual. Draft. OSWER Directive 9234.1-01. U.S. Environmental Protection Agency, Washington, DC.

USEPA, 1988c. Guidance on remedial actions for contaminated ground water at Superfund sites. Draft. OSWER Directive No. 9283.1-2, Office of Emergency and Remedial Response, U.S. Environmental Protection Agency Washington, DC.

USEPA, 1989a. Risk Assessment Guidance for Superfund, Vol. I; Human Health Evaluation Manual (Part A); Interim final. EPA/540/1-89/002, Office of Emergency and Remedial Response, U.S. Environmental Protection Agency, Washington, DC.

USEPA, 1989b. *Treatability Studies under CERCLA: An Overview*. Publication No. 9380.3-02FS. Office of Solid Waste and Emergency Response, United States Environmental Protection Agency, 6 p.

USEPA, 1989c. Evaluation of Ground-Water Extraction Remedies, Vol. 1, Summary Report (EPA/540/2-89/054, NTIS PB90-183583, 66 pp.); Vol. 2, Case Studies 1-19 (EPA/540/2-89/054b); and Vol. 3, General Site Data Base Reports (EPA/540/2-89/054c). Office of Solid Waste and Emergency Response, Washington, DC.

USEPA, 1991a. Risk Assessment Guidance for Superfund, Vol. I: Human Health Evaluation Manual (Part B, Development of Risk-Based Preliminary Remediation Goals); Interim. EPA/540/R-92/003, Office of Emergency and Remedial Response, U.S. Environmental Protection Agency, Washington, DC.

USEPA, 1991b. Ecological Assessment of Superfund Sites: An Overview. Publication

9345.0-05I. Office of Solid Waste and Emergency Response, U.S. Environmental Protection Agency, 8 p.

USEPA, 1991c. Risk Assessment Guidance for Superfund, Vol. I: Human Health Evaluation Manual (Part C, Risk Evaluation of Remedial Alternatives); Interim. Publication 9285.7-01C, Office of Emergency and Remedial Response, U.S. Environmental Protection Agency, Washington, DC.

USEPA, 1992. Guidance for conducting treatability studies under CERCLA; final. EPA/540/R-92/071a, Office of Solid Waste and Emergency Response, U.S. Environmental Protection Agency, Washington, DC, 74 p.

USEPA, 1993. Guidance for evaluating the technical impracticability of ground-water restoration. OSWER Directive 9234.2-25, EPA/540-R-93-080, September 1993.

USEPA, 1994. Methods for monitoring pump-and-treat performance, EPA/600/R-94/123, U.S. Environmental Protection Agency, Office of Research and Development, R.S. Kerr Environmental Research Laboratory, Ada, OK.

USEPA, 1996a. A citizen's guide to soil vapor extraction and air sparging. EPA 542-F-96-008, Office of Solid Waste and Emergency Response, U.S. Environmental Protection Agency, Washington, DC, 4 p.

USEPA, 1996b. Pump-and-treat ground-water remediation; a guide for decision makers and practitioners. EPA/625/R-95/005, Office of Research and Development, U.S. Environmental Protection Agency, Washington, DC, 74 p.

USEPA, 1996c. A Citizen's guide to bioremediation. EPA 542-F-96-007, Office of Solid Waste and Emergency Response, 4 p.

USEPA 1996d. A citizen's guide to natural attenuation. EPA 542-F-96-015, Office of Solid Waste and Emergency Response, 4 p.

USEPA, 1997. Analysis of selected enhancements for soil vapor extraction. EPA/542/R-97/007, Office of Solid Waste and Emergency Response, U.S. Environmental Protection Agency, Washington, DC, various pages.

USEPA, 1998a. Evaluation of subsurface engineered barriers at waste sites. EPA 542-R-98-005, Office of Solid Waste and Emergency Response, U.S. Environmental Protection Agency, Washington, DC, 102 p. + appendices.

USEPA, 1998b. Steam injection for soil and aquifer remediation. EPA/540/S-97/505, Office of Solid Waste and Emergency Response, U.S. Environmental Protection Agency, Washington, DC, 16 p.

USEPA, 1998c. Permeable reactive barrier technologies for contaminant remediation. EPA/600/R-98/125, Office of Solid Waste and Emergency Response, U.S. Environmental Protection Agency, Washington, DC, 94 p.

USEPA, 1999a. Use of monitored natural attenuation at Superfund, RCRA corrective action, and underground storage tank sites, Final. Office of Solid Waste and Emergency Response, Directive Number 9200.4-17P, 32 p.

USEPA, 1999b. Groundwater cleanup: Overview of operating experience at 28 Sites. EPA/542/R-99/006, Office of Solid Waste and Emergency Response, Washington, DC.

USEPA, 2000a. Presenter's manual for: Superfund risk assessment and how you can help; A 40-minute videotape. EPA/540/R-99/013, Office of Solid Waste and Emergency Response, U.S. Environmental Protection Agency, Washington, DC.

USEPA, 2000b. *Engineered approaches to in situ bioremediation of chlorinated solvents: fundamentals and field applications.* EPA 542-R-00-008, Office of Solid Waste and Emergency Response, U.S. Environmental Protection Agency, Washington, DC.

USEPA, 2002. Elements for effective management of operating pump and treat systems. EPA 542-R-02-009, Office of Solid Waste and Emergency Response, U.S. Environmental Protection Agency, Washington, DC, 18 p.

USEPA, 2003. Atrazine interim reregistration eligibility decision (IRED), Q&A's - January 2003. Available at: http://www.epa.gov/pesticides/factsheets/atrazine.htm#q1. Accessed January 23, 2008.

USEPA, 2004. In situ thermal treatment of chlorinated solvents; Fundamentals and field applications. EPA 542/R-04/010. Office of Solid Waste and Emergency Response, U.S. Environmental Protection Agency, Washington, DC, various pages.

USEPA, 2006a. Off-gas treatment for soil vapor extraction systems: State of the practice. EPA 542/R-05/028. Office of Solid Waste and Emergency Response, U.S. Environmental Protection Agency, Washington, DC, 129 p.

USEPA, 2006b. In situ treatment technologies for contaminated soil; Engineering Forum Issue Paper. EPA 542/F-06/013, U.S. Environmental Protection Agency, Office of Solid Waste and Emergency Response, Washington, DC, 35 p.

USEPA, 2007a. Treatment technologies for site cleanup: Annual status report (12th ed.). EPA-542-R-07-012, U.S. Environmental Protection Agency, Office of Solid Waste and Emergency Response, Washington, DC, various pages.

USEPA, 2007b. Optimization strategies for long-term ground water remedies (with particular emphasis on pump and treat systems). EPA 542-R-07-007, U.S. Environmental Protection Agency, Office of Solid Waste and Emergency Response, Washington, DC, 15 p. + appendices.

USEPA, 2007c. Options for discharging treated water from pump and treat systems. EPA 542-R-07-006, U.S. Environmental Protection Agency, Office of Solid Waste and Emergency Response, Washington, DC, 14 p. + appendices.

USEPA, 2008. A systematic approach for evaluation of capture zones at pump and treat systems; Final project report. EPA 600/R-08/003, U.S. Environmental Protection Agency, Office of Research and Development, Washington, DC, 38 p. + appendices.

USGS (United States Geological Survey), 1999. Deep Aquifer Remediation Tools (DARTs): A new technology for ground-water remediation. USGS Fact Sheet 156-99, U.S. Geological Survey, 2 p.

Vinegar, H., and Stegemeier, G., 1998. In situ thermal desorption. Paper presented at the In Situ Thermal Treatment Conference, USEPA TIO, Atlanta, GA. December 1998.

Waddill, D.W., and Widdowson, M.A., 2000. *SEAM3D: A Numerical Model for Three-Dimensional Solute Transport and Sequential Electron Acceptor-Based Bioremediation in Groundwater*. Engineer Research and Development Center, Vicksburg MS Environmental Laboratory, U.A. Army Corps of Engineers, Vicksburg, MS.

White, M.D., and Oostrom, M., 2000. *STOMP Subsurface Transport over Multiple Phases*, Version 2.0, Theory Guide. PNNL-12030. Pacific Northwest National Laboratory, Richland, WA.

White, M.D., and Oostrom, M., 2004. *STOMP Subsurface Transport over Multiple Phases*, Version 3.1, User's Guide. PNNL-14478. Pacific Northwest National Laboratory, Richland, WA.

White, M.D., Oostrom, M., and Lenhard, R.J., 2004. A practical model for mobile, re-sidual, and entrapped NAPL in water-wet porous media. *Ground Water*, vol. 42, pp. 734–746.

Wiedemeier, T.H., et al., 1998. Technical Protocol for Evaluating Natural Attenuation of Chlorinated Solvents in Ground Water. EPA/600/R-98/128, U.S. Environmental Protection Agency, Office of Research and Development, Washington, DC.

Wiedemeier, T.H., et al., 1999. *Technical Protocol for Implementing Intrinsic Remediation with Long-Term Monitoring for Natural Attenuation of Fuel Contamination Dissolved in Ground-water*, Vol. I (Revision 0). Air Force Center for Environmental Excellence (AFCEE), Technology Transfer Division, Brooks Air Force Base, San Antonio, TX.

Williams, P.R.D., Benton, L., and Sheehan, P.J., 2004. The risk of MTBE relative to other VOCs in public drinking water in California. *Risk Analysis*, vol., 24, no. 3, pp. 621–634.

Wilkin, R.T., and Puls, R.W., 2003. Capstone report on the application, monitoring, and performance of permeable reactive barriers for groundwater remediation: Volume 1—Performance evaluation at two sites. EPA/600/R-03/045a. U.S. Environmental Protection Agency, Washington, DC.

Wilson, R.D., and Mackay, D.M., 1997. Arrays of unpumped wells: An alternative to permeable walls for in situ treatment. In: *International Contaminant Technology Conference Proceedings*, February 9–12, 1997, St. Petersburg, FL, pp. 888–894.

Wolfe, W.J., and Haugh, C.J., 2001. Preliminary conceptual models of chlorinated-solvent accumulation in karst aquifers. In: *2001 U.S. Geological Survey Karst Interest Group Proceedings*, Kuniansky, E.L., editor. U.S. Geological Survey Water-Resources Investigations Report 01-4011, pp. 157–162.

Zhang, W., 2003. Nanoscale iron particles for environmental remediation. *Journal of Nanoparticle Research*, vol. 5, pp. 323–332.

Zhu, J., and Sykes, J.F., 2004. Simple screening models of NAPL dissolution in the subsurface. *Journal of Contaminant Hydrology*, vol. 72, pp. 245–258.

Values of $W(u)$ for Fully Penetrating Wells in a Confined, Isotropic Aquifer

u	$W(u)$	u	$W(u)$	u	$W(u)$	u	$W(u)$	u	$W(u)$	u	$W(u)$
1.0E-11	24.7512	1.0E-09	20.1460	1.0E-07	15.5409	1.0E-05	10.9357	1.0E-03	6.3315	1.0E-01	1.8229
1.5E-11	24.3458	1.5E-09	19.7406	1.5E-07	15.1354	1.5E-05	10.5303	1.5E-03	5.9266	1.5E-01	1.4645
2.0E-11	24.0581	2.0E-09	19.4529	2.0E-07	14.8477	2.0E-05	10.2426	2.0E-03	5.6394	2.0E-01	1.2227
2.5E-11	23.8349	2.5E-09	19.2298	2.5E-07	14.6246	2.5E-05	10.0194	2.5E-03	5.4167	2.5E-01	1.0443
3.0E-11	23.6526	3.0E-09	19.0474	3.0E-07	14.4423	3.0E-05	9.8317	3.0E-03	5.2349	3.0E-01	0.9057
3.5E-11	23.4985	3.5E-09	18.8933	3.5E-07	14.2881	3.5E-05	9.6830	3.5E-03	5.0813	3.5E-01	0.7942
4.0E-11	23.3649	4.0E-09	18.7598	4.0E-07	14.1546	4.0E-05	9.5495	4.0E-03	4.9482	4.0E-01	0.7024
4.5E-11	23.2471	4.5E-09	18.6420	4.5E-07	14.0368	4.5E-05	9.4317	4.5E-03	4.8310	4.5E-01	0.6253
5.0E-11	23.1418	5.0E-09	18.5366	5.0E-07	13.9314	5.0E-05	9.3263	5.0E-03	4.7261	5.0E-01	0.5598
5.5E-11	23.0465	5.5E-09	18.4413	5.5E-07	13.8361	5.5E-05	9.2310	5.5E-03	4.6313	5.5E-01	0.5034
6.0E-11	22.9595	6.0E-09	18.3543	6.0E-07	13.7491	6.0E-05	9.1440	6.0E-03	4.5448	6.0E-01	0.4544
6.5E-11	22.8794	6.5E-09	18.2742	6.5E-07	13.6691	6.5E-05	9.0640	6.5E-03	4.4652	6.5E-01	0.4115
7.0E-11	22.8053	7.0E-09	18.2001	7.0E-07	13.5950	7.0E-05	8.9899	7.0E-03	4.3916	7.0E-01	0.3738
7.5E-11	22.7363	7.5E-09	18.1311	7.5E-07	13.5260	7.5E-05	8.9209	7.5E-03	4.3231	7.5E-01	0.3403
8.0E-11	22.6718	8.0E-09	18.0666	8.0E-07	13.4614	8.0E-05	8.8563	8.0E-03	4.2591	8.0E-01	0.3106
8.5E-11	22.6112	8.5E-09	18.0060	8.5E-07	13.4008	8.5E-05	8.7957	8.5E-03	4.1990	8.5E-01	0.2840
9.0E-11	22.5540	9.0E-09	17.9488	9.0E-07	13.3437	9.0E-05	8.7386	9.0E-03	4.1423	9.0E-01	0.2602
9.5E-11	22.4999	9.5E-09	17.8948	9.5E-07	13.2896	9.5E-05	8.6845	9.5E-03	4.0887	9.5E-01	0.2387

1.0E-10	22.4486	1.0E-08	17.8435	1.0E-06	13.2383	1.0E-04	8.6332	1.0E-02	4.0379	1.0	0.2194
1.5E-10	22.0432	1.5E-08	17.4380	1.5E-06	12.8328	1.5E-04	8.2278	1.5E-02	3.6374	1.5	0.1000
2.0E-10	21.7555	2.0E-08	17.1503	2.0E-06	12.5451	2.0E-04	7.9402	2.0E-02	3.3547	2.0	0.04890
2.5E-10	21.5323	2.5E-08	16.9272	2.5E-06	12.3220	2.5E-04	7.7172	2.5E-02	3.1365	2.5	0.02491
3.0E-10	21.3500	3.0E-08	16.7449	3.0E-06	12.1397	3.0E-04	7.5348	3.0E-02	2.9591	3.0	0.01305
3.5E-10	21.1959	3.5E-08	16.5591	3.5E-06	11.9855	3.5E-04	7.3807	3.5E-02	2.8099	3.5	0.00698
4.0E-10	21.0623	4.0E-08	16.4572	4.0E-06	11.8520	4.0E-04	7.2472	4.0E-02	2.6813	4.0	0.00378
4.5E-10	20.9446	4.5E-08	16.3394	4.5E-06	11.7342	4.5E-04	7.1295	4.5E-02	2.5684	4.5	0.00207
5.0E-10	20.8392	5.0E-08	16.2340	5.0E-06	11.6289	5.0E-04	7.0242	5.0E-02	2.4679	5.0	0.00115
5.5E-10	20.7439	5.5E-08	16.1387	5.5E-06	11.5336	5.5E-04	6.9289	5.5E-02	2.3775	5.5	0.000641
6.0E-10	20.6569	6.0E-08	16.0517	6.0E-06	11.4465	6.0E-04	6.8420	6.0E-02	2.2953	6.0	0.000360
6.5E-10	20.5768	6.5E-08	15.9717	6.5E-06	11.3665	6.5E-04	6.7620	6.5E-02	2.2201	6.5	0.000203
7.0E-10	20.5027	7.0E-08	15.8976	7.0E-06	11.2924	7.0E-04	6.6879	7.0E-02	2.1508	7.0	0.000116
7.5E-10	20.4337	7.5E-08	15.8286	7.5E-06	11.2234	7.5E-04	6.6190	7.5E-02	2.0867	7.5	6.58E-05
8.0E-10	20.3692	8.0E-08	15.7640	8.0E-06	11.1589	8.0E-04	6.5545	8.0E-02	2.0269	8.0	3.77E-05
8.5E-10	20.3086	8.5E-08	15.7034	8.5E-06	11.0982	8.5E-04	6.4939	8.5E-02	1.9711	8.5	2.16E-05
9.0E-10	20.2514	9.0E-08	15.6462	9.0E-06	11.0411	9.0E-04	6.4368	9.0E-02	1.9187	9.0	1.25E-05
9.5E-10	20.1973	9.5E-08	15.5922	9.5E-06	10.9870	9.5E-04	6.3828	9.5E-02	1.8695	9.5	7.19E-06

Adapted from Ferris, J.G., Knowles, D.B., Brown, R.H., and Stallman, R.W., 1962. Theory of Aquifer Tests, *Geological Survey Water-Supply Paper 1536-E*. US Government Printing Office, Washington, DC, 174 p.

Unit Conversion Table for Length, Area, and Volume

To Convert From	To	Multiply By
feet (ft)	meters (m)	0.3048
ft	centimeters (cm)	30.48
ft	millimeters (mm)	304.8
ft	inches (in)	12
ft	yards (yd)	0.333
inches (in)	ft	0.083
in	m	0.0254
in	cm	2.54
miles (mi)	kilometers (km)	1.609
mi	m	1609
mi	ft	5280
mi	yd	1760
meters (m)	ft	3.281
m	in.	39.37
m	yd	1.094
m	mm	1000
m	cm	100
m	km	0.001
kilometers (km)	m	1000
km	mi	0.6215
km	ft	3281
square feet (ft^2)	square meters (m^2)	0.0929
ft^2	hectares (ha)	9.29×10^{-6}
ft^2	square inches (in^2)	144
acres	ft^2	43560
acres	m^2	4046.86

To Convert From	To	Multiply By
acres	ha	0.4047
square miles (mi^2)	acres	640
mi^2	ft^2	2.788×10^7
mi^2	ha	259
mi^2	square kilometers (km^2)	2.59
square meters (m^2)	ft^2	10.764
m^2	square yards (yd^2)	1.196
m^2	in^2	1550
m^2	square centimeters (cm^2)	10000
hectares (ha)	acres	2.471
ha	m^2	10000
square kilometers (km^2)	ft^2	1.076×10^7
km^2	acres	247.1
km^2	m^2	1×10^6
km^2	mi^2	0.3861
cubic feet (ft^3)	cubic meters (m^3)	0.02832
ft^3	liters (L)	28.32
ft^3	gallons (gal)	7.48
ft^3	cubic inches (in^3)	1728
acre-feet (acre-ft)	ft^3	4.354×10^4
acre-ft	m^3	1233.48
gallons (gal)	m^3	0.003785
gal	L	3.785
gal	ft^3	0.134
liters (L)	ft^3	0.035
L	gal	0.2642
L	in^3	61.02
L	cubic centimeter (cm^3)	1000
L	milliliter (mL)	1000
cubic centimeters (cm^3)	mL	1
cubic meters (m^3)	gal	264.2
m^3	ft^3	35.31
cubic kilometers (km^3)	m^3	1×10^9
km^3	cubic mile (mi^3)	0.24

Unit Conversion Table for Flow Rate

To Convert From	To	Multiply By
cubic feet per second (ft^3/s; cfs)	cubic meters per second (m^3/s)	0.0283
ft^3/s	liters per second (L/s)	28.32
ft^3/s	cubic meters per day (m^3/d)	2446.6
ft^3/s	cubic feet per day (ft^3/d)	8.64×10^4
ft^3/s	gallons per minute (gal/min)	448.8
ft^3/s	gallons per day (gal/d)	6.46×10^5
ft^3/s	acre-feet per day (acre-ft/d)	1.984
gallons per minute (gal/min)	m^3/s	6.3×10^{-5}
gal/min	m^3/d	5.451
gal/min	L/s	0.0631
gal/min	ft^3/s	0.00223
gal/min	ft^3/d	192.5
gal/min	acre-ft/d	0.00442
acre-feet per day (acre-ft/d)	m^3/s	0.0143
acre-ft/d	m^3/d	1233.5
acre-ft/d	ft^3/s	0.5042
acre-ft/d	ft^3/d	43,560
cubic meters per second (m^3/s)	ft^3/s	35.31
m^3/s	ft^3/d	3.051×10^6
m^3/s	gal/min	1.58×10^4
m^3/s	L/s	1000
m^3/s	m^3/d	8.64×10^4
m^3/s	acre-ft/d	70.05
liters per second (L/s)	ft^3/s	0.0353
L/s	ft^3/d	3051.2
L/s	acre-ft/d	0.070
L/s	gal/min	15.85
L/s	m^3/s	0.001
L/s	m^3/d	86.4

Unit Conversion Table for Hydraulic Conductivity and Transmissivity

To Convert From	To	Multiply By
feet per day (ft/d)	centimeters per second (cm/s)	3.53×10^{-4}
ft/d	meters per second (m/s)	3.53×10^{-6}
ft/d	meters per day (m/d)	0.305
centimeters per second (cm/s)	ft/d	2835
cm/s	m/d	864
meters per day (m/d)	ft/d	3.28
m/d	cm/s	0.00116
square feet per day (ft²/d)	square meters per day (m²/d)	0.0929
ft²/d	liters per meter per day (L/m d)	92.903
ft²/d	gallons per foot per day (gal/ft d)	7.4805
square meters per day (m²/d)	ft²/d	10.764

Temperature

To convert degrees of Fahrenheit (°F) to degrees of Celsius (°C)

$$°C = (°F - 32)/1.8$$

To convert degrees of Celsius (°C) to degrees of Fahrenheit (°F)

$$°F = °C \times 1.8 + 32$$

Index

Note: Page numbers referencing figures are italicized and followed by an *"f"*. Page numbers referencing tables are italicized and followed by a *"t"*.

▬ Numerics ▬

1,1-dichloroethylene, *385t*, *415t*, *417t*
1,1,1-trichloroethane, *390t*, *415t*, *417t*
1,1,2-trichloroethane, *390t*, *415t*, *417t*
1,2-dichloroethane, *385t*, *415t*, *417t*
1,2-dichloropropane, *385t*
1,2-dibromo-3-chloropropane (DBCP), *385t*, *415t*, *417t*
1,2,4-trichlorobenzene, *390t*, *415t*, *417t*
1,4-dioxane plume, 770–771
2,4-D, *384t*
2,4,5-TP (Silvex), *390t*
3H-3He (tritium-helium-3), 273
1996 Amendments to Safe Drinking Water Act, 378

▬ A ▬

abiotic degradation, 758
absorption, 409
access to water, 34
Acevedo, W., 260
acid, 503
acid treatment, 515
acid washing, 513
acrylamide, *382t*
activated alumina, 451, 468
active biobarriers, 783–784
actual recharge, 99, 236
actual renewable water resources, 10–11
acute exposure to contaminants, 343
acute toxicity, 706
Ada Ciganlija wellfield, 522–524
adsorption, 409–410, 468
advection, 401–403
advection-dispersion equation, 422–425
aeration, 439, 458, 466
aerobic conditions, 333, 419–422

Africa
  anthropogenic climate change, 321
  earth-fill dams in, 529–530
  freshwater use, 24–26
  nonrenewable aquifers in Sahara Desert, *81f*, 82
  Nubian Sandstone Aquifer System, 163–165
  sand storage dams in, *528f*
  subsurface dams in, *527f*, 529
  water disputes in, 36
  water scarcity in, 31
age, groundwater, 269, 275–278
Agency for Toxic Substances & Disease Registry (ATSDR), 344
agricultural contaminants, 369–374
agricultural drought, 307
agricultural products, 56, *57t*
agricultural water use
  adjusting prices of, 52–54
  center-pivot irrigation systems, *32f*
  in China, 27
  contamination from, 351–352
  dominance of, 17
  in Eastern Snake River Plain, 104–107
  in India, 28–29
  in Lake Chad Basin, 25
  overview, 13–14
  and recharge, 261–262
  subsidized, 40–41
  in U.S., 19–21
  and water scarcity, 31
Agua Fria Recharge Project, 673
Ahmed, S., 268
Air Force Plant 4, Fort Worth, Texas, 761
air jetting, 502
air sparging, 741–745

air stripping, 466
air temperature, 257
airlift pumping, 501–502
Alabama, 37
alachlor, *382t, 415t, 417t*
Alaska, *7f*
Albian-Neocomian aquifer system, 553–554
Alicante Declaration, 63–66
alkalinity, *422t*
Allen, R. G., 242
alpha particles, *382t*
alpha radiation, 358–359
alumina, activated, 451
aluminum, 333–334
AlWasia Treatment Plant, 464
Amarasinghe, U., 32–33
ambient monitoring, 581–583
Amendments to Safe Drinking Water Act, 378
American rule, 562
American Society for Testing and Materials
    (ASTM), 414, 649
American Water Works Association
    (AWWA), 48–49
ammonium, 359–360
anaerobic bioremediation, 780–781
anaerobic conditions, 333, 419–422
anaerobic dechlorination, 781, 785
analytes, *415t–417t, 422t*
analytical contaminant detection
    methods, *593t*
analytical data, 609, *610t*
analytical models, 635–636, 730
Andrews, C. B., 425
angle of incidence, 127, *128f*
angle of refraction, 127, *128f*
animal wastes, 361, 373–374
anion exchange, 465. *See also* ion exchange
anisotropy, 126–132, *137f*, 183–184, *743t*
Ann Arbor, Michigan, 770–771
ANSI/AWWAA100 standard, 488
antecedent moisture condition, 237
anthracite filters, 458
anthropogenic climate change
    impacts on surface water and groundwater
        Africa, 321
        Asia, 321–322
        Australia and New Zealand, 322
        Europe, 322
        Latin America, 322
        North America, 323
        overview, 317–321
        polar regions, 323
        small islands, 323
    overview, 313–317
anthropogenic contaminants, 340
anthropogenic radionuclides, 358–359
antibiotic resistance, 374
antimony, *382t*, 474
apertures, screen, 495
aphelion, 298–299
applicable or relevant and appropriate
    requirements (ARARs), 702, 708
AQUASTAT database, 10
aqueducts, 194–195
aqueous solubility, 398, *399t*
aquicludes, 77–78
aquifer mining, 95
aquifer properties datasets, *601t*
aquifer storage and recovery (ASR), 665,
    673–675
aquifers. *See also* artificial aquifer recharge;
        Blue Ridge Province case study;
        confined aquifers; karst aquifers; sand
        and gravel aquifers; unconfined
        aquifers
    basaltic and other volcanic rock, 174–175
    changes in due to extensive withdrawals,
        96–97
    coastal, 650
    comparative features of, 577
    compressibility in, 90
    defined, 75–77
    fractured rock, 175–176, 728
    heterogeneous, 497, *712f*, 750
    homogeneous, 496–497
    international, 39, 571–575
    monitoring, 546
    overview, 149–150
    pumping from, *550f*
    recharging, 521–522
    sandstone, 160–167
    storage properties of, 90–92, *93f–94f*
    transboundary, 559–560
    types of, 78–79, 81–82
    withdrawals from in United States, 184–186
aquitards, 77, 186–194

AR (autoregressive) models, 659

Aral Sea, 6, 8–9

ARARs (applicable or relevant and appropriate requirements), 702, 708

area, unit conversion tables for, 811–812

areal recharge, 100

arid land, 3, *4f*

Arizona, 43, *156f*

ARMA (autoregressive moving average) models, 659

arsenic

  health effects of, 344

  overview, 355–357, *382t*

  removal from water

    activated alumina, 468

    additional technologies for, 472

    coagulation/membrane filtration, 470–471

    developing countries, 473

    electrodialysis reversal, 469

    granular ferric hydroxide, 471

    greensand filtration, 472

    ion exchange, 467–468

    modified coagulation/filtration, 469

    modified lime softening, 469

    nanofiltration, 471

    oxidation/filtration, 469–470

    POU and POE treatment devices, 472–473

    RO membrane technologies, 468–469

  as secondary constituent, 329

  sources of, 372–373

artesian aquifers, 79

artesian springs, 198–199

artesian wells, *78f*, 79, *153f–154f*, 165, 167

artificial aquifer recharge

  aquifer storage and recovery, 673–675

  methods of

    Central Arizona Project, 672–673

    dams, 670–671

    overview, 667–668

    spreading structures and trenches, 668–670

    wells, 671–672

  overview, 662–667

  source water quality and treatment, 675–680

artificial contaminants, 340

asbestos, *382t*, 474

ascending springs, 198–199

Asia, 321–322

asiatic clams, *454t*

Asmussen, M. P., 281

ASR (aquifer storage and recovery), 665, 673–675

"Assessment framework for ground-water model applications" USEPA directive, 649

assessment monitoring, 569

Associated Press, 37

ASTM (American Society for Testing and Materials), 414, 649

at risk supplies, 437

atrazine, *382t*, *415t*, *417t*, 696

ATSDR (Agency for Toxic Substances & Disease Registry), 344

attributes, 601, 603–604

Aurelius, L. A., 373

Austin, Texas, 612–613

Australia

  anthropogenic climate change, 322

  drought policy in, 305–306

  Great Artesian Basin, 165, 167

  IWCM principles, 578

  water disputes in, 38

  water prices in, 47, *48f*

Australian Bureau of Statistics, 48

Austria, *196f*

autocorrelation analysis, 206–210

autocovariance, 207

automated calibration, 646–647

autoregressive (AR) models, 659

autoregressive moving average (ARMA) models, 659

availability

  of freshwater, 10–11

  in Water Poverty Index, 33–34

  of world water resources, *15t*

average annual growth, *16t*

average contour maps, 132, 134

average spring discharges, 201–202

avoided costs, 52

Avra Valley Recharge Project, 673

awareness, groundwater management, 547, 558

AWWA (American Water Works Association), 48–49
axial tilt and precession, 297–299
azaarenes, 364

━━━ **B** ━━━

Bachman Road site, Michigan, 785–787
backwash process, 443, 501
bacteria, 375–377, 627
bag and cartridge filters, 446
Baisan, C. A., 313
Baker, J. R., 414, *415t–417t, 418f*
Balmorhea, Texas, *613f*
Bangladesh, 355, 473
bank filtration, 150–151, 570–571
barium, *382t*, 474
Barrett, M. H., 361
barrier springs, 198–199
barrier technologies, 734–735
barriers, hydraulic, 665
basal streams, 315
basaltic aquifers, 174–175
baseflow separation, 278–282
baseline groundwater quality, 328
baseline risk assessment, 701–702, *703f*
basic water requirement (BWR), 17
basin-fill aquifers, 150–154
basin-fill basins, 141–143
basins, spring, 530–531
basin-scale management models, 634
BATs (best available technologies), *470t*
BBC (British Broadcasting Corporation), 38
Becquerel (Bq) units, 358
bedrock wells, 491, *494f*
Beijing, China, 46
Belgrade waterworks, 522–524
benzene, *382t, 416t–417t*
benzene, toluene, ethylbenzene, and xylenes (BTEX), 349, 412, 754–756
benzo(a)pyrene, *383t*
beryllium, *383t*, 474
best available technologies (BATs), *470t*
beta particles, *383t*
beta radiation, 358–359
beverage industry, 38
Big Chino aquifer, 43
Big Lost River, 105

Big Spring, MS, 212–213
bioaugmentation, 420, 785
biobarriers, 777–778, 783–784
biodegradation
    contaminants, 419–422
    equations for, 788–790
    functions of, 779–781
    NAPL source-zone treatment with, 756
    of organic substrates, 784
biological activated carbon filters, 472
biological clogging, 514
biological degradation pathways, 759
biological growth control, *454t*
biological treatment, 451–452, 459
bioremediation, 756–757, 778–787
BIOSCREEN AT analytical model, 425, 711
biostimulation, 778, 785
biotic degradation, 758
bioventing, 740
Biscayne aquifer, 168
blanket sand and gravel aquifers, 150, 154–156
blasting, well, 503
Blue Ridge Province case study
    high-yield well sitting, 182–183
    investigation of anisotropy, 183–184
    monitoring program, 178–182
    overview, 177–178
    wells database, 178
Boethling, R. S., 413–414
Bölke, J. K., 361
Bonninn, J., 361
Borchardt, M. A., 188
Borden Landfill test, 405
borings, pilot, 488
bottled water, 38
Botucatu aquifer, 163
boundary conditions, 82, 137–149
Boussinesq equation, 204, 206
Bouzelboudjen, M., 208
Box-and-Whisker plots, 334, *336f*
Bq (Becquerel) units, 358
brackish groundwater
    desalination, 463–464
    inland brackish water, 220–223
    overview, 213–218
    saltwater intrusion, 218–220
Bradbury, K. R., 188, 192–193

brines, 220, 351

Briscoe, John, 30

British Broadcasting Corporation (BBC), 38

British Geologic Survey and Environment Agency, 328, 335–338

bromate, *383t*

Brooks and Corey equation, 251–253

Brown, Lester, 31

Brown, P., 242

Brundtland Report, 61–62

Brunswick aquifers, 75, *76f*

BTEX (benzene, toluene, ethylbenzene, and xylenes), 349, 412, 754–756

budgets, water. *See* water budget

Bunker-Hill Basin, 145, *146f*

buried infrastructure datasets, *601t*

Burkina Faso, 529–530

BWR (basic water requirement), 17

━━ C ━━

cadmium, *383t*, 474

CAFOs (concentrated animal feeding operations), 352, 373–374

caissons, 518

calcite saturation index ($SI_{calcite}$), 212–213

calibrated recharge rates, 236

calibration, numeric groundwater model, 642–648

California

  basin-fill aquifers, *155f*

  Bunker Hill dike, *632f*

  cancer risk slope factors, 707

  Department of Water Resources, 662–664

  faults, 145, *146f*

  Fresno Water Division, 589

  groundwater management, *562f*

  groundwater storage projects, 664

  Health Effects Study, 678

  numeric model of Antelope Valley, 640–641

  Orange County Water District, 575–576

  Pierce Service Station site, 754–756

  recharge areas in, 629–630

  Salinas Valley project, 665

  San Bernardino Basin, 655

  San Joaquin Valley land subsidence, 97, *98f*

  urbanization and runoff, 261, *262f–263f*

  water supply legislation, 549

  Yucca Valley, 698–700

Cambrian-Ordovician aquifer system, 160–162

Camden County, Georgia, 75, *76f–77f*

Canada, 47–48, 627, *628f*

CAP (Central Arizona Project), 672–673

Cape Cod, Massachusetts, 405–406, 642

capital investments, 44, *54t*

Cappaert v. United States, 1976, 560–561

caps, 734

capture, rule of, 562

capture zones, 127, *128f*, 766, *767f*

carbofuran, *383t*

carbon adsorption, 450

carbon dioxide (CO_2), *296f*, 314–315

carbon filters, 472

carbon tetrachloride, *383t*

carbon-14, 277–282

carbonate aquifers. *See* karst aquifers

carbonate hardness, 459

carcinogens, 343, 346, 706–707

case studies. *See* Blue Ridge Province case study

casings, well, 489–491

cation exchangers, 465

caustic soda, 460

Cavazos, T., 312–313

cave systems, 299–300

CCL (contaminant candidate list), USEPA, 378–379, 381, *392t–393t*

CDI (chronic daily intake), 705

cement-bentonite slurry walls, 737

Center for Biological Diversity, 43

center-pivot irrigation systems, *32f*, *630f–631f*

Central Arizona Project (CAP), 672–673

Central Basin karst region, Tennessee, 587–588

Central Corridor region, Nevada, 124

Central Ground Water Authority, India, 29

ceramic filters, 446–447, 472

CERCLA (Comprehensive Environmental Response, Compensation, and Liability Act), 566–567

CFCs (chlorofluorocarbons), 275–277

CFP (Conduit Flow Process), 772

CFR (Code of Federal Regulations), 563

Chapagain, A. K., 56, 59

Charleston, South Carolina, *675f*

CHEMFATE database, 414, *415t–417t*, *418f*

chemical analyses of groundwater, 334–338

chemical composition of aquitard water, 193–194
chemical corrosion, 512–513
chemical grouts, 738
chemical oxidants, 439–441, 458
chemical pollution, 594
chemical precipitation, 441–442, 460
chemicals. *See also* nonaqueous phase liquids (NAPLs)
    contaminants, 344–345, *392t–393t*
    for iron bacteria removal, 515
    prices of, 477–478
    in well development, 500
    in well rehabilitation, 513–514
Chemicals cause changes in fish and raise concerns for humans article, 377–378
chemicals of potential concern (COPCs), 703
Chen, J.-S., 282
Cherry, J. A., 188, 193–194, 361
Chery, L., 361
Chicago, Illinois, 540–541
China
    freshwater use, 26–30
    karst in, 172–173, *174t*
    water prices in, 46
chloramines, *383t*, *453t*, 454–455
chlordane, *383t*, *415t*, *417t*
chloride, 269–271, *422t*, *653f*
chlorinated aliphatic hydrocarbons, *781t*
chlorinated compounds, 419–420, *421f*, *422t*
chlorinated DNAPLs, 749
chlorinated solvents, 758, 780
chlorinated volatile organic compounds (CVOCs), 720
chlorine, *384t*, 439, *440f*, 452–454, 515
chlorine dioxide (ClO₂), *384t*, 441, *453t*, 455
chlorite, *384t*
chlorobenzene, *384t*, *415t*, *417t*
chlorofluorocarbons (CFCs), 275–277
chromium, *384t*, 474
chronic daily intake (CDI), 705
chronic exposure to contaminants, 343
chronic toxicity, 706
Ci (Curie) units, 358
circular radial flow clarifiers, *442f*
cis-1,2-dichloroethylene, *385t*, *416t–417t*
cities, 14, 17
clarification, 441–443, *475t*

clarifiers, circular radial flow, *442f*
Clark, I. D., 272–273
Class V injection wells, 672
classification of springs, 197
clay filters, 446–447
Clean Water Act, 563–564
Clean Water Action Plan, 562–563
cleanup projects, 563, 700–701
clear-cutting forests, 260
Clement, T. P., 425
climate. *See also* anthropogenic climate change; cycles, climatic
    change, 5–10, 295–296
    recharge process and, 253–257
    variability, 633
Clinton, T., 677
ClO₂ (chlorine dioxide), *384t*, 441, *453t*, 455
clogging, 513–514, 669–670
closed hydrogeologic structures, *80f*, 81
CN (curve number) method, 237–241, 260–261
CO₂ (carbon dioxide), *296f*, 314–315
COAG (Council of Australian Governments), 38
coagulant aids, 478
coagulation, 441–442, 458, 465, 470–471
coal tar, 367
coastal aquifers, 650
coastal areas
    inland brackish water, 220–223
    overview, 213–218
    saltwater intrusion, 218–220
coastal estuaries, 485
coastal land subsidence, 485
coastal lowlands aquifer system, 157
Coca-Cola Company, 38
COCs (contaminants of concern), 591, *592–593t*, 703–705
Code of Federal Regulations (CFR), 563
coefficients of cross-correlation, 208–209
coefficients of discharge, 204
coefficients of storage, 90–92
Cohen, Robert M. *See* Blue Ridge Province case study
Colborn, T., 345
coliform bacteria, 377, *389t*
coliphage viruses, 376–377
collection systems, *579t*

collector wells
Belgrade waterworks, 522–524
design and construction, 517–521
maintenance, 521
other applications, 522
overview, 515–517
riverbank filtration, 521–522
colloidal particles, 330
colloids, manganese, 458
color removal, *454t*
Colorado River, 295, 301–304
Colorado River Compact, 300
Colorado State University, 242
Columbia Plateau, 255–257
Columbia River, 147
Comal Springs, Texas, 201–203
cometabolism, 419
commercial water use, 12
common-law rule, 562
community water systems (CWSs), 564
compaction, aquifer system, 97
competent aquitards, 77
compliance monitoring, 569, 583–586
Comprehensive Environmental Response,
Compensation, and Liability Act
(CERCLA), 566–567
Comprehensive State Ground Water
Protection Program (CSGWPP), 566
compressed air pumping, 501
compressibility in aquifers, 90
Comrie, A. C., 312–313
concentrated animal feeding operations
(CAFOs), 352, 373–374
concentrated recharge, 100
conceptual definitions of drought, 305–306
conceptual site model (CSM), 639–640, 709,
711f
concrete slurry walls, 737
condensate, *761f*
Conduit Flow Process (CFP), 772
confined aquifers
defined, 79
flow in, 114–115
screening for, 497
storage properties of, 90–92, *93f–94f*
unrenewability of, 557
values of W(u) for fully penetrating wells
in, 807–809

confining beds, 77–78
Conger, R. W., 281
conjunctive-use optimization model, 654
consolidated sedimentary rock porosity,
85–87
constant springs, 197–198
constituents of groundwater, 328–329
constrained optimization model, 653, 655
consumed water, 11–13, 22
contact springs of descending type, 198, *199f*
containment
hydraulic, 766
physical
deep soil mixing, 737
grouted barriers, 738
overview, 732–735
sheet-pile walls, 738–739
slurry walls, 735–737
contaminant candidate list (CCL), USEPA,
378–379, 381, *392t–393t*
contaminants. *See also* restoration,
groundwater; synthetic organic
chemicals (SOCs)
advection, 401–403
agricultural, 369–374
biodegradation, 419–422
in delineation of wellhead protection
areas, 623
dispersion and diffusion, 403–409
dissolution, 394–398
emerging, 377–380
fate and transport, 82, 422–425
and heterogeneity, 126–127
IAS applicability on, *743t*
ISCO treatment, 752
leachability studies, 721–722
and leaky aquitards, 186–188
mass flux, 793–794
microbiological, 374–377
migration through soil, 327
naturally occurring, 355–359
nitrogen and nitrate, 359–362
recharge rates and, 236
in recharge systems, 677–680
saltwater intrusion, 218–220
sorption and retardation
organic solutes, 412–418
overview, 409–412

contaminants (*Cont.*)
  tailing of, 765
  volatilization, 398–401
contaminants of concern (COCs), 591,
    592–593t, 703–705
contamination. *See also* restoration,
    groundwater
  aquifer restoration, 541–542
  compliance monitoring for, 583
  cost of, 52
  detecting, 589–594
  fecal, 437, 568–569
  hazard maps, 615–616
  health effects, 342–346
  of karst systems, 626–627
  lack of monitoring, 578
  nitrate, 461
  overview, 340–342
  sources of
    agricultural practices, 351–352
    fuel storage practices, 348–349
    industrial practices, 352–353
    mining operations, 353
    overview, 347–348
    waste disposal practices, 349–351
    wells as contamination conduits,
      353–355
  of surface waters, 485–486
continental-scale sandstone aquifers, 163
continuous time series models, 657–659
continuous-slot screens, 495, *496f*, 519–520
contour maps, 607, *608f*
contributing areas, 83, 624
Convention on the Protection and Use of
    Transboundary Watercourses and
    International Lakes, 574
COPCs (chemicals of potential concern),
    703
copper, *384t*, 474
correlograms, 207–208
corrosion, 456, 512–513
cosolvent flushing, 748–750
costs
  of contamination, 52
  of deep well drilling, 489
  distribution of water systems, *580t*
  of irrigation, 52–54
  of ISCO, 754

of PRBs, 778
  remediation, 719
  of wells, 509–510
Council of Australian Governments (COAG),
    38
countries, water footprints of, 56, 59
covariance, 207
covered slow sand filters, *445f*
Coyote Spring Valley aquifer, 124
creosote, 367
Croatia, 209–210
crop evaporation, reference, 241–242
crop type and evapotranspiration, 242
cross correlation analysis, 206–210
cross-sectional area of flow, 108, 114
cross-stratification, *130f*
Cryptosporidium, 376, *384t*, *453t*, 570
CSGWPP (Comprehensive State Ground
    Water Protection Program), 566
CSM (conceptual site model), 639–640, 709,
    711f
Cub Run Watershed, Virginia, 260–261
cumulative frequency diagrams, 335–338
cumulative sediment load, *263f*
Curie (Ci) units, 358
Curriero, F. C., 374
curve number (CN) method, 237–241,
    260–261
CVOCs (chlorinated volatile organic
    compounds), 720
CWMI Map Viewer, *574f*
CWSs (community water systems), 564
cyanide, *384t*, 474
cycles, climatic
  droughts
    drought indices, 309
    occurrence of droughts, 309–313
    overview, 304–309
  overview, 296–304
cycles in time series, 658

━━━ **D** ━━━

daily water balance model, INFILv3, 285
dalapon, *385t*
DAMA (Data Management Association),
    596
dams, 483–484, 526, 670–671
Darcy-Buckingham equation, 249

Darcy's law
    hydraulic conductivity and permeability, 111–114
    hydraulic head and hydraulic gradient, 109–111
    infiltration and water movement through vadose zone, 247–250
    overview, 108–109
Darcy's velocity, 125
DART (deep aquifer remediation tool), 775–776
data collection forms, standardized, 599
data management
    data quality, 609–610
    geographic information systems and, 607–609
    metadata, 610
    overview, 594–600
    tools, 604–607
    types of data, 600–604
Data Management Association (DAMA), 596
datasets, groundwater management, *601t*
dating groundwater, 275–278
Dayton, Ohio, WHPP, 51–52
DBCP (1,2-dibromo-3-chloropropane), *385t, 415t, 417t*
DBPs (disinfection by-products), 362–363, 439, 452, 455
Death Valley, 243–245, 285–286
decay series, radionuclide, 358–359
dechlorination, reductive, 420, *421f, 422t*
decision support system (DSS), 634–635
deep aquifer remediation tool (DART), 775–776
deep percolation, 236
Deep Percolation Model (DPM), 285–287
deep soil mixing, 737
deep well injection, 351
deep wells, 489, *494f*
degradation rates, 646, 789
Delin, G. N., 268
delineation of potential aquifer zones, *727f*
Delmarva Peninsula, 276–277
demand management, 543, 548–549
demand, supply and, 45
denitrification, 359
dense nonaqueous phase liquids. *See* DNAPLs

density flow simulation, variable, *652f*
Denver, J. M., 361
Department of Agriculture, U.S. (USDA), 237–241, 414
Department of Water Resources (DWR), 662–664
depression springs, 198, *199f*
desalination, 221, 223, 463–464, 522
descending springs, 198, *199f*
desert wash, recharge through, 282–285
design, well, 488, 491
desorption, 409–410
detached plumes, *763f*
detection of potential contaminants, *593t*
deterministic approach, 790
deterministic connectivity, 123
deterministic models, 635
Dettinger, M. D., 124
deuterium, 273–275
development of groundwater. *See also* wells
    overview, 483–487
    springs, 530–535
    subsurface dams, 524–530
development stages, groundwater resource, 545
deviation of residuals, 648
Dewandel, B., 268
diagnostic profiles, 544
diet, changes in, 56
diffuse recharge, 100
diffusion
    and dispersion of contaminants, 403–409
    molecular, 194
Diocletian's Palace, 194–195
dioxin, *386t*
dipole moment for water, 329–330
diquat, *386t*
direct injection wells, *668f*, 783
direct-push drilling, *752f*
discharge
    classification of springs based on, 197
    in Eastern Snake River Plain aquifer, *106f*, 107, *110f*
    spring hydrograph analysis, 201–206
    submarine, 214–218
    urbanization and runoff, 261, *262f*
discharge areas, 82–83
discrete time series models, 657–659

disease, waterborne, 569

disinfectants, 515

disinfection
chloramines, 454–455
chlorine, 452–454
chlorine dioxide, 455
cost of, *476t*
by oxidants, 439–441
ozone, 455
of springs, 532
ultraviolet light, 455–456

disinfection by-products (DBPs), 362–363,
439, 452, 455

dispersion of contaminants, 403–409

dispersivity, 125–126, 425

disputes, freshwater, 36–39

dissolution, 329–330, 727–728

dissolved air flotation, 441

dissolved oxygen, *422t*

dissolved phase remediation
bioremediation, 778–787
monitored natural attenuation, 787–792
overview, 762–765
permeable reactive barriers, 773–778
pump and treat
1,4-dioxane plume, 770–771
fractured rock and karst aquifers,
772–773
optimization, 769–770
overview, 765–769

distillation, 452, 462, 465, 473

distributed-parameter areal recharge
models, 285–287

distribution of water systems costs, *580t*

distribution systems, *579t*

DNAPLs (dense nonaqueous phase liquids)
air sparging, 742
in bedrock, 731–732
emulsified system for, 777
ERH treatment of, 761
indications of, 724–727
mobile zones of, 748
overview, 364–369
performance data of contamination by, *757f*
SEAR technology for, 749

documentation, numeric groundwater
models, 649–650

Doesken, N. J., 309

Domenico solution, 423–425

domestic water use, 12, 14, 21, 27, *29t*

double water table fluctuation method, 268

DPE (dual-phase extraction) technology, 747

DPM (Deep Percolation Model), 285–287

drainage areas, 83

DRASTIC index method, 617

drawdown, 117–118, 504–509

drawdown-pumping rate ratio, *506f*

drilling, direct push, *752f*

drilling technology, 488–489, *492t*

drinking water. *See also* treatment
chlorine in treatment of, *454t*
disinfection, 362
standards, 380–381
TDS levels in, 463
treatment costs, 474–478
treatment plants, *438f*
USEPA standards for, 564–565

droughts
Colorado River Basin, 301–304
indices, 309
long-term, 553
occurrence of, 309–313
overview, 304–309
underground water storage, 664

DSS (decision support system), 634–635

dual-phase extraction (DPE) technology, 747

dual-porosity media, 86, *124f*, 407

dual-pump systems, 746

Dublin Principles, 39–40

Dugan, J. T., 254

Dumanoski, D., 345

dune filtration systems, 666

Dupuits radius of well influence, 507–508

Dust Bowl drought, 304, *305f*

DWR (Department of Water Resources),
662–664

dye tracing tests, 126

dynamic storage, 93, 95–96

━━━ **E** ━━━

E. Coli, *453t*

early-warning monitoring wells, 583–584

earth-fill dams, 529–530

Eaton, T. T., 188

ebb-and-flow springs, 198

eccentricity of earth's orbit, 297–298

Eckstein, Y., 406–407
economic instruments, 630–631
economic water scarcity, 33
economics, freshwater
    overview, 39–44
    price and value of, 44–55
    virtual water and global water trade,
        55–61
economies of scale, 50
ED (electrodialysis), 449
EDR (electrodialysis reversal), 450, 462, 465,
    469
education
    component of WHPPs, 51
    groundwater management, 547, 577
    groundwater protection, 613
    public, 620
Edwards aquifer, 670–671
Edwards, D. C., 301–304
EEB (European Environmental Bureau), 540
effective infiltration, 99, 236
effective porosity, 86–90, 125, 401–402
effective rainfall, 99
Eisenlohr, L., 208
El Niño–Southern Oscillation (ENSO),
    300–302
El Paso Water Utility (EPWU), 220–223,
    463–464, *549f*
electrical resistive heating (ERH), 760–761
electrical resistivity (ER), 182–183, *675f*
electrical subsidies, 46
electrochemical corrosion, 513
electrodialysis (ED), 449
electrodialysis reversal (EDR), 450, 462, 465,
    469
electrolytes, 330
electron donors and acceptors, 419–422,
    778–779
emerging contaminants, 377–380
empirical approach, 790–791
empirical models, 635
emulsified system, 777
endocrine disruptors, 344–345
endothall, *386t*
endrin, *386t, 415t, 417t*
energy expenses for irrigation water, *54t*
energy water use, *29t*
engineered bioremediation, 778, 785

engineering methods, artificial aquifer
    recharge, 667–668
English rule, 562
enhanced anaerobic bioremediation, 781–785,
    786t
enhanced bioremediation, 756–757
ENSO (El Niño–Southern Oscillation),
    300–302
entrance velocity, screen, 495
entry pressure, DNAPL TCE, *726f*
Environment Canada, 47–48
environmental costs, *44f*, 53
environmental factors in Water Poverty
    Index, 34
environmental impact, *579t*
Environmental Protection Agency. *See* U.S.
    Environmental Protection Agency
environmental reforms, *579t*
environmental risk assessment, 702–703
environmental tracers
    carbon-14, 277–282
    CFCs and sulfur hexafluoride, 275–277
    chloride, 269–271
    overview, 268–269
    oxygen and deuterium, 273–275
    tritium, 272–273
    tritium-helium-3, 273
epichlorohydrin, *386t*
episodic variation, 658
EPM (equivalent porous medium), 123
EPWU (El Paso Water Utility), 220–223,
    463–464, *549f*
equipotential boundaries, 140–141
equipotential lines, 134–135
equivalent porous medium (EPM), 123
ER (electrical resistivity), 182–183, *675f*
ergs, 3
ERH (electrical resistive heating), 760–761
erosion, 485
errors, numeric groundwater model, 642–648
estavelles, 198
estimates, exposure, 704
estimating recharge process
    baseflow separation, 278–282
    environmental tracers, 268–278
    lysimeters, 264–266
    numeric modeling, 282–287
    overview, 262–264

estimating recharge process (*Cont.*)
soil moisture measurements, 266
water table fluctuations, 266–268
estuaries, coastal, 485
ET (evapotranspiration), 236, 241–245
ethylbenzene, 349, *386t*, 412, *416t–417t*, 754–756
ethylene dibromide, *386t*
Europe, 22–24, 322
European Environmental Bureau (EEB), 540
European Water Framework Directive, 610–611
evaluation, quality, 546
evaporation, 24
evaporative loss, 486
evapotranspiration (ET), 236, 241–245
expanding plumes, *763f*
experimental substrates, *782t*
export, water, 59, *60t*
exposure assessment, 704–705
extent, system, 83
external societal costs, *44f*
external system boundaries, 138–139
external water footprints, 56, 59
extinction of species, 317
extraction from basin-fill aquifers, 154
extraction rights, 559
extraction wells, 759

▬▬ **F** ▬▬

F&T (fate and transport), 82, 646, *763f*
fabric, geologic, 129
failed capture, 767
Falkenmark Index Measure, 35
Falteisek, J. D., 268
FAO (Food and Agriculture Organization), 10–11
Farm and Ranch Irrigation Survey, 53
Farvolden, R. N., 193–194
fate and transport (F&T), 82, 646, *763f*
faucet treatments, 438–439
faults, 144–145, *146f*
Faust, Charles R. *See* Blue Ridge Province case study
feasibility studies (FS), 567, 702, 708–718, *711f*
fecal contamination, 437, 568–569
federal role in water disputes, 36–37
feed systems, *440f*, *441f*

FEFLOW numerical model, *676f*
Feinstein, D. T., 192–193
Fenske, B. A., 51–52
Fenton's reagent, 754–756
ferric coagulants, 470
fertilizers, 352, 369–370, 696
Fick's law, 404–405, 408
field data, Theis equation, 119–120
filters
anthracite/sand dual, 458
biological activated carbon, 472
material for, 493, 497–498
oxidation, 465
pressure manganese greensand, 458
rapid sand, *444f*
slow sand, 472
filtration
arsenic, 469–470
bag and cartridge filters, 446
bank, 150–151
ceramic filters, 446–447
cost of media, *475t*
iron and manganese removal, 458–459
natural, 521–522
precoat filters, 446
pressure filters, 445–446
rapid sand filters, 443–444
slow sand filters, 444–445
Filtron systems, 446–447
Finding and estimating chemical property data for environmental assessment article, 413–414
finite element models, 637
finite-difference models, 637
first-order degradation, 788–789
fish, 485
fissure springs, 196
fixed-head, 140–141
flat file type databases, 605
Flint, A. L., 285–286
Flint, L. E., 285–286
flocculation, 441, 458
flood plains, 150–151, *152f*
floodwater retention dams, 670
Florida
ambient monitoring, 581–583
Biscayne aquifer, 168
Kissengen Spring, 551–553

monitoring of springs, 583
monitoring wells, *602f*
springs in, 196
water disputes in, 37
Florida State University Department of
    Anthropology, 170
Floridan aquifer, 75, *76f–77f, 133f,* 170, 214,
    353–355
flow
    anisotropy and heterogeneity, 126–132
    Darcy's law
        hydraulic conductivity and permeability,
        111–114
        hydraulic head and hydraulic gradient,
        109–111
        overview, 108–109
    defined, 82
    effect of boundary conditions on, 141–143
    for fractured rock and karst aquifers, *139f*
    in karst aquifers, 168–169
    measuring in aquitards, 187–194
    overview, 107–132
    types and calculations of
        fractured rock and karst aquifers,
        123–125
        overview, 114–118
        Theis equation, 118–123
    unit conversion tables for rate of, 813
    in vadose zone, 247–252
    variably saturated flow models, 282–285
    velocity, 125–126
flow clarifiers, *442f*
flow diagrams, 599–600
flow lines, 134–135
flow meters, 512
flow modeling, 621–622, 650
flow nets, 134–137
flowing wells, *78f,* 79, *153f–154f,* 165, 167
fluid removal technologies
    air sparging, 741–745
    free-phase and multiphase extraction,
    745–748
    soil vapor extraction and bioventing,
    739–740
    surfactant and cosolvent flushing, 748–750
fluoride, 339, *386t,* 474
flushed zones, *674f*
flushing, 513, 521

flux
    contaminant mass, 793–794
    of contaminants, 408
    equations, 790–791
    water, 138–141
Food and Agriculture Organization (FAO),
    10–11
food consumption patterns, 56
food trade, 59–61
footprints, water, 55–56
forests, clear-cut, 260
formation grains size distribution curves, 499
formation loss, 504, *506f*
fossil aquifers, 81–82
Foster, S. S. D., 369
fracture springs, 196
fractured aquitards, 188–192
fractured rock
    anisotropy and heterogeneity, *131f*
    contour maps of, 137
    flow patterns, *139f*
    groundwater flow, 123–125
    groundwater velocity in, 126
    pump and treat, 772–773
fractured rock aquifers, 175–176, 728. *See also*
    Blue Ridge Province case study
Franke, O. L., 97–98, 403
free-phase dense nonaqueous phase liquids,
    368–369
free-phase extraction, 745–748
French, M. J., 361
frequency, monitoring, *586t*
freshwater
    availability, 10–11
    in coastal areas, 213–218
    conversion of saltwater to, 2
    disputes, 36–39
    economics
        overview, 39–44
        price and value of, 44–55
        virtual water and global water trade,
        55–61
    overview, 1–10
    scarcity, 30–35
    sustainability, 61–66
    use of
        in Africa, 24–26
        in China and India, 26–30

freshwater (*Cont.*)
in Europe, 22–24
overview, 11–17
in United States, 17–22
volume of, 1–3
freshwater lenses, 217–218
Fresno Water Division, California, 589
Freundlich isotherm, 410–412
Fritz, P., 272–273
FS (feasibility studies), 567, 702, 708–718, *711f*
fuel hydrocarbons, 419, *422t*
fuel storage practices, 348–349
full water price, 44–45
Fuller, M. L., 620, *621f*
fully penetrating wells, 807–809
functionality, data management system, 598
funding, 51–52, *579t*, 701
Funkhouser, G., 312–313

━━━ G ━━━

GAC filters, 450
Galloway, D., 97
gamma radiation, 358–359
gaseous chlorine feed systems, *440f*
gases, 398, 400–401
gasoline products, 365–366
GE (General Electric) electrodialysis reversal, *462t*
Gelhar, L. W., 406–407
General Electric (GE) electrodialysis reversal, *462t*
geographic information systems (GIS), 607–609, 641
geologic fabric, 129
geologic formations, *492t*
geology
datasets, *601t*
factors in IAS, *743t*
recharge process and, 257–259
geometry, system, 82
Georgia, 37, 75, *76f*
German Research Center for Environmental Health (GmbH), *265f*
Germany, *196f*
Geyh, M., 275
geysers, 196–197
GFH (granular ferric hydroxide), 471
GHGs (greenhouse gases), 314–315

Giardia lamblia, 375–376, *386t*, *453t*
Gibson, K. E., 374
GIS (geographic information systems), 607–609, 641
glacial drift, 157
glacial-deposit aquifers, 150, 157–160
Glacier Bay National Park and Preserve, Alaska, *7f*
glaciers, melting of, *7f*, *320f*
glauconite, 458
Glen Canyon dam, 484
global temperature, *296f*
Global Water Partnership, 560
global water trade, 55–61
glyphosate, *386t*
GmbH (German Research Center for Environmental Health), *265f*
goals
aquifer management, 596–598
NAPL source-zone cleanup, 792
treatabililty study performance, 718
Gollehon, N., 53
Gotkowitz, M. B., 188
governance, water, 539–540, 542–543, 560
governmental programs, 627–628
graded gravel packs, 498
Grand Challenges in Environmental Sciences report, 633–634
granular ferric hydroxide (GFH), 471
granular filters, 443
granular iron, 774, 776–777
granulated active carbon, 451
Grapevine Springs, 244–245
graph pumping rate, *506f*
Grasso, D. A., 208
gravel deposits, 150, *151f*
gravel packs, 493, *494f*, 497–500
gravel-packed screens, *520f*
gravity springs, 198, *199f*
gray water, 13
Great Artesian Basin, 165, 167
Great Britain, 33–35
Green, C., 40
greenhouse gases (GHGs), 314–315
Greenland ice sheet, 317
greensand filtration, 458, 472
Ground Water Atlas of the United States, The, *148f*, 149–150

Ground Water Supply Survey, 362
Ground Water Task Force (GWTF), 611
Ground Water Use, Value, and Vulnerability as Factors in Setting Cleanup Goals discussion paper, 695
groundwater chemistry datasets, *601t*
groundwater contamination hazard maps, 615–616
Groundwater Directive, European Parliament, 539–540
"Groundwater in international law. Compilation of treaties and other legal instruments" publication, 573
Ground-Water Management Process (GWM) program, 656
Groundwater Rule (GWR), 174, 568–569
grouted barriers, 738
grouting casings, 489
growth, future, *579t*
Grza spring, Serbia, 210
Guaraní Aquifer System, 163
Guidelines for Evaluating Ground-Water Flow Models, 650
Guinamant, J.-L., 361
Gurr, C., 412–414, *415t–417t, 418f*
Guyonnet, D., 425
GWM (Ground-Water Management Process) program, 656
GWR (Groundwater Rule), 174, 568–569
GWTF (Ground Water Task Force), 611

▬ H ▬

H2S (hydrogen sulfide), 466
haa5 (haloacetic acids), *386t*
Haefner, R. J., 403
Haitjema, H. M., 620–621
Halford, K. J., 281
Hall, A. W., 560
haloacetic acids (haa5), *386t*
halogenated aliphatic hydrocarbons, 419–420, *421f, 422t*
halogenated solvents, 366–367
Hanford, Washington, 255–257
hardness, water, 459–461
Hart, D. J., 188, 192–193
hazard ranking system (HRS), 567
Hazardous and Solid Waste Amendments (HSWA), 568

hazardous substances, 706, 709
HBSLs (health-based screening levels), 345–346
HDWD (Hi-Desert Water District), 698–699
head-dependent flux, 140
health effects of contamination, 342–346
health effects studies, 678
health-based screening levels (HBSLs), 345–346
Healy, R. W., 97–98
heat conductivities, soil, 762
heavy metals, 338–339, 343–344
Hebei province, China, 28
Helmholtz Center Munich, *265f*
Helsinki convention, 574
Henry's law, *399t*, 400–401, 758
heptachlor, *386t, 415t, 417t*
heptachlor epoxide, *387t*
herbicides, 370
heterogeneity, 126–132, *136f*, 403, 497, 750
heterotrophic plate count (HPC), *387t*
Hevesi, J. A., 285–286
hexachlorocyclopentadiene, *387t*
Hickey, J. P., 414, *415t–417t, 418f*
Hi-Desert Water District (HDWD), 698–699
Hieroglyphic Mountains Recharge Project, 673
High Plains aquifer, 101–104, 154–156, 268–269
high rate clarification, 441
high-frequency measurements, *587f*
high-yield well sitting, 182–183
Hiscock, K. M., 361
Hoekstra, A. Y., 56, 59
homogeneity, 126, *131f, 136f*, 496–497
horizontal capture zones, *766f*
horizontal collector wells. *See* collector wells
horizontal hydraulic gradients, *112f*
horizontal wells, *771f*
hot springs, 200
Howard, P. H., 413–414
HPC (heterotrophic plate count), *387t*
HRS (hazard ranking system), 567
HRUs (hydrologic response units), DPM, 287
HSWA (Hazardous and Solid Waste Amendments), 568
Hueco Bolson, 220–221

Hughes, M. K., 312–313
human activity and climate change, 299–300.
        See also anthropogenic climate change
human health risk assessment, 702–703
humidity, 242–243
Hung, P. Q., 59
Hussain, I., 32–33
hydraulic barriers, 665
hydraulic conductivity
    anisotropy, 129
    in aquitards, 192–193
    contour maps, *136f–137f*
    Darcy's law, 108–109
    heterogeneity of, 127, *128f*
    permeability and, 111–114
    in situ air sparging, *743t*
    soil water retention and, 247, *248f*
    unit conversion tables for, 815
hydraulic containment, 766
hydraulic gradients, 108–111, *112f*
hydraulic heads
    in aquifers, 78–79
    boundary conditions, 141–143, *144f*
    calibration, *644f*
    and hydraulic gradient, 109–111
    initial conditions and, 132–137
    measurements of and aquitard flow,
        187–188
    residuals, 647
    in water budget, 101–104
hydrochemical separation
    of spring hydrographs, 211–213
    of streamflow hydrographs, 281
hydrofracturing, 503
hydrogen isotopes, 273–274
hydrogen peroxide, 751
hydrogen sulfide (H2S), 466
hydrogen-ion activity (pH), 331–332,
        456
hydrogeologic mapping, 625–626
hydrogeologic structures, 79–81
hydrographs
    baseflow separation, 278–281
    pumping rate, 503–504
    spring
        autocorrelation and cross correlation,
            206–210
        hydrochemical separation of, 211–213

overview, 201–203
    potentially exploitable reserves, *534f*
    recession analysis, 203–206
  streamflow, 281
hydrologic forecasting, 633
hydrologic response units (HRUs), DPM,
        287
hydrologic years, 197
hydrological characteristics, 577
hydrological drought, 307–308
hydrology datasets, *601t*
hydrolysis, 758
hydrophobicity, 418
hypochlorite feed systems, *441f*

IAEA (International Atomic Energy Agency),
        269, *270t*
IAH (International Association of
        Hydrogeologists), 39
ice melting, 315
ice-contact deposits, 159
Idaho National Laboratory (INL), Radiation
        Control Division, 104–107
Idaho Water Resources Research Institute,
        University of Idaho, 104–107
IGRAC (International Groundwater
        Resources Assessment Centre), *574f*
IHE Delft, 55–56
ILC (International Law Commission), 574
Imes, J. L., 212
immobilization
    contaminant, 409
    nitrogen, 359
impermeability, dam, 526
import, water, 59, *60t*
impounding effect of faults, 145
impoundments, 350–351
in situ cathodic protection, 513
in situ chemical oxidation (ISCO)
    BTEX treatment with Fenton's reagent,
        754–756
    overview, 750–754
in situ remediation, 716–717
incentives, user, 630
incidence, angle of, 127, *128f*
income poverty, 33
incrustation, 513–514

index mapping methods, 616–619

India, 26–30

indirect estimates of recharge, 263–264

individual water footprints, 55–56

industrial practices, 352–353

industrial products, virtual water content of, 56, *58t*

industrial water use, 12, 14, 21–22, 27–28, *29t*

infiltration, 99

   rainfall-runoff relationship, 236–241

   rates of and spreading, 669

   riverbank filtration, 521

   urbanization and runoff, 261

   and water movement through vadose zone

      Brooks and Corey equation, 251–253

      Darcy's law, 247–250

      overview, 245–247

      Richards equation, 251–253

      soil water retention and hydraulic conductivity, 247

      van Genuchten equation, 251–253

INFILv3 model, 285–286

infrastructure

   major, *579t*

   wastewater, 13

   water, 40

Ingebritsen, S. E., 97

inhalation slope factor, 707

initial conditions, 82, 132–137

injection grouting, 738

injection wells, 351, 671–672, 759

INL (Idaho National Laboratory), Radiation Control Division, 104–107

inland brackish water, 220–223

inner protection zones, 624

inorganic adsorption, 450–451

inorganic compounds, 473–474

inorganic fertilizers, 369

instrumentation for ET computation, 244–245

instruments, groundwater management, 546–547

intake dose estimates, 705

integrated groundwater resource management, *544f*

integrated pumping test, 794

Integrated Risk Information System (IRIS), 705

integrated water cycle management (IWCM), 578

Integrated Water Resources Management (IWRM), 575–578

interception, 100

interflow, 99

Intergovernmental Panel on Climate Change (IPCC)

   anthropogenic climate change, 313–317

   impact of global warming, 317–323

intergranular porosity, 86

intermittent springs, 198

internal system boundaries, 139

internal water footprints, 56, 59

internally renewable freshwater resources (IRWR), 11

international aquifers, 39, 571–575

International Association of Hydrogeologists (IAH), 39

International Atomic Energy Agency (IAEA), 269, *270t*

International Groundwater Resources Assessment Centre (IGRAC), *574f*

International Law Commission (ILC), 574

International Symposium on Groundwater Sustainability (ISGWAS), 63–66

International Union of Pure and Applied Chemistry (IUPAC), *365t*, *399t*

international water cost comparison, *47t*

International Water Management Institute (IWMI), 17, 59

Internationally Shared Aquifer Resources Management (ISARM), 39, 571–572

interpretation of test data, 122

interstices, rock, 86–87

intrinsic permeability, 111–114

intrinsic values of water sources, 50–51

intrinsic vulnerability, 615

inverse modeling, 397

in-well vapor stripping process, 744–745

ion exchange, 450–451, 459, 461, 467–468

IPCC (Intergovernmental Panel on Climate Change)

   anthropogenic climate change, 313–317

   impact of global warming, 317–323

IRIS (Integrated Risk Information System), 705

iron
    bacteria, 491, 514–515
    biodegradation, *422t*
    coagulation, 470
    granular, 774, 776–777
    and manganese removal
        biological treatment, 459
        coagulation/precipitation, 458
        filtration, 458–459
        ion exchange, 459
        oxidation, 457–458
        sequestration, 459
    oxidation, *438f, 454t*
    as primary constituent, 333–334
irrigation
    adjusting prices of, 52–54
    center-pivot systems, *32f*
    in Eastern Snake River Plain, 104–107
    efficient, 630–631
    in European Union, 23–24
    in India, 28–29
    irrigation water use, 13
    subsidized, 40–41
    in U.S., 19–21
    and water scarcity, 31
IRWR (internally renewable freshwater
    resources), 11
ISARM (Internationally Shared Aquifer
    Resources Management), 39,
    571–572
ISCO. *See* in situ chemical oxidation
ISGWAS (*International Symposium on
    Groundwater Sustainability*), 63–66
islands, 217–218, 323
isotherms, sorption, 410–412
isotopic effects, 361–362
isotropic aquifers, *131f*, 807–809
Issaouane Erg, 3, *5f*
IUPAC (International Union of Pure and
    Applied Chemistry), *365t, 399t*
IWCM (integrated water cycle management),
    578
IWMI (International Water Management
    Institute), 17, 59
IWRM (Integrated Water Resources
    Management), 575–578
Izbicki, J. A., 270–271

J

Jabal Tuwayq, Saudi Arabia, *32f*
Jadar karst spring, *195f*
Jadro Spring, Croatia, 661–662
Jeannin, Y., 208
jetting, air, 502
Johnson Muni-Pak prepacked well screen,
    520f
Johnston, C. D., 270
Jones, D. R., 97

K

Karanovic, M., 425
karst aquifers
    chlorinated DNAPL in, 731
    contour maps of, 137
    flow patterns, *139f*
    groundwater contamination
        complications, *733f*
    groundwater flow, 123–125
    groundwater velocity in, 126
    overview, 167–174
    potential DNALP accumulation sites in,
        731f
    pump and treat, 772–773
    source water protection zones, 626–627
    vulnerability of, 618–619
karst recharge rate, 258, *259f*
karst springs, 532, 534, 661–662
Keele University, Great Britain, 33–35
Kelson, V. A., 620–621
Kentucky, 170–171
Kenya, *527f–528f*
kinematic viscosity, 112–113
Kiraly, L., 208
Kissengen Spring, Florida, 551–553
Kleeschulte, M. J., 212
Kleist, J., 309
Kloppmann, W., 361
Knight, T., 313
known flux, 138–139
known-head, 140–141
K_{oc} (organic carbon), 412–418
Koïchiro Matsuura, 1
K_{ow} (octanol-water partition coefficient), 418
Kraemer, S. R., 620–621
Kueper, B. H., 425

L

LaBaugh, J. W., 97–98
LACDPW (Los Angeles County Department of Public Works), *641f*
Lake Baikal, 2
Lake Chad, 9–10
Lake Chad Basin, 25–26
Lake Mead, 302–304, 377–378
Lake Michigan, 540–541
lake sediments, 310
Lake Superior, 6
lakes, 2
land cover, 237, 239–241, 259–262, *263t*
land subsidence, 92–97, *98f*, 485
land use, 259–262, *263t*, *341f*, 633
land use and land cover (LULC), DPM, 287
landfills, 349–350, 590
landslides, 526
Langmuir isotherm, 410–411
LARWQCB (Los Angeles Regional Water Quality Control Board), 754
Las Vegas, Nevada, 63, *64f–65f*
Las Vegas Sun, 377–378
lateral well screens, 520–521
Latin America, 322
Lawrence, P., 33, 35
lead, *387t*, 474
leaking underground storage tanks, 348
leaks, industrial, 353
leaky aquitards, 77, 186–187
Lee, K. K., 425
legacy data, 598
Legionella, *387t*
legislation, water, 546
length, unit conversion tables for, 811–812
Lerner, D. N., 361
light nonaqueous phase liquids (LNAPLs), 364–367, 724–725, 742, 745–746
lime, 460
lime softening, 465
lime-soda ash, 460
lindane, *387t*, *415t*, *417t*
line type data, 602
lineament analysis, 182–183
linear groundwater velocity, 125
linear isotherm, 410–412
linear losses, 504–505

linear regression (LR), 312
lithology, vadose zone, 250
livestock, 13, 352
Lloyd aquifer, 554–555
LNAPLs (light nonaqueous phase liquids), 364–367, 724–725, 742, 745–746
log-log curves, *121f*
Long Island groundwater system, *554f*
longitudinal dispersivity, 404–407, 425
long-term ambient monitoring, 581–583
long-term models, *626f*
long-term monitoring (LTM) program, 585
Los Angeles County Department of Public Works (LACDPW), *641f*
Los Angeles Regional Water Quality Control Board (LARWQCB), 754
loss of commodity value, 52
Lower Santa Cruz Recharge Project, 673
low-viscosity fluid substrates, *782t*
LR (linear regression), 312
LTM (long-term monitoring) program, 585
Luehrs, D. C., 414, *415t–417t*, *418f*
Lukas, J. J., 313
LULC (land use and land cover), DPM, 287
lysimeters, 262, 264–266

M

macro constituents of groundwater, 328
macroporosity, 86, 264
macroscopic heterogeneity, 403
magmatic rock porosity, *88f*
Maillet equation, 204
maintenance
   collector wells, 521
   expenses for irrigation water, *54t*
   vertical wells, 511–515
major constituents of groundwater, 328
Mammoth Cave area, 170–171
managed aquifer recharge (MAR), 665
management, groundwater
   artificial aquifer recharge
      aquifer storage and recovery, 673–675
      methods of, 667–673
      overview, 662–667
      source water quality and treatment, 675–680

management, groundwater (*Cont.*)
  data management
    data quality, 609–610
    geographic information systems, 607–609
    metadata, 610
    overview, 594–600
    tools, 604–607
    types of data, 600–604
  Integrated Water Resources Management, 575–578
  modeling and optimization
    numeric groundwater models, 635–656
    time series models, 657–662
  monitoring
    ambient, 581–583
    compliance, 583–586
    detecting contamination, 589–594
    overview, 578–581
    performance, 586–589
  overview, 539–549
  protecting resources
    overview, 610–614
    source water protection zones, 619–627
    strategies, 627–634
    vulnerability maps, 615–619
  regulatory framework
    groundwater quality, 562–571
    groundwater quantity, 560–562
    transboundary aquifers, 571–575
  sustainability
    nonrenewable groundwater resources, 557–560
    overview, 549–560
manganese, *422t, 438f, 454t,* 472
manganese colloids, 458
Mangin, A., 208
manual calibration, 644
manual modeling optimization, 652
manure, 361, 373–374
maps
  contour, 132–137, 607, *608f*
  groundwater contamination hazard, 615–616
  risk, 615–616
MA(*q* (moving average models)), 659
Maquoketa Formation, 193
MAR (managed aquifer recharge), 665

Maréchal, J.-C., 268
MAROS program, 585
mass balance, 269–270
Mass Flux Toolkit, 794
mass transfer rates, 730
Massachusetts, 405–406, 642
match points, 120, 253
materials, reinventing use of, 633–634
matric potential, 247, *248f*
matrix porosity, 86, 407
maximum contaminant level goals (MCLGs), 380
maximum contaminant levels (MCLs), 345, 358–359, 380–381, *382t–391t*
Mayer, G. C., 281
MBR (mountain block recharge), 255
McDade, J. M., 719–720
McGuire, T. M., 719–720
McKee, T. B., 309
MCLGs (maximum contaminant level goals), 380
MCLs (maximum contaminant levels), 345, 358–359, 380–381, *382t–391t*
measured drawdown, 507–509
measured hydraulic heads, *648f*
mechanical corrosion, 513
mechanical dispersion, 125–126
media filtration, *475t*
megacities, 14, 17
Meigh, J., 33, 35
Meinzer, O. E.
  artificial aquifer recharge, 665
  classification of springs based on discharge, 197–198
  faults, 145
  rock porosity, 85
  thermal springs, 200
  total versus effective porosity, 90
  U.S. springs, 196, 199–200
Meko, D. M., 313
melting, *7f,* 315, *320f*
membrane filtration, 447–450, *476t*
memory, 207–208
mercury, *387t,* 474
Mesozoic carbonate aquifers, 169–170
metadata, 610
metals, 343–344
metamorphic rock porosity, *88f*

methane, *422t*

methemoglobinemia, 360

methoxychlor, *387t*

Mexico, *171f*

Meyers, J.-P., 345

Meylan, W. M., 413–414

MF (microfiltration) membrane filtration, 448

mho units, 331

Michigan, 770–771, 785–787

microbial contaminant candidates, *392t–393t*

microbiological contaminants, 374–377

microfiltration (MF) membrane filtration, 448

microorganism reduction ability, *453t*

microregimes of discharge, 204–206

migration of dense nonaqueous phase
     liquids, 368–369

Mihelcic, J. R., 414, *415t–417t, 418f*

Milankovitch, Milutin, 297–299

Millennium Declaration 2000, 575–576

mineral springs, 200–201

mineralization, nitrogen, 359

mining, aquifer, 95

mining operations, 353

Ministry of Water Resources, India, 28–29

minor constituents of groundwater, 329

Mississippi, 212–213

Mississippi Embayment aquifer system, 157

Mississippi River Valley alluvial aquifer, 156

MNA (monitored natural attenuation),
    787–792

model-calculated hydraulic heads, *648f*

modeling. *See also* numeric models
    predictive groundwater, 559
    programs, 607, *608f*
    standards, 649

MODFLOW program, 637, 772, *773f*

modified coagulation/filtration, 469

modified lime softening, 469

moisture
    Darcy's law, 249–250
    soil, 237, 242
    variably saturated flow models, 282–285

moisture characteristic curve, 247, *248f*

Molden, D., 32–33

molecular diffusion, 194, 404–405

molecular weight cutoff (MWCO), 448

monitored natural attenuation (MNA),
    787–792

monitoring
    ambient, 581–583
    compliance, 569, 583–586
    detecting contamination, 589–594
    networks, 584–585
    overview, 578–581
    performance, 586–589
    water, 569
    of wells, 511–512

monitoring program, Blue Ridge Province,
    178–182

monitoring wells, 117–118, 188, *189f*

Montebello Forebay Project, 678

mountain block recharge (MBR), 255

mountain-front recharge, 153

moving average models (MA(q)), 659

Mozley, P. S., 145

Mualem, Y., 252–253

multiphase extraction, 745–748

municipal water systems, 12

Murray-Darling Basin Commission,
    Australia, 38

MWCO (molecular weight cutoff), 448

m-xylene, *416t–417t*

N

Nace, R. L., 3, 5

Namibia, 529

nanofiltration (NF), 448–449, 471

nanomaterials, 379

NAPLs (nonaqueous phase liquids)
    air sparging, 742
    overview, 364–369
    solubility of, 397–398
    source-zone depletion models, 721
    source-zone remediation, 719, 724–732
    volatilization, 400–401
    zone cleanup goals for, 792

National Climatic Data Center (NCDC),
    313–314, 323–324

National Contingency Plan (NCP), 567,
    700

National Drought Mitigation Center
    (NDMC), 307–308

National Geographic News, 6

National Institutes of Health (NIH), 414

*National Oil and Hazardous Substances
    Pollution Contingency Plan*, 700

National Primary Drinking Water
Regulations (NPDWRs), 380–381,
382t–391t, 564–565
National Priorities List (NPL), 567, *764f*
National Research Council (NRC), 297, 632
National Secondary Drinking Water
Regulations (NSDWRs), 381, *393t*, 565
National Water-Quality Assessment
(NAWQA) program, 376–377
natural groundwater constituents
hydrogen-ion activity, 331–332
overview, 328–329
primary constituents, 333–338
Redox potential, 332–333
secondary constituents, 338–340
TDSs, specific conductance, and salinity,
329–331
naturally developed wells, 493, *494f*
naturally occurring contaminants, 340,
355–359
NAWQA (National Water-Quality
Assessment) program, 376–377
NCDC (National Climatic Data Center),
313–314, 323–324
NCP (National Contingency Plan), 567, 700
NDMC (National Drought Mitigation
Center), 307–308
net infiltration, 99
net recharge, 100
network design for performance monitoring,
584f
Neuman, S. P., 407
neural network (NN), 312
Nevada, 63, *64f–65f*, 124
Neville, C., 425
Neville, C. J., 425
New York, 589–590
New York Times, The, 555
New Zealand, 322
Newell, C. J., 719–720
NF (nanofiltration), 448–449, 471
NF/RO (spiral-wound) module pressure
vessels, *449f*
Ni, F., 312–313
nickel, 474
NIH (National Institutes of Health), 414
nitrate
ambient monitoring, 583

biodegradation, *422t*
contamination from, 359–362, 461–462
groundwater dating, 276–277
groundwater restoration, 696, 698–699
overview, *388t*
nitrification, 359
nitrite, 359–360, *388t*
nitrogen, 359–362, 369–370
NN (neural network), 312
nonaqueous phase liquids. *See* NAPLs
noncarbonate hardness, 459–460
noncommunity water systems, 564
nonelectrolytes, 330–331
nonpoint sources of pollution, 347
nonpolar arid land, 3, *4f*
nonrenewable aquifers, 81–82
nonrenewable groundwater resources, 557–560
nontarget analyte monitoring, 594
nontransient non-CWS, 564
North America, 323. *See also* United States
North Carolina, 334–336, 461
North Western Sahara Aquifer System
(NWSAS), 165
Northern Atlantic Coastal Plain aquifer
system, 157, *158f–159f*
NPDWRs (National Primary Drinking Water
Regulations), 380–381, *382t–391t*,
564–565
NPL (National Priorities List), 567, *764f*
NRC (National Research Council), 297, 632
NSDWRs (National Secondary Drinking
Water Regulations), 381, *393t*, 565
Nubian Sandstone Aquifer System (NSAS),
163–165, *166t*
numeric models
calibration, sensitivity analysis, and error,
642–648
contour maps, 135, *136f–137f*
distributed-parameter areal recharge
models, 285–287
documentation and standards, 649–650
hydrogeologic mapping, 625–626
optimization, 652–656
overview, 122–123, 635–639
in RI/FS process, 710, 712
saltwater intrusion, 650–652
setup, 639–642
variably saturated flow models, 282–285

NUS Consulting Group, 46–47
Nutzotin Mountains, *320f*
NWSAS (North Western Sahara Aquifer System), 165

O

O&M (operations and maintenance) costs, 44
obliquity cycle, 297–299
observed hydraulic heads, *644f*
Oceanic Niño Index (ONI), *301f*
octanol-water partition coefficient (K_{ow}), 418
OCWD (Orange County Water District), 575–576
o-dichlorobenzene, *385t*
odors, *454t*, 514
off-farm water suppliers, 53, *54t*
Ogallala aquifer, 101–104, 154–156, 268–269
Ohio, 51–52
oil field brines, 351
Ombla spring, Croatia, 209–210
ONI (Oceanic Niño Index), *301f*
open hydrogeologic structures, 80
open soil evaporation, 243
operational definitions of drought, 306
operations and maintenance (O&M) costs, 44
opportunity cost, 53, 55
optimization
    numeric groundwater models, 652–656
    pump and treat, 769–770
optimum remediation systems, 653
oral slope factor, 707
Orange County Water District (OCWD), 575–576
orbital variations, 297–299
organic carbon (K_{oc}), 412–418
organic chemicals, 341
organic compounds, 450
organic fertilizers, 369–370
organic matter in water, 339–340, 361
organic solute sorption, 412–418
organic substance solubility, 397–398
orientation, axial, 297, 299
Ortega-Guerrero, A., 192
Our Common Future report, 61–62
Our Stolen Future book, 345
outwash, 159
Ouyang, Y., 282
overexploitation, 578

overflowing springs, 198, *199f*
overpumping, 501, 534–535
oxamyl, *388t*
oxidation
    arsenic removal, 467, 469–470
    chemical oxidants, 439–441
    cost of, *475t*
    filters, 465
    groundwater treatment, 439–441
    iron and manganese removal, 457–458
    redox potential, 332–333, *422t*
    in treatment plants, *438f*
oxygen, 273–275, 333, *422t*
o-xylene, *416t–417t*
ozone, 440, *453t*, 455

P

P&T (pump and treat)
    1,4-dioxane plume, 770–771
    fractured rock and karst aquifers, 772–773
    optimization, 769–770
    overview, 765–769
PAHs (polycyclic aromatic hydrocarbons), 364, *383t*
paleoclimate, 310–313
Paleozoic limestone, *169f*
Pall Aria system, 470–471
Palmer Drought Severity Index (PDSI), 309
parameter estimation, 660
parameter sensitivity, 644–645
parametric mapping methods, 616
parasites, 375–376
Parker, B. L., 188
PART program, 281
participatory approach to water management, 39
particles of rock and porosity, 86–87
particulate grouts, 738
partitioning, 409
passive biobarriers, 783
passive bioremediation, 787
passive flux meter (PFM), 793–794
pasteurization, 515
pathogenic bacteria, 627
pathogens, 374–377, *453t*, 678–679
PCBs (polychlorinated biphenyls), 367, *388t*
PCE (tetrachloroethylene), *389t*, *416t–417t*, 707

p-dichlorobenzene, *385t*

PDSI (Palmer Drought Severity Index), 309

Peckenpaugh, J. M., 254

Pecos River Basin alluvial aquifer, 156

Pedley, S., 361

pentachlorophenol, *388t*

perched aquifers, 79

Pereira, L. S., 242

performance goals, 718

performance monitoring, *584f*, 586–589

perihelion, 298–299

periodic springs, 198

periodicity, 658

permanent injection wells, 783

permanganate, 439–440, 751, 753

permeability, 111–114

permeable reactive barriers (PRB), 773–778

permits, extraction, 559

Perry, C. J., 41

persistence, 207

persulfate, 753

Perth, Australia, *48f*

pesticides, 352, 370–373, 696

Pesticides in Ground Water Database, 371

PET (potential evapotranspiration), 241–243

petroleum products, 348–349

PFM (passive flux meter), 793–794

pH (hydrogen-ion activity), 331–332, 456

pharmaceuticals and personal care products (PPCPs), 349, 379–380, 594

phosphate, 372–373

phosphorous, 369

photon emitters, *383t*

PHREEQC program, 212–213, 396–397

physical containment

deep soil mixing, 737

grouted barriers, 738

overview, 732–735

sheet-pile walls, 738–739

slurry walls, 735–737

physical water scarcity, 33

picloram, *388t*

Pierce Service Station site, 754–756

piezometric level. *See* hydraulic heads

pilot borings, 488

pilot tests, 717

Pima Mine Road Recharge Project, 673

Pipe diagrams, 334–335

Plan B 2.0 Rescuing a Planet Under Stress and a Civilization in Trouble book, 31

planar flow, 115

planned nonrenewable groundwater management scenarios, 557

plants

desalination, 463–464

drinking-water treatment, *438f*

and recharge rate, 257

type of and evapotranspiration, 242

uptake of nitrogen by, 359–360

Pleasant Grove Spring basin, 697–698

plume remediation. *See* dissolved phase remediation

Plummer, L. N., 212

POE (point of entry) devices

arsenic, 472–473

nitrate removal in, 462

for radon removal, 466

treatment systems with, 438–439

point of use devices. *See* POU devices

point type data, 602

polar regions, 323

pollution

chemical, 594

control of

CERCLA, 566–567

Clean Water Act, 563

in groundwater management, 547

integrated water systems, *580t*

programs, 620

defined, 13

polychlorinated biphenyls (PCBs), 367, *388t*

polycyclic aromatic hydrocarbons (PAHs), 364, *383t*

polygon data, 602

polymer additives, 442

polymer beads, 472

polyphosphates, 459

Ponderosa Cave, Yucatan, Mexico, *171f*

ponding time, 246

pools, NAPL, 397–398

population

and average annual growth, *16t*

trends in U.S., *18f*

and water resources in regions of world, 24t

Population Action International, 31
pore-water concentration, 727–728
porosity
    diffusion, 407–409
    effective, 86–90, 125, 401–402
    infiltration rates and, 246
    in karst aquifers, 167–168
    permeability, 111–114
    storage and, 83–86
potable water, 12
potassium, 369
potential evapotranspiration (PET), 241–243
potential recharge, 99, 236
potentially responsible parties (PRPs), 700
potentially useable water resources (PUWR),
        11, 33
POTWs (publicly owned treatment works),
        13
POU (point of use) devices
    arsenic, 472–473
    nitrate removal in, 462
    for radon removal, 466
    treatment systems with, 438–439
PPCPs (pharmaceuticals and personal care
        products), 349, 379–380, 594
PRB (permeable reactive barriers), 773–778
precession, axial, 297, 299
precipitation
    chemical, 441–442
    classifications of climate based on,
        253–254
    in global warming scenarios, 320–321
    rainfall-runoff-recharge relationship,
        236–241
    and recharge rate, 257
precipitative softening, 460
precoat filters, 446
predevelopment conditions, 132, *133f*
predictive modeling, 559, 638, 651
prefabricated slow sand filters, 444–445
prepacked well screens, *520f*
pre-ROD treatability investigation, 717–718
Prescott Valley, Arizona, 43
pressure filters, 445–446, 458
pressure vessels, *449f*
pressure-driven membrane process
        classification, *447f*
Preussag method, 516–517

prevention
    of pollution, 628
    of side effects, *547t*
pricing, water, 40–43
primary constituents of groundwater, 329,
        333–338
primary drinking water standards, 380–381
primary porosity, 85
primary springs, 531
primary water supply (PWS), 11, 33, 570–571
prior appropriation doctrine, 561
priority pollutants, 594
privatization of water supply, 41
probabilistic models, 635
probability graphs, 202–203
programs, data management, 604
projection pipe method, 518–519
protecting resources
    overview, 610–614
    protection planning, 563
    recharge areas, 629
    regional water supplies, 52
    source water protection zones
        hydrogeologic mapping and
            groundwater modeling, 625–626
        karst aquifers, 626–627
        overview, 619–625
    strategies, 627–634
    vulnerability maps, 615–619
protozoa, 375–376
proxies, 310–313
PRPs (potentially responsible parties), 700
public awareness campaigns, 558
public education, 619–620
public property resources, 559
public supply, 12, 21–22, *23t*
public water systems (PWSs), 12, 564
public water use, 12
publicly owned treatment works (POTWs),
        13
Puerto Rican karst, 170, *171f*
pump and treat. *See* P&T
pump assembly, 512
pump efficiencies, 508–509
pumping rates, *493t*, 510, *524f*
pumping tests, 503–508
pumping wells, 116–118
purveyors, water, 12

PUWR (potentially useable water resources), 11, 33

PWS (primary water supply), 11, 33, 570–571

PWSs (public water systems), 12, 564

p-xylene, *416t–417t*

pyroclastic material, 174

Q

quality control (QC), 609, *610t*

quality of groundwater
  bank filtration, 570–571
  CERCLA, 566–567
  Clean Water Act, 563–564
  contaminants
    advection, 401–403
    agricultural, 369–374
    analytical equations of fate and
      transport, 422–425
    biodegradation, 419–422
    dispersion and diffusion, 403–409
    dissolution, 394–398
    emerging, 377–380
    health effects, 342–346
    microbiological, 374–377
    naturally occurring, 355–359
    nitrogen and nitrate, 359–362
    overview, 340–342
    sorption and retardation, 409–418
    sources of contamination, 347–355
    synthetic organic contaminants, 362–369
    volatilization, 398–401
  defined, 82
  drinking water standards, 380–381
  evaluating, *546t*
  Groundwater Rule, 568–569
  natural groundwater constituents
    hydrogen-ion activity, 331–332
    overview, 328–329
    primary constituents, 333–338
    Redox potential, 332–333
    secondary constituents, 338–340
    TDSs, specific conductance, and salinity,
      329–331
  overview, 327–328, 562–563
  Resource Conservation and Recovery Act,
    567–568
  Safe Drinking Water Act, 564–565

Sole Source Aquifer Protection Program,
    565–566
  standards for, 12
quantity, groundwater, 560–562
quasi-hydrogeologic delineation methods,
    624–625
quasi-steady-state flow, 114
Queen Inlet, Alaska, *7f*
Quinby, W., 53

R

radial collector wells. *See* collector wells
radial flow, 118
radial flow clarifiers, circular, *442f*
Radiation Control Division, Idaho National
    Laboratory, 104–107
radiocarbon, 277–278
radionuclides, 357–359, 465–466
radium 226 and radium 228, *388t*
radius of well influence, 766, *767f*
radon, 465–466
Raes, D., 242
Raftelis Financial Consultants, 48–49
rainfall
  in Africa, 24
  excess, 99–100
  in global warming scenarios, 320–321
rainfall-runoff-recharge relationship,
    236–241
rainwater tanks, *580t*
Rake, Launce, 377–378
random fluctuations, 658
Ranney method, 516, 518
Ranney wells. *See* collector wells
Raoult's law, 400–401
rapid sand filters, 443–444
RASA (Regional Aquifer-System Analysis),
    149–150
raster datasets, 602
rate of mass transfer, 730
rates, pumping, 510
rationalization scenarios, *558f*
Ravi, V., 282
RBF (riverbank filtration), 517–518, 666
RCRA (Resource Conservation and Recovery
    Act), 567–568, 591
RD/RA (remedial design/remedial action)
    treatability studies, 717–718

reactions, oxidant, 751–752

reactive materials, PRB, 774–775

real price of water, 44–45

real water savings, 59

reasonable maximum exposure (RME), 705

reasonable use rule, 562

rebound, 720

recession analysis, 203–206

recession coefficient, 204

recharge. *See also* artificial aquifer recharge
  to basin-fill aquifers, 153
  in Blue Ridge Province, 182
  in Eastern Snake River Plain aquifer,
    105–106, *110f*
  evapotranspiration, 241–245
  factors influencing
    climate, 253–257
    geology and topography, 257–259
    land cover and land use, 259–262
  infiltration and water movement through
    vadose zone
    Brooks and Corey equation, 251–253
    Darcy's law, 247–250
    overview, 245–247
    Richards equation, 251–253
    soil water retention and hydraulic
      conductivity, 247
    van Genuchten equation, 251–253
  methods for estimating
    baseflow separation, 278–282
    environmental tracers, 268–278
    lysimeters, 264–266
    numeric modeling, 282–287
    overview, 262–264
    soil moisture measurements, 266
    water table fluctuations, 266–268
  overview, 235–236
  rainfall-runoff-recharge relationship,
    236–241
  riverbank filtration, 521–522
  in water budget, 99–101

recharge areas, 82–83, 629

recharge basins, *668f*

recharge trenches, 670

recirculation systems, 783

reclaimed wastewater, 13

record of decision (ROD), 710, 716–718,
  764–765

recycled water supply, 11

Redmond, K. T., 301–304

redox reactions, 779, *780f*

reduction-oxidation (redox) potential (Eh),
  332–333, *422t*

reductive dechlorination, 420, *421f*, *422t*

reference crop evaporation, 241–242

refraction, angle of, 127, *128f*

regional ambient monitoring, 581

Regional Aquifer-System Analysis (RASA),
  149–150

regional protection of water supplies, 52

regolith, 731

regulation, spring, 534–535

regulatory framework
  development of, *546t*
  groundwater quality
    bank filtration, 570–571
    CERCLA, 566–567
    Clean Water Act, 563–564
    Groundwater Rule, 568–569
    overview, 562–563
    Resource Conservation and Recovery
      Act, 567–568
    Safe Drinking Water Act, 564–565
    Sole Source Aquifer Protection Program,
      565–566
  groundwater quantity, 560–562
  transboundary aquifers, 571–575

Rehfeldt, K. R., 406–407

Reilly, T. E., 403

relational databases, 605–607

remedial design/remedial action (RD/RA)
  treatability studies, 717–718

remedial investigation/feasibility study
  (RI/FS), 567, 702, 708–718, *711f*

remediation. *See also* restoration,
  groundwater
  costs of, 700–701
  federal resources for, 563
  resource protection, 613–614
  trends, *764f*

remedy selection considerations, 708

remote sensing technology, 630

repair expenses for irrigation water, *54t*

representative elementary volume (REV),
  129–130

residence-time criterion, 623

residential water prices in U.S., 48–50
residual disinfectants, 452
residual mean, 647
residual phase, NAPL, 397–398
residual saturation, 726
residual water content, 252
residuals, hydraulic head, 647
resins, ion exchange, 450, 461
resource allocation, 547
resource assessment, 546
Resource Conservation and Recovery Act
        (RCRA), 567–568, 591
resource dependency, 579t
resource-oriented strategies, 614
resources, water, 15t, 24t, 33–34, 633
responsibilities, 580t
restoration, groundwater. See also dissolved
        phase remediation; source-zone
        remediation
    measuring success of remediation,
        792–794
    overview, 695–701
    remedial investigation and feasibility
        study, 708–718
    risk assessment
        characterization, 706–708
        data collection and evaluation, 703–704
        exposure assessment, 704–705
        overview, 701–703
        toxicity assessment, 705–706
retardation, 409–412
RETC program, 253
retention of water, 236–241, 251–253
retention, specific, 88–89
return flow, 12
returned water, 767–768
REV (representative elementary volume),
        129–130
revenue collection, 579t
reverse osmosis membrane filtration. See RO
        membrane filtration
reverse-circulation method, 500
Rialto-Colton Basin, 145
Richards equation, 251–253
Riebek, H., 316
RI/FS (remedial investigation/feasibility
        study), 567, 702, 708–718, 711f
rights, water, 546, 559

riparian doctrine, 561
riparian habitats, 483–485
riser pipe diameters, 489
rising springs, 198–199
risk assessment
    characterization, 706–708
    data collection and evaluation, 703–704
    exposure assessment, 704–705
    overview, 701–703
    toxicity assessment, 705–706
risk maps, 615–616
Risser, D. W., 281
riverbank filtration (RBF), 517–518, 666
rivers
    development of, 483–484
    horizontal collector wells, 517
RME (reasonable maximum exposure), 705
RO (reverse osmosis) membrane filtration
    of arsenic, 468–469
    in membrane filtration process, 448–449
    in nitrate removal, 461–462
    in POE/POU arsenic removal, 473
    for removal of radionuclides, 465
Robbins, G. A., 423–425
Rochdi, H., 361
Rock, M., 41
rock permeability, 113–114
rock porosity, 83–88
ROD (record of decision), 710, 716–718,
        764–765
Rogers, P., 560
RORA computer program, 281
Rorabaugh's method, 281
Rosetta program, 253
Rossier, I., 208
rule of capture, 562
runoff, 24, 259–261, 278–281
rural populations, 17
Rutledge, A. T., 281

——— S ———

Safe Drinking Water Act (SDWA), 378,
        564–565
safe yield, 550
Sahara Desert, 81f, 82
Salinas Valley project, 665
salinity, 331, 352
salinization, 651, 652f

salt lakes, 2
saltwater, 2, 215–218
saltwater intrusion, 218–220, 650–652
Salzer, M. W., 313
sampling frequency, *586t*
San Bernardino Basin, California, 655
San Joaquin Valley, California, 97, *98f*
sand and gravel aquifers
    basin-fill, 151–154
    blanket sand and gravel, 154–156
    glacial-deposit, 157–160
    overview, 150–151
    semiconsolidated sand, 156–157
sand content, well, 501
sand dunes, 310–311
sand filters, *444f*, 472
sand seas, 3, *5f*
sand storage dams, 526–528
sanding, well, 498
sandstone and carbonate-rock aquifers, 174
sandstone aquifers, 160–167
sandstone wells, 503
sanitary surveys, 569
sanitary water, 13
sanitary wells, 341, *342f*
sanitary zones, 623–624
sanitation, 13–14
Santa Ana River, 575–576
Sapkota, A. R., 374
SARA (Superfund Amendments and
        Reauthorization Act), 567
saturated aquifer thickness, 101, *102f–103f*
saturated water content, 252
saturated zone of aquifers, 79
saturated-zone injection wells, 671
Saudi Arabia, 31, *32f*, 557–558
Sava River flood plain, 151, *152f*
Savenije, H. H. G., 40–43
savings, real water, 59
SCADA (Supervisory Control And Data
        Acquisition), 588–589
scale economies, 50
scale formation, 460
scarcity, freshwater, 30–35
Schiff, S. L., 361
schmutzdecke, 444
Schumacher, J. G., 212
Schwab, K. J., 374

screens, well, 489, 491–497, 519–521
SCS (Soil Conservation Service), 237–241,
        260–261
SDG (submarine discharge of groundwater),
        214–218
SDWA (Safe Drinking Water Act), 378,
        564–565
sea-level rise, 315, 317, 321
seasonal fluctuations, 486
seasonal time series models, 659–660
seawater intrusion, 665
Seckler, D., 41
secondary constituents of groundwater, 329,
        338–340
secondary drinking water standards, 381,
        393t
secondary porosity, 85
secondary sewage effluent, 675
secondary springs, 198, 531
second-order statistics, 647–648
sediment load, cumulative, *263f*
sedimentary rock porosity, 85–87
sedimentation, 465
sediments, lake, 310
SEE (steam-enhanced extraction), 759–760
seepage springs, 196, 532, *533f*
seeps, 196
Seiler, R. L., 361
selenium, *389t*, 474
self-reporting, 630–631
self-supplied water use, 21–22
semiclosed hydrogeologic structures, 80
semiconfined aquifers, 79
semiconfining beds, 78
semiconsolidated sand aquifers, 156–157
semilog curves, *121f*
semimetals, 338
semiopen hydrogeologic structures, 80
semipassive biobarriers, 783–784
semivolatile organic compounds (SVOCs),
        364
sensitivity analysis, 642–648
sentinel wells, 583–584
septage, 698–699
septic systems, 349, 360–361
sequestration, 459
Serbia, 210
serial correlation coefficient, 207

sewage, 349
Seymour aquifer, 156
sharp interface model, *650f*
sheet-pile walls, 738–739
shock chlorination, 532
shortages, water, 540
Shoshone Falls, *107f–108f*
shrinking plumes, *763f*
SI$_{calcite}$ (calcite saturation index), 212–213
side effects, 547
Siemen units, 331
Sierra Nevada Mountains, 317–320
silica, 334
silicic lavas, 174
silver, 474
Silver Spring, Florida, 196
Silvex (2,4,5-TP), *390t*
simazine, *389t*
Simmons, D. L., 403
simulated hydraulic heads, *644f*
simulation-optimization approach, 768–769
simulators, seawater intrusion, 651
single-pump systems, 745–746
sites, contamination, *592t*
Skipp, David C. *See* Blue Ridge Province case
    study
slow sand filters, 444–445, 472
slurry walls, 735–737
small islands, 323
smearing, 366
Smith, Allan, 355
Smith, M., 242
Smith, Z. A., 36
Snake River, 106–107
Snake River Plain, 175
Snow, Lester, 295
snowmelt recharge, 255–256
snowpack, 317–321
societal cost, 53
socioeconomic drought, 307–308
socioeconomic factors, 577
SOCs (synthetic organic chemicals)
    health effects of, 343–345
    nonaqueous-phase liquids, 364–369
    overview, 362–363
    prevalence of, 340–341
    removing, 466
    VOCs and SVOCs, 363–364

softening process
    cost of, *475t*
    for iron and manganese removal, 458
    lime in, 465
    methods for, 460
    precipitation through, 442
Soil Conservation Service (SCS), 237–241,
    260–261
soil vapor extraction (SVE), 739–740, *741f*
soil water balance models, 254
soil-bentonite slurry walls, 736
soils
    heat conductivities, 762
    infiltration and water movement through
        vadose zone, 245–247
    migration of contaminants through, 327
    moisture and evapotranspiration, 242
    moisture measurements, 266, *267f*
    open evaporation, 243
    rainfall-runoff relationship, 237, 239–241
    remediation, 722
    water retention, 247, 251–253
Sole Source Aquifer (SSA) Protection
    Program, 565–566
solid substrates, *782t*
solidification stabilization, 737
solubility
    dissolution, 394–397
    of organic substances, 397–398
solubilization, NAPL, 756
soluble substrates, *782t*
solutes, 330
solvents, halogenated, 366–367
Somalia, 36
sorption
    of organic solutes, 412–418
    overview, 409–412
sounding lines, 512
source protection costs, *44f*
source water
    protection zones
        hydrogeologic mapping and
            groundwater modeling, 625–626
        karst aquifers, 626–627
        overview, 619–625
    quality and treatment, 675–680
"Source Water Assessment and Protection"
    memorandum, 619

Source Water Assessment Programs (SWAPs), 564
source water classification, *463t*
source-oriented approaches, 614
source-zone cleanup goals, NAPL, 792
source-zone remediation
    enhanced bioremediation, 756–757
    fluid removal technologies
        air sparging, 741–745
        free-phase and multiphase extraction, 745–748
        soil vapor extraction and bioventing, 739–740
        surfactant and cosolvent flushing, 748–750
    non-aqueous-phase liquids, 724–732
    overview, 718–724
    physical containment
        deep soil mixing, 737
        grouted barriers, 738
        overview, 732–735
        sheet-pile walls, 738–739
        slurry walls, 735–737
    in situ chemical oxidation
        BTEX treatment with Fenton's reagent, 754–756
        overview, 750–754
    thermal technologies
        electrical resistive heating, 760–761
        overview, 757–759
        steam-enhanced extraction, 759–760
        thermal conductive heating, 761–762
South American Guaraní Aquifer System, 163
South Carolina, *675f*
Southeastern Coastal Plain aquifer system, 157
Southern Californian basin-fill aquifers, *155f*
spatial correlation, 648
spatial data, 601–603
speciation modeling, 396–397
species extinction, 317
specific conductance, 331
specific retention, 88–89
specific storage, 90–92
specific vulnerability, 615
specific yield, 88–90
speleothems, 299–300
SPI (Standardized Precipitation Index), 309

spills, 353
spiral-wound (NF/RO) module pressure vessels, *449f*
spreading structures, 668–670
spring boxes, 530, *532f–533f*
spring flow hydrographs, 281–282
springs
    development and regulation of, 530–535
    discharging into reservoir, *172f*
    flow rates, 124–125
    general use of, 170
    hydrograph analysis
        autocorrelation and cross correlation, 206–210
        hydrochemical separation of, 211–213
        overview, 201–203
        recession analysis, 203–206
    overview, 194–196
    thermal and mineral, 200–201
    types and classifications of, 196–200
SRC (Syracuse Research Corporation), 414
Srinivasan, V., 425
SSA (Sole Source Aquifer) Protection Program, 565–566
St. Marys well cluster, *76f–77f*
St. Peter-Prairie du Chien-Jordan aquifer, *626f*
stable isotopes, 274
stable plumes, *763f*
staining, mineral, 457
stakeholder participation, 546
standard deviation, 207
standardized data collection forms, 599
Standardized Precipitation Index (SPI), 309
state parks, 613
static storage, 93, 95
steady springs, 197–198
steady state flow, 114–118
steady-state calibration, 643
steady-state models, *626f*
steam injection, 742
steam-enhanced extraction (SEE), 759–760
stiff diagrams, 335, *337f*
stimulation, well, 503
stochastic interconnectivity, 123
stochastic time series models, 657–662
storage
    defined, 82
    droughts, 553

storage (*Cont.*)
  fuel storage practices, 348–349
  land subsidence and, 92–97
  leaking tanks, 595
  porosity and effective porosity, 83–88
  specific, 90–92
  specific yield and coefficient of, 88–92
  sustainable pumping, 551
  underground, 664
storativity of aquifers, 90–92
stormwater management, *580t*
Strahler, A. H., 253–254
Strahler, A. N., 253–254
stratigraphic layers, *84f*
streamflow, 182, 654–655
streamlines, 134–135
stream-valley aquifers, 150
Streeten, P., 34–35
strength, oxidant, *751t*
stress, water, 30–31
structures, spreading, 668–670
styrene, *389t*
submarine discharge of groundwater (SDG),
    214–218
subsidence, land, 92–97, *98f*, 485
subsidies, 40–41, 45–46
substrates, 781–784
subsurface barriers, 734–735
subsurface dams, 524–530
suction pressure of soil, 247, *248f*
sulfate, *422t*
sulfur hexafluoride, 275–277
Sullivan, C., 33, 35
Superfund Amendments and
    Reauthorization Act (SARA), 567
Superfund program, 566–567, 591, 708,
    764–765
Superstition Mountains, 673
Supervisory Control And Data Acquisition
    (SCADA), 588–589
supply and demand, 45
supply-driven groundwater development,
    543f
supply-side management, 543
Surface Impoundment Assessment, 351
surface impoundments, 350–351
surface loading of contaminants, 722–723,
    724f

surface runoff, 278–281
surface water
  comparative features of, 577
  cost of irrigation water from, 53
  development of, 483–486
  and groundwater in contour maps, 135,
    137, *138f*
  horizontal collector wells, 517
  including groundwater in management
    of, 540
  intrinsic value of, 50–51
  prices, *49f–50f*
  pumping from reservoir, *550f*
  as single resource with groundwater,
    327–328
  United States rights to, 560–561
  withdrawals, 18–19, 22–23
surfactants, 748–750
sustainability
  freshwater, 61–66
  of MNA, 790–792
  nonrenewable groundwater resources,
    557–560
  overview, 549–560
sustainability cost, 53, 55
sustainable pumping, 551
"Sustainable water use in Europe. Part 2:
    Demand management" report,
    548–549
sustainable yield, 655
SUTRA program, 637, 651–652
SVE (soil vapor extraction), 739–740, *741f*
SVOCs (semivolatile organic compounds),
    364
SWAPs (Source Water Assessment
    Programs), 564
swine production practices, 373–374
Switzerland, *196f*
synthetic organic chemicals. *See* SOCs
Syracuse Research Corporation (SRC), 414
systems, groundwater. *See also* aquifers;
    flow; springs
  aquitards, 186–194
  boundary conditions, 137–149
  coastal areas and brackish groundwater
    inland brackish water, 220–223
    overview, 213–218
    saltwater intrusion, 218–220

definitions, 75–82
geometry, 82–83
initial conditions and contouring of
    hydraulic head, 132–137
overview, 42, 75
storage
    land subsidence and, 92–97
    porosity and effective porosity, 83–88
    specific yield and coefficient of storage,
        88–92
system extent, 83
vulnerability, 82
water budget
    Eastern Snake River Plain Aquifer,
        104–107
    overview, 97–107

━━━ T ━━━

TAGD (Texas Alliance of Groundwater
    Districts), 556
tailing effect, *771f*
Tarawa atoll, 217–218
target analyte monitoring, 594
TARM (Transboundary Aquifer Resources
    Management), 39, 571–575
TARWR (Total Actual Renewable Water
    Resources), 10–11
taste control, *454t*
Taylor, J. L., 260
TCE (trichloroethylene), *390t, 413t, 416t–417t,*
    726f–727f, 729, 761f
TDS (total dissolved solids), 220–221,
    329–331, 462–464
technical impracticability (TI) waivers, 700
Technical Release 55 (TR-55), USDA,
    237–241
technical tools, 546
technologies, screening of, *714t–715t*
telescoping casing, 489
Tellam, J. H., 361
temperature
    of air and recharge rate, 257
    and evapotranspiration, 242–243
    global, and carbon dioxide levels,
        296f
    and hydraulic conductivity, 113
temporal correlation, 648
Tennessee, 587–588

tensiometers, 264, 266
tetrachloroethylene (PCE), *389t, 416t–417t,*
    707
Texas
    Air Force Plant 4, Fort Worth, 761
    Balmorhea, *613f*
    Comal Springs, 201–203
    Edwards aquifer, 670–671
    El Paso Water Utility water management,
        220–223
    groundwater in storage in, 5
Texas Alliance of Groundwater Districts
    (TAGD), 556
Texas Groundwater Protection Committee
    (TGPC), 555–556
texture, rock, 86–87
TGPC (Texas Groundwater Protection
    Committee), 555–556
thallium, *389t*, 474
Theis equation, 118–123, 507
Theis type curve, 119
theoretical drawdown, 504–509
thermal conductive heating, 761–762
thermal springs, 198, 200–201
thermal technologies
    electrical resistive heating, 760–761
    overview, 757–759
    steam-enhanced extraction, 759–760
    thermal conductive heating, 761–762
thermocouple psychrometers, 266
thermoelectric power water use, 18
Thiem equation, 116–118
THMs (trihalomethanes), 363
Thomas-Fiering models, 659
Thornthwaite, C. W., 253
Thousand Springs, Idaho, *106f, 109f*
three-dimensional sequential electron
    acceptor model, 791–792
TI (technical impracticability) waivers,
    700
till, 157, 159
tilt, axial, 297–299
time series models
    Jadro Spring, Croatia, 661–662
    overview, 657–661
time-dependent boundary conditions, 143,
    145, 147–149
titanium-based media, 472

Title 40-Protection of Environment, 563
TLDP (Turkana Livestock Development Program), 529
TMDL (total maximum daily load), 564
TOC (total organic carbon), 467
toluene, 349, *389t*, 412, *416t–417t*, 754–756
Tonopah Desert Recharge Project, 673, *674f*
topography, recharge process and, 257–259
tortuosity, 408–409
Total Actual Renewable Water Resources (TARWR), 10–11
total dissolved solids (TDS), 220–221, 329–331, 462–464
total maximum daily load (TMDL), 564
total measured drawdown, 504
total organic carbon (TOC), 467
total porosity, 86–90
total trihalomethanes (TTHMs), *389t*
toxaphene, *389t*
toxicity, 342–343, 705–706
TPE (two-phase extraction) technology, 747–748
TR-55 (Technical Release 55), USDA, 237–241
trace constituents of groundwater, 329
trace elements, 679
trace metals, 473–474
trading systems, groundwater remediation, 701
trans-1,2-dichloroethylene, *385t*, *416t–417t*
Transboundary Aquifer Resources Management (TARM), 39, 571–575
transboundary water, 39, 559–560
transient calibration, 643–644
transient flow, 114–115
transient non-CWS, 564
transition zones, *674f*
transmissivity, 116–117, 815
transpiration, 241
transport of contaminants, 82
treated wastewater, 13
treatment. *See also* arsenic
  alternatives, 713, 716
  biological treatment, 451–452
  carbon adsorption, 450
  clarification, 441–443
  corrosion control, 456
  costs of, 474–478

disinfection, 452–456
distillation, 452
filtration, 443–447
hardness, 459–461
hydrogen sulfide removal, 466
inorganic compound removal, 473–474
and intrinsic value, 50–51
ion exchange and inorganic adsorption, 450–451
iron and manganese removal, 457–459
membrane filtration, 447–450
nitrate removal, 461–462
overview, 437–439
oxidation, 439–441
radionuclide removal, 465–466
in situ projects, *765f*
total dissolved solid removal, 462–464
total organic carbon removal, 467
trace metal removal, 473–474
VOC and SOC removal, 466
treatment plants, *438f*
treatment techniques (TTs), 380–381, *382t–391t*
tree rings, 310–313
tremie pipes, 499–500
trench emplacement, PRB, 775
trenches, 668–670
trends, 657, *764f*
trial-and-error calibration, 644
triazines, 697–698
trichloroethylene (TCE), *390t*, *413t*, *416t–417t*, *726f–727f*, 729, *761f*
triggered monitoring, 569
trihalomethanes (THMs), 363
tritium, *271f*, 272–273
tritium-helium-3 (3H-3He), 273
TTHMs (total trihalomethanes), *389t*
TTs (treatment techniques), 380–381, *382t–391t*
Tucson Basin, Arizona, *156f*
Tully, Andrew, 38
turbidity, *390t*, 443, 661–662, *663f*, 677
turbulent losses, 504, 506
Turkana Livestock Development Program (TLDP), 529
Twin Falls County, Idaho, *105f*
two-phase extraction (TPE) technology, 747–748

═══ U ═══

Ulrich, J. E., 281
ultrafiltration (UF) membrane filtration, 448
ultraviolet (UV) light disinfection, *453t*, 455–456
unconfined aquifers
  flow in, 114–115
  radial flow toward, 118
  screening for, 496–497
  storage properties of, 90–92, *93f–94f*
  unrenewability of, 557
  vertical wells in, *511f*
unconsolidated sand and gravel aquifers, 150–151
unconsolidated sedimentary rock porosity, 85–87
underground dams. *See* subsurface dams
underground source of drinking water (USDW), 327
underground storage tanks (USTs), 348
Ungs, M. J., 425
UNICEF, 355
uniform gravel packs, 498–499
unit conversion tables, 811–813, 815
United Kingdom Environment Agency, 696
United Nations Economic Commission for Europe (UN/ECE), 572
United Nations Educational, Scientific and Cultural Organization (UNESCO), 11
United Nations Environment Programme (UNEP), 13
United Nations General Assembly, 61–62
United Nations International Law Commission, 574
United States
  arsenic concentrations, 357
  cost of groundwater treatment plants, 474
  delineated wellhead protection areas in, 622
  freshwater use, 17–22
  groundwater development in Midwest, *484f*
  historical droughts in, 310–313
  impact of global warming, 317–320
  karst in, 170–171, *172f*
  land use in, 260
  pathogens, 376–377
  pesticides, 371–372
  regional PET patterns, 243
  replenishment of groundwater in, 3, 5
  sources of contamination, 347–348
  springs in, 195–196, 199–200, *531f*
  subsurface dams in, 525
  volatile organic chemicals, 363–364
  water disputes in, 36–37
  water prices in, 48–49
  water rights in, 560–562
  wellhead protection programs in, 51–52
University of Idaho, 104–107
University of Wisconsin–Madison, 345
unplanned nonrenewable groundwater management scenarios, 558
upper Brunswick aquifer, 75, *76f*
Upper Floridan aquifer, 75, *76f–77f, 133f*, 353–355
upscaling, 130
uptake of nitrogen by plants, 359–360
uranium, 358–359, *390t*
urbanization, 14, 17, 259–261, 629
U.S. Department of Agriculture (USDA), 237–241, 414
U.S. Department of Health and Human Services, 344
U.S. Environmental Protection Agency (USEPA)
  analytical methods, 282
  carcinogens, 343
  contaminant candidate list, 378–379, 381, *392t–393t*
  dispersivity, 405
  distributed-parameter areal recharge models, 285–287
  endocrine disruptors, 344
  Ground Water Supply Survey, 362
  health-based screening levels, 345–346
  MCL standards for radionuclides, 359
  pesticides, 371
  Primary Standards, 380–381
  remediation of contaminated water, 613–614
  sorption
    of organic solutes, 414, *415t–417t, 418f*
    overview, 412
  sources of contamination, 347–348
  Surface Impoundment Assessment, 351
  synthetic organic chemicals, 341

U.S. Geological Survey (USGS)
arsenic concentrations, 357
emerging contaminants, 377–378
ET rates in Death Valley, 243–245
health-based screening levels, 345–346
impact of global warming, 317–320
lysimeters, 266
MCL standards for radionuclides, 359
monitoring network, 581–582
National Water-Quality Assessment
program, 376–377
pesticides, 371–372
pharmaceuticals and personal care
products, 349
RORA and PART programs, 281
volatile organic chemicals, 363–364
U.S. National Academy of Sciences, 297
USDW (underground source of drinking
water), 327
use, water, 18f, 34, 611
USTs (underground storage tanks), 348
utilities, U.S. water, 48–50
UV (ultraviolet) light disinfection, 453t,
455–456

━━ **V** ━━
vacuum-enhanced recovery, 746–747
vadose zones
infiltration and water movement through
Brooks and Corey equation, 251–253
Darcy's law, 247–250
overview, 245–247
Richards equation, 251–253
soil water retention and hydraulic
conductivity, 247
van Genuchten equation, 251–253
injection wells, 668f, 671
modeling, 723
variably saturated flow models, 282–285
validation, data, 609–610
value, groundwater, 611
Van der Zaag, P., 40–43
van Genuchten equation, 251–253
van Genuchten-Mualem equation, 253
vapor pressure deficit (VPD), 242–243
vapor stripping process, in-well, 744–745
vaporization, 758
Vargas, C., 192

variability, spring discharge, 197
variable density flow simulation, 652f
variable-density groundwater modeling, 676f
variably saturated flow models, 282–285, 722f
variance, 207
vauclusian springs, 534
vegetative cover, 257
velocity
in aquitards, 188–192
groundwater flow, 125–126
screen entrance, 495
Verde River, 43
vertical barrier technologies, 734–735
vertical capture zones, 766f
vertical hydraulic gradients, 112f
vertical wells
completion, 510–511
gravel pack, 497–500
maintenance and rehabilitation, 511–515
overview, 488–491
screens, 491–497
testing and performance
overview, 503–508
well efficiency, 508–510
well development, 500–503
Village of White Oak Water System, North
Carolina, 461
vinyl chloride, 390t, 416t–417t
Virginia, 260–261. See also Blue Ridge
Province case study
virtual water, 55–61
viruses, 375–377, 390t, 453t, 678–679
viscosity
kinematic, 112–113
thermal technologies, 757–758
viscous fluid substrates, 782t
Visual Modflow, 607, 608f
vitrification, 757
voids, 84–85
volatile organic compounds (VOCs),
362–364, 466
volatilization, 398–401
volcanic rock aquifers, 174–175
volume
of freshwater in world, 1–3
unit conversion tables for, 811–812
Vordzorgbe, S. D., 31
VPD (vapor pressure deficit), 242–243

VS2DT model, 282–285
vulnerability, 82, 611–612, 614
vulnerability maps, 615–619

═══ W ═══

Walkerton, Ontario, 627, *628f*
walls, barrier, 734–735, 773–774
warm springs, 200
Warrenton, Virginia, *541f*
Washington, 255–257
Washington Post, The, 36–37
waste disposal, 349–351, 567–568
wastes, animal, 361, 373–374
wastewater, 13
water bottling companies, 38
water budget
  calibration, 646
  defined, 82
  Eastern Snake River Plain Aquifer, 104–107
  overview, 97–104
water consumption, 11–13, 22
water demand management, 548–549
water export, 59, *60t*
water footprints, 55–56
Water Framework Directive (WFD), 573
water governance, 539–540, 542–543, 560
water import, 59, *60t*
Water Information System Europe (WISE), 573
water jetting, 502
water legislation, 546
water management solutions, *656f*
water monitoring, 569
Water Poverty Index (WPI), 33–35
water purveyors, 12
water quality, 12, 82. *See also* quality of groundwater
water quality standards (WQS), 563–564
Water Resource Monitoring Program (WRMP), 178, 182
water resources, *15t*, *24t*, 33–34
water retention, 236–241, 251–253
water rights, 546, 559
water savings, real, 59
Water Services Association of Australia (WSAA), 47
water shortages, 540
water stress, 30–31

water supply, 12, 14
water systems, 12. *See also* systems, groundwater
water table, 79, 101–104, 266–268
water treatment. *See* treatment
water use, *18f*, 34, 611
water withdrawal. *See* withdrawals
waterborne diseases, 374–377, 569
waterfall aerators, 439
Waterloo Barrier, 738–739
watershed-based approach, 577
waves, groundwater, 101
Wax, Emily, 36
weather, 296–297
weighted residual mean, 647
Weizhou island, 652–653
well function W(u), 118–120, 507, 807–809
wellhead protection areas (WPAs), 563, 615
wellhead protection laws, 619
wellhead protection programs (WHPPs), 51–52
wells. *See also* collector wells; flowing wells; vertical wells
  in American Midwest, *485f*
  artificial aquifer recharge, 671–672
  in Blue Ridge Province, 178, *179f–181f*, *182t*
  boundary conditions and estimates of yield, 143, *144f*
  casings, 489–491
  chloride concentration in, *653f*
  as contamination conduits, 353–355
  contamination in, 589
  diameter of, 489, *490f*, *493t*
  gravel pack, 497–500
  monitoring, 117–118, 188, *189f*
  monitoring networks for, *585t*
  overview, 487–488
  pathogens in, 377
  performance testing, 511–512
  pesticides in, 373
  pumping, 116–118
  saltwater intrusion, 219–220
  sand content, 501
  sanding, 498
  sanitary, 341, *342f*
  screens, 489, 491–497, 519–521
  SEE, 759–760
  specific capacity, 510

wells (*Cont.*)
  stimulation methods, 503
  treatment in, 438
  use of in Lake Chad Basin, 25–26
  well efficiency, 508–510
  well loss, 117–118
  well properties datasets, *601t*
Welty, C., 406–407
West, M., 425
West, M. R., 425
Westchester County, New York, 589–590
wetlands, 2, 483
WFD (Water Framework Directive), 573
WhAEM2000 program, 621–622
WHPPs (wellhead protection programs),
    51–52
WHYMAP (World-wide Hydrogeological
    Mapping and Assessment
    Programme), 572
Widory, D., 361
Wiedemeier, T. H., 419, 420–422
Wilhelm, S. R., 361
Williams, J. R., 282
Williams, M. B., 51–52
windows in aquitards, 193
Winter, T. C., 97–98
wire-wrapped well screens, 519–520
Wisconsin, 193
WISE (Water Information System Europe),
    573
withdrawals
  in Africa, 25
  Cambrian-Ordovician aquifer system,
    161–162
  effect of boundary conditions on, 141–143
  in European Union, 22–24
  impact of on water system storage, 96–97
  Nubian Sandstone Aquifer System, 163,
    165
  trends, 11–13
  in U.S., 17–22
women in water management, 39–40

Woodhouse, C. A., 313
world availability of water resources, *15t*
World Bank, 30, 46
World Commission on Environment and
    Development, 61–62
World Water Council (WWC), 40, 56
World-wide Hydrogeological Mapping and
    Assessment Programme (WHYMAP),
    572
WPAs (wellhead protection areas), 563, 615
WPI (Water Poverty Index), 33–35
WQS (water quality standards), 563–564
WRMP (Water Resource Monitoring
    Program), 178, 182
WSAA (Water Services Association of
    Australia), 47
WWC (World Water Council), 40, 56

X

Xu, M., 406–407
xylenes, 349, *390t*, 412, *416t*, 754–756

Y

yields, well, 178, *181f*, *182t*, 489, *490f*
young carbonate sediments, 173–174
Younger Dryas, 324
Yucatan, Mexico, *171f*

Z

Zaidi, F. K., 268
zebra mussels, *454t*
zero flux plane, 266, *267f*
zero-flux boundaries, 139
zero-order degradation, 789
zero-valent iron (ZVI), 774
zinc, 474
zoning, groundwater
  inner protection, 624
  overview, 614
  of strict protection, 623–624
  wellhead protection, 623–624
ZVI (zero-valent iron), 774